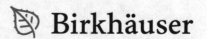

Birkhäuser Advanced Texts Basler Lehrbücher

Series Editors

Steven G. Krantz, Washington University, St. Louis, USA

Shrawan Kumar, University of North Carolina at Chapel Hill, Chapel Hill, USA

Jan Nekovář, Sorbonne Université, Paris, France

Shouchuan Hu • Nikolaos S. Papageorgiou

Research Topics in Analysis, Volume I

Grounding Theory

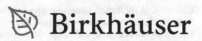

Shouchuan Hu
Department of Mathematics
Missouri State University
Springfield, MO, USA

Nikolaos S. Papageorgiou (iD)
Department of Mathematics, Zografou
Campus
National Technical University
Athens, Attiki, Greece

ISSN 1019-6242   ISSN 2296-4894   (electronic)
Birkhäuser Advanced Texts Basler Lehrbücher
ISBN 978-3-031-17839-9   ISBN 978-3-031-17837-5   (eBook)
https://doi.org/10.1007/978-3-031-17837-5

Mathematics Subject Classification: 28-XX, 46-XX, 46Bxx, 49-XX

This book is published under the imprint Birkhäuser, www.birkhauser-science.com by the registered company Springer Nature Switzerland AG
The registered company address is: Gewerbestrasse 11, 6330 Cham, Switzerland

To the memory of our parents

Preface

The aim of this two-volume work is to present in a comprehensive way the main ingredients of some of the active research topics in nonlinear analysis and of its applications. The two volumes will equip young researchers with a good theoretical background and a taste of how these tools can be used to deal with more advanced studies on various topics and applications. The preparatory nature of Volume 1 is emphasized by the presence of relevant (and sometimes challenging) problems at the end of each chapter, which will help the reader digest and test his/her understanding of the materials. Some of the problems also provide additional notions and results. The flow of the materials is "linear" in the sense that each chapter sets the ground in order for the next ones to develop and the topics to expand naturally. In Volume 2, the chapters are more independent, and the reader can choose a particular chapter of interest to focus on without much difficulty.

Volume 1 presents the foundations of the subject and provides the necessary introductory materials for analysis in order for young researchers to approach the subject at the research level equipped with the necessary knowledge and background, with which they will be able to identify the areas where new contributions and advancements can be made and know the tools that need to be deployed. For this reason, the first three chapters are devoted to the presentation of the main aspects of topology, measure theory, and functional analysis. Our aim is to familiarize the reader with the basic background knowledge of these subjects without overwhelming them with details and topics which are not relevant to the study of modern nonlinear analysis and its applications. Our presentation focuses only on the essential aspects, which are used in the applications. Therefore, the reader will gain a good overview of the subjects without getting carried away in details that are not relevant. Chapter 4 continues in this spirit and is devoted to a comprehensive analysis of the main function spaces, which one encounters in applications. So, we present Lebesgue spaces (also with variable exponents), Lebesgue–Bochner spaces (needed in evaluation equations), Sobolev spaces (an indispensable tools in the study of boundary value problems, a modern treatment of partial differential equations), and spaces of measures (used in relaxation theory as well as in diverse applications such as optimal control, mathematical economics, and game theory).

Specifically, we have

In Chap. 1, we present a compact but comprehensive introduction of topology examined from the point of view of an analyst.

In Chap. 2, we do the same approach for the subject of measure theory. We also present how topology and measure theory interact. Such an interaction is very common in modern analysis and its applications. The theory of Young measures, which is important in calculus of variations and in optimal control theory, is a characteristic example of such an interaction of the two subjects.

Chapter 3 deals with functional analysis and gives a detailed account of Banach space theory and of the spaces of linear operators between them.

Chapter 4 deals with various function spaces that one encounters in applications. So, we cover Lebesgue spaces including those with variable exponents, Lebesgue–Bochner spaces (an indispensable tool in evolution equations), Sobolev spaces (the heart of the modern treatment of partial differential equations), and spaces of measures (important in probability theory and applications such as the transport theory).

In Chap. 5, we pass to more concrete subjects, which are closely related to applications and more difficult to find in compact form. The first such topic to be examined is that of "Multivalued Analysis" (set-valued functions). Set-valued maps were originally introduced to address the needs of subjects such as optimal control, optimization, mathematical economics, game theory, and calculus of variations. We will have a complete overview of the subjects, examining them from both topological and measure-theoretic viewpoints.

In Chap. 6, we present the smooth and nonsmooth calculus of general maps, with the latter developed in parallel and in symbiotic relation with multivalued analysis. Our treatment covers many aspects of the theory of convex analysis and of locally Lipschitz functions. The latter are central in the modern approach to the geometric measure theory and also provide a natural framework to extend convex analysis to nonconvex functions. Furthermore, Lipschitz functions are important since they are the counterpart of smooth functions in the study of equations in general metric spaces. We establish remarkable continuity properties that convex functions exhibit and then introduce the notion of "convex subdifferential" that extends the concept of derivative to nonsmooth convex functions and develops the corresponding calculus.

In Chap. 7, we introduce certain important classes of nonlinear operators such as compact and Fredholm operators, maximal monotone operators, and operators of monotone type. So, we deal with compact operators and with monotone operators and the generalizations. These classes arise naturally from nonsmooth analysis (Chap. 6), the link being the notion of the convex subdifferential. Compact operators are important because of their spectral properties and are the main ingredient in the infinite-dimensional extensions of degree theory (Leray–Schauder degree). On the other hand, operators of monotone type have many applications in nonlinear boundary value problems.

Complementing Chaps. 5–7, Chapter 8 presents some topics of variational analysis. So, we study convergence of sets, a topic useful in "Shape-Optimization" and "Sensitivity Analysis" of optimization and optimal control problems (a guide for the development of efficient numerical algorithms). Then we present the Γ-convergence of functions and the G-convergence of operators, two topics crucial in problems of the calculus of variations and of homogenization theory. Finally, we present some important variational principles such as the Ekeland variational principle, which are useful in applications.

Each chapter is followed by a collection of related exercise problems, which are of different levels of difficulty and are chosen so that the reader can check and confirm his understanding of the concepts introduced in the chapter. Also, in some occasions, new notions and results are introduced in the exercise problems.

The flow of the materials in this volume is natural in that each chapter leads naturally and smoothly to the next one. There are also listings of detailed indices of terminologies and mathematical symbols. We believe that the volume will equip the reader with necessary tools and theoretical background to address more specialized research topics and applications.

Volume 2 consists of various applications using tools and techniques developed in Volume 1. Volume 2 will be oriented to have the following eight chapters.

Chapter 1: "Degree Theory" The material of this chapter is useful to everyone applying the topological methods in the study of boundary value problems. So, we will present the Brouwer degree, the Leray–Schauder degree, and the degree theory for operator type and for multifunctions.

Chapter 2: "Fixed Point Theory" This is a topic used in all kinds of applications beyond mathematical. We will cover metric, topological, and order fixed point theories and also discuss the Fixed Point Index.

Chapter 3: "Critical Point Theory" This is the main tool in the variational methods in the study of boundary value problems. Together with critical groups (Morse theory), it leads to the powerful existence and multiplicity results for elliptic equations.

Chapter 4: "Spectra of Differential Operators" We will discuss about the spectrum under different boundary conditions of the Laplacian and the p-Laplacian including the scalar case, and of anisotropic and fractional ones.

Chapter 5: "Boundary Value Problems" In this chapter, we will examine concrete equations and will show how the material from the previous chapters can be used to obtain existence and multiplicity results for different classes of problems (variational, singular, problems with convection, anisotropic problems, etc.).

Chapter 6: "Evolution Equations" In this chapter, we will deal with dynamic equations and show how the formalism of evolution triples, and the theory of monotone operators can be used to treat certain parabolic and hyperbolic equations. We will also address the issue of the structure of the solution set and its stability using the notion of G-convergence.

Chapter 7: "Optimization and Optimal Control" We will study problems in calculus of variations and of optimal control, focusing on their relaxation properties and on sensitivity analysis.

Chapter 8: "Mathematical Economics and Game Theory" We will present both equilibrium and dynamic economic models and also discuss about Nash equilibrium.

Springfield, MO, USA   Shouchuan Hu
Athens, Greece   Nikolaos S. Papageorgiou

Contents

Contents   xiii

Chapter 1
Topology

In this chapter we present some basic notions and results from Topology that are standard in Analysis. We start with the basic definitions and then discuss continuous functions and nets, since in general spaces such as infinite dimensional Banach spaces with the weak topology, sequences are not sufficient to describe topological properties. We then present classifications of topological spaces based on their separation and countability properties. In Sect. 1.4 we present new topologies which can be generated from pre-existing ones. One of them is the "weak topology," which will resurface in Chap. 3, in the study of Banach spaces. A particular case of a weak topology is the product topology which is a standard item in many applications. In Sect. 1.5 we discuss the fundamental notions of compactness and locally compact spaces. The latter provide the right framework for the interaction between Topology and Analysis, and are used in the study of topological groups. In Sect. 1.6 we deal with the other fundamental topological notions: namely connected, path-connected, and locally connected spaces. In Sect. 1.7 we turn our attention to Polish, Souslin, and Baire spaces, which are often encountered in Measure Theory and Functional Analysis. We conclude with a brief discussion of semicontinuous functions that arise in applications, such as Calculus of Variation, Optimization, Optimal Control and Mathematical Economics.

1.1 Basic Notions and Facts

We start with spaces whose topology is defined by a distance, which are known as "metric spaces" and are important in modern analysis. Having a metric is a very desirable property of the space since it implies some important properties. In particular, we can study convergence in the space by reducing it to convergence in $\mathbb{R}$.

© The Author(s), under exclusive license to Springer Nature Switzerland AG 2022
S. Hu, N. S. Papageorgiou, *Research Topics in Analysis, Volume I*, Birkhäuser
Advanced Texts Basler Lehrbücher, https://doi.org/10.1007/978-3-031-17837-5_1

Definition 1.1 A "metric" on a set X is a function: $d : X \times X \to \mathbb{R}_+ = [0, \infty)$ such that:

(a) $d(x, u) = 0 \iff x = u$.
(b) $d(x, u) = d(u, x)$ for all $x, u \in X$.
(c) $d(x, u) \le d(x, v) + d(v, u)$ for all $x, u, v \in X$ (triangle inequality).

We say that (X, d) is a "metric space."

Remark 1.2 If (a) is replaced by the weaker condition of $(a)' : d(x, x) = 0$ for all $x \in X$, then it is a "semimetric" or "pseudometric."

Example 1.3 For $X = \mathbb{R}^N$, $x = (x_k)$, $u = (u_k) \in X$, and $1 \le p < \infty$, we have

$$d_\varrho(x, u) = \left[\sum_{k=1}^{N} |x_k - u_k|^p \right]^{1/p},$$

$$d_\infty(x, u) = \max\left[|x_k - u_k| : 1 \le k \le N \right]$$

are both metrics on $\mathbb{R}^N$. If X is any set and we define

$$d(x, u) = \begin{cases} 0 & \text{if } x = u \\ 1 & \text{if } x \ne u, \end{cases}$$

then we have the "discrete metric" on X. If (X, d) is a metric space and $V \subseteq X$, then $d|_{V \times V}$ is a metric on V.

For a metric space (X, d), $x \in X$ and $\varrho > 0$, set

$$B_\varrho(x) = \{u \in X; d(x, u) < \varrho\}, \qquad \overline{B}_\varrho(x) = \{u \in X : d(x, u) \le \varrho\},$$

the open and closed balls with center x and radius ϱ, respectively.

Definition 1.4 Let (X, d) be a metric space and

$$\tau_d = \{U \subseteq X : U \text{ is the union of a family of open balls in } X\},$$

which is the "topology" on X induced by d and compatible with d.

Remark 1.5 $U \in \tau_d$ iff for any $x \in U$ we can find $\varrho > 0$ such that $B_\varrho(x) \subseteq U$. We also have (a) $\emptyset, X \in \tau_d$, (b) τ_d is closed under arbitrary unions, (c) τ_d is closed under finite intersection.

Definition 1.6 Let (X, d) be a metric space and $\{x_n\} \subseteq X$.

(a) $x \in X$ is called a "cluster point" of $\{x_n\}$ if $B_\epsilon(x)$ contains infinitely many terms of the sequence for every $\epsilon > 0$.
(b) $\{x_n\}$ "converges" to x if $\lim_{n \to \infty} d(x_n, x) = 0$.
(c) $\{x_n\}$ is a "Cauchy sequence" if $\lim_{n,m \to \infty} d(x_n, x_m) = 0$.
(d) The metric space is "complete" if every Cauchy sequence converges to some element in X.

Motivated by these observations we can extend the notion of topology to nonmetrizable spaces.

Definition 1.7 Let X be a set and τ a collection of subsets of X. τ is called a topology on X and (X, τ) a "topological space" if

(a) $\emptyset, X \in \tau$.
(b) τ is closed under arbitrary unions.
(c) τ is closed under finite intersections.
     Elements of τ are called "open sets" and their complements are called "closed sets."

Remark 1.8 If (X, τ) is a topological space, whose topology is induced by a metric d, i.e., $\tau = \tau_d$, then we say that X is "metrizable." In the metric space we have fixed a metric, while in a metrizable space the choice is still open.

Definition 1.9 Let (X, τ) be a topological space and $A \subseteq X$.

(a) The "interior" of A is the set $\operatorname{int} A = \bigcup \{U : U \subseteq A \text{ is open }\}$.
(b) The "closure" of A is the set $cl A = \overline{A} = \{C : A \subseteq C \text{ is closed}\}$.
(c) The "boundary" of A is the set $bd A = \partial A = cl A \cap cl(X \setminus A)$.

Remark 1.10 By definition, $\operatorname{int} A$ is open and is the largest open set contained in A, while $cl A$ is the smallest closed set containing A. A is open iff $A = \operatorname{int} A$, and closed iff $A = cl A$. Also, $bd A$ is closed. In fact, $cl A = A \cup bd A$, $\operatorname{int} A = A \setminus bd A$, and $X = \operatorname{int} A \cup bd A \cup \operatorname{int}(X \setminus A)$.

In general topological spaces, the abstract notion of closeness is expressed using the concept of neighborhood. The language of neighborhoods is that of point set topology.

Definition 1.11 Let (X, τ) be a topological space and $x \in X$. We say that A is a neighborhood of x if we can find $U \in \tau$ such that $x \in U \subseteq A$. Let $\mathcal{N}(x)$ denote the collection of all neighborhoods of x, which is called the "filter of neighborhoods" of x. A "neighborhood basis" at $x \in X$ is a subcollection $\mathcal{B}(x)$ of $\mathcal{N}(x)$ such that

$$\mathcal{N}(x) = \{A : E \subseteq A, E \in \mathcal{B}(x)\}.$$

As we did in the above definition, locally at a point, we can do it globally for the whole topology. We can specify the topology on X without describing every open set by giving a "basis" for the topology.

Definition 1.12 Let (X, τ) be a topological space. A "basis" for the topology τ is a subcollection $\mathcal{B} \subseteq \tau$ such that $\tau = \cup\{B : B \in \mathcal{B}\}$. A "subbasis" is a subcollection $\mathcal{S} \subseteq \tau$ such that the family of all finite intersections of elements from $\mathcal{S}$ forms a basis for τ.

Remark 1.13 It is clear that $\mathcal{B}$ is a basis for τ iff for any $x \in U \in \tau$ we can find $B \in \mathcal{B}$ such that $x \in B \in U$. Moreover, $\mathcal{B}$ is a basis iff for any $x \in X$, $\mathcal{B}(x) = \{B \in \mathcal{B} : x \in B\}$ is a neighborhood basis at x. Finally, any collection of subsets of X is a subbasis of some topology on X.

Example 1.14 (a) If (X, d) is a metric space, then $\mathcal{B} = \big\{B_\varrho(x) : x \in X, \varrho > 0\big\}$ is a subbasis for the metric topology τ_d. (b) The collection $\{\{x\} : x \in X\}$ is a basis for the discrete topology on X.

A subset of a topological space inherits a topology from the ambient space.

Definition 1.15 Let (X, τ) and $A \subseteq X$ be given. $\tau_A = \{A \cap U : U \in \tau\}$, the trace of τ on A, is a topology on A. We call τ_A the "relative or subspace topology" on A and in the sequel, every subset is considered with the corresponding subspace topology.

Remark 1.16 Evidently, $C \subseteq A$ is τ_A-closed iff $C = A \cap K$ with some $K \subseteq X$ τ-closed. Let $x \in A$ and $\mathcal{N}_A(x)$ (respectively, $\mathcal{B}_A(x)$) be the filter of neighborhoods of x (respectively, a neighborhood basis at x) for the τ_A-topology, then

$$\mathcal{N}_A(x) = A \cap \mathcal{N}(x) \quad (\text{respectively, } \mathcal{B}_A(x) = A \cap \mathcal{B}(x)).$$

Similarly, if $\mathcal{B}$ is basis for τ, then $\mathcal{B}_A = A \cap \mathcal{B}$ is a basis for τ_A.

1.2   Continuous Functions: Nets

Definition 1.17 Let X, Y be topological spaces and $f : X \to Y$. f is said to be "continuous at x_0" if for any $V \in \mathcal{N}(f(x_0))$ we can find $U \in \mathcal{N}(x_0)$ such that $f(U) \subseteq V$. f is continuous if it is continuous at every point.

Remark 1.18 It is clear that we can replace neighborhoods by basic neighborhoods.

Proposition 1.19 *If X, Y are topological spaces and $f : X \to Y$, the statements below are equivalent.*

(a) f is continuous.
(b) $f^{-1}(V)$ is open in X for every open $V \subseteq Y$.
(c) $f^{-1}(C)$ is closed in X for every closed $C \subseteq Y$.
(d) $f(\overline{A}) \subseteq \overline{f(A)}$ for every $A \subseteq X$.

Proof (a) $\Rightarrow$ (b): Let $V \subseteq Y$ be open and $x \in f^{-1}(V)$. We have $V \in \mathcal{N}(f(x))$ and thus we can find $U \in \mathcal{N}(x)$ such that $f(U) \subseteq V$. Hence $U \subseteq f^{-1}(V)$. Therefore $f^{-1}(V)$ is open.

(b) $\Rightarrow$ (c): Let $C \subseteq Y$ be closed. Then $Y \setminus C$ is open and thus $f^{-1}(Y \setminus C) \subseteq X$ is open. Therefore, $f^{-1}(C)$ is closed in X since $f^{-1}(Y \setminus C) = X \setminus f^{-1}(C)$.

(c) $\Rightarrow$ (d): Let $C \supseteq f(A)$ be closed. Then $f^{-1}(C) \subseteq X$ is closed and $A \subseteq f^{-1}(C)$. Hence $\overline{A} \subseteq f^{-1}(C)$ and $f(\overline{A}) \subseteq C$. Therefore, $f(\overline{A}) \subseteq \overline{f(A)}$.

(d) $\Rightarrow$ (a): Let V be an open neighborhood of $f(x)$. Set $A = X \setminus f^{-1}(V)$ and $U = X \setminus \overline{A}$. We claim that $x \in U$. Suppose that $x \notin U$. Then $x \in \overline{A}$ and so $f(x) \in f(\overline{A}) \cap V$. On the other hand, $f(A) \subseteq Y \setminus V$ and the latter is closed in Y, hence $\overline{f(A)} \subseteq Y \setminus V$. By (d), $f(\overline{A}) \subseteq Y \setminus V$, thus $f(\overline{A}) \cap V = \emptyset$, a contradiction. Thus, $x \in U$ and $f(U) \subseteq V$, which implies that f is continuous. $\qquad\square$

In the next proposition we gather some basic facts about continuous functions, whose proofs are straightforward consequences of the definitions.

Proposition 1.20 *Given X, Y, Z topological spaces. (a) If $f : X \to Y$ and $g : Y \to Z$ are continuous, then $g \circ f$ is continuous.*
(b) If $f : X \to Y$ is continuous and $A \subseteq X$, then $f|_A : A \to Y$ is continuous.
(c) $f : X \to B \subseteq Y$ is continuous iff it is continuous as a map into Y.

The next result is known as the "pasting lemma."

Proposition 1.21 *For topological spaces X and Y, where $X = A \cup C$ with A, C closed, and continuous $f : A \to Y$ and $g : C \to Y$ with $f \equiv g$ on $A \cap C$, we have*

$$h(x) = \begin{cases} f(x) & \text{if } x \in A \\ g(x) & \text{if } x \in C \end{cases}$$

is continuous.

Proof For $K \subseteq Y$ closed, we have $h^{-1}(K) = f^{-1}(K) \cup g^{-1}(K)$ and $f^{-1}(K) \subseteq A, g^{-1}(K) \subseteq C$ are all closed. Thus, h is continuous. $\qquad\square$

If X and Y are metric spaces, then we have the following more familiar form of continuity.

Definition 1.22 Let (X, d) and (Y, ϱ) be two metric spaces and $f : X \to Y$. We say that f is "continuous" at $x_0 \in X$ if for any $\epsilon > 0$ we can find $\delta > 0$ such that $\varrho(f(x), f(x_0)) < \epsilon$ whenever $d(x, x_0) \leq \delta$. f is continuous if it is continuous at every point of X.

Remark 1.23 Translated in the language of sequences, the above definition says that $\varrho(f(x_n), f(x_0)) \to 0$ whenever $d(x_n, x_0) \to 0$. However, such sequential characterization of continuity is valid only for topologies having a countable system of basic neighborhoods at each point. For more general spaces we need to consider generalized sequences, known as "nets."

Definition 1.24 D is a "directed set" if there is a relation $\preceq$ on D satisfying

(1) $\vartheta \preceq \vartheta$ for each $\vartheta \in D$ (reflexive).
(2) $\vartheta_1 \preceq \vartheta_2$ and $\vartheta_2 \preceq \vartheta_3 \Rightarrow \vartheta_1 \preceq \vartheta_3$ (transitive).
(3) For any $\vartheta_1, \vartheta_2 \in D$ there is $\vartheta \in D$ such that $\vartheta_1 \preceq \vartheta, \vartheta_2 \preceq \vartheta$ (directed).

Example 1.25

(a) The set $\mathbb{N}$ of natural numbers with the relation (direction) $\preceq$ defined by $m \preceq n$ if $m \leq n$.
(b) The sets $(0, \infty)$ and $(0, 1)$ are directed sets with $\preceq$ being the usual ordering.
(c) The filter of neighborhoods at $x \in X, \mathcal{N}(x)$ with the relation (direction) $\preceq$ defined by $V \preceq U$ whenever $U \subseteq V$ (inverse inclusion).

When D, E are directed sets, there is a natural direction on the product $D \times E$, defined by $(a, b) \preceq (c, d)$ if $a \preceq_D c$ and $b \preceq_E d$.

Definition 1.26 A "net" in a set X is a function $u : D \to X$, where D is a directed set. We write $u(\vartheta) = u_\vartheta$ and refer to the net $\{u_\vartheta\}_{\vartheta \in D}$. A net $\{v_i\}_{i \in I}$ is a "subnet" of a net $\{u_\vartheta\}_{\vartheta \in D}$ if there is a function $p : I \to D$ satisfying

(a) $p(i) \preceq_D p(j)$ if $i \preceq_I j$, namely p is increasing.
(b) For each $\vartheta \in D$ there is $i \in I$ such that $\vartheta \preceq_D p(i)$; (p is cofinal in D).
(c) $v_i = u_{p(i)}$ for all $i \in I$.
   $\{u_\vartheta\}_{\vartheta \, in \, D}$ "converges" to $x \in X$ if for every $U \in \mathcal{N}(x)$ we can find $\vartheta_U \in D$ such that $u_\vartheta \in U$ for all $\vartheta \succeq \vartheta_U$.

We write $u_\vartheta \to u$ or $u_\vartheta \xrightarrow{\tau} u$ when we wish to emphasize the topology in which the convergence occurs.

Remark 1.27 Evidently, sequences are nets, and every subsequence is a subnet. The converse is not true, for instance, consider the sequence $\{x_n = n^2 + 1\}$ and the net $\{v_{m,n} = m^2 + 2mn + n^2 + 1\}$ that is a subnet of $\{x_n\}$ but not a subsequence. Note also that a subnet of a subnet of $\{x_\vartheta\}_{\vartheta \in D}$ is still a subnet of $\{x_\vartheta\}$.

Nets perform many of the tasks that sequences do in a metric space.

Definition 1.28 Let (X, τ) be a topological space and $A \subseteq X$. $x \in X$ is called a "cluster point (accumulation point or limit point)" of A if for any $U \in \mathcal{N}(x)$ we have $A \cap (U \setminus \{x\}) \neq \emptyset$. The set, A', of all the cluster points of A is called the "derived" set of A".

Remark 1.29 It is clear that $\overline{A} = A \cup A'$.

Proposition 1.30 *If (X, τ) is a topological space and $A \subseteq X$, then $x \in \overline{A}$ iff there exists a net $\{x_\vartheta\}_{\vartheta \in D} \subseteq A$ such that $x_\vartheta \to x$ in X.*

Proof $\Rightarrow$: Given $U \in \mathcal{N}(x)$, then $A \cap U \neq \emptyset$. So, let $x_U \in A \cap U$. This way we produce a net $\{x_U\}_{U \in \mathcal{N}(x)}$ and we have $x_U \to x$ in X.

$\Leftarrow$: If $\{x_\vartheta\}_{\vartheta_D} \subseteq A$ and $x_\vartheta \to x$, then $A \cap U \neq \emptyset$ for all $U \in \mathcal{N}(x)$ and so $x \in \overline{A}$. $\square$

Proposition 1.31 *For any topological space* (X, τ), x *is a cluster point of a net* $\{x_\vartheta\}_{\vartheta \in D} \subseteq X$ *iff there exists a subnet that converges to* x.

Proof $\Rightarrow$: Consider $D \times \mathcal{N}(x)$ with the product direction and let $M \subseteq D \times \mathcal{N}(x)$ be defined by $M = \{(\vartheta, U) \in D \times \mathcal{N}(x) : x_\vartheta \in U\}$. Then M is a directed set. We introduce $p : M \to D$ defined by $p(\vartheta, U) = \vartheta$. Evidently, p is increasing and cofinal. Thus, it defines a subnet. Let $V \in \mathcal{N}(x)$. There is $\vartheta_0 \in D$ such that $x_{\vartheta_0} \in V$. Then $(\vartheta_0, V) \in M$ and if $(\vartheta_0, V) \preccurlyeq (\vartheta, V') \in M$, we have $V' \subseteq V$ and so $x_\vartheta \in V' \subseteq V$. This means the subnet converges to x.

$\Leftarrow$: Suppose that a subnet $\{v_i\}_{i \in I}$ of a net $\{x_\vartheta\}_{\vartheta \in D}$ converges to some point x. Fix $\vartheta_0 \in D$, $V \in \mathcal{N}(x)$ and $p : I \to D$ be the map defining the subnet. Choose $i_0 \in I$ such that $v_i \in V$ for all $i \succcurlyeq i_0$. Next we choose $i_1 \in I$ such that $\vartheta_0 \preccurlyeq_D p(i)$ for all $i \succcurlyeq i_1$. Take $i_2 \in I$ such that $i_0, i_1 \preccurlyeq i_2$. Then $\vartheta = p(i_2)$ satisfies $\vartheta_0 \preccurlyeq_D \vartheta$ and $x_\vartheta = x_{p(i_2)} = v_{i_2} \in V$, which implies x is a cluster point of $\{x_\vartheta\}_{\vartheta \in D}$. $\square$

Proposition 1.32 *In a topological space* X, *a net converges to a point* x *iff every subnet converges to* x.

Proof $\Rightarrow$: This is immediate from the definition of subnet.

$\Leftarrow$: Arguing by contradiction. Suppose that every subnet converges to x but the net itself does not. Then there exists $V \in \mathcal{N}(x)$ such that for any $\vartheta \in D$ we can find $p(\vartheta) \succcurlyeq_D \vartheta$ with $x_{p(\vartheta)} \notin V$. Set $v_\vartheta = x_{p(\vartheta)}$, then the subnet $\{v_\vartheta\}_{\vartheta \in D}$ fails to converge to x, a contradiction. $\square$

Remark 1.33 Note that the limits need not be unique in the above result.

Proposition 1.34 *Let* X, Y *be topological spaces,* $f : X \to Y$ *and* $x \in X$. *Then,* f *is continuous at* x *iff* $f(x_\vartheta) \to f(x)$ *in* Y *for every net* $x_\vartheta \to x$ *in* X.

Proof $\Rightarrow$: Let $V \in \mathcal{N}(f(x))$. Then $f^{-1}(V) \in \mathcal{N}(x)$ and so, there is ϑ_0 such that $x_\vartheta \in f^{-1}(V)$ for $\vartheta_0 \preccurlyeq \vartheta$, thus $f(x_\vartheta) \in V$ for $\vartheta_0 \preccurlyeq \vartheta$. So, $f(x_\vartheta) \to f(x)$ in Y.

$\Leftarrow$: Arguing by contradiction. Suppose that f is not continuous at x. Then there is $V \in \mathcal{N}(f(x_0))$ such that $f(U) \nsubseteq V$ for every $U \in \mathcal{N}(x)$. For each $U \in \mathcal{N}(x)$ let $x_U \in U$ such that $f(x_U) \notin V$. Then we have a net $\{x_U\}_{U \in \mathcal{N}(x)} \subseteq X$ such that $x_U \to x$ in X, but $f(x_U) \nrightarrow f(x)$, a contradiction. $\square$

Remark 1.35 Sequences are used if X is a metric space.

1.3   Separation and Countability Properties

We will introduce a classification of topological spaces according to their separation properties.

Definition 1.36 A topological space (X, τ) is (a) "Hausdorff" if for any $x, u \in X$, $x \neq u$, there are open $U \in \mathcal{N}(x)$ and $V \in \mathcal{N}(u)$ such that $U \cap V = \emptyset$.

(b) "Regular" if it is Hausdorff and for every closed $C \subseteq X$ and $x \notin C$, there are open $U \supseteq C$ and $V \in \mathcal{N}(x)$ such that $U \cap V = \emptyset$.

(c) "Completely regular" if it is Hausdorff and for any closed $C \subseteq X$ and $x \notin C$, there exists continuous $f : X \to [0, 1]$ such that $f(x) = 0$ and $f|_C = 1$.

(d) "Normal" if it is Hausdorff and for any closed $C, D \subseteq X$ with $C \cap D = \emptyset$ there exist open sets $U \supseteq C$ and $V \supseteq D$ such that $U \cap V = \emptyset$.

Remark 1.37 It is clear that the limit of a convergent net in a Hausdorff space is unique. For this reason the Hausdorff property in Analysis is the minimum structure of a space. Thus, in the sequel we always consider Hausdorff spaces.

Proposition 1.38 *For a Hausdorff space (X, τ), the following statements are equivalent:*

(a) X is regular.
(b) For any $x \in X$, $U \in \mathcal{N}(x)$ there exists $V \in \mathcal{N}(x)$ such that $x \in \overline{V} \subseteq U$.
(c) For any closed $C \subseteq X$ and $x \in X \setminus C$, there exists $U \in \mathcal{N}(x)$ such that $C \cap \overline{U} = \emptyset$.

Proof (a) $\Rightarrow$ (b): For $U \in \mathcal{N}(x)$, $C = X \setminus U$ is closed, thus the regularity of X implies that there are open sets $W \supseteq C$ and $V \in \mathcal{N}(x)$ such that $W \cap V = \emptyset$. Hence, $\overline{V} \subseteq X \setminus W$. Note that $\overline{V} \cap C \subseteq \overline{V} \cap W = \emptyset$. Therefore, $V \subseteq \overline{V} \subseteq U$.

(b) $\Rightarrow$ (c): For closed $C \subseteq X$ and $x \in X \setminus C$, $U = X \setminus C \in \mathcal{N}(x)$. Thus, we can find $V \in \mathcal{N}(x)$ such that $V \subseteq \overline{V} \subseteq U$. Clearly, $C \cap \overline{V} = \emptyset$.

(c) $\Rightarrow$ (a) Consider closed $C \subseteq X$ and $x \in X \setminus C$. We can find $V \in \mathcal{N}(x)$ such that $C \cap \overline{V} = \emptyset$. Let $W \in X \setminus \overline{V}$. Then $W \supseteq C$ is open and $W \cap V = \emptyset$, which proves the regularity of X. $\qquad\qquad\square$

Corollary 1.39 *A Hausdorff space is regular iff each $x \in X$ admits a neighborhood basis consisting of closed sets.*

Proposition 1.40 *A completely regular space is regular.*

Proof Let $C \subseteq X$ be closed and $x \in X \setminus C$. There exists a continuous $f : X \to [0, 1]$ such that $f(x) = 0$ and $f|_C = 1$. Thus the sets $U = f^{-1}([0, 1/2)) \in \mathcal{N}(x)$ and $V = f^{-1}((1/2, 1]) \supseteq C$ are open and disjoint. Therefore, X is regular. $\qquad\square$

We can have a result characterizing normal spaces, similar to Proposition 1.38, whose proof is similar.

Proposition 1.41 *The following statements are equivalent for a Hausdorff space* X:

(a) X *is normal.*
(b) *For any closed* $C \subseteq X$ *and open* $U \supseteq C$, *there exists open* $V \subseteq X$ *such that* $C \subseteq V \subseteq \overline{V} \subseteq U$.
(c) *For any disjoint closed sets* $C, D \subseteq X$, *there exists open* $U \supseteq C$ *such that* $\overline{U} \cap D = \emptyset$.

The next theorem is a deep result, which shows why normal spaces are important. The result is known in the literature as "Urysohn's Lemma."

Theorem 1.42 *A Hausdorff space* (X, τ) *is normal iff for any disjoint closed pair* $C, D \subseteq X$ *there exists a continuous* $f : X \to [0, 1]$ *such that* $f|_C = 0$ *and* $f|_D = 1$.

Proof $\Rightarrow$: A rational dyadic number $\tau \in [0, 1]$ can be expressed as

$$r = \frac{m}{2^n}, \ m = 0, 1, 2, \ldots, 2^n, \ n \in \mathbb{N}_0 = \{0, 1, \cdots\}.$$

We will construct a family $\{U_r\}$ of open sets such that

$$C \subseteq U_0, \ U_1 = X \setminus D, \ \text{and} \ \overline{U}_r \subseteq U_s \ \text{if } r < s.$$

The normality of X implies that there exists open $U_0 \supseteq C$ such that $\overline{U}_0 \subseteq X \setminus D$ (Proposition 1.41). So, for $m = n = 0$ we choose this U_0. For $m = 1$ and $n = 0$ we choose $U_1 = X \setminus D$.

For $m = 0, 1, 2$ and $n = 1$ we have three open sets $U_0, U_{1/2}, U_1$, where $U_{1/2}$ is chosen such that

$$\overline{U}_0 \subseteq U_{1/2} \subseteq \overline{U}_{1/2} \subseteq U_1.$$

For $m = 0, 1, 2, 3, 4$ and $n = 2$, we have open sets $U_{\frac{m}{2^n}}$ of which only $U_{1/4}$ and $U_{3/4}$ have to be defined. By the normality of X we can find open sets $U_{1/4}$ and $U_{3/4}$ such that

$$\overline{U}_0 \subseteq U_{1/4} \subseteq \overline{U}_{1/4} \subseteq U_{1/2} \ \text{and} \ \overline{U}_{1/2} \subseteq U_{3/4} \subseteq \overline{U}_{3/4} \subseteq U_1.$$

We continue inductively. Suppose that we have produced $U_{\frac{m}{2^{n-1}}}$. Now we choose open sets $U_{\frac{m}{2^n}}$ as follows. For even m the sets have already been selected. For odd m, as before this can be done using Proposition 1.41. We can find open set $U_{\frac{m}{2^n}}$ such that

$$\overline{U}_{\frac{m-1}{2^n}} \subseteq U_{\frac{m}{2^n}} \subseteq \overline{U}_{\frac{m}{2^n}} \subseteq U_{\frac{m+1}{2^n}}.$$

Then define $f : X \to [0, 1]$ by

$$f(x) = \begin{cases} 1 & \text{if } x \in D \\ \inf\{f(x) : x \in U_r\} & \text{if } x \in X \setminus D. \end{cases}$$

Clearly, $f|_C = 0$ and $f|_D = 1$. It remains to show that f is continuous. For every $\lambda \in (0, 1]$, we have

$$\{f < \lambda\} = \bigcup_{r < \lambda} U_r,$$

thus $\{f < \lambda\}$ is open. Since $\overline{U}_r \le U_s$ for $r < s$, we have

$$\{f \le \lambda\} = \bigcap_{r > \lambda} \overline{U}_r,$$

thus $\{f \le \lambda\}$ is closed. We conclude that f is continuous.

$\Leftarrow$: Let $U = f^{-1}([0, 1/2))$ and $V = f^{-1}((1/2, 1])$. These are open sets, with $C \subseteq U, D \subseteq V$ and $U \cap V = \emptyset$. So, X is normal. $\square$

Remark 1.43 The function f is called a "Urysohn function" for the sets C and D. It is not true in general that $f^{-1}(0) = C$ and $f^{-1}(1) = D$. Spaces for which this is true are called "perfectly normal." When (X, d) is a metric space, it is easy to construct a Urysohn function for a pair (C, D) of disjoint closed sets by letting

$$f(x) = \frac{d(x, C)}{d(x, C) + d(x, D)}$$

for all $x \in X$. Notice that $d(x, E) = \inf[d(x, e) : e \in E]$ for $E \subseteq X$. Therefore, a metric space is perfectly normal.

Before passing to the second fundamental theorem characterizing normal spaces, we introduce a notion central in topology.

Definition 1.44 Let X, Y be topological spaces. $f : X \to Y$ is called a "homeomorphism" if it is $1 - 1$ (injective) and onto (surjective), continuous and f^{-1} is continuous. Then we say the two spaces are "homeomorphic" and denote it by $X \sim Y$. If f has all the other properties except surjectivity, then we say that X is embedded in Y and it is denoted by $X \hookrightarrow Y$.

Remark 1.45 A homeomorphism maps open (respectively, closed) sets to open (respectively, closed) sets and $f(\overline{A}) = \overline{f(A)}$. The relation $\sim$ is an equivalence relation. A property of a topological space is a "topological property" if it is preserved by homeomorphisms. Hausdorff, regular, and normal are all topological properties.

Example 1.46 It is clear that $(a, b) \subseteq \mathbb{R}$ is homeomorphic to $(0, 1)$, and homeomorphic to all half lines of the form (a, ∞). All half lines (a, ∞) and $(-\infty, -a)$ are homeomorphic via the homeomorphism $f(x) = -x$. Finally, $\mathbb{R}$ is homeomorphic to $(-\frac{\pi}{2}, \frac{\pi}{2})$ via $f(x) = \sin^{-1} x$. Thus, all intervals, half lines and $\mathbb{R}$ itself, are homeomorphic. Similarly, all bounded closed intervals are homeomorphic.

There is a second characterization of normal spaces in terms of extensions of continuous functions. The result is known in the literature as the "Tietze Extension Theorem."

Theorem 1.47 *A Hausdorff space (X, τ) is normal iff for any closed $C \subseteq X$ and continuous $f : C \to \mathbb{R}$ there exists a continuous extension of f, namely a continuous $\widehat{f} : X \to \mathbb{R}$ such that $\widehat{f}|_C = f$.*

Proof $\Rightarrow$: First assuming $f : C \to [-1, 1]$. The continuity of f implies that the sets $\left\{ f \geq \frac{1}{3} \right\}$ and $\left\{ f \leq -\frac{1}{3} \right\}$ are closed and disjoint. By Theorem 1.42 there exists a continuous function $g_1 : X \to [-\frac{1}{3}, \frac{1}{3}]$ such that

$$g_1|_{\left\{ f \geq \frac{1}{3} \right\}} = \frac{1}{3} \quad \text{and} \quad g_1|_{\left\{ f \leq -\frac{1}{3} \right\}} = -\frac{1}{3}.$$

Then we have $|f(x) - g_1(x)| \leq \frac{2}{3}$ for all $x \in C$. $h_1 = f - g_1$ is a bounded continuous function on C. The sets $\left\{ h_1 \geq \frac{2}{9} \right\}$ and $\left\{ h_1 \leq -\frac{2}{9} \right\}$ are closed and disjoint. So, again by Theorem 1.42 we can find a continuous function $g_2 : X \to [-\frac{2}{9}, \frac{2}{9}]$ such that

$$g_2|_{\left\{ h_1 \geq \frac{2}{9} \right\}} = \frac{2}{9} \quad \text{and} \quad g_2|_{\left\{ h_1 \leq -\frac{2}{9} \right\}} = -\frac{2}{9}.$$

Evidently, for $x \in C$ we have

$$|h_1(x) - g_2(x)| = |f(x) - (g_1 + g_2)(x)| \leq \frac{4}{9}.$$

Inductively we generate a sequence of continuous functions $g_n : X \to \mathbb{R}$ such that for $x \in X$

$$|g_n(x)| \leq \frac{1}{3} \left(\frac{2}{3} \right)^{n-1} \tag{1.1}$$

and for $x \in C$ we have

$$|f(x) - g_n(x)| \leq \left(\frac{2}{3} \right)^n. \tag{1.2}$$

By (1.1), $f_n(x) = \sum_{k=1}^{2} g_k(x)$ converges uniformly to $\widehat{f}(x) = \sum_{k \geq 1} g_k(x)$ and $|\widehat{f}(x)| \leq 1$ for all $x \in X$. By (1.2), $\widehat{f}_C = f$. This proves the theorem when $|f(x)| \leq 1$ for all $\in C$. If $|f(x)| \leq M$ on C, then the conclusion is valid by replacing f by $\frac{1}{M} f$. Finally, suppose that f is unbounded. Recall that $\mathbb{R}$ and $(-1, 1)$ are homeomorphic. Thus we may assume that $f : C \to (-1, 1)$. By the first part of the proof it admits a continuous extension $f_0 : X \to [-1, 1]$. Then the sets C and $C_0 = \{x \in X : |f_0(x)| = 1\}$ are disjoint and closed. Thus by Theorem 1.42 we can find a Urysohn function $h : X \to [0, 1]$ such that $h|_{C_0} = 0$ and $h|_C = 1$. Set $\widehat{f}(x) = h(x) f_0(x)$. Then $\widehat{f} : X \to (-1, 1)$ is continuous and $\widehat{f}|_C = f$.

$\Leftarrow$: For disjoint closed sets $C, D \subseteq X$, $f : C \cup D \to [0, 1]$ defined by $f|_C = 0$ and $f|_D = 1$ is continuous. Thus, f admits a continuous extension $\widehat{f}$, a Urysohn function for the pair $\{C, D\}$. By Theorem 1.42 we conclude that X is normal. □

Definition 1.48 A property of a topological space is called "hereditary" if is inherited by every subset with the subspace topology.

Remark 1.49 The Hausdorff and regularity properties are hereditary, but normality is not. However, closed subsets of a normal space with the subspace topology are normal too.

We now introduce new classes of spaces based on countability properties.

Definition 1.50 A Hausdorff topological space (X, τ) is

(a) "First countable" if it has countable neighborhood basis at each $x \in X$.
(b) "Second countable" if it has a countable basis.
(c) "Separable" if it has a countable dense subset, namely there exists a countable set $D \subseteq X$ such that $X = \overline{D}$.
(d) An "open cover" of a set A is a collection of open sets whose union includes A. X is said "Linderlöf" if every open cover of X has a countable subcover.

The first result is a straightforward consequence of the above definitions.

Proposition 1.51 *(a) A second countable space is separable. (b) An open subset of a second countable space furnished with the subspace topology is Lindelöf.*

Corollary 1.52 *Every second countable space is Lindelöf, thus every open basis has a countable subfamily which is also an open basis.*

Remark 1.53 In general a separable space is not second countable, but for metric spaces the two properties are equivalent.

Proposition 1.54 *A metric space (X, d) is second countable iff it is separable.*

Proof $\Rightarrow$: This follows from Proposition 1.51 (a).
    $\Leftarrow$: Let D be a countable dense subset of X, and

$$\mathcal{Y} = \{B_r(x) : x \in D, r \in \mathcal{Q}, r > 0\}.$$

Then, $\mathcal{Y}$ is countable. We claim that $\mathcal{Y}$ is an open basis for the metric topology of X. Let $U \subseteq X$ be open and $x \in U$. Choose $r > 0$ such that $B_r(x) \subseteq U$, $u \in B_{r/3}(x) \cap D$ (possible since D is dense in X) and $s \in (r/3, 2r/3) \cap \mathcal{Q}$. Then, $x \in B_s(u) \subseteq B_r(x) \subseteq U$. Therefore, $\mathcal{Y}$ is a countable open basis for X. □

Remark 1.55 Every metric space is first countable. In first countable spaces, sequences suffice to describe all topological notions and there is no need to consider nets.

Proposition 1.56 *If (X, τ) is a regular Lindelöf topological space, then X is normal.*

Proof Let C, D be disjoint closed subsets of X. Since X is regular, by Proposition 1.41 for any $x \in C$ there exists $U_x \in \mathcal{N}(x)$ such that $\overline{U_x} \cap D = \emptyset$. Similarly, for any $u \in D$ there exists $V_u \in \mathcal{N}(u)$ such that $\overline{V_u} \cap C = \emptyset$. Then $\{U_x\}_{x \in C} \cup \{V_u\}_{u \in D} \cup (X \setminus (C \cup D))$ is an open cover of X. Since X is Lindelöf we can find countable subcollections $\{U_n\}_{n \geq 1}$ of $\{U_x\}_{x \in X}$ and $\{V_n\}_{n \geq 1}$ of $\{V_u\}_{u \in D}$ such that $C \subseteq \cup_{n \geq 1} U_n$ and $D \subseteq \cup_{n \geq 1} V_n$ (Proposition 1.51 (b)). For any $n \in \mathbb{N}$, define the open sets

$$U_n^* = U_n \setminus \cup_{k=1}^n \overline{V_k} \text{ and } V_n^* = V_n \setminus \cup_{k=1}^n \overline{U_k}.$$

We have $U_n^* \cap V_m^* = \emptyset$ for all $n, m \in \mathbb{N}$, $C \subseteq \cup_{n \geq 1} U_n^* = U$ and $D \subseteq \cup_{n \geq 1} V_n^* = V$. Finally, observe that $U \cap V = \emptyset$. This proves the normality of X. □

1.4 Weak, Product, and Quotient Topologies

Let X be a set and τ_1, τ_2 two topologies on X such that $\tau_1 \subseteq \tau_2$. Then we say that τ_1 is "weaker" than τ_2 or alternatively that τ_2 is "stronger" than τ_1. Evidently, the trivial topology $\{\emptyset, X\}$ is the weakest topology on X, while the discrete topology, 2^X, is the strongest topology on X. The intersection of any two topologies is clearly a topology on X. In fact the intersection of any nonempty family of topologies on X is still a topology on X. This topology is the greatest lower bound of the family. On the other hand, the union of two topologies need not to be a topology. However, if $\{\tau_i\}$ is a family of topologies, then the intersection of all topologies that are stronger than each τ_i is itself a topology stronger than each τ_i. It is the least upper bound of the family. So, the set of topologies is a complete lattice, with a least element (the trivial topology) and a greatest element (the discrete topology).

Next let X be a set and $\{(Y_i, \tau_i)\}_{i \in I}$ a family of topological spaces. Let a function $f_i : X \to Y_i$ be given for each $i \in I$. Then all f_i are continuous if X is equipped with the discrete topology. However, we may find some weaker topologies on X

which do the same job. In fact, there is a unique topology of this kind, which is the intersection of all topologies on X which make all the functions f_i continuous. This is the "weak topology" on X induced by the family $\{f_i\}_{i \in I}$.

Definition 1.57 Let X be a set, $\{(Y_i, \tau_i)\}_{i \in I}$ a family of topological spaces and $f_i : X \to Y_i, i \in I$ are functions. The "weak topology" or "initial topology" on X induced by the family $\{f_i\}$ is the weakest topology on X which makes each f_i continuous. Therefore, it is the topology generated by the subbasis

$$\left\{ f_i^{-1}(V) : i \in I, V \in \tau_i \right\}.$$

In fact, if $\Im_i \subseteq \tau_i$ is a subbasis for each $i \in I$, then the weak topology is generated by the even smaller subbasis

$$\left\{ f_i^{-1}(V) : i \in I : V \in \Im_i \right\}.$$

All the finite intersections of these sets form a basis for the weak topology. We denote the weak topology by w or $w(\{f_i\})$ if we wish to emphasize the generating family of functions.

Remark 1.58 For $Y_i = \mathbb{R}$ for each $i \in I$, a subbasis for $w(\{f_i\})$ is given by the sets

$$U_i(x, \epsilon) = \{u \in X : |f_i(u) - f_i(x)| < \epsilon\}$$

with $i \in I, x \in X, \epsilon > 0$.

Lemma 1.59 *A net $\{x_\alpha\}_{\alpha \in A} \subseteq X$ converges to x for $w(\{f_i\})$ iff for each $i \in I$*

$$f_i(x_\alpha) \stackrel{\tau_i}{\to} f_i(x).$$

Proof $\Rightarrow$: Since each f_i is w-continuous and $x_\alpha \stackrel{w}{\to} x$, we have $f_i(x_\alpha) \to f_i(x)$ for each $i \in I$ (see Proposition 1.34).

$\Leftarrow$: Let $U = \cap_{k=1}^n f_{i_k}^{-1}(V_{i_k})$ be a basic w-neighborhood of x, with $V_{i_k} \in \tau_{i_k}$ for all $k \in \mathbb{N}$. We can find $\alpha_{i_k} \in A$ such that $x_\alpha \in f_{i_k}^{-1}(V_{i_k})$ for $\alpha \succcurlyeq \alpha_{i_k}$. Choose $\widehat{\alpha} \succcurlyeq \alpha_{i_k}$ for all $k = 1, \cdots, n$. Then $x_\alpha \in U$ for $\alpha \succcurlyeq \widehat{\alpha}$, and $x_\alpha \stackrel{w}{\to} x$ □

Proposition 1.60 *Let $(X, w(\{f_i\}_{i \in I})$ and (Y, τ) topological spaces and $g : Y \to X$. Then g is continuous iff $f_i \circ g : Y \to X_i$ is continuous for each $i \in I$.*

Proof $\Rightarrow$: This implication is true since composition of continuous functions is continuous.

$\Leftarrow$: Let $U = \cap_{k=1}^n f_{i_k}^{-1}(V_{i_k})$, with $V_{i_k} \in \tau_{i_k}$, be basic elements for the w-topology on X. Then,

$$g^{-1}(U) = \bigcap_{k=1}^{n} g^{-1}(f_{i_k}^{-1}(V_{i_k})) = \bigcap_{k=1}^{n} (f_{i_k} \circ g)^{-1}(V_{i_k}) \in \tau$$

and so, g is continuous.   □

Definition 1.61 We say that a family of functions $f_i : X \rightarrow \mathbb{R}$, $i \in I$, is "separating" if for any $x \neq u$, there exists $i \in I$ such that $f_i(x) \neq f_i(u)$.

The following result is an easy consequence of the above definition.

Proposition 1.62 *The weak topology $w(\{f_i\}_{i\in I})$ on X is Hausdorff iff the defining family $\{f_i\}_{i\in I}$ is separating.*

Let X be a set, $f_i : X \rightarrow \mathbb{R}$, $i \in I$, and $A \subseteq X$. Then, on A we can consider two topologies. The first, w_1, is the restriction of $w(\{f_i\})$ on A; while the second, w_2, is induced by $f_i|_A$ for $i \in I$. Since the two topologies on A have the same convergent nets by Lemma 1.59, we have

Proposition 1.63 $w_1 = w_2$.

Definition 1.64 Let (X, τ) be a topological space. Define

$$C(X) = \{f : X \rightarrow \mathbb{R} \text{ is } \tau - \text{continuous}\}$$

and

$$C_b(X) = \{f \in C(X) : f \text{ is bounded}\}.$$

Remark 1.65 Both are vector spaces and $C_b(X) \subseteq C(X)$.

Proposition 1.66 *Let (X, τ) is a topological space, $w(X, C(X))$ the weak topology induced by $C(X)$, and $w(X, C_b(X))$ the weak topology induced by $C_b(X)$. Then, $w(X, C(X)) = w(X, C_b(X))$.*

Proof Obviously, we have $w(X, C_b(X)) \subseteq w(X, C(X))$. Let U be a subbasic element for the $w(X, C(X))$-topology. Then by Remark 1.58 we have

$$U = U(f, x, \epsilon) = \{u \in X : |f(u) - f(x)| < \epsilon\}$$

with $f \in C(X), x \in X, \epsilon > 0$. Consider $g : X \rightarrow \mathbb{R}$ defined by

$$g(u) = \min\{f(x) + \epsilon, \max\{f(x) - \epsilon, f(u)\}\}.$$

Evidently, $g \in C_b(X)$ and $U(g, x, \epsilon) = U(f, x, \epsilon)$, thus $w(X, C(X)) \subseteq w(X, C_b(X))$, hence $w(X, C(X)) = w(X, C_b(X))$.   □

Proposition 1.67 *A Hausdorff space* (X, τ) *is completely regular iff* $\tau = w(X, C(X)) = w(X, C_b(X))$.

Proof $\Rightarrow$: Let $U \in \tau$ and $x \in U$. Then we can find $f \in C(X)$ such that $f(x) = 0$ and $f|_{X \setminus U} = 1$ (see Definition 1.36(c)). Let $V = \{u \in X : f(u) < 1\}$. Then $x \in V \in \tau, V \subseteq U$ and so $\tau \subseteq w(X, C(X)) = w(X, C_b(X))$. But by Definition 1.61 we have $w(X, C(X)) \subseteq \tau$. Therefore, we conclude that $\tau = w(X, C(X)) = w(X, C_b(X))$.

$\Leftarrow$: Assume that $\tau = w(X, C(X)) = w(X, C_b(X))$. Let $C \subseteq X$ be closed and $x \notin C$. We have $V = X \setminus C \in \tau$ and so we can find

$$U = \bigcap_{k=1}^{n} \{u \in X : |f_k(u) - f_k(x)| \leq 1\} \subseteq V, \ \{f_k\}_{k=1}^{n} \subseteq C(X).$$

Let $g_k(u) = \min\{|f_k(u) - f_k(x)|, 1\}$ for all $u \in X$ and $g = \max_{q \leq k \leq n} g_k$. Then $g : X \to [0, 1]$ and is continuous. Note that $g(x) = 0$ and $g|_C = 1$. Therefore, X is completely regular. $\square$

Corollary 1.68 *In a completely regular topological space* (X, τ), $x_i \to x$ *in* X *iff* $f(x_i) \to f(x)$ *for all* $f \in C(X)$ *or* $f \in C_b(X)$.

An important particular case of weak topology is the so-called product topology, defined on Cartesian products of sets.

Definition 1.69 For a family of sets: $\{X_i\}_{i \in I}$, their "Cartesian product" is the set

$$\prod_{i \in I} X_i = \left\{ x : I \to \bigcup_{i \in I} X_i : x(i) \in X_i \text{ for each } i \in I \right\}.$$

Usually, instead of $x(i)$ we write x_i and this element is referred to as the "ith-coordinate" of x. The space X_i is known as the "ith-factor space" and the map $p_j : \prod_{i \in I} X_i \to X_j$ defined by $p_j(x) = x_j$ is known as the "jth-projection map."

Remark 1.70 We need the axiom of choice to prove that $\prod_{i \in I} X_i \neq \emptyset$.

Evidently, every projection map is surjective.

Definition 1.71 Let $\{(X_i, \tau_i)\}_{i \in I}$ be a family of topological spaces and $X = \prod_{i \in I} X_i$. The "product topology," τ, is the weak topology on X induced by the family $\{p_j\}_{i \in I}$ of projections.

Remark 1.72 A subbasis for the product topology is given by the sets of the form: $p_j^{-1}(V_j) = \prod_{i \in I} V_i$, where $V_i = X_i$ for all $i \neq j$ and $V_j \in \tau_j$. A basis for the product topology consists of all sets of the form: $V = \prod_{i \in I} V_i$, where $V_i \in \tau_i$ and $V_i = X_i$ for all $i \in J$ with a finite $J \subseteq I$.

Next we introduce a topology dual to the notion of the weak topology. Now we have topological spaces $\{(Y_i, \tau_i)\}_{i \in I}$, Y is a set and $g_i : Y_i \to Y$. We introduce the notion of "strong topology" induced on Y by the family of functions $\{g_i\}_{i \in I}$, which is the strongest topology on Y making all these functions continuous. When the family consists of only one function $g : X \to Y$, the strongest topology induced by g is called the "quotient topology."

Definition 1.73 Let (X, τ) be a topological space, Y a set and $g : X \to Y$ surjective. We declare that $U \subseteq Y$ is open if $g^{-1}(U) \in \tau$. This collection of open sets in Y is a topology on Y, called the "quotient topology induced by g." Then we say that Y is a "quotient space" of X and the inducing map g is a "quotient map."

Remark 1.74 Evidently, the quotient topology on Y induced by g is the biggest topology on Y making g continuous. Also, the quotient topology induced by g can be defined as: $C \subseteq Y$ is closed in the quotient topology induced by g iff $g^{-1}(C) \subseteq X$ is closed.

Suppose that (X, τ) is a topological space and $\sim$ is an equivalent relation on X. Let $X/\sim$ denote the set of all equivalence classes $[x] = \{u \in X : x \sim u\}$, determined by $\sim$. A collection $\mathcal{D}$ of equivalence classes is an "open" set in $X/\sim$ if the union of the members of $\mathcal{D}$ is an open set in X. These open sets form the quotient topology of $X/\sim$ and the set $X/\sim$ with this topology is called the "quotient space modulo $\sim$" of X. A comparison with Definition 1.73 reveals that the quotient topology on $X/\sim$ coincides with the quotient topology induced by the quotient map $p : X \to X/\sim$, where $p(x) = [x]$ for $x \in X$.

Now let (X, τ_X) and (Y, τ_Y) be topological spaces and $f : X \to Y$ a continuous map. Under what conditions on f will the preassigned topology τ_Y on Y coincide with the quotient topology τ_f on Y induced by f?

Definition 1.75 $f : X \to Y$ is said to be "open" (respectively, "closed") if for every $U \in \tau_X$ (respectively, closed $C \subseteq X$), we have $f(U) \in \tau_Y$ (respectively, $f(C)$ is closed in Y).

Remark 1.76 An open (respectively, closed) function need not to be continuous. In fact, a function which is both open and closed need not to be continuous. Also, an open function need not be closed and conversely. A continuous function need not be open nor closed. However, a homeomorphism is both open and closed and so is its inverse.

Proposition 1.77 *Let $(X, \tau_X), (Y, \tau_Y)$ be topological spaces and $f : X \to Y$ continuous, which is onto and is either open or closed. Then, $\tau_Y = \tau_f$.*

Proof Let $U \in \tau_Y$, The continuity of f implies that $f^{-1}(U) \in \tau_X$ and so, $U \in \tau_f$. Therefore, $\tau_Y \subseteq \tau_f$. Suppose that f is open and let $V \in \tau_f$. Then $f^{-1}(V) \in \tau_X$. Since f is open, we have $f(f^{-1}(V)) = V \in \tau_Y$. Thus, $\tau_f \subseteq \tau_Y$. We conclude that $\tau_Y = \tau_f$. Similarly if f is closed. $\square$

The next proposition is an analog of Proposition 1.60 for the quotient topology.

Proposition 1.78 *Let* $(X, \tau_X), (Z, \tau_Z)$ *be topological spaces,* $f : X \rightarrow Y$ *a surjection,* Y *a set furnished with the quotient topology* τ_f *induced by* f. *Let* $g : Y \rightarrow Z$. *Then,* g *is continuous iff* $g \circ f : X \rightarrow Z$ *is continuous.*

1.5   Compact and Locally Compact Spaces

Compactness is one of the fundamental notions of Analysis.

Definition 1.79 Let (X, τ) be a Hausdorff space and $A \subseteq X$.

(a) An "open cover" of A is a family $\mathcal{S} \subseteq \tau$ such that $A \subseteq \bigcup_{U \in \mathcal{S}} U$. A "subcover" originating from an open cover $\mathcal{S}$ is a subfamily $\mathcal{S}'$ of $\mathcal{S}$ with $A \subseteq \bigcup_{U' \in \mathcal{S}'} U'$.
(b) X is "compact" if every open cover of X has a finite subcover. A subset of X is "compact" if it is a compact topological space in its subspace topology.

Remark 1.80 Many authors do not require that X be Hausdorff. Here we follow the Bourbaki school. From the definition of the subspace topology (Definition 1.15) we see that $A \subseteq X$ is compact iff every open cover of A by open sets in X has a finite subcover. Dually to the above notions, a family of subsets of $2^X \setminus \{\emptyset\}$ has the "finite intersection property" if every finite subfamily has a nonempty intersection.

The next proposition provides some useful characterizations of compactness.

Proposition 1.81 *For a Hausdorff space* (X, τ), *the following statements are equivalent:*

(a) X is compact.
(b) Every family of closed subsets having the finite intersection property has a nonempty intersection.
(c) Every net in X has a convergent subnet.

Proof (a) $\Leftrightarrow$ (b): Let us assume that X is compact and let $\mathcal{A} \subseteq 2^X \setminus \{\emptyset\}$ be a family of closed sets with the finite intersection property. If $\bigcap \{C : C \in \mathcal{A}\} = \emptyset$, then $X = \bigcup \{X \setminus C : C \in \mathcal{A}\}$. But this is an open cover of the compact X, it has a finite subcover. So, we have $X = \bigcup_{k=1}^n (X \setminus C_k)$, with $C_k \in \mathcal{A}$. Then $\bigcap_{k=1}^n C_k = \emptyset$, contradicting the fact that $\mathcal{A}$ has the finite intersection property.

Now we assume (b). Let $\mathcal{S} \subseteq \tau$ be an open cover of X. Then $\bigcap_{U \in \mathcal{S}} (X \setminus U) = \emptyset$, which means $\mathcal{S}$ does not have the finite intersection property. Thus, there exist $U_1, \cdots, U_k \in \mathcal{S}$ such that $\bigcap_{k=1}^n (X \setminus U_k) = \emptyset$, hence $X = \bigcup_{k=1}^n U_k$ and therefore we conclude that X is compact.

(a) $\Leftrightarrow$ (c): Suppose X is compact and $\{x_\alpha\}_{\alpha \in I}$ is a net. According to Proposition 1.31. it suffices to show that $\{x_\alpha\}$ has a cluster point. If no cluster point exists, then for any $x \in X$ there exists $U_x \in \mathcal{N}(x)$ open and $\alpha_x \in I$ such that $x_\alpha \notin U$ for all $\alpha \succcurlyeq \alpha_x$. Then the family $\mathcal{S} = \{U_x\}_{x \in X}$ is an open cover of X. The compactness of X implies that we can find $x_1, \cdots, x_n \in X$ such that $X = \cup_{k=1}^n U_{x_k}$. Since I is directed we can find $\alpha_0 \succcurlyeq \alpha_{x_k}$ for all $k = 1, \cdots, n$. But then $x_{\alpha_0} \notin U_{x_k}$ for all $k = 1, \cdots, n$, which is a contradiction since $\{U_{x_k}\}_{k=1}^n$ is a finite subcover. So, $\{x_\alpha\}_{\alpha \in I}$ has a cluster point, thus a convergent subset.

Now suppose that every net has a convergent subnet. Arguing by contradiction, suppose that X is not compact. Then we can find an open cover S with no finite subcover. Let $\mathcal{D}$ be the collection of all finite subfamilies of S. We order $\mathcal{D}$ by inclusion to obtain a directed set. For each $\mathcal{F} = \{U_k\}_{k=1}^n \in \mathcal{D}$ we choose $u_{\mathcal{F}} \notin \cup_{k=1}^n U_k$. Then we have generated a net $\{u_{\mathcal{F}}\}$ and by hypothesis it has a cluster point u. Let $u \in U \in S$. We can find $\mathcal{F} \in \mathcal{D}$ such that $\{U\} \preccurlyeq \mathcal{F}$ and $u_{\mathcal{F}} \in U$. Then $U \subseteq \cup_{V \in \mathcal{F}} V$ and $u_{\mathcal{F}} \in \cup_{V \in \mathcal{F}} V$, which is a contradiction. $\qquad\square$

Proposition 1.82 *In a compact topological space (X, τ), $A \subseteq X$ is compact iff it is closed.*

Proof $\Rightarrow$: Consider a net $\{x_\vartheta\}_{\vartheta \in D} \subseteq A$ and assume that $x_\vartheta \to x$ in X. Then Propositions 1.81 and 1.32 imply that $x \in A$ and so A is closed.

$\Leftarrow$: Let S be a cover of A by open sets of X. Then $S^* = S \cup \{X \setminus A\}$ is an open cover of X. We can find a finite subcover. This will consist of $\{U_k\}_{k=1}^n \subseteq S$ and may contain also $X \setminus A$. Then,

$$X = \left(\cup_{k=1}^n U_k\right) \cup (X \setminus A),$$

thus $A \subseteq \cup_{k=1}^n U_k$ and A is compact. $\qquad\square$

Proposition 1.83 *Let (X, τ) be a Hausdorff space, $A, C \subseteq X$ disjoint compact sets. Then there are disjoint $U, V \in \tau$ such that $A \subseteq U$ and $C \subseteq V$.*

Proof Let $x \in C$. For $u \in A$ there exist disjoint $U_u, V_u \in \tau$ such that $u \in U_u$ and $x \in V_u$. $\{U_u\}_{u \in A}$ is an open cover of A and by compactness there is a finite subcover $\{U_{u_k}\}_{k=1}^n$. Let

$$U_x = \cup_{k=1}^n U_{u_k} \in \tau \text{ and } V_x = \cap_{k=1}^n V_{u_k} \in \tau.$$

Evidently, these are disjoint sets and $A \subseteq U_x$ and $x \in V_x$. The family $\{V_x\}_{x \in C}$ is an open cover of C and so, there is a finite subcover $\{V_{x_k}\}_{k=1}^m$. We set

$$U = \cap_{k=1}^m U_{x_k} \in \tau \text{ and } V = \cup_{k=1}^m V_{x_k} \in \tau.$$

These sets are disjoint, $A \subseteq U$ and $C \subseteq V$. $\qquad\square$

Corollary 1.84 *Let (X, τ) be a compact topological space, and $A, C \subseteq X$ disjoint closed sets. Then there exist disjoint $U, V \in \tau$ such that $A \subseteq U$ and $C \subseteq V$.*

Compactness is important in analysis because it is closely related to the notion of continuity.

Proposition 1.85 *Let (X, τ_Y) and (Y, τ_Y) be Hausdorff spaces, $f : X \to Y$ continuous and $A \subseteq X$ compact. Then, $f(A) \subseteq Y$ is compact.*

Proof Let $\{V_i\}_{i \in I} \subseteq \tau_Y$ be an open cover of $f(A)$. Then $\{f^{-1}(V_i)\}_{i \in I} \subseteq \tau_X$ is an open cover of A. There exist $i_1, \cdots, i_n \in I$ such that $A \subseteq \cup_{k=1}^n f^{-1}(V_{i_k})$. Then we have

$$f(A) \subseteq f(\cup_{k=1}^n f^{-1}(V_{i_k})) = \cup_{k=1}^n f(f^{-1}(V_{i_k})) \subseteq \cup_{k=1}^n V_{i_k},$$

therefore, $f(A)$ is compact.   □

Corollary 1.86 *Every continuous function on a compact topological space achieves its maximum and minimum values.*

Corollary 1.87 *Compactness is a topological property (Remark 1.45).*

Proposition 1.88 *Let (X, τ_X) and (Y, τ_Y) be Hausdorff spaces, X compact, and $f : X \to Y$ is continuous and bijective. Then f is a homeomorphism.*

Proof Let $C \subseteq X$ be closed, hence compact and $f(C) \subseteq Y$ is compact. Therefore, f^{-1} is continuous and so, f is a homeomorphism.   □

Definition 1.89 A Hausdorff topological space has the "Bolzano–Weierstrass property" if every infinite subset has a cluster point.

Proposition 1.90 *All of compact topological spaces have the Bolzano–Weierstrass property.*

Remark 1.91 The converse is not true in general. However, for metric spaces the two notions are equivalent (see Theorem 1.96).

Definition 1.92 Let (X, d) be a metric space. $A \subseteq X$ is said to be "totally bounded" if for any $\epsilon > 0$ there exists a finite "ϵ-dense" set $\{x_k\}_{k=1}^n \subseteq A$. Namely, $\{B_\epsilon(x_k) = \{x \in X : d(x, x_k) < \epsilon\}\}_{k=1}^n$ is an open cover of A.

Remark 1.93 Apparently, a compact metric space is totally bounded. Moreover, a totally bounded metric space is separable. Therefore, a compact metric space is separable.

The next result is known as "Cantor's Intersection Theorem." Recall that for a metric space (X, d) and $A \subseteq X$, the diameter of A is defined as $\operatorname{diam} A = \sup\{d(x, u) : x, u \in A\}$.

Proposition 1.94 *Let (X, d) be a complete metric space and $\{C_n\}_{n \geq 1}$ a decreasing sequence of closed subsets of X such that $\operatorname{diam} C_n \to 0$ as $n \to \infty$. Then, $\cap_{n \geq 1} C_n$ is a singleton.*

Proof Let $x_n \in C_n$. Then $d(x_n, x_m) \le \operatorname{diam} C_n$ for $m \ge n$. So, $\{x_n\}_{n \ge 1}$ is Cauchy and the completeness of X implies $x_n \xrightarrow{d} x \in X$. Clearly, $x \in \cap_{n \ge 1} C_n \neq \emptyset$. If $x, u \in \cap_{n \ge 1} C_n$, with $x \neq u$, then $0 < d(x, u) \le \operatorname{diam} C_n$ for all $n \ge 1$, a contradiction since $\operatorname{diam} C_n \to 0$. Therefore, $\cap_{n \ge 1} C_n$ is a singleton. $\square$

Proposition 1.95 *Let (X, d) be a compact metric space and $S = \{U_\alpha\}_{\alpha \in I}$ an open cover of X. Then there exists $\varrho > 0$ such that each ball $B_\varrho(y)$ is contained in some U_α. This $\varrho > 0$ is called the "Lebesgue number" of the cover.*

Proof For $x \in X$, choose $r(x) > 0$ such that $B_{r(x)}(x) \subseteq U_\alpha$ for some $\alpha \in I$. Then we obtain the open cover $\{B_{r(x)/2}(x)\}_{x \in X}$ of X and by compactness we can find a finite subcover $\{B_{r(x_k)/2}(x_k)\}_{k=1}^n$. Let $\varrho = \min\left\{\frac{r(x_k)}{2}\right\}_{k=1}^n > 0$. For $u \in B_\varrho(x)$, $u \in B_{r(x_k)/2}(x_k)$ for some $k \in \{1, \cdots, n\}$ and then for any $v \in B_\varrho(u)$ we have

$$d(v, x_k) \le d(v, u) + d(u, x_k) < \varrho + \frac{r(x_k)}{2} \le r(x_k).$$

Thus, $B_\varrho(x) \subseteq B_{r(x_k)/2}(x_k) \subseteq U_\alpha$ for some $\alpha \in I$. $\square$

Theorem 1.96 *For a metric space (X, d), the following statements are equivalent:*

(a) X is compact.
(b) X is complete and totally bounded.
(c) X is sequentially compact, namely every sequence has a convergent subsequence.

Proof (a) $\Rightarrow$ (b): We already know that X is totally bounded. Let $\{x_n\} \subseteq X$ be Cauchy and $\epsilon > 0$. There exists $n_0 \in \mathbb{N}$ such that $d(x_n, x_m) \le \epsilon$ for all $n, m \ge n_0$. By Proposition 1.90, $\{x_n\}_{n \ge 1}$ has cluster point x. Choose $m \ge n_0$ such that $d(x_m, x) < \epsilon$. Then for $n \ge n_0$ we have

$$d(x_n, x) \le d(x_n, x_m) + d(x_m, x) < 2\epsilon,$$

thus $x_n \xrightarrow{d} x$. So, (X, d) is complete.

(b) $\Rightarrow$ (c): Consider $\{x_n\}_{n \ge 1}$. Since X is totally bounded, there is a closed ball of radius $1/2$ containing infinitely many elements of the sequence. Since this closed ball is also totally bounded, in it there is a closed $\frac{1}{4}$-ball containing infinitely many elements of the sequence. Inductively we generate a sequence of closed balls with vanishing radii. Then Proposition 1.94 implies a convergent subsequence.

(c) $\Rightarrow$ (a): Let $\{U_\alpha\}_{\alpha \in I}$ be an open cover of X and $\varrho > 0$ the Lebesgue number of the cover. We claim that there exist $x_1, \cdots, x_n \in X$ such that $X = \cup_{k=1}^n B_\varrho(x_k)$, which of course implies the open cover has a finite subcover. If the claim is not true, then for fixed $x_1 \in X$ we can find $x_2 \in X$ such that $d(x_1, x_2) \ge \varrho$. Since $X \neq B_\varrho(x_1) \cup B_\varrho(x_2)$, there is $x_3 \in X$ such that $d(x_1, x_3) \ge \varrho$ and $d(x_2, x_3) \ge \varrho$.

Inductively, we can produce a sequence $\{x_n\}_{n\geq 1} \subseteq X$ such that $d(x_n, x_m) \geq \varrho$ for all $n \neq m$. However, such a sequence cannot have a convergent subsequence, a contradiction.   □

Definition 1.97 Let (X, d_X) and (Y, d_Y) be metric spaces and $f : X \to Y$. We say that f is "uniformly continuous" if for any $\epsilon > 0$ there is a $\delta > 0$ such that $d_Y(f(x), f(u)) < \epsilon$ whenever $d_X(x, u) < \delta$.

Remark 1.98 As uniformly continuous function is continuous, while the converse is not true. Uniform continuity is a global notion, while continuity is a local one.

Proposition 1.99 *Let (X, d_X) be a compact metric space, (Y, d_Y) a metric space and $f : X \to Y$ is continuous. Then f is uniformly continuous.*

Proof Let $\epsilon > 0$ and $x \in X$. Then $U_x = f^{-1}(B_{\epsilon/2}(f(x)))$ is open in X. For $u, v \in U_x$, $d_Y(f(u), f(v)) < \epsilon$. $\{U_x\}_{x\in X}$ is an open cover of X. Let $\varrho > 0$ be the Lebesgue number of this cover. Then for each $u \in X$ we have $B_\varrho(u) \subseteq U_x$ for some $x \in X$. Therefore, $v \in B_\delta(u)$ implies $d_Y(f(v), f(u)) < \epsilon$, which shows that f is uniformly continuous.   □

Remark 1.100 Though compactness is a topological property, for metric spaces the notions of completeness and total boundedness are not.

The next theorem is one of the most important theorems in mathematical analysis, known as "Tychonov's theorem." First, we have a lemma.

Lemma 1.101 *Let (X, τ) be a Hausdorff space and $S \subseteq 2^X$ has the finite intersection property. Then there exists a maximal collection $S^* \subseteq 2^X$ with the finite intersection property and $S \subseteq S^*$. Moreover, S^* is closed under finite intersections and if $C \subseteq X$ intersects every element in S^*, then $C \in S^*$.*

Proof Consider the family $\mathcal{F}$ of all collections in 2^X, which contain S and have the finite intersection property. We partially order this family by inclusion. By the Hausdorff Maximal Principle (see Hewitt–Stromberg [137], p.14), there is a maximal chain $\mathcal{F}_0$ in $\mathcal{F}$. Let $S^* = \cup_{S'\in\mathcal{F}_0} S'$, which is the desired maximal collection containing S and closed under finite intersections. Finally, let $C \subseteq X$ intersect every element in S'. Then the collection $S^* \cup \{C\}$ has the finite intersection property and contains S. Therefore, $S^* \cup \{C\} = S^*$, that is, $C \in S^*$.   □

Now we are ready for the Tychonov Theorem.

Theorem 1.102 *Let $\{X_\alpha\}_{\alpha\in I}$ is a family of Hausdorff topological spaces. Then $X = \prod_{\alpha\in I} X_\alpha$, furnished with the product topology, is compact iff X_α is compact for each $\alpha \in I$.*

Proof $\Rightarrow$: Recall that the projection map, $p_\alpha : X \to X_\alpha$, is continuous (see Definition 1.71). By Proposition 1.85, $X_\alpha = p_\alpha(X)$ is compact.

$\Leftarrow$: Let $\mathcal{S}$ be the collection of closed sets of X with the finite intersection property. Let $\mathcal{S}^*$ be the maximal collection containing $\mathcal{S}$ as postulated by Lemma 1.101. The elements of $\mathcal{S}^*$ need not be closed. Define

$$\mathcal{S}^*_\alpha = \left\{ p_\alpha(A) : A \in \mathcal{S}^* \right\}.$$

The collection has the finite intersection property in X_α. So, the set

$$\overline{\mathcal{S}^*_\alpha} = \left\{ \overline{p_\alpha(A)} : A \in \mathcal{S}^* \right\}$$

has a nonempty intersection for each $\alpha \in I$. Let

$$u_\alpha \in \bigcap_{A \in \cap \mathcal{S}^*} \overline{p_\alpha(A)} \subseteq X_\alpha, \quad u = (u_\alpha)_{\alpha \in I} \in \prod_{\alpha \in I} X_\alpha.$$

We claim that $u \in \overline{A}$ for all $A \in \mathcal{S}^*$. To this end, let $U \in \mathcal{N}(u)$ open. Then we can find $\{\alpha_k\}_{k=1}^n \subseteq I$ and open $U_{\alpha_k} \subseteq X_{\alpha_k}$ such that $u \in \cap_{k=1}^n p_{\alpha_k}^{-1}(U_{\alpha_k}) \subseteq U$. Since $x_{\alpha_k} \in \overline{C}$ for all $C \in \mathcal{S}^*_{\alpha_k}$, we have

$$U_{\alpha_k} \cap C \neq \emptyset$$

for all $C \in \mathcal{S}^*_{\alpha_k}$, thus $p_{\alpha_k}^{-1}(U_{\alpha_k}) \cap A \neq \emptyset$ for all $A \in \mathcal{S}^*$. Hence, $p_{\alpha_k}^{-1}(U_{\alpha_k}) \in \mathcal{S}^*$ for all $k = 1, \cdots, n$ (see Lemma 1.101).

Also, the finite intersection $\cap_{k=1}^n p_{\alpha_k}^{-1}(U_{\alpha_k})$ intersects every element of $\mathcal{S}^*$ and so, it belongs to $\mathcal{S}^*$ (see Lemma 1.101). Thus, $U \cap \mathcal{S}^* \neq \emptyset$ and so $u \in \overline{A}$ for all $A \in \mathcal{S}^*$, which means that X is compact (see Proposition 1.81). $\qquad\square$

It is useful to know when a given Hausdorff space can be embedded in a compact space, which is closely related to a local compactness property of the space.

Definition 1.103 A Hausdorff space (X, τ) is "locally compact" if for each $x \in X$ there exists an open set $U \in \mathcal{N}(x)$ such that $\overline{U}$ is compact.

Remark 1.104 Evidently, local compactness is a topological property. Also, a compact space is automatically locally compact, but the converse is not true (think of $X = \mathbb{R}^N$). In fact, for a topological vector space (that is, a vector space equipped with a Hausdorff topology for which vector addition and scalar multiplication are continuous), local compactness is equivalent to finite dimensionality.

Proposition 1.105 *Let (X, τ) be a Hausdorff space. The following statements are equivalent:*

(a) *X is locally compact.*
(b) *For any $x \in X$ and $U \in \mathcal{N}(x)$, there exists open $V \subseteq X$ with $\overline{V}$ compact such that $x \in V \subseteq \overline{V} \subseteq U$.*

(c) *For any compact $K \subseteq X$ and open $U \supseteq K$, there exists open $V \subseteq X$ with $\overline{V}$*
     compact such that $K \subseteq V \subseteq \overline{V} \subseteq U$.
(d) *There exists a basis of X consisting of open sets with compact closures.*

Proof (a) $\Rightarrow$ (b): By Definition 1.103 we can find open $W \in \mathcal{N}(x)$ with $\overline{W}$ compact hence regular (see Corollary 1.84) and $\overline{W} \cap U$ is a neighborhood of x in $\overline{W}$. By Proposition 1.38 we can find open $E \subseteq \overline{W}$ such that $x \in E \subseteq \overline{E}^{\tau_{\overline{W}}} \subseteq \overline{W} \cap U$. We know that $E = D \cap \overline{W}$ for some $D \in \tau$. Set $V = D \cap W$, which is the desired open neighborhood of X with compact closure.

   (b) $\Rightarrow$ (c): For each $x \in K$ we can find open $V_x \in \mathcal{N}(x)$ with compact closure. Then $\{V_x\}_{x \in K}$ is an open cover of K and so we can find $\{V_{x_k}\}_{k=1}^{n}$ a finite subcover. Let $V = \cup_{k=1}^{n} V_{x_k}$. Then $V \in \tau$, $\overline{V}$ is compact and $K \subseteq V \subset \overline{V} \subseteq U$.

   (c) $\Rightarrow$ (d): Let $\mathcal{B}$ be the family of all open sets in X with compact closure. Since the singleton $\{x\}$ is compact, from (c) we infer that $\mathcal{B}$ is a basis for the topology τ.

   (d) $\Rightarrow$ (a): Immediate from Definition 1.103. $\square$

Remark 1.106 A set with compact closure is called "relatively compact."

The following proposition is an easy consequence of Definition 1.103.

Proposition 1.107 *Every open and every closed subset of a locally compact space, furnished with the subspace topology, is locally compact.*

Let X be a Hausdorff space. A "compactification" of X is a compact space Y such that X is homeomorphic to a dense subset of Y. In fact, we can say that X itself is a dense subset of Y. Locally compact spaces admit a particularly simple compactification, known as the "Alexandrov one-point compactification."

Definition 1.108 Let (X, τ) be a locally compact space and $\infty \notin X$. Define $Y = X \cup \{\infty\}$, furnished with a topology, τ_Y, consisting of the following sets:

(a) Every $U \in \tau$ belongs in τ_Y.
(b) Every set $Y \setminus K$, with $K \subseteq X$ compact, belongs in τ_Y.

   (Y, τ_Y) is called the "Alexandrov one-point compactification" of X.

We now verify that indeed this is a compactification of X.

Theorem 1.109 *For a locally compact space (X, τ), the above space (Y, τ_Y) is a compactification of X.*

Proof It is easy to see that τ_Y is actually a topology on Y and that $\tau_Y|_X = \tau$. We show that Y is Hausdorff. Clearly we have only to check the separation of each $x \in X$ from ∞. Since X is locally compact, there exists open $U \in \mathcal{N}(x)$ with $\overline{U}$ compact. Then U and $V = Y \setminus \overline{U}$ separate x and ∞. So, Y is Hausdorff. To show that Y is compact, let $\{V_\alpha\}_{\alpha \in I}$ be an open cover of Y. Then there exists some $\alpha_0 \in I$ such that $\infty \in V_{\alpha_0}$. Hence $V_{\alpha_0} = Y \setminus K$ for some compact $K \subseteq X$. Thus, K has a cover $\{V_\alpha \cap X\}_{\alpha \in I}$ and by compactness it admits a finite subcover $\{V_{\alpha_k} \cap X\}_{k=1}^{n}$.

Then $\{V_{\alpha_k}\}_{k=0}^{n}$ is a finite open over of Y and so (Y, τ_Y) is compact. Finally, note that X is dense in Y. Therefore, (Y, τ_Y) is a compactification of X. □

Remark 1.110 Evidently for any $C \subseteq X$, $C \cup \{\infty\}$ is closed in Y iff C is closed in X.

Example 1.111 Consider $\mathbb{R}$ with the standard metric topology, which is locally compact and the Alexandrov one-point compactification is homeomorphic to the circle. One such homeomorphism is the stereographic projection. Similarly, the Alexandrov one-point compactification of $\mathbb{R}^2$ is the sphere S^1, known to the mapmakers.

The embedding of a locally compact space in a compact space has the following consequence.

Proposition 1.112 *Every locally compact space is completely regular.*

Definition 1.113 A Hausdorff space (X, τ) is "σ-compact" if

$$X = \cup_{n \geq 1} K_n, \quad \text{with} \quad K_n \subseteq X \text{ compact}.$$

Remark 1.114 Every Euclidean space $\mathbb{R}^N$ is σ-compact.

Proposition 1.115 *A second countable and locally compact space X is σ-compact.*

Proof Let $\mathcal{B}$ be a countable basis for X. Let $\mathcal{B}^* = \{W \in \mathcal{B} : \overline{W} \text{ is compact}\}$ and consider $x \in U$ with $U \in \tau$. By Proposition 1.105, there exists open $V \in \mathcal{N}(x)$ with $\overline{V}$ compact such that $x \in V \subseteq \overline{V} \subseteq U$. Since $\mathcal{B}$ is a basis we can find $W \in \mathcal{B}$ such that $x \in W \subseteq V$, hence $\overline{W}$ is compact. Therefore, $W \in \mathcal{B}^*$. This proves that $\mathcal{B}^*$ is a countable basis for X and from this we infer that X is σ-compact. □

In fact, the spaces of Proposition 1.115 are metrizable, which is a consequence of the so-called Urysohn Metrization Theorem (see Dugundji [91], p.195).

Theorem 1.116 *For a Hausdorff space (X, τ), the following statements are equivalent:*

(a) X is regular and second countable.
(b) X is metrizable and separable.
(c) X is homeomorphic to a subset of the Hilbert cube I^N, where $I = [0, 1]$.

Remark 1.117 Recall that every locally compact space is completely regular, it is regular in particular.

Proposition 1.118 *For any locally compact and σ-compact space (X, τ), there exists a sequence $\{C_n\}$ of compact sets such that $C_n \subseteq \text{int } C_{n+1}$ for all $n \in \mathbb{N}$, and $X = \cup_{n \geq 1} C_n$.*

Proof The proof is by induction. Since X is σ-compact, we have $X = \cup_{n \geq 1} K_n$ with compact $K_n \subseteq X$ for all $n \in \mathbb{N}$. Let $C_1 = K_1$ and suppose that C_m has been constructed. Set $D_m = C_m \cup K_m$. If $x \in D_m$, then by local compactness we can find $V_x \in \mathcal{N}(x)$ open with $\overline{V}_x$ compact. Then $\{V_x\}_{x \in D_m}$ is an open cover and since D_m is compact, we can find a finite subcover $\{V_{x_i^m}\}_{i=1}^{l_m}$.

This way we can produce points $x_i^m \in D_m$ for $i \in \{1, \cdots, l_m\}$ such that

$$C_m \subseteq \cup_{i=1}^{l_m} V_{x_i^m}.$$

Let $C_{m+1} = \cup_{i=1}^{l_m} \overline{V}_{x_i^m}$, which is a compact set in X. Also, we have

$$C_m \subseteq \bigcup_{i=1}^{l_m} V_{x_i^m} \subseteq \text{int}\, C_{m+1}, \quad X = \bigcup_{n \in \mathbb{N}} K_n = \bigcup_{n \in \mathbb{N}} C_n.$$

$\square$

1.6 Connectedness

Now we turn to another basic topological notion: connectedness. It is an intuitive concept that in essence implies that the space is all in one piece.

Definition 1.119 A topological space (X, τ) is "disconnected" if it is the union of two disjoint nonempty open sets. Namely, $X = U \cup V$, with $U, V \in \tau$ and $U \cap V = \emptyset$. Such a pair of open sets (U, V) is called a "separation" of X. If X is not disconnected, then it is "connected." A subset $A \subseteq X$ is connected, provided it is connected when furnished with the subspace topology.

Remark 1.120 In the above definition we can replace "open" by "closed." Thus, X is connected iff it has no subset that is both closed and open (clopen for short) other than $\emptyset$ and X. Apparently, X is connected iff $\emptyset$ and X are the only subsets with empty boundary. The property of being connected is definitely not hereditary.

Proposition 1.121 *A topological space (X, τ) is disconnected   iff   there is a continuous function from X onto a discrete two-point space $\{a, b\}$.*

Proof $\Rightarrow$: Let (U, V) be a separation of X. The f defined below is continuous:

$$f(x) = \begin{cases} a & \text{if } x \in U \\ b & \text{if } x \in V. \end{cases}$$

$\Leftarrow$: If $f : X \to \{a, b\}$ is continuous, then $U = f^{-1}(a)$ and $V = f^{-1}(b)$ are disjoint open sets in X and $X = U \cup V$. So, (U, V) is a separation of X, which is disconnected. $\qquad\square$

Proposition 1.122 *Let (X, τ_X) and (Y, τ_Y) be topological spaces, $f : X \to Y$ continuous and $A \subseteq X$ connected. Then $f(A)$ is connected in Y.*

Proof Arguing by contradiction, suppose that $f(A)$ is disconnected and let (U, V) be a separation of $f(A)$. Then, both sets $f^{-1}(U)$ and $f^{-1}(V)$ are open on account of the continuity of f. Clearly, they are disjoint and nonempty. Finally, we have

$$A \subseteq f^{-1}(f(A)) = f^{-1}(U \cup V) = f^{-1}(U) \cup f^{-1}(V),$$

hence A is disconnected, a contradiction. $\qquad\square$

Definition 1.123 Let (X, τ) be a topological space. Nonempty $A, B \subseteq X$ are "separated sets" if $\overline{A} \cap B = \emptyset$ and $A \cap \overline{B} = \emptyset$.

Proposition 1.124 *Let (X, τ) be a topological space and $C \subseteq X$. Then C is connected iff there are no separated sets $\{A, B\}$ such that $C = A \cup B$.*

Proof $\Rightarrow$: Suppose that $C = A \cup B$ with $\{A, B\}$ separated sets. Then,

$$\overline{A}^{\tau_C} = (A \cup B) \cap \overline{A}^{\tau}$$
$$= (A \cap \overline{A}^{\tau}) \cup (B \cap \overline{A}^{\tau})$$
$$= A,$$

thus $A \subseteq C$ is closed. Similarly, we can show that $B \subseteq C$ is closed. Therefore, A is disconnected (see Remark 1.120), a contradiction.

$\Leftarrow$: If C is disconnected, then $C = A \cup B$ for some disjoint clopen sets A and B, a contradiction to the hypothesis. $\qquad\square$

Corollary 1.125 *Let (X, τ) be a topological space, $\{A, B\}$ separated subsets of X, and $C \subseteq A \cup B$ connected. Then $C \subseteq A$ or $C \subseteq B$.*

It is clear that connectedness is not preserved by unions of sets. However, we have the following result.

Proposition 1.126 *Let (X, τ) be a topological space and $\{C_\alpha\}_{\alpha \in I}$ a family of connected subsets of X with $\cap_{\alpha \in I} C_\alpha \neq \emptyset$. Then, $\cup_{\alpha \in I} C_\alpha$ is connected.*

Proof Arguing by contradiction, suppose that $C = \cup_{\alpha \in I} C_\alpha$ is disconnected. Then by Proposition 1.124, we have $C = A \cup B$ with $\{A, B\}$ separated sets in C. We have $C_\alpha \subseteq A \cup B$ for all $\alpha \in I$. It follows from Corollary 1.125 that $C_\alpha \subseteq A$ for all $\alpha \in I$ or $C_\alpha \subseteq B$ for all $\alpha \in I$. Then $C \subseteq A$ or $C \subseteq B$. Hence, either $B = \emptyset$ or $A = \emptyset$, a contradiction. This proves that $C = \cup_{\alpha \in I} C_\alpha$ is connected. $\qquad\square$

Corollary 1.127 *Let (X, τ) be a topological space, $\{C_\alpha\}_{\alpha \in I}$ a family of connected subsets of X, $B \subseteq X$ connected such that $C_\alpha \cap B \neq \emptyset$ for every $\alpha \in I$. Then $((\cup_{\alpha \in I} C_\alpha) \cup B$ is connected.*

Proof By Proposition 1.126, $C_\alpha \cup B$ is connected for each $\alpha \in I$. Note that

$$\emptyset \neq B \subseteq \cap_{\alpha \in I}(C_\alpha \cup B).$$

Thus, $(\cup_{\alpha \in I} C_\alpha) \cup B$ is connected by Proposition 1.129.   □

Proposition 1.128 *Let (X, τ) be a topological space, $\{C_n\}$ a sequence of connected subsets of X such that C_n has nonempty intersection with one of the previous sets for $n \geq 2$. Then, $\cup_{n \geq 1} C_n$ is connected.*

Proof C_1 is connected. If $\cup_{k=1}^{n-1} C_k$ is connected, then so is $\cup_{k=1}^{n} C_k$ by Proposition 1.126. Therefore, $\{D_n = \cup_{k=1}^{n} C_k\}$ is a sequence of connected sets such that $\emptyset \neq C_1 \subseteq \cap_{k \geq 1} D_k$. Proposition 1.126 implies that $\cup_{k \geq 1} C_k$ is connected.   □

Proposition 1.129 *Let (X, τ) be a topological space, $B \subseteq X$ connected and $B \subseteq A \subseteq \overline{B}$. Then A is connected too.*

Proof It is enough to consider $\overline{B}$. Indeed, if $B \subseteq A \subseteq \overline{B}$, we have $A = \overline{B}^{\tau_A}$ and so X may be replaced by A. Arguing by contradiction, suppose that $\overline{B}$ is disconnected. Then $\overline{B} = U \cup V$, with U, V disjoint nonempty open in $\overline{B}$. Thus $B = (U \cap B) \cup (V \cap B)$ and the latter are disjoint, nonempty open sets in B. Therefore, B is disconnected, a contradiction.   □

Example 1.130 The "topologist's sine curve." Let $A = \{(0, y) \in \mathbb{R}^2 : |y| \leq 1\}$ and $B = \{(x, y) \in \mathbb{R}^2 : 0 < x \leq 1, y = \sin(\pi/x)\}$. Set $S = A \cup B$, which is called the "topologist's sine curve." The set B is connected since it is the image of $(0, 1]$ under the continuous map $x \to \sin(\pi/x)$. We mention that in $\mathbb{R}$ all the connected sets are intervals. Note that $S = \overline{B}$ and hence connected according to Proposition 1.129.

If a space is not connected, then we may try to decompose it into a family of disjoint maximal connected subspaces. We will show that this can be done, always.

Definition 1.131 Let (X, τ) be a topological space. A maximal connected subspace of X is a connected subspace that is not a proper subset of any connected subspace, called a connected "component" of X.

Remark 1.132 Let $x \in X$ and C_x be the largest connected subspace containing x. It exists since it is the union of all connected subspaces of X containing x. For $x \neq y$, either $C_x = C_y$ or $C_x \cap C_y = \emptyset$. Indeed, otherwise $C_x \cup C_y$ would be a connected set containing $x, y \in X$ strictly larger than both C_x and C_y, a contradiction. Evidently, every connected subset of X is contained in some connected component. Furthermore, every component is closed. To see this, let C be a component of X. According to Proposition 1.129, $\overline{C}$ is connected and so $\overline{C} \subseteq C$. Hence, $C = \overline{C}$ and

C is thus closed. However, the connected components need not be open. To see this, consider the space $X = \mathbb{Q}$. Then the components are singletons $\{q\}$, $q \in \mathbb{Q}$, which are closed sets but not open. Clearly, X is connected iff it has only one component. Finally, if C is a component of X and $\{U, V\}$ is a separation of X, then either $C \subseteq U$ or $C \subseteq V$.

It is clear from the previous discussion that the notion of connectedness is negative in nature. It requires the nonexistence of a particular splitting of the space. A more positive approach is provided by the stronger notion of path-connectedness.

Definition 1.133 A topological space (X, τ) is "path-connected" if for any two points $x, u \in X$, there exists a continuous function $p : [0, 1] \to X$ such that $p(0) = x$ and $p(1) = u$. The function p, as well as its range, is known as the "path" joining x and u.

Proposition 1.134 *A path-connected topological space X is connected.*

Proof Arguing by contradiction, suppose that X is not connected and let $\{U, V\}$ be a separation of X. Let $x \in U$ and $u \in V$. We can find a path: $p : [0, 1] \to X$ joining x and u. Then $\{f^{-1}(u), f^{-1}(v)\}$ is a separation of $[0, 1]$, which is connected, a contradiction. So, X is connected. □

Remark 1.135 The converse of the above proposition is not true. The topologist's sine curve from Example 1.130 is connected but not path-connected. In fact, there is no path joining $(0, 0)$ and $(\frac{1}{n}, 0)$ (see Dugundji [91], p.115). Path-connectedness is a topological property. In fact it is easy to see that the continuous image of path-connected space is path-connected. However, in contrast to connectedness, the closure of a path-connected space need not be path-connected. Let $B = \{(x, y) \in pr^2 : 0 < x \leq 1, y = \sin(\frac{\pi}{x})\}$. Clearly, B is path-connected, but $\overline{B} = S$ (the topologist's sine curve, see Example 1.130) is not, as we just mentioned.

Proposition 1.136 *Let (X, τ) be a topological space and $x_0 \in X$. Then X is path-connected iff each point $u \in X$ can be joined with x_0 by a path.*

Proof $\Rightarrow$: It follows from Definition 1.133.
$\Leftarrow$: Let $x, u \in$. By assumption we can find paths $p_1, p_2 : [0, 1] \to X$ such that

$$p_1(0) = x, \, p_2(1) = x_0, \text{ and } p_2(0) = x_0, \, p_2(1) = u.$$

Define $\widehat{p} : [0, 1] \to X$ by

$$\widehat{p}(t) = \begin{cases} p_1(2t) & \text{if } 0 \leq t \leq 1/2 \\ p_2(2t - 1) & \text{if } 1/2 < t \leq 1. \end{cases}$$

Clearly, $\widehat{p}$ is a path connecting x and u. Therefore, X is path-connected. □

Corollary 1.137 *Let (X, τ) be a topological space and $\{C_\alpha\}_{\alpha \in I}$ a family of path-connected subsets of X with $\cap_{\alpha \in I} C_\alpha \neq \emptyset$. Then $\cup_{\alpha \in I} C_\alpha$ is path-connected.*

Remark 1.138 The corollary allows the introduction of path-connected components of a topological space as maximal path-connected subsets. As before, the path-connected components partition the space. However, the path-connected components need not be closed, in contrast to the connected components (see Remark 1.132). Also, every path-connected subset is contained in a path-connected component and the space is path-connected iff it has a single path-connected component.

Definition 1.139 A topological space is called "locally connected" (respectively, "locally path-connected") if every point has a neighborhood basis consisting of open connected sets (respectively, of open path-connected sets).

The notion of local path-connectedness provides a partial converse to Proposition 1.134.

Proposition 1.140 *A connected and locally path-connected topological space is path-connected.*

Proof Let $x \in X$ and $D = \{u \in X : $ it can be joined to x by a path in $X\}$.

(a) D is open. To prove it, let $u \in D$ and choose $U \in \mathcal{N}(u)$, which is path-connected. So, if $v \in U$ we can find a path joining v and u. Concatenate the path joining u and x to generate a path joining v and x. Therefore, $v \in D$ and so D is open.

(b) D is closed. To prove it, let $u \in \overline{D}$ and choose $U \in \mathcal{N}(u)$, which is path-connected. Then $U \cap D \neq \emptyset$ and let $v \in U \cap D$. We can find a path joining u and v, and a path joining v and x. Concatenating the two produces a path joining u and x and so, $u \in D$ and D is closed.

From (a) and (b) and since $D \neq \emptyset$ and X is connected, we concluded that $D = X$. $\square$

Corollary 1.141 *An open connected subset of $\mathbb{R}^N$ is path-connected.*

Remark 1.142 The conclusion in the corollary fails for nonopen subsets of $\mathbb{R}^N$ (see Example 1.130).

Proposition 1.143 *A topological space (X, τ) is locally connected iff each connected component of every open set is open.*

Proof $\Rightarrow$: Let $U \in \tau$ and C a component of U. Given $x \in C$, we can find $V \in \tau$ connected such that $x \in V \subseteq U$. Then $V \subseteq C$ (see Remark 1.132) and so, C is open.

$\Leftarrow$: Let $U \in \mathcal{N}(x)$ be open. By assumption the connected component of U containing x is open and in U, which means that X is locally connected. $\square$

Corollary 1.144 *(a) The connected components of a locally connected space are clopen; (b) A compact locally connected space has a finite number of connected components.*

1.7 Polish, Souslin, and Baire Spaces

Polish spaces are generalizations of complete separable metric spaces.

Definition 1.145 A topological space (X, τ) is "completely metrizable" if it admits a compatible metric d, namely $\tau = \tau_d$, such that (X, d) is complete. A separable completely metrizable space is a "Polish space."

Remark 1.146 There are Polish spaces that have no natural (simple) compatible metric. However, what matters is the existence of a compatible metric and not its particular choice. So working with Polish spaces we have a more general framework of analysis.

Proposition 1.147 *Every closed or open subset of a Polish space is itself Polish.*

Proof Let $C \subseteq X$ be closed and d a compatible complete metric. Then $d_C = d|_{C \times C}$ is a compatible complete metric on C. Therefore, C is Polish.

Let $U \subseteq X$ be open. Assume that $U \neq X$. Consider the distance function from $U^c = X \setminus U$, defined by $d(x, U^c) = \inf[d(x, u) : u \in U^c]$. Define

$$d_0(x, u) = d(x, u) + \left[\frac{1}{d(x, U^c)} - \frac{1}{d(u, U^c)} \right].$$

This is well defined. It is easy to check that d_0 is a metric on U.

Claim: d_0 is compatible with the subspace topology of U and complete.

Note that

$$|d(x, U^c) - d(u, U^c)| \leq d(x, u).$$

Thus, $u_n \xrightarrow{d} u$ iff $u_n \xrightarrow{d_0} u$. Therefore, d_0 is compatible with the subspace topology on U.

Next, we show that U is d_0-complete. So, let $\{u_n\} \subseteq U$ be d_0-Cauchy, hence d-Cauchy too. Thus we have $u_n \xrightarrow{d} u \in X$. If $u \notin U$, then $d(u_n, U^c) \to 0$ and so, $\limsup_{n,m \to \infty} d_0(u_n, u_m) = \infty$, a contradiction to the assumption that $\{u_n\}$ is d_0-Cauchy. So, $u \in U$ and this proves the d_0-completeness of U.

From the Claim and since U is separable, we concluded that U is Polish. $\square$

Note that for a metric d on X, $d_0(x, u) = \min\{1, d(x, u)\}$ is also a metric, which is bounded. We have $\tau_d = \tau_{d_0}$, namely the two metrics generate the same topology thus, topologically equivalent. Moreover, X is d-complete iff it is d_0-complete.

Proposition 1.148 $X = \prod_{n \geq 1} X_n$ *with product topology is Polish if each X_n is Polish.*

Proof Let d_n be a compatible metric on X_n. From the comments preceding the proposition, we may assume that $d_n(x, u) \leq 1$ for all $x, u \in X_n$. Define, for $\widehat{x} = (x_n)_{n \geq 1}$ and $\widehat{u} = (u_n)_{n \geq 1}$,

$$\widehat{d}(\widehat{x}, \widehat{u}) = \sum_{n \in \mathbb{N}} \frac{1}{2^n} d_n(x_n, u_n).$$

Then it is easy to see that $\widehat{d}$ is a metric on X and the product topology on X equals the $\tau_{\widehat{d}}$-topology. It remains to show that the separability of X. According to Proposition 1.54 we need to show that X is second countable. So, let $\mathcal{B}_n$ be a countable basis for X_n. Consider the family

$$\mathcal{B} = \{U_1 \times \cdots, \times U_m \times X_{m+1} \times X_{m+2} \times \cdots : U_k \in \mathcal{B}_k, k \in \{1, \cdots, m\}\}.$$

Then $\mathcal{B}$ is countable and it is a basis for the product topology on X, which is then second countable, thus separable. Therefore, X is Polish. □

For locally compact metric spaces, we have the Alexandrov one-point compactification (see Definition 1.108 and Theorem 1.109). It is natural to ask when such a compactification is metrizable.

Theorem 1.149 *Let (X, d) be a locally compact metrizable space and Y its one-point compactification. Then Y is metrizable iff X is separable.*

Proof $\Rightarrow$: Since Y is metrizable and compact, it is separable and then so is X.

$\Leftarrow$: By Propositions 1.115 and 1.118, it follows that $X = \cup_{n \geq 1} K_n$, with K_n compact and $K_n \subseteq \text{int } K_{n+1}$. Then the collection of $\{Y \setminus K_n\}_{n \geq 1}$ is a countable neighborhood basis for ∞. Therefore, Y is second countable and hence metrizable (see Theorem 1.116). □

Corollary 1.150 *Every locally compact σ-compact metrizable space is Polish.*

Proposition 1.151 *An intersection of countable Polish spaces is Polish.*

Proof Let $\{X_n\}$ be the Polish spaces and set

$$\Delta = \left\{\widehat{x} = (x, \cdots, x, \cdots) \in \prod_{n \geq 1} X_n\right\}.$$

Then Δ is closed and so Polish (see Propositions 1.147, 1.148). Clearly, Δ is homeomorphic to $\cap_{n \geq 1} X_n$, which is therefore Polish.   □

Corollary 1.152 $X = \mathbb{R} \setminus \mathbb{Q}$ *with the subspace topology is Polish.*

Proof Let $\{q_n\}$ be an enumeration of the rationals and $X_n = \mathbb{R} \setminus \{q_n\}$. Then X_n is open in $\mathbb{R}$ and hence Polish (see Proposition 1.147). Note that $X = \cap_{n \geq 1} X_n$ and apply Proposition 1.151 to conclude that $X = \mathbb{R} \setminus \mathbb{Q}$ is Polish.   □

Definition 1.153 A subset of a topological space is a "G_δ-set" (respectively, "F_σ-set") if it is the intersection of countable open sets (respectively, the union of countable closed sets).

We wish to characterize those subspaces of a Polish space that are Polish spaces themselves when furnished with the subspace topology. To this end, we will need the following notion.

Definition 1.154 Let (X, τ) be a Hausdorff space, (Y, d) a metric space, and $f : A \subseteq X \to Y$. The "oscillation" of f at $x \in X$ is defined by

$$o_f(x) = \inf\{\text{diam } f(A \cap U) : U \in \mathcal{N}(x) \cap \tau\},$$

where for $C \subseteq Y$, diam $C = \sup_{u, v \in Y} d(u, v)$, with the convention that $diam \emptyset = 0$.

Remark 1.155 f is continuous at $x \in A$ iff $o_f(x) = 0$.

Proposition 1.156 *Let (X, τ) be a Hausdorff topological space, Y a metrizable space and $f : X \to Y$. Then the set $C_f \subseteq X$ of points of continuity of f is G_δ.*

Proof For $x \in X$ and $\epsilon > 0$, we have $o_f(x) < \epsilon$ iff we can find $U \in \mathcal{N}(x) \cap \tau$ such that diam $f(A \cap U) < \epsilon$. Hence,

$$\{x \in X : o_f(x) < \epsilon\} = \cup\{U : U \in \mathcal{N}(x) \cap \tau, \text{ diam } f(A \cap U) < \epsilon\}$$

thus $\{x \in X : o_f(x) < \epsilon\} = A_\epsilon$ is open.
    Note that $C_f = \{x \in X : o_f(x) = 0\} = \cap_{n \geq 1} A_{1/n}$ and so, C_f is G_δ.   □

Proposition 1.157 *In a metrizable space every closed subset is G_δ, hence every open subset is F_σ.*

Proof Let d be a compatible metric on X. Recall that $d(x, A) = \inf[d(x, a) : a \in A]$ for $x \in X$ and $A \subseteq X$. Then $|d(x, A) - d(u, A)| \leq d(x, u)$ for $x, u \in X$. Define the ϵ-enlargement of A by

$$A_\epsilon = \{x \in X : d(x, A) < \epsilon\},$$

which is clearly open. If A is closed, we have

$$A = \{x \in X : d(x, A) = 0\} = \cap_{n \geq 1} A_{1/n},$$

thus A is G_δ. □

We will use the previous two propositions to prove the following fundamental extension theorem due to Kuratowski [167] (p.422).

Theorem 1.158 *Let X be metrizable, Y completely metrizable, and $f : A \subseteq X \to Y$ continuous. Then there exists a G_δ-set C such that $A \subseteq C \subseteq \overline{A}$ and a continuous extension $\widehat{f} : C \to Y$ of f.*

Proof Let $C = \overline{A} \cap \{x \in X : o_f(x) = 0\}$. Using Propositions 1.156 and 1.157, we see that C is G_δ and $A \subseteq C \subseteq \overline{A}$. Let $x \in C$. Since $x \in \overline{A}$, there exists $\{x_n\} \subseteq A$ such that $x_n \to x$. Since $o_f(x) = 0$, $\{f(x_n)\}$ is Cauchy in Y, hence convergent. Define

$$\widehat{f}(x) = \lim f(x_n).$$

It is easy to see that $\widehat{f}$ is well defined, namely $\widehat{f}(x)$ is independent of the choice of the sequence $\{x_n\}$. Also, $\widehat{f}|_A = f$. Finally, we check the continuity of $\widehat{f}$. We need to show that $o_{\widehat{f}}(x) = 0$ for all $x \in C$. If $U \subseteq X$ is open, then $\widehat{f}(C \cap U) \subseteq \overline{f(A \cap U)}$ and so, diam $\widehat{f}(C \cap U) \leq$ diam $f(A \cap U)$, hence $o_{\widehat{f}}(x) \leq o_f(x) = 0$. □

The next theorem is an important application of the previous result and is known as the "Lavrentiev Extension Theorem."

Theorem 1.159 *Let X, Y be completely metrizable spaces, and $f : A \subseteq X \to B \subseteq Y$ a homeomorphism. Then, there exist G_δ-sets $C \supseteq A$ and $D \supseteq B$ and a homeomorphism $\widehat{f} : C \to D$ which extends f.*

Proof Let $h = f^{-1}$. By Theorem 1.158 we can find $\widetilde{A} \supseteq A$, a G_δ-set, and a continuous extension $\widetilde{f} : \widetilde{A} \to Y$ of f. Similarly, we can find $\widetilde{B} \supseteq B$, a G_δ-set, and a continuous extension $\widetilde{h} : \widetilde{B} \to X$ of h. Set

$$E = gr\,\widetilde{f} \quad \text{and} \quad F = (gr\,\widetilde{h})^{-1} = \{(x, y) \in X \times \widetilde{B} : x = \widetilde{y}\}.$$

Let $C = p_1(E \cap F)$ and $D = p_2(E \cap F)$, where p_1, p_2 are the corresponding projection maps. Since $\widetilde{f}$ is continuous on G_δ-set $\widetilde{A}$, Proposition 1.156 implies that C is G_δ. Since F is closed and $\widetilde{B}$ is G_δ, it follows that F is G_δ (see Proposition 1.157). Therefore, as above by exploiting the continuity of $\widetilde{h}$ we conclude that D is G_δ. Then we see that $\widehat{f} = \widetilde{f}|_C$ is a homeomorphism of C onto D which extends f. □

Theorem 1.160 *(a) Let X be metrizable and $A \subseteq X$ completely metrizable with the subspace topology. Then, A is G_δ in X. (b) Let X be completely metrizable and $A \subseteq X$ a G_δ-set. Then, A is completely metrizable with the subspace topology.*

Proof (a) Consider the identity map $i_A : A \to A$. By Theorem 1.158 we can find C, a G_δ-set, with $A \subseteq C \subseteq \overline{A}$, and a continuous extension $\widehat{f} : C \to Y$ of i_A. Since A is dense in C, we have $\widehat{f} = i_C$. Therefore, $A = C$.

(b) By assumption, $A = \cap_{n \geq 1} U_n$, with $U_n \subseteq X$ open. Let $C_n = X \setminus U_n$ and d be a complete compatible metric on X. Set, for $x, u \in A$,

$$d_0(x, u) = d(x, u) + \sum_{n \geq 1} \min \left\{ \frac{1}{2^n}, \left| \frac{1}{d(x, C_n)} - \frac{1}{d(u, C_n)} \right| \right\}.$$

Reasoning as in the proof of Proposition 1.147 we can show that d_0 is a metric on A, compatible with the subspace topology and (A, d_0) is complete.   □

As a consequence we have the following characterization.

Theorem 1.161 *A subset of a Polish space is Polish iff it is G_δ.*

Definition 1.162 A Hausdorff space X is a "Souslin" space if there exists a Polish space Y and a continuous surjection $f : Y \to X$.

Remark 1.163 Evidently every Polish space is Souslin. Also, a Souslin space is separable but need not be metrizable. For example, anticipating material from Sect. 3.5, an infinite dimensional separable Banach space furnished with the weak topology. A Souslin subset of a Polish space is known as "analytic set." For a Souslin X, Hausdorff Z and $g : X \to Z$ continuous, we always have that $g(X) \subseteq Z$ is Souslin.

Proposition 1.164 *All closed sets and all open sets in a Souslin space are Souslin.*

Proof Let (X, τ) be Souslin. There exist a Polish space Y and continuous surjective $f : Y \to X$. Let $A \subseteq X$ be closed (respectively, open). Then $f^{-1}(A)$ is closed (respectively, open) in Y. Thus, $f^{-1}(A)$ is Polish (see Proposition 1.147). Since $f(f^{-1}(A)) = A$ (due to the surjectivity of f), we conclude that A is Polish.   □

Proposition 1.165 *Let (X, τ_X) be Souslin, (Z, τ_Z) Hausdorff, $g : X \to Z$ continuous, and $A \subseteq Z$ Souslin. Then $g^{-1}(A)$ is Souslin in X.*

Proof Since X and A are Souslin, we can find Polish spaces Y_1 and Y_2, and continuous surjections $f_1 : Y_1 \to X$ and $f_2 : Y_2 \to A$. Consider $S = \{(u, v) \in Y_1 \times Y_2 : g(f_1(u)) = f_2(v)\}$. Then S is closed in $Y_1 \times Y_2$ and thus Polish (see Propositions 1.147 and 1.148). Now consider the projection $p_1 : Y_1 \times Y_2 \to Y_1$ and let $\vartheta = p_1|_S$. Then $g^{-1}(A) = (f_1 \circ \vartheta)(S)$ and thus $g^{-1}(A)$ is Souslin in X (see Remark 1.163).   □

Proposition 1.166 $X = \prod_{n \geq 1} X_n$ *with the product topology is Souslin when all X_n are.*

Proof For each $n \in \mathbb{N}$, we can find a Polish space Y_n and a continuous surjection $f_n : Y_n \to X_n$. Then $Y = \prod_{n \geq 1} Y_n$ is Polish (see Proposition 1.148) and $f = (f_n) : Y \to X$ is a continuous surjection. Hence X is Souslin. $\square$

In a similar fashion we also can show the following;

Proposition 1.167 *Let (X, τ) be a Hausdorff space and $\{A_n\}$ are Souslin subsets of X. Then $\cap_{n \geq 1} A_n$ is Souslin.*

We have not examined what happens with countable unions of such spaces (Polish and Souslin). It turns out that the families of Polish and Souslin spaces are closed under countable unions, which can be seen using the notion of "disjoint union" of spaces.

Definition 1.168 Let $\{(X_\alpha, \tau_\alpha)\}_{\alpha \in I}$ be Hausdorff spaces and

$$X_\alpha^* = \{(x, \alpha) : x \in X_\alpha\}.$$

Furnished with the topology $\tau_\alpha^* = \tau_\alpha \times \{\alpha\}$, X_α^* is homeomorphic with X_α, and the collection $\{X_\alpha^*\}_{\alpha \in I}$ differs from the collection $\{X_\alpha\}_{\alpha \in I}$ only in that $X_\alpha^* \cap X_\beta^* = \emptyset$ when $\alpha \neq \beta$.

Now consider $X = \cup_{\alpha \in I} X_\alpha^*$, equipped with a topology as follows: $U \subseteq X$ is open iff $U \cap X_\alpha^* \in \tau_\alpha^*$ for each $\alpha \in I$. Then X is called the "disjoint union" of the spaces and denoted by $\sum_{\alpha \in I} X_\alpha$.

Remark 1.169 In practice we almost always ignore the distinction between X_α and X_α^* and treat X_α as a subset of the disjoint union, which does not cause any trouble.

Using this notion we obtain the following result.

Proposition 1.170 *Countable unions of Polish (respectively, Souslin) spaces is Polish (respectively, Souslin).*

The two theorems of Topology with the greatest value in Analysis are the "Tychonov Theorem" and "Baire Category Theorem." While we presented the former, we will present the latter now. Roughly speaking, the Baire Category Theorem identifies "thick" topological spaces.

Definition 1.171 Let (X, τ) be a Hausdorff space and $A \subseteq X$. A is said to be "nowhere dense" if int $\overline{A} = \emptyset$. A is of "first category" (or "meager") if it is a countable union of nowhere dense sets; and of "second category" if it is not of the first category. Finally, a Hausdorff space is a "Baire space" if the open sets are not of first category.

Remark 1.172 If $U \subseteq X$ is open dense, then $U^c = X \setminus U$ is nowhere dense. Indeed, note that U^c is closed. Thus, we need to show that int $U^c = \emptyset$. However, int $U^c = X \setminus \overline{U}$, which is empty on account of the density of U.

The next proposition characterizes Baire spaces.

Proposition 1.173 *For a Hausdorff space* (X, τ), *we have the following equivalent statements:*

(a) X is a Baire space.
(b) Countable intersections of open dense sets are dense.
(c) If $X = \cup_{n \geq 1} C_n$ with each C_n closed, then $\cup_{n \geq 1}$ int C_n is dense.

Proof (a) $\Rightarrow$ (b): Let $\{U_n\} \subseteq \tau$, which are dense in X, and set $A = \cap_{n \geq 1} U_n$. Let $\emptyset \neq V \in \tau$, and suppose that $A \cap V = \emptyset$. Then $X = (X \setminus A) \cup (X \setminus V)$ and so we obtain

$$V = (X \setminus A) \cap V = (X \setminus \cap_{n \geq 1} U_n) \cap V = \cup_{n \geq 1}((X \setminus U_n) \cap V),$$

hence V is of first category (see Remark 1.172), a contradiction to the assumption that X is Baire. So, $A \cap V \neq \emptyset$ for all $V \in \tau$ nonempty, thus A is dense in X.

(b) $\Rightarrow$ (c): Let $\{C_n\}$ be a sequence of closed sets such that $X = \cup_{n \geq 1} C_n$. Then $U = \cup_{n \geq 1}$ int $C_n \in \tau$. Set $F_n = C_n \setminus$ int C_n. Then F_n is nowhere dense and so, $F = \cup_{n \geq 1} F_n$ is of the first category. Notice that $X \setminus F_n$ is open dense, thus $D = \cap_{n \geq 1}(X \setminus F_n)$ is dense by assumption. We have

$$X \setminus U = \bigcup_{n \geq 1} C_n \setminus \bigcup_{n \geq 1} \text{int } C_n \subseteq \bigcup_{n \geq 1}(C_n \setminus \text{int } C_n) = F,$$

thus $X \setminus F = D = \cap_{n \geq 1}(X \setminus F_n) \subseteq U$ and therefore $U = \cup_{n \geq 1}$ int C_n is dense in X.

(c) $\Rightarrow$ (a): Let $U \in \tau$ nonempty. If U is of first category, then $U = \cup_{n \geq 1} E_n$, with int $\overline{E} = \emptyset$ for each $n \in \mathbb{N}$. Then

$$X = (X \setminus U) \cup (\cup_{n \geq 1} \overline{E}_n).$$

By assumption, we must have that int$(X \setminus U)$ is dense in X. Hence $X \setminus U$ is dense in X. Thus, we must have $U \cap (X \setminus U) \neq \emptyset$, a contradiction. Therefore, U is of second category and X is a Baire space. $\square$

The following theorem is known as the "Baire Category Theorem."

Theorem 1.174 *Locally compact spaces and completely metrizable spaces are Baire spaces.*

Proof We will do the proof for completely metrizable spaces. The proof for locally compact spaces can be found in Dugundji [91], p.249.

Let d be a compatible complete metric on X. Suppose that $\{U_n\}$ is a sequence of dense open sets in X and $A = \cap_{n \geq 1} U_n$. According to Proposition 1.173 we need to show that A is dense in X, which is equivalent to saying that for every $x \in X$ and $r > 0$ we have $A \cap B_r(x) \neq \emptyset$.

Since U_1 is open dense in X, we can find $u_1 \in X$ and $0 < r_1 < 1$ such that

$$\overline{B}_{r_1}(u_1) \subseteq U_1 \cap B_r(x).$$

Similarly since U_2 is open dense in X we can find $u_2 \in X$ and $0 < r_2 < 1/2$ such that

$$\overline{B}_{r_2}(u_2) \subseteq U_2 \cap B_{r_1}(u_1).$$

By induction we can produce two sequences $\{u_n\} \subseteq X$ and $\{r_n\} \subseteq (0, 1)$ such that for $n \geq \mathbb{R}^N$,

$$\overline{B}_{r_{n+1}}(u_{n+1}) \subseteq U_n \cap B_{r_n}(u_n), \text{ and } 0 < r_n < \frac{1}{n}.$$

From Cantor's Intersection Theorem (see Proposition 1.94), $\cap_{n \geq 1} \overline{B}_{r_n}(u_n)$ is a singleton. Since $\cap_{n \geq 1} \overline{B}_{r_n}(u_n) \subseteq A \cap B_r(x)$, we conclude that $A \cap B_r(x) \neq \emptyset$. □

Remark 1.175 In fact every G_δ-set of a compact topological space is a Baire space.

1.8 Semicontinuous Functions

Semicontinuity (lower and upper) extends the notion of continuity and is important in applications.

Definition 1.176 Let (X, τ) be a Hausdorff space and $f : X \to \mathbb{R}^* = \mathbb{R} \cup \{\pm\infty\}$.

(a) f is called "lower semicontinuous" at $x \in X$ if for any $\lambda \in \mathbb{R}$ with $\lambda < f(x)$ there exists $U \in \mathcal{N}(x)$ such that $\lambda < f(u)$ for all $u \in U$. f is lower semicontinuous if it is lower semicontinuous at every $x \in X$.
(b) In (a) if $<$ is replaced by $>$, then f is called "upper semicontinuous" at $x \in X$. f is upper semicontinuous if it is upper semicontinuous at every $x \in X$.

Remark 1.177 It is clear from the definition that f is lower semicontinuous iff the set $L_\lambda = \{x \in X : f(x) \leq \lambda\}$ is closed for every $\lambda \in \mathbb{R}$; and f is upper semicontinuous iff the set $C_\lambda = \{x \in X : f(x) \geq \lambda\}$ is closed for every $\lambda \in \mathbb{R}$. Note that f is lower semicontinuous iff $-f$ is upper semicontinuous. According to the definition, f is lower semicontinuous at $x \in X$ iff

$$f(x) = \sup_{U \in \mathcal{N}(x)} \inf_{u \in U} f(u).$$

And similarly, f is upper semicontinuous at $x \in X$ iff

$$f(x) = \inf_{U \in \mathcal{N}(x)} \sup_{u \in U} f(u).$$

Proposition 1.178 *Let (X, τ) be a Hausdorff space and $f : X \to \mathbb{R}^*$. Then,*

(a) f is lower semicontinuous iff for any net $\{x_\vartheta\}_{\vartheta \in D} \subseteq X$ such that $x_\vartheta \to x \in X$ we have

$$f(x) \le \liminf_{\vartheta \in D} f(x_\vartheta) = \sup_{\vartheta \in D} \inf_{\vartheta \preccurlyeq \eta} f(x_\eta).$$

(b) f is upper semicontinuous iff for every net $\{x_\vartheta\}_{\vartheta \in D} \subseteq X$ such that $x_\vartheta \to x \in X$ we have

$$f(x) \ge \limsup_{\vartheta \in D} f(x_\vartheta) = \inf_{\vartheta \in D} \sup_{\vartheta \preccurlyeq \eta} f(x_\eta).$$

Proof (a) $\Rightarrow$: We assume that $f(x) > -\infty$. Fix $\lambda < f(x)$. Then $U_\lambda = \{x \in X : \lambda < f(x)\}$ is open (see Remark 1.177). Since $x \in U_\lambda$, we can find $\vartheta_0 \in D$ such that $x_\vartheta \in U_\lambda$ for all $\vartheta \succcurlyeq \vartheta_0$. Hence

$$\lambda < f(x_\vartheta) \quad \text{for all} \ \vartheta \succcurlyeq \vartheta_0,$$

thus $\lambda \le \liminf_\vartheta f(x_\vartheta)$. Since $\lambda < f(x)$ is arbitrary, we conclude that $f(x) \le \liminf_\vartheta f(x_\vartheta)$.

$\Leftarrow$: For $\lambda \in \mathbb{R}$ let $L_\lambda = \{x \in X : f(x) \le \lambda\}$. Consider a net $\{x_\vartheta\}_{\vartheta \in D} \subseteq L_\lambda$ such that $x_\vartheta \to x$ in X. Since $f(x_\vartheta) \le \lambda$ for all $\vartheta \in D$, by assumption we have $f(x) \le \liminf_\vartheta f(x_\vartheta) \le \lambda$, hence $x \in L_\lambda$ and so L_λ is closed. This implies that f is lower semicontinuous.

(b) It follows from (a) since f is upper semicontinuous iff $-f$ is lower semicontinuous. $\qquad\square$

Remark 1.179 Nets can be replaced by sequences if X is first countable.

Proposition 1.180 *Let X be a compact space, and $f : X \to \overline{\mathbb{R}} = \mathbb{R} \cup \{\infty\}$ lower semicontinuous with $f \neq \infty$. Then there exists $x_0 \in X$ such that*

$$f(x_0) = \inf_{x \in X} f(x)$$

and the set of all such minimizers is compact.

Proof Let $S = f(X)$ and $s \in S$. Then $L_s = \{x \in X : f(x) \le s\}$ is nonempty and closed. The family $\{L_s\}_{s \in S}$ has the finite intersection property. Thus, by Proposition 1.81, the set $C = \cap_{s \in S} L_s$, which is the set of minimizers of f, is nonempty compact. $\qquad\square$

Remark 1.181 Similarly, an upper semicontinuous $\mathbb{R}$-valued function on a compact space attains its supremum and the set of maximizers is compact.

Proposition 1.182 *Let (X, τ) be a Hausdorff space and S a family of lower semicontinuous (respectively, upper semicontinuous) functions. Then*

$$\widehat{f} = \sup_{S} f \quad (\text{respectively, } \widetilde{f} = \inf_{S} f)$$

is lower semicontinuous (respectively, upper semicontinuous).

Proof We only do the proof for the lower semicontinuous case. For each $\lambda \in \mathbb{R}$, set

$$\{x \in X : \widehat{f}(x) \le \lambda\} = \cap_S \{x \in X : f(x) \le \lambda\},$$

which is closed, hence $\widehat{f}$ is lower semicontinuous.   □

In metric spaces, we have a very useful approximation result for semicontinuous functions.

Proposition 1.183 *Let (X, d) be a metric space. (a) A bounded below $f : X \to \mathbb{R}$ is lower semicontinuous iff it is the pointwise limit of an increasing sequence of Lipschitz functions. (b) A bounded above $f : X \to \mathbb{R}$ is upper semicontinuous iff it is the pointwise limit of a decreasing sequence of Lipschitz functions.*

Proof (a) $\Rightarrow$: For each $n \in \mathbb{R}^N$, we define

$$f_n(x) = \inf[f(u) + nd(x, u) : u \in X].$$

Clearly, $f_n(x) \le f_{n+1}(x) \le f(x)$ for all $x \in X$ and $n \in \mathbb{N}$. Also, we have $|f_n(x) - f_n(y)| \le nd(x, y)$ for every $n \in \mathbb{N}$. Thus, f_n is Lipschitz. We have $f_n(x) \uparrow \widehat{f}(x) \le f(x)$ as $n \to \infty$ for every $x \in X$.

For fixed $x \in X$ and $\epsilon > 0$, we can find $u_n \in X$ such that

$$f(u_n) \le f(u_n) + nd(x, u_n) \le f_n(x) + \epsilon.$$

Let $\eta \in \mathbb{R}$ be a lower bound for f. From the above inequalities we have

$$d(x, u_n) \le \frac{f_n(x) + \epsilon - f(u_n)}{n} \le \frac{f(x) + \epsilon - \eta}{n}.$$

Thus $d(x, u_n) \to 0$ as $n \to \infty$. Since $f(u_n) \le f_n(x) + \epsilon$, we have $f(x) \le \widehat{f}(x) + \epsilon$ (see Proposition 1.178(a)). Therefore, $f \le \widehat{f}$, and $f = \widehat{f}$.

$\Leftarrow$: It follows from Proposition 1.182.

(b) It follows from (a) since $-f$ is lower semicontinuous.   □

A continuous function is both lower and upper semicontinuous. Thus we have

Corollary 1.184 *For any* $f \in C_b(X)$, *with* (X, d) *a metric space, there exist sequences* $\{k_n\}$ *and* $\{h_n\}$ *of bounded Lipschitz functions such that for any* $x \in X$,

$$k_n(x) \uparrow f(x) \ \text{ and } \ h_n(x) \downarrow f(x).$$

We conclude this section with a well-known fact about metric spaces.

Definition 1.185 Two metric spaces (X, d_x) and (Y, d_Y) are said to be "isometric" if there exists a bijection, namely a 1-1 and onto map, $f : X \to Y$ such that for all $x, u \in X$,

$$d_Y(f(x), f(u)) = d_X(x, u).$$

The function f is called an "isometry."

The next well-known theorem is usually proved by working on equivalence classes of Cauchy sequences. In Dugundji [90], p.304, we find a more elegant proof.

Theorem 1.186 *Every metric space can be isometrically embedded as a dense subset of a complete metric space, known as the "completion." The completion is unique up to an isometry.*

1.9 Remarks

The origin of topology can be traced back to the work of Riemann. However, the first abstract axiomatic definition based on neighborhoods was given by Hausdorff [136]. Alexandrov [6] produced the definition that is now used for topology. The notion of subbasis is due to Bourbaki. Nets and subnets were introduced by Moore [195] and developed further by Birkhoff [33] and Kelley [156] (in fact, Kelley is the one that coined the term "net"). There is an alternative approach based on filters due to Bourbaki [41]. Regular spaces were introduced by Vietoris [276] and normal spaces by Tietze [260]. The characterizations of normality in Theorems 1.42 and 1.47 are due to Urysohn [271] and Tietze [265], respectively. The countability axioms are due to Hausdorff [136]. The Lindelöf property was established for Euclidean spaces by Lindelöf [183] and extended to general topological spaces by Kuratowski-Sierpinski [170]. The notion of separability goes back to Frechet. Weak topologies were first considered by Bourbaki [41], who used the name "initial topologies." The quotient topology was first studied by Moore [197] and Alexandrov [6]. The term compact space was first used by Frechet [116] for metric spaces (sequential compactness). The general definition in terms of open covers (see Definition 1.79) is due to Alexandrov-Urysohn [7]. Theorem 1.102 (one of the most important results in topology) was proved by Tychonov [270]. In the definition of compactness, we can use only basic elements. There is a remarkable extension of this fact, known as the "Alexander Subbasis Theorem" (see, for example, Kelley [157], p.139).

Theorem 1.187 *A Hausdorff space is compact iff there exists a subbasis such that every open cover of the space by members of the subbasis has a finite subcover.*

Locally compact spaces were introduced independently by Alexandrov [6] and Tietze [266]. The one-point compactification (Theorem 1.109) is due to Alexandrov [6]. Theorem 1.116 (Urysohn Metrization Theorem) was proved by Urysohn [272].

There is another notion worth mentioning since it is associated with the useful tool of partition of unity, which is the notion of paracompactness introduced by Dieudonné [86].

Definition 1.188

(a) For covers S and S' of X, we say that S "refines" S' if for each $U \in S$ there exists $V \in S'$ such that $U \subseteq V$. In such a case we say S is a "refinement" of S' and write $S \prec S'$.
(b) A collection of subsets, $\mathcal{D}$, of a Hausdorff space is "locally finite" if for each x there exits $U \in \mathcal{N}(x)$ which intersects with only finite number of sets in $\mathcal{D}$.
(c) A Hausdorff space is "paracompact" if every open cover has an open locally finite refinement.

Theorem 1.189 *(a) Metric spaces are paracompact; (b) paracompact spaces are normal.*

As mentioned, paracompact spaces are associated with partitions of unity, which are "moving convex combinations" and a basic analytical tool in many other areas such as multivalued analysis (see Chap. 5) and fixed point theory. For a topological space (X, τ), the "support" of $f : X \to \mathbb{R}$ is the closed set $\{x \in X : f(x) \neq 0\}$, which is denoted by supp f. Notice that $u \notin$ supp f iff there exists $U \in \mathcal{N}(u) \cap \tau$ such that $f|_U \equiv 0$.

Definition 1.190 Given a Hausdorff space (X, τ), a family of continuous maps $f_\alpha : X \to [0, 1], \alpha \in I$, is called "partition of unity" of X, provided that

(a) $\{\text{supp } f_\alpha\}_{\alpha \in I}$ is a locally finite cover of X.
(b) $\sum_{\alpha \in I} f_\alpha(x) = 1$ (the sum is well defined due to (a)).

Let S be an open cover of X. We say that the partition of unity $\{f_\alpha\}_{\alpha \in I}$ is "subordinated" to S if each supp f_α is a subset of some $U_\alpha \in S$.

Theorem 1.191 *A Hausdorff space is paracompact iff every open cover of the space admits a locally finite partition of unity subordinated to it.*

The notion of connected spaces has its origins to the works of Cantor (1879) and Jordan (1893), while path-connectedness was used by Weierstrass (1885). Locally connected spaces are due to Hahn (1914).

Polish spaces were given the name to honor the famous school of Polish mathematicians, who in the interwar period (1919–1939) produced many ground breaking results in Analysis and Topology. Souslin spaces, also known as analytic sets, were named after the Soviet mathematician M.Y. Souslin, who initiated their study. We will return to them in Chap. 2. Polish and Souslin spaces are discussed in Bourbaki [43], Cohn [71], Dudley [89], Schwartz [252]. Theorem 1.158 is due to Kuratowski [167] (p.422) and Theorem 1.160 is due to Lavrentiev [171]. In Theorem 1.159, part (a) is due to Mazurkievicz [189] and part (b) is due to Alexandrov [6]. Spaces of first and second category were introduced by Baire [20]. Theorem 1.174 was proved by Baire [20] (for completely metrizable spaces) and Moore [196] (for locally compact spaces). As we will see in Chap. 3, two very basic results of Functional Analysis (the Open Mapping Theorem and the Uniform Boundedness Principle) are proved using Theorem 1.174. Another remarkable consequence of this theorem is the following result proved by Banach [23].

Theorem 1.192 *There exists a continuous function, $f : [0, 1] \to \mathbb{R}$, which is nowhere differentiable.*

Lower semicontinuous functions were introduced by Tonelli [267] and used in problems of Calculus of Variations (see also Fan [107]). More detailed discussion of these and other topics of Topology can be found in the books of Bourbaki [43], Choquet [66], Dugundji [91], Engelking [104], Kelley [157], Kuratowski [167, 168], Munkres [202], Nagata [203], and Willard [285].

We conclude with another result known as the "double limit lemma."

Lemma 1.193 *Let X be a metric space and $\{x_{m,n}\}_{mn \in \mathbb{N}} \subseteq X$. Assume that $x_m = \lim_{n \to \infty} x_{mn}$ and $x = \lim_{m \to \infty} x_m$. Then, there exist sequences $\{n_m\}_{m \in \mathbb{N}}$ and $\{m_n\}_{n \in \mathbb{N}}$ such that*

$$\lim_{m \to \infty} x_{mn_m} = \lim_{n \to \infty} x_{m_n n} = x.$$

1.10 Problems

Problem 1.1 Let (X, τ) be a Hausdorff space, $\{x_n\} \subseteq X$ such that $x_n \to x$. Show that $\{x_n, x\}_{n \geq 1} \subseteq X$ is compact. Show that the result fails for nets.

Problem 1.2 In a topological space X, does $\emptyset \neq A \subseteq X$ imply $\operatorname{bd} A = \operatorname{bd} \overline{A}$? Explain.

Problem 1.3 Given Hausdorff spaces X and Y, with Y compact, show that $f : X \to Y$ is continuous iff $\operatorname{Gr} f = \{(x, y) \in X \times Y : y = f(x)\}$ is closed.

Problem 1.4 Let τ_1 and τ_2 be comparable compact topologies on a set. Show that $\tau_1 = \tau_2$.

Problem 1.5 Let (X, τ) be a compact topological space and (Y, d) a metric space. Equip the space $C(X, Y) = \{f : X \to Y \text{ continuous}\}$ with the supremum metric:

$$\widehat{d}(f, h) = \max_{x \in X} d(f(x), h(x)).$$

Show that $(C(X, Y), \widehat{d})$ is complete iff (Y, d) is complete.

Problem 1.6 Let (X, d) be a metric space and

$$U_b(X, \mathbb{R}) = \{f : X \to \mathbb{R} \text{ uniformly continuous and bounded}\},$$

furnished with the supremum metric:

$$d_\infty(f, h) = \sup_{x \in X} |f(x) - h(x)|.$$

Show that X can be embedded isometrically into $U_b(X\mathbb{R})$.

Problem 1.7 Let (X, τ) be a compact topological space, Y is a metrizable space with two compatible metrics d_1 and d_2. Let $\widehat{d}_1$ and $\widehat{d}_2$ be the corresponding supremum metrics (see Problem 1.5). Show that $\widehat{d}_1$ and $\widehat{d}_2$ are topologically equivalent.

Problem 1.8 Show that a connected normal topological space is either a singleton or uncountable.

Problem 1.9 Let (X, τ) be a topological space, A and C are subsets of X with C connected. Assume that $A \cap C \neq \emptyset$ and $(X \setminus A) \cap C \neq \emptyset$. Show that bd $A \cap C \neq \emptyset$.

Problem 1.10 Let (X, τ_X) and (Y, τ_Y) be Hausdorff spaces, $f : X \to Y$ a closed function (see Definition 1.75) and assume that $f^{-1}(y) \subseteq X$ is compact for every $y \in Y$. Show that $f^{-1}(K)$ is compact in X for every compact $K \subseteq Y$.

Problem 1.11 Show that $S^n = \{x \in \mathbb{R}^{n+1} : |x| = 1\}$ is path-connected for each $n \in \mathbb{N}$.

Problem 1.12 Show that a Baire space cannot be expressed as a union of a sequence of nowhere dense subsets.

Problem 1.13 Let X be a Baire space and D_1, D_2 two dense subsets. Is it true that $D = D_1 \cap D_2$ must be dense in X? Explain.

Problem 1.14 Let (X, d) be a compact metric space, (Y, τ) a Hausdorff space, and $f : X \to Y$ a continuous surjection. Show that Y is metrizable.

Problem 1.15 Show that a locally compact dense subset in a metrizable space is open.

Problem 1.16 Show that every locally compact metrizable space is completely metrizable.

Problem 1.17 Show that a self-isometry in a compact metric space is surjective.

Problem 1.18 Let X be a regular Souslin space and $\mathcal{F}$ is a family of some continuous $\mathbb{R}$-valued functions which is separating (see Definition 1.61). Show that there is a countable subfamily $\mathcal{F}'$ of $\mathcal{F}$ which is separating too.

Problem 1.19 Show that every locally compact Souslin space is Polish.

Problem 1.20 Show that $\mathbb{Q}$ is not completely metrizable.

Problem 1.21 Is there an $\mathbb{R}$-valued function, which is continuous on the rationals and discontinuous on the irrationals? Explain.

Problem 1.22 Let $A \subseteq \mathbb{R}^N$ be connected and $A_\epsilon = \{u \in \mathbb{R}^N : d(u, A) \leq \epsilon\}$ for $\epsilon > 0$. Show that A_ϵ is path-connected.

Problem 1.23 Let $\chi_{\mathbb{Q}}(x) = 1$ if $x \in \mathbb{Q}$; and 0 if $x \in \mathbb{R} \setminus \mathbb{Q}$, which is the characteristic function of the set of the rationals. Show that $\chi_{\mathbb{Q}}$ is not the pointwise limit of a sequence of continuous functions on $\mathbb{R}$.

Problem 1.24 Let X and Y be compact metric spaces and $C(X, Y)$ be equipped with the supremum metric $\widehat{d}$ (see Problem 1.5). Is $(C(X, Y), \widehat{d})$ compact? Explain.

Problem 1.25 A Hausdorff space X is a "k-space," provided that $C \subseteq X$ is closed iff $C \cap K$ is closed for every compact $K \subseteq X$. Show that every locally compact space and every first countable space is a k-space.

Problem 1.26 Let (X, τ) be a countably compact Hausdorff topological space, $\varphi_n : X \to \mathbb{R}$ an increasing (respectively, decreasing) sequence of lower (respectively, upper) semicontinuous functions such that $\varphi_n(u) \to \varphi(u)$ for all $u \in X$ and $\varphi : X \to \mathbb{R}$ is upper (respectively, lower) semicontinuous. Show that φ is continuous and $\varphi_n \to \varphi$ uniformly. The result is known as the Dini's theorem.

Chapter 2
Measure Theory

This chapter is devoted to the presentation of the essential parts of "Measure Theory," which will be needed in the study of function spaces and in the applications. In the first section we introduce the basic language of measure theory and show how we can construct interesting examples of measures, starting from minimal basic ingredients. We also introduce the Lebesgue measure on $\mathbb{R}$, historically the starting point of the theory. In the second section with an eye on the integration theory of Lebesgue, we introduce and discuss measurable functions. In that process we also introduce product σ-algebras, product measures, and Caratheodory functions. In Sect. 2.3 we return to Polish and Souslin spaces and examine them from a measure theoretic viewpoint. We also discuss Borel spaces. In Sect. 2.4, starting from simple functions we define the Lebesgue integral of measurable functions. We establish the main properties of this integral and prove the basic convergence theorems, which are essentially different from those for the Riemann integral. In Sect. 2.5 we introduce and discuss measures with values in $\mathbb{R}^* = [-\infty, \infty]$ (signed measures), which lead to the fundamental "Lebesgue–Radon–Nikodym Theorem." It is one of the main theorems of measure theory and provides conditions for a signed measure to be expressed as an integral. It is also related to L^p-spaces, which are studied in Sect. 2.6. The Radon–Nikodym Theorem makes it possible to identify the dual of the Banach spaces L^p. In Sect. 2.7 we discuss various modes of convergence of sequences of measurable functions and introduce the notion of "Uniformly Integrable" set in L^1, which leads to a generalization of the Lebesgue dominated convergence theorem (Vitali's theorem). Finally, in Sect. 2.8 we examine the interplay between Measure Theory and Topology. The mixing of the two structures leads to new stronger results.

© The Author(s), under exclusive license to Springer Nature Switzerland AG 2022   47
S. Hu, N. S. Papageorgiou, *Research Topics in Analysis, Volume I*, Birkhäuser
Advanced Texts Basler Lehrbücher, https://doi.org/10.1007/978-3-031-17837-5_2

2.1   Algebras of Sets and Measures

Length, area and volume, are particular familiar instances of the notion of measure. In general, a measure is a set function, namely a function defined on a family of sets. Of course, some structure must be imposed on this family of sets. So, we start by discussing possible such structures.

Definition 2.1 Let X be a set and $\mathcal{S} \subseteq 2^X$.

(a) $\mathcal{S}$ is called an "algebra" (or a "field") if $X \in \mathcal{S}$ and $A \setminus B = A \cap (X \setminus B) \in \mathcal{S}$ for all $A, B \in \mathcal{S}$.

(b) $\mathcal{S}$ is called a "σ-algebra" (or a "σ-field") if it is an algebra and $\cup_{n \geq 1} A_n \in \mathcal{S}$ for any $\{A_n\}_{n \geq 1} \subseteq \mathcal{S}$.

(c) $(X, \mathcal{S})$ is called a "measurable space" if X is a set and $\mathcal{S} \subseteq 2^X$ is a σ-algebra.

Remark 2.2 It is easy to see that if $\mathcal{S}$ is an algebra, it contains the empty set and it is closed under finite unions and intersections. Moreover, an algebra $\mathcal{S} \subseteq 2^X$ is a σ-algebra if it is closed under countable "disjoint" unions. Indeed, for $\{A_n\}_{n \geq} \subseteq \mathcal{S}$, set $B_m = A_m \setminus \cup_{n=1}^{m-1} A_n$. Then these sets, B_m, are disjoint and $\cup_{n \geq 1} A_n = \cup_{m \geq 1} B_m$.

Proposition 2.3 *For any given family $\mathcal{F} \subseteq 2^X$, there exists a unique smallest σ-algebra containing $\mathcal{F}$ and it is denoted by $\sigma(\mathcal{F})$.*

Proof Let $\mathcal{S} = \cap \{\mathcal{D} : \mathcal{D}$ is a σ-algebra such that $\mathcal{F} \subseteq \mathcal{D}\}$. Note that 2^X is a σ-algebra. Thus $\mathcal{S} \neq \emptyset$. It is routine to check that $\mathcal{S}$ is a σ-algebra. Therefore, $\mathcal{S} = \sigma(\mathcal{F})$. $\square$

The σ-algebra $\sigma(\mathcal{F})$ is characterized as follows.

Proposition 2.4 *For $X \in \mathcal{F} \subseteq 2^X$, $\sigma(\mathcal{F})$ is the smallest family in 2^X containing $\mathcal{F}$ and such that*

(a) $A^c = X \setminus A \in \sigma(\mathcal{F})$ for any $A \in \mathcal{F}$.

(b) $\sigma(\mathcal{F})$ is closed under countable intersections.

(c) $\sigma(\mathcal{F})$ is closed under countable disjoint unions.

Proof Let $\mathcal{A}$ be the smallest family of sets in X such that $\mathcal{F} \subseteq \mathcal{A}$ and it satisfies (a)-(b)-(c). Then $\mathcal{A} \subseteq \sigma(\mathcal{F})$. Let $\mathcal{M} = \{A \in \mathcal{A} : A^c \in \mathcal{A}\}$. By (a) we have $\mathcal{F} \subseteq \mathcal{M} \subseteq \mathcal{A}$.

Claim: $\mathcal{M}$ is a σ-algebra.

First, we show that $A \setminus B \in \mathcal{M}$ if $A, B \in \mathcal{M}$. Since $A^c, B^c \in \mathcal{M}$, we have $A \setminus B = A \cap B^c \in \mathcal{A}$ (see (6)) and also $(A \setminus B)^c = A^c \cup B \in \mathcal{A}$ (see (b)). Therefore, $A \setminus B \in \mathcal{M}$ and so $\mathcal{M}$ is an algebra. Suppose $\{A_n\} \subseteq \mathcal{M}$. Since $\mathcal{M}$ is an algebra, we may assume that the sequence $\{A_n\}$ is disjoint (see Remark 2.2) and so by (c) we have $\cup_{n \geq 1} A_n \in \mathcal{A}$. We have $(\cup_{n \geq 1} A_n)^c = \cap_{n \geq 1} A_n^c \in \mathcal{A}$ (see (b)). We infer that $\cup_{n \geq 1} A_n \in \mathcal{M}$ and so $\mathcal{M}$ is a σ-algebra. Hence we have $\sigma(\mathcal{F}) \subseteq \mathcal{M} \subseteq \mathcal{A}$, from which we conclude that $\sigma(\mathcal{F}) = \mathcal{A}$. $\square$

Definition 2.5 Let X be a set and $\mathcal{M} \subseteq 2^X$. Then,

(a) $\mathcal{M}$ is a "monotone class" if it is closed under countable increasing unions and countable decreasing intersections, namely if $A_n \in \mathcal{M}$ and $A_n \uparrow A$ or $A_n \downarrow A$, then $A \in \mathcal{M}$.
(b) $\mathcal{M}$ is a "Dynkin system," provided that (1) $X \in \mathcal{M}$; (2) $A \setminus B \in \mathcal{M}$ for any $A, B \in \mathcal{M}$ with $B \subseteq A$; (3) $A \in \mathcal{M}$ if $A_n \uparrow A$ with $A_n \in \mathcal{M}$.

Remark 2.6 If $\mathcal{M}$ is a Dynkin system, then (1) and (2) imply that $\mathcal{M}$ is closed under complementation. Hence by (3), a Dynkin system is a monotone class. Moreover, if $\mathcal{M}$ is also closed under finite unions or intersections, then it is a σ-algebra.

Proposition 2.7 *Let X be a set, $\mathcal{S} \subseteq 2^X$ an algebra, $\mathcal{M}$ a monotone class and $\mathcal{S} \subseteq \mathcal{M}$. Then $\sigma(\mathcal{S}) \subseteq \mathcal{M}$.*

Proof Let $\mathcal{M}_0$ be the smallest monotone class containing $\mathcal{S}$. We will show that $\mathcal{M}_0 = \sigma(\mathcal{S})$ (the smallest monotone class and the smallest σ-algebra generated by an algebra coincide).

Let $A \in \mathcal{M}_0$ and $\mathcal{M}_A = \{B \in \mathcal{M}_0 : A \cap B, A \cap B^c, A^c \cap B \in \mathcal{M}_0\}$. Then $\mathcal{M}_A$ is a monotone class. Actually we have $\mathcal{M}_A = \mathcal{M}_0$. Indeed, if $A \in \mathcal{S}$ then $\mathcal{S} \subseteq \mathcal{M}_A$ since $\mathcal{S}$ is an algebra and so $\mathcal{M}_0 = \mathcal{M}_A$. But then for any $B \in \mathcal{M}_0$ we have $A \cap B, A \cap B^c, A^c \cap B \in \mathcal{M}_0$ for any $A \in \mathcal{S}$ and so $\mathcal{S} \subseteq \mathcal{M}_B$ and thus $\mathcal{M}_0 = \mathcal{M}_B$ for all $B \in \mathcal{M}_0$.

Note that $\mathcal{M}_0$ is an algebra. However, a monotone class that is also an algebra is in fact a σ-algebra. Therefore, $\sigma(\mathcal{S}) = \mathcal{M}_0 \subseteq \mathcal{M}$. $\square$

An interesting byproduct of the above proof is that the smallest monotone class and the smallest σ-algebra over an algebra coincide. A companion to Proposition 2.7 is the so-called Dynkin System Theorem.

Theorem 2.8 *Let $\mathcal{S} \subseteq 2^X$ be closed under finite intersections, $\mathcal{M}$ a Dynkin system and $\mathcal{S} \subseteq \mathcal{M}$. Then $\sigma(\mathcal{S}) \subseteq \mathcal{M}$.*

Proof Let $\mathcal{M}_0$ be a the smallest Dynkin system containing $\mathcal{S}$. We will show that $\mathcal{M}_0 = \sigma(\mathcal{S})$, namely the smallest Dynkin system and the smallest σ-algebra over a family closed under finite intersections, coincide.

Since $\sigma(\mathcal{S})$ is a Dynkin system (see Definition 2.5(b)), we have $\mathcal{M}_0 \subseteq \sigma(\mathcal{S})$. Let $\mathcal{D} = \{A \in \mathcal{M}_0 : A \cap B \in \mathcal{M}_0 \text{ for all } B \in \mathcal{S}\}$. Since $\mathcal{S}$ is closed under finite intersections, we see that $\mathcal{S} \subseteq \mathcal{D}$. Because $\mathcal{M}_0$ is a Dynkin system, then so is $\mathcal{D}$ and we have $\mathcal{M}_0 \subseteq \mathcal{D}$. Therefore, $\mathcal{M}_0 = \mathcal{D}$.

Let $\mathcal{D}' = \{C \in \mathcal{M}_0 : C \cap D \in \mathcal{M}_0 \text{ for all } D \in \mathcal{M}_0\}$. Since $\mathcal{M}_0 = \mathcal{D}$, we have $\mathcal{S} \subseteq \mathcal{D}'$. Because $\mathcal{D}'$ is a Dynkin system, it follows that $\mathcal{M}_0 \subseteq \mathcal{D}'$ and thus $\mathcal{M}_0 = \mathcal{D}'$. So, $\mathcal{M}_0$ is closed under finite intersections and hence a σ-algebra (see Remark 2.6). We conclude that $\mathcal{M}_0 = \sigma(\mathcal{S}) \subseteq \mathcal{M}$. $\square$

Remark 2.9 Again, a byproduct of the above proof is that if $S \subseteq 2^X$ is closed under finite intersections, then the smallest Dynkin system and the smallest σ-algebra over S coincide.

One of the most important σ-algebras arises when the set X admits a topological structure.

Definition 2.10 Let (X, τ) be a topological space. The "Borel σ-algebra" of X is the σ-algebra $\sigma(\tau)$, denoted by $B(X)$. Elements of $B(X)$ are known as "Borel sets."

Remark 2.11 $B(X)$ is the smallest Dynkin system containing the open sets (or the closed sets), see Remark 2.9.

Also, from Proposition 2.4 we have

Proposition 2.12 *For a topological space (X, τ), $B(X)$ is the smallest family in 2^X containing all open sets and all closed sets, and it is closed under countable intersections and countable disjoint unions.*

In a metric space, every closed set is G_δ (see Proposition 1.157 and Definition 1.153). So, the above proposition becomes

Proposition 2.13 *For a metrizable space X, $B(X)$ is the smallest family in 2^X containing the open sets and closed under countable intersections and under countable disjoint unions.*

Remark 2.14 Since every open set is F_σ (see Proposition 1.157 and Definition 1.153), in order to have the above proposition with open sets replaced by closed ones we have to require closedness under all countable unions, not just disjoint ones. For $X = \mathbb{R}$, $B(\mathbb{R})$ is generated by each of the following families: (a) open intervals; (b) closed intervals; (c) half-open intervals; (d) open rays; (e) closed rays.

In the sequel, σ-algebras will be denoted by Σ.

Definition 2.15 Let Σ be a σ-algebra on a set X. $\mu : \Sigma \to [0, \infty]$ is a "measure" if the following properties hold true:

(a) $\mu(\emptyset) = 0$.
(b) $\mu(\cup_{n \geq 1} A_n) = \Sigma_{n \geq 1} \mu(A_n)$ for any mutually disjoint $A_n \in \Sigma$ (countable additivity).

The triple (X, Σ, μ) is called a "measure space." We say that μ is "finite" if $\mu(X) < \infty$; "σ-finite" if $X = \cup_{n \geq 1} A_n$ for some $A_n \in \Sigma$ with $\mu(A_n) < \infty$ for all $n \geq 1$.

Remark 2.16 If μ is finite, then $\mu(A) < \infty$ for every $A \in \Sigma$ since $\mu(A) + \mu(A^c) = \mu(X) < \infty$.

The next proposition summarizes the main properties of a measure.

Proposition 2.17 *In a measure space* (X, Σ, μ) *we have*

(a) $\mu(A) \leq \mu(B)$ *for* $A, B \in \Sigma$ *with* $A \subseteq B$ *(monotonicity).*
(b) $\mu(\cup_{n\geq1} A_n) \leq \Sigma_{n\geq1} \mu(A_n)$ *for any* $\{A_n\} \subseteq \Sigma$ *(subadditivity).*
(c) $\mu(A_n) \uparrow \mu(A)$ *for* $A_n \subseteq \Sigma$ *with* $A_n \uparrow A$ *(continuity from below).*
(d) $\mu(A_n) \downarrow \mu(A)$ *for* $\{A_n\} \subseteq \Sigma$ *with* $\{A_n\} \downarrow A$ *and* $\mu(A_1) < \infty$ *(continuity from above).*

Proof

(a) We have $\mu(B) = \mu(A) + \mu(B \setminus A) \geq \mu(A)$.
(b) As in Remark 2.2 we consider a sequence $\{B_n\} \subseteq \Sigma$ of mutually disjoint sets such that $B_n \subseteq A_n$ and $\cup_{n\geq1} A_n = \cup_{n\geq1} B_n$. By (a) we

$$\mu(\cup_{n\geq1} A_n) = \mu(\cup_{n\geq1} B_n) = \Sigma_{n\geq1} \mu(B_n) \leq \Sigma_{n\geq1} \mu(A_n).$$

(c) Let $A_0 = \emptyset$. We have

$$\mu(\cup_{n\geq1} A_n) = \Sigma_{n\geq1} \mu(A_n \setminus A_{n-1}) = \lim_{m\to\infty} \Sigma_{n=1}^{m} \mu(A_n \setminus A_{n-1}) = \lim_{n\to\infty} \mu(A_n).$$

(d) We have $A_1 \setminus A_n \uparrow A_1 \setminus A$ since $A_n \downarrow A$. Hence from (c) we have

$$\mu(A_1 \setminus A_n) \uparrow \mu(A_1 \setminus A).$$

But since $\mu(A_1) < \infty$, we have $\mu(A_1 \setminus A_n) = \mu(A_1) - \mu(A_n)$ and $\mu(A_1 \setminus A) = \mu(A_1) - \mu(A)$. Therefore, $\mu(A_n) \downarrow \mu(A)$. □

Definition 2.18

(a) $A \in \Sigma$ is called an "μ-null set" if $\mu(A) = 0$.
(b) A property that is true for all $x \in X$, except on a μ-null set, is said to hold "almost everywhere" or "for almost all" x (abbreviated by a.e. or a. a.).
(c) A measure μ on a σ-algebra Σ is said to be "complete" iff every subset of a null set belongs to Σ.

Completeness of a measure is often desirable and it can be easily realized by simply enlarging the domain of μ, as stated in the proposition below. The proof of this result is straightforward and so it is omitted.

Proposition 2.19 *Let* (X, Σ, μ) *be a measure space, Define*

$$\Sigma_\mu = \{A \cup D : A \in \Sigma, D \subseteq B \in \Sigma \text{ for some } B \text{ with } \mu(B) = 0, \}$$

and $\widehat{\mu} : \Sigma_\mu \to \mathbb{R}_+$ *by* $\widehat{\mu}(A \cup D) = \mu(A)$. *Then* Σ_μ *is a* σ-*algebra and* $\widehat{\mu}$ *is complete on* Σ_μ.

Remark 2.20 The measure space $(X, \Sigma_\mu, \widehat{\mu})$ is called the "completion" of (X, Σ, μ) and $\widehat{\mu}$ the "completion" of μ.

The notion of outer measure is essential in constructing measures. It generalizes the notion of outer area from elementary geometry.

Definition 2.21 An "outer measure" is a set function $\mu : 2^X \to [0, \infty]$ such that $\mu(\emptyset) = 0$ and it is monotone (namely, $\mu(A) \leq \mu(B)$ if $A \subseteq B$) and σ-subadditive (namely, $\mu(\cup_{n \geq 1} A_n) \leq \Sigma_{n \geq 1} \mu(A_n)$ for $\{A_n\} \subseteq 2^X$).

Proposition 2.22 *Let Σ be a σ-algebra on X, μ a measure on Σ. Define*

$$\mu^*(B) = \inf[\mu(A) : B \subseteq A, A \in \Sigma].$$

Then μ^* is an outer measure on X.

Proof Clearly, $\mu^*(\emptyset) = 0$. Also, for $B_1 \subseteq B_2 \subseteq X$ we have $\{A \in \Sigma : B_2 \subseteq A\} \subseteq \{A \in \Sigma : B_1 \subseteq A\}$, thus $\mu^*(B_1) \leq \mu^*(B_2)$. Finally, for a given sequence $\{B_n\} \subseteq 2^X$ we consider $B_n \subseteq A_n \in \Sigma$. Then $\mu(\cup_{n \geq 1} A_n) \leq \Sigma_{n \geq 1} \mu(A_n)$. By taking infima we conclude that $\mu^*(\cup_{n \geq 1} B_n) \leq \Sigma_{n \geq 1} \mu^*(B_n)$. $\square$

In fact we can produce outer measures with less information.

Proposition 2.23 *Let $\mathcal{S} \subseteq 2^X$ and $\mu : \mathcal{S} \to [0, \infty]$ such that $\emptyset, X \in \mathcal{S}$ and $\mu(\emptyset) = 0$. For any $B \subseteq X$, define*

$$\mu^*(B) = \inf[\sum_{n \geq 1} \mu(A_n) : A_n \in \mathcal{S} \text{ and } B \subseteq \bigcup_{n \geq 1} A_n].$$

Then μ^* is an outer measure on X.

Proof μ^* is well defined since $X \in \mathcal{S}$. Also, $\mu^*(\emptyset) = 0$ since $\emptyset \in \mathcal{S}$. Moreover, as in the previous proof we have $\mu^*(B_1) \leq \mu^*(B_2)$ for $B_1 \subseteq B_2 \subseteq X$. Finally, suppose $\{B_n\} \subseteq X$ and $\epsilon > 0$. For each $n \geq 1$ there exists $\{A_n^m\} \subseteq \mathcal{S}$ such that $B_n \subseteq \cup_{m \geq 1} A_n^m$ and $\sum_{m \geq 1} \mu(A_n^m) \leq \mu^*(B_n) + \frac{\epsilon}{2^m}$. Then $\cup_{n \geq 1} B_n \subseteq \cup_{m,n \geq 1} A_n^m$ and $\sum_{m,n \geq 1} \mu(A_n^m) \leq \sum_{n \geq 1} \mu^*(B_n) + \epsilon$. Thus, $\mu^*(\cup_{n \geq 1} B_n) \leq \sum_{n \geq 1} \mu^*(B_n) + \epsilon$. We are done by letting $\epsilon \to 0^+$. $\square$

The basic step that leads from outer measures to measures is based on the following notion.

Definition 2.24 Let μ^* be an outer measure on X. $A \subseteq X$ is called "μ^*-measurable" if $\mu^*(B) = \mu^*(B \cap A) + \mu^*(B \cap A^c)$ for all $B \subseteq X$, namely A splits additively over every set in X.

Remark 2.25 To motivate the definition, note that if A is μ^*-measurable and $A \subseteq B$, then the outer measure $\mu^*(A)$ equals the inner measure $\mu^*(B) - \mu^*(B \cap A^c)$ of A. Also note that on account of the subadditivity of μ^*, in order to check the μ^*-

measurability of a set A, it suffices to show that $\mu^*(B \cap A) + \mu^*(B \cap A^c) \leq \mu^*(B)$ for all $B \subseteq X$.

The next theorem justifies Definition 2.24.

Theorem 2.26 *Let μ^* be an outer measure on X and Σ_{μ^*} the collection of all μ^*-measurable sets. Then Σ_{μ^*} is a σ-algebra and $\mu^*|_{\Sigma_{\mu^*}}$ is a complete measure.*

Proof Since Definition 2.24 is symmetric in A and A^c, the collection Σ_{μ^*} is closed under complementation. Also, $\mu^*(B) = \mu^*(B) + \mu^*(\emptyset)$ for all $B \subseteq X$, hence $X, \emptyset \in \Sigma_{\mu^*}$. Now let $A, D \in \Sigma_{\mu^*}$ and $B \subseteq X$, we have

$$\mu^*(B) = \mu^*(B \cap A) + \mu^*(B \cap A^c)$$
$$= \mu^*(B \cap A \cap D) + \mu^*(B \cap A \cap D^c)$$
$$+ \mu^*(B \cap A^c \cap D) + \mu^*(B \cap A^c \cap D^c).$$

Note that $A \cup D = (A \cap D) \cup (A \cap D^c) \cup (A^c \cap D)$. By subadditivity we have

$$\mu^*(B \cap (A \cup D)) \leq \mu^*(B \cap A \cap D) + \mu^*(B \cap A \cap D^c) + \mu^*(B \cap A^c \cap D),$$

thus $\mu^*(B) \geq \mu^*(B \cap (A \cup D)) + \mu^*(B \cap (A \cup D)^c)$ and so, $A \cup D \in \Sigma_{\mu^*}$. This proves that Σ_{μ^*} is an algebra. Moreover, for $A, D \in \Sigma_{\mu^*}$ with $A \cap D = \emptyset$, we have

$$\mu^*(A \cup D) = \mu^*((A \cup D) \cap A) + \mu^*((A \cup D) \cap A^c) = \mu^*(A) + \mu^*(D),$$

hence μ^* is additive (by induction).

According to Remark 2.2, to show that Σ_{μ^*} is a σ-algebra it suffices to show that it is closed under countable disjoint unions. So, let $\{A_n\} \subseteq \Sigma_{\mu^*}$ be a pairwise disjoint sequence. Let $D_m = \bigcup_{n=1}^{m} A_n \in \Sigma_{\mu^*}$, $A = \bigcup_{n \geq 1} A_n$ and $B \subseteq X$. We have

$$\mu^*(B) = \mu^*(B \cap D_m) + \mu^*(B \cap D_m^c)$$
$$\geq \mu^*(B \cap D_m) + \mu^*(B \cap A^c) \text{ (by monotonicity)}$$
$$= \sum_{n=1}^{m} \mu^*(B \cap A_n) + \mu^*(B \cap A^c) \text{ (by additivity).}$$

This is true for all $m \in \mathbb{N}$. Let $m \to \infty$ and use the subadditivity of μ^*. We obtain

$$\mu^*(B) \geq \mu^*(B \cap A) + \mu^*(B \cap A^c),$$

thus $A \in \Sigma_{\mu^*}$ and so Σ_{μ^*} is a σ-algebra.

For each $m \in \mathbb{N}$ we have

$$\sum_{n=1}^{m} \mu^*(A_n) = \mu^*(\bigcup_{n=1}^{m} A_n) \le \mu^*(\bigcup_{n \ge 1} A_n) \le \sum_{n \ge 1} \mu^*(A_n),$$

thus μ^* is a measure on Σ_{μ^*}.

Finally, if $\mu^*(A) = 0$ then we have for any $B \subseteq X$,

$$\mu^*(B) \le \mu^*(B \cap A) + \mu^*(B \cap A^c) \le \mu^*(B \cap A^c) \le \mu^*(B),$$

thus $A \in \Sigma_{\mu^*}$ and so, $\mu^*|_{\Sigma_{\mu^*}}$ is a complete measure.   □

Let $S \subseteq 2^X$ be an algebra and $\mu : S \to [0, \infty]$ a measure in the sense that $\mu(\emptyset) = 0$ and $\mu(\cup_{n \ge 1} A_n) = \Sigma_{n \ge 1} \mu(A_n)$ for any $\{A_n\} \subseteq S$ pairwise disjoint such that $\cup_{n \ge 1} A_n \in S$. As before for $B \subseteq X$ we set

$$\mu^*(B) = \inf \left[\sum_{n \ge 1} \mu(A_n) : A_n \in S, B \subseteq \cup_{n \ge 1} A_n \right]. \tag{2.1}$$

Proposition 2.27 (a) $\mu^*|_S = \mu$; (b) $S \subseteq \Sigma_{\mu^*}$.

Proof

(a) Let $B \in S$, with $B \subseteq \cup_{n \ge 1} A_n$ for some $A_n \in S$, Set $C_n = B \cap (A_n) \setminus \cup_{k=1}^{n-1} A_k \in S$, which are pairwise disjoint with $\cup_{n \ge 1} C_n = B$. Hence, $\mu(B) = \Sigma_{n \ge 1} \mu(C_n) \le \Sigma_{n \ge 1} \mu(A_n)$ and consequently $\mu(B) \le \mu^*(B)$. The opposite inequality is obvious since $B \in S$. Therefore, $\mu^*|_S = \mu$.

(b) Let $D \in S, B \subseteq X$ and $\epsilon > 0$. Find $\{A_n\} \subseteq S$ such that $B \subseteq \cup_{n \ge 1} A_n$ and $\Sigma_{n \ge 1} \mu(A_n) \le \mu^*(B) + \epsilon$ (see (2.1)). The additivity of μ on S implies

$$\mu^*(B \cap D) + \mu^*(B \cap D^c) \le \Sigma_{n \ge 1} \mu(A_n \cap D) + \Sigma_{n \ge 1} \mu(A_n \cap D^c) \le \mu^*(B) + \epsilon,$$

thus $\mu^*(B \cap D) + \mu^*(B \cap D^c) \le \mu^*(B)$ (letting $\epsilon \downarrow 0$), hence $D \in \Sigma_{\mu^*}$, namely $S \subseteq \Sigma_{\mu^*}$.   □

Next we can state the theorem that gives a unique measure out of an outer measure.

Theorem 2.28 Let $S \subseteq 2^X$ be an algebra, $\mu : S \to [0, \infty]$ a measure as above, which is σ-finite and $\Sigma = \sigma(S)$. Then, there exists a unique measure $\widehat{\mu} : \Sigma \to [0, \infty]$ such that

$$\widehat{\mu}|_S = \mu \text{ and } \mu^*|_\Sigma = \widehat{\mu} \text{ (see (1)).}$$

Proof By Theorem 2.26 there exists this extension $\widehat{\mu}$ of μ (note that $S \subseteq \Sigma \subseteq \Sigma_{\mu^*}$). Next we show the uniqueness of this extension. So, suppose that $\widetilde{\mu} : \Sigma \to$

$[0, \infty]$ is another such extension of μ. Let $B \in \Sigma$ and $B \subseteq \cup_{n \geq 1} A_n$ with $A_n \in \Sigma$. Then $\widetilde{\mu}(B) \leq \Sigma_{n \geq 1} \widetilde{\mu}(A_n) = \Sigma_{n \geq 1} \mu(A_n)$, hence $\widetilde{\mu}(B) \leq \widehat{\mu}(B)$. If $D = \cup_{n \geq 1} A_n$, then

$$\widetilde{\mu}(D) = \lim_{m \to \infty} \widetilde{\mu}(\cup_{n=1}^m A_n) = \lim_{m \to \infty} \widehat{\mu}(\cup_{n=1}^m A_n) = \widehat{\mu}(D). \tag{2.2}$$

If $B \in \Sigma$ and $\widehat{\mu}(B) < \infty$, then we choose $\{A_n\} \subseteq \mathcal{S}$ such that $\widehat{\mu}(D) \leq \widehat{\mu}(B) + \epsilon$. Thus $\widehat{\mu}(D \setminus B) < \epsilon$. We have

$$\widehat{\mu}(B) \leq \widehat{\mu}(D) \text{ (see (2.2))}$$
$$= \widetilde{\mu}(B) + \widetilde{\mu}(D \setminus B) \leq \widetilde{\mu}(B) + \widehat{\mu}(D \setminus B)$$
$$\leq \widetilde{\mu}(B) + \epsilon,$$

thus $\widehat{\mu}(B) = \widetilde{\mu}(B)$.

By assumption, $X = \cup_{k \geq 1} C_k$, with $C_k \in \mathcal{S}$ and $\widehat{\mu}(C_k) = \mu(C_k) < \infty$. Then for any $B \in \Sigma$ we have

$$\widehat{\mu}(B) = \sum_{k \geq 1} \widehat{\mu}(B \cap C_k) = \sum_{k \geq 1} \widetilde{\mu}(B \cap C_k) = \widetilde{\mu}(B),$$

namely $\widehat{\mu} = \widetilde{\mu}$. □

Using this theorem we can produce the Lebesgue measure on $\mathbb{R}$. We start with the half-open intervals of the form: $(a, b]$ or $[a, b)$ or $\emptyset$, with $-\infty \leq a < b < \infty$. Clearly, this family is closed under finite intersections and if A, C are sets in this family, then $A \setminus C$ is the disjoint union of elements from the family (such a collection of sets is usually called "semiring"). Then the finite disjoint unions of such sets form an algebra $\mathcal{S}$ and $\sigma(\mathcal{S}) = B(\mathbb{R})$ (see Remark 2.14). For $B \subseteq \mathbb{R}$ we set

$$\mu^*(B) = \inf \left[\sum_{n \geq 1} (b_n - a_n) : B \subseteq \bigcup_{n \geq 1} (a_n, b_n] \right].$$

Then, this is an outer measure and Σ_{μ^*} is the σ-algebra of "Lebesgue measurable sets," usually denoted by $\mathcal{L}_{\mathbb{R}}$ and $B(\mathbb{R}) \subseteq \mathcal{L}_{\mathbb{R}}$. The unique measure λ on $\mathbb{R}$, generated by Theorem 2.28, is known as the "Lebesgue measure" on $\mathbb{R}$. The most important properties of λ are the translation invariance and its simple behavior under dilations. These properties follow easily from the above construction of λ.

Proposition 2.29 *For all $B \in \mathcal{L}_{\mathbb{R}}$ and $a, b \in \mathbb{R}$, we have*

$$\lambda(B) = \lambda(B + a) \text{ and } \lambda(bB) = |b| \lambda(B).$$

Remark 2.30 Up to a multiplication with a positive constant, λ is the unique measure exhibiting these two properties.

2.2 Measurable Functions

In this section we discuss measurable functions, on which we define the Lebesgue integral (see Sect. 2.4).

Definition 2.31

(a) Let X, Y be topological spaces. $f : X \to Y$ is "Borel measurable, " if $f^{-1}(B) \in B(X)$ for each $B \in B(Y)$.
(b) Let (X, Σ) and $(Y, \mathcal{M})$ be measurable spaces. $f : X \to Y$ is "$(\Sigma, \mathcal{M})$-measurable" (or simply "measurable" if the σ-algebras are clearly understood), if $f^{-1}(B) \in \Sigma$ for each $B \in \mathcal{M}$.
(c) Let $X = \mathbb{R}$ and Y a topological space. $f : \mathbb{R} \to Y$ is "Lebesgue measurable" if $f^{-1}(B) \in \mathcal{L}_{\mathbb{R}}$ for each $B \in B(Y)$.

Remark 2.32 Note that in both the Borel and Lebesgue measurability the Borel σ-algebra is used on the range space. Indeed, there exists a continuous nondecreasing $f : [0, 1] \to [0, 1]$ and $B \in \mathcal{L}_{\mathbb{R}}$ such that $f^{-1}(B) \notin \mathcal{L}_{\mathbb{R}}$, assuming as usual the axiom of choice. This function is known as the "Cantor function" and its construction can be found in Dudley [85], p. 124. It is easy to see that in Definition 2.31(b) if $\mathcal{M} = \sigma(\mathcal{S})$ and $f : X \to Y$ is such that $f^{-1}(B) \in \Sigma$ for each $B \in \mathcal{S}$, then f is measurable. Composition of measurable functions is measurable; continuous functions between topological spaces are Borel measurable.

Combining Remarks 2.14 and 2.32 we have

Proposition 2.33 *Let (X, Σ) be a measurable space and $f : X \to \mathbb{R}$. Then the following statements are equivalent.*

(a) f is measurable.
(b) $f^{-1}((a, \infty)) \in \Sigma$ for any $a \in \mathbb{R}$.
(c) $f([a, \infty)) \in \Sigma$ for any $a \in \mathbb{R}$.
(d) $f^{-1}((-\infty, a)) \in \Sigma$ for any $a \in \mathbb{R}$.
(e) $f^{-1}((-\infty, a]) \in \Sigma$ for any $a \in \mathbb{R}$.

Remark 2.34 If $f : X \to \mathbb{R}^* = \mathbb{R} \cup \{\pm\infty\}$ and $V = f^{-1}(\mathbb{R})$, then f is measurable iff $f^{-1}(\infty) \in \Sigma$, $f^{-1}(-\infty) \in \Sigma$, and $f : V \to \mathbb{R}$ is measurable (on V we consider the trace of Σ, namely the σ-algebra $\Sigma_V = \Sigma \cap V$).

Proposition 2.35 *Let (X, Σ) be a measurable space, (Y, d) a metric space, $f_n : X \to Y$ measurable for each $n \in \mathbb{N}$ with $f_n(x) \to f(x)$ in Y for each $x \in X$. Then $f : X \to Y$ is measurable.*

Proof Let $U \subseteq Y$ be open and define $C_m = \{u \in U : B_{1/m}(u) \subseteq U\}$, which is closed. To see this, let $\{u_n\}_{n \geq 1} \subseteq C_m$ and assume that $u_n \to u$. Consider $y \in Y$ such that $d(u, y) < 1/m$. We can find $n_0 \in \mathbb{R}$ such that $d(u_n, y) < 1/m$ for all $n \geq n_0$, hence $y \in U$ and so, $B_{1/m}(u) \subseteq U$, which means that $u \in C_m$. Note that

$$f^{-1}(U) = \bigcup_{m \in \mathbb{N}} \bigcup_{k \in \mathbb{N}} \bigcap_{n \geq k} f_n^{-1}(C_m) \in \Sigma,$$

which proves the measurability of f. $\qquad\square$

Proposition 2.36 *Let* (X, Σ) *be a measurable space and* $f_n : X \to \mathbb{R}$ *is measurable for all* $n \in \mathbb{N}$. *Then* $f^*(x) = \sup_{n \geq 1} f_n(x)$ *and* $f_*(x) = \inf_{n \geq 1} f_n(x)$ *are both measurable.*

Proof For $a \in \mathbb{R}$, $f_n^{-1}((a, \infty)) \subseteq (f^*)^{-1}((a, \infty))$ and so, $\cup_{n \geq 1} f_n^{-1}((a, \infty)) \subseteq (f^*)^{-1}((a, \infty))$. Suppose that $x \notin \cup_{n \geq 1} f_n^{-1}((a, \infty))$. Then $f_n(x) \leq a$ for all $n \in \mathbb{N}$ and so $f^*(x) \leq a$, hence $x \notin (f^*)^{-1}((a, \infty))$. Thus, $\cup_{n \geq 1} f_n^{-1}((a, \infty)) \supseteq (f^*)^{-1}((a, \infty))$ and we conclude that $(f^*)^{-1}((a, \infty)) = \cup_{n \geq 1} f_n^{-1}((a, \infty))$ and so $(f^*)^{-1}((a, \infty)) \in \Sigma$. By Proposition 2.33 this proves the measurability of f^*. Since $f_*(x) = - \sup(-f_n(x))$, we also have the measurability of f_*. $\qquad\square$

Corollary 2.37 *Let* (X, Σ) *be a measurable space and* $f_n : X \to \mathbb{R}$ *measurable for each* $n \in \mathbb{N}$. *Then* $x \to \liminf_{n \to \infty} f_n(x)$ *and* $x \to \limsup_{n \to \infty} f_n(x)$ *are both measurable.*

Remark 2.38 So, we can say that the set of measurable functions is a linear space which is also an algebra, namely if f, g are both measurable, then so is $x \to (fg)(x)$.

To continue the discussion of measurable functions we need to introduce the so-called product σ-algebra.

Definition 2.39 Let $\{(X_\alpha, \Sigma_\alpha)\}_{\alpha \in I}$ be a family of measurable spaces, $X = \prod_{\alpha \in I} X_\alpha$ and $p_\alpha : X \to X_\alpha$ the corresponding projection (coordinate) map. Set $S = \{p_\alpha^{-1}(A_\alpha) : A_\alpha \in \Sigma_\alpha, \alpha \in I\}$. Then $\sigma(S)$ is the "product σ-algebra" and is denoted by $\bigotimes_{\alpha \in I} \Sigma_\alpha$.

Proposition 2.40 *If* I *is countable, then* $\bigotimes_{\alpha \in I} \Sigma_\alpha = \sigma(\{\prod_{\alpha \in I} A_\alpha : A_\alpha \in \Sigma_\alpha\})$.

Proof Let $A_\alpha \in \Sigma_\alpha$. Then $p_\alpha^{-1}(A_\alpha) = \prod_{\gamma \in A} A_\gamma$, with $A_\gamma = X_\gamma$ for all $\gamma \neq \alpha$. Also, $\prod_{\alpha \in I} A_\alpha = \bigcap_{\alpha \in I} p_\alpha^{-1}(A_\alpha)$. This proves the proposition. $\qquad\square$

Similarly we show the following result.

Proposition 2.41 *Let* I *be countable,* $\Sigma_\alpha = \sigma(S_\alpha)$, $X_\alpha \in S_\alpha$ *for each* $\alpha \in I$ *and*

$$\mathcal{S}^* = \{\prod_{\alpha \in I} A_\alpha : A_\alpha \in \mathcal{S}_\alpha\},$$

then $\bigotimes_{\alpha \in I} \Sigma_\alpha = \sigma(\mathcal{S}^*)$.

Proposition 2.42 *Let* $\{(X_k, \tau_k)\}_{k=1}^n$ *be topological spaces. Then*

(a) $\bigotimes_{k=1}^n B(X_k) \subseteq B(\prod_{k=1}^n X_k)$; *and*

(b) $\bigotimes_{k=1}^n B(X_k) = B(\prod_{k=1}^n X_k)$, *provided that each* X_k *is second countable.*

Proof

(a) From Proposition 2.41 we have $B(X_k) = \sigma(\{p_k^{-1}(U_k) : U_k \in \tau_k\})$. The generating sets are open in $X = \prod_{k=1}^n X_k$ and so we conclude that $\bigotimes_{k=1}^n B(X_k) \subseteq B(X)$.

(b) Let $\mathcal{B}_k$ be a countable basis for X_k. Then $B(X_k) = \sigma(\mathcal{B}_k)$ and $B(X) = \sigma(\mathcal{B})$, where $\mathcal{B} = \{\prod_{k=1}^n B_k : B_k \in \mathcal{B}_k\}$. We conclude that $\bigotimes_{k=1}^n B(X_k) = B(X)$.

$\square$

There is a variant of Proposition 2.42(b), in which we can replace the second countability requirement by another property.

Proposition 2.43 *Let* $\{(X_k, \tau_k)\}_{k=1}^n$ *be Hausdorff topological spaces and* $\prod_{k=1}^n X_k$ *and all of its subsets have the Lindelöf property (we say that* $\prod_{k=1}^n X_k$ *is "hereditary Lindelöf"). Then*

$$\bigotimes_{k=1}^n B(X_k) = B(\prod_{k=1}^n X_k).$$

Proof By assumption, every open subset of $\prod_{k=1}^n X_k$ can be represented as a countable union of products of open sets in the spaces X_k. This fact and Proposition 2.42(a) establish the desired equality. $\square$

We mention that both Propositions 2.42 and 2.43 remain true also for countable products.

Proposition 2.44 *Let* X, Y *be Hausdorff topological spaces and*

$$\Delta_Y = \{(y, y) \in Y \times Y : y \in Y\} \in B(Y) \bigotimes B(Y).$$

Then for any Borel measurable function $f : X \to Y$ *we have*

$$\text{Gr } f \in B(X) \bigotimes B(Y).$$

Proof Consider the map $g : X \times Y \to Y \times Y$ defined by $g(x, y) = (f(x), y)$. Then g is $(B(X) \otimes B(Y), B(Y) \otimes B(Y))$-measurable. Note that $\operatorname{Gr} f = g^{-1}(\Delta_Y) \in B(X) \otimes B(Y)$. $\qquad \square$

Proposition 2.45 *Let* (X, Σ) *and* $\{(Y_\alpha, S_\alpha)\}_{\alpha \in I}$ *be measurable spaces,* $Y = \prod_{\alpha \in I} Y_\alpha$ *and* $S = \bigotimes_{\alpha \in I} S_\alpha$. *Then,* $f : X \to Y$ *is* (Σ, S)-*measurable iff* $f_\alpha = p_\alpha \circ f : X \to Y_\alpha$ *is* (Σ, S_α)-*measurable for all* $\alpha \in I$.

Proof $\Rightarrow$: As the composition of two measurable functions, $f_\alpha = p_\alpha \circ f$ is (Σ, S_α)-measurable.

$\Leftarrow$: For each $A_\alpha \in S_\alpha$ we have $f^{-1}(p_\alpha^{-1}(A_\alpha)) = f_\alpha^{-1}(A_\alpha) \in \Sigma$ and so f is measurable (see Remark 2. 32 and Definition 2.39). $\qquad \square$

Next we turn our attention to a special class of functions, which arise in many applications of analysis.

Definition 2.46 Let (Ω, Σ) be a measurable space, X and Y two topological spaces. $f : \Omega \times X \to Y$ is a "Caratheodory function" if

(a) $\omega \to f(\omega, x)$ is $(\Sigma, B(Y))$-measurable for every $x \in X$.
(b) $x \to f(\omega, x)$ is continuous from X into Y for every $\omega \in \Omega$.

When X, Y have rich structures, then Caratheodory functions are jointly measurable.

Theorem 2.47 *If* (Ω, Σ) *is a measurable space,* X *is separable metrizable and* Y *is a metrizable space, then every Caratheodory function* $f : \Omega \times X \to Y$ *is jointly measurable (that is* $(\Sigma \otimes B(X), B(Y))$-*measurable).*

Proof Let d be a compatible metric on X and e a compatible metric on Y. Since X is separable, we can find $\mathcal{D} = \{x_m\}_{m \in \mathbb{N}} \subseteq X$ dense. Then for $C \subseteq Y$ closed, we have "$f(\omega, x) \in C$ if and only if for every $n \in \mathbb{N}$, there is $x_m \in \mathcal{D}$ such that $d(x, x_m) < \frac{1}{n}$ and $e(f(\omega, x_m), C) < \frac{1}{n}$." It follows that

$$f^{-1}(C) = \bigcap_{n \geq 1} \bigcup_{m \geq 1} \{\omega \in \Omega : f(\omega, x_m) \in C_{\frac{1}{n}}\} \times B_{\frac{1}{n}}(x_m) \tag{2.3}$$

with $C_{\frac{1}{n}} = \{y \in Y : e(y, C) < \frac{1}{n}\}$. From (2.3), the measurability of $f(\cdot, x)$ and since $C_{\frac{1}{n}} \subseteq Y$ is open, we infer that $\{\omega \in \Omega : f(\omega, x_m) \in C_{\frac{1}{n}}\} \in \Sigma$ for all $m, n \in \mathbb{N}$. Therefore we conclude that $f^{-1}(C) \in \Sigma \otimes B(X)$, that is, f is jointly measurable. $\qquad \square$

Remark 2.48 The result fails if X is not separable. Also, if $Y = \mathbb{R}$ and $f(\omega, x)$ is Σ-measurable in $\omega \in \Omega$ and lower (or upper) semicontinuous in $x \in \mathbb{R}$, then again the result fails (see Hu-Papageorgiou [145], Example 7.1, p 226).

Let X, Y be metrizable spaces and $C(X, Y) = \{f : X \to Y \text{ continuous}\}$. Let d be a compatible metric on Y and define

$$\widehat{d_d}(f, g) = \sup_{y \in Y} d(f(y), g(y)) \quad for \ all \ f, g \in C(X, Y). \tag{2.4}$$

This is a metric on $C(X, Y)$ and the metric topology is the "topology of uniform convergence" (see Problem 1.5). We know that if d' is another compatible metric on Y, then $\widehat{d_d}$ and $\widehat{d_{d'}}$ are equivalent. (see Problem 1.7).

Definition 2.49 For every $x \in X$, the map $e_x : C(X, Y) \to Y$ defined by $e_x(f) = f(x)$ is known as the "evaluation at x functional."

Remark 2.50 The functional e_x is continuous, hence Borel measurable. We will see that in certain circumstances these functionals generate the Borel σ-algebra of $C(X, Y)$.

Proposition 2.51 *If X is compact, metrizable, Y is separable, metrizable and*

$$\mathcal{S} = \{e_x^{-1}(C) : x \in X, C \subseteq Y \ closed\},$$

then $B(C(X, Y)) = \sigma(\mathcal{S})$.

Proof The continuity of e_x implies that $e_x^{-1}(C) \subseteq C(X, Y)$ is closed. Therefore we have $\sigma(\mathcal{S}) \subseteq B(C(X, Y))$. We need to show that the opposite inclusion also holds. Let d be a compatible metric on Y and let $\widehat{d_d}$ be the corresponding metric on $C(X, Y)$ (see (2.4)). The separability of Y implies the separability of $(C(X, Y), \widehat{d_d})$ (see Kuratowski [164], Theorem 1, p 244). Let $D \subseteq C(X, Y)$ be a countable dense subset. Every open set in $C(X, Y)$ is the countable union of closed balls $\overline{B}_{\frac{1}{n}}(f) = \{g \in C(X, Y) : \widehat{d_d}(g, f) \leq \frac{1}{n}\}, f \in D, n \in \mathbb{N}$. So, we need to show that each $\overline{B}_{\frac{1}{n}}(f) \in \sigma(\mathcal{S})$.

We fix $f \in D, x \in X$ and $\varepsilon > 0$. We have

$$\{g \in C(X, Y) : d(g(x), f(x)) \leq \varepsilon\}$$

$$= \{g \in C(X, Y) : e_x(g) \in \overline{B}_{\varepsilon}^{Y}(f(x))\} \quad (\overline{B}_{\varepsilon}^{Y}(f(x)) = \{y \in Y : d(y, f(x)) \leq \varepsilon\})$$

$$= e_x^{-1}(\overline{B}_{1}^{Y}(f(x)) \in \mathcal{S}.$$

Let $\{x_n\}_{n \geq 1} \subseteq X$ be dense. Then

$$\overline{B}_{\varepsilon}(f) = \bigcap_{n \geq 1}\{g \in C(X, Y) : d(g(x_n), f(x_n)) \leq \varepsilon\} \in \mathcal{S}$$

$$\Rightarrow B(C(X, Y)) = \sigma(\mathcal{S}). \qquad \square$$

Corollary 2.52 *If X is compact metrizable, Y is separable metrizable, $\mathcal{D} \subseteq 2^Y$ is such that $\sigma(\mathcal{D}) = B(Y)$ and $\mathcal{S} = \{e_x^{-1}(C) : x \in X, C \in \mathcal{D}\}$, then $B(C(X, Y)) = \sigma(\mathcal{S})$.*

If $f : \Omega \times X \to Y$, then we can define $\widehat{f} : \Omega \to Y^X$ by $\widehat{f}(\omega) = f(\omega, \cdot)$. Conversely, if $\widehat{h} : \Omega \to Y^X$, then we define $h : \Omega \times X \to Y$ by setting $h(\omega, x) = \widehat{h}(\omega)(x)$. Note that $f \to \widehat{f}$ and $\widehat{h} \to h$ are inverse maps.

Theorem 2.53 *If (Ω, Σ) is a measurable space, X is compact metrizable, (Y, d) is a separable metric space and $C(X, Y)$ is furnished with the $\widehat{d}_d$-metric topology (see (2.4)), then*

(a) *for every Caratheodory function $f : \Omega \times X \to Y$, $\widehat{f}(\omega) \in C(X, Y)$ for all $\omega \in \Omega$ and $\widehat{f} : \Omega \to C(X, Y)$ is Borel measurable;*
(b) *for every $\widehat{h} : \Omega \to C(X, Y)$ Borel measurable, $h(\omega, x) = \widehat{h}(\omega)(x)$ is Caratheodory.*

Proof

(a) Clearly for all $\omega \in \Omega$, $\widehat{f}(\omega) \in C(X, Y)$. Let $x \in X$, $C \subseteq Y$ closed and set $A = e_x^{-1}(C)$. According to Proposition 2.51, it suffices to show that $\widehat{f}^{-1}(B) \in \Sigma$. So, let $\xi : \Omega \times X \to \mathbb{R}_+$ be defined by $\xi(\omega, x) = d(f(\omega.x), C)$. This is an $\mathbb{R}_+$-valued Caratheodory function. Hence by Theorem 2.47, ξ is jointly measurable. Then

$$\widehat{f}^{-1}(A) = \{\omega \in \Omega : \xi(\omega, x) = 0\} \in \Sigma,$$

thus $\widehat{f} : \Omega \to C(X, Y)$ is Σ-measurable.
(b) Clearly for every $\omega \in \Omega$, $h(\omega, \cdot)$ is continuous. Let $x \in X$, $U \subseteq Y$ open and let $V = \{g \in C(X, Y) : g(x) \in U\}$. Clearly this is open in $C(X, Y)$. We have

$$\{\omega \in \Omega : h(\omega, x) \in U\} = \{\omega \in \Omega : h(\omega, \cdot) \in V\} = \widehat{h}^{-1}(V) \in \Sigma,$$

hence $\omega \to h(\omega, x)$ is Σ-measurable, hence $h(\cdot, \cdot)$ is Caratheodory. □

2.3 Polish, Souslin, and Borel Spaces

In this section we return to the subject of Polish and Souslin spaces (see Sect. 1.7) and examine them from the point of view of measure theory. We also introduce Borel spaces.

Recall that a Polish space is a topological space which is homeomorphic to a complete metric space and a Souslin space is a Hausdorff topological space which is the continuous image of a Polish space. Souslin sets are also called "Analytic Sets."

Let X be a Hausdorff topological space. By $\alpha(X)$ we denote the family of Souslin subsets of X.

Proposition 2.54 *If X is a Souslin space, then $B(X) \subseteq \alpha(X)$.*

Proof First we assume that X is a Polish space. We know that $\alpha(X)$ is closed under countable intersections and countable unions (see Propositions 1.167 and 1.170). Also the open (or closed) subsets of X are Polish, hence Souslin (see Proposition 1.147). Therefore $B(X) \subseteq \alpha(X)$.

Now suppose that X is a Souslin space. Then $X = f(P)$ with P a Polish space and $f : P \to X$ a continuous map. If $A \in B(X)$, then $f^{-1}(A) \in B(P) \subseteq \alpha(P)$ (from the first part of the proof). Since f is surjective, we have $A = f(f^{-1}(A)) \in \alpha(X)$ (see Remark 1.163). We conclude that $B(X) \subseteq \alpha(X)$.   □

Corollary 2.55 *If X and Y are Souslin spaces and $C \in B(X \times Y)$, then $p_X(C) \in \alpha(X)$.*

The next theorem is a fundamental technical fact about Souslin sets and its proof can be found in Cohn [71] (Corollary 8.32, p.274)

Theorem 2.56 *If X is a Hausdorff topological space, $\{A_n\}_{n \in \mathbb{N}} \subseteq \alpha(X)$ are pairwise disjoint, then there exist $\{B_n\}_{n \in \mathbb{N}} \subseteq B(X)$ pairwise disjoint such that $A_n \subseteq B_n$ for all $n \in \mathbb{N}$.*

Corollary 2.57 *If X is a Hausdorff topological space and $A \in \alpha(X)$ such that $A^c \in \alpha(X)$ too, then $A \in B(X)$.*

Proof By Theorem 2.55, there are disjoint sets $B_1, B_2 \in B(X)$ such that $A \subseteq B_1$ and $A^c \subseteq B_2$. Clearly then $A = B_1$ and $A^c = B_2$ and so $A \in B(X)$.   □

Definition 2.58 Let X be a topological space. We say that X is a "Borel space," if there exists a Polish space Y and a set $B \in B(Y)$ such that X is homeomorphic to B.

Remark 2.59 From the above definition we see that every Borel space is metrizable and separable. Clearly Polish spaces are Borel spaces. Any countable set X, equipped with the discrete topology τ_d (that is, $\tau_d = 2^X$), is a Borel space.

The next result is a simple observation about the Borel σ-algebra of a topological space. Its proof is straightforward and so it is omitted.

Lemma 2.60 *If X is a topological space and $A \subseteq X$ is furnished with the relative topology, then $B(A) = B(X) \cap A$.*

Another simple observation concerning Borel σ-algebras is stated in the next lemma.

Lemma 2.61 *If X and Y are topological spaces and $f : X \to Y$ is an embedding into Y, then $f(B(X)) = B(f(X))$.*

Proof Let τ_X be the topology on X. Then since f is an embedding, we infer that $f(\tau_X)$ is the topology of $f(X)$. Since f is one-to-one, we have

$$f(B(X)) = f(\sigma(\tau_X)) = \sigma(f(\tau_X)) = B(f(X)). \qquad \square$$

Theorem 2.62 *If X is a Borel space and $A \in B(X)$, then A is a Borel space too.*

Proof Since X is a Borel space, there exist Y a Polish space and an embedding $f : X \to Y$ such that $f(X) \in B(Y)$. From Lemma 2.61, we have $f(A) \in B(f(X))$. Finally Lemma 2.60 implies that $f(A) \in B(Y)$. $\qquad \square$

A model Polish space is the so-called Baire space $\mathcal{N} = \mathbb{N}^{\mathbb{N}}$ (sometimes also denoted by $\mathbb{N}^{\infty}$). The elements of this space are sequences of positive integers. That this is a Polish space follows from Proposition 1.148 (clearly $\mathbb{N}$ with the discrete topology is Polish). In fact we have the surprising result that $\mathcal{N}$ is homeomorphic to the space of irrational numbers in $(0, 1)$ (with the usual metric topology; see Cohn [71], p.255). Moreover, every Polish space is the continuous image of $\mathcal{N}$ (see Cohn [71], Proposition 8.27, p.263). As a consequence of these facts, we have the following result.

Proposition 2.63 *If X is a Hausdorff topological space and $A \subseteq X$, then A is Souslin if and only if it is the continuous image of $\mathcal{N}$ (respectively of the irrationals in $(0, 1)$).*

Now we can give a more precise characterization of Souslin sets.

Theorem 2.64 *If X is a Polish space and $A \subseteq X$, then the following statements are equivalent*

(a) $A \in \alpha(X)$.
(b) A is the projection of a closed set in $X \times \mathcal{N}$.
(c) A is the projection of a Borel set in $X \times \mathbb{R}$.

Proof (a)$\Rightarrow$(b): By Proposition 2.63, there exists a continuous map $f : \mathcal{N} \to X$ such that $f(\mathcal{N}) = A$. Let $(\mathrm{Gr}\, f)^{-1} = \{(x, u) \in X \times \mathcal{N} : f(u) = x\}$. Then $(\mathrm{Gr}\, f)^{-1}$ is a closed subset of $X \times \mathcal{N}$ and its projection on X is A.

(b)$\Rightarrow$(a): The space $X \times \mathcal{N}$ is Polish (see Proposition 1.148) and the projection map is continuous. Therefore A is Souslin (see Remark 1.163).

(a)$\Rightarrow$(c): By Proposition 2.63 there exists a continuous map $f : (0, 1) \setminus \mathbb{Q} \to X$ such that $f((0, 1) \setminus \mathbb{Q}) = A$. Let $x \in A$ and define $\widehat{f} : \mathbb{R} \to X$ by

$$\widehat{f}(y) = \begin{cases} f(y) & \text{if } y \in (0, 1) \setminus \mathbb{Q} \\ x & \text{otherwise.} \end{cases}$$

Then $\widehat{f}(\mathbb{R}) = A$ and $\widehat{f}(\cdot)$ is Borel measurable. Hence $\mathrm{Gr}\, \widehat{f} \in B(\mathbb{R} \times X) = B(\mathbb{R}) \otimes B(X)$ (see Proposition 2.44). Thus, so is $(\mathrm{Gr}\, f)^{-1} = \{(x, y) \in X \times \mathbb{R} : \widehat{f}(y) = x\}$ and A is the projection on X of $(\mathrm{Gr}\, \widehat{f})^{-1}$.

(c)$\Rightarrow$(a): Since $X \times \mathbb{R}$ is a Polish space, any Borel subset of $X \times \mathbb{R}$ is Souslin (see Proposition 2.54). Since the projection is continuous, we conclude that $A \in \alpha(X)$. $\qquad \square$

To proceed further, we will need the following simple observation concerning Souslin spaces.

Lemma 2.65 *Every Souslin space X is hereditary Lindelöf.*

Proof Since X is Souslin, there is a Polish space P and a continuous surjection $f : P \to X$. Let $U_\alpha \subseteq X, \alpha \in I$ be open sets. Then $f^{-1}(\bigcup_{\alpha \in I} U_\alpha) = \bigcup_{\alpha \in I} f^{-1}(U_\alpha)$ is an open cover in P which is hereditary Lindelöf being Polish. So, there is a countable subcollection $\{f^{-1}(U_{\alpha_k})\}_{k \in \mathbb{N}}$ such that $f^{-1}(\bigcup_{k \in \mathbb{N}} U_{\alpha_k}) = f^{-1}(\bigcup_{\alpha \in I} U_\alpha)$. Hence $\bigcup_{k \in \mathbb{N}} U_{\alpha_k} = \bigcup_{\alpha \in I} U_\alpha$, which proves that X is hereditary Lindelöf.   □

Corollary 2.66 *If $\{X_k\}_{k=1}^n$ are Souslin spaces, then $\bigotimes_{k=1}^n B(X_k) = \prod_{k=1}^n B(X_k)$.*

Theorem 2.67 *If X and Y are Souslin spaces, $f : X \to Y$ is Borel measurable $A \in \alpha(X)$ and $B \in \alpha(Y)$, then $f(A) \in \alpha(Y)$ and $f^{-1}(B) \in \alpha(X)$; moreover, if f is injective, then $f^{-1} : f(X) \to X$ is Borel measurable.*

Proof From Proposition 2.44 and Corollary 2.66, we have that $\mathrm{Gr}\, f \in B(X) \otimes B(Y)$. The sets $A \times Y$ and $X \times B$ are Souslin subsets of $X \times Y$. Note that $f(A)$ is the projection of the Souslin set $(A \times Y) \cap \mathrm{Gr}\, f$ on X. Hence $f(A) \in \alpha(Y)$. Similarly $f^{-1}(B)$ is the projection of the Souslin set $(X \times B) \cap \mathrm{Gr}\, f$ on X. Hence $f^{-1}(B) \in \alpha(X)$. Finally suppose that $f(\cdot)$ is injective and let $A \in B(X)$. Then from the first part of the proof we have $f(A) \in \alpha(Y)$. Since $A^c \in B(X)$, it follows that $f(X) \setminus f(A) = f(A^c) \in \alpha(Y)$. Then Corollary 2.57 implies that $f(A) \in B(f(X))$.   □

2.4 Integration

In this section we introduce the Lebesgue integral which is more powerful than the Riemann integral. In particular, the Lebesgue integral has a better behavior under pointwise limits of functions, than the Riemann integral which requires uniform convergence. Moreover, the Lebesgue integral also applies to functions on spaces much more general than $\mathbb{R}$ and with respect to general measures.

The building blocks for the Lebesgue integral are the simple functions.

Definition 2.68 Let (X, Σ) be a measurable space.

(a) Given $A \subseteq X$, the "characteristic function," χ_A of A, is defined by

$$\chi_A(x) = \begin{cases} 1 & if \ x \in A \\ 0 & if \ x \notin A \end{cases}$$

(b) A "simple function" is a finite linear combination with real coefficients of characteristic functions of sets in Σ.

Remark 2.69 So, a simple function is a measurable function $s : X \to \mathbb{R}$ with finite range. We usually write $s(\cdot)$ in its standard representation, namely

$$s(x) = \sum_{k=1}^{n} a_k \mathcal{X}_{A_k}(x) \quad with \ \{a_k\}_{k=1}^{n} = range \ s(\cdot), \ A_k = s^{-1}(a_k).$$

So, in this expression, the linear combination has distinct coefficients (one of them may be zero) and the sets in the characteristic function are disjoint and their union is all of X. Sums and products of simple functions are again a simple function. Of course simple functions can be defined with values in any metric space. We will need them, when we will discuss the integration theory for vector valued functions (see Sect. 4.4).

Proposition 2.70 *If (X, Σ) is a measurable space and $f : X \to \mathbb{R}_+ = [0, \infty)$ measurable, then there exists a sequence $\{s_n\}_{n \geq 1}$ of simple functions with values in $\mathbb{R}_+$ such that $s_n(x) \uparrow f(x)$ for all $x \in X$ and the convergence is uniform on any set on which $f(\cdot)$ is bounded.*

Proof For every $n \in \mathbb{N}$ and for $k \in \mathbb{N}_0$ with $0 \leq k \leq 2^{2n} - 1$ we define

$$A_n^k = \{x \in X : \frac{k}{2^n} \leq f(x) \leq \frac{k+1}{2^n}\} \ and \ B_n = \{x \in X : 2^n < f(x)\}.$$

Evidently $A_n^k, B_n \in \Sigma$. We set $s_n(x) = \sum_{k=0}^{2^{2n}-1} \frac{k}{2^n} \mathcal{X}_{A_n^k}(x) + 2^n \mathcal{X}_{B_n}(x)$. Then $\{s_n\}_{n \geq 1}$ is increasing and $0 \leq f - s_n \leq \frac{1}{2^n}$ on the set where $f \leq 2^n$. This proves the proposition. □

Since $f = f^+ - f^-$ with $f^{\pm} = max\{\pm f, 0\}$, we have:

Corollary 2.71 *If (X, Σ) is a measurable space and $f : X \to \mathbb{R}$, then $f(\cdot)$ is Σ-measurable if and only if there exists a sequence $\{s_n\}_{n \geq 1}$ of simple functions with $|s_n| \leq |f|$ for all $n \in \mathbb{N}$ and $s_n(x) \to f(x)$ for every $x \in X$.*

The next result tells us that we need not worry too much whether the underlying measure space is complete.

Proposition 2.72 *If (X, Σ, μ) is a measure space, $(X, \Sigma_\mu, \widehat{\mu})$ its completion (see Proposition 2. 19) and $f : X \to \mathbb{R}$ is Σ_μ-measurable, then there is a function $g : X \to \mathbb{R}$ which is Σ-measurable and $f = g$ for $\widehat{\mu} - a.a, x \in X$.*

Proof From the definition of $\widehat{\mu}(\cdot)$ (see Proposition 2.19), the result is true if $f = \chi_A$ with $A \in \Sigma_\mu$. Consequently it is also true for f being a Σ_μ-measurable simple function. For the general case, using Corollary 2.71 we can find $\{\widehat{s}_n\}_{n \geq 1}$ a sequence of Σ_μ-simple functions which converge pointwise to f. For each $n \in \mathbb{N}$, we can find a Σ-measurable simple function s_n such that $\widehat{s_n}(x) = s_n(x)$ for all $x \in X \setminus \widehat{A_n}$, $\widehat{\mu}(\widehat{A_\mu}) = 0$. Let $D \in \Sigma$ such that $\mu(D) = 0$ and $\bigcup_{n \geq 1} \widehat{A_n} \subseteq D$ and set $g = \lim_{n \to \infty} \chi_{X \setminus D} s_n$. Then $g(\cdot)$ is Σ-measurable and $f = g$ on $X \setminus D$.   □

Using simple functions as our building blocks, we will define the "Lebesgue integral."

Definition 2.73 Let (X, Σ, μ) be a measure space.

(a) Given a simple function $s(x) = \sum_{k=1}^{n} a_k \chi_{A_k}(x)$ and $B \in \Sigma$, we define

$$\int_B s d\mu = \sum_{k=1}^{n} a_k \mu(A_k \cap B),$$

using the convention that $0 \cdot \infty = 0$.

(b) Given $f : X \to \mathbb{R}_+$, Σ-measurable, and $B \in \Sigma$, we define

$$\int_B f d\mu = \sup[\int_B s d\mu : 0 \leq s \leq f, s = simple].$$

Evidently $\int_B f d\mu \in [0, \infty]$ and it is known as the "Lebesgue integral" of f on B.

Remark 2.74 In this definition, we see the basic difference between the Lebesgue integral and the Riemann integral of elementary calculus. In the Riemann integral we partition the domain, while in the Lebesgue integral with the use of simple functions, we partition the range. Note that for every simple function $s \geq 0$, $A \to \int_A s d\mu$ is a measure on Σ.

The following properties of the Lebesgue integral are straightforward consequences of Definition 2.73. In this proposition all functions and sets involved are assumed to be measurable.

Proposition 2.75

(a) $0 \leq g \leq f \Rightarrow \int_A g d\mu \leq \int_A f d\mu$.
(b) $A \subseteq B$ and $f \geq 0 \Rightarrow \int_A f d\mu \leq \int_B f d\mu$.
(c) $f \geq 0$ and $\lambda \in [0, +\infty] \Rightarrow \int_A \lambda f d\mu = \lambda \int_A f d\mu$.
(d) $f|_A = 0 \Rightarrow \int_A f d\mu = 0$.
(e) $\mu(A) = 0 \Rightarrow \int_A f d\mu = 0$ *even if* $f \equiv +\infty$.
(f) $f \geq 0 \Rightarrow \int_A f d\mu = \int_X \chi_A f d\mu$.

Using the positive and negative parts of f, we can extend the Lebesgue integral to Σ-measurable functions $f : \Sigma \to \mathbb{R}^* = \mathbb{R} \cup \{\pm\infty\}$.

Definition 2.76 Let $f : X \to \mathbb{R}^* = \mathbb{R} \cup \{\pm\infty\}$ be Σ-measurable and suppose that at least one of $\int_X f^+ d\mu$ and $\int_X f^- d\mu$ is finite. Then the "Lebesgue integral" of f is defined to be $\int_X f d\mu = \int_X f^+ d\mu - \int_X f^- d\mu$. We say that f is "(Lebesgue) integrable," if both $\int_X f^+ d\mu$, $\int_X f^- d\mu$ are finite.

Remark 2.77 Since $|f| = f^+ + f^-$, we see that f is integrable if and only if $|f|$ is. Also on account of Proposition 2.75, the set of integrable functions is a vector space and the integral is a linear functional on it. We denote the space of integrable functions by $\mathcal{L}^1(X)$.

Proposition 2.78 *If $f \in \mathcal{L}^1(X)$, then $|\int_X f d\mu| \leq \int_X |f| d\mu$.*

Proof We have $|\int_X f d\mu| = |\int_X f^+ d\mu - \int_X f^- d\mu| \leq \int_X f^+ d\mu + \int_X f^- d\mu = \int_X |f| d\mu$. $\square$

Proposition 2.79

(a) If $f \in \mathcal{L}^1(X)$, then $\mu(\{f = \pm\infty\}) = 0$ and $\{x \in X : f(x) \neq 0\}$ is σ-finite.
(b) If $f, g \in \mathcal{L}^1(X)$, then $\int_A f d\mu = \int_A g d\mu$ for all $A \in \Sigma \iff \int_X |f - g| d\mu = 0 \iff f = g$ $\mu - a.e.$

Proof

(a) Let $A_{\pm} = \{x \in X : f(x) = \pm\infty\} \in \Sigma$. Suppose that $\mu(A_+) > 0$. Then $\int_{A_+} f^+ d\mu = \int_X \chi_{A_+} f^+ d\mu = +\infty$. Similarly if $\mu(A_-) > 0$, then $\int_{A_-} f^- d\mu = +\infty$. Both contradict the fact that $f \in \mathcal{L}^1(X)$. Therefore $A = \{x \in X : f(x) = \pm\infty\} \in \Sigma$ is μ-null. For every $\vartheta > 0$ we have

$$\vartheta \mu(\{|f| \geq \vartheta\}) \leq \int_{\{f \geq \vartheta\}} |f| d\mu \leq \int_X |f| d\mu < \infty.$$

Finally note that $\{f \neq 0\} = \bigcup_{n \geq 1} \{|f| \geq \frac{1}{n}\}$.

(b) From Proposition 2.75(d) we see that $\int_X |f - g| d\mu = 0 \iff f = g$ $\mu - a.e.$ Also $f = g$ $\mu - a.e$ implies $\int_A f d\mu = \int_A g d\mu$ for all $A \in \Sigma$. Conversely, suppose that $\int_A f d\mu = \int_A g d\mu$ for all $A \in \Sigma$. If it is not true that $f = g$ $\mu - a.e$, then at least one of $(f - g)^+$ and $(f - g)^-$ is nonzero. Suppose $(f - g)^+ \neq 0$ and let $A_+ = \{(f - g)^+ > 0\} \in \Sigma$. Then $\mu(A_+) > 0$ and we have $0 = \int_{A_+} (f - g) d\mu = \int_{A_+} (f - g)^+ d\mu > 0$, a contradiction. Finally suppose that $\int_X |f - g| d\mu = 0$ and let $A \in \Sigma$. Then $|\int_A f d\mu - \int_A g d\mu| \leq \int_X \chi_A |f - g| d\mu \leq \int_X |f - g| d\mu = 0$. $\square$

The next result is very important in the theory of integration of nonnegative functions. It is known as the "Monotone Convergence Theorem."

Theorem 2.80 *If (X, Σ, μ) is a measure space, $f_n : X \to \overline{\mathbb{R}}_+ : [0, +\infty]$ $n \in \mathbb{N}$ is a sequence of Σ-measurable functions, $f_n \leq f_{n+1}$ for all $n \in \mathbb{N}$ and $f_n \to f$, then $\int_X f d\mu = \lim_{n \to \infty} \int_X f_n d\mu$.*

Proof From Proposition 2.75(a), we know that $\{\int_X f_n d\mu\}_{n \geq 1}$ is increasing. So, the limit $\lim_{n \to \infty} \int_X f_n d\mu$ exists (possibly equals ∞). Also, we have $\int_X f_n d\mu \leq \int_X f d\mu$ for all $n \in \mathbb{N}$, hence $\lim_{n \to \infty} \int_X f_n d\mu \leq \int_X f d\mu$. We need to show that the opposite inequality is also true. So, fix $\lambda \in (0, 1)$ and let $s(\cdot)$ be a simple function such that $0 \leq s \leq f$. We set $A_n = \{x \in X : \lambda s(x) \leq f_n(x)\}$. Evidently $\{A_n\}_{n \geq 1} \in \Sigma$ is increasing and $\bigcup_{n \geq 1} A_n = X$. We have $\lambda \int_{A_n} s d\mu \leq \int_{A_n} f_n d\mu \leq \int_X f_n d\mu$. Recall that $\Sigma \ni B \to \int_B s d\mu$ is a measure (see Remark 2.74). So, from Proposition 2.17(c) we have that $\lim_{n \to \infty} \int_{A_n} s d\mu = \int_X s d\mu$. Hence $\lambda \int_X s d\mu \leq \lim_{n \to \infty} \int_X f_n d\mu$. But $\lambda \in (0, 1)$ is arbitrary. So it follows that $\int_X s d\mu \leq \lim_{n \to \infty} \int_X f_n d\mu$ for every simple function $0 \leq s \leq f$. By Definition 2.73(b), we infer that $\int_X f d\mu \leq \lim_{n \to \infty} \int_X f_n d\mu$. So, we conclude that $\int_X f d\mu = \lim_{n \to \infty} \int_X f_n d\mu$. $\square$

Remark 2.81 The Monotone Convergence Theorem fails in general for decreasing sequences. To see this, let $X = \mathbb{R}$ equipped with the Lebesgue measure λ and let $f_n = \frac{1}{n} \chi_{[n, +\infty)}$. Then, $\{f_n\}_{n \geq 1}$ is decreasing, but $\lim_{n \to \infty} \int_{\mathbb{R}} f_n d\lambda = +\infty \neq 0 = \int_{\Omega} \lim_{n \to \infty} f_n d\lambda$. (see Proposition 2.17(d)).

Corollary 2.82 *If (X, Σ, μ) is a measure space and $f_n : X \to \overline{\mathbb{R}}_+ = [0, +\infty]$, $n \in \mathbb{N}$, are Σ-measurable functions, then $\sum_{n \geq 1} \int_X f_n d\mu = \int_X (\sum_{n \geq 1} f_n) d\mu$.*

Proof Follows from the Monotone Convergence Theorem applied to the sequence of partial sums and the linearity of the integral. $\square$

Remark 2.83 An interesting application of this corollary is the following known fact about double sums. So, let $\{a_{nk}\}_{n,k \in \mathbb{N}} \subseteq \mathbb{R}_+$. We have $\sum_{n \geq 1} \sum_{k \geq 1} a_{nk} = \sum_{k \geq 1} \sum_{n \geq 1} a_{nk}$. To see this, let $X = \mathbb{N}$ furnished with the counting measure μ. If we consider the functions $f_n : \mathbb{N} \to \mathbb{R}_+$ defined by $f_n(k) = a_{nk}$, then $\int_X f_n d\mu = \sum_{k \geq 1} a_{nk}$ and so the double sum equality follows from Corollary 2.82.

Then next result is known as "Fatou's Lemma."

Theorem 2.84 *If (X, Σ, μ) is a measure space and $f_n : X \to \overline{\mathbb{R}}_+ = [0, +\infty]$, $n \in \mathbb{N}$ are Σ-measurable functions, then*

$$\int_X \liminf_{n\to\infty} f_n d\mu \le \liminf_{n\to\infty} \int_X f_n d\mu.$$

Proof For every $m \in \mathbb{N}$, let $g_m = \inf_{n\ge m} f_n \le f_n$ for $n \ge m$. We have $\int_X g_m d\mu \le \int_X f_n d\mu$ for all $n \ge m$, hence $\int_X g_m d\mu \le \inf_{n\ge m} \int_X f_n d\mu$. Since $\{g_m\}_{m\ge 1}$ is increasing, applying Theorem 2.80, we obtain $\int_X \lim_{n\to\infty} \inf f_n d\mu \le \lim_{n\to\infty} \inf \int_X f_n d\mu$. $\qquad\square$

Remark 2.85 Evidently we can have $f : X \to \overline{\mathbb{R}} = \mathbb{R} \cup \{+\infty\}, n \ge 1$ Σ-measurable such that $h(x) \le f_n(x)$ $\mu - a.e$ with $h \in \mathcal{L}^1(x)$. Simply apply Fatou's lemma on $\widehat{f_n} = f_n - h, n \in \mathbb{N}$. Without such a condition, Fatou's lemma fails for $\overline{\mathbb{R}}$-valued functions. To see this let $X = \mathbb{R}$ with the Lebesgue measure λ and let $f_n = -\frac{1}{n}\chi_{[0,n]}$. Then $\lim_{n\to\infty} \inf \int_\mathbb{R} f_n d\lambda = -1 < 0 = \int_\mathbb{R} \lim_{n\to\infty} \inf f_n d\lambda$.

Now we present the main convergence theorem, which reveals the power of the Lebesgue integral. The result is known as "Lebesgue Dominated Convergence Theorem."

Theorem 2.86 *If (X, Σ, μ) is a measure space and $f_n : X \to \mathbb{R}^* = \mathbb{R} \cup \{\pm\infty\}, n \in \mathbb{N}$ a sequence of Σ-measurable functions such that $f_n(x) \to f(x)$ $\mu - a.e$ and $|f_n(x)| \le h(x)$ $\mu - a.e$, for all $n \in \mathbb{N}$, with $h \in \mathcal{L}^1(X)$, then $\int_X f_n d\mu \to \int_X f d\mu$ and $\int_X |f_n - f| d\mu \to 0$.*

Proof Note that $f(\cdot)$ is Σ-measurable (possibly by redefining it on μ-null set). Also, we have $|f(x)| \le h(x)$ $\mu - a.e$, hence $f \in \mathcal{L}^1(X)$. Note that $h + f_n \ge 0, h - f_n \ge 0$ $\mu - a.e$ for all $n \in \mathbb{N}$. By Fatou's lemma (see Theorem 2.84), we have

$$\int_X h d\mu + \int_X f d\mu \le \liminf_{n\to\infty} \int_X [h + f_n] d\mu = \int_X h d\mu + \liminf_{n\to\infty} \int_X f_n d\mu;$$
$$(2.5)$$

$$\int_X h d\mu - \int_X f d\mu \le \liminf_{n\to\infty} \int_X [h - f_n] d\mu = \int_X h d\mu - \limsup_{n\to\infty} \int_X f_n d\mu$$
$$(2.6)$$

From (2.5) and (2.6), we obtain

$$\limsup_{n\to\infty} \int_X f_n d\mu \le \int_X f d\mu \le \liminf_{n\to\infty} \int_X f_n d\mu,$$

$$\Rightarrow \int_X f_n d\mu \to \int_X f d\mu.$$

Since $|(f_n - f)(x)| \le 2h(x)$ $\mu - a.e$ for all $n \in \mathbb{N}$ and $|(f - f_n)(x)| \to 0$ $\mu - a.e$ as $n \to \infty$, we also have that $\int_X |f_n - f| d\mu \to 0$. $\qquad\square$

Remark 2.87 It is important that $h \in \mathcal{L}^1(X)$. Otherwise the result fails. To see this, let $X = [0, 1]$ equipped with the Lebesgue measure λ and let $f_n = n\chi_{[0,\frac{1}{n}]}$. Then

$$\lim_{n\to\infty} \int_0^1 f_n d\lambda = 1 \neq 0 = \int_0^1 \lim_{n\to\infty} f_n d\lambda.$$

Proposition 2.88 *If (X, Σ, μ) is a measure space and $\{f_n\}_{n\geq 1} \subseteq \mathcal{L}^1(X)$ such that*

$$\sum_{n\geq 1} \int_X |f_n| d\mu < \infty,$$

then $f = \sum_{n\geq 1} f_n \in \mathcal{L}^1(X)$ and $\int_X f d\mu = \sum_{n\geq 1} \int_X f_n d\mu$.

Proof From Corollary 2.82 we have $\sum_{n\geq 1} \int_X |f_n| d\mu = \int_X (\sum_{n\geq 1} |f_n|) d\mu < \infty$. Hence $\sum_{n\geq 1} |f_n| = h \in \mathcal{L}^1(X)$ and $|\sum_{k=1}^n f_k(x)| \leq h(x) \ \mu - a.e.$ Apply Theorem 2.86 to get the desired equality. $\square$

Also using Corollary 2.71 and Theorem 2.86, we have the following approximation result.

Proposition 2.89 *If (X, Σ, μ) is a measure space, $f \in \mathcal{L}^1(X)$ and $\varepsilon > 0$, then we can find an integrable simple function $s(\cdot)$ such that $\int_X |f - s| d\mu \leq \varepsilon$.*

We conclude this section with a result which leads to the topic of the next section.

Proposition 2.90 *If (X, Σ, μ) is a measure space, $f : X \to \overline{\mathbb{R}}_+ = [0, +\infty]$ is Σ-measurable and $m(A) = \int_A f d\mu$ for every $A \in \Sigma$, then $m : \Sigma \to \overline{\mathbb{R}}_+$ is a measure and for every Σ-measurable function $h : X \to \overline{\mathbb{R}}_+$, $\int_X h dm = \int_X h f d\mu$.*

Proof Let $\{A_n\}_{n\geq 1} \subseteq \Sigma$ be mutually disjoint sets and let $A = \bigcup_{n\geq 1} A_n$. We have

$$\chi_A f = \sum_{n\geq 1} \chi_{A_n} f,$$

$$\Rightarrow m(A) = \int_X \chi_A f d\mu, \ m(A_n) = \int_X \chi_{A_n} f d\mu \ for \ all \ n \in \mathbb{N}.$$

From Proposition 2.88 we have $m(A) = \sum_{n\geq 1} m(A_n)$. Since $m(\emptyset) = 0$, we conclude that $m(\cdot)$ is a measure.

When $h = \chi_A$ with $A \in \Sigma$, then $\int_X h dm = m(A) = \int_A f d\mu = \int_X h f d\mu$. Therefore the equality $\int_X h dm = \int_X h f d\mu$ holds for all $h = simple$. The general case follows from Proposition 2. 70 and the Monotone Convergence Theorem (see Theorem 2.80). $\square$

Remark 2.91 The second conclusion of Proposition 2.90, can be expressed as $dm = f d\mu$.

The converse of this assertion is known as the "Radon–Nikodym Theorem" which is one of the main topics of the next section.

2.5 Signed Measures and the Lebesgue–Radon–Nikodym Theorem

Until now we have considered measures with values in $\overline{\mathbb{R}}_+ = [0, \infty]$. However, for the purpose of differentiating a measure with respect to another measure on the same σ-algebra, it is useful to generalize the notion of measure and allow also negative values.

Definition 2.92 Let (X, Σ) be a measurable space and $\mu : \Sigma \to [-\infty, +\infty]$. We say that $\mu(\cdot)$ is a "signed measure" if

(a) $\mu(\emptyset) = 0$.
(b) $\mu(\cdot)$ takes at most one of the values $+\infty$ and $-\infty$ (that is, $\mu : \Sigma \to (-\infty, +\infty]$ or $\mu : \Sigma \to [-\infty, +\infty)$).
(c) for every sequence $\{A_n\}_{n \geq 1} \subseteq \Sigma$ of pairwise disjoint sets, $\mu(\bigcup_{n \geq 1} A_n) = \sum_{n \geq 1} \mu(A_n)$ (σ-additivity).

Proposition 2.93 *If* $\mu : \Sigma \to \mathbb{R}^* = [-\infty, +\infty]$ *is a signed measure,* $A, B \in \Sigma, |\mu(A)| < \infty$ *and* $B \subseteq A$, *then* $|\mu(B)| < \infty$.

Proof Note that $\mu(A) = \mu(B) + \mu(A \setminus B)$. Then $\mu(A)$ is finite if and only if both items in the right hand side are finite. □

The next result is the analogue of Proposition 2.17 for signed measures. The proof is essentially the same and so it is omitted.

Proposition 2.94

(a) *If* $\mu : \Sigma \to \mathbb{R}^* = [-\infty, +\infty]$ *is a signed measure an* $\{A_n\}_{n \geq 1} \subseteq \Sigma$ *is increasing, then* $\mu(\bigcup_{n \geq 1}) A_n = \lim_{n \to \infty} \mu(A_n)$.
(b) *If* $\mu : \Sigma \to \mathbb{R}^* = [-\infty, +\infty]$, $\{A_n\}_{n \geq 1} \subseteq \Sigma$ *is decreasing and* $|\mu(A_1)| < \infty$, *then* $\mu(\bigcap_{n \geq 1} A_n) = \lim_{n \to \infty} \mu(A_n)$.

Using the next notion we will be able to write a signed measure as the difference of two measures.

Definition 2.95 Let (X, Σ) be a measurable space and $\mu : \Sigma \to \mathbb{R}^* = [-\infty, +\infty]$ is a signed measure. We say that $A \in \Sigma$ is "positive" (respectively, "negative," "null"), if $\mu(B) \geq 0$ (respectively, $\mu(B) \leq 0$, $\mu(B) = 0$) for all $B \in \Sigma, B \subseteq A$.

Remark 2.96 A set $A \in \Sigma$ may have positive μ-measure without being positive. To see this, let $X = \mathbb{R}$, λ =the Lebesgue measure and $f \in \mathcal{L}^1(\mathbb{R})$ odd with $f(x) > 0$ for a.a $x > 0$. Set $\mu(A) = \int_A f d\lambda$, $A \in B(\mathbb{R})$. This is a signed measure and any set of the form $[-a, b]$ with $0 < a < b$ has positive measure without being positive. Also note that alternatively we can say that $A \in \Sigma$ is positive (respectively, negative) for μ, if $\mu(A \cap B) \geq 0$ (respectively, $\mu(A \cap B) \leq 0$) for all $B \in \Sigma$ and it is null for μ, if it is both positive and negative.

Proposition 2.97 *Every Σ-subset of a positive (respectively, negative, null) set for μ is itself positive (respectively, negative, null). Also countable unions of positive (respectively, negative, null)sets is again positive (respectively, negative, null).*

Proof We do the proof for positive sets. The other two cases can be done similarly. The first assertion is an immediate consequence of Definition 2.95. For the second assertion, let $\{A_n\}_{n\geq 1} \subseteq \Sigma$ be positive sets and set $A = \bigcup_{n\geq 1} A_n \in \Sigma$. If $B \in \Sigma, B \subseteq A$, we let $B_n = B \cap (A_n \setminus \bigcup_{k=1}^{n-1} A_k), n \in \Sigma$. Then $B_n \in \Sigma, B_n \subseteq A_n$ and so $\mu(B_n) \geq 0$. The sets $\{B_n\}_{n\geq 1}$ are disjoint and $B = \bigcup_{n\geq 1} B_n$. So, we have $\mu(B) = \sum_{n\geq 1} \mu(B_n) \geq 0$. Therefore A is a positive set. $\square$

Proposition 2.98 *If $\mu : \Sigma \to \mathbb{R}^* = [-\infty, +\infty]$ is a signed measure, $C \in \Sigma$ and $0 < \mu(C) < \infty$, (respectively, $-\infty < \mu(C) < 0$), then there exists a positive (respectively, negative) set $A \in \Sigma, A \subseteq C$ such that $\mu(A) > 0$ (respectively, $\mu(A) < 0$).*

Proof Again we do the proof for the positive case, the negative case being done similarly. If $C \in \Sigma$ is not positive, then it contains a Σ-set of negative μ-measure. Let $n_1 \in \mathbb{N}$ be the smallest positive integer such that there exists $B_1 \in \Sigma, B_1 \subseteq C$ such that $\mu(B_1) < -\frac{1}{n_1}$. Inductively, let $n_k \in \mathbb{N}$ be the smallest positive integer for which we can find $B_k \in \Sigma, B_k \subseteq C \setminus \bigcup_{i=1}^{k-1} B_i$ such that $\mu(B_k) < -\frac{1}{n_k}$. We set $A = C \setminus \bigcup_{k\geq 1} B_k$. Then $C = A \cup (\bigcup_{k\geq 1} B_k)$. We have

$$\mu(C) = \mu(A) + \sum_{k\geq 1} \mu(B_k). \tag{2.7}$$

The series converges absolutely and so $\sum_{k\geq 1} \frac{1}{n_k} < \infty$, hence $n_k \to +\infty$. Since $\mu(C) > 0$, $\mu(B_k) \leq 0$, we have $\mu(A) > 0$ (see (2.7)).

We show that $A \in \Sigma$ is positive for μ. Given $\varepsilon > 0$, we can find $k \in \mathbb{N}$ big such that $\frac{1}{n_k - 1} < \varepsilon$. We have $A \subseteq C \setminus \bigcup_{i=1}^{k-1} B_k$ and so A cannot contain Σ-sets with μ-measure less than $-\frac{1}{n_k - 1} > -\varepsilon$. So, A contains no Σ-sets of μ-measures less than $-\varepsilon$. But $\varepsilon > 0$ is arbitrary. Let $\varepsilon \to 0^+$ to conclude that A contains no Σ-sets of negative measure. Hence $A \in \Sigma$ is positive for μ.   □

Now we show that X can be partitioned by a positive and a negative set. This partition is not unique. The result is known as "Hahn Decomposition theorem."

Theorem 2.99 *If (X, Σ) is a measurable space and $\mu : \Sigma \to \mathbb{R}^* = [-\infty, +\infty]$ is a signed measure, then there exist $A, B \in \Sigma$ with A positive and B negative for μ such that $X = A \cup B$, $A \cap B = \emptyset$.*

Proof Suppose that $\mu(\cdot)$ has values in $(-\infty, +\infty]$. Let

$$\eta = \inf[\mu(A) : A \in \Sigma, A \text{ is negative for } \mu].$$

We consider $\{A_n\}_{n \geq 1} \subseteq \Sigma$ negative sets such that $\mu(A_n) \downarrow \eta$ and let $A = \bigcup_{n \geq 1} A_n$, $B = X \setminus \bigcup_{n \geq 1} A_n$. We define $C_1 = A_1$ and for $n \geq 2$ $C_n = A_n \setminus \bigcup_{k=1}^{n-1} A_k$. Then for every $D \in \Sigma$, $D \subseteq A$ since the sets $\{C_n\}_{n \geq 1}$ are mutually disjoint, we have

$$\mu(D) = \sum_{n \geq 1} \mu(D \cap C_n) \leq 0,$$

thus $A \in \Sigma$ is negative for μ. Hence, for every $n \in \mathbb{N}$, we have

$$\mu(A) = \mu(A_n) + \mu(A \setminus A_n) \leq \mu(A_n), \Rightarrow \mu(A) = \eta. \tag{2.8}$$

Suppose that B is not positive for μ. Then we can find $D \in \Sigma, D \subseteq B$ with $\mu(D) < 0$. Then from Proposition 2.98, we know that we can find $E \in \Sigma, E \subseteq D$ such that E is negative for μ and $\mu(E) < 0$. Then $A \cup E$ is negative for μ (see Proposition 2.97) and we have

$$\mu(A \cup E) = \mu(A) + \mu(E) = \eta + \mu(E) < \eta \quad (see \ (2.8)),$$

a contradiction to the definition of η.

Now suppose that $\mu : \Sigma \to [-\infty, +\infty)$. Then $-\mu : \Sigma \to (-\infty, +\infty]$ and of course is also a signed measure. Then from the first part of the proof we can find $A \in \Sigma$ negative for $(-\mu)$ and $B \in \Sigma$ positive for $(-\mu)$ such that $X = A \cup B$, $A \cap B = \emptyset$. Then B is negative for μ and A is positive for μ.   □

Remark 2.100 The pair (A, B) produced in Theorem 2.99, is called "Hahn decomposition" of X. This decomposition is not unique since we can replace A by $A \cup N$

with $N \in \Sigma$ null set for μ and B by $B \setminus N = B \cap N^c$. In this direction we have the following result.

Proposition 2.101 *If (A, B) and (A_0, B_0) are Hahn decompositions of X, then $A \triangle A_0 = B \triangle B_0$ is null for μ.*

Proof We have $A \setminus A_0 \subseteq A$ and $A \setminus A_0 \subseteq B_0$. So $A \setminus A_0$ is both positive and negative for μ, therefore is null. Similarly for $A_0 \setminus A$. So, finally $A \triangle A_0$ is null for μ (see Proposition 2.97). $\qquad \square$

The Hahn decomposition may not be unique, but it leads to a canonical representation of a signed measure as the difference of two measures. First a definition.

Definition 2.102 Let (X, Σ) be a measurable space and $\mu_1, \mu_2 : \Sigma \to \mathbb{R}^* = [-\infty, +\infty]$ two signed measures. We say that μ_1 and μ_2 are "mutually singular," denoted by $\mu_1 \perp \mu_2$, if there exist sets $A, B \in \Sigma$ with $X = A \cup B$, $A \cap B = \varnothing$ and A is null for μ_1, B is null for μ_2 (that is, the two signed measures live on disjoint sets).

The next theorem gives the canonical decomposition of a signed measure as the difference of two measures. The result is known as the "Jordan Decomposition Theorem."

Theorem 2.103 *If (X, Σ) is a measurable space and $\mu : \Sigma \to \mathbb{R}^* = [-\infty, +\infty]$ is a signed measure, then there exist unique measures $\mu^+, \mu^- : \Sigma \to [0, +\infty]$ one of which is finite such that $\mu^+ \perp \mu^-$ and $\mu = \mu^+ - \mu^-$.*

Proof Let (A, B) be a Hahn decomposition for μ (see Theorem 2.99). We define $\mu^+(C) = \mu(C \cap A)$ and $\mu^-(C) = -\mu(C \cap B)$ for all $C \in \Sigma$.

Evidently $\mu = \mu^+ - \mu^-$ and $\mu^+ \perp \mu^-$.

Suppose $\mu = \mu_1^+ - \mu_1^-$ and $\mu_1^+ \perp \mu_1^-$. Let $D, E \in \Sigma$ with $X = D \cup E$, $D \cap E = \varnothing$ and $\mu_1^+(E) = \mu_1^-(D) = 0$. Then (D, E) is another Hahn decomposition for μ. By Proposition 2.101, we have that $A \triangle D$ is null for μ. Therefore for any $F \in \Sigma$ we have

$$\mu_1^+(F) = \mu_1^+(F \cap D) = \mu(F \cap D) = \mu(F \cap A) = \mu^+(F),$$

$$\Rightarrow \mu_1^+ = \mu^+.$$

Similarly we show that $\mu^- = \mu_1^-$. $\qquad \square$

Proposition 2.104 *If (X, Σ) is a measurable space and $\mu : \Sigma \to \mathbb{R}^* = [-\infty, +\infty]$ is a signed measure, then there exist $A_*, A^* \in \Sigma$ such that $\mu(A_*) = \inf[\mu(A) : A \in \Sigma] \le 0 \le \sup[\mu(A) : A \in \Sigma] = \mu(A^*)$*

Proof Let (P, N) be a Hahn decomposition for $\mu(\cdot)$ and let (μ^+, μ^-) be the Jordan decomposition for $\mu(\cdot)$ (see Theorem 2.99 and Theorem 2.103, respectively). Then for every $A \in \Sigma$, we have

$$\mu^+(A) = \mu(A \cap P) \leq \mu(P) \quad and \quad \mu^-(A) = -\mu(A \cap N) \leq \mu^-(N) = -\mu(N),$$
$$\Rightarrow \mu(N) \leq -\mu^-(A) \leq \mu^+(A) - \mu^-(A) = \mu(A) \leq \mu^+(A) \leq \mu(P)$$

So, take $A_* = N$ and $A^* = P$. □

Corollary 2.105 *Every signed measure* $\mu : \Sigma \to \mathbb{R}^* = [-\infty, +\infty]$ *is either bounded from above or bounded from below.*

Definition 2.106 Let (X, Σ) be a measurable space, $\mu : \Sigma \to [-\infty, +\infty]$ a signed measure, and (μ^+, μ^-) its Jordan decomposition. The measures $\mu^+(\cdot)$ and $\mu^-(\cdot)$ are called the "positive variation" of $\mu(\cdot)$ and the "negative variation" of $\mu(\cdot)$. The measure $|\mu| = \mu^+ + \mu^-$ is called the "total variation" of $\mu(\cdot)$.

The next proposition provides characterizations of the above items.

Proposition 2.107 *If* (X, Σ) *is a measurable space and* $\mu : \Sigma \to [-\infty, +\infty]$ *a signed measure, then*

(a) *for all* $A \in \Sigma, \mu^+(A) = \sup[\mu(B) : B \in \Sigma, B \subseteq A, B \text{ positive}] = \sup[\mu(B) : B \in \Sigma, B \subseteq A]$ *and* $\mu^-(A) = -\inf[\mu(B) : B \in \Sigma, B \subseteq A, B \text{ negative}] = -\inf[\mu(B) : B \in \Sigma, B \subseteq A]$.

(b) *For all* $A \in \Sigma, |\mu(A)| = \sup[\sum_{k=1}^{n} |\mu(A_k)| : n \in \mathbb{N}, \{A_k\}_{k=1}^{n} \subseteq \Sigma \text{ are disjoint}$
and $A = \bigcup_{k=1}^{n} A_k, \text{ that is, } \{A_k\}_{k=1}^{n} \text{is a } \Sigma - \text{partition of } A]$.

Proof

(a) Without any loss of generality, we may assume that $\mu : \Sigma \to [-\infty, +\infty)$. Let (P, N) be a Hahn decomposition for μ. Since P is positive, we have

$$\mu^+(A) = \mu(A \cap P)$$
$$= sup[\mu(B) : B \in \Sigma, B \subseteq A, B \text{ positive}]$$
$$= sup[\mu(B) : B \in \Sigma, B \subseteq A, \mu^-(B) = 0]$$
$$= sup[\mu(B) : B \in \Sigma, B \subseteq A]$$

.

Similarly, since N is negative,

$$\mu^-(A) = -\mu(A \cap N)$$
$$= -\inf[\mu(B) : B \in \Sigma, B \subseteq A, B \text{ negative}].$$

If $|\mu(A)| < +\infty$, then

$$\mu^-(A) = \mu^+(A) - \mu(A)$$
$$= sup[\mu(B) - \mu(A) : B \in \Sigma, B \subseteq A]$$
$$= sup[-\mu(A \setminus B) : B \in \Sigma, B \subseteq A]$$
$$= -inf[\mu(B) : B \in \Sigma, B \subseteq A].$$

If $\mu(A) = -\infty$, then since $\mu^+(A) < +\infty$, we have

$$\mu^-(A) = \mu^+(A) - \mu(A) = +\infty,$$

hence $\mu^-(A) = -inf[\mu(B) : B \in \Sigma, B \subseteq A] = +\infty$.

(b) Let η be the right hand side of the equality that we want to prove. Then

$$\sum_{k=1}^{n} |\mu(A_k)| = \sum_{k=1}^{n} |\mu^+(A_k) - \mu^-(A_k)|$$

$$\leq \sum_{k=1}^{n} [\mu^+(A_k) + \mu^-(A_k)] = \sum_{k=1}^{n} |\mu|(A_k) = |\mu|(A)$$

$$\Rightarrow \eta \leq |\mu|(A). \tag{2.9}$$

On the other hand, let (P, N) be a Hahn decomposition for $\mu(\cdot)$ and consider the Σ-partition of A $\{A \cap P, A \cap N\}$. We have

$$|\mu|(A) = \mu^+(A) + \mu^-(A) = \mu(A \cap P) + |\mu(A \cap N)| \leq \eta. \tag{2.10}$$

From (2.9) and (2.10) we conclude that $|\mu|(A) = \eta$. $\qquad\qquad\qquad\square$

Definition 2.108 Let (X, Σ) be a measurable space, $\lambda : \Sigma \to \mathbb{R}^* = [-\infty, \infty]$ a signed measure and $\mu : \Sigma \to \mathbb{R}^* = [0, +\infty]$ a measure. We say that λ is "absolutely continuous " with respect to μ, denoted by $\lambda \ll \mu$, if $\lambda(A) = 0$ for all $A \in \Sigma$, with $\mu(A) = 0$.

Remark 2.109 It is easy to check that $\lambda \ll \mu$ if and only if $|\lambda| \ll \mu$ if and only if $\lambda^+ \ll \mu$ and $\lambda^- \ll \mu$. Absolute continuity is the opposite of mutual singularity in the sense that if $\lambda \perp \mu$ and $\lambda \ll \mu$, then $\lambda = 0$. Indeed suppose $A, B \in \Sigma, X = A \cup B, A \cap B = \emptyset$ and $\mu(A) = |\lambda|(B) = 0$. Then since $\lambda \ll \mu$ is equivalent to $|\lambda| \ll \mu$, we have $|\lambda|(A) = 0$, which means that $\lambda \equiv 0$. If λ_1, λ_2 are signed measures and μ is a measure, then we can easily show the following facts. (a) $\lambda_1 \perp \mu$, $\lambda_2 \perp \mu \Rightarrow \lambda_1 + \lambda_2 \perp \mu$; (b) $\lambda_1 \ll \mu, \lambda_2 \ll \mu \Rightarrow \lambda_1 + \lambda_2 \ll \mu$; (c) $\lambda_1 \ll \mu, \lambda_2 \perp \mu$

$\Rightarrow \lambda_1 \perp \lambda_2$. Statements (a) and (b) remain true also for sequences of signed measures and measures.

Proposition 2.110 *If (X, Σ) is a measurable space, $\lambda : \Sigma \to \mathbb{R}$ is a signed measure and $\mu : \Sigma \to \overline{\mathbb{R}} = [0, +\infty]$ is a measure, then $\lambda \ll \mu$ if and only if for every $\varepsilon > 0$, there exists $\delta > 0$ such that "for $A \in \Sigma$, $\mu(A) < \delta \Rightarrow |\lambda|(A) < \varepsilon$."*

Proof Recall that $\lambda \ll \mu \iff |\lambda| \ll \mu$ and $|\lambda(A)| \leq |\lambda|(A)$ for all $A \in \Sigma$. Hence we may assume that $\lambda = |\lambda|$, that is, λ is a measure.
$\Rightarrow$: Suppose that this implication is not true. Then we can find $\varepsilon > 0$ such that for all $n \in \mathbb{N}$, there exists $A_n \in \Sigma$ with $\mu(A_n) < \frac{1}{2^n}$ and $\lambda(A_n) \geq \varepsilon$. Let $B_k = \bigcup_{n \geq k} A_n \in \Sigma$
and $B = \bigcap_{k \geq 1} B_k \in \Sigma$. We have $\mu(B) < \sum_{n \geq k} 2^{-n} = \frac{1}{2^{k-1}}$ for all $k \in \mathbb{N}$ and so
$\mu(B) = 0$ (see Proposition 2.17(d)). But $\lambda(B_k) \geq \lambda(A_n) \geq \varepsilon$ for all $n \geq k$, all $k \in \mathbb{N}$. Since λ is finite, $\lambda(B) = \lim_{k \to \infty} \lambda(B_k) \geq \varepsilon$ (again Proposition 2.17(d)), a contradiction to our hypothesis that $\lambda \ll \mu$ (see Definition 2.108).
$\Leftarrow$: It is clear from Definition 2.108. $\square$

Corollary 2.111 *If (X, Σ, μ) is a measure space and $f \in \mathcal{L}^1(X)$, then given $\varepsilon > 0$, we can find $\delta > 0$ such that*

$$\text{"}A \in \Sigma \text{ and } \mu(A) < \delta \Rightarrow |\int_A f d\mu| < \varepsilon.\text{"}$$

The next lemma will be used in the proof of the Lebesgue–Radon–Nikodym Theorem (see Theorem 2.113).

Lemma 2.112 *If (X, Σ) is a measurable space and $\lambda, \mu : \Sigma \to \mathbb{R}$ are finite measures, then either $\lambda \perp \mu$ or there exist $\varepsilon > 0$ and $A \in \Sigma$ such that*

$$\mu(A) > 0 \text{ and } \varepsilon\mu \leq \lambda \text{ on } A$$

(so A is a positive set for $\lambda - \varepsilon\mu$).

Proof Let $\vartheta_n = \lambda - \frac{1}{n}\mu$, $n \in \mathbb{N}$ and let (P_n, N_n) be a Hahn decomposition for $\vartheta_n(\cdot)$. We set $P = \bigcup_{n \geq 1} P_n$ and $N = \bigcap_{n \geq 1} N_n = P^c$. Note that N is negative set for
ϑ_n for all $n \in \mathbb{N}$, hence $0 \leq \lambda(N) \leq \frac{1}{n}\mu(N)$ for all $n \in \mathbb{N}$ and so $\lambda(N) = 0$. If $\mu(P) = 0$, then $\lambda \perp \mu$. If $\mu(P) > 0$, then for some $n \in \mathbb{N}$, we have $\mu(P_n) > 0$ and P_n is a positive set for $\vartheta_n = \lambda - \frac{1}{n}\mu$. $\square$

We already know that if $f \in \mathcal{L}^1(X)$, then $\Sigma \ni A \to \lambda(A) = \int_A f d\mu$ is a signed measure (see Proposition 2.90 and recall that $f = f^+ - f^-$). If μ is a σ-finite, then this gives a complete characterization of all signed measures λ which are absolutely continuous with respect to $\mu(\cdot)$. The result is known as the "Lebesgue–

Radon–Nikodym Theorem." We mention that a signed measure $\lambda(\cdot)$ is said to be "finite"(respectively, "σ-finite"), if $|\lambda|(\cdot)$ is "finite"(respectively, "σ-finite").

Theorem 2.113 *If (X, Σ) is a measurable space, $\lambda : \Sigma \to \mathbb{R}^* = [-\infty, \infty]$ is a σ-finite signed measure and $\mu : \Sigma \to \overline{\mathbb{R}} = [0, +\infty]$ is a σ-finite measure, then there exist unique σ-finite signed measures ϑ and ξ on (X, Σ) such that*

$$\vartheta \ll \mu, \xi \perp \mu \text{ and } \lambda = \vartheta + \xi.$$

Moreover, we can find $f \in \mathcal{L}^1(X, \mu)$ such that

$$\vartheta(A) = \int_A f d\mu \text{ for all } A \in \Sigma.$$

Proof First suppose that λ and μ are finite measures. We set $S = \{f : X \to \overline{\mathbb{R}} = [0, +\infty] : \int_A f d\mu \leq \lambda(A) \text{ for all } A \in \Sigma\}$. Note that $0 \in S$ and so $S \neq \emptyset$. If $f, h \in S$, then $u = max\{f, h\} \in S$. Indeed, let $D = \{h < f\}$. Then for any $A \in \Sigma$ we have

$$\int_A u d\mu = \int_{A \cap D} f d\mu + \int_{A \backslash D} h d\mu \leq \lambda(A \cap D) + \lambda(A \backslash D) = \lambda(A).$$

Let $\eta = \sup[\int_X f d\mu : f \in \mathcal{S}] \leq \lambda(X)$ and consider a sequence $\{f_n\}_{n \geq 1} \subseteq \mathcal{S}$ such that $\int_X f_n d\mu \to \eta$. We set $h_n = max\{f_k\}_{k=1}^n$ and $f = \sup_{n \geq 1} f_n$. From the previous part of the proof, we have $h_n \in S$. Also $h_n \uparrow f$ and $\int_X f_n d\mu \leq \int_X h_n d\mu$ and so $\int_X h_n d\mu \to \eta$. Then by the Monotone Convergence theorem (see Theorem 2.80), we have that $\int_X f d\mu = \eta < \infty$. Hence $f(x) < +\infty$ for a. a $x \in X$ and so we can say that f is $\mathbb{R}$-valued.

Let $\xi = \lambda - \vartheta$ ($\vartheta(A) = \int_A f d\mu$ for all $A \in \Sigma$). Then $\xi \perp \mu$. If not, then by Lemma 2.112 we can find $\varepsilon > 0$ and $A \in \Sigma$ such that $\varepsilon \mu \leq \xi$ on A. Then we have $\varepsilon m_A \leq \xi = \lambda - m$ with $m_A(B) = \int_B \chi_A d\mu$ for all $B \in \Sigma$. So $m + \varepsilon m_A \leq \lambda$. Note that $(m + \varepsilon m_A)(B) = \int_B [f + \varepsilon \chi_A] d\mu$ for all $B \in \Sigma$. Hence $f + \varepsilon \chi_A \in S$ and $\int_X [f + \varepsilon \chi_A] d\mu = \eta + \varepsilon \mu(A) > \eta$, a contradiction (since $f + \varepsilon \chi_A \in S$). Therefore $\xi \perp \mu$.

So, we have proved the existence of ϑ, ξ and f. Next we show the uniqueness of these items. Suppose that $\widehat{\vartheta}, \widehat{\xi}$ and $\widehat{f}$ form another such triple. We have $\lambda = \vartheta + \xi = \widehat{\vartheta} + \widehat{\xi}$, hence $(\vartheta - \widehat{\vartheta})(A) = \int_A [f - \widehat{f}] d\mu = (\widehat{\xi} - \xi)(A)$ for all $A \in \Sigma$. Since $\xi \perp \mu, \widehat{\xi} \perp \mu$, we have $(\widehat{\xi} - \xi) \perp \mu$ and $(\vartheta - \widehat{\vartheta}) \ll \mu$ (see Remark 2.109). Therefore we infer that $\vartheta - \widehat{\vartheta} = 0 = \widehat{\xi} - \xi$, hence $\vartheta = \widehat{\vartheta}, \widehat{\xi} = \xi$.

Next suppose that λ, μ are σ-finite measures. Then we can find $\{A_n\} \subseteq \Sigma$ mutually disjoint such that $\lambda(A_n)$ and $\mu(A_n)$ are finite and $X = \cup_{n \geq 1} A_n$. Let $\lambda_n(A) = \lambda(A \cap A_n)$ and $\mu_n(A) = \mu(A \cap A_n)$ for all $A \in \Sigma$. From the first part of the proof, $\lambda_n = \vartheta_n + \xi_n$ with $\vartheta_n \ll \mu_n$ and $\xi_n \perp \mu_n$ for all $n \in \mathbb{N}$. Note that

$\lambda_n(A_n^c) = \mu_n(A_n^c) = 0$. So, we have $\xi_n(A_n^c) = \lambda_n(A_n^c) - \vartheta_n(A_n^c) = 0$ and we may assume that $f_n|_{A_n^c} = 0$, where $f_n = \frac{d\mu_n}{d\vartheta_n}$. Set

$$\xi = \sum_{n \geq 1} \xi_n, \quad f = \sum_{n \geq 1} f_n, \quad \text{and} \quad \vartheta(A) = \int_A f \, d\mu \quad \text{for } A \in \Sigma.$$

We have $\lambda = \vartheta + \xi$, $\vartheta \ll \mu$, $\xi \perp \mu$, with ϑ and ξ both σ-finite. Uniqueness can be shown as before.

Finally, suppose λ is a signed measure. Then apply the previous results on λ^+ and on λ^- to yield the results. $\qquad \square$

Remark 2.114 The decomposition $\lambda = \vartheta + \xi$ with $\vartheta \ll \mu$ and $\xi \perp \mu$ is called the "Lebesgue Decomposition" of λ with respect to μ. If $\lambda \ll \mu$, then $\xi = 0$ and we have $\lambda = \vartheta$. Using more suggestive notation, we write $d\lambda = f \, d\mu$ for some f with f^+ or f^- integrable. This result is known as the "Radon–Nikodym Theorem" and the function f is the "Radon–Nikodym derivative" of λ with respect to μ. We write $f = \frac{d\lambda}{d\mu}$.

We have a final remark about the total variation which is useful when dealing with spaces of measures.

Proposition 2.115 *If (X, Σ) is a measurable space and $\mu : \Sigma \to \mathbb{R}^* = [-\infty, +\infty]$ is a signed measure, then $\sup[|\mu(B)| : B \in \Sigma, B \subseteq A] \leq |\mu|(A) \leq 2 \sup[|\mu(B)| : B \in \Sigma, B \subseteq A]$.*

Proof Let $B \in \Sigma, B \subseteq A$. Then $\{B, A \setminus B\}$ is a Σ-partition of A. So, from Proposition 2.107(b) we have

$$|\mu(B)| \leq |\mu(B)| + |\mu(A \setminus B)| \leq |\mu|(A),$$

$\Rightarrow \sup[|\mu(B)| : B \in \Sigma, B \subseteq A] \leq |\mu|(A)$.

Let $\{A_k\}_{k=1}^n \subseteq \Sigma$ be a Σ-partition of A. Assume that $\mu(A_1), \cdots, \mu(A_m) \geq 0$ and $\mu(A_{m+1}), \cdots, \mu(A_n) < 0$. So, we have

$$\sum_{k=1}^n |\mu(A_k)| = \sum_{k=1}^m \mu(A_k) - \sum_{k=m+1}^n \mu(A_k),$$

$$= \mu(\bigcup_{k=1}^m A_k) - \mu(\bigcup_{k=m+1}^n A_k),$$

$$= |\mu(\bigcup_{k=1}^m A_k)| + |\mu(\bigcup_{k=m+1}^n A_k)|,$$

$$\leq 2 \sup[|\mu(B)| : B \in \Sigma, B \subseteq A]. \qquad \square$$

2.6 L^p-Spaces

Let (X, Σ, μ) be a measure space. Given two Σ-measurable functions $f, h : X \to \mathbb{R}^* = [-\infty, +\infty]$, we say that f and h are "equivalent," denoted by $f \sim h$, if and only if $f(x) = h(x)$ μ—a.e. Evidently $\sim$ is an equivalence relation. Given $1 \leq p \leq \infty$, we define

$$\mathcal{L}^p(X) = \{f : X \to \mathbb{R}^* = [-\infty, +\infty] \ is \ \Sigma - measurable, \int_X |f|^p d\mu < \infty\}.$$

For $p = +\infty$, we define

$$\mathcal{L}^\infty(X) = \{f : X \to \mathbb{R}^* = [-\infty, +\infty] \ is \ \Sigma - measurable, \operatorname{esssup}_\Omega |f| < \infty\},$$

where $\operatorname{esssup}_\Omega |f| = \inf[M > 0 : |f(x)| \leq M \ \mu - a.e]$.

Definition 2.116 For $1 \leq p \leq \infty$, $L^p(X) = \mathcal{L}^p(X)/\sim$. For $f \in L^p(X)$ we define

$$\|f\|_p = (\int_X |f|^p d\mu)^{\frac{1}{p}}, if \ 1 \leq p < \infty$$

$$\|f\|_\infty = \operatorname{esssup}_\Omega |f|, if \ p = +\infty.$$

Remark 2.117 For $1 \leq p < \infty$, we have $|f + h|^p \leq 2^p(\max\{|f|, |h|\})^p \leq 2^p[|f|^p + |h|^p]$. Therefore the spaces $L^p(X)$ $1 \leq p \leq \infty$ are vector spaces and $\| \cdot \|_p$ are norms. Only the triangle inequality needs verification and as we will see below it is valid precisely when $p \geq 1$. For $0 < p < 1$ the triangle inequality fails. Finally we mention that if $X = \mathbb{N}$ and μ is the counting measure, then $L^p(\mathbb{N}) = l^p$ for all $1 \leq p \leq \infty$.

In what follows if $1 \leq p \leq \infty$, then $1 \leq p' \leq \infty$ is the "conjugate" exponent defined by

$$p' = \begin{cases} \dfrac{p}{p-1} & \text{if } 1 < p < \infty, \\[2mm] \infty & \text{if } p = 1, \\[2mm] 1 & \text{if } p = \infty. \end{cases}$$

Note that we have $\frac{1}{p} + \frac{1}{p'} = 1$.

The main inequality in the theory of L^p-spaces, is the so-called Hölder inequality. It is based on the following elementary inequality.

Lemma 2.118 *If* $1 \leq p \leq \infty$ *and* $a, b \in \mathbb{R}$, *then* $|ab| \leq \frac{1}{p}|a|^p + \frac{1}{p'}|b|^{p'}$.

Proof The inequality is clear if $a = 0$ or $b = 0$. So, suppose that $|a| > 0$, $|b| > 0$. Again the inequality is clear if $p = 1$ or $p = \infty$. Therefore we may assume that $1 < p < \infty$. Consider the function $\xi(x) = \frac{x^p}{p} + \frac{1}{p'} - x$ for $x \geq 0$. Then $\xi(\cdot)$ has an absolute minimum at $x = 1$. Hence $x \leq \frac{x^p}{p} + \frac{1}{p'}$ for all $x \geq 0$ and equality holds for $x = 1$. Let $x = \frac{|a|}{|b|^{p'/p}}$. Then $\frac{|a|}{|b|^{p'/p}} \leq \frac{1}{p}\frac{|a|^p}{|b|^{p'}} + \frac{1}{p'} \Rightarrow |ab| \leq \frac{1}{p}|a|^p + \frac{1}{p'}|b|^{p'}$. $\qquad\square$

Using this lemma we can prove the "Hölder inequality."

Theorem 2.119 *If* $1 \leq p \leq \infty$, $f \in L^p(X)$ *and* $h \in L^{p'}(X)$, *then* $fh \in L^1(X)$ *and* $\|fh\|_1 \leq \|f\|_p\|h\|_{p'}$; *moreover, equality holds if and only if* $|f(x)|^p = c|h(x)|^{p'}$ $\mu - a.e$ *for some* $c \in \mathbb{R}$.

Proof Without any loss of generality we may assume that $f(x), h(x) \geq 0$ $\mu - a.e.$ Also, we can have that $f \neq 0$, $h \neq 0$ and $1 < p < \infty$ (otherwise the inequality is clear). Let $a = \frac{f}{\|f\|_p}$ and $b = \frac{h}{\|h\|_{p'}}$. Using Lemma 2.118, we have

$$\frac{fh}{\|f\|_p\|h\|_{p'}} \leq \frac{1}{p}\frac{f^p}{\|f\|_p^p} + \frac{1}{p'}\frac{h^{p'}}{\|h\|_{p'}^{p'}},$$

$$\Rightarrow \int_X fh\,d\mu \leq \|fh\|_1 \leq \|f\|_p\|h\|_{p'}.$$

From the proof of Lemma 2.118, we see that equality holds only if

$$f(x)^p = \frac{\|f\|_p^p}{\|h\|_{p'}^{p'}}h(x)^{p'} \quad \mu - a.e.$$

$\qquad\square$

Corollary 2.120 *If* $1 < p_1, \cdots, p_n < \infty$, $\sum_{k=1}^{n}\frac{1}{p_k} = 1$ *and* $f_k \in L^{p_k}(X)$ *with* $k = 1, \cdots, n$, *then* $\|\prod_{k=1}^{n} f_k\|_1 \leq \prod_{k=1}^{n}\|f_k\|_{p_k}$.

Proof Follows by applying Theorem 2.119 and an induction argument. $\qquad\square$

In fact more generally we can have:

Corollary 2.121 *If* $1 < p_1, \cdots, p_n < \infty$, $\frac{1}{p} = \sum_{k=1}^{n}\frac{1}{p_k} \leq 1$ *and* $f_k \in L^{p_k}(\Omega)$ *with* $k = 1, \cdots, n$, *then* $\prod_{k=1}^{n} f_k \in L^p(X)$ *and* $\|\prod_{k=1}^{n} f_k\|_p \leq \prod_{k=1}^{n}\|f_k\|_{p_k}$.

Next we present some important consequences of the "Hölder inequality." The first is the so-called Minkowski Inequality. It shows that $L^p(X)$ $1 \leq p \leq \infty$ is a normed space.

Theorem 2.122 *If* $1 \le p \le \infty$ *and* $f, h \in L^p(X)$, *then* $\|f + h\|_p \le \|f\|_p + \|h\|_p$; *moreover equality holds if* $f(x) = ch(x)$ $\mu - a.e.$ *for some* $c \in \mathbb{R}$.

Proof The inequality is clear if $p = 1$ and if $p = +\infty$. So, we assume that $1 < p < \infty$. Then we have

$$\|f + h\|_p^p = \int_X |f + h|^p d\mu$$

$$= \int_X |f + h|^{p-1} |f + h| d\mu,$$

$$\le \int_X |f + h|^{p-1} |f| d\mu + \int_X |f + h|^{p-1} |h| d\mu,$$

$$\le \|f + g\|_p^{p-1} [\|f\|_p + \|h\|_p] \quad (using \ Theorem \ 2.119),$$

$$\Rightarrow \|f + h\|_p \le \|f\|_p + \|h\|_p.$$

Also, from Theorem 2.119 we see that equality holds if $f = ch$ with $c \in \mathbb{R}$. $\square$

Another consequence of the Hölder inequality is the so-called Interpolation Inequality.

Proposition 2.123 *If* $1 \le p \le q \le \infty$ *and* $f \in L^p(X) \cap L^q(X)$, *then* $f \in L^r(X)$ *for all* $r \in [p, q]$ *and* $\|f\|_r \le \|f\|_p^{1-t} \|f\|_q$ *with* $t \in [0, 1]$ *such that* $\frac{1}{r} = \frac{t}{p} + \frac{1-t}{q}$.

Proof If $q = +\infty$, then $t = \frac{p}{r}$ and $|f|^r \le \|f\|_\infty^{r-p} |f|^p$. Hence $\|f\|_r \le \|f\|_\infty^{1-\frac{p}{r}} \|f\|_p^{\frac{p}{r}} = \|f\|_p^t \|f\|_\infty^{1-t}$. So, suppose that $q < +\infty$. Also, we assume that $r \ne p, r \ne q$, hence $t \in (0, 1)$. We consider the conjugate exponents $\frac{p}{tr}, \frac{q}{(1-t)r}$. Using Hölder inequality (see Theorem 2.119), we have

$$\|f\|_r^r = \int_X |f|^r d\mu = \int_X |f|^{tr} |f|^{(1-t)r} d\mu \le \|f\|_p^{tr} \|f\|_q^{(1-t)r},$$

$$\Rightarrow \|f\|_r \le \|f\|_p^t \|f\|_q^{1-t}.$$ $\square$

The Hölder inequality leads to a relation between different L^p-spaces.

Proposition 2.124 *If* (X, Σ, μ) *is a finite measure space and* $1 \le p < q \le \infty$, *then* $L^q(X) \subseteq L^p(X)$ *and* $\|f\|_p \le \|f\|_q \mu(X)^{\frac{1}{p} - \frac{1}{q}}$.

Proof If $q = +\infty$, then the assertions of the proposition are clear. So suppose $q < \infty$. Let $f \in L^q(X)$ and set $r = \frac{q}{p} > 1$, $s = \frac{q}{q-p} > 1$. Note that $\frac{1}{r} + \frac{1}{s} = 1$ (that is, $s = r'$). Since the constant function 1 belongs in $L^s(X)$, using Theorem 2.119 we have

$$\int_X |f|^p d\mu \le (\int_X |f|^{pr} d\mu)^{\frac{1}{r}} (\int_X 1^s d\mu)^{\frac{1}{s}} \le \|f\|_q^p \mu(X)^{\frac{1}{s}},$$

$$\Rightarrow \|f\|_p \le \|f\|_q \mu(X)^{\frac{1}{p} - \frac{1}{q}} < \infty.$$ □

Remark 2.125 The inclusion $L^q(X) \subseteq L^p(X)$ is true if $0 < p < q \le \infty$.

The Minkowski inequality (see Theorem 1.222) implies that for $1 \le p \le \infty$, $L^p(X)$ is a normed space. In fact we can say more.

Theorem 2.126 *If* $1 \le p \le \infty$, *then* $L^p(X)$ *is a Banach space.*

Proof First suppose that $1 \le p < \infty$. Let $\{f_n\}_{n \ge 1} \subseteq L^p(X)$ be a Cauchy sequence. We extract a subsequence $\{f_{n_k}\}_{k \in \mathbb{N}}$ $n_1 < n_2 < \cdots < n_k < \cdots$ such that

$$\|f_{n_{k+1}} - f_{n_k}\|_p < \frac{1}{2^k} \quad for \ all \ k \in \mathbb{N}. \tag{2.11}$$

We set $h_m = \sum_{k=1}^m |f_{n_{k+1}} - f_{n_k}|$ and $h = \sum_{k \ge 1} |f_{n_{k+1}} - f_{n_k}|$. From (2.11) and the Minkowski inequality, we have $\|h_m\|_p < 1$, $m \in \mathbb{N}$. Applying Fatou's lemma (see Theorem 2.84) on the sequence $\{h_m^p\}_{m \in \mathbb{N}}$, we obtain $\|h\|_p \le 1$. Hence $h(x) < +\infty$ $\mu - a.e.$ Then the series $f_{n_1}(x) + \sum_{k \ge 1} [f_{n_{k+1}}(x) - f_{n_k}(x)]$ converges absolutely for $\mu - a.a$ $x \in X$. Let $f(x) = f_{n_1}(x) + \sum_{k \ge 1} [f_{n_{k+1}}(x) - f_{n_k}(x)]$ (we set $f(x) = 0$ on the μ-null set where we do not have convergence). Then $f_{n_1} + \sum_{k=1}^{m-1} [f_{n_{k+1}} - f_{n_k}] = f_{n_k} \to f$ $\mu - a.e.$ Given $\varepsilon > 0$, we can find $n_0 \in \mathbb{N}$ such that

$$\|f_n - f_m\| \le \varepsilon \quad for \ all \ n, m \ge n_0,$$

$$\Rightarrow \int_X |f - f_m|^p d\mu \le \lim_{k \to \infty} \inf \int_X |f_{n_k} - f_m|^p d\mu \le \varepsilon^p,$$

$$\Rightarrow f - f_m \in L^p(X),$$

$$\Rightarrow f \in L^p(X) \ and \ \|f - f_m\|_p \to 0 \ as \ m \to \infty.$$

This proves that $L^p(X)$ is a Banach when $1 \le p < \infty$.

Next let $p = +\infty$ and let $\{f_n\}_{n \ge 1} \subseteq L^\infty(X)$ be a Cauchy sequence. Let $A_k = \{x \in X : |f_k(x)| > \|f_k\|_\infty\} \in \Sigma$ and $B_{m,n} = \{x \in X : |f_n(x) - f_m(x)| > \|f_n - f_m\|_\infty\} \in \Sigma$, $k, m, n \in \mathbb{N}$. Let $C = \bigcup_{k,m,n \ge 1} (A_k \cup B_{m,n}) \in \Sigma$. Evidently $\mu(C) = 0$ and on C^c we have uniform convergence of $\{f_n\}_{n \ge 1}$ to a bounded function $f(\cdot)$. Set $f(x) = 0$ for $x \in C$. Then $f \in L^\infty(X)$ and $\|f_n - f\|_\infty \to 0$ as $n \to \infty$. □

An interesting byproduct of the above proof is the following result.

Proposition 2.127 *If* $1 \leq p \leq \infty$ *and* $\{f_n\}_{n\geq 1} \subseteq L^p(X)$ *satisfies* $\|f_n - f\|_p \to 0$, *then* $f \in L^p(X)$ *and we can find a subsequence* $\{f_{n_k}\}_{k\in\mathbb{N}}$ *of* $\{f_n\}_{n\in\mathbb{N}}$ *such that* $f_{n_k}(x) \to f(x)$ $\mu - a.e.$

Next we indicate some useful dense subsets of $L^p(X)$, $1 \leq p \leq \infty$.

Proposition 2.128 *If* $1 \leq p < \infty$ *and* S *is the set of all integrable simple functions (that is,* $\mu(\{x \in X : s(x) \neq 0\}) < \infty$ *for* $s \in S$*), then* S *is dense in* $L^p(X)$.

Proof Let $f \in L^p(X)$ and first assume that $f \geq 0$. Then according to Proposition 2.70, we can find a sequence of simple functions $\{s_n\}_{n\geq 1}$ such that $0 \leq s_n \leq f$ and $s_n(x) \uparrow f(x)$ $\mu - a.e.$ Evidently $s_n \in L^p(X)$ for all $n \in \mathbb{N}$ and $0 \leq (f - s_n)^p \leq f^p$. So, by the Dominated Convergence Theorem (see Theorem 2.86), we have $\|f - s_n\|_p \to 0$. For general, $f \in L^p(X)$ simply write $f = f^+ - f^-$.   □

Directly from Proposition 2.70, we have:

Proposition 2.129 *Simple functions are dense in* $L^\infty(X)$.

Anticipating a notion from Sect. 2.8 we make the following definition:

Definition 2.130 Let (X, τ) be a Hausdorff topological space and $\mu : B(X) \to \overline{\mathbb{R}} = [0, +\infty]$ a measure. We say that $\mu(\cdot)$ is a "Radon measure," if it has the following properties:

(a) $\mu(\cdot)$ is outer regular, that is, for all $A \in B(X)$ we have

$$\mu(A) = \inf[\mu(U) : U \in \tau, A \subseteq U].$$

(b) $\mu(\cdot)$ is tight, that is, for all $A \in B(X)$ we have

$$\mu(A) = \sup[\mu(K) : K \ compact, K \subseteq A].$$

(c) $\mu(K) < +\infty$ for all compact $K \subseteq X$.

Proposition 2.131 *If* X *is a locally compact topological space and* $\mu : B(X) \to \overline{\mathbb{R}} = [0, +\infty]$ *a Radon measure, then* $C_c(X)$ =*space of continuous* $\mathbb{R}$-*valued functions on* X *with compact support is dense in* $L^p(X, \mu)$ *for all* $1 \leq p < \infty$.

Proof From Proposition 2.128 we know that the set S of integrable simple function is dense in $L^p(X, \mu)$. So, it suffices to show that if $A \in B(X)$ and $\mu(A) < +\infty$, then given $\varepsilon > 0$, we can find $f \in C_c(X)$ such that $\|\chi_A - f\|_p \leq \varepsilon$.

Since $\mu(\cdot)$ is Radon, we can find $K \subseteq A \subseteq U$ with K compact, U open such that $\mu(U \setminus K) \leq (\frac{\varepsilon}{2})^p$. Since X is locally compact, according to Proposition 1.105(c), we can find $V \subseteq X$ open with $\overline{V}$ compact such that $K \subseteq V \subseteq \overline{V} \subseteq U$. Then by Proposition 1.112 we can find $f \in C(K)$ such that $f \mid_K = 1$ and $f \mid_{X\setminus V} = 0$. So $f \in C_c(X)$ and we have

$$(\int_A |\chi_A - f|^p d\mu)^{\frac{1}{p}} \le (\int_X |\chi_A - \chi_K|^p d\mu)^{\frac{1}{p}} + (\int_X |\chi_K - f|^p d\mu)^{\frac{1}{p}},$$

$$\le (\int_X \chi_{A \setminus K} d\mu)^{\frac{1}{p}} + (\int_X \chi_{U \setminus K} d\mu)^{\frac{1}{p}},$$

$$\le (\frac{\varepsilon}{2})^{\frac{1}{p}} + (\frac{\varepsilon}{2})^{\frac{1}{p}}$$

$$\le 2(\frac{\varepsilon}{2})^{1/p},$$

$$\Rightarrow C_c(X) \text{ is dense in } L^p(X), \ 1 \le p < \infty. \qquad \square$$

For $X = \mathbb{R}^N$ and using classical approximations by mollification, we can replace $C_c(\mathbb{R}^N)$ by the smaller set $C_c^\infty(\mathbb{R}^N)$.

Proposition 2.132 *If* $1 \le p < \infty$ *and on* $\mathbb{R}^N$ *we consider the Lebesgue measure, then* $C_c^\infty(\mathbb{R}^N)$ *is norm dense in* $L^p(\mathbb{R}^N)$.

Now we will identify the continuous linear functionals of the Banach space $L^p(X), 1 \le p \le \infty$, that is, we will identify the dual of $L^p(X)$ (see Chap. 3). The result is known as the "Riesz Representation Theorem."

Theorem 2.133 *If* (X, Σ, μ) *is a* σ-*finite measure space,* $1 \le p < \infty$ *and* φ *is a bounded linear functional on* $L^p(X)$, *then there exists a unique* $h \in L^{p'}(\Omega)(\frac{1}{p} + \frac{1}{p'} = 1)$ *such that*

$$\varphi(f) = \int_X f h d\mu \quad \text{for all } f \in L^p(X) \tag{2.12}$$

$$\|\varphi\|_* = \|h\|_{p'} \ (\| \cdot \|_* \text{ denoting the dual norm}). \tag{2.13}$$

Proof The uniqueness of h is easy. Indeed for any other $h' \in L^{p'}(X)$ for which (2.12) is true, we have $\int_A [h - h'] d\mu = 0$ for all $A \in \Sigma, \mu(A) < \infty$ (just take $f = \chi_A$). Then on account of Proposition 2.79 and the σ-finiteness of the measure space, we have $h = h' \ \mu - a.e.$

If (2.12) holds, then via Hölder's inequality (see Theorem 2.119), we have $\|\varphi\|_* \le \|h\|_{p'}$. Clearly we may assume that $\varphi \neq 0$.

First we assume that $\mu(X) < \infty$. For every $A \in \Sigma$, we set $\lambda(A) = \varphi(\chi_A)$. Then it is easy to see that $\lambda(\cdot)$ is a signed measure and $\lambda \ll \mu$. So, by Theorem 2.113 there exists unique function $h \in L^1(X)$ such that

$$\varphi(\chi_A) = \int_A h d\mu = \int_X \chi_A h d\mu \quad \text{for all } A \in \Sigma,$$

thus $\varphi(s) = \int_X s h d\mu$ for every simple function s. Hence, for every $f \in L^\infty(X)$

$$\varphi(f) = \int_X f h d\mu \quad \text{(see Proposition 2.129).} \tag{2.14}$$

First assume $p = 1$. We have

$$\left| \int_A h d\mu \right| \leq \|\varphi\|_* \|\chi_A\|_1 = \|\varphi\|_* \mu(A) \quad \text{for all } A \in \Sigma,$$

$$\Rightarrow |h(x)| \leq \|\varphi\|_* \quad \mu - a.e,$$

$$\Rightarrow \|h\|_\infty \leq \|\varphi\|_*.$$

Next suppose $1 < p < \infty$. Let ξ be a Σ-measurable function with $|\xi(x)| = 1$ for $\mu - a.a \ x \in X$ and $\xi h = |h|$. Let $A_n = \{x \in X : |h(x)| \leq n\} \in \Sigma$ and $f = \chi_{A_n} |h|^{p'-1} \xi$. We have $|f|^p = |h|^{p'}$ on A_n. From (2.14) we have

$$\int_{A_n} |h|^{p'} d\mu = \int_X f h d\mu = \varphi(f) \leq \|\varphi\|_* \left[\int_{A_n} |h|^{p'} d\mu \right]^{\frac{1}{p'}},$$

$$\Rightarrow \int_X \chi_{A_n} |h|^{p'} d\mu \leq \|\varphi\|_*^{p'} \quad n \in \mathbb{N}.$$

$$\Rightarrow \|h\|_{p'} \leq \|\varphi\|_* \quad \text{(by the Monotone Convergence Theorem)}$$

$$\Rightarrow \|\varphi\|_* = \|h\|_{p'}$$

Now in (2.14) both sides are continuous functions and coincide on $L^\infty(X)$ which is dense in $L^p(X)$. So, they coincide on $L^p(X)$.

Now we assume that (X, Σ, μ) is σ-finite. Then $X = \bigcup_{n \geq 1} A_n, A_n \in \Sigma, 0 < \mu(A_n) < \infty$ for all $n \in \mathbb{N}$. Let $g : X \to (0, \infty)$ be defined by $g(x) = \frac{1}{n^2} \mu(A_n)$ if $x \in A_n$. Then $g \in L^1(X)$ and $\vartheta(A) = \int_A g d\mu \ A \in \Sigma$ is a finite measure on Σ (see Proposition 2.90). The map $u \to g^{\frac{1}{p}} u$ is a linear isometry of $L^p(X, \vartheta)$ onto $L^p(X, \mu)$. Therefore, if $\psi(u) = \varphi(g^{\frac{1}{p}} u)$ for all $u \in L^p(X, \vartheta)$, then $\psi(\cdot)$ is a continuous linear functional on $L^p(X, \vartheta)$ and $\|\psi\|_* = \|\varphi\|_*$. Since $\vartheta(\cdot)$ is finite, we can use the first part of the proof and find $v \in L^{p'}(X, \vartheta)$ such that

$$\psi(u) = \int_X u v d\vartheta \quad \text{for all } u \in L^p(X, \vartheta).$$

Let $h = g^{\frac{1}{p'}} v$ (if $p = 1$, then $h = v$). We have

$$\int_X |h|^{p'} d\mu = \int_X |v|^{p'} d\vartheta = \|\psi\|_*^{p'} = \|\varphi\|_*^{p'}, \quad \text{if } 1 < p,$$

$$\|h\|_\infty = \|v\|_\infty = \|\psi\|_* = \|\varphi\|_* \quad \text{if } p = 1.$$

Therefore, (2.13) holds and we have

$$\varphi(f) = \psi(g^{-\frac{1}{p}} f) = \int_X g^{-\frac{1}{p}} f v d\vartheta = \int_X f h d\mu. \quad \text{for all } f \in L^p(X, \mu).$$

$\square$

Remark 2.134 Based on this theorem, we can write

$$L^p(X)^* = L^{p'}(X) \quad \text{for all } 1 \le p < \infty.$$

Note that if $p = 2$, then $p' = 2$ and so $L^2(X)^* = L^2(X)$.

2.7 Modes of Convergence: Uniform Integrability

In this section we will introduce and compare some modes of convergence for measurable functions. We also introduce and discuss the important notion of uniform integrability, which leads to an extension of the Dominated Convergence Theorem.

Definition 2.135 Let (X, Σ, μ) be a measure space and $\{f_n\}_{n \ge 1}$ μ-measurable functions.

(a) We say that $f_n \to f$ $\mu - a.e$, if there exists $B \in \Sigma$ with $\mu(B) = 0$ such that $f_n(x) \to f(x)$ for all $x \in X \setminus B$.
(b) We say that $f_n \to f$ in $L^p(X)$ $1 \le p \le \infty$, if $\|f_n - f\|_p \to 0$ as $n \to \infty$.
(c) We say that $f_n \xrightarrow{\mu} f$ (converges in μ-measure), if for every $\varepsilon > 0$, we have $\mu(\{x \in \Omega : |f_n(x) - f(x)| \ge \varepsilon\}) \to 0$ as $n \to \infty$.

Remark 2.136 We also write $f_n \xrightarrow{a.e} f$ for the almost everywhere convergence, $f_n \xrightarrow{L^p} f$ for convergence in the L^p-norm and $f_n \xrightarrow{\mu} f$ for the convergence in μ-measure. If $\mu(X) = 1$, then (c) is called "convergence in probability."

Example 2.137 Let $X = \mathbb{R}$ with the Lebesgue measure and let

(a) $f_n = \frac{1}{n} \chi_{(0,n)}$.
(b) $f_n = \chi_{(n,n+1)}$.
(c) $f_n = n \chi_{[0,\frac{1}{n}]}$.
(d) $f_n = \chi_{[\frac{m}{2^k}, \frac{m+1}{2^k}]}$ $n = 2^k + m, 0 \le m < 2^k$.

Then the sequences (a), (b), (c) converge to 0 pointwise (in fact uniformly), but not in L^1. For (d) we have $f_n \xrightarrow{L^1} 0$ but not pointwise. In fact $\{f_n(x)\}_{n \ge 1}$ does not converge for any $x \in \mathbb{R}$. Also, (a), (c), (d) converge to zero in measure, but not (b).

The following facts about the convergence in μ-measure, can be easily deduced from Definition 2.135(c).

Proposition 2.138

(a) The limit in the convergence in μ-measure, is unique.

(b) $f_n \xrightarrow{\mu} f$ and $g_n \xrightarrow{\mu} g \Rightarrow \vartheta f_n + \eta g_n \xrightarrow{\mu} \vartheta f + \eta g$ for all $\vartheta, \eta \in \mathbb{R}$.

(c) $f_n \xrightarrow{\mu} f \Rightarrow f_n^{\pm} \xrightarrow{\mu} f^{\pm}$ and $|f_n| \xrightarrow{\mu} |f|$.

In general pointwise convergence does not imply convergence in μ-measure (see Example 2.137, sequence (b)). However, if we restrict $\mu(\cdot)$, we can have the implication.

Proposition 2.139 *If (X, Σ, μ) is a finite measure space, then $f_n \xrightarrow{a.e} f \Rightarrow f_n \xrightarrow{\mu} f$.*

Proof Let $A_{n\varepsilon} = \{x \in X : |f_n(x) - f(x)| \geq \varepsilon\}$ and let $A_{\varepsilon} = \lim\sup_{n\to\infty} A_{n\varepsilon} = \bigcap_{k\geq 1}\bigcup_{n\geq k} A_{n\varepsilon}$. We see that $\bigcup_{n\geq k} A_{n\varepsilon} \downarrow A_{\varepsilon}$ and so by Proposition 2.17(d), we have $\mu(\bigcup_{n\geq k} A_{n\varepsilon}) \to \mu(A_{\varepsilon})$ as $k \to \infty$.

Note that $\{x \in X : f_n(x) \not\to f(x)\} = \bigcup_{\varepsilon>0} A_{\varepsilon} = \bigcup_{m\geq 1} A_{\frac{1}{m}}$ (since A_{ε} is decreasing with $\varepsilon > 0$). Therefore $f_n \xrightarrow{a.e} f \iff \mu(A_{\varepsilon}) = 0$ for all $\varepsilon > 0 \iff \mu(\bigcup_{n\geq k} A_{n\varepsilon}) \to 0$ as $k \to \infty$ for all $\varepsilon > 0$. Hence $f_n \xrightarrow{a.e} f \Rightarrow f_n \xrightarrow{\mu} f$.   □

Corollary 2.140 *If (X, Σ, μ) is a finite measure space, then $f_n \xrightarrow{a.e} f \iff \mu(\bigcup_{n\geq k} \{x \in X : |f_n(x) - f(x)| \geq \varepsilon\}) \to 0$ as $k \to \infty$ for all $\varepsilon > 0$.*

Proposition 2.141 *If $f_n \xrightarrow{\mu} f$ as $n \to \infty$, then there exists a subsequence $\{f_{m_n}\}_{n\geq 1}$ of $\{f_n\}$ such that $f_{m_n} \xrightarrow{a.e} f$ as $n \to \infty$.*

Proof On account of Definition 2.135, we can find $\{m_n\}_{n\geq 1}$ strictly increasing such that $\mu(\{x \in X : |f_m(x) - f(x)| \geq \frac{1}{n}\}) < \frac{1}{2^n}$ for all $m \geq m_n$. We set $A_n = \{x \in X : |f_{m_n}(x) - f(x)| \geq \frac{1}{n}\}$ and let $A = \lim\sup_{n\to\infty} A_n = \bigcap_{n\geq 1}\bigcup_{m\geq n} A_m$. We have

$$\mu(A) \leq \mu(\bigcup_{m\geq n} A_m) \leq \sum_{m\geq n} \mu(A_m) < \frac{1}{2^{n-1}} \quad for\ all\ n \in \mathbb{N},$$

$$\Rightarrow \mu(A) = 0.$$

From Corollary 2.140 we have that $f_{m_n}(x) \to f(x)$ for all $x \notin A$.   □

Proposition 2.142 *If $1 \leq p \leq \infty$ and $f_n \xrightarrow{L^p} f$, then $f_n \xrightarrow{\mu} f$.*

Proof The case $p = +\infty$ is clear. So, we assume that $1 \leq p < \infty$. Let $\varepsilon > 0$ and set $A_n = \{x \in X : |f_n(x) - f(x)| \geq \varepsilon\}$. Then $\varepsilon^p \chi_{A_n}(x) \leq |f_n(x) - f(x)|^p$ for all $x \in X$. Hence

$$\mu(A_n) \leq \frac{1}{\varepsilon^p} \|f_n - f\|_p^p \to 0 \quad as \ n \to \infty,$$

$$\Rightarrow f_n \xrightarrow{\mu} f.$$

$\square$

Let $L^0(X)$ be the vector space of all equivalence classes of measurable functions. The next theorem shows that on finite measure spaces convergence in measure is a metric convergence.

Theorem 2.143 *If (X, Σ, μ) is a finite measure space, then in the vector space $L^0(X)$ convergence in measure is equivalent to convergence with respect to the translation invariant metric $d_0(f, h) = \int_X \frac{|f-h|}{1+|f-h|} d\mu$ (so $f_n \xrightarrow{\mu} f$ if and only if $d_0(f_n, f) \to 0$ as $n \to \infty$).*

Proof First note that $0 \leq \frac{|f-h|}{1+|f-h|} \leq 1$ on X. So $0 \leq d_0(f, h) < \infty$ for all $f, h \in L^0(X)$. Then, we can easily check that d_0 is a metric on $L^0(X)$. For the triangle inequality use the fact that $x \to \frac{x}{1+x}, x \geq 0$ is nondecreasing. This leads to the elementary inequality that $\frac{a}{1+a} \leq \frac{b}{1+b} + \frac{c}{1+c}$ for all $a, b, c \geq 0$ such that $a \leq b+c$. Clearly d_0 is translation invariant.

Let $\{f_n\}_{n \geq 1} \subseteq L^0(X)$ and $\varepsilon > 0$. We define

$$A_n = \{x \in X : |f_n(x) - f(x)| \geq \varepsilon\} = \{x \in X : \frac{|f_n(x) - f(x)|}{1 + |f_n(x) - f(x)|} \geq \frac{\varepsilon}{1 + \varepsilon}\}.$$

Suppose that $d_0(f_n, f) \to 0$ as $n \to \infty$. We have $\frac{\varepsilon}{1+\varepsilon} \chi_{A_n} \leq \frac{|f_n-f|}{1+|f_n-f|}$, hence $\mu(A_n) \leq \frac{1+\varepsilon}{\varepsilon} d_0(f_n, f)$ and so $\mu(A_n) \to 0$, therefore $f_n \xrightarrow{\mu} f$.

Conversely, suppose that $f_n \xrightarrow{\mu} f$. Then we can find $n_0 \in \mathbb{N}$ such that $\mu(A_n) < \varepsilon$ for all $n \geq n_0$. We have

$$d_0(f_n, f) = \int_{A_n} \frac{|f_n - f|}{1 + |f_n - f|} d\mu + \int_{X \setminus A_n} \frac{|f_n - f|}{1 + |f_n - f|} d\mu,$$

$$\leq \mu(A_n) + \frac{\varepsilon}{1 + \varepsilon} \mu(A_n^c) \leq \varepsilon[1 + \mu(X)] \quad for \ all \ n \geq n_0,$$

$$\Rightarrow d_0(f_n, f) \to 0.$$

$\square$

Next we introduce a notion which will lead to an extension of the Dominated Convergence Theorem (see Theorem 2.86).

Definition 2.144 Let (X, Σ, μ) be a measure space and $\mathcal{F} \subseteq L^0(X)$.

(a) We say that $\mathcal{F}$ is "uniformly integrable," if for every $\varepsilon > 0$ we can find $\delta > 0$ such that $\mu(A) \leq \delta \Rightarrow \int_A |f| d\mu \leq \varepsilon$ for all $f \in \mathcal{F}$.

(b) We say that $\mathcal{F}$ is "p-uniformly integrable" $(1 \leq p < \infty)$, if $\{|f|^p : f \in \mathcal{F}\}$ is uniformly integrable.

Proposition 2.145 *If (X, Σ, μ) is a measure space, $\{f_n\}_{n \geq 1} \subseteq L^0(X)$ and there exists $h \in L^p(X)$ $1 \leq p < \infty$ such that $|f_n(x)| \leq h(x)$ $\mu - a.e$ for all $n \in \mathbb{N}$, then*

(a) *$\{f_n\}_{n \geq 1}$ is p-uniformly integrable.*

(b) *For every $\varepsilon > 0$, we can find $A \in \Sigma$ such that $\mu(A) < \infty$ and $\int_{X \setminus A} |f_n|^p d\mu \leq \varepsilon$ for all $n \in \mathbb{N}$.*

Proof Let $\lambda(A) = \int_A h^p d\mu$. Then $\lambda(\cdot)$ is a finite measure on Σ and $\lambda \ll \mu$. So, given $\varepsilon > 0$, we can find $\delta > 0$ such that $\mu(A) \leq \delta \Rightarrow \int_A h^p d\mu \leq \varepsilon$ (see Proposition 2.110). Also, using Proposition 2.17(c), we have

$$\lim_{\vartheta \to +\infty} \int_{\{\frac{1}{\vartheta} < h < \vartheta\}} h^p d\mu = \int_{\{0 < h < \infty\}} h^p d\mu = \int_X h^p d\mu \tag{2.15}$$

(see Proposition 2.79(a)).

For every $A \in \Sigma$ with $\mu(A) \leq \delta$, we have $\int_A |f_n|^p d\mu \leq \int_A h^p d\mu \leq \varepsilon$ for all $n \in \mathbb{N}$ and so $\{f_n\}_{n \geq 1}$ is p-uniformly integrable.

On account of (2.15), we can find $\vartheta > 0$ big so that

$$\int_X h^p d\mu - \int_{\{\frac{1}{\vartheta} < h < \vartheta\}} h^p d\mu = \int_{X \setminus \{\frac{1}{\vartheta} < h < \vartheta\}} h^p d\mu \leq \varepsilon. \tag{2.16}$$

Let $A = \{x \in X : \frac{1}{\vartheta} < h(x) < \vartheta\} \in \Sigma$. Then $\mu(A) \leq \vartheta^p \int_X h^p d\mu < \infty$ and

$$\int_{X \setminus A} |f_n|^p d\mu \leq \int_{X \setminus A} h^p d\mu \leq \epsilon$$

for all $n \in \mathbb{N}$ (see (2.16)). $\qquad\square$

The next result is known as "Egorov's Theorem" and says that pointwise convergence is "almost" uniform.

Theorem 2.146 *Let (X, Σ, μ) be a finite measure space, (Y, d) a separable metric space, $f_n : X \to Y$ Σ-measurable functions and $f_n(x) \to f(x)$, μ—a.e. Then, for any $\epsilon > 0$ there exists $A_\epsilon \in \Sigma$ such that $\mu(X \setminus A_\epsilon) \leq \epsilon$ and*

$$f_n \to f \quad \text{uniformly on } A_\epsilon.$$

Proof Given $\epsilon > 0$ and $m \in \mathbb{N}$, choose $n_m \in \mathbb{N}$ such that $A_m = \cup_{n \geq n_m}\{|f_n - f| \geq \frac{1}{m}\}$ has μ-measure less that $\epsilon/2^m$ (see Corollary 2.140). Let $A = \cup_{m \geq 1} A_m$. Then $\mu(A) \leq \Sigma_{m \geq 1} \mu(A_m) < \epsilon$. Moreover, for $\delta > 0$ and $m \in \mathbb{N}$ with $0 < 1/m < \delta$, for any $n \geq n_m$ and $x \in X \setminus A$ we have $|f_n(x) - f(x)| \leq \frac{1}{m} < \delta$. Therefore, $f_n \to f$ uniformly on $A^c = X \setminus A$. $\qquad\square$

Now we are ready for the extension of Theorem 2.86, known as "Vitali's Theorem" or as the "Extended Dominated Convergence Theorem."

Theorem 2.147 *Let (X, Σ, μ) be a measure space, $\{f_n\} \subseteq L^1(X)$ a uniformly integrable sequence satisfying (b) of Proposition 2.145. Assume that $f_n \xrightarrow{\text{a.e.}} f$, with $f(x) \in \mathbb{R}$, μ—a.e. Then $f \in L^1(X)$ and $\|f_n - f\|_1 = \int_X |f_n - f| d\mu \to 0$ as $n \to \infty$.*

Proof First we assume that all f_n and f are $\mathbb{R}$-valued. Let $\epsilon > 0$ be given. Then by assumption we can find $\delta > 0$ and $C_\epsilon \in \Sigma$ such that

$$\sup_{n \geq 1} \int_A |f_n| d\mu \leq \epsilon, \text{ provided that } \mu(A) \leq \delta, \tag{2.17}$$

and also, $\mu(C_\epsilon) < \infty$ and $\sup_{n \geq 1} \int_{C_\epsilon^c} |f_n| d\mu \leq \epsilon$.

Also by Theorem 2.146 we can find $C_\epsilon \supseteq B_\epsilon \in \Sigma$, with $\mu(B_\epsilon) \leq \delta$ such that $f_n \to f$ uniformly on $C_\epsilon \setminus B_\epsilon$. We have

$$\int_{C_\epsilon} |f_n - f| d\mu = \int_{B_\epsilon} |f_n - f| d\mu + \int_{C_\epsilon \setminus B_\epsilon} |f_n - f| d\mu$$

$$\leq \int_{B_\epsilon} |f_n| d\mu + \int_{B_\epsilon} |f| d\mu + \mu(C_\epsilon) \sup_{C_\epsilon \setminus B_\epsilon} |f_n - f| \tag{2.18}$$

$$\leq 2\epsilon + \mu(C_\epsilon) \sup_{C_\epsilon \setminus B_\epsilon} |f_n - f| \text{ (see (2.17))}.$$

Since $\int_{C_\epsilon^c} |f| d\mu \leq \epsilon$, by Fatou's lemma we obtain

$$\int_X |f_n - f| d\mu = \int_{C_\epsilon} |f_n - f| d\mu + \int_{C_\epsilon^c} |f_n - f| d\mu$$

$$\leq \int_{C_\epsilon} |f_n - f| d\mu + 2\epsilon$$

$$\leq 4\epsilon + \mu(C_\epsilon) \sup_{C_\epsilon \setminus B_\epsilon} |f_n - f| \text{ (see (2.18))}.$$

Thus, $\int_X |f_n - f| d\mu \to 0$ as $n \to \infty$. Since $f = (f - f_n) + f_n$, it follows that $f \in L^1(X)$.

Now we remove the extra hypothesis that f_n and f are all $\mathbb{R}$-valued. Let

$$\widehat{N}_n = \{x \in X : |f_n(x)| = \infty\} \text{ and } \widehat{N}_0 = \{x \in X : |f(x)| = \infty\}.$$

We know that $\mu(\widehat{N}_n) = \mu(\widehat{N}_0) = 0$ (see Proposition 2.79 (a)). Let $\widehat{N} = \cup_{n \geq 1} \widehat{N}_n$. Then $\mu(\widehat{N}) = 0$ and from the first part of the proof the result holds on $\widehat{N}^c$ Finally, note that $\int_X |f_n - f| d\mu = \int_{\widehat{N}^c} |f_n - f| d\mu \to 0$ as $n \to \infty$. $\qquad \square$

Remark 2.148 On account of Proposition 2.141 we can replace the hypothesis $f_n \xrightarrow{a.e.} f$ by $f_n \xrightarrow{\mu} f$.

Also, there is an extended version of Fatou's lemma. We leave the details of the proof to the reader.

Proposition 2.149 *Let (X, Σ, μ) be a measure space and $\{f_n\}_{n \geq 1} \subseteq L^1(X)$ uniformly integrable. Then*

$$\int_X \liminf_{n \to \infty} f_n d\mu \leq \liminf_{n \to \infty} \int_X f_n d\mu \leq \limsup_{n \to \infty} \int_X f_n d\mu \leq \int_X \limsup_{n \to \infty} f_n d\mu.$$

We close this section with the statement of the Fubini-Tonelli Theorem. For a proof we refer to G.B. Folland [115], Theorem 2.39, p. 68.

Theorem 2.150 *Let (X, Σ, μ) and $(Y, \mathcal{S}, \lambda)$ be complete measure spaces and $f \in L^1(X \times Y, \mu \otimes \lambda)$. Then, we have*

(a) $f(x, \cdot) \in L^1(Y)$ for μ-a. a. $x \in X$.
(b) $f(\cdot, y) \in L^1(X)$ for λ-a. a. $y \in Y$.
(c) $x \to \int_Y f(x, y) d\lambda(y)$ is integrable on X.
(d) $y \to \int_X f(x, y) d\mu(x)$ is integrable on Y.
(e) $\int_X (\int_Y f d\lambda) d\mu = \int_{X \times Y} f d(\mu \otimes \lambda) = \int_Y (\int_X f d\mu) d\lambda$.

Remark 2.151 The measure space $(X \times Y, \Sigma \otimes \mathcal{S}, \mu \otimes \lambda)$ is seldom complete even if (X, Σ, μ) and $(Y, \mathcal{S}, \lambda)$ are both complete.

2.8 Measures and Topology

In most applications the measures are defined in a nice topological or metric space. Then by combining the topological and measure theoretic structures, we can have a more robust and richer theory.

Definition 2.152 Let (X, τ) be a topological space and $\mu : B(X) \to \overline{\mathbb{R}}$ a measure

(a) μ is "outer regular" if for all $A \in B(X)$ we have

$$\mu(A) = \inf[\mu(U) : U \in \tau, A \subseteq U].$$

(b) μ is "inner regular" if for all $A \in B(X)$ we have

$$\mu(A) = \sup[\mu(C) : C \text{ is closed and } C \subseteq A].$$

(c) μ is "regular" if it is both inner and outer regular.
(d) μ is "tight" if it is inner regular for compact sets, i.e., for every $A \in \Sigma$ we have

$$\mu(A) = \sup[\mu(K) : K \text{ is compact and } K \subseteq A].$$

(e) μ is "Radon" if $\mu(K) < \infty$ for all compact $K \subseteq X$ and it is both outer regular and tight.
(f) We say a signed measure μ has any of the above properties if both μ^+ and μ^- have it.

Remark 2.153 Every Radon measure is tight. Also, note that a finite Borel measure is regular iff for any $A \in B(X)$ and $\epsilon > 0$ we can find an open set U and a closed set C such that $C \subseteq A \subseteq U$ and $\mu(U \setminus C) = \mu(U) - \mu(C) < \epsilon$.

Theorem 2.154 *Every finite Borel measure on a metrizable space is regular.*

Proof Let X be the space. We introduce the following family of subsets of X.

$$\mathcal{L} = \{A \in B(X) : \mu(A) = \inf[\mu(U) : U \subseteq X \text{ open and } A \subseteq U]$$
$$= \sup[\mu(C) : C \subseteq X \text{ close and } C \subseteq A]\}.$$

If we show that $\mathcal{L}$ is a σ-algebra containing the open sets, then $\mathcal{L} = B(X)$ and so we are done.

Clearly, $A^c = X \setminus A \in \mathcal{L}$ if $A \in \mathcal{L}$. Let $\{A_n\} \subseteq \mathcal{L}$ such that $A = \cup_{n \geq 1} A_n$. Find open U_n and closed C_n such that

$$C_n \subseteq A_n \subseteq U_n \text{ and } \mu(U_n) \leq \mu(C_n) + \frac{\epsilon}{2^n}.$$

Set $U = \cup_{n \geq 1} U_n$ and $C = \cup_{n \geq 1} C_n$. Then U is open, $A \subseteq U$ and since $U \setminus A \subseteq \cup_{n \geq 1}(U_n \setminus A_n)$, we have

$$\mu(U) - \mu(A) = \mu(U \setminus A) \leq \sum_{n \geq 1} \mu(U_n \setminus A_n)$$
$$= \sum_{n \geq 1}(\mu(U_n) - \mu(A_n)) \leq \sum_{n \geq 1} \frac{\epsilon}{2^n} = \epsilon.$$

Thus, $\mu(A) = \inf[\mu(U) : U \subseteq X \text{ open and } A \subseteq U]$. Similarly, we can show that $\mu(A) \leq \mu(C) + \epsilon$. Note that $\mu(E_n) \uparrow \mu(C)$ if $E_n = \cup_{k=1}^n C_k$ (see Proposition 2.17(c)). So, we can find closed $C_n \subseteq X$ such that $\mu(A) \leq \mu(C_n) + \frac{\epsilon}{2}$.

It follows that

$$\mu(A) = \sup[\mu(C) : C \subseteq X \text{ closed and } C \subseteq A],$$

hence $A \in \mathcal{L}$.

Finally, we show that $\mathcal{L}$ contains the open sets. So, consider an open set $U \subseteq X$. From Proposition 1.157, we know that U is F_σ. Hence we can find $\{C_n\}$, an increasing sequence of closed sets such that $U = \cup_{n \geq 1} C_n$. Then $\mu(C_n) \uparrow \mu(U)$ (see Proposition 2.17(c)) and so

$$\mu(U) = \sup[\mu(C) : C \subseteq C \text{ closed and } C \subseteq A],$$

hence $U \in \mathcal{L}$ and so $\mathcal{L} = B(X)$.   □

Proposition 2.155 *Let X be a Hausdorff topological space and μ a finite Borel measure on X that is tight. Then μ is regular.*

Proof Since X is Hausdorff, every compact set is closed. So, for any $A \in B(X)$ we have

$$\mu(A) = \sup[\mu(K) : K \subseteq X \text{ compact and } K \subseteq A]$$

$$\leq \sup[\mu(C) : C \subseteq X \text{ closed and } C \subseteq A]$$

$$\leq \mu(A).$$

Then we have

$$\mu(A^c) = \sup[\mu(C) : C \subseteq X \text{ closed and } C \subseteq A^c]$$

$$= \mu(X) - \inf[\mu(U) : U \subseteq X \text{ open and } A \subseteq U],$$

thus $\mu(A) = \inf[\mu(U) : U \subseteq X \text{ open and } A \subseteq U]$, and μ is regular.   □

Tightness of a finite Borel measure on a metrizable space can be verified by checking inner regularity for compact sets.

Proposition 2.156 *Let X be a metrizable space and μ a finite Borel measure on X. Then μ is tight iff for any $\epsilon > 0$ there exists a compact $K_\epsilon \subseteq X$ such that $\mu(X \setminus K_\epsilon) < \epsilon$.*

Proof $\Rightarrow$: Immediate from Definition 2.152(d).

$\Leftarrow$: From Theorem 2.154 we know that μ is regular. So, it suffices to show that for any closed $C \subseteq X$ we have

$$\mu(C) = \sup[\mu(K) : K \subseteq X \text{ compact and } K \subseteq C].$$

Arguing by contradiction, suppose there exists $\epsilon > 0$ such that

$$\sup[\mu(K) : K \subseteq X \text{ compact and } K \subseteq X] \leq \mu(C) - \epsilon.$$

Let $E \subseteq X$ be compact. Then so is $C \cap E$ and we have

$$\mu(E) = \mu(E \cap C) + \mu(E \cap C^c) \leq \mu(C) + \mu(C^c) - \epsilon = \mu(X) - \epsilon,$$

a contradiction. □

Radon measures arise in the duality theory of certain spaces of continuous functions (see Theorem 4.228). Next we show that on Polish spaces every finite Borel measure is Radon.

Theorem 2.157 *Every finite Borel measure on a Polish space is Radon.*

Proof Let μ be the finite Borel measure on the Polish space X, d a comparable metric on X. By Proposition 2.155 it suffices to show that μ is tight. Let $\{x_n\}$ be dense in X and consider the closed set $C_n^k = \{x \in X : d(x, x_n) \leq 1/k\}$. Then $X = \cup_{n \geq 1} C_n^k$ for each $k \in \mathbb{N}$. Given $\epsilon > 0$, choose $m_k \in \mathbb{N}$ such that

$$\mu(X \setminus \bigcup_{n=1}^{m_k} C_n^k) < \frac{\epsilon}{2^k}, \quad k \in \mathbb{N}.$$

Set $C = \bigcap_{k \geq 1} \bigcup_{n=1}^{m_k} C_n^k$. Then C is closed and totally bounded, thus compact. We have

$$\mu(X \setminus C) = \mu(\bigcup_{k \geq 1} (X \setminus \bigcup_{n=1}^{m_k} C_n^k))$$

$$\leq \sum_{k \geq 1} \mu(X \setminus \bigcup_{n=1}^{m_k} C_n^k)$$

$$\leq \sum_{k \geq 1} \frac{\epsilon}{2^k} = \epsilon.$$

Thus, μ is tight (see Proposition 2.156) and so μ is Radon. □

The next result is a topological counterpart of Theorem 2.146. The result is known as "Lusin Theorem" and shows that a measurable function is "almost" continuous.

Theorem 2.158 *Let X be a Hausdorff topological space, μ a Radon measure on X, Y is a Polish space and $f : X \to Y$ measurable. Then for any $\epsilon > 0$ there exists compact $K_\epsilon \subseteq X$ such that*

$$\mu(X \setminus K_\epsilon) < \epsilon \quad \text{and} \quad f|_{K_\epsilon} \text{ is continuous.}$$

Proof First we prove the result for simple functions. Let $\{a_k\}_{k=1}^{m} \subseteq Y$ be the values of the simple functions and $A_k = f^{-1}(\{a_k\})$. We can find compact $K_k \subseteq A_k$ such that $\mu(A_k \setminus K_k) < \epsilon/m$. Since $\{K_k\}$ are mutually disjoint and X is Hausdorff, compact sets can be separated by open sets and so, $f|_{\cup_{k=1}^{m} K_k}$ is continuous and $\mu(X \setminus \cup_{k=1}^{m} A_k) < \epsilon$.

Next suppose f has countably may values. We can find simple functions $\{s_n\}$ such that $s_n(x) \to f(x)$ for all $\in X$. From the first part of the proof, we know that we can find compact $K_n \subseteq X$ such that $f|_{K_n}$ is continuous and $\mu(X \setminus K_n) < \frac{\epsilon}{2^{n+1}}$. On the other hand, by Egorov's theorem (see Theorem 2.146) we can find compact $K \subseteq X$ such that $s_n \to f$ uniformly on K and $\mu(X \setminus K) \leq \epsilon/2$. Therefore, $f|_{\cap_{n \geq 1} K_n}$ is continuous (being the uniform limit of continuous functions).

Finally, we consider the general case of a measurable function. Let $\{y_n\}_{n \in \mathbb{N}}$ be dense in Y and inductively we define $B_n = B_{1/m}(y_n) \setminus \cup_{k=1}^{n-1} B_k$, $B_1 = B_1(y_1)$. Evidently, $B_n \in B(X)$ and diam $B_n \leq 2/m$. Let $f_m : X \to Y$ be defined by

$$f_m(x) = y_n \text{ on } f^{-1}(B_n).$$

Note that each f_m takes countably many values and $f_m \to f$ uniformly on X. Then as before, using the previous part of the proof we can find compact $K \subseteq X$ such that $\mu(X \setminus K) < \epsilon$ and $f|_K$ is continuous. $\qquad \square$

We will close this section with a brief discussion of the notion of the support of a measure.

Definition 2.159 Let X be a topological space and μ a Borel measure on X. $C \subseteq X$ is called the "support" of μ iff C is the smallest closed set whose complement is μ-null. We denote it by supp μ.

Remark 2.160 Not every Borel measure has a support.

Proposition 2.161 *A Radon measure on a topological space always has a support.*

Proof Let X be the topological space and μ the measure. Let $\{C_i\}_{i \in I}$ be the collection of all closed subsets of X such that $\mu(C_i) = \mu(X)$ for all $i \in I$. Set $C = \cap_{i \in I} C_i$. Suppose that $\mu(X \setminus C) = \epsilon > 0$. Since μ is Radon, we can find compact $K \subseteq X \setminus C$ such that $\mu((X \setminus C) \setminus K) < \epsilon$. Note that $\{U_i = C_i^c = X \setminus C_i\}_{i \in I}$ is an open cover of K. Since it is compact, we have $K \subseteq \cup_{k=1}^{m} U_{i_k}$. Then

$$\mu(C^c) = \mu(X \setminus C) \leq \mu((X \setminus C) \setminus K) + \sum_{k=1}^{m} \mu(U_{i_k}) < \epsilon,$$

a contradiction. Therefore supp μ exists. $\qquad \square$

2.9 Remarks

Measure theory and the modern theory of integration started with the thesis of Lebesgue [173] (see also Lebesgue [176]). Caratheodory [57] introduced Definitions 2.2 and 2.24, namely the notions of outer measure and μ^*-measurable sets, and proved Theorem 2.26. Note that Definition 2.24 is not intuitive and in fact it is rather a mystery how Caratheodory came up with this definition. However, his definition turned out to be very efficient and has many remarkable implications. Theorem 2.28 is due to Frechet [118] and Hahn [132]. In connection with measures, we should mention the "Banach-Tarski Paradox," due to Banach-Tarski [26]. They show using the Axiom of Choice that the unit ball B_1 in $\mathbb{R}^3$ can be partitioned into two disjoint sets A and C such that each set can be partitioned into five disjoint sets which can be put back together using only translations and rotations and produce a copy of B_1. So, the unit ball can be broken into pieces, which are regrouped to generate two balls B_1! Evidently, these pieces cannot not be Lebesgue measurable. This result reveals that we cannot define a finitely additive, translation invariant measure on $2^{\mathbb{R}^N}$ for all $N \geq 3$.

The construction of the integral starts from measures. There is an approach which goes the opposite way, namely start with the integral as a positive linear functional defined on a suitable space of functions and from it to produce the measure. This approach was first developed by Daniell [75]. Using the Axiom of Choice, Sierpinski [255] proved the existence of a nonmeasurable subset of $S \subseteq [0, 1] \times [0, 1]$ such that its intersection with any straight line contains at most two points (see also Hewitt–Stromberg [139], p. 393). Using this fact we can show that in Theorem 2.47 we cannot relax the continuity of $f(\omega, \cdot)$ to just lower or upper semicontinuity (see Hu-Papageorgiou [145], p. 226). Moreover, in the direction of Theorem 2.53 we can find additional material in Aumann [17, 19]

The theory of Polish and Souslin sets was developed in the interwar period of 1919–1939, primarily by Polish and Soviet mathematicians. Corollary 2.55 was in fact the starting point of the theory of Souslin (Analytic) spaces. Lebesgue [174] claimed that a Borel set in $\mathbb{R}^2$ has a Borel projection on the x-axis. Souslin [259] realized that this is not true and in the process invented the Souslin (Analytic) sets. Applications of Souslin (Analytic) sets can be found in Arveson [12] (operator theory), Dellacherie [80] (probability theory) and Mackey [187] (group representation theory). The books of Cohn [71], Dudley [89], and Schwartz [252] contain further information on the subject. For Borel spaces we refer to the book of Srivastava [261].

In Sect. 2.4 the main result is Theorem 2.86 (the Lebesgue Dominated Convergence Theorem), which also illustrates the power of the Lebesgue integral compared to the Riemann integral. In the latter in order to pass the limit inside the integral, we need to have uniform convergence of the sequence.

In Sect. 2.5 the main result is Theorem 2.113 (the Lebesgue–Radon–Nikodym Theorem). It was proved first by Lebesgue [175] for μ being the Lebesgue measure on $\mathbb{R}^N$. Under the hypothesis that $\lambda \ll \mu$ it was generalized by Radon [223] to arbitrary regular Borel measures on $\mathbb{R}^N$ and by Nikodym [211] to measures on abstract spaces.

Another important inequality for integrable functions is the so-called Jensen's Inequality.

Proposition 2.162 *Let* (X, Σ, μ) *be a finite measure space,* $f \in L^1(X)$, *and* $\varphi : \mathbb{R} \to \mathbb{R}$ *a convex function, with* $(\varphi \circ f) \in L^1(X)$. *Then,*

$$\varphi\left(\frac{1}{\mu(x)} \int_X f d\mu\right) \le \frac{1}{\mu(x)} \int_X (\varphi \circ f) d\mu.$$

The spaces $L^p[a, b]$ were first investigated by Riesz [226], who also proved Theorem 2.133, the main result of Sect. 2.5.

Convergence in measure (initially called "asymptotic convergence") can be found in the early works of Borel and Lebesgue, but a more systematic study was conducted by Riesz [228], who also introduced the metric in Theorem 2.113. Alternative metrics metrizing the convergence in measure were proposed by Frechet and KyFan (see Dudley [89], p. 259).

Additional formulations of uniform integrability (see Definition 2.144) valid also for arbitrary measure spaces can be found in Gasinski-Papageorgiou [122] (see Problems 1.7, 1.15-1.17). Theorem 2.146 was proved by Egorov [98]. In the next proposition we state a version of the result without the hypothesis that $\mu(X) < \infty$.

Proposition 2.163 *Let* (X, Σ, μ) *be a measure space,* $f_n : \Sigma \to \mathbb{R}$ *a sequence of* Σ-*measurable functions such that*

$$f_n(x) \to f(x), \mu - a.e. \text{ and } |f_n(x)| \le h(x), \mu - a.e.$$

for all $n \in \mathbb{N}$ *and some* $h \in L^1(X)$. *Then, for any* $\epsilon > 0$ *there exists* $A_\epsilon \in \Sigma$ *with* $\mu(X \setminus A_\epsilon) < \epsilon$ *such that* $f_n \to f$ *uniformly on* A_ϵ.

Theorem 2.147 is a very useful result and has many interesting applications. It is due to Vitali [278]. The terminology in Definition 2.152 is not universal. In the literature there are also other names used for these notions. So, the reader should be alert. Radon measures have their origin in the seminal paper of Radon [223]. A classical reference for Radon measures is the book of Schwartz [252]. Theorem 2.158 is due to Lusin [185], though it was first stated without proof by Lebesgue [175].

There is a useful regularity result for functions which are integrable with respect to a Radon measure. The result is known as the "Vitali-Caratheodory Theorem" and can be found in Rudin [243], p. 57.

Theorem 2.164 *Let* X *be a locally compact topological space,* $\mu : B(X) \to \overline{\mathbb{R}}_+ = [0, \infty]$ *a Radon measure,* $f \in L^1(X, \mu)$ *and* $\epsilon > 0$. *Then we can find* $g_1 : X \to \mathbb{R}$ *upper semicontinuous bounded above, and* $g_2 : X \to \mathbb{R}$ *lower semicontinuous bounded below, such that*

$$g_1(x) \leq f(x) \leq g_2(x), \mu - a.e., \quad and \quad \int_X [g_2 - g_1] d\mu < \epsilon.$$

Definition 2.165 Let (X, Σ, μ) be a measure space. $A \in \Sigma$ is called an "atom" if $0 < \mu(A) < \infty$ and for any $\Sigma \ni B \subseteq A$ we have either $\mu(B) = 0$ or $\mu(B) = \mu(A)$. A measure without atoms is called "nonatomic" (for example, the Lebesgue measure on $\mathbb{R}^N$).

The next result is known as Saks' Lemma and can be found in Dunford–Schwartz [95], p. 308.

Lemma 2.166 *Let (X, Σ, μ) be a finite measure space and $\epsilon > 0$. Then there exists a finite partition $\{A_k\}_{k=1}^n \subseteq \Sigma$ of X by mutually disjoint sets such that each A_k is either an atom or $\mu(A_k) \leq \epsilon$.*

Concerning nonatomic measures there is the so-called Lyapunov's Convexity Theorem which has important applications in optimal control, mathematical economics and game theory. For two different proofs of this theorem, we refer to Halmos [133] and Lindenstrauss [184].

Theorem 2.167 *Let (X, Σ) be a measurable space and $\mu_k : \Sigma \to \mathbb{R}$ are finite nonatomic measures on X for $k = 1, \ldots, n$. Then*

$$R\left(\{\mu_k\}_{k=1}^n\right) = \left\{(\mu_k(A))_{k=1}^n : A \in \Sigma\right\} \subseteq \mathbb{R}^n$$

is compact and convex.

Definition 2.168

(a) Let (X, Σ) be a measurable space. The "universal σ-algebra" of $\overline{\Sigma}$ is defined by

$$\overline{\Sigma} = \bigcap\{\Sigma_\mu : \mu \text{ is probability measure on } \Sigma\}.$$

(b) A σ-algebra, Σ, of subsets of X is called "countably generated" if there exists a countable subfamily $\mathcal{L}$ of Σ such that $\Sigma = \sigma(\mathcal{L})$.
(c) Σ is called "separable" if it is countably generated and separates points, namely for any $x, x' \in X$ we can find $A \in \Sigma$ such that $\chi_A(x) \neq \chi_A(x')$.

Finally, we mention some books dealing with measure theory which the reader can consult for additional topics and information: Aliprantis-Border [10], Ash [13], Bogachev [35], Bourbaki [42], Cohn [71], Denkowski-Migorski-Papageorgiou [81], Dudley [89], Folland [115], Halmos [135], Hewitt–Stromberg [139], Papageorgiou-Winkert [216], Royden [241], and Rudin [243]

2.10 Problems

Problem 2.1 Let $f : \mathbb{R}^N \to \mathbb{R}_+$ and set $S(f) = \{(x, \lambda) \in \mathbb{R}^N \times \mathbb{R} : 0 \leq \lambda \leq f(x)\}$. Show that (a) f is Borel measurable iff $S(f) \in B(\mathbb{R}^N \times \mathbb{R}) = B(\mathbb{R}^N) \bigotimes B(\mathbb{R})$; and (b) If f is Borel measurable and $\vartheta > 0$, then $\int_{\mathbb{R}^N} f(x)^\vartheta dx = \vartheta \int_0^\infty \lambda^{\vartheta-1} |\{f > \lambda\}|_N d\lambda$, where $|\cdot|_N$ denotes the Lebesgue measure on $\mathbb{R}^N$.

Problem 2.2 Given $\epsilon > 0$, find an open dense set $U \subseteq \mathbb{R}$ such that $|U|_1 \leq \epsilon$.

Problem 2.3 Let (Ω, Σ, μ) be a finite measure space and $\{A_n\} \subseteq \Sigma$ satisfy $\Sigma_{n \geq 1} \mu(A_n) < \infty$. Show that

$$\mu(\limsup_{n \to \infty} A_n) = 0,$$

with $\limsup_{n \to \infty} A_n = \cap_{k \geq 1} \cup_{n \geq k} A_n$. This result is known as the "Borel–Cantelli Lemma."

Problem 2.4 Let (X, Σ, μ) be a finite nonatomic measure space and $\{\vartheta_n\} \subseteq (0, \infty)$ such that $\Sigma_{n \geq 1} \vartheta_n \leq \mu(X)$. Show that there exist $\{A_n\} \subseteq \Sigma$, pairwise disjoint, such that $\mu(A_n) = \vartheta_n$ for all $n \in \mathbb{R}^N$.

Problem 2.5 Let (X, Σ, μ) be a σ-finite measure space and $f, g : X \to \mathbb{R}_+$ measurable functions such that $\int_A f d\mu \leq \int_A g d\mu$ for all $A \in \Sigma$ with $\mu(A) < \infty$. Show that $f(x) \leq g(x)$, μ—a.e.

Problem 2.6 Let (X, Σ, μ) be a nonatomic measure space and $f : X \to \mathbb{R}_+$ a measurable function. Set $\lambda(A) = \int_A f d\mu$. Show that λ is nonatomic iff $\mu(\{f = \infty\}) = 0$.

Problem 2.7 Let (X, Σ, μ) be a measure space and $\{A_n\} \subseteq \Sigma$ satisfy $\mu(\cup_{n \geq 1} A_n) < \infty$ and $\inf_{n \geq 1} \mu(A_n) \geq \eta > 0$. Set

$$A = \{x \in X : x \text{ belongs to an infinite number of the sets } A_n\}.$$

Show that $A \in \Sigma$ and $\mu(A) \geq \eta > 0$.

Problem 2.8 Let $A \subseteq \mathbb{R}$ be Lebesgue measure with finite Lebesgue measure. Show that $f(x) = |A \cap (-\infty, x]|_1$ is continuous.

Problem 2.9 Let (X, Σ, μ) be a measure space, $f \in L^1(X)$, $f(x) \geq 0$ μ—a.e. and $g : X \to \mathbb{R}$ is Σ-measurable such that $a \leq g(x) \leq b$, μ—a.e. Show that there exits $c \in [a, b]$ such that $\int_X f g d\mu = c \int_X f d\mu$.

Problem 2.10 Let (X, Σ, μ) be a measure space and $f_n : X \to \mathbb{R}^* = [-\infty, \infty]$ a sequence of Σ-measurable functions such that $\sup_{n \geq 1} \|f_n\|_1 < \infty$. Show that $\{f_n\}$ is uniformly integrable iff

$$\lim_{\substack{k \to \infty \\ k \in \mathbb{N}}} \sup_{n \in \mathbb{N}} \int_{\{|f_n| > k\}} (|f_n| - k) d\mu = 0.$$

Problem 2.11 Let (X, Σ, μ) be a σ-finite measure space, and $f : X \to \mathbb{R}^* = [-\infty, \infty]$ a Σ-measurable function such that $fg \in L^1(X)$ for $g \in L^{p'}(X)$, where $1 \leq p \leq \infty$ and $1/p + 1/p' = 1$. Show that $f \in L^p(X)$.

Problem 2.12 Let (X, Σ, μ) be a measure space and $1 \leq q < p < r \leq \infty$. Show that

$$L^p(X) \subseteq L^q(X) + L^r(X).$$

Problem 2.13 Let (X, Σ, μ) be a measure space, $f : X \to \mathbb{R}^* = [-\infty, \infty]$ a Σ-measurable function and $d_f : (0, \infty] \to [0, \infty]$ is defined by $d_f(\eta) = \mu(\{|f| > \eta\})$. Show that (a) d_f is decreasing and right continuous; (b) $d_f \leq d_g$ if $|f| \leq |g|$; (c) $d_{f_n} \uparrow d_f$ if $|f_n| \uparrow |f|$; (d) if $f = g + h$, then for any $0 < \eta < \infty$,

$$d_f(\eta) \leq d_g(\frac{1}{2}\eta) + d_h(\frac{1}{2}\eta).$$

Problem 2.14 Let (X, Σ, μ) be a measure space, $1 \leq p < \infty$, $\{f_n - f\} \subseteq L^p(X)$ and assume that $\|f_n\|_p \to \|f\|_p$, $f_n \xrightarrow{\mu-a.e.} f$. Show that $\|f_n - f\| \to 0$ as $n \to \infty$.

Problem 2.15 Let (X, Σ, μ) be a measure space, $1 \leq p < \infty$, $\{f_n\} \subseteq L^p(X)$ and assume that $\sup_{n \geq 1} \|f_n\|_p < \infty$ and $f_n \xrightarrow{\mu-a.e} f$. Show that $f \in L^p(X)$ and

$$\lim_{n \to \infty} [\|f_n\|_p^p - \|f_n - f\|_p^p] = \|f\|_p^p.$$

The result is known as the "Brezis-Lieb Lemma," see [50].

Problem 2.16 Let $A \subseteq \mathbb{R}$ be Lebesgue measurable with positive measure. Show that $A - A$ is a neighborhood of zero.

Problem 2.17 Let X be a metric space and μ, λ are finite Borel measures on X. Show that (a) $\mu = \lambda$ if $\mu = \lambda$ on open sets; (b) $\mu = \lambda$ if X is σ-compact and $\mu = \lambda$ on compact sets.

Problem 2.18 Let X be a metric space, μ a σ-finite Borel measure on X and $A \in B(X)$ with $\mu(A) < \infty$. Show that A is inner regular.

Problem 2.19 Let (X, Σ, μ) be a measure space and $f_n, f : X \to \mathbb{R}_+$ functions Σ-measurable such that $f_n \overset{\mu}{\to} f$. Show that $f_n^\vartheta \overset{\mu}{\to} f^\vartheta$ for any $\vartheta > 0$.

Problem 2.20 Let (X, Σ, μ) be a measure space, $C \subseteq \mathbb{R}^N$ closed, $f \in L^1(X, \mathbb{R}^N)$ such that $\frac{1}{\mu(A)} \int_A f(x) d\mu \in C$ for all $A \in \Sigma$ with $\mu(A) > 0$. Show that $f(x) \in C$, μ—a.e.

Problem 2.21 Let (X, Σ, μ) be a measure space and $f \in L^1(X) \cap L^2(X)$. Show that (a) $f \in L^r(X)$ for all $r \in [1, 2]$; and (b) $\|f\|_p \to \|f\|_1$ as $p \to 1^+$.

Problem 2.22 Let I be an interval in $\mathbb{R}$, $A \subseteq I$ with finite Lebesgue measure and $f : I \to \mathbb{R}$ measurable and differentiable at every point in A. Show that

$$\lambda^*[f(A)] \le \int_A |f'(t)| dt,$$

with λ^* being the Lebesgue outer measure on $\mathbb{R}$.

Problem 2.23 Show that any nonempty complete metric space without isolated points has a Souslin (Analytic) set which is not Borel.

Problem 2.24 Let (X, Σ, μ) be a finite measure space. Show that there exists an at most countable family $\{A_n\}$ of atoms such that $X \setminus \cup_{n \ge 1} A_n$ is atom free.

Problem 2.25 Let X be a metric space, μ a finite Borel measure on X and $f : X \to \mathbb{R}$ continuous. Show that there exist at most countably many $\vartheta \in \mathbb{R}$ such that $\mu(\{f = \vartheta\}) > 0$.

Problem 2.26 Let (X, Σ_1) and (Y, Σ_2) be measurable spaces and $f : X \to Y$ be (Σ_1, Σ_2)-measurable. Show that f is also $(\widehat{\Sigma_1}, \widehat{\Sigma_2})$-measurable.

Problem 2.27 Let (X, Σ) be a measurable space, V a Hausdorff topological space and $G \in \Sigma \bigotimes B(V)$. Show that there exists countably generated sub σ-algebra, $\Sigma_0 \subseteq \Sigma$, such that $G \in \Sigma_0 \bigotimes B(V)$.

Chapter 3
Banach Space Theory

3.1 Introduction

In this chapter we present some basic concepts and results from Functional Analysis (Banach Space Theory). In Sect. 3.2 we introduce the terminology of general locally convex spaces, we define normed and Banach spaces, and then study linear operators between them. We also prove some distinguishing facts about finite dimensional normed spaces. In Sect. 3.3 we present the analytic and geometric forms of the Hahn-Banach theorem. The latter refer to the separation theorems for convex sets. We also prove some of their useful consequences. In Sect. 3.4 we prove the three fundamental theorems of linear functional analysis. These are the Open Mapping Theorem, the Closed Graph Theorem and the Uniform Boundedness Principle (Banach-Steinhaus Theorem). All of the three are outgrowths of Baire's Category Theory. In Sect. 3.5 we introduce and discuss the weak and weak* topologies. These are locally convex, nonmetrizable (in infinite dimensions) topologies, which are strictly weaker than the norm (metric) topology and exhibit many interesting features. In Sect. 3.6 we focus on the special classes of separable and reflexive normed spaces. Most of the spaces we encounter in applications are separable and/or reflexive. In Sect. 3.7 we discuss dual operators, compact operators and projection operators. Compact operators exhibit properties similar to those of a finite dimensional operator. In Sect. 3.8 we present the main features of Hilbert spaces. In fact, Hilbert spaces are infinite dimensional generalizations of Euclidean spaces. We also outline the spectral theory for compact self-adjoint operators. Finally, in Sect. 3.9 we have a brief encounter with unbounded linear operators.

S. Hu, N. S. Papageorgiou, *Research Topics in Analysis, Volume I*, Birkhäuser
Advanced Texts Basler Lehrbücher, https://doi.org/10.1007/978-3-031-17837-5_3

3.2   Locally Convex Spaces: Banach Spaces

In the sequel all vector spaces are over the field of real numbers unless otherwise stated. We start with a notion that combines the algebraic (vector) and topological structures on a set.

Definition 3.1

(a) Let X be a vector space and τ a Hausdorff topology on X for which vector addition and scalar multiplication are both continuous. Then (X, τ) is a "topological vector space" and τ is a "vector topology" on X.

(b) Let (X, τ) be a topological vector space and $A \subseteq X$. A is "bounded" if for any $U \in \mathcal{N}(0)$ we can find $\lambda_0 > 0$ such that $A \subseteq \lambda U$ for all $\lambda > \lambda_0$.

Remark 3.2 Given $u \in X$ and $\lambda \in \mathbb{R}$, we can define the translation map: $x \rightarrow t_u(x) = u + x$ and the scalar multiplication map: $x \rightarrow m_\lambda(x) = \lambda x$. It is easy to see that both maps are homeomorphisms of X onto X. Hence, the vector topology τ is translation invariant, namely $U \in \mathcal{N}(0)$ iff $x + U \in \mathcal{N}(x)$. Thus, τ is completely determined by a local basis at the origin. A metric d on X is called translation invariant if $d(u + y, x + y) = d(u, x)$ for all $u, x, y \in X$.

Next we recall the basic notion of a norm and list the main classes of topological vector spaces.

Definition 3.3

(a) A map $\| \cdot \| : X \rightarrow \mathbb{R}_+$ is a "norm" on X if it satisfies (1) $\|x\| = 0$ iff $x = 0$; (2) $\|\lambda x\| = |\lambda| \|x\|$ for all scalar λ; (3) $\|u + x\| \leq \|u\| + \|x\|$ for all $u, x \in X$. Then $(X, \| \cdot \|)$ is a "normed space" and $\| \cdot \|$ generates a metric on X by $d(u, x) = \|u - x\|$. Evidently, this metric is translation invariant.

(b) Let (X, τ) be a topological vector space.

   (i) X is "locally convex" if it has a local basis consisting of convex sets.
   (ii) X is "locally bounded" if there is a bounded $U \in \mathcal{N}(0)$.
   (iii) X is a "Frechet space" if it is locally convex and τ is generated by a complete translation invariant metric.
   (iv) X is "normable" if there is a norm on X so that the metric corresponding to it generates the vector topology.
   (v) X is a "Banach Space" if it is a complete normed space.

Proposition 3.4 *Every topological vector space (X, τ) is regular.*

Proof Let $\mathcal{B}(0)$ be a local basis for X and $U \in \mathcal{B}(0)$. The continuity of the vector operations implies that there exists $V \in \mathcal{N}(0)$ such that $V - V \subseteq U$. Hence, $V \cap [(X \setminus U) + V] = \emptyset$. Note that $(X \setminus U) + V = \cup_{y \in X \setminus U}(y + V)$ is open and so $\overline{V} \cap [(X \setminus U) + V] = \emptyset$. Since $0 \in V$ it follows that $\overline{V} \cap (X \setminus U) = \emptyset$. Therefore, $\overline{V} \subseteq U$. By Proposition 1.38 we conclude that (X, τ) is regular. $\square$

Proposition 3.5 *Let (X, τ) be a topological vector space.*

(a) *If $U \in \mathcal{N}(0)$, $\{\lambda_n\}_{n \in \mathbb{N}} \subseteq (0, +\infty)$ is strictly increasing and $\lambda_n \to +\infty$, then*
$$X = \bigcup_{n \in \mathbb{N}} \lambda_n U.$$
(b) *If $K \subseteq X$ is compact, then K is bounded.*
(c) *If X is locally bounded, then it is first countable.*

Proof

(a) Fix $x \in X$. The continuity of the map $\lambda \to \lambda x$ implies that $V = \{\lambda \in \mathbb{R} : \lambda x \in U\}$ is open and contains 0. So $\frac{1}{\lambda_n} \in V$ for all $n \geq n_0$, therefore $x \in \lambda_n U$ for $n \geq n_0$.

(b) Let $U \in \mathcal{N}(0)$. The continuity of the scalar multiplication, implies that we can find $\delta > 0$ and $V \in \mathcal{N}(0)$ such that $\lambda V \subseteq U$ for all $|\lambda| < \delta$. Let $W = \bigcup_{|\lambda| < \delta} \lambda V$.
Then $W \in \mathcal{N}(0)$, $\vartheta W \subseteq W$ for all $|\vartheta| \leq 1$ (such a neighborhood of the origin is said to be "balanced") and $W \subseteq U$. From (a) we have $K \subseteq \bigcup_{n \in \mathbb{N}} nW$.
The compactness of K implies that we can find a finite set $F \subseteq \mathbb{N}$ such that $K \subseteq n_0 W$ with $n_0 = \max F$ (recall that W is balanced). If $\lambda > n_0$, then $K \subseteq \lambda W \subseteq \lambda U$.

(c) Let $U \subseteq \mathcal{N}(0)$ be bounded and $V \subseteq \mathcal{N}(0)$. Then we can find $\lambda_0 > 0$ such that $U \subseteq \lambda V$ for all $\lambda > \lambda_0$. Suppose $\{\delta_n\}_{n \in \mathbb{N}} \subseteq (0, +\infty)$ is strictly decreasing and $\delta_n \to 0^+$. Then for $n \geq n_0$, we have $\lambda_0 \delta_n < 1$ and so $U \subseteq \frac{1}{\delta_n} V$ for all $n \geq n_0$. Therefore (X, τ) is first countable. $\quad\square$

A byproduct of the proof of part (b) is the following corollary.

Corollary 3.6 *If (X, τ) is a topological vector space, then every $U \in \mathcal{N}(0)$ (convex) contains $W \in \mathcal{N}(0)$ balanced (convex).*

Proposition 3.7 *If (X, τ) is a locally compact topological vector space, then X is finite dimensional.*

Proof Let $U \in \mathcal{N}(0)$ be relatively compact. Then from Proposition 3.5(c), we have that $\{\frac{1}{2^n} U\}_{n \in \mathbb{N}}$ is a local basis for X. The compactness of $\overline{U}$ implies that we can find a finite set $F \subseteq X$ such that $\overline{U} \subseteq F + \frac{1}{2} U$. Let $Y = \operatorname{span} F$. Then Y is finite dimensional and so a closed subspace of X. We have $U \subseteq Y + \frac{1}{2} U$, hence $\frac{1}{2} U \subseteq Y + \frac{1}{4} U$ and so $U \subseteq Y + \frac{1}{2} U \subseteq Y + Y + \frac{1}{4} U = Y + \frac{1}{4} U$. Inductively we have $U \subseteq Y + \frac{1}{2^n} U$ for all $n \in \mathbb{N}$ and so $U \subseteq \bigcap_{n \in \mathbb{N}} (Y + \frac{1}{2^n} U) = \overline{Y} = Y$. Since $\{\frac{1}{2^n} U\}_{n \in \mathbb{N}}$ is a local basis of X, it follows that $X \subseteq Y$, hence $X = Y$. $\quad\square$

Proposition 3.8 *If (X, τ) is a topological vector space and $A \subseteq X$, then the following two statements are equivalent*

(a) *A is bounded.*
(b) *If $\{u_n\}_{n \geq 1} \subseteq X$ and $\{\lambda_n\}_{n \geq 1} \subseteq \mathbb{R}$ such that $\lambda_n \to 0$, then $\lambda_n u_n \to 0$.*

Proof (a) $\Rightarrow$ (b) Let $U \in \mathcal{N}(0)$ be balanced. We have $A \subseteq \widehat{\lambda} U$ for some $\widehat{\lambda} > 0$. Then $|\lambda_n|\widehat{\lambda} < 1$ for all $n > n_0$ and since $\frac{1}{\lambda} A \subseteq U$, we have $\lambda_n x_n \in U$ for all $n > n_0$. This means that $\lambda_n x_n \to 0$ as $n \to \infty$.

(b) $\Rightarrow$ (a) Arguing by contradiction, suppose that A is not bounded. Then we can find $U \in \mathcal{N}(0)$ and $\lambda_n \to +\infty$ such that no $\lambda_n U$ contains A. Let $u_n \in A$ such that $u_n \notin \lambda_n U$, $n \in \mathbb{N}$ and so $\frac{1}{\lambda_n} u_n \not\to 0$, a contradiction. $\square$

In Analysis we often encounter spaces X with a function $p : X \to \mathbb{R}$ which satisfies the requirements of a norm except $(\alpha)_1$ (see Definition 3.3(a)).

Definition 3.9 Let X be a vector space and $p : X \to \mathbb{R}$. We say that $p(\cdot)$ is a "seminorm."

If (a) $p(u + x) \leq p(u) + p(x)$ for all $u, x \in X$ (subadditivity);

(b) $p(\lambda u) = |\lambda| p(u)$ for all $\lambda \in \mathbb{R}$, $u \in X$ (positive homogeneity). A family $\{p_\alpha\}_{\alpha \in I}$ of seminorm is said to be "separating," if for each $x \neq 0$ we can find at least one $\alpha \in I$ such that $p_\alpha(x) \neq 0$.

Remark 3.10 It is an easy consequence of the positive homogeneity property that $p(0) = 0$. Every norm is a seminorm and a seminorm $p(\cdot)$ is a norm if $x \neq 0 \Rightarrow p(x) \neq 0$. So, a seminorm $p(\cdot)$ is not a norm if $\ker p$ is a nontrivial subspace of X.

Proposition 3.11 *If X is a vector space and $p(\cdot)$ a seminorm on X, then*

(a) $|p(u) - p(x)| \leq p(u - x)$ and $p \geq 0$.
(b) $\ker p = \{u \in X : p(u) = 0\}$ is a subspace of X.
(c) $B_1 = \{u \in X : p(u) < 1\}$ is convex, balanced and absorbing (that is, for every $u \in X$, we can find $t_u > 0$ such that $u \in t_u B_1$).

Proof

(a) We have $p(u) = p(u - x - x) \leq p(u - x) + p(x)$, hence $p(u) - p(x) \leq p(u - x)$. Similarly, we also have $p(x) - p(u) \leq p(x - u)$. Since by positive homogeneity we have $p(u - x) = p(x - u)$, we conclude that $|p(u) - p(x)| \leq p(u - x)$ for all $u, x \in X$. Taking $x = 0$, we see that $p \geq 0$.

(b) Let $u, x \in \ker p$ and $\lambda \in \mathbb{R}$. Then $0 \leq p(\lambda u + x) \leq |\lambda| p(u) + p(x) = 0$. Hence $\lambda u + x \in \ker p$.

(c) Clearly B_1 is balanced. If $u, x \in B_1$ and $0 < t < 1$, then $p(tu + (1 - t)x) \leq tp(u) + (1 - t)p(x) < 1 \Rightarrow B_1$ is convex. Moreover, if $u \in X$ and $\lambda > p(u)$, then $p(\frac{1}{\lambda} u) = \frac{1}{\lambda} p(u) < 1$ and so B_1 is also absorbing. $\square$

Remark 3.12 Evidently every absorbing set contains the origin.

To have a description of a locally convex topology, we need the following technical devise.

Definition 3.13 Let X be a vector space and $A \subseteq X$. The "Minkowski functional" (or "gauge") of A is the function $p_A : X \to \overline{\mathbb{R}}_+ = \mathbb{R}_+ \cup \{+\infty\}$ defined by $p_A(x) = \inf\{\lambda > 0 : x \in \lambda A\}$.

As always, we use the convention that $\inf \emptyset = +\infty$. Minkowski functionals are useful in the context of convex, absorbing sets.

Proposition 3.14 *If X is a vector space and $A \subseteq X$ is convex, absorbing, then*

(a) *p_A is subadditive and $p_A(\lambda u) = \lambda p_A(u)$ for all $u \in X$, $\lambda > 0$.*
(b) *If A is also balanced, then $p_A(\cdot)$ is a seminorm.*
(c) *If $B_1 = \{u \in X : p_A(u) < 1\}$ and $\widehat{B}_1 = \{u \in X : p_A(u) \leq 1\}$, then $B_1 \subseteq A \subseteq \widehat{B}_1$ and $p_A = p_{B_1} = p_{\widehat{B}_1}$.*

Proof

(a) Let $u \in X$ and let $L_A(u) = \{\lambda > 0 : \frac{1}{\lambda}u \in A\}$. If $\lambda \in L_A(u)$ and $\mu > \lambda$, then $\mu \in L_A(u)$ (recall that A is convex and $0 \in A$). Therefore $L_A(u)$ is a half line starting at $p_A(u)$. Let $p_A(u) < \lambda$, $p_A(x) < \mu$ and $\vartheta = \lambda + \mu$. We have $\frac{1}{\lambda}u \in A$, $\frac{1}{\mu} \in A$.

The convexity of A implies that

$$\frac{1}{\vartheta}(u + x) = \frac{\lambda}{\vartheta}(\frac{1}{\lambda}u) + \frac{\mu}{\vartheta}(\frac{1}{\mu}x) \in A$$

$$\Rightarrow p_A(u + x) \leq \vartheta$$

and so $p_A(\cdot)$ is subadditive. Clearly, we also have $p_A(\lambda u) = \lambda p_A(u)$ for all $u \in X$, $\lambda > 0$.

(b) Since A is balanced, we have $p_A(\lambda u) = |\lambda| p_A(u)$ for all $\lambda \in \mathbb{R}$. Hence $p_A(\cdot)$ is a seminorm.

(c) Note that $p_A(u) < 1 \Rightarrow 1 \in L_A(u)$ and so $u \in A$. On the other hand if $u \in A$, then $p_A(u) \leq 1$. Hence we have $B_1 \subseteq A \subseteq \widehat{B}_1$ from which we infer that $L_{B_1}(u) \subseteq L_A(u) \subseteq L_{\widehat{B}_1}(u)$ for all $u \in X$. It follows that $p_{\widehat{B}_1} \leq p_A \leq p_{B_1}$. Let $p_{\widehat{B}_1}(u) < \lambda < \mu$. Then $\frac{1}{\lambda}u \in \widehat{B}_1$ and so $p_A(\frac{1}{\lambda}u) \leq 1$, from which we infer that $p_A(\frac{1}{\mu}u) \leq \frac{\lambda}{\mu} < 1$. Hence $\frac{1}{\mu}u \in B_1$ and so $p_{B_1}(\frac{1}{\mu}u) \leq 1$, from which we obtain $p_{B_1}(u) \leq \mu$. Therefore $p_{B_1} = p_A = p_{\widehat{B}_1}$. $\square$

Proposition 3.15 *If X is a topological vector space and $A \subseteq X$ is convex, absorbing balanced, then $p_A(\cdot)$ is continuous if and only if $A \in \mathcal{N}(0)$.*

Proof $\Rightarrow$ The continuity of $p_A(\cdot)$ implies that $B_1 = \{u \in X : p_A(u) < 1\}$ is open. But by Proposition 3.14(c) we have $B_1 \subseteq A$ and so we conclude that $A \in \mathcal{N}(0)$.
$\Leftarrow$ Given $\epsilon > 0$, we have $p_A(u) \leq \epsilon$ for all $u \in \epsilon A$. Since $A \in \mathcal{N}(0)$, we infer that $p_A(\cdot)$ is continuous at the origin. But from Proposition 3.11(a) we have $|p_A(u) - p_A(x)| \leq p_A(u - x)$ for all $u, x \leq X$ and this implies the continuity of $p_A(\cdot)$. $\square$

Therefore, in a locally convex space X with a local basis $\mathbb{B}(0)$ consisting of convex, balanced sets (see Corollary 3.6), then $\{p_U\}_{U \in \mathbb{B}(0)}$ is a family of continuous seminorms with the following additional property.

Proposition 3.16 *If X is a locally convex vector space with local basis $\mathbb{B}(0)$, then the family of seminorms $\{p_u\}_{u \in \mathbb{B}(0)}$ is separating.*

Proof We need to show that if $u \in X$ and $p_U(u) = 0$ for all $U \in \mathbb{B}(0)$, then $u = 0$. Indeed, if $u \neq 0$ and $p_U(0) = 0$ for all $U \in \mathbb{B}(0)$, then $u \in \bigcap_{U \in \mathbb{B}(0)} \{x \in X : p_U(x) < 1\} \subseteq \bigcap_{U \in \mathbb{B}(0)} U = \{0\}$, a contradiction.   □

So, there is a very convenient representation of locally convex spaces (see Rudin [229], Theorem 1.37, p.26).

Theorem 3.17 *If X is a vector space and $\mathcal{P}$ is a separating family of seminorms on X, then the weak topology on X corresponding to the family $\mathcal{P}$ is a locally convex topology on X.*

Then next theorem characterizes normable spaces and is known in the literature as the "Kolmogorov Normability Criterion."

Theorem 3.18 *If X is a locally convex vector space, then X is normable if and only if it is locally bounded.*

Proof $\Rightarrow B_1 = \{u \in X : \|u\| < 1\} \in \mathcal{N}(0)$.
$\Leftarrow$ Let $U \in \mathcal{N}(0)$ be bounded, convex, balanced. Let $\|u\| = p_U(u)$ for all $u \in X$. From Proposition 3.5(c) we know that $\{\frac{1}{n}U\}_{n \in \mathbb{N}}$ is a local basis for the topology of X. If $u \neq 0$, then $u \notin \frac{1}{n}U$ for some $n \in \mathbb{N}$ and so $\|u\| \geq \frac{1}{n}$. Therefore by Proposition 3.14(b), $\|\cdot\|$ is a norm on X. We have $\{u \in X : \|u\| < r\} = rU$ for all $r > 0$. So, the topology of X and the norm topology coincide.   □

Example 3.19 Let $\Omega \subseteq \mathbb{R}^N$ be an open set. We can find $\{K_n\}_{n \in \mathbb{N}}$ a sequence of compact sets in Ω such that $K_n \subseteq \text{int } K_{n+1}$ for all $n \in \mathbb{N}$ and $\Omega = \bigcup_{n \in \mathbb{N}} K_n$ (we can always take $K_n = \{x \in \mathbb{R}^N : d(x, \mathbb{R}^N \setminus \Omega) > \frac{1}{n}, |x|_N \leq n\}$, with $|\cdot|_N$ being the Euclidean norm on $\mathbb{R}^N$). Consider the vector space $C(\Omega)$ of continuous functions on Ω with the topology τ determined by the local basis $\{U_n\}_{n \in \mathbb{N}}$ with

$$U_n = \{g \in C(\Omega) : \sup_{x \in K_n} |g(x)| < \frac{1}{n}\}.$$

Then $(C(\Omega), \tau)$ is a locally convex space and τ does not depend on the choice of $\{K_n\}_{n \in \mathbb{N}}$. Also τ is the topology of uniform convergence on compact sets, that is, $g_n \overset{\tau}{\to} g$ if and only if $\sup_{x \in K} |(g_n - g)(x)| \to 0$ as $n \to \infty$ for every compact $K \subseteq \Omega$, Finally $(C(\Omega), \tau)$ is a Frechet space which is not normable (every U_n contains functions g for which $\sup\limits_{K_{n+1}} |g|$ can be arbitrarily big and so no U_n is bounded; see Theorem 3.18).

Now we turn our attention to normed spaces. First we consider a particular case of the quotient space and quotient mapping introduced in Definition 1.73 and Remark 1.74. So, let X be a normed space and $V \subseteq X$ a closed subspace. We introduce the equivalence relation: $x \sim u$ if and only if $x - u \in V$. Then for every $x \in X$, the coset $\hat{x}$ is given by $\hat{x} = \{u \in X : x - u \in V\} = \{x + v : v \in V\}$ and we write the quotient space $X/\sim$ as X/V. This is a linear space with vector addition $\hat{x} + \hat{u} = \widehat{(x+u)}$ and scalar multiplication $\lambda\hat{x} = \widehat{(\lambda x)}$. Also if we set $\|\hat{x}\| = \inf[\|u\| : u \in \hat{x}]$, then we can easily verify that this is a norm on X/V.

Definition 3.20 Let X be a normed space and $V \subseteq X$ a closed subspace. The space X/V furnished with the norm $\|\hat{x}\| = \inf[\|u\| : u \in \hat{x}]$ is called the "quotient space" of X with respect to V.

Remark 3.21 For every $u \in \hat{x}$, we have $\|\hat{x}\| = \inf[\|u - v\| : v \in V] = d(u, V)$.

Proposition 3.22 *If X is Banach space and $V \subseteq X$ is a closed subspace, then X/V is a Banach space too.*

Proof Let $\{\hat{x}_n\}_{n \geq 1} \subseteq X/V$ be a Cauchy sequence. So, we can find $n_0(k) \in \mathbb{N}$ such that $\|\hat{x}_n - \hat{x}_m\| = \|(x_n + V) - (x_m + V)\| < \frac{1}{2^k}$ for all $n, m > n_0(k)$.

We consider a subsequence $\{\hat{x}_{n_k}\}_{k \in \mathbb{N}}$ such that $\|\hat{x}_{n_k} - \hat{x}_{n_{k+1}}\| = \|(x_{n_k} + V) - (x_{n_{k+1}} + V)\| < \frac{1}{2^k}$ with $n_k > n_0(k)$. Let $x_k \in \hat{x}_{n_k}$ $k \in \mathbb{N}$ such that $\|x_k - x_{k+1}\| < \frac{1}{2^k}$ (see Remark 3.21). Then for $n > k$ we have $\|x_k - x_n\| \leq \|x_k - x_{k+1}\| + \ldots + \|x_{n-1} - x_n\| < \frac{1}{2^{k-1}}$ for all $k \in \mathbb{N}$. Hence $\{x_k\}_{k \in \mathbb{N}} \subseteq X$ is Cauchy. Therefore $x_k \to x \in X$. We have $\|\hat{x}_{n_k} - \hat{x}\| \leq \|x_k - x\|$ for every $k \in \mathbb{N}$, $\Rightarrow \hat{x}_{n_k} \to \hat{x}$ and this proves the completeness of X/V. □

Proposition 3.23 *If X, Y are normed spaces and $A : X \to Y$ is a linear operator, then $A(\cdot)$ is continuous if and only if $A(\overline{B}_1)$ is bounded, where $\overline{B}_1 = \{u \in X : \|u\|_X \leq 1\}$.*

Proof $\Rightarrow$ We claim that $\|A(u)\|_Y \leq M\|u\|_X$ for some $M > 0$, all $u \in X$. If we prove this the implication follows. Arguing by contradiction, suppose that for every $n \in \mathbb{N}$ we can find $u_n \in X$ such that $\|A(u_n)\|_Y > n\|u_n\|_X$. So $\|A(\frac{1}{n}\frac{u_n}{\|u_n\|_X})\|_Y > 1$ for all $n \in \mathbb{N}$. But $\frac{1}{n}\frac{u_n}{\|u_n\|_X} \to 0$, a contradiction.
$\Leftarrow$ Evidently $\|A(u)\|_Y \leq M\|u\|_X$ for some $M > 0$, all $u \in X$. This means that $A(\cdot)$ is continuous at the origin and so by linearly it is continuous everywhere. □

Remark 3.24 In fact $A(\cdot)$ is Lipschitz continuous. Also $A(\cdot)$ maps bounded sets to bounded sets. Such an operator is said to be bounded. So, according to Proposition 3.23 a linear operator $A : X \to Y$ is continuous if and only if it is bounded.

Definition 3.25 Let X, Y be normed spaces. By $\mathcal{L}(X, Y)$ we denote the space of linear continuous (equivalently, bounded) operators from X to Y. This is a normed space with norm $\|A\|_{\mathcal{L}} = \sup\{\|A(u)\|_Y : \|u\|_X \leq 1\}$. If $X = Y$, then we write $\mathcal{L}(X)$ instead of $\mathcal{L}(X, X)$.

Proposition 3.26 *If X, Y are normed spaces and $A : X \to Y$ is linear, surjective, then $A^{-1} \in \mathcal{L}(Y, X)$ if and only if there exists $\vartheta > 0$ such that $\vartheta \|u\|_X \leq \|A(u)\|_Y$ for all $u \in X$.*

Proof $\Rightarrow$ By Proposition 3.23 we have $\|A^{-1}(y)\|_X \leq M\|y\|_Y$ for some $M > 0$, all $y \in Y = A(X)$. We have $y = A(u)$ and so $\|A^{-1}(A(u))\|_X = \|u\|_X \leq M\|A(u)\|_Y$, hence $\vartheta \|u\|_X \leq \|A(u)\|_Y$ for all $u \in X$, with $\vartheta = \frac{1}{M}$.

$\Leftarrow$ If $\vartheta \|u\|_X \leq \|A(u)\|_Y$ for all $u \in X$, then $\ker A = \{0\}$ and so A^{-1} exists and it is linear. Let $u = A^{-1}(y)$ for any $y \in Y$.

We have $\vartheta \|A^{-1}(y)\|_X \leq \|A(A^{-1}(y))\|_Y = \|y\|_Y \Rightarrow \|A^{-1}(y)\|_X \leq \frac{1}{\vartheta}\|y\|_Y$ and so $A^{-1} \in \mathcal{L}(Y, X)$ (see Proposition 3.23). $\square$

Definition 3.27 Let X, Y be normed spaces and $A \in \mathcal{L}(X, Y)$. We say that $A(\cdot)$ is an "isomorphism" (or "linear isomorphism" or "topological isomorphism"), if it is bijective and $A^{-1} \in \mathcal{L}(Y, X)$. In this case we say that X and Y are isomorphic (or "topologically isomorphic'). We say that $A \in \mathcal{L}(X, Y)$ is an "isometric isomorphism" if $\|A(u)\|_Y = \|u\|_X$ for all $u \in X$. Then we say that the spaces X, Y are "isometrically isomorphic."

Remark 3.28 Clearly an isometric isomorphism is an isomorphism and the relation "X and Y are isomorphic (respectively, isometrically isomorphic)" is an equivalence relation. If $A \in \mathcal{L}(X)$ and $\|A\|_{\mathcal{L}} < 1$, then $I - A$ is an isomorphism and $(I - A)^{-1} = \sum_{n \geq 0} A^n$ (the series is absolutely convergent in $\mathcal{L}(X)$). It follows that the set $\mathcal{D} \subseteq \mathcal{L}(X)$ of all isomorphisms is open and the map $A \to A^{-1}$ is a homomorphism of $\mathcal{D}$ onto $\mathcal{D}$ (for details, see Brown and Pearcy [55], pp.257–258).

From Propositions 3.23 and 3.26, we have

Proposition 3.29 *If X, Y are normed spaces and $A \in \mathcal{L}(X, Y)$, then A is an isomorphism if and only if $\vartheta \|u\|_X \leq \|A(u)\|_Y \leq M\|u\|_X$ for some $0 < \vartheta < M$, all $u \in X$.*

For general metrizable spaces completeness is metric dependent, that is, it can happen that we have two equivalent metrics (they generate the same topology) and one is complete the other is not. However, completeness is preserved by linear isomorphisms.

Proposition 3.30 *If X, Y are isomorphic normed spaces and one is a Banach space, then so is the other.*

Proof Let $A : X \to Y$ be an isomorphism and suppose X is a Banach space. By Proposition 3.29, we have

$$\vartheta \|u\|_X \le \|A(u)\|_Y \le M\|u\|_X \tag{3.1}$$

for all $u \in X$. Let $\{y_n\}_{n\ge 1} \subseteq Y$ be a Cauchy sequence. We have $y_n = A(u_n)$ with $u_n \in X$ and $\vartheta \|u_n - u_m\| \le \|y_n - y_m\|$ for all $n, m \in \mathbb{N}$ (see (3.1)). $\Rightarrow \{u_n\} \subseteq X$ is Cauchy. So, we have $u_n \to u$ in X and so $y_n = A(u_n) \to A(y) = y$ in Y. Hence Y is Banach too. $\qquad \square$

Remark 3.31 Suppose that on the vector space X there are two norms $\|\cdot\|_1, \|\cdot\|_2$. Then we say that

(i) "$\|\cdot\|_1$ is weaker than $\|\cdot\|_2$," if the identity map id: $(X, \|\cdot\|_2) \to (X, \|\cdot\|_1)$ is continuous.
(ii) "$\|\cdot\|_1$ and $\|\cdot\|_2$ are equivalent" if id: $(X, \|\cdot\|_1) \to (X, \|\cdot\|_2)$ is an isomorphism.

In the study of continuous linear functionals on a topological vector space, the following algebraic notion is useful.

Definition 3.32 Let X be a vector space and V a proper subspace of X. We say that V is a "hyperplane" (or "subspace of codimension 1") if given $u_0 \in X\backslash V$, every $u \in X$ can be expressed as $u = \vartheta u_0 + v$ for some $\vartheta \in \mathbb{R}, v \in V$.

In what follows by a linear functional, we mean a linear function $f : X \to \mathbb{R}$. Next we show that hyperplanes correspond to kernels of linear functionals.

Proposition 3.33 *If X is a vector space and $V \subseteq X$ a subspace, then V is a hyperplane if and only if $V = \ker f$ with f a linear functional.*

Proof $\Rightarrow$ Let $u_0 \in X\backslash V$ and $u \in X$. We know that we can write u uniquely as $u = \vartheta u_0 + v$ for some $\vartheta \in \mathbb{R}, v \in V$. Let $f(u) = \vartheta$. Then f is a linear functional and $\ker f = V$.
$\Leftarrow$ Let $u_0 \in X\backslash \ker f$, then $f(u_0) \ne 0$ and so for every $u \in X$, we can write that $u = \frac{f(u)}{f(u_0)}u_0 + v$ with $v \in \ker f$. $\qquad \square$

Corollary 3.34 *If X is a normed space and $f : X \to \mathbb{R}$ is a nontrivial linear functional, then $\ker f$ is either closed in X or dense in X.*

Proof The nontriviality of $f(\cdot)$ implies $\ker f \ne X$. Now assume that $\ker f$ is not closed. So, we can find $u_0 \in \overline{\ker f}\backslash \ker f$. Then by Proposition 3.33, $X = span\{u_0, \ker f\}$. Since $u_0 \in \overline{\ker f}$, we conclude that $X = \overline{\ker f}$. $\qquad \square$

For linear functionals we have the following convenient criterion of continuity in terms of its kernel.

Proposition 3.35 *If X is a normed space and $f : X \to \mathbb{R}$ a nontrivial linear functional, then f is continuous if and only if $\ker f$ is closed.*

Proof $\Rightarrow$ Clear.
$\Leftarrow$ Arguing by contradiction, suppose f is not continuous. According to Proposition 3.23, $f(\overline{B}_r)$ is not bounded ($\overline{B}_r = \{u \in X : \|u\| \le r\}$) and so by linearity $f(\overline{B}_r) =$

$\mathbb{R}$. Therefore for any $x \in X$ we can find $u \in \overline{B}_r$ such that $f(x) = -f(u)$. Hence $x + u \in \ker f \cap (x + B_r)$, which means that $\ker f$ is dense in X, hence $f \equiv 0$, a contradiction. $\square$

Remark 3.36 Combining Corollary 3.34 and Proposition 3.35, we see that every discontinuous linear functional has dense kernel. Also Proposition 3.35 is not true for general linear operators.

Definition 3.37 Let X be an infinite dimensional normed space. A "Schauder basis" for X, is a sequence $\{e_n\}_{n \in \mathbb{N}} \subseteq X$ such that $span\{e_n\}_{n \in \mathbb{N}}$ is dense in X, that is, for every $u \in X$ there exists a unique sequence such $\{\lambda_n\}_{n \in \mathbb{N}} \subseteq \mathbb{R}$ such that $\|u - \sum_{n=1}^{m} \lambda_n e_n\| \to 0$ as $m \to \infty$. We write $u = \sum_{n \in \mathbb{N}} \lambda_n e_n$ (the series converges in norm) and call $\{\lambda_n\}_{n \in \mathbb{N}}$ the coordinates of u.

Remark 3.38 Evidently the basis $\{e_n\}_{n \in \mathbb{N}} \subseteq X$ is a linearly independent set. A Banach space with a Schauder basis, is separable (finite rational combinations of its elements form a countable dense subset). The converse is not true. This was proved by Enflo [99]. If X is finite dimensional, then the Schauder basis coincides with the algebraic basis.

For the rest of this section, we focus on finite dimensional spaces. In what follows by $| \cdot |_2$ we denote the Euclidean norm on $\mathbb{R}^N$.

Proposition 3.39 *Every N-dimensional normed space X is isomorphic to* $(\mathbb{R}^N, | \cdot |_2)$.

Proof Let $\{e_k\}_{k=1}^{N}$ be a basis for X and consider the linear map $A : \mathbb{R}^N \to X$ defined by $A(\{\lambda_k\}_{k=1}^{N}) = \sum_{k=1}^{N} \lambda_k e_k$. Using the Cauchy–Schwarz inequality, we see that $\|A(\{\lambda_k\}_{k=1}^{N})\| \leq M |(\{\lambda_k\}_{k=1}^{N})|_2$ with $M = (\sum_{k=1}^{N} \|e_k\|^2)^{1/2}$. Let $g = \| \cdot \| \circ A : \mathbb{R}^N \to \mathbb{R}$. This is a continuous map and so by the Weierstrass theorem we can find $\{\widehat{\lambda}_k\}_{k=1}^{N} \in \mathbb{R}^N$ such that $g(\{\widehat{\lambda}_k\}_{K=1}^{N}) = \min_{\partial B_1} g = \vartheta > 0$ (since A is $1 - 1$). Then $\vartheta \leq \|A(\{\lambda_k\}_{k=1}^{N})\|$ for all $\{\lambda_k\}_{k=1}^{N} \in \partial B_1$. Hence by Proposition 3.26, A is an isomorphism. $\square$

Remark 3.40 It follows that all finite dimensional normed spaces with the same dimension are isomorphic. Moreover, from Propositions 3.30 and 3.39, we see that all finite dimensional normed spaces are Banach spaces. So, we recover the well-known fact that if V is a finite dimensional subspace of a normed space X, then V is closed. Also, in a finite dimensional normed space, the compact sets are the closed and bounded ones (Heine-Borel property).

Corollary 3.41 *If X is a finite dimensional vector space, then any two norms on X are equivalent.*

To characterize finite dimensional normed spaces, we need the following result, known in the literature as "Riesz Lemma."

Lemma 3.42 *If X is a normed space and $V \subseteq X$ a proper closed subspace, then for every $\epsilon \in (0, 1)$, we can find $\widehat{u} \in X \setminus V$ with $\|\widehat{u}\| = 1$ such that $d(\widehat{u}, V) \geq 1 - \epsilon$.*

Proof Let $u \in X \setminus V$ and set $\eta = d(u, V) > 0$. We can find $v \in V$ such that $\eta \leq \|u - v\| \leq \frac{\eta}{1-\epsilon}$. Let $\widehat{u} = \frac{u-v}{\|u-v\|}$, Then $\|\widehat{u}\| = 1$ and for any $x \in V$ we have
$$\|\widehat{u} - x\| = \frac{\|u - (v + \|u-v\|x)\|}{\|u-v\|} \geq \frac{\eta(1-\epsilon)}{\eta} = 1 - \epsilon. \qquad \square$$

Using this lemma we can characterize finite dimensional normed spaces.

Theorem 3.43 *A normed space X is finite dimensional if and only if $\overline{B}_1$ is compact.*

Proof $\Rightarrow$ See Remark 3.40

$\Leftarrow$ Arguing by contradiction, suppose that X is infinite dimensional. Starting with $u_0 \in X$ and using Lemma 3.42, inductively we can produce $u_n \in X$ such that $\|u_n\| = 1$ and $d(u_n, span\{u_k\}_{k=0}^{n-1}) \geq \frac{1}{2}$ for all $n \in \mathbb{N}$, $\Rightarrow \|u_n - u_m\| \geq \frac{1}{2}$ for all $n \neq m$. So, $\{u_n\}_{n \geq 1} \subseteq \overline{B}_1$ has no convergent subsequence, a contradiction. Therefore X is finite dimensional. $\qquad \square$

Remark 3.44 So, for a normed space, finite dimensionality and local compactness are equivalent notions.

An important consequence of Proposition 3.39, is the following result about linear operators.

Proposition 3.45 *If X and Y are normed spaces and X is finite dimensional then every linear operator $A : X \to Y$ is continuous.*

Proof Let $\{e_k\}_k^N = 1$ be a basis for X. Given $u \in X$, we have $u = \sum_{k=1}^{N} \lambda_k e_k$, $\lambda_k \in \mathbb{R}$. Then $A(u) = \sum_{k=1}^{N} \lambda_k A(e_k)$, hence $\|A(u)\|_Y \leq M |\{\lambda_k\}_{k=1}^{N}|_2$ with $M = (\sum_{k=1}^{N} \|A(e_k)\|_Y^2)^{\frac{1}{2}}$. Using Proposition 3.39, we have $|\{\lambda_k\}_{k=1}^{N}|_2 \leq M_0 \|u\|_X$ for some M_0. Hence $\|A(u)\|_Y \leq M M_0 \|u\|_X$ for all $u \in X$, $\Rightarrow A \in \mathcal{L}(X, Y)$. $\qquad \square$

Remark 3.46 In fact the converse of the above proposition is also true, namely if every linear operator $A : X \to Y$ is continuous, then $dim X < \infty$ (see Kubrusly [158], p.234).

As a consequence of Propositions 3.45 and of Remark 3.46, we have the following corollary.

Corollary 3.47 *A normed space X is finite dimensional if and only if every linear functional $f : X \to \mathbb{R}$ is continuous.*

In Proposition 3.35, we have seen that for linear functionals continuity is characterized using their kernel and we mentioned that in general this does not extend to linear operators (see Remark 3.36). However, we can have the following partial generalization.

Proposition 3.48 *If X, Y are normed spaces, Y is finite dimensional and $A : X \to Y$ is linear, then A is continuous if and only if $\ker A \subseteq X$ is closed.*

Proof $\Rightarrow$ Clear.
$\Leftarrow$] Let $N = \dim A(X)$ and let $\{e_k\}_{k=1}^N$ be a basis for $A(X)$. For $u \in X$, we have

$$A(u) = \sum_{k=1}^{N} c_k(u) e_k \tag{3.2}$$

with $c_k(\cdot)$ linear functionals on X. Let $x_k \in X$ such that $A(x_k) = e_k$ and let $V_N = \text{span}\{x_k\}_{k=1}^N$. We have $X = V_N \oplus \ker A$ and so $d(x, \ker A)$ is a norm on V_N (see Remark 3.21). On account of Corollary 3.41, we can find $M > 0$ such that

$$\|x\| \leq M d(x, V_N) \text{ for all } x \in V_N. \tag{3.3}$$

The linear functionals $c_k|_{V_N}$ are continuous (see Corollary 3.47). So, we can find $\eta_k > 0$ such that

$$|c_k(x)| \leq \eta_k \|x\| \text{ for all } x \in V_N. \tag{3.4}$$

If $u \in X$, then $u = x + y$ with $x \in V_N$, $y \in \ker A$. and we have $|c_k(u)| = |c_k(x)| \leq \eta_k \|x\|$ (see (3.4)) $\leq M \eta_k d(x, V_N)$ (see (3.3)) $\leq M \eta_k \|x + y\| = M \eta_k \|u\|$, $\Rightarrow c_k(\cdot)$ is continuous on X, $\Rightarrow A(\cdot)$ is continuous on X (see (3.2)). $\square$

Finally recall that $\mathcal{L}(X, Y)$ is a normed space with norm

$$\|A\|_{\mathcal{L}} = \sup\{\|A(u)\|_Y : \|u\|_X \leq 1\}, A \in \mathcal{L}(X, Y).$$

Using this definition, it is straightforward to prove the following result.

Proposition 3.49 *If X is a normed space and Y is a Banach space, then $(\mathcal{L}(X, Y), \|\cdot\|_{\mathcal{L}})$ is a Banach space.*

Remark 3.50 We write $X^* = \mathcal{L}(X, \mathcal{R})$. This is the "dual space." According to Proposition 3.49, the dual of a normed spaces is always a Banach space.

3.3   Hahn-Banach Theorem-Separation Theorems

In this section, first we prove the so-called Hahn-Banach Theorem. This is one of the fundamental results of linear functional analysis, with far reaching implications. It guarantees the existence of many continuous linear functionals, enough to develop a rich duality theory. An important part of this duality theory are the "separation theorems" for convex sets, which occupy the second half of this section.

Recall that if X is a vector space, then we say that $p : X \to \mathbb{R}$ is a "sublinear functional," if (a) $p(x + u) \leq p(x) + p(u)$ for all $x, u \in X$ (subadditive); (b) $p(\lambda x) = \lambda p(x)$ for all $\lambda \geq 0$, all $x \in X$ (positively homogeneous).

Theorem 3.51 *If X is a vector space, $V \subseteq X$ is a subspace, $p : X \to \mathbb{R}$ is a sublinear functional, $f : V \to \mathbb{R}$ is a linear functional and $f(x) \leq p(x)$ for all $x \in V$, then there exists a linear functional $\widehat{f} : X \to \mathbb{R}$ such that $\widehat{f}(x) \leq p(x)$ for all $x \in X$ and $\widehat{f}|_V = f$.*

Proof The proof uses transfinite induction. Let $x \in X \setminus V$. First we show that $f(\cdot)$ can be extended to $V + \mathbb{R}x$ and preserve the domination by $p(\cdot)$. Every element $u \in V + \mathbb{R}x$ can be written in a unique way as $u = v + \lambda x$ with $v \in V, \lambda \in \mathbb{R}$. Indeed, if $u = v_1 + \lambda_1 x = v_2 + \lambda_2 x$, then $v_1 - v_2 = (\lambda_2 - \lambda_1)x \in V$, hence $\lambda_1 = \lambda_2$ and so $v_1 = v_2$. Then let $f_1 : V + \mathbb{R}x \to \mathbb{R}$ be defined by $f_1(v + \lambda x) = f(x) + \lambda \vartheta$. Evidently $f_1(\cdot)$ is linear, $f_1|_V = f$ and now we have to choose $\vartheta \in \mathbb{R}$ so that $\widehat{f_1} \leq p$. To this end, note that for $\lambda > 0$, we have
$f_1(v \pm \lambda x) = \lambda f_1(\frac{v}{\lambda} \pm x) \leq \lambda p(\frac{v}{\lambda} \pm x) = p(v \pm \lambda v), \Rightarrow \vartheta \leq p(v + x) - f_1(x)$
and $-\vartheta \leq p(w - x) - f_1(w)$ for all $v, w \in V$, $\Rightarrow f_1(w) - p(w - x) \leq \vartheta \leq p(v + x) - f_1(v)$ for all $v, w \in V$. So, to find a desired $\vartheta \in \mathbb{R}$, we need to verity that

$$\sup_{w \in V} [f_1(w) - p(w - x)] \leq \inf_{v \in V} [p(v + x) - f_1(v)]. \qquad (3.5)$$

Note that (3.5) is equivalent to saying that for every $v, w \in V$, we have $f_1(w) - p(w - x) \leq p(v + x) - f_1(v), \Leftrightarrow f_1(v + w) \leq p(v + x) + p(w - x)$ and the last inequality is true since $f_1 \leq p$ and $p(\cdot)$ is subadditive. So, (3.5) holds and the choice of $\vartheta \in \mathbb{R}$ is possible. Now consider the family $\mathcal{P}$ of all pairs (h, Y) with Y a subspace of X such that $V \subseteq Y$ and $h : Y \to \mathbb{R}$ is a linear functional such that $h|_V = f, h \leq p$. We introduce on $\mathcal{P}$ the partial order $\preceq$, defined by

$$(h, Y) \preceq (\widehat{h}, \widehat{Y}) \text{ if and only if } Y \subseteq \widehat{Y} \text{ and } \widehat{h}|_Y = h.$$

If $\{(h_\alpha, Y_\alpha)\}_{\alpha \in J}$ is a chain in $\mathcal{P}$, then we set $Y = \bigcup_{\alpha \in J} Y_\alpha$ and $h|_{Y_\alpha} = h_\alpha$ for all $\alpha \in J$. Evidently $(h, Y) \in \mathcal{P}$ is an upper bounded for the chain and so by Zorn's lemma $\mathcal{P}$ has a maximal element $(\widehat{f}, \widehat{Y})$. By the first part of the proof $\widehat{Y} = X$. □

Corollary 3.52 *If X is a vector space, $V \subseteq X$ is a subspace, $p : X \to \mathbb{R}$ is a seminorm, $f : V \to \mathbb{R}$ is a linear functional and $|f(x)| \leq p(x)$ for all $x \in V$, then there exists a linear functional $\widehat{f} : X \to \mathbb{R}$ such that $|\widehat{f}(x)| \leq p(x)$ for all $x \in X$ and $\widehat{f}|_V = f$.*

As we already mentioned in the beginning of the section, the Hahn-Banach theorem leads to a rich duality theory for normed spaces. The next results illustrate this fact.

Definition 3.53 Let X be a normed space and X^* its dual (that is, $X^* = \mathcal{L}(X, \mathbb{R})$). The dual norm of X^* is defined by $\|u^*\|_* = \sup[|u^*(x)| : \|x\| \leq 1]$.

Proposition 3.54 *If X is a normed space, $V \subseteq X$ is a subspace and $v^* \in V^*$, then there exists $u^* \in X^*$ such that $u^*|_V = v^*$ and $\|u^*\|_{X^*} = \|v^*\|_{V^*}$.*

Proof Let $\widehat{c} = \sup[|v^*(v)| : v \in V, \|v\| \leq 1]$. Then $|v^*(v)| \leq \widehat{c}\|v\|$ for all $v \in V$. The function $p(x) = \widehat{c}\|x\|$ for all $x \in X$ is a norm on X. Hence by Corollary 3.52, we can find $u^* \in X^*$ such that $u^*|_{V^*} = v^*$ and $|u^*(x)| \leq \widehat{c}\|x\|$ for all $x \in X$. Hence $\|u^*\|_{X^*} \leq \widehat{c} = \|v^*\|_{V^*}$. On the other hand since u^* is an extension of v^*, we always have $\|v^*\|_{V^*} \leq \|u^*\|_{X^*}$. Therefore $\|u^*\|_{X^*} = \|v^*\|_{V^*}$. $\square$

Proposition 3.55 *If X is a normed space and $x \in X \setminus \{0\}$, then we can find $u^* \in X^*$, $\|u^*\|_* = 1$ such that $u^*(x) = \|x\|$.*

Proof Let $V = \mathbb{R}x$ and consider $v^* \in V^*$ defined by $v^*(\lambda x) = \lambda\|x\|$ for all $\lambda \in \mathbb{R}$. Then $\|v^*\|_{V^*} = 1$ and $v^*(x) = \|x\|$. Invoking Proposition 3.54, we can find $u^* \in X^*$ such that $u^*|_V = v^*$ and $\|u^*\|_{X^*} = \|v^*\|_{V^*} = 1$. $\square$

Corollary 3.56 *If X is a normed space, then $\|x\| = \sup[|u^*(x)| : u^* \in X^*, \|u^*\|_* \leq 1]$.*

Remark 3.57 The formula in this corollary is dual to the one in Definition 3.53. However, we point out that for $\| \cdot \|_*$ is a norm, while for $\| \cdot \|$ is a consequence of the Hahn-Banach theorem.

This is the right time to introduce the "canonical embedding." If X is a normed space, set $X^{**} = (X^*)^*$. We furnish X^{**} with the dual norm (that is $\|f\|_{X^{**}} = \sup[|f(u^*)| : \|u^*\|_{X^*} \leq 1]$, see Definition 3.53).

Definition 3.58 Let X be a normed apace. The "canonical embedding" is the linear map $\widehat{i} : X \to X^{**}$ defined by $\widehat{i}(x)(u^*) = u^*(x)$ for all $x \in X, u^* \in X^*$. It is easy to see that $\widehat{i}(\cdot)$ is an isometry.

Remark 3.59 Using the canonical embedding, we can view X as a subspace of X^{**}, and as such we can think of its elements as continuous linear functionals on X^*. For this reason, in the sequel given $x \in X, u^* \in X^*$ instead of $u^*(x)$, we write $\langle u^*, x \rangle$ which permits us to think that x acts on u^*. We call $\langle \cdot, \cdot \rangle$ the duality brackets for the dual pair (X, X^*). Note that the completion $\widehat{X}$ of X is given by $\widehat{X} = \overline{\widehat{i}(X)}^{\|\cdot\|_{X^{**}}}$.

Proposition 3.60 *If X is a Banach space, $V \subseteq X$ a closed subspace and $x_0 \in X \backslash V$, then there exists $u^* \in X^*$ such that $\|u^*\|_* = 1$, $u^*|_V = 0$ and $\langle u^*, x_0 \rangle = d(x_0, V)$.*

Proof Let $\vartheta = d(x_0, V) > 0$ and $Y = span\{V, x_0\}$. We introduce that linear functional $y^* : Y \to \mathbb{R}$ defined by $y^*(v + \lambda x_0) = \lambda \vartheta$ for all $v \in V, \lambda \in \mathbb{R}$. Then $y^*|_V = 0$. If $y = v + \lambda x_0 \in Y$, then

$$\langle y^*, y \rangle = \lambda \vartheta = \frac{\lambda \|y\|}{\|y\|} \vartheta = \frac{\lambda \|y\|}{\|v + \lambda x_0\|} \vartheta = \frac{\|y\|}{\|\frac{v}{\lambda} + x_0\|} \vartheta \leq \frac{\|y\| \vartheta}{d(x_0, V)} = \|y\|$$

hence $\|y^*\|_{Y^*} \leq 1$. Let $\{v_n\}_{n \in \mathbb{N}} \subseteq V$ such that $\|v_n - x_0\| \downarrow d(x_0, v)$. We have $\vartheta = |\langle y^*, v_n - x_0 \rangle| \leq \|y^*\|_{Y^*} \|v_n - x_0\|$, $\Rightarrow \vartheta \leq \|y^*\|_{Y^*} \vartheta$ and so $\|y^*\|_{Y^*} = 1$, $\langle y^*, x_0 \rangle = \vartheta$. Apply Proposition 3.54, to produce $u^* \in X^*$ such that $u^*|_V = 0$ and $\|u^*\|_{x^*} = \vartheta$ $\square$

Definition 3.61 Let X be a normed space and $C \subseteq X$. The "annihilator of C" is the set $C^\perp = \{x^* \in X^* : \langle x^*, x \rangle = 0 \text{ for all } x \in C\}$.

Remark 3.62 Evidently $C^\perp$ is a closed subspace of X^*.

Proposition 3.63 *If X is a normed space and $V \subseteq X$ is a subspace, then (a) V^* is isometrically isomorphic to $X^*/V^\perp$. (b) If V is closed, then $(X/V)^*$ is isometrically isomorphic to $V^\perp$.*

Proof

(a) Let $v^* \in V^*$. By Proposition 3.54 there exists $u^* \in X^*$ such that $u^*|_V = v^*$ and $\|u^*\|_{X^*} = \|v^*\|_{V^*}$. Then we consider the map $\xi : V^* \to X^*/V^\perp$ defined by $\xi(u^*) = u^* + V^\perp$. This map is well defined since if $\widehat{u}^* \in X^*$ is another extension of v^*, then $u^* - \widehat{u}^* \in V^\perp$. Also $\xi(\cdot)$ is linear. Each element of $u^* + V^\perp$ is an extension of v^*, hence

$$\|v^*\|_{V^*} \leq \inf\{\|u^* + x^*\|_{X^*} : x^* \in V^\perp\} = \|u^* + V^\perp\|_{X^*/V^\perp} = \|\xi(v^*)\|_{X^*/V^\perp}$$

(see Definition 3.20). Also we have

$$\|v^*\|_{V^*} = \|u^*\|_{X^*} \geq \inf[\|u^* + x^*\|_{X^*} : x^* \in V^\perp]$$

$$= \|u^* + V^\perp\|_{X^*/V^\perp} = \|\xi(v^*)\|_{X^*/V^\perp}.$$

Therefore ξ is an isometric isomorphism.

(b) Consider the map $\widehat{\xi} : V^\perp \to (X/V)^*$ defined by $\widehat{\xi}(x^*)(\widehat{x}) = \langle x^*, x \rangle$. Since $\langle x^*, y \rangle = \langle x^*, z \rangle$ for all $y, z \in \widehat{x} = x + V$, we see that $\widehat{\xi}(\cdot)$ is well defined. Also it is linear. Let $y^* \in (X/V)^*$ and define $x^* \in X^*$ by $\langle x^*, x \rangle = y^*(\widehat{x})$. Then $x^* \in V^\perp$ and $\widehat{\xi}(x^*)(\widehat{x}) = \langle x^*, x \rangle = y^*(\widehat{x})$. So $\widehat{\xi}(\cdot)$ is surjective. Also we have

$$\|\widehat{\xi}(x^*)\|_{(X/V)^*} = \sup[|\xi(x^*)(\widehat{x}) : \|\widehat{x}\|_{X/V} \le 1]$$

$$= \sup[|\langle x^*, x \rangle| : \|x\| \le 1]$$

$$= \|x^*\|_{X^*},$$

thus $\widehat{\xi}$ is an isometric isomorphism.   □

Next we will present the geometric interpretations of the Hahn-Banach theorem, which are the separation theorems for convex sets. We start with some simple but useful facts about convex sets.

Proposition 3.64 *If X is a topological vector space and $C \subseteq X$ is a convex set then (a) $0 < \lambda \le 1 \Rightarrow \lambda \operatorname{int} C + (1 - \lambda)\overline{C} \subseteq \operatorname{int} C$; (b) $\overline{\operatorname{int} C} = \overline{C}$; (c) $\operatorname{int} \overline{C} = \operatorname{int} C$.*

Proof

(a) Evidently we may assume $0 < \lambda < 1$. Let $x \in \operatorname{int} C$, $u \in \overline{C}$. We can find $U \in \mathcal{N}(0)$ open such that $x + U \subseteq C$. Then $u - \frac{\lambda}{1-\lambda}U \in \mathcal{N}(u)$ and so we can find $y \in C \cap [u - \frac{\lambda}{1-\lambda}U]$ and then $(1 - \lambda)(u - y) \in \lambda U$. Let $D = \lambda[x + U] + (1 - \lambda)y \subseteq C$ (since $x + U \subseteq C$, $y \in C$ and C is convex). Also we have $\lambda x + (1-\lambda)u = \lambda x + (1-\lambda)(u-y) + (1-\lambda)y \in \lambda x + \lambda U + (1-\lambda)y = D \subseteq C$.

(b) In (a) we let $\lambda \to 0^+$ and obtain $\overline{C} = \overline{\operatorname{int} C}$.

(c) Let $x \in \operatorname{int} C$ and $u \in \operatorname{int} \overline{C}$. Choose $U \in \mathcal{N}(0)$ such that $u + U \subseteq \overline{C}$. Then we can find $\lambda \in (0, 1)$ such that $\lambda(u - x) \in U$, hence $u + \lambda(u - x) \in \overline{C}$. Moreover, on account at (a) we have $u - \lambda(u - x) = \lambda x + (1 - \lambda)u \in \operatorname{int} C$. Now note that $u = \frac{1}{2}[u - \lambda(u - x)] + \frac{1}{2}[u + \lambda(u - x)] \in \frac{1}{2} \operatorname{int} C + \frac{1}{2}\overline{C} \subseteq \operatorname{int} C_+$ (again by (a)), $\Rightarrow \operatorname{int} \overline{C} = \operatorname{int} C$.   □

Definition 3.65 Let X be a vector space and $A \subseteq X$. The "convex hull" of A, denoted by $\operatorname{conv} A$, is the set of all convex combinations of elements in A, that is,

$$\operatorname{conv} A = \{u = \sum_{k=1}^{n} \lambda_k x_k : x_k \in A, \lambda_k \ge 0, \sum_{k=1}^{n} \lambda_k = 1\}.$$

Remark 3.66 The convex hull of A is the smallest convex set containing A. Hence it is the intersection of all convex sets which contain A. If X is a topological vector space, then the "closed convex hull" of A, denoted by $\overline{\operatorname{conv}} A$, is the smallest closed convex set which contains A. We have $\overline{\operatorname{conv}} A = \overline{\operatorname{conv} A}$.

The next result is known in the literature as "Caratheodory's Convexity Theorem."

Theorem 3.67 *If X is an N-dimensional vector space, $A \subseteq X$ and $x \in \operatorname{conv} A$, then x is a convex combination of at most $N + 1$ points from A.*

Proof Let $D = \{m \in \mathbb{N} : x$ is a convex combination of m vectors in $A\}$ Let $\vartheta = \min D$. We claim that $\vartheta \le N + 1$. Arguing by contradiction, suppose that $\vartheta > N + 1$. Let $\{x_k\}_{k=1}^{\vartheta} \subseteq A$ and $\{\lambda_k\}_{k=1}^{\vartheta} \subseteq \mathbb{R}_+$ with $\sum_{k=1}^{\vartheta} \lambda_k = 1$ such that

$x = \sum_{k=1}^{\vartheta} \lambda_k x_k$. Since $\vartheta - 1 > N$, then the vectors $\{x_k - x_1\}_{k=2}^{\vartheta}$ are linearly dependent.

So, we can find $\{\mu_k\}_{k=2}^{\vartheta} \subseteq \mathbb{R}$ not all zero such that $\sum_{k=2}^{\vartheta} \mu_k(x_k - x_1) = 0$. Set

$\beta_1 = -\sum_{k=2}^{\vartheta} \mu_k$ and $\beta_k = \mu_k$ for $k = 2, \ldots, \vartheta$. Then not all the β_k's are zero and we have

$$\sum_{k=1}^{\vartheta} \beta_k x_k = 0 \text{ and } \sum_{k=1}^{\vartheta} \beta_k = 0. \tag{3.6}$$

Let $m \in \{1, \ldots, \vartheta\}$ such that $|\frac{\lambda_k}{\beta_k}| \leq |\frac{\lambda_m}{\beta_m}|$ for all $k \in \{1, \ldots, \vartheta\}$ and define $\gamma_k = \lambda_k - \frac{\beta_k \lambda_m}{\beta_m}$ for $k \in \{1, \ldots, \vartheta\}$. Then $\gamma_k \geq 0$, $\sum_{k=1}^{\vartheta} \gamma_k = \sum_{k=1}^{\vartheta} \lambda_k = 1$. (see (3.6))

$x = \sum_{k=1}^{\vartheta} \gamma_k x_k$ and $\gamma_m = 0$. This contradicts the definition of ϑ. $\square$

Corollary 3.68 *If X is an N-dimensional vector space and $A \subseteq X$ is compact, then $\text{conv}A$ is compact too.*

As for normed spaces, given X a topological vector space, by X^* we denote the vector space of all continuous linear functionals $f : X \to \mathbb{R}$. Now we present the two main separation theorems for convex sets.

Theorem 3.69 *If X is a topological vector space, then (a) given $A \subseteq X$ nonempty, open convex and $x_0 \in X \setminus A$, we can find $\widehat{x}^* \in X^*$ such that $\widehat{x}^*(a) < \widehat{x}^*(x_0)$ for all $\alpha \in A$; (b) If X is locally convex, $C \subseteq X$ is nonempty, closed, convex $x_0 \in X \setminus C$, we can find $\widehat{x}^* \in X^*$ such that $\sup[\widehat{x}^*(c) : c \in C] < \widehat{x}^*(x_0)$.*

Proof

(a) Shifting A and x_0 if necessary, we may assume that $0 \in A$. Then by Propositions 3.14 and 3.15, the Minkowski functional $p_A(\cdot)$ is sublinear, continuous and $A = \{x \in X : p_A(x) < 1\}$. Let $x^* : \mathbb{R}x_0 \to \mathbb{R}$ be defined by $x^*(\lambda x_0) = \lambda p_A(x_0)$ for all $\lambda \in \mathbb{R}$, then $x^*(\lambda x_0) \leq p_A(\lambda x_0)$ for all $\lambda \in \mathbb{R}$. By the Hahn-Banach Theorem (see Theorem 3.51), we can find $\widehat{x}^*$ a linear extension of x^* on X and $\widehat{x}^* \leq p_A$. The continuity of $p_A(\cdot)$ implies that $\widehat{x}^* \in X^*$ and we have $\widehat{x}^*(a) \leq p_A(a) < 1 \leq p_A(x_0) = \widehat{x}^*(x_0)$ for all $a \in A$.

(b) Suppose that X is locally convex. Let $U \in \mathcal{N}(0)$ and we may assume that U is open, convex, balanced and $[x_0 + U] \cap C = \emptyset$. Note that $C + U \subseteq X$ is open (see Proposition 3.64(a)) and $x_0 \notin C + U$. So, we can apply part (a) and find $\widehat{x}^* \in X^*$ such that $\widehat{x}^*(c + u) < \widehat{x}^8(x_0)$ for all $c \in C$ and all $u \in U$. Then $\sup[\widehat{x}^*(c) : c \in C] < x^*(x_0) - x^*(u)$ for all $u \in U$. Since $U \in \mathcal{N}(0)$ is balanced choose $u \in U$ such that $\widehat{x}^*(u) > 0$. Then $\sup[\widehat{x}^*(c) : c \in C] < \widehat{x}^*(x_0)$. $\square$

This theorem guarantees that X^* is rich for a locally convex space X.

Corollary 3.70 *If X is locally convex and $x,, u \in X$ with $x \neq u$, then we can find $\widehat{x}^* \in X^*$ such that $\widehat{x}^*(x) \neq \widehat{x}^*(u)$.*

The second separation theorem is the following one:

Theorem 3.71 *If X is a topological vector space and $A, C \subseteq X$ are convex sets such that $A \cap C = \emptyset$, then (a) when A is open we can find $\widehat{x}^* \in X^*$ such that $\widehat{x}^*(a) < \inf[\widehat{x}^*(c) : c \in C]$ for all $a \in A$; (b) when X is locally convex, A is closed and C is compact, then we can find $\widehat{x} \in X^*$ and $\vartheta \in \mathbb{R}$ such that $\sup[\widehat{x}^*(a) : a \in A] < \vartheta < \inf[\widehat{x}^*(c) : c \in C]$.*

Proof

(a) The set $A - C$ is open convex and $0 \notin A - C$. So, by Theorem 3.69(a) we can find $\widehat{x}^* \in X^*$ such that $\widehat{x}^*(a - c) < \widehat{x}^*(0) = 0$ for all $a \in A, c \in C$. Hence we have that $\widehat{x}^*(a) \leq \inf[\widehat{x}^*(c) : c \in C] = m_C$ for all $a \in A$. Suppose that we can find $a_0 \in A$ such that $\widehat{x}^*(a_0) = m_C$. Then since A is open and $a_0 \in A$, we can find $a \in A$ such that $\widehat{x}^*(a) > m_C$, a contradiction. Therefore $\widehat{x}^*(a) < \inf[\widehat{x}^*(c) : c \in C]$ for all $a \in A$.

(b) If X is locally convex, then given $c \in C$, we can find $U_c \in \mathcal{N}(0)$ open convex, balanced such that $(c + U_c) \cap A = \emptyset$. Then the family $\{c + U_c\}_{c \in C}$ is an open cover of C. So, by compactness we can find $F \subseteq C$ finite such that $C \subseteq \bigcup_{c \in F} (c + U_c)$. Let $U = \bigcap_{c \in F} U_c$. Then $U \in \mathcal{N}(0)$ is open, convex, balanced and $(A + U) \cap C = \emptyset$. We apply Theorem 3.69(b) on the sets $A + U$ and C and produce $\widehat{x}^* \in X^*$ such that $\widehat{x}^*(a + u) < \min[\widehat{x}^*(c) : c \in C]$. Again it suffices to choose $u \in U$ with $\widehat{x}^*(u) > 0$.   □

Remark 3.72 Usually we call Theorem 3.71(a) "Weak Separation Theorem" and Theorem 3.71(b) "Strong Separation Theorem."

Corollary 3.73 *If X is a locally convex space, $C \subseteq X$ is nonempty closed convex and $x \notin C$, then we can find $\widehat{x}^* \in X^*$ such that $\sup[\widehat{x}^*(c) : c \in C] < \widehat{x}^*(x)$.*

Corollary 3.74 *If X is a locally convex space and $V \subseteq X$ is a subspace, then V is not dense in X if and only if we can find $\widehat{x}^* \in X^* \setminus \{0\}$ such that $\widehat{x}^*|_V = 0$.*

Given $\widehat{x}^* \in X^*$ and $\lambda \in \mathbb{R}$, the sets $L_\lambda = \{x \in X : \widehat{x}^*(x) \leq \lambda\}$ and $U_\lambda = \{x \in X : \widehat{x}^*(x) \geq \lambda\}$ are said to be "closed half spaces."

Corollary 3.75 *If X is a locally convex space and $C \subseteq X$ is a proper, closed, convex set, then C is the intersection of all closed half spaces which contain C.*

Definition 3.76 (a) Let X be a vector space and $P : X \to X$ a linear operator. We say that P is a "projection" onto a subspace $V \subseteq X$, if $P(X) = V$ and $P|_V = id$ (that is, $P^2 = P$). (b) Let X be a vector space. We say that X is the "direct sum" of two subspaces V and Y, if every $x \in X$ has a unique decomposition of the form $x = v + y$ with $v \in V$, $y \in Y$. We write $X = V \bigoplus Y$. Let X be a Banach space, and $V \subseteq X$ a subspace. We say that V is "complemented" in X, if we can find a continuous projection of X onto V. (c) Let X be a Banach space and $V \subseteq X$ a closed subspace. We say that Y is a "topological complement" of V, if $Y \subseteq X$ is a closed subspace and $X = V \bigoplus Y$. (d) Let X be a Banach space and $V \subseteq X$ a subspace. The "codimension" of V is $\dim(X/V)$.

Proposition 3.77 *If X is a Banach space, $V \subseteq X$ is a subspace and V is either finite dimensional or it is closed and has finite codimension, then V admits a topological complement.*

Proof First we assume that V is finite dimensional. Let $\{e_k\}_{k=1}^{N}$ be a basis for V. Then for every $v \in V$ we have $v = \sum\limits_{k=1}^{N} \lambda_k e_k$. Let $p_k(v) = \lambda_k$. Evidently $p_k(\cdot)$ is linear continuous. So, we can find $\widehat{p}_k \in X^*$ such that $\widehat{p}_k|_V = p_k$. Let $Y = \bigcap_{k=1}^{N} \widehat{p}_k^{-1}(0)$. Then this is the desired topological complement of V. Next assume that V is closed and has finite codimension. Choose any finite dimensional subspace Y such that $V \cap Y = \{0\}$, $X = V + Y$. Note that Y is closed being finite dimensional. $\qquad \square$

Remark 3.78 A subspace of finite codimension, need not be closed. For closed subspaces, the properties of being complemented (see Definition 3.76(c)) and of having a topological complement (see Definition 3.76(d)), are equivalent.

3.4 Three Basic Theorems

In this section we present the three fundamental results of linear functional analysis, which are closely related to each other and are all essentially variants of the Baire Category Theorem (see Sect. 1.7). These results are the "Open Mapping Theorem," the "Closed Graph Theorem" and the "Uniform Boundedness Principle" (or "Banach-Steinhaus Theorem"). We start with the Open Mapping Theorem. To prove it we will need the following two Lemmata.

Lemma 3.79 *If X, Y are Banach spaces and $A : X \to Y$ is linear and surjective, then $\overline{A(B_1^X)} \in \mathcal{N}(0)$ with $B_1^X = \{x \in X : \|x\|_X < 1\}$.*

Proof The surjectivity of A implies that $Y = \bigcup\limits_{n \in \mathbb{N}} nA(B_1^X)$. According to Proposition 1.173, we can find $n_0 \in \mathbb{N}$ such that $\operatorname{int} \overline{n_0 A(B_1^X)} \neq \emptyset$. So, we can find $u_0 \in n_0 A(B_1^X)$ such that $u_0 \in \operatorname{int} \overline{A(B_1^X)}$. Recall that translations are homeomorphisms (see Remark 3.2). Hence $0 \in \operatorname{int}[\overline{n_0 A(B_1^X)} - u_0]$. Let $x_0 \in n_0 B_1^X$ such that

$u_0 = A(x_0)$. Hence $n_0 A(B_1^X) - u_0 = A(n_0 B_1^X - x_0)$. For any $x \in n_0 B_1^X$ we have $\|x - x_0\| < 2n_0$ and so $n_0 A(B_1^X) - u_0 \subseteq 2n_0 A(B_1^X)$. Exploiting once again the fact that translations are homeomorphisms, we have $\overline{n_0 A(B_1^X)} - u_0 = \overline{n_0 A(B_1^X) - u_0}$. Therefore $0 \in \text{int}[n_0 \overline{A(B_1^X)} - u_0] \subseteq \text{int } 2n_0 \overline{A(B_1^X)}$. Scalar multiplication is also a homeomorphism and so we conclude that $0 \in \text{int } \overline{A(B_1^X)}$, which means that $\overline{A(B_1^X)} \in \mathcal{N}(0)$ (= filter of neighborhoods of the origin, see Definition 1.11). $\quad\square$

Remark 3.80 We point out that in the above lemma the only restriction on the linear map A is its surjectivity. No continuity is required. In case A is also continuous, we can improve the result.

Lemma 3.81 *If X, Y are Banach spaces, $A \in \mathcal{L}(X, Y)$ and $\overline{A(B_1^X)} \in \mathcal{N}(0)$, then $A(B_1^X) \in \mathcal{N}(0)$.*

Proof We have $0 \in \text{int } \overline{A(B_1^X)}$ and so we can find $\delta > 0$ such that $\delta B_1^Y \subseteq \overline{A(B_1^X)}$ (here $B_1^Y = \{y \in Y : \|y\|_Y < 1\}$). Let $y \in \delta B_1^Y$. Then we can find $x_1 \in B_1^X$ with $y_1 = A(x_1)$ and $\|y - y_1\|_Y < \frac{\delta}{2}$. We have $\frac{\delta}{2} B_1^Y \subseteq \overline{A(\frac{1}{2} B_1^X)}$ and so as before we can find $x_2 \in B_1^X$ with $y_2 = A(x_2)$ and $\|(y - y_1) - y_2\|_Y < \frac{\delta}{2}$. Inductively, we generate a sequence $\{x_n\}_{n \in \mathbb{N}} \subseteq X$ such that $x_n \in \frac{1}{2^{n-1}} B_1^X$ with $y_n = A(x_n)$ and $\|y - \sum_{k=1}^n y_k\|_Y < \frac{\delta}{2^n}$. Let $v_n = \sum_{k=1}^n x_k$ $n \in \mathbb{N}$. We have

$$\|v_n - v_m\| \leq \sum_{k=m+1}^n \|x_k\|_X < \frac{1}{2^{m-1}} \text{ for all } n > m,$$

hence $\{v_n\}_{n \in \mathbb{N}} \subseteq X$ is Cauchy and $\|v_n\|_X < 2$ for all $n \in \mathbb{N}$, thus,

$$v_n \to v \in X \quad \text{and} \quad \|v\|_X \leq 2. \tag{3.7}$$

Also we have

$$\|A(v) - y\|_Y \leq \|A(v) - A(v_n)\|_Y + \|A(v_n) - \sum_{k+1}^n y_k\|_Y + \|\sum_{k=1}^n y_k - y\|_Y,$$

hence $A(v) = y$ (see (3.7) and recall $A \in \mathcal{L}(X, Y)$), $\Rightarrow y \in 3A(B_1^X)$. Therefore $\delta B_1^Y \subseteq 3A(B_1^X)$ and so $\frac{\delta}{3} B_1^Y \subseteq A(B_1^X)$, hence $A(B_1^X) \in \mathcal{N}(0)$. $\quad\square$

We can now prove the "Open Mapping Theorem."

Theorem 3.82 *If X, Y are Banach spaces and $A \in \mathcal{L}(X, Y)$ is surjective then A is an open map.*

Proof Let $U \subseteq X$ be open and $y \in A(U)$. We can find $x \in U$ such that $y = A(x)$. Let $\lambda > 0$ small such that $x + \lambda B_1^X \subseteq U$. From Lemmata 3.80 and 3.81 it follows that for some $\delta > 0$ we have $\delta B_1^Y \subseteq A(B_1^X)$. Then $y + \delta B_1^Y \subseteq y + A(B_1^X) = A(x + B_1^X) \subseteq A(U)$. Therefore $A(U) \subseteq Y$ is open. $\quad\square$

Corollary 3.83 *If X, Y are Banach spaces and $A \in \mathcal{L}(X, Y)$ is 1–1 and onto then A is an isomorphism.*

Proof Evidently $A^{-1} : Y \to X$ exists and it is linear. Also, from Theorem 3.82. $A(\cdot)$ is open and so $A^{-1} \in \mathcal{L}(Y, X)$. Therefore A is an isomorphism. $\quad\square$

Remark 3.84 If in the above corollary $A \in \mathcal{L}(X, Y)$ is only surjective (that is not 1–1), then Y is isomorphic to $X/\ker A$.

Corollary 3.85 *If X is a vector space, $\| \cdot \|$ and $| \cdot |$ are two norms on X such that $(X, \| \cdot \|)$, $(X, | \cdot |)$ are Banach spaces and there exists $c > 0$ such that $\|x\| \leq c|x|$ for all $x \in X$ then $\| \cdot \|$ and $| \cdot |$ are equivalent norms on X.*

Proof Consider the identity operator id: $(X, | \cdot |) \to (X, \| \cdot \|)$. By hypothesis this operator is continuous. Apply Corollary 3.83 to infer that id is an isomorphism. Therefore the two norms are equivalent. $\quad\square$

The second basic result of Linear Functional Analysis is the so-called Closed Graph Theorem. It provides a way to decide whether a linear operator between two Banach spaces is continuous. Its proof is in fact a simple application of the previous corollary.

Theorem 3.86 *If X, Y are Banach spaces and $A : X \to Y$ is a linear operator, then $A \in \mathcal{L}(X, Y)$ if and only if $\operatorname{Gr} A \subseteq X \times Y$ is closed.*

Proof $\Rightarrow$ Graph of any continuous map between topological spaces is closed.

$\Leftarrow$ On X we consider a new norm $| \cdot |$ given by $|x| = \|x\|_X + \|A(x)\|_Y$ for all $x \in X$. Evidently A is continuous from $(X, | \cdot |)$ into Y. Also we have $\|x\|_X \leq \|x\|_X + \|A(x)\|_Y = |x|$ for all $x \in X$. To apply Corollary 3.85, it remains to show that $(X, | \cdot |)$ is a Banach space. So, let $\{x_n\}_{n \in \mathbb{N}} \subseteq X$ be $| \cdot |$-Cauchy. For $n, m \in \mathbb{N}$ we have $|x_n - x_m| = \|x_n - x_m\|_X + \|A(x_n) - A(x_m)\|_Y$, $\Rightarrow \{x_n\}_{n \in \mathbb{N}} \subseteq X$ is $\| \cdot \|_X$-Cauchy and $\{A(x_n)\}_{n \in \mathbb{N}} \subseteq Y$ is $\| \cdot \|_Y$-Cauchy. Since $(X, \| \cdot \|_X)$ and $(Y, \| \cdot \|_Y)$ are Banach spaces, we have

$$x_n \xrightarrow{\| \cdot \|_X} x \in X \text{ and } A(x_n) \xrightarrow{\| \cdot \|_Y} y \in Y.$$

By hypothesis $\operatorname{Gr} A$ is closed. Hence $(x, y) \in \operatorname{Gr} A$ and so $y = A(x)$. Then $|x_n - x| = \|x_n - x\|_X + \|A(x_n) - A(x)\|_Y \to 0$ as $n \to \infty$, $\Rightarrow (X, | \cdot |)$ is Banach, $\Rightarrow | \cdot |$ and $\| \cdot \|_X$ are equivalent norms (see Corollary 3.85), $\Rightarrow A \in \mathcal{L}(X, Y)$. $\quad\square$

Remark 3.87 Combining Theorems 3.82 and 3.86, we can state the following useful facts. In what follows X and Y are Banach spaces. (a) If $A : X \to Y$ is linear, surjective and Gr $A \subseteq X \times Y$ is closed, then A is an open map. (b) If $A : X \to Y$ is linear, bijective and Gr $A \subseteq X \times Y$ is closed, then A is a homeomorphism.

In Problem 1.26 we saw a situation where the pointwise limit of a sequence of continuous functions, is a continuous function (Dini's Theorem). The "Uniform Boundedness Principle" (or "Banach-Steinhaus Theorem"), is a result in this spirit.

Definition 3.88 Let X, Y be normed spaces and $\mathscr{S} \subseteq \mathcal{L}(X, Y)$. (a) We say that $\mathscr{S}$ is "pointwise bounded," if for all $x \in X$ the set $\{A(x) : A \in \mathscr{S}\} \subseteq Y$ is bounded. (b) We say that $\mathscr{S}$ is "uniformly bounded," if $\mathscr{S}$ is bounded in the normed space $\mathcal{L}(X, Y)$, that is $\sup\{\|A\|_{\mathcal{L}} : A \in \mathscr{S}\} < \infty$.

Remark 3.89 Evidently uniform boundedness implies pointwise boundedness. The converse is not in general true. To see this consider the vector space ℓ^{∞} of all bounded real sequences $\widehat{x} = (x_n)_{n \in \mathbb{N}}$ with norm

$$\|\widehat{x}\|_{\infty} = \sup\{|x_n| : n \in \mathbb{N}\} < \infty.$$

Then ℓ^{∞} is a Banach space. We consider the subspace $c_{00} \subseteq \ell^{\infty}$ consisting of all real sequences which have only a finite number of nonzero elements. Then $(c_{00}, \| \cdot \|_{\infty})$ is not complete. Let $\{f_n\}_{n \in \mathbb{N}} \subseteq c_{00}^*$ be defined by $f_n(\widehat{x}) = nx_n \, n \in \mathbb{N}$. Then for any $\widehat{x}^0 \in c_{00}$ we have $|f_n(\widehat{x}^0)| \leq n_0 \|\widehat{x}^0\|_{\infty}$ for some $n_0 \in \mathbb{N}$ ($x_n^0 = 0$ for all $n > n_0$). So $\{f_n\}_{n \in \mathbb{N}}$ is pointwise bounded. But $\|f_n\|_* = n$ for all $n \in \mathbb{N}$ and so $\{f_n\}_{n \in \mathbb{N}}$ is not uniformly bounded.

The next result, known as the "Uniform Boundedness Principle" (or "Banach-Steinhaus Theorem"), says that if X is complete (a Banach space), then pointwise boundedness implies uniform boundedness.

Theorem 3.90 *If X is a Banach spaces, Y is normed space and $\mathscr{S} \subseteq \mathcal{L}(X, Y)$ is pointwise bounded, then $\mathscr{S}$ is uniformly bounded.*

Proof For every $n \in \mathbb{N}$, define

$$C_n = \{x \in X : \|A(x)\|_Y \leq n \text{ for all } A \in \mathscr{S}\}.$$

which is closed as the intersection of the closed sets $A^{-1}(\overline{B})_n^Y)(\overline{B}_n^Y = \{y \in Y : \|y\|_Y \leq n\})$. Moreover, since $\mathscr{S}$ is by hypothesis pointwise bounded, we have $X = \bigcup_{n \in \mathbb{N}} C_n$. Then by Proposition 1.173, there exists $n_0 \in \mathbb{N}$ such that int $C_{n_0} \neq \emptyset$. Hence we can find $x_0 \in X$ and $\varrho > 0$ such that $x_0 + \overline{B}_{\varrho}^X \subseteq C_{n_0}$ ($\overline{B}_{\varrho}^X = \{x \in X : \|x\|_X \leq \varrho\}$). Then

$$\|A(x_0 + \overline{B}_{\varrho})\|_Y \leq n_0 \text{ for all } A \in \mathscr{S}. \tag{3.8}$$

On account of the linearity of $A \in \mathscr{S}$ we have $A(\overline{B}_{\varrho}) = A(x_0 + \overline{B}_{\varrho}) - A(x_0)$ and so from (3.8) we have $\|A(x)\|_Y \leq 2n_0$ for all $x \in \overline{B}_{\varrho}$, all $A \in \mathscr{S}$, thus,

$$\|A(x)\|_Y \le \frac{2n_0}{\varrho} \text{ for all } x \in \overline{B}_1 \text{ and all } A \in \mathcal{L},$$

hence $\|A\|_{\mathcal{L}} \le \frac{2n_0}{\varrho}$ for all $A \in \mathcal{L}$. □

An important consequence of this theorem is the following result.

Proposition 3.91 *If X, Y are Banach spaces, $\{A_n\}_{n \in \mathbb{N}} \subseteq \mathcal{L}(X, Y)$ and for every $x \in X$ we have $A_n(x) \to A(x)$ in Y (pointwise convergence), then $A \in \mathcal{L}(X, Y)$ and $\|A\|_{\mathcal{L}} \le \liminf\limits_{n \to \infty} \|A_n\|_{\mathcal{L}}$.*

Proof Evidently $A : X \to Y$ is linear. Since for every $x \in X$, $A_n(x) \to A(x)$ in Y, it follows that $\{\|A_n(x)\|_Y\}_{n \in \mathbb{N}} \subseteq Y$ is bounded for every $x \in X$. Then Theorem 3.90 implies that $\{A_n\}_{n \in \mathbb{N}} \subseteq \mathcal{L}(X, Y)$ is bounded. Let $\eta = \liminf\limits_{n \to \infty} \|A_n\|_{\mathcal{L}}$. Then for every $x \in X$, we have

$$\|A(x)\|_Y = \lim_{n \to \infty} \|A_n(x)\|_Y \le \|x\| \liminf_{n \to \infty} \|A_n\|_{\mathcal{L}} = \eta \|x\|,$$

hence $A \in \mathcal{L}(X, Y)$ and $\|A\|_{\mathcal{L}} \le \eta$. □

3.5 Weak and Weak* Topologies

In this section we return to the notion of weak topology introduced in Definition 1.57 and examine it in the context of normed spaces and linear functionals. For finite dimensional spaces the resulting topology coincides with the norm (strong) topology. However, for infinite dimensional spaces, this new topology is distinct (strictly coarser) than the norm topology and important properties which fail for the richer norm topology, are true for the weak topology.

Definition 3.92 Let X be a normed space and X^* its dual. For every $x^* \in X^*$, let $f_{x^*} : X \to \mathbb{R}$ be the linear functional defined by $f_{x^*}(x) = \langle x^*, x \rangle$ for all $x \in X$. The "weak topology" $w = w(X, X^*)$ on X is the weakest (coarsest) topology on X for which all the functionals $\{f_{x^*}\}_{x^* \in X^*}$ are continuous. So, in the notation of Definition 1.57, we have

$$w = w(\{f_{x^*}\}_{x^* \in X^*}).$$

Then at every $x_0 \in X$ a local basis for the weak topology consists of the sets

$$U = \left\{ x \in X : |f_{x_k^*}(x - x_0)| < \epsilon, \ k = 1, \cdots, n \right\},$$

for all $\epsilon > 0$ and all finite choices $x_1^*, \ldots, x_n^* \in X^*$. Of course the weak topology is coarser than norm (strong) topology. By X_w we denote the normed space X furnished with the weak topology. The space X_w is completely regular.

When we are in a dual space X^* (which is a Banach space), we can also have the weak* topology, since we can consider topologies τ on X^* such that $(X_\tau^*)^* = X$.

Definition 3.93 Let X be a normed space and X^* its dual. For every $x \in X$, let $f_x : X^* \to \mathbb{R}$ be the linear functional defined by $f_x(x^*) = \langle x^*, x \rangle$ for all $x^* \in X^*$. The "weak* topology" $w^* = w(X^*, X)$ on X^*, is the weakest (coarsest) topology on X^* for which all the functionals $\{f_x\}_{x \in X}$ are continuous. So, in the notation of Definition 1.57, we have $w^* = w(\{f_x\}_{x \in X})$. Then at every $x_0^* \in X^*$ a local basis for the weak* topology, consists of the sets $U = \{x^* \in X^* : |f_{x_k}(x^* - x_0^*)| < \epsilon, k = 1, \ldots, n\}$, for all $\epsilon > 0$ and all finite sets $x_1, \ldots, x_n \in X$. On X^* we can also consider the weak topology w for which we have $(X_w^*)^* = X^{**}$. Evidently, we have $w^* \subseteq w \subseteq s$ (= the norm (strong) topology). By $X_{w^*}^*$ we denote the space X^* furnished with weak* topology. The space $X_{w^*}^*$ is completely regular.

In the sequel by $\xrightarrow{w^*}$ (respectively, $\xrightarrow{w}$, $\to$) we denote convergence in the weak* (respectively, weak, norm) topology. The next proposition is a straightforward consequence of the above definitions.

Proposition 3.94 *If X is a normed space, X^* is its dual and $\{x_\vartheta\}_{\vartheta \in D} \subseteq X$ and $\{x_i^*\}_{i \in I} \subseteq X^*$ are nets, then*

(a) $x_\vartheta \to x \Rightarrow x_\vartheta \xrightarrow{w} x$ *in X; and $x_i^* \xrightarrow{w} x^* \Rightarrow x_i^* \xrightarrow{w^*} x^*$ in X^*.*

(b) $x_\vartheta \xrightarrow{w} x \Leftrightarrow \langle x^*, x_\vartheta \rangle \to \langle x^*, x \rangle$ *for all $x^* \in X^*$;*

   $x_i^* \xrightarrow{w^*} x^* \Leftrightarrow \langle x_i^*, x \rangle \to \langle x^*, x \rangle$ *for all $x \in X$.*

(c) $x_\vartheta \xrightarrow{w} x$ *in X and $x_\vartheta^* \to x^*$ in $X^* \Rightarrow \langle x_\vartheta^*, x_\vartheta \rangle \to \langle x^*, x \rangle$;*

   $x_i^* \xrightarrow{w^*} x^*$ *in X^* and $x_i \to x^*$ in $X \Rightarrow \langle x_i^*, x \rangle \to \langle x^*, x \rangle$.*

(d) $A \in \mathcal{L}(X, Y)$ *if and only if $A \in \mathcal{L}(X_w, Y_w)$. The result fails for X, Y dual spaces furnished with the corresponding w^*-topologies.*

In finite dimensional normed spaces, both these topologies coincide with the norm topology. Note that the dual of an N-dimensional normed space is also an N-dimensional normed space. This is an immediate consequence of Proposition 3.39.

Proposition 3.95 *If X is a finite dimensional normed space and X^* is its dual, then (a) the weak and norm topologies on X coincide; and (b) the weak* and norm topologies on X^* coincide.*

Proof (a) Let $V \subseteq X^*$ be norm open. Given $x_0 \in V$ we can find $\epsilon > 0$ such that $x_0 + B_\epsilon \subseteq V$. Since X^* is finite dimensional, we can have a basis $\{e_k^*\}_{k=1}^N$ for it. Then for $x \in X$ we define $|x| = \max_{1 \le k \le N} |\langle e_k^*, x \rangle|$. Evidently $|\cdot|$ is a norm on X and since X is finite dimensional, on account of Corollary 3.41 $|\cdot|$ and $\|\cdot\|$ are equivalent. Hence we can find $\delta > 0$ such that for every $x \in X$

$|x| < \delta \Rightarrow \|x\| < \epsilon$. It follows that the w-open set in X

$$U = \{x \in X : |\langle e_k^x, x - x_0 \rangle| < \delta\}$$

is contained in $B_\epsilon(x_0) = \{x \in X : \|x - x_0\| < \epsilon\}$. We conclude that $w -$ int $B_\epsilon(x)) \neq \emptyset$ and so $w = s$.

(b) The proof of this part is similar to the proof of (a). □

However, for infinite dimensional normed spaces, the situation changes. Then the weak and weak* topologies do not coincide with the norm topology. To show this we will need two simple results from linear algebra.

Lemma 3.96 *If X, Y, V are vector spaces and $f : X \rightarrow V$ and $g : X \rightarrow Y$ are linear maps, then there exists a linear map $h : Y \rightarrow V$ such that $f = h \circ g$ if and only if $\ker g \subseteq \ker f$.*

Proof $\Rightarrow$ Clear.

$\Leftarrow$ Let $Z = g(X) \subseteq Y$. For every $z \in Z$ we have $z = g(x)$ with $x \in X$. We consider $\widehat{h} : Z \rightarrow V$ linear defined by $\widehat{h}(z) = \widehat{h}(g(x)) = f(x)$. This is well defined. Indeed if $g(x_1) = g(x_2)$, then $x_1 - x_2 \in \ker g \subseteq \ker f$ and so $f(x_1) = f(x_2)$. We extend $h(\cdot)$ to a linear map h on Y. Then we have $f = h \circ g$. □

Lemma 3.97 *If X is a vector space and $f, g_1, \ldots, g_m : X \rightarrow \mathbb{R}$ are linear functionals satisfying $\bigcap_{k=1}^{m} \ker g_k \subseteq \ker f$, then $f = \sum_{k=1}^{m} \lambda_k g_k$ for some $\lambda_k \in \mathbb{R}$.*

Proof Let $Y = \mathbb{R}^m$, $V = \mathbb{R}$ and $g = (g_k)_{k=1}^m : X \rightarrow \mathbb{R}^m$ and apply Lemma 3.96. Then we can find a linear map $h : \mathbb{R}^m \rightarrow \mathbb{R}$ such that $f = h \circ g$. Then $h(y) = \sum_{k=1}^{m} \lambda_k y_k$ for every $y \in \mathbb{R}^m$ with $\lambda_k \in \mathbb{R}$. Hence $f(x) = \sum_{k=1}^{m} \lambda_k g_k(x)$ for every $x \in X$. □

Using this lemma we can now show that in an infinite dimensional context the weak and weak* topologies do not coincide with the norm topology.

Proposition 3.98 *If X is an infinite dimensional normed space and $U \subseteq X$ is nonempty and w-open, or $U \subseteq X^*$ is nonempty and w^*-open, then U is unbounded.*

Proof We do the proof for $U \subseteq X$, the proof for $U \subseteq X^*$ being similar. We may assume that $0 \in U$. Then according to Definition 3.92, we can find $\epsilon > 0$ and $\{x_k^*\}_{k=1}^n \subseteq X^*$ such that

$$\{x \in X : |\langle x_k^*, x \rangle| < \epsilon, k = 1, \ldots, n\} \subseteq U.$$

We see that $D = \bigcap_{k=1}^{n} \ker x_k^* \subseteq U$. We claim $D \neq \{0\}$. Arguing by contradiction suppose that $D = \{0\}$. Then $D \subseteq \ker x^*$ for every $x^* \in X^*$. So, by Lemma 3.97, $x^* \in span\{x_k^*\}_{k=1}^n$ which means that X^* is finite dimensional, a contradiction. So, we can find $x \in D$ with $x \neq 0$. Then $\mathbb{R}x \subseteq U$ and so U is unbounded. Similarly for $U \subseteq X^*$.   □

Corollary 3.99 *If X is an infinite dimensional normed space, then the weak topology on X and the weak* topology on X^* do not coincide with the corresponding norm topologies.*

Definition 3.100 Let X be a normed space and $C \subseteq X, D \subseteq X^*$. (a) We say that C is "w-bounded" if for every $x^* \in X^*$ we have

$$\sup[|\langle x^*, u \rangle| : u \in C] < \infty.$$

(b) We say that D is "w^*-bounded," if for every $x \in X$ we have

$$\sup[|\langle u^*, x \rangle| : u^* \in D] < \infty.$$

Proposition 3.101 *(a) If X is a normed space and $C \subseteq X$, then C is bounded if and only if it is w-bounded. (b) If X is a Banach space and $D \subseteq X^*$, then D is bounded if and only if it is w^*-bounded.*

Proof

(a) $\Rightarrow$ For $x^* \in X^*$ and $u \in C$, we have $|\langle x^*, u \rangle| \leq \|x^*\|_* \|u\| \leq M \|x^*\|_*$ for some $M > 0$ (C is bounded). It follows that C is w-bounded.

$\Leftarrow$ Let $\widehat{i} : X \to X^{**}$ be the canonical embedding (see Definition 3.58). For $x \in X$, define $f_{\widehat{i}(x)} \in X^{**}$ by $f_{\widehat{i}(x)}(x^*) = \langle x^*, x \rangle$. By hypothesis $\{f_{\widehat{i}(u)}\}_{u \in C}$ is pointwise bounded. Since X^* and X^{**} are both Banach spaces, from Theorem 3.90 it follows that $\widehat{i}(C) \subseteq X^{**}$ is bounded. Since $\widehat{i}(\cdot)$ is an isometry, we conclude that C is bounded.

(b) $\Rightarrow$ Some as in (a).

$\Leftarrow$ For $x^* \in X^*$, let $f_{x^*}(x) : \langle x^*, x \rangle$ for all $x \in X$. Then $\{f_{u^*}\}_{u^* \in D}$ is pointwise bounded and since X is a Banach space, we can use Theorem 3.90 to conclude that D is bounded.   □

Remark 3.102 In part (b) if X is not complete, then the result is no longer true (see the example in Remark 3.89).

Proposition 3.103 *If X is a Banach space and $\xi \in X^{**}$ is w^*-continuous, then $\xi = \widehat{i}(x)$ for some $x \in X$ ($\widehat{i}$ is canonical embedding).*

Proof Since $\xi(\cdot)$ is w^*-continuous, we can find

$$U = \{x^* \in X^* : |\langle x^*, x_k \rangle| < \epsilon, \, \epsilon > 0, k = 1, \dots, n\}$$

such that $|\xi(U)| \leq 1$. Using the Hahn-Banach Theorem, we can find $x_0^* \in X^*$ such that $\langle x_0^*, x_k \rangle = 0$ for all $k = 1, \ldots, n$. Then $\lambda x_0^* \in U$ for all $\lambda \in \mathbb{R}$ and so $|\xi(\lambda x_0^*)| \leq 1$ for all $\lambda \in \mathbb{R}$. Hence $\xi(x_0^*) = 0$ and this implies that $\bigcap_{k=1}^{n} \ker \widehat{i}(x_k) \subseteq \ker \xi$. We apply Lemma 3.97 to conclude that $\xi = \sum_{k=1}^{n} \lambda_k \widehat{i}(x_k) \in X$. $\qquad \square$

Combining Propositions 3.91 and 3.94, we have:

Proposition 3.104 *If X is a normed space, then $\varphi(x) = \|x\|$ is w-lower semicontinuous for $x \in X$, and $\psi(x^*) = \|x^*\|_*$ is w*-lower semicontinuous for $x^* \in X^*$.*

The next result, known in the literature as "Mazur's theorem," is an interesting one because an algebraic condition on the set (convexity) leads to a topological conclusion.

Theorem 3.105 *If X is a normed space and $C \subseteq X$ is convex, then C is closed if and only if it is w-closed.*

Proof $\Rightarrow$ Let $x_0 \in X \setminus C$. Invoking Theorem 3.69(b) we can find $\widehat{x}^* \in X^*$ such that

$$\widehat{m} = \sup[\langle \widehat{x}^*, c \rangle : c \in C] < \langle \widehat{x}^*, x_0 \rangle.$$

Consider the set $U = \{x \in X : \langle \widehat{x}^*, x \rangle > \widehat{m}\}$, which is w-open, $x_0 \in U$ and $U \cap C = \emptyset$. This proves that C is w-closed.

$\Leftarrow$ The weak topology is weaker than the norm topology. Therefore every w-closed set is closed. $\qquad \square$

Remark 3.106 Mazur's theorem says that for convex sets the norm and weak closures coincide. In general for a set $C \subseteq X$, we have $\overline{C} \subseteq \overline{C}^w$ and the inclusion can be strict, Mazur's Theorem fails for the w^*-topology.

Proposition 3.107 *If X is normed space and $x_n \overset{w}{\to} x$ in X, then we can find a sequence $\{u_n\}_{n \in \mathbb{N}} \subseteq X$, $u_n = \sum_{k=n}^{N} \lambda_k x_k$ with $\lambda_k \geq 0$ and $\sum_{k=n}^{N} \lambda_k = 1$ such that $u_n \to x$.*

Proof For every $n \in \mathbb{N}$, we have

$$x \in \overline{\bigcup_{k \geq n} \{x_k\}}^w \subseteq \overline{\text{conv}}^w \bigcup_{k \geq n} \{x_k\} = \overline{\text{conv}} \bigcup_{k \geq n} \{x_k\}$$

(see Theorem 3.105). So, we choose $u_n \in \text{conv} \bigcup_{k \geq n} \{x_k\}$ such that $\|u_n - x\| < \frac{1}{n}$. This is the desired sequence of convex combinations. $\qquad \square$

Now we come to one of the main results of Banach space theory, which illustrates the importance of the weak topologies. It can be viewed as an infinite dimensional analog of the Heine-Borel property. Namely, in a finite dimensional space, closed and bounded sets are compact, see Theorem 3.43. The result is known in the literature as "Alaoglu's theorem."

Theorem 3.108 *If X is a normed space and $\overline{B}_1^* = \{x^* \in X^* : \|x^*\|_* \leq 1\}$, then $\overline{B}_1^*$ is w^*-compact.*

Proof For $x \in X$, let $E_x = \{\vartheta \in \mathbb{R} : |\vartheta| \leq \|x\|\}$ and let $E = \prod_{x \in X} E_x$ furnished with the product topology. By Tychonov's Theorem (see Theorem 1.102), E is compact. We consider the map $\gamma : (\overline{B}_1^*, w^*) \to E$ defined by

$$\gamma(x^*) = \{\langle x^*, x \rangle\}_{x \in X}.$$

Evidently, γ is continuous and injective (see Corollary 3.70). Also, again from the definitions of the product and weak* topologies, we see that

$$\gamma^{-1} : \gamma(\overline{B}_1^*) \to \overline{B}_1^*$$

is continuous too. Therefore $\overline{B}_1^*$ is homeomorphic to the subset $\gamma(\overline{B}_1^*)$ of E. So, to finish the proof, it suffices to show that $\gamma(\overline{B}_1^*)$ is a closed subset of the compact set E (see Proposition 1.82).

Let $\widehat{\vartheta} = (\vartheta_x)_{x \in X} \in E$ be in the closure of $\gamma(\overline{B}_1^*)$. We introduce the functional $f : X \to \mathbb{R}$ defined by $f(x) = \vartheta_x$ for all $x \in X$. We claim that $f \in X^*$. First we show the linearity of f. Since $\widehat{\vartheta} \in \overline{\gamma(\overline{B}_1^*)}$, for every $n \in \mathbb{N}$, we can find $x_n^* \in \overline{B}_1^*$ such that for all $x, u \in X$ and all $\lambda, \mu \in \mathbb{R}$, we have

$$\frac{1}{n} > |f(x) - \langle x_n^*, x \rangle| + |f(u) - \langle x_1^*, u \rangle|$$

$$+ |f(\lambda x + \mu u) - \langle x_n^*, \lambda x + \mu u \rangle|.$$

Thus, $f(\lambda x + \mu y) = \lambda f(x) + \mu f(u)$, and $f(\cdot)$ is linear. Also, we have $|f(x)| = |\vartheta_x| \leq \|x\|$ for every $x \in X$, hence f is continuous. Therefore $f \in X^*$ and $\|f\|_* \leq 1$, that is $f \in \overline{B}_1^*$. Then from the definition of f we see that $\widehat{\vartheta} = \gamma(f) \in \gamma(\overline{B}_1^*)$ and so $\gamma(\overline{B}_1^*) \subseteq E$ is closed, thus compact. $\square$

Remark 3.109 In the above proof, Tychonov's theorem was a basic tool. This is not a surprise given that both the weak* and product topologies are weak topologies.

Alaoglu's theorem, provides additional information concerning the canonical embedding. The result is known as "Goldstine's Theorem." In what follows $\overline{B}_1 = \{x \in X : \|x\| \leq 1\}$ and $\overline{B}_1^{**} = \{x^{**} \in X^{**} : \|x^{**}\|_{X^{**}} \leq 1\}$.

Theorem 3.110 *If X is a normed space, then we have (a) $\overline{\widehat{i(B_1)}}^{w^*} = \overline{B}_1^{**}$; and (b) $X^{**} = \overline{\widehat{i(X)}}^{w^*}$.*

Proof

(a) We have $\widehat{i}(\overline{B}_1) \subseteq \overline{B}_1^{**}$ and the latter is w^*−compact (see Theorem 3.108). Therefore $K = \overline{\widehat{i}(\overline{B}_1)}^{w^*} \subseteq \overline{B}_1^{**}$. Suppose that the inclusion is strict, we can find $x_0^{**} \in \overline{B}_1^{**} \setminus \overline{\widehat{i}(\overline{B}_1)}^{w^*}$. According to Theorem 3.69(b), we can find $\widehat{x}^* \in X^*$ such that

$$0 < m_K = \sup[\langle u^{**}, \widehat{x}^* \rangle : u^{**} \in K] < \langle x_0^{**}, \widehat{x}^* \rangle.$$

By scaling if necessary we may assume that $m_K = 1$. Since $\widehat{i}(\overline{B}_1) \subseteq K$, we obtain $\|\widehat{x}^*\|_* \leq 1$. Also we have $\|x_0^{**}\|_{X^{**}} \leq 1$. Therefore

$$1 = m_K < \langle x_0^{**}, \widehat{x}^* \rangle \leq \|x_0^{**}\|_{X^{**}} \|\widehat{x}^*\|_* \leq 1,$$

a contradiction. We conclude that $\overline{\widehat{i}(\overline{B}_1)}^{w^*} = \overline{B}_1^{**}$.

(b) Follows at once from (a). □

Remark 3.111 From now on we drop the use of the isometry $\widehat{i}(\cdot)$ (the canonical embedding) and view X as a subspace of X^{**}. We have

$$w(X^{**}, X^*)|_X = w(X, X^*).$$

Proposition 3.112 *If X is an infinite dimensional normed space and $\partial B_1 = \{x \in X : \|x\| = 1\}$, then ∂B_1 is a dense, G_δ-subset of $(\overline{B}_1, w)$.*

Proof Let $x_0 \in B_1 = \{x \in X : \|x\| < 1\}$ and $U \in \mathcal{N}_w(x_0)$ (= filter of weak neighborhoods of x_0). W can take

$$U = \{x \in X : |\langle x_k^*, x - x_0 \rangle| < \epsilon, \quad k = 1, \ldots, n\}$$

(see Definition 3.92). Let $\gamma : X \to \mathbb{R}^n$ be defined by $\gamma(x) = (\langle x_k^*, x \rangle)_{k=1}^n$. This map is not injective or otherwise dim $X = n$, a contradiction. So, we can find $u_0 \neq 0$ such that $\gamma(u_0) = 0$. Then $\langle x_k^*, u_0 \rangle = 0$ for every $k \in \{1, \ldots, n\}$. It follows that

$$x_0 + \lambda u_0 \in U \quad \text{for all } \lambda \in \mathbb{R}.$$

We introduce the map $\eta(\lambda) = \|x_0 + \lambda u_0\|$. Then $\eta(0) = \|x_0\| < 1$ and $\eta(\lambda) \to +\infty$ as $\lambda \to +\infty$. By Bolzano's theorem, we can find $\lambda_0 > 0$ such that $\|x_0 + \lambda_0 u_0\| = 1$. We have $x_0 + \lambda_0 u_0 \in \partial B_1 \cap U \neq \emptyset$ and so $x_0 \in \overline{\partial B_1}^w$. Since $x \in B_1$, is arbitrary we conclude that $\overline{\partial B_1}^w = \overline{B}_1$.

Finally let $U_n = \{x \in \overline{B}_1 : \|x\| > 1 - \frac{1}{n}\}$. By Proposition 3.104, U_n is open in $(\overline{B}_1, w)$ and $\partial B_1 = \bigcap_{n \geq 1} U_n$. Hence ∂B_1 is G_δ. □

Remark 3.113 It is clear from Proposition 3.98, that in an infinite dimensional normed space, the weak interior of a closed or open ball is empty. Similarly, if $K \subseteq X$ is compact, then from Theorem 3.43 we infer that the (norm) interior of K is empty.

We conclude this section by introducing one more locally convex topology on X^*. Recall that w^* is the weakest topology τ on X^* such that $(X^*_\tau)^* = X$. There is also a strongest such topology.

Definition 3.114 There is a strongest topology m on X^* for which we have $(X^*_m)^* = X$. This is the topology of uniform convergence on w-compact sets, that is $x^*_\vartheta \xrightarrow{m} x^*$ in X^* if and only if for all $K \subseteq X$ w-compact

$$\sup_{u \in K} \langle x^*_\vartheta - x^*, u \rangle \to 0.$$

This topology is locally convex, it is known as the "Mackey topology" and is denoted by $m(X^*, X)$.

Remark 3.115 If we consider X instead of X^*, then the Mackey topology coincides with the norm (strong) topology.

3.6   Separable and Reflexive Normed Spaces

Recall that a topological space is separable, if it has a countable dense subset (see Definition 1.50(c)).

Proposition 3.116 *A w-separable Banach space X is separable.*

Proof Let $D \subseteq X$ be countable and $\overline{D}^w = X$, $V = span\, D$ and $S = span_\mathbb{Q} D$ (rational linear combinations of elements in D). Then S is countable and dense in V. So, V is norm separable and w-dense in X. On account of Theorem 3.105 (Mazur's Theorem), we have $X = \overline{V}^w = \overline{V}$. We conclude that X is separable. □

Separability of X turns out to be equivalent to metrizability of $(\overline{B}^*_1, w^*)$.

Theorem 3.117 *If X is normed space, then X is separable if and only if $(\overline{B}^*_1, w^*)$ is metrizable.*

Proof $\Rightarrow$ Since X is separable, so is $\overline{B}_1$. Let $\{x_n\}_{n \in \mathbb{N}}$ be dense in $\overline{B}_1$. We set

$$d(x^*, u^*) = \sum_{n \in \mathbb{N}} \frac{1}{2^n} |\langle x^* - u^*, x_n \rangle|$$

for all $x^*, u^* \in X^*$. It is easy to see that $d(\cdot, \cdot)$ is a metric on X^* and $d(x^*, u^*) \leq \|x^*\|_* + \|u^*\|_*$. We claim that $d(\cdot, \cdot)$ generates the w^*-topology on $\overline{B}^*_1$ (and for that matter on any bounded subset of X^*). To this end, we consider the identity map $i : (\overline{B}_1^*, w^*) \to (\overline{B}_1^*, d)$. From Theorem 3.108 (Alaoglu's Theorem), we know that $(\overline{B}_1^*, w^*)$ is compact. So, according to Proposition 1.88 it suffices to show that $i(\cdot)$ is continuous. For this purpose consider a net $\{x_\vartheta^*\}_{\vartheta \in D} \subseteq \overline{B}_1^*$ such that $x_\vartheta^* \xrightarrow{w^*} x^*$. Given $\epsilon > 0$, we can find $n_0 \in \mathbb{N}$ such that $\sum_{n > n_0} \frac{1}{2^n} < \frac{\epsilon}{2}$. Note that $|\langle x_\vartheta^*, x_n \rangle - \langle x^*, x_n \rangle| \leq 2$ (recall $\|x_n\|, \|x_\vartheta^*\|_*, \|x^*\|_* \leq 1$ for all $n \in \mathbb{N}$, all $\vartheta \in D$). Hence we have

$$d(x_\vartheta^*, x^*) < \sum_{n=1}^{n_0} |\langle x_\vartheta^*, x_n \rangle - \langle x^*, x_n \rangle| + \epsilon$$

and $\limsup_{\vartheta \in D} d(x_\vartheta^*, x^*) \leq \epsilon$. Since $\epsilon > 0$ is arbitrary, we infer that $d(x_\vartheta^*, x^*) \to 0$ and so $i(\cdot)$ is continuous, thus a homeomorphism.

$\Leftarrow$ Since $(\overline{B}_1^*, w^*)$ is metrizable, it is first countable and so we can find a local basis of w^* open neighborhoods $\{U_n\}_{n \in \mathbb{N}}$ of 0 in $\overline{B}_1^*$ such that $\bigcap_{n \geq 1} U_n = \{0\}$. We can always take

$$U_n = \{x^* \in \overline{B}_1^* : |\langle x^*, x_k \rangle| < \epsilon \text{ for } k = 1, \dots, n\}$$

(see Definition 3.93). Let $A_n = \{x_k\}_{k=1}^n$ and $A = \bigcup_{n \in \mathbb{N}} A_n$. The set $A \subseteq X$ is countable. If $x^*|_A = 0$, then $x^* \in U_n$ for all $n \in \mathbb{N}$ and so $x^* = 0$. Therefore $x^*|_{\overline{span}A} = 0$ implies $x^* = 0$. The space $\overline{span}A$ is separable. Using Proposition 3.60, we infer $X = \overline{span}A$. □

Similarly, we may also show the following "dual version" of Theorem 3.117.

Theorem 3.118 *If X is a Banach space, Then X^* is separable if and only if $(\overline{B}_1, w)$ is metrizable.*

Another useful metrizability result is the following one.

Theorem 3.119 *If X^* is w^*-separable and $K \subseteq X$ is w-compact, then (K, w) is metrizable.*

Proof Since X^* is w^*-separable, we can find $\{x_n^*\}_{n \in \mathbb{N}}$ separating sequence. We can always assume that $\|x_n^*\|_* \leq \frac{1}{2^n}$ for all $n \in \mathbb{N}$. We define

$$d(x, u) = \sum_{n \in \mathbb{N}} |\langle x_n^*, x - u \rangle| \text{ for all } x, u \in X.$$

This is a metric on X. We will show that this metric induces the w^*-topology on K. As before (see the proof of Theorem 3.117), we consider the identity map $i : (K, w) \to (K, d)$ and it suffices to show that it is continuous. So, let $x_\vartheta \xrightarrow{w} x$ in K, which is w-compact, thus w-bounded (see 3.100(a)) and so bounded (see

Proposition 3.101(a)). So, given $\epsilon > 0$, we can find $n_0 \in \mathbb{N}$ such that $\sum\limits_{n>n_0} |\langle x_n^*, x_\vartheta -$

$x \rangle| < \epsilon$. Therefore $d(x_\vartheta, x) \leq \sum\limits_{n=1}^{n_0} |\langle x_n^*, x_\vartheta - x \rangle| + \epsilon. \Rightarrow \limsup\limits_{\vartheta \in D} d(x_\vartheta, x) \leq \epsilon$.

Since $\epsilon > 0$ is arbitrary, we infer that $d(x_0, x) \to 0$ and so $i(\cdot)$ is continuous, hence a homeomorphism. $\square$

Corollary 3.120 *If X^* is separable and $K \subseteq X$ is w-compact, then (K, w) is metrizable.*

In general for a w-compact set K, (K, w) is not metrizable and so sequences do not suffice to describe topological notions and properties. Nevertheless, there is the following remarkable result, known as the "Eberlein–Smulian Theorem." For a proof of this result, we refer to Dunford–Schwartz [91] (p.430). In a topological space (V, τ) a set $C \subseteq V$ is said to be "relatively compact" if $\overline{C}^\tau \subseteq V$ is compact.

Theorem 3.121 *If X is a normed space and $K \subseteq X$, then K is (relatively) w-compact if and only if K is (relatively) sequentially weakly compact.*

Another important class of Banach spaces are the reflexive spaces. They are the next best thing to Hilbert spaces.

Definition 3.122 A Banach space X is said to be "reflexive," if the canonical embedding $\widehat{i} : X \to X^{**}$ (see Definition 3.58) is surjective.

Remark 3.123 It is important that for the definition of reflexivity we use the canonical embedding. There are examples of nonreflexive Banach spaces X which are isometrically isomorphic to X^{**} but not using $\widehat{i}(\cdot)$ (the so-called James space, see James [150]).

Proposition 3.124 *A Banach space X is reflexive if and only if $\overline{B}_1 = \{x \in X : \|x\| \leq 1\}$ is w-compact.*

Proof $\Rightarrow$ By reflexivity we have $X = X^{**}$ (for notational simplicity we drop the use of the map $\widehat{i}(\cdot)$). Then $\overline{B_1} = \overline{B}_1^{**} = \{x^{**} \in X^{**} : \|x^{**}\|_{X^{**}} \leq 1\}$ and the latter is w-compact (see Theorem 3.108). Since $w^*(X^{**}, X^*)|_X = w(X, X^*)$, we conclude that $\overline{B}_1$ is w-compact.

$\Leftarrow$ Since $\overline{B}_1$ is w-compact, viewed as a subset of X^{**} it is w^*-closed. From Theorem 3.110 (Goldstine's Theorem), we know that $\overline{B}_1 = \overline{\overline{B}_1}^{w^*} = \overline{B}_1^{**}$. Therefore $X = X^{**}$. $\square$

Proposition 3.125 *A Banach space X is reflexive if and only if X^* is reflexive.*

Proof $\Rightarrow$ Since X is reflexive, we have $X = X^{**}$ and so on X the weak and weak* topologies coincide. Then by Theorem 3.108 (Alaoglu's Theorem), $\overline{B}_1^*$ is w-compact and this by Proposition 3.124 means that X^* is reflexive.

$\Leftarrow$ Since X^* is reflexive, by the first part of the proof we have that X^{**} is reflexive. So, by Proposition 3.124 $\overline{B}_1^{**}$ is w-compact. We know that $\overline{B}_1$ is closed and convex, thus w-closed (see Theorem 3.105). Because $\overline{B}_1 \subseteq \overline{B}_1^{**}$, we infer that $\overline{B}_1$ is w-compact, hence X is reflexive. $\qquad\qquad\square$

Reflexivity is a hereditary property on closed subspaces.

Proposition 3.126 *If X is a reflexive Banach space and V is a closed subspace, then V is a reflexive Banach space.*

Proof Using Proposition 3.54, we see that $w(V, V^*) = w(X, X^*)|_V$. Then if $\overline{B}_1^V = \{u \in V : \|u\| \leq 1\}$, we have that $\overline{B}_1^V$ is a w-closed subset of $\overline{B}_1$ which is w-compact (see Proposition 3.124). Therefore $\overline{B}_1^V$ is w-compact and this implies the reflexivity of V. $\qquad\qquad\square$

A separable Banach space need not have a separable dual. (a typical example is $L^1[0, 1]$ which is separable, but $L^\infty[0, 1] = L^1[0, 1]^*$ is not). However, a Banach space X is separable and reflexive if and only if X^* is separable and reflexive. Then using Theorem 3.118, we have.

Proposition 3.127 *If X is a separable and reflexive Banach space, then $(\overline{B}_1, w)$ is a compact metrizable space.*

There is a beautiful criterion for w-compactness of a set due to James [146]. This result of James is one of the deepest and most influential results in Banach space theory. For a proof we refer to Floret [114], p.77.

Theorem 3.128 *If X is a Banach space and $C \subseteq X$ is w-closed, then X is w-compact if and only if every $x^* \in X^*$ attains its supremum on C.*

This theorem can be used to show that w-compactness is preserved if we take the closed convex hull of the set. The result is known in the literature as the "Krein-Smulian Theorem."

Theorem 3.129 *If X is a Banach space and $C \subseteq X$ is w-compact, then $\overline{\mathrm{conv}}C$ is w-compact too.*

Proof Let $x^* \in X^* \setminus \{0\}$. We have

$$m = \sup[\langle x^*, u\rangle : u \in C] = \sup[\langle x^*, v\rangle : v \in \overline{\mathrm{conv}}C].$$

Since C is w-compact, there exists $\widehat{u} \in C \subseteq \overline{\mathrm{conv}}C$ such that $m = \langle x^*, \widehat{u}\rangle$. Invoking Theorem 3.128 (James Theorem), we conclude that $\overline{\mathrm{conv}}C$ is w-compact. $\qquad\square$

The same is also true for norm compact sets. The result is sometimes called "Mazur's Theorem" (a second one, see also Theorem 3.105).

Theorem 3.130 *If X is a Banach space and $C \subseteq X$ is compact, then $\overline{\mathrm{conv}}C$ is compact too.*

Proof Since C is compact, there exits $F \subseteq C$ finite such that with $\epsilon > 0$

$$C \subseteq F + \frac{1}{2}B_\epsilon \quad \text{(recall } B_\epsilon = \{x \in X : \|x\| < \epsilon\}). \tag{3.9}$$

Let $x \in \text{conv}\, C$. Then $x = \sum_{k=1}^{n} \lambda_k x_k$, $\{x_k\}_{k=1}^{n} \subseteq C$, $\{\lambda_k\}_{k=1}^{n} \subseteq [0, +\infty)$, $\sum_{k=1}^{n} \lambda_k = 1$. On account of (3.9), we can find $\{u_k\}_{k=1}^{n} \subseteq F$ such that $x_k \in u_k + \frac{1}{2}B_\epsilon$ for all $k = 1, \ldots, n$. We have

$$x = \sum_{k=1}^{n} \lambda_k u_k + \sum_{k=1}^{n} \lambda_k (x_k - u_k) \subseteq \text{conv}\, F + \frac{1}{2}B_\epsilon$$

thus

$$\text{conv}\, C \subseteq \text{conv}\, F + \frac{1}{2}B_\epsilon. \tag{3.10}$$

The set $\text{conv}\, F$ is compact (see Corollary 3.68). So, we can find $F_0 \subseteq \text{conv}\, F$ finite such that $\text{conv}\, F \subseteq F_0 + \frac{1}{2}B_\epsilon$. Using this in (3.10) we obtain $\text{conv}\, C \subseteq F_0 + B_\epsilon$, $\Rightarrow$ conv C is totally bounded and so $\overline{\text{conv}}C$ is compact (see Theorem 1.96). $\qquad \square$

Proposition 3.131 *If X is a Banach space and $C \subseteq X^*$ is w^*-compact, then $\overline{\text{conv}}^{w^*} C$ is w^*-compact too.*

Proof The set C is bounded (see Proposition 3.101(b)). So, $C \subseteq \varrho \overline{B}_1^*$ for some $\varrho > 0$. Hence $\overline{\text{conv}}^{w^*} C \subseteq \varrho \overline{B}_1^*$ and the latter is w^*-compact (see Theorem 3.108). Therefore, we conclude that $\overline{\text{conv}}^{w^*} C$ is w^*-compact. $\qquad \square$

Another important theorem dealing with w-compactness, is the so-called Grothendieck's Theorem.

Theorem 3.132 *If X is a Banach space and $C \subseteq X$, then C is relatively w-compact if and only if for every $\epsilon > 0$, there exists $C_\epsilon \subseteq X$, w-compact, such that $C \subseteq C_\epsilon + \epsilon \overline{B}_1$.*

Proof $\Rightarrow$ Obvious.

$\Leftarrow$ We view C as a subset of X^{**} and consider $\overline{C}^{w^*}$. We have

$$\overline{C}^{w^*} \subseteq w^* - cl[C_\epsilon + \epsilon \overline{B}_1] = \overline{C_\epsilon}^{w^*} + \epsilon \overline{\overline{B}_1}^{w^*} = C_\epsilon + \epsilon \overline{B}_1^{**}$$

(see Theorem 3.110), thus,

$$\overline{C}^{w^*} = \bigcap_{\epsilon>0}(C_\epsilon + \epsilon\overline{B}_1^{**}) \subseteq X. \tag{3.11}$$

The set C is bounded and so $\overline{C}^{w^*}$ is w^*-compact. From (3.11) we infer that C is w-compact. □

In an infinite dimensional normed space, the weak topology is never metrizable. So, sequences do not suffice to describe topological notions and properties. However, in certain cases this is possible, see Theorem 3.121 (the Eberlein–Smulian Theorem). We present one more such situation.

Definition 3.133 A Hausdorff topological space X is said to be "angelic," if for every $C \subseteq X$ relatively countably compact, the following two properties hold: (a) C is relatively compact: (b) for every $x \in \overline{C}$ we can find a sequence $\{x_n\}_{n\in\mathbb{N}} \subseteq C$ such that $x_n \to x$.

Remark 3.134 In angelic spaces, we have "compact $\Leftrightarrow$ countably compact $\Leftrightarrow$ sequentially compact."

The next theorem is a special case of a more general result, which can be found in Floret [114], p.37).

Theorem 3.135 *A normed space with the weak topology is angelic.*

On account of this theorem, we can state the following result.

Proposition 3.136 *If X is a reflexive Banach space, $D \subseteq X$ is bounded and $x \in \overline{D}^w$, then we can find a sequence $\{x_n\}_{n\geq1} \subseteq D$ such that $x_n \overset{w}{\to} x$ in X.*

3.7 Dual Operators —Compact Operators— Projections

A continuous linear operator between two normed spaces, generates in a natural way a continuous linear operator between their duals.

Definition 3.137 Let X, Y be normed spaces and $A \in \mathcal{L}(X,Y)$. The "dual operator" $A^* \in \mathcal{L}(Y^*, X^*)$ is defined by $A^*(y^*) = y^* \circ A$, that is,

$$\langle A^*(y^*), x\rangle_X = \langle y^*, A(x)\rangle_Y,$$

where by $\langle \cdot, \cdot \rangle_X$ *(respectively, $\langle \cdot, \cdot \rangle_Y$)* we denote the duality brackets for the pair (X, X^*) (respectively, (Y, Y^*)).

Remark 3.138 The name "adjoint" is also used for A^*. We have decided to reserve this name for the operators defined on a Hilbert space. It is clear from the above definition that $A^* \in \mathcal{L}(Y^*, X^*)$. Note that $(B \circ A)^* = A^* \circ B^*$ for all $A \in \mathcal{L}(X, Y)$, $B \in \mathcal{L}(Y, V)$.

Proposition 3.139 *If X, Y are normed spaces and $A \in \mathcal{L}(X, Y)$, then $\|A\|_{\mathcal{L}} = \|A^*\|_{\mathcal{L}}$.*

Proof Given $\epsilon > 0$, we can find $x_\epsilon \in X$ with $\|x_\epsilon\|_X \leq 1$ such that $\|A\|_{\mathcal{L}} - \epsilon \leq \|A(x_\epsilon)\|_Y$. Proposition 3.55 says that we can find $\widehat{y}^* \in Y^*$, $\|\widehat{y}^*\|_{Y^*} = 1$ such that $\langle \widehat{y}^*, A(x_\epsilon) \rangle = \|A(x_\epsilon)\|_Y$. We have

$$\|A^*\|_{\mathcal{L}} \geq \|A^*(\widehat{y}^*)\|_{X^*} \geq |\langle A^*(\widehat{y}^*), x_\epsilon \rangle_X|$$
$$= |\langle \widehat{y}^*, A(x_\epsilon) \rangle_Y| = \|A(x_\epsilon)\|_Y$$
$$\geq \|A\|_{\mathcal{L}} - \epsilon.$$

Since $\epsilon > 0$ is arbitrary, we infer that $\|A\|_{\mathcal{L}} \leq \|A^*\|_{\mathcal{L}}$. On the other hand directly from Definition 3.137 we have that

$$\|A^*(y^*)\|_{X^*} = \|y^* \circ A\|_{X^*} \leq \|A\|_{\mathcal{L}} \|y^*\|_{Y^*}$$

for all $y^* \in Y^*$. Thus, $\|A^*\|_{\mathcal{L}} \leq \|A\|_{\mathcal{L}}$ and so $\|A\|_{\mathcal{L}} = \|A^*\|_{\mathcal{L}}$. □

In analogy to Definition 3.61, we introduce the following notion.

Definition 3.140 For a normed space X and $D \subseteq X^*$, the "pre-annihilator of D" is the set $^\perp D = \{x \in X : \langle x^*, x \rangle = 0 \text{ for all } x^* \in D\}$.

Remark 3.141 Evidently $^\perp D$ is a closed subspace of X.

Proposition 3.142 *If X, Y are normed spaces and $A \in \mathcal{L}(X, Y)$, then*

(a) $R(A)^\perp = \ker A^*$ and $^\perp R(A^*) = \ker A$.
(b) $R(A) \subseteq Y$ is dense if and only if A^* is injective.
(c) A is injective if and only if $R(A^*) \subseteq X^*$ is w^*-dense.

Proof

(a) We have

$$y^* \in R(A)^\perp \Leftrightarrow \langle y^*, A(x) \rangle_Y = 0 \text{ for } x \in X \Leftrightarrow \langle A^*(y^*), x \rangle_X = 0 \text{ for } x \in X.$$

It follows that $A^*(y^*) = 0$ and so $y^* \in \ker A^*$. Hence $R(A)^\perp = \ker A^*$. Similarly we have $x \in {}^\perp R(A^*) \Leftrightarrow \langle A^*(y^*), x \rangle_X = 0$ for all $y^* \in Y^* \Leftrightarrow \langle y^*, A(x) \rangle_Y = 0$ for all $y^* \in Y^*$. It follows that $A(x) = 0$ and so $x \in \ker A$. Hence $^\perp R(A^*) = \ker A$.
(b) A^* is injective $\Leftrightarrow \ker A^* = \{0\} \Leftrightarrow R(A)^\perp = \{0\}$ (see (a)) $\Leftrightarrow R(A)$ is dense in Y (see Corollary 3.74).

(c) A is injective $\Leftrightarrow \ker A = \{0\} \Leftrightarrow {}^{\perp}R(A^*) = \{0\}$ (see (a)) $\Leftrightarrow R(A^*)$ is w^*-dense in X^*. $\square$

Proposition 3.143 *If X, Y are normed spaces and $A \in \mathcal{L}(X, Y)$ is an isomorphism, then $A^* \in \mathcal{L}(Y^*, X^*)$ is an isomorphism and $(A^*)^{-1} = (A^{-1})^*$.*

Proof From Proposition 3.142(a), (b) we see that A^* is an isomorphism. We have

$$id_{Y^*} = id_Y^* = (A \circ A^{-1})^* = (A^{-1})^* \circ A^*,$$

and $id_{X^*} = id_X^* = (A^{-1} \circ A)^* = A^* \circ (A^{-1})^*$. Therefore $(A^*)^{-1} = (A^{-1})^*$. $\square$

Proposition 3.144 *If X is Banach space, Y a normed space, $A \in \mathcal{L}(X, Y)$ and $A^* \in \mathcal{L}(Y^*, X^*)$ is an isomorphism, then A is an isomorphism.*

Proof From Proposition 3.143 we know that $A^{**} = (A^*)^* \in \mathcal{L}(X^{**}, Y^{**})$ is an isomorphism. So, we can find $0 < c_1 < c_2$ such that

$$c_1 \|x^{**}\|_{X^{**}} \leq \|A^{**}(x^{**})\|_{Y^{**}} \leq c_2 \|x^{**}\|_{X^{**}} \text{ for all } x^{**} \in X^{**}.$$

Since $A^{**}|_X = A$, we have $c_1 \|x\|_X \leq \|A(x)\|_Y \leq c_2 \|x\|_X$ for all $x \in X$. To show that A is an isomorphism too, we need to show that $A(\cdot)$ is surjective. The completeness of X, implies the completeness of $R(A)$. So $R(A) \subseteq Y$ is closed, which means that $R(A) = Y$ (see Proposition 3.142(b)). Therefore A is an isomorphism. $\square$

Proposition 3.145 *If X, Y are Banach spaces and $A \in \mathcal{L}(X, Y)$, then $R(A) \subseteq Y$ is closed if and only if $R(A^*) \subseteq X^*$ is closed.*

Proof $\Rightarrow$ Note that $R(A) \subseteq Y$ being closed is a Banach space and

$$\widehat{A} \in \mathcal{L}(X/\ker A, R(A)),$$

defined by $\widehat{A}([x]) = A(x)$ is an isomorphism. Proposition 3.143 implies that $\widehat{A}^* \in \mathcal{L}(R(A^*), (X/\ker A)^*)$ is an isomorphism. According to Proposition 3.63 $(X/\ker A)^* = (\ker A)^{\perp} = R(A^*)$ (see Proposition 3.142). Therefore $R(A^*)$ is closed.

$\Leftarrow$ We consider $A_0 : X \to \overline{R(A)}$. Then $A_0^*(Y^*/\ker A^*) = R(A^*)$ which is closed since A_0^* is surjective, from Proposition 3.142, it follows that A_0 is surjective and so $R(A) = \overline{R(A)}$. $\square$

On account of Proposition 3.139 and the previous results, if X, Y are Banach spaces and $A \in \mathcal{L}(X, Y)$, then A is an isometric isomorphism if and only if A^* is an isometric isomorphism. Moreover, it is easy to check that $R(A) \subseteq Y$ is closed (respectively, $R(A^*) \subseteq X^*$ is closed) if and only if there exists $c > 0$ such that

$$\inf[\|x + u\|_X : u \in \ker A] \leq c\|A(x)\|_X$$

(respectively, $\inf[\|y^* + v^*\|_{Y^*} : v^* \in \ker A^*] \leq c\|A^*(y^*)\|_{X^*}$).

When we pass from finite to infinite dimensional normed spaces, we want to consider operators which preserve some of the basic properties of maps between finite dimensional spaces. This is the case with so-called compact operators.

Definition 3.146 Let X, Y be Banach spaces, $D \subseteq X$ and $f : D \rightarrow Y$ a map. We say that $f(\cdot)$ is "compact," if $f(\cdot)$ is continuous and for every $A \subseteq D$ bounded, $\overline{f(A)} \subseteq Y$ is compact. We denote the set of compact maps by $K(D, Y)$. We say that $f(\cdot)$ is "finite rank," if the range of f is in a finite dimensional subspace of Y. We denote the set of finite rank maps by $K_f(D, Y)$. Evidently $K_f(D, Y) \subseteq K(D, Y)$. We write

$$\mathcal{L}_c(X, Y) = K(X, Y) \cap \mathcal{L}(X, Y)$$

and $\mathcal{L}_f(X, Y) = K_f(X, Y) \cap \mathcal{L}(X, Y)$. Both $K(D, Y)$ and $K_f(D, Y)$ are vector spaces. We say that $f : D \rightarrow Y$ is "completely continuous" if

$$x_n \xrightarrow{w} x \text{ in } X, \{x_n, x\}_{n \in \mathbb{N}} \subseteq D \Rightarrow f(x_n) \rightarrow f(x) \text{ in } Y.$$

In general the classes of compact and completely continuous maps are distinct.

Example 3.147 We know that the sequence space, which is a Banach space,

$$l^1 = \{\widehat{x} = (x_n)_{n \in \mathbb{N}} : \|\widehat{x}\|_{l^1} = \sum_{n \geq 1} |x_n| < +\infty\}$$

has the so-called Schur property, namely every weakly convergent sequence is norm convergent (see Megginson [192], pp. 219–220). Then the identity map $i : l^1 \rightarrow l^1$ is completely continuous but not compact (see Theorem 3.43).

Proposition 3.148 *If X is a reflexive Banach space, Y is a Banach space, $D \subseteq X$ is closed, convex and $f : D \rightarrow Y$ is completely continuous, then $f \in K(D, Y)$.*

Proof Evidently $f(\cdot)$ is continuous. Let $A \subseteq D$ be bounded. Let $\{y_n\}_{n \in \mathbb{N}} \subseteq f(A)$. Then $y_n = f(x_n)$ with $x_n \in A$ for all $n \in \mathbb{N}$. The reflexivity of X implies that $\{x_n\}_{n \in \mathbb{N}}$ is relatively w-compact and so by the Eberlein–Smulian Theorem (see Theorem 3.121) we may assume that $x_n \xrightarrow{w} x \in D$ (see Theorem 3.105). Then the complete continuity of $f(\cdot)$ implies that $y_n = f(x_n) \rightarrow f(x) = y$ in Y. We conclude that $f(A)$ is relatively compact and so $f \in K(D, Y)$. $\square$

Corollary 3.149 *If X is a reflexive Banach space, Y is a Banach space and $A \in \mathcal{L}(X, Y)$, then $A \in \mathcal{L}_c(X, Y)$ if and only if $A(\cdot)$ is completely continuous.*

Proposition 3.150 *If X, Y are Banach spaces, $\{f_n\}_{n \in \mathbb{N}} \subseteq K(X, Y)$ and $f_n(x) \to f(x)$ in Y for every $x \in X$, uniformly on bounded subsets of X, then $f \in K(X, Y)$.*

Proof Since the convergence is uniform on bounded sets, $f(\cdot)$ is continuous. Let $A \subseteq X$ be bounded. Then given $\epsilon > 0$, we can find $n \in \mathbb{N}$ such that $\|f_n(x) - f(x)\|_Y \leq \epsilon/3$ for all $x \in A$. The set $f_n(A) \subseteq Y$ is relatively compact. So, we can find $\{x_m\}_{m=1}^{M} \subseteq A$ such that $f_n(A) \subseteq \bigcup_{m=1}^{N} B_{\epsilon/3}(f_n(x_m))$. Then for $x \in A$, we can find $m \in \{1, \ldots, M\}$ such that

$$\|f(x) - f(x_m)\|_Y \leq \|f(x) - f_n(x)\|_Y + \|f_n(x) - f_n(x_m)\|_Y$$
$$+ \|f_n(x_m) - f(x_m)\|_Y$$
$$\leq \frac{\epsilon}{3} + \frac{\epsilon}{3} + \frac{\epsilon}{3} = \epsilon,$$

hence $\overline{f(A)} \subseteq Y$ is compact (see Theorem 1.96), and so $f \in K(X, Y)$. □

Next we show that $K(D, Y)$ consists of those maps, which can be approximated uniformly by elements in $K_f(D, Y)$. It is this approximation property that makes compact maps, a very useful class of maps.

Theorem 3.151 *If X, Y are Banach spaces and $D \subseteq X$ is bounded, then $f \in K(D, Y)$ if and only if f is the uniform limit of elements in $K_f(D, Y)$.*

Proof $\Rightarrow$: We have that $f(D) \subseteq Y$ is relatively compact. So, given $n \in \mathbb{N}$, we can find $\{y_k\}_{k=1}^{m_n} \subseteq f(D)$ such that

$$\min_{1 \leq k \leq m_\varrho} \|f(x) - y_k\|_Y < \frac{1}{n} \text{ for all } x \in D. \tag{3.12}$$

We define $\vartheta_k(x) = \max\{\frac{1}{n} - \|f(x) - y_k\|_Y, 0\}$. Clearly $\vartheta_k \in C(D)$ and the $\vartheta_k's$ do not vanish simultaneously at $x \in D$ see (3.12). So, we can define

$$f_n(x) = \frac{\sum_{k=1}^{m_n} \vartheta_k(x) y_k}{\sum_{k=1}^{m_n} \vartheta_k(x)} \text{ for } x \in D.$$

We have $f_n \in K_f(D, Y)$ and for all $x \in D$ (see (3.12)),

$$\|f(x) - f_n(x)\| = \left\| \frac{\sum_{k=1}^{m_n} \vartheta_k(x)(f(x) - y_k)}{\sum_{k=1}^{m_n} \vartheta_k(x)} \right\|_Y < \frac{1}{n}.$$

$\Leftarrow$: Follows from Proposition 3.150. $\square$

This theorem raises the following question. "If $K \in \mathcal{L}_c(X, Y)$, then can we find a sequence $\{K_n\}_{n \in \mathbb{N}} \subseteq \mathcal{L}_f(X, Y)$ such that $\|K - K_n\|_{\mathcal{L}} \to 0$ as $n \to \infty$?". The answer to this question was an open problem in Banach space theory. It turns out that the answer depends on the Banach space Y.

Definition 3.152 We say that a Banach space Y has the "approximation property," if for every Banach space X, every $K \in \mathcal{L}_c(X, Y)$ and every $\epsilon > 0$, we can find $F \in \mathcal{L}_f(X, Y)$ such that $\|K - F\|_{\mathcal{L}} < \epsilon$.

Remark 3.153 A Banach space Y with a Schauder basis (see Definition 3.37) has the approximation property. However, not every Banach space has the approximation property. The first example at a Banach space without the approximation property, was constructed by Enflo [103] and it was a separable and reflexive Banach space. For compact maps, we have an extension theorem due to Dugundji [90] (see also Granas and Dugundji [128], p.163). We state the result in its most general form. Thee theorem generalizes the Tietze Extension Theorem (see Theorem 1.47).

Theorem 3.154 *If X is a metrizable space, Y is a locally convex topological vector space, $D \subseteq X$ is closed and $f \in K(D, Y)$, then there exists $\widehat{f} \in K(X, Y)$ such that $\widehat{f}|_D = f$ and $\widehat{f}(X) \subseteq \text{conv} f(D)$.*

Now we focus on linear compact operators.

Proposition 3.155 *If X,Y,V are Banach spaces, $A \in \mathcal{L}(X, Y)$, $T \in \mathcal{L}(Y, V)$ and one of them is compact, then $T \circ A \in \mathcal{L}_c(X, V)$.*

Proof First suppose that $A \in \mathcal{L}_c(X, Y)$. Let $\{x_n\}_{n \in \mathbb{N}} \subseteq X$ be bounded. Then we can find a subsequence $\{x_{n_k}\}_{k \in \mathbb{N}}$ such that $A(x_{n_k}) \to y$ in Y. Hence $T(A(x_{n_k})) \to T(y)$ in V and so $T \circ A \in \mathcal{L}_c(X, Y)$.

Next suppose that $T \in \mathcal{L}_c(Y, V)$. If $\{x_n\}_{n \in \mathbb{N}} \subseteq X$ is bounded, then $\{A(x_n)\}_{n \in \mathbb{N}} \subseteq Y$ is bounded. So, there exists a subsequence $\{A(x_{n_k})\}_{k \in \mathbb{N}}$ $\subseteq Y$ such that $T(A(x_{n_k})) \to v$ in V. Hence $T \circ A \in \mathcal{L}_c(X, V)$. $\square$

Directly from Proposition 3.150, we have:

Proposition 3.156 *If X, Y are Banach space, then $\mathcal{L}_c(X, Y)$ is a closed subspace of $\mathcal{L}(X, Y)$.*

Corollary 3.157 *If $\{A_n\}_{n \in \mathbb{N}} \subseteq \mathcal{L}_f(X, Y)$ and $\|A_n - A\|_{\mathcal{L}} \to 0$, then $A \in \mathcal{L}_c(X, Y)$.*

The next result shows that the property of compactness is duality invariant. The result is often called "Schauder's Theorem."

Theorem 3.158 *If X, Y are Banach spaces and $A \in \mathcal{L}(X, Y)$, then $A \in \mathcal{L}_c(X, Y)$ if and only if $A^* \in \mathcal{L}_c(Y^*, X^*)$.*

Proof $\Rightarrow$: Let $D = A(\overline{B}_1^X)$ (recall $\overline{B}_1^X = \{x \in X : \|x\|_X \le 1\}$). This is a compact metric space. For every $y^* \in Y^*$, let $h_{y^*} = y^*|_D$. We define

$$\mathscr{S} = \{h_{y^*} : y^* \in \overline{B}_1^{Y^*}\} \subseteq C(D)$$

(recall $\overline{B}_1^{Y^*} = \{y^* \in Y^* : \|y^*\|_{Y^*} \le 1\}$). We have

$$
\begin{aligned}
\|h_{y^*}\|_{C(D)} &= \sup[|\langle y^*, y\rangle_Y| : y \in D] \\
&= \sup[|\langle y^*, A(x)\rangle_Y| : x \in \overline{B}_1^X] \\
&= \sup[|\langle A^*(y^*), x\rangle_X| : x \in \overline{B}_1^X] \\
&= \|A^*(y^*)\|_{X^*},
\end{aligned}
\tag{3.13}
$$

thus, $\|h_{y^*}\|_{C(D)} \le \|A^*\|_{\mathcal{L}}\|y^*\|_{Y^*} \le \|A\|_{\mathcal{L}}$ (Proposition 3.139), $\Rightarrow \mathscr{S} \subseteq C(D)$ is bounded. Also we have

$$|h_{y^*}(y) - h_{y^*}(\widehat{y})| = |\langle y^*, y - \widehat{y}\rangle_Y| \le \|y - \widehat{y}\|_Y$$

for all $y, \widehat{y} \in D$, $\Rightarrow \mathscr{S}$ is equicontinuous. By the Arzela–Ascoli Theorem $\mathscr{S} \subseteq C(D)$ is relatively compact. So, if $\{y_n^*\}_{n\in\mathbb{N}} \subseteq \overline{B}_1^{Y^*}$, then there exists a subsequence $\{h_{y_{n_k}^*}\}_{k\in\mathbb{N}} \subseteq C(D)$ which converges in $C(D)$. From (3.13),

$$\|h_{y_{n_k}^*} - h_{y_{n_m}^*}\|_{C(D)} = \|A^*(y_{n_k}^*) - A^*(y_{n_m}^*)\|_{X^*}$$

for all $k, m \in \mathbb{N}$, $\Rightarrow \{A^*(y_{n_k}^*)\}_{k\in\mathbb{N}} \subseteq X^*$ is Cauchy, thus it converges. This shows that $A^* \in \mathcal{L}_c(Y^*, X^*)$.

$\Leftarrow$: If $A^* \in \mathcal{L}_c(Y^*, X^*)$, then from the first part we have that $A^{**} = (A^*)^* : X^{**} \to Y^{**}$ is compact. So, if $\{x_n\}_{n\ge 1} \subseteq X$ is bounded, then $\{\widehat{i}_X(x_n)\}_{n\in\mathbb{N}} \subseteq X^{**}$ is bounded (recall that $\widehat{i}_X(\cdot)$ is the canonical embedding of X into X^{**}). The compactness of A^{**} implies that we can find a subsequence $\{\widehat{i}_X(x_{n_k})\}_{k\in\mathbb{N}}$ such that $\{A^{**}(\widehat{i}_X(x_{n_k})) = \widehat{i}_Y(A(x_{n_k}))\}_{k\in\mathbb{N}}$ converges in Y^{**} (recall that $\widehat{i}_Y(\cdot)$ is the canonical embedding of Y into Y^{**}). Therefore $\{A(x_{n_k})\}_{k\in\mathbb{N}} \subseteq Y$ is Cauchy, which means that $A \in \mathcal{L}_c(X, Y)$. $\qquad\square$

Next we turn our attention to operators of the form I-A with $A \in \mathcal{L}_c(X)$ (compact perturbations of identity). We encounter such operators in degree theory. Such operators are important because they exhibit properties similar to those of linear operators on finite dimensional spaces. The main theorem concerning such operators, is the so-called Fredholm Alternative Theorem. This result has significant applications in the study of boundary value problems.

Theorem 3.159 *If X is a Banach space and $A \in \mathcal{L}_c(X)$, then (a) $\ker(I - A)$ is finite dimensional; (b) $R(I - A)$ is closed and $R(I - A) = {}^{\perp}\ker(I - A^*)$; (c) $\ker(I - A) = \{0\}$ if and only if $R(I - A) = X$; (d) $\dim \ker(I - A) = \dim \ker(I - A^*)$.*

Proof

(a) Let $V = \ker(I - A)$ and $\overline{B}_1 = \{x \in X : \|x\| \leq 1\}$. Let

$$\overline{B}_1^V = \{u \in V : \|u\| \leq 1\}.$$

If $u \in \overline{B}_1^V$, then $u = A(u)$ hence $\overline{B}_1^V \subseteq A(\overline{B}_1)$. Since $A \in \mathcal{L}_c(X)$, it follows that $\overline{B}_1^V$ is compact and so V is finite dimensional (see Theorem 3.43).

(b) Let $\{u_n\}_{n \in \mathbb{N}} \subseteq R(I - A)$ and suppose that $u_n \to u$. Then $u_n = x_n - A(x_n)$ for some $x_n \in X$, $n \in \mathbb{N}$. From (a) we know that $V = \ker(I - A)$ is finite dimensional. So, we can find $y_n \in V$ such that $\|x_n - y_n\| = d(x_n, V)$ for all $n \in \mathbb{N}$. Since $y_n \in V$, we have

$$u_n = x_n - y_n - A(x_n - y_n) \text{ for all } n \in \mathbb{N}. \tag{3.14}$$

Claim: $\{x_n - y_n\}_{n \in \mathbb{N}} \subseteq X$ is bounded.

Arguing by contradiction, we may say by passing to a subsequence if necessary, that $\|x_n - y_n\| \to \infty$ as $n \to \infty$. Let $v_n = \frac{x_n - y_n}{\|x_n - y_n\|}$, $n \in \mathbb{N}$. Then $\|v_n\| = 1$ for all $n \in \mathbb{N}$. Also, from (3.14) we have

$$v_n - A(v_n) = \frac{u_n}{\|x_n - y_n\|} \text{ for all } n \in \mathbb{N},$$

thus

$$v_n - A(v_n) \to 0.$$

Since $A(\cdot)$ is compact, we can find a subsequence $\{v_{n_k}\}_{k \in \mathbb{N}}$ such $A(v_{n_k}) \to w$ in X, hence $v_{n_k} \to w$ in X. Hence $A(v_{n_k}) \to A(w)$ in X and so $w = A(w)$, that is, $w \in \ker(I - A)$. Since $y_{n_k} \in V = \ker(I - A)$ we have

$$d(v_{n_k}, V) = \frac{d(x_{n_k}, V)}{\|x_{n_k} - y_{n_k}\|} = 1 \text{ for all } k \in \mathbb{N},$$

thus $d(w, V) = 1$, which contradicts $w \in V$. This proves the Claim.

On account of the Claim and since $A \in \mathcal{L}_c(X)$, we have that $\{A(x_n - y_n)\}_{n \in \mathbb{N}} \subseteq X$ is relatively compact. Hence we may assume that

$$A(x_n - y_n) \to h \text{ in } X \text{ as } n \to \infty. \tag{3.15}$$

From (3.14) and (3.15), we obtain $x_n - y_n \to u + h$ in X, thus

$$A(x_n - y_n) \to A(u + h) \text{ in } X,$$

hence $h = A(u + h)$ (see (3.15)), If $g = u + h \in X$, then $g - h = u$, hence $g - \lim_{n \to \infty} A(x_n - y_n) = u$. Therefore $u \in R(I - A)$ and so $R(I - A)$ is closed. From Proposition 3.142, we know that $R(I - A)^\perp = \ker(I - A^*)$, $\Rightarrow^\perp [R(I - A)^\perp] = {}^\perp \ker(I - A^*)$, $\Rightarrow R(I - A) = {}^\perp \ker(I - A^*)$ (since $R(I - A)$ is closed).

(c) $\Rightarrow$ Suppose $Y_1 = R(I - A) \neq X$. By part (b) Y_1 is a Banach space and if $y \in Y_1$, then $y = x - A(x)$ for some $x \in X$ and so

$$A(y) = (I - A)(A(x)),$$

hence $A(Y_1) \subseteq Y_1$. Then $A \in \mathcal{L}_c(Y_1)$ and so $(I - A)|_{Y_1}$ has closed range (see (b)). Next let $Y_2 = (I - A)(Y_1)$. This is a closed subspace of Y_1 and $Y_2 \neq Y_1$. Indeed, if $Y_2 = Y_1$, then for every $x \in X$, we can find $u \in X$ such that

$$(I - A)(x) = (I - A)^2(u),$$

$\Rightarrow x = (I - A)(u)$ (since by hypothesis $(I - A)(\cdot)$ is injective), $\Rightarrow X = Y_1$, a contradiction to our hypothesis. So, by induction we generate a decreasing sequence $\{Y_n\}_{n \in \mathbb{N}}$ of closed subspaces of X such that $Y_n = (I - A)^n(X)$ and $Y_{n+1} \neq Y_n$ for all $n \in \mathbb{N}$. Using Lemma 3.42 (the Riesz Lemma), we can find $\widehat{x}_n \in Y_n$ such that $\|\widehat{x}_n\| = 1$ and $d(\widehat{x}_n, Y_{n+1}) \geq \frac{1}{2}$. For $m < n$, we have

$$A(\widehat{x}_m - \widehat{x}_n) = (\widehat{x}_n - A(\widehat{x}_n)) - (\widehat{x}_m - A(\widehat{x}_m)) + (\widehat{x}_m - \widehat{x}_n). \tag{3.16}$$

Note that $\widehat{x}_n - A(\widehat{x}_n) \in Y_{n+1}, \widehat{x}_m - A(\widehat{x}_m) \in Y_{m+1}, \widehat{x}_n \in Y_n$ and $Y_{n+1} \subseteq Y_n \subseteq Y_{m+1} \subseteq Y_m$. Therefore

$$A(\widehat{x}_m - \widehat{x}_n) = \widehat{x}_m - v,$$

with $v \in Y_{m+1}$ (see (3.16). $\Rightarrow \|A(\widehat{x}_m) - A(x_n)\| \geq \frac{1}{2}$, $\Rightarrow \{A(\widehat{x}_n)\}_{n \in \mathbb{N}}$ cannot have a convergent subsequence, which contradicts the fact that $A \in \mathcal{L}_c(X)$. This proves that $R(I - A) = X$.

$\Leftarrow$ Using Proposition 3.142 we have

$$\ker(I - A^*) = R(I - A)^\perp = \{0\},$$

thus $R(I - A^*) = X^*$ (see Theorem 3.158 and part (b)), $\Rightarrow \ker A :^\perp R(I - A^*) = \{0\}$ (see again Proposition 3.142)

(d) Let $d = \dim \ker(I - A)$ and $d^* = \dim \ker(I - A^*)$. Suppose that $d < d^*$. The subspace $\ker(I - A)$ is finite dimensional and so by Proposition 3.77 it is complemented. So, there exists a continuous projection $P : X \to V = \ker(I - A)$ (see Definition 3.76). Also $R(I - A) = {}^\perp \ker(I - A^*)$ and so $R(I - A)$ is finite codimensional with codimension d^*. Let Z be a d^* dimensional subspace of X which is a complement to $R(I - A)$. Recall that we have assumed that $d < d^*$, so we can find a linear operator $T : \ker(I - A) \to Z$ which is injective but not surjective. We define

$$S = A + T \circ P.$$

Note that $T \circ P$ is of finite rank. Therefore $S \in \mathcal{L}_c(X)$. Let $x \in \ker(I - S)$. Then

$$0 = u - A(u) + T(P(u)).$$

We have $u - A(u) \in R(I - A)$ and $T(P(u)) \in Z$ which is the topological complement of $R(I - A)$. It follows that $u - A(u) = 0$ and $T(P(u)) = 0$, that is $u \in \ker(I - A)$ and $T(u) = 0$. Then $u = 0$, hence $\ker(I - S) = \{0\}$. Apply part (c) on the operator S to infer that $R(I - S) = X$. But this is not possible since there exists $z \in Z$ with $z \notin R(T)$ and so the equation $x - S(x) = z$ has no solution. This proves that $d^* \leq d$. Applying this fact to A^*, we obtain

$$\dim \ker(I - A^{**}) \leq \dim \ker(I - A^*) \leq \dim \ker(I - A).$$

But since $\ker(I - A) \subseteq \ker(I - A^{**})$, we conclude that $d = d^*$.   □

Remark 3.160 In fact (c) is usually called the "Fredholm Alternative." It says that either the homogeneous problem $x - A(x) = 0$ has only the trivial solution and so for every $u \in X$ the equation $x - A(x) = u$ has a unique solution or otherwise $x - A(x) = 0$ has d-linearly independent solutions and so the equation $x - A(x) = u$, $u \in X$, is solvable if and only if u satisfies d-orthogonality relations, that is $u \in {}^\perp \ker(I - A^*)$.

Theorem 3.159 is the starting point of the theory of Fredhelm operators.

Definition 3.161 Let X, Y be Banach spaces and $A \in \mathcal{L}(X, Y)$. We say that $A(\cdot)$ is a Fredholm operator, if $R(A)$ is closed with finite codimension and $\ker A$ is finite dimensional. The "Fredholm index" of A is given by

$$\mathrm{ind}(A) = \dim \ker A - \mathrm{codim}\, R(A).$$

In fact in the above definition it is enough to require that $A(\cdot)$ has finite codimension (that is, $\dim(Y/R(A)) < \infty$) and then $R(A)$ is necessarily closed.

Proposition 3.162 *If X, Y are Banach spaces and $A \in \mathcal{L}(X, Y)$ has range with finite codimension, then $R(A) \subseteq Y$ is closed.*

Proof Let $d = \mathrm{codim}\, R(A)$ and let $y_1, \dots y_d \in Y$ such that

$$\{[y_k] = y_k + R(A)\}_{k=1}^d \subseteq Y/R(A)$$

form a basis for the finite dimensional space $Y/R(A)$. We consider the Banach space $V = X \times R^d$ with $\|(x, \xi)\|_V = \|x\|_X + |\xi|, |\cdot|$ being the Euclidean norm on $\mathbb{R}^d$. Let $\widehat{A} \in \mathcal{L}(V, Y)$ be defined by

$$\widehat{A}((x, \xi)) = A(x) + \sum_{k=1}^d \xi_k y_k$$

$(\xi = (\xi_k)_{k=1}^d \in \mathbb{R}^d)$. The operator $\widehat{A}$ is surjective and

$$\ker \widehat{A} = \{(x, \xi) \in X \times \mathbb{R}^d : A(x) = 0, \xi = 0\} = \ker A \times \{0\}.$$

So, we can find $c > 0$ such that

$$\inf[\|x + u\|_X + |\xi| : u \in \ker A] \le c\|A(x) + \sum_{k=1}^d \xi_k y_k\|_Y$$

for all $x \in X, \xi \in \mathbb{R}^d$, (see Remark 3.21), thus

$$\inf[\|x + u\|_X : u \in \ker A] \le c\|A(x)\|_Y$$

for all $x \in X$. Hence $R(A) \subseteq Y$ is closed (see Proposition 3.145). □

Recall that if X is a vector space, a linear operator $P : X \to X$ is called a "projection," if $P^2 = P$, that is, P is idempotent (see Definition 3.76(a)). Projections are closely related to direct sum decompositions (see Definition 3.76(b)). This is illustrated in the next proposition, the proof of which is a straightforward consequence of Definitions 3.76(a), (b).

Proposition 3.163 *Let X be a vector space. We have (a) if $P : X \to X$ is a projection, then $X = R(P) \oplus \ker P$; (b) if $X = V \oplus W$, then the operator $P : X \to X$ defined by $P(x) = v$ for all $x = v + w$ with $v \in V, w \in W$, is a projection.*

Corollary 3.164 *If X is a vector space and $P : X \to X$ is a linear operator, then P is a projection if and only if $I - P$ is a projection.*

In a normed space, we are interested in continuous projections.

Theorem 3.165 *(a) If X is a normed space and $P \in \mathcal{L}(X)$ is a projection, then $x = R(P) \oplus \ker P$ and $R(A)$, $\ker A \subseteq X$ are closed subspaces. (b) If X is a Banach space, $X = V \oplus W$ with $V, W \subseteq X$ closed subspaces, then the operator $P : X \to X$ defined by $P(x) = v$ for all $x = v + w$ with $v \in V, w \in W$, is a continuous projection.*

Proof

(a) From Proposition 3.163, we have $X = R(P) \oplus \ker P$. Since $P \in \mathcal{L}(X)$, we know that $\ker P$ is closed. Since $I - P \in \mathcal{X}$, we have $\ker(I - P) \subseteq X$ is closed. But $R(P) = \ker(I - P)$. So $R(P)$ is closed too.

(b) For every $x = v + w$, $v \in V$, $w \in W$, we define $|x| = \|v\| + \|w\|$. Evidently this is a norm on X. Since X is a Banach space and $V, W \subseteq X$ are closed subspaces (hence Banach spaces too), we infer that $(X, |\cdot|)$ is a Banach space. Note that $\|\cdot\| \leq |\cdot|$. So, from Corollary 3.85 we have that $\|\cdot\|$, $|\cdot|$ are equivalent norms. Also for all $x \in X$, $\|P(x)\| = \|v\| \leq \|v\| + \|w\| = |x|$, hence $P : (X, \|\cdot\|) \to (X, |\cdot|)$ is continuous, which on account of the equivalence of the two norms implies that P is a continuous projection (see Proposition 3.163(b)). $\square$

3.8 Hilbert Spaces

Hilbert spaces are a natural generalization of Euclidean spaces to infinite dimensional settings. For such spaces the notion of orthogonal vectors can be defined and this leads to a better geometrical insight concerning the structure of the space. In what follows H is a real vector space unless otherwise stated.

Definition 3.166

(a) An "inner product" on H is a map $(\cdot, \cdot) : H \times H \to \mathbb{R}$ which satisfies

   (i) $(x, x) \geq 0$ for all $x \in H$ and $(x, x) = 0$ if and only $x = 0$.
   (ii) $(x, u) = (u, x)$ for all $x, u \in X$.
   (iii) $(\vartheta x + \eta u, y) = \vartheta(x, y) + \eta(u, y)$ for all $x, u, y \in H$, all $\vartheta, \eta \in \mathbb{R}$. A vector space H equipped with an inner product, is called an "inner product space." Such a space is a normed with norm $\|x\| = (x, x)^{1/2}$ for all $x \in H$.

(b) An inner product space H which is complete for the norm defined by the inner product, is said to be a "Hilbert space."

Remark 3.167 If H is a complex vector space, then (ii) in Definition 3.166(a) becomes $(x, u) = \overline{(u, x)}$ for all $x, u \in H$, with the bar indicating complex conjugate.

The next proposition, presents one of the most fundamental inequalities in mathematics. It is known as the "Cauchy–Bunyakovsky–Schwarz Inequality."

Proposition 3.168 *If H is an inner product space, then $|(x, u)| \leq \|x\| \, \|u\|$ for all $x, u \in H$ and the equality holds if and only if $x = \lambda u$ for some $\lambda \in \mathbb{R}$.*

Proof We may assume that $u \neq 0$. We have

$$0 \le \|x - \frac{(x,u)}{\|u\|^2}u\|^2 = \|x\|^2 - \frac{(x,u)^2}{\|u\|^2}, \tag{3.17}$$

thus $|(x,u)| \le \|x\|\|u\|$. If $x = \lambda u$, then clearly $|(x,u)| = \|x\|\|u\|$. Conversely, if $(x,u) = \pm\|x\|\|u\|$ and $u \ne 0$, then from (3.17) we have $\|x - \frac{(x,u)}{\|u\|^2}u\| = 0$ and so $x = \lambda u$ with $\lambda = \frac{(x,u)}{\|u\|^2}$. $\qquad\qquad\qquad\qquad\qquad\qquad\qquad\qquad\qquad\qquad\qquad\qquad\qquad\square$

The following identities follow from Definition 3.166.

Proposition 3.169 *If H is an inner product space, then for all $x, u \in H$*

$$\|x + u\|^2 + \|x - u\|^2 = 2\|x\|^2 + 2\|u\|^2 \text{ (“parallelogram law”);}$$

and the following “polarization identity” formulas:

$$(x,u) = \frac{1}{4}[\|x + u\|^2 - \|x - u\|^2] \text{ (real case);}$$

$$(x,u) = \frac{1}{4}[\|x + u\|^2 - \|x - u\|^2 + i\|x + iu\|^2 - i\|x - iu\|^2] \text{ (complex case).}$$

Remark 3.170 The parallelogram law characterizes those norms which come from an inner product. So, if a norm satisfies the parallelogram law, then it is generated by an inner product (see Weidmann [281], p.267).

The presence of an inner product, means that we can have the notion of orthogonality between elements in the space and this endows the space with a geometric structure, similar to that of Euclidean spaces.

Definition 3.171 Let H be an inner product space and $x, u \in H$. We say that x and u are “orthogonal” and write $x \perp u$, if $(x,u) = 0$. Also, if $C \subseteq X$ and $x \in X$, then we say that x is “orthogonal to C” and we write $x \perp C$, if $x \perp u$ for every $u \in C$. If $C \subseteq H$, then we define

$$C^\perp = \{x \in H : x \perp u \text{ for all } u \in C\}$$

and call $C^\perp$ the “orthogonal complement of C.”
  Clearly $C^\perp$ is a closed subspace of H and $(\overline{span}C)^\perp = C^\perp$.

Remark 3.172 If $x \perp C$, then $x \perp spanC$ and if $x \perp u$ for every $u \in X$, then $x = 0$.

Proposition 3.173 *If H is an inner product space and $x, u \in H$, then $(x,u) = 0$ if and only if $\|x\| \le \|x + \lambda u\|$ for all $\lambda \in \mathbb{R}$.*

Proof $\Rightarrow$: For every $\lambda \in \mathbb{R}$, we have

$$\|x\|^2 = \|\|x\|^2 + \lambda(u, x)\| = |(x + \lambda u, x)| \leq \|x + \lambda u\| \|x\|.$$

Hence $\|x\| \leq \|x + \lambda u\|$ for all $\lambda \in \mathbb{R}$
$\Leftarrow$: For all $\lambda \in \mathbb{R}$, we have

$$0 \leq \|x + \lambda u\|^2 - \|x + \lambda u\| \|x\| \leq (x, x + \lambda u) + \lambda(u, x + \lambda u) - (x, x + \lambda u)$$

(see Proposition 3.168). Hence $\lambda(u, x + \lambda u) \geq 0$ for all $\lambda \in \mathbb{R}$. Then for $\lambda > 0$, we have $\lambda \|u\|^2 + (x, u) \geq 0$ and for $\lambda < 0$, we have $\lambda \|u\|^2 + (x, u) \leq 0$. Letting $\lambda \to 0^{\pm}$, we obtain $(x, u) = 0$. $\qquad \square$

A remarkable consequence of this proposition is the following corollary.

Corollary 3.174 *If H is an inner product space and $x, u \in H$, then*

$$\|x + \lambda u\| \geq \|x\| \quad \text{for all } \lambda \in \mathbb{R}$$

if and only if $\|u + \lambda x\| \geq \|u\|$ for all $\lambda \in \mathbb{R}$.

The next theorem is a basic result about closed convex sets in a Hilbert space.

Theorem 3.175 *If H is a Hilbert space, $C \subseteq H$ is nonempty closed convex and $x \in H$, then there exists a unique $\widehat{u} \in C$ such that*

$$\|x - \widehat{u}\| = \min[\|x - u\| : u \in C] = m.$$

Proof Let $\{u_n\}_{n \geq 1} \subseteq C$ such that $\|x - u_n\| \downarrow m$. For $n, m \in \mathbb{N}$, from the parallelogram law (see Proposition 3.169), we have

$$\|(x - u_n) + (x - u_m)\|^2 + \|(x - u_n) - (x - u_m)\|^2 = 2\|x - u_n\|^2 + 2\|x - u_m\|^2,$$

thus $\|u_n - u_m\|^2 = 2\|x - u_n\|^2 + 2\|x - u_m\|^2 - 4\|x - \frac{u_n + u_m}{2}\|^2$. Since C is convex, $\frac{u_n + u_m}{2} \in C$ and so

$$\|u_n - u_m\|^2 \leq 2\|x - u_n\|^2 + 2\|x - u_m\|^2 - 4m^2,$$

hence $\|u_n - u_m\|^2 \to 0$ as $n, m \to \infty$, $\Rightarrow \{u_n\}_{n \in \mathbb{N}} \subseteq H$ is Cauchy. So, $u_n \to \widehat{u}$ in H, $\widehat{u} \in C$. We have $\|x - \widehat{u}\| = m$.

Next we show the uniqueness of this best approximation to $x \in H$. If $\widehat{v} \in C$ is another nearest point, then from the parallelogram we have

$$\|\widehat{u} - \widehat{v}\|^2 = 2\|x - \widehat{u}\|^2 + 2\|x - \widehat{v}\|^2 - 4\|x - \frac{\widehat{u} + \widehat{v}}{2}\|^2 \leq 0,$$

therefore, $\widehat{u} = \widehat{v}$. $\qquad \square$

So, we can define the map $p_C : H \to C$ by setting $p_C(x) = \widehat{u}$. This map is known as the "metric projection on C." The properties of this map are stated in the next proposition. Evidently if $x \in C$, then $p_C(x) = x$.

Proposition 3.176 *If H is a Hilbert space and $C \subseteq H$ is nonempty, closed, convex, then*

(a) $(x - p_C(x), u - p_C(x)) \leq 0$ *for all $x \in X$ and all $u \in C$.*
(b) $p_C(\cdot)$ *is nonexpansive, that is $\|p_C(x) - p_C(v)\| \leq \|x - v\|$ for all $x, v \in X$.*
(c) *If $C \subseteq H$ is a closed subspace, then $(x - p_C(x), u) = 0$ for all $u \in C$.*
(d) *If $x \notin C$, then $p_C(x) \in \partial C$.*

Proof

(a) Let $x \in H$, $u \in C$ and $0 < t < 1$. We have

$$\|x - p_C(x)\|^2 \leq \|x - (tu + (1-t)p_C(x))\|^2$$
$$= \|(x - p_C(x)) - t(u - p_C(x))\|^2$$
$$= \|x - p_C(x)\|^2 - 2t(x - p_C(x), u - p_C(x))$$
$$+ t^2\|u - p_C(x)\|^2,$$

thus, $2(x - p_C(x), u - p_C(x)) \leq t\|u - p_C(x)\|$. We let $t \to 0^+$ and obtain

$$(x - p_C(x), u - p_C(x)) \leq 0$$

for all $x \in H$, all $u \in C$.
(b) Let $x, v \in H$. From (a) we have

$$(x - p_C(x), p_C(v) - p_P(x)) \leq 0. \tag{3.18}$$

Interchanging x and v, we also have

$$(v - p_C(v), p_C(x) - p_C(v)) \leq 0. \tag{3.19}$$

We add (3.18) and (3.19) and obtain

$$(x - v + p_C(v) - p_C(x), p_C(v) - p_C(x)) \leq 0,$$

thus $\|p_C(x) - p_C(x)\| \leq \|v - x\|$ for all $x, v \in H$.
(c) Since C is a vector space $p_C(x) + \lambda(\pm u) \in V$ for all $u \in C$, all $\lambda \in \mathbb{R}$. Then

$$\|x - p_C(x)\|^2 \leq \|x - (p_C(x) + \lambda(\pm u))\|^2$$
$$= \|x - p_C(x)\|^2 \mp 2\lambda(x - p_C(x), u) + \lambda^2\|u\|^2,$$

thus $\pm 2\lambda(x - p_C(x), u) \leq \lambda^2 \|u\|^2$ for all $\lambda \in \mathbb{R}$. If $\lambda > 0$, then

$$\pm 2(x - p_C(x), u) \leq \lambda \|u\|^2$$

and letting $\lambda \to 0^+$, $\pm 2(x - p_C(x), u) \leq 0$. If $\lambda < 0$, then

$$\mp 2(x - p_C(x), u) \leq |\lambda| \|u\|^2$$

and letting $\lambda \to 0^-$, $\mp 2(x - p_C(x), u) \leq 0$. We conclude that

$$(x - p_C(x), u) = 0$$

for all $x \in H$, all $u \in C$.

(d) For $t \in (0, 1)$ let $x_t = tx + (1 - t)p_C(x)$. Then

$$\|x - x_t\| = (1 - t)\|x - p_C(x)\| < \|x - p_C(x)\| \text{ for all } t \in (0, 1). \qquad (3.20)$$

Suppose $p_C(x) \in \text{int} \, C$. Then for $t \in (0, 1)$ small $x_t \in C$ and so

$$\|x - p_C(x)\| \leq \|x - x_t\|. \qquad (3.21)$$

Comparing (3.20) and (3.21), we have a contradiction. Hence $p_C(x) \in \partial C$.

$\square$

The next result is an abstract generalization of the famous Pythagorean theorem from Euclidean geometry.

Theorem 3.177 *If H is an inner product space and $\{x_k\}_{k=1}^n$ are pairwise orthogonal, then $\|\sum_{k=1}^n x_k\|^2 = \sum_{k=1}^n \|x_k\|^2$.*

Proof If $n = 2$, then from the definition of the norm we see that $\|x_1 + x_2\|^2 = \|x_1\|^2 + \|x_2\|^2$. The general case follows by induction. $\square$

Proposition 3.178 *If H is a Hilbert space and $V \subseteq H$ is a proper closed subspace, then $V^\perp \setminus \{0\} \neq \emptyset$.*

Proof Let $x \in H \setminus V$. Then by Theorem 3.176, there exists unique $u \in V$ such that $\|x - u\| = d(x, V)$. Let $h = x - u$, then for any $v \in V$ and $\lambda \in \mathbb{R}$, we have

$$\|h + \lambda v\| = \|x - (u - \lambda v)\| \geq d(x, V) = \|h\|,$$

thus $(h, v) = 0$ (see Proposition 3.173), hence $h \perp V$ and so $h \in V^\perp$. $\square$

The next theorem justifies the name "orthogonal complement."

Theorem 3.179 *If H is a Hilbert space and $V \subseteq H$ is a closed subspace, then $H = V \oplus V^{\perp}$.*

Proof Evidently $V \cap V^{\perp} = \{0\}$ and so $V + V^{\perp} = V \oplus V^{\perp}$ (see Definition 3.76). Next we show that $V \oplus V^{\perp}$ is closed. So, let $\{x_n\}_{n \in \mathbb{N}} \subseteq V \oplus V^{\perp}$ and assume that $x_n \to x$ in H. We have

$$x_n = v_n + u_n \text{ with } v_n \in V, u_n \in V^{\perp}.$$

Since $(v_n, u_m) = 0$, we see that

$$\|x_n - x_m\|^2 = \|v_n - v_m\|^2 + \|u_n - u_m\|^2.$$

Hence $\{v_n\}_{n \in \mathbb{N}} \subseteq V$ and $\{u_n\}_{n \in \mathbb{N}} \subseteq V^{\perp}$ are both Cauchy. Since V and $V^{\perp}$ are closed, we have $v_n \to v \in V$ and $u_n \to u \in V^{\perp}$ and so $x = v + u \in V \oplus V^{\perp}$. Therefore $V \oplus V^{\perp}$ is closed. If $V \oplus V^{\perp}$ is a proper subspace, then by Proposition 3.178 we can find $h \in (V \oplus V^{\perp})^{\perp}$, $h \neq 0$. Then $h \perp V$, $h \perp V^{\perp}$, hence $h = 0$, a contradiction. We conclude that $H = V \oplus V^{\perp}$ □

Corollary 3.180 *If H is a Hilbert space and $V \subseteq H$ is a closed subspace, then $V^{\perp\perp} = (V^{\perp})^{\perp} = V$*

If H is an inner product space, then for any $x \in H$, we can define the linear functional $f_x : H \to \mathbb{R}$ by setting $f_x(u) = (x, u)$ for all $u \in H$. On account of Proposition 3.168, $f_x \in H^*$ and $\|f_x\|_* = \|x\|$ (just note that $f_x(x) = \|x\|^2$). Therefore the map $j : H \to H^*$ defined by $j(x) = f_x$ is an isometric injection. It turns out that in a Hilbert space, all continuous linear functionals can be obtained this way, that is, $j(\cdot)$ is an isometric isomorphism. (usually called the "Riesz isomorphism" for H). This result is known as the "Riesz-Frechet Representation Theorem."

Theorem 3.181 *If H is a Hilbert space and $\widehat{f} \in H^*$, then there exists a unique $\widehat{x} \in H$ such that $\widehat{f}(u) = (\widehat{x}, u)$ for all $u \in H$ and $\|\widehat{f}\|_* = \|\widehat{x}\|$ (that is, $j(\widehat{x}) = \widehat{f}$).*

Proof Clearly we may assume that $\widehat{f} \neq 0$. Let $x_0 \in (\ker \widehat{f})^{\perp}$, $\|x_0\| = 1$. We define

$$\xi(u) = \widehat{f}(x_0)u - \widehat{f}(u)x_0 \text{ for all } u \in H$$

and have $\xi(u) = \ker \widehat{f}$ for all $u \in H$. Hence

$$0 = (\xi(u), x_0) = \widehat{f}(x_0)(u, x_0) - \widehat{f}(u) = (u, \widehat{f}(x_0)x_0) - \widehat{f}(u)$$

for all $u \in H$. We set $\widehat{x} = \widehat{f}(x_0)x_0$ and have $(\widehat{x}, u) = (\widehat{f}(x_0)x_0, u) = \widehat{f}(u)$ for all $u \in H$. Moreover, if $(\widehat{x}, u) = (\widetilde{x}, u)$ for all $u \in H$, then $\widehat{x} - \widetilde{x} \in H^{\perp} = \{0\}$, that is, $\widehat{x} = \widetilde{x}$, which proves the uniqueness of $\widehat{x}$. Finally from Proposition 3.168 we see that $\|\widehat{x}\| \leq \|\widehat{f}\|_*$, while $\widehat{f}(\frac{\widehat{x}}{\|\widehat{x}\|}) = \|\widehat{x}\|$. Therefore $\|\widehat{f}\|_* = \|\widehat{x}\|$. □

Remark 3.182 The space H^*, furnished with the inner product $(x^*, u^*)_* = (j^{-1}(x^*), j^{-1}(u^*))$ for all $x^*, u^* \in H^*$, is a Hilbert space too.

Corollary 3.183 *Every Hilbert space is reflexive.*

Proof Let $\widehat{h} \in H^{**}$. Since H^* is a Hilbert space, according to Theorem 3.181 we can find $\widehat{x}^* \in H^*$ such that $\widehat{h}(u^*) = (\widehat{x}^*, u^*)_* = (j^{-1}(\widehat{x}^*), j^{-1}(u^*))$ for all $u^* \in H^*$. From Proposition 3.103 it follows that $\widehat{h} \in H$ and so $H = H^{**}$, that is, H is reflexive.   □

Let H be a Hilbert space and $A \in \mathcal{L}(H)$. Using the Riesz isomorphism $j(\cdot)$, we are able to associate to A another operator still defined on H and related to the dual operator of A (see Definition 3.137). In the sequel for Hilbert space operators by A^d we denote the dual operator and by A^* this new operator defined on H.

Definition 3.184 Let H be a Hilbert space and $A \in \mathcal{L}(H)$. The "adjoint operator" to A, denoted by A^*, is defined by $A^* = j^{-1} \circ A^d \circ j \in \mathcal{L}(H)$. It is easy to see that this definition is equivalent to saying that for all $x, u \in H$

$$(A(x), u) = (x, A^*(u)).$$

Proposition 3.185 *If H is a Hilbert space and $A \in \mathcal{L}(H)$, then (a) $\|A\|_{\mathcal{L}} = \|A^*\|_{\mathcal{L}}$; (b) $A^{**} = (A^*)^* = A$; (c) $\|A^*A\|_{\mathcal{L}} = \|A\|_{\mathcal{L}}^2$.*

Proof

(a) By Definition 3.184, Proposition 3.139 and the fact that $j(\cdot)$ is an isometric isomorphism, we have

$$\|A^*\|_{\mathcal{L}} = \|j^{-1} \circ A^d \circ j\|_{\mathcal{L}} = \|A^d\|_{\mathcal{L}} = \|A\|_{\mathcal{L}}.$$

(b) For every $x, u \in H$, we have

$$(u, (A^*)^*(x)) = (A^*(u), x) = (x, A^*(u)) = (Ax, u) = (u, A(x)),$$

thus, $A^{**} = A$.

(c) Using (a) we have

$$\|A\|_{\mathcal{L}}^2 = \sup[\|A(x)\|^2 : \|x\| \leq 1] = \sup[(A(x), A(x)) : \|x\| \leq 1]$$

$$= \sup[(x, A^*(A(x)) : \|x\| \leq 1]$$

$$\leq \|A^*A\|_{\mathcal{L}} \leq \|A^*\|_{\mathcal{L}}\|A\|_{\mathcal{L}} = \|A\|_{\mathcal{L}}^2,$$

thus $\|A^*A\|_{\mathcal{L}} = \|A\|_{\mathcal{L}}^2$.   □

In a similar fashion, we can establish the following computation rules for the adjoint.

Proposition 3.186 *If it is a Hilbert space, then (a)* $(\lambda A + T)^* = \lambda A^* + T^*$ *for all* $A, T \in \mathcal{L}(H)$, $\lambda \in \mathbb{R}$; *(b)* $(T \circ A)^* = A^* \circ T^*$ *for all* $A, T \in \mathcal{L}(H)$.

Remark 3.187 If H is a complex Hilbert space and $\lambda \in \mathbb{C}$, then $(\lambda A)^* = \bar{\lambda} A^*$.

Definition 3.188 Let H be a Hilbert space and $A \in \mathcal{L}(H)$. We say that A is "self-adjoint" if $A = A^*$, that is, $(A(x), u) = (x, A(u))$ for all $x, u \in H$.

Remark 3.189 Note that for any $A \in \mathcal{L}(H)$, AA^* and A^*A are self-adjoint. So, there exist plenty of self-adjoint operators. Also $(A(x), x) \in \mathbb{R}$ when H is complex.

Similar to the polarization identities for the norm in an inner product space (see Proposition 3.169), we have corresponding identities for operators.

Proposition 3.190 *If H is an inner product space and $A \in \mathcal{L}(H)$, then*

$$(A(x), u) = \frac{1}{4}[(A(x + u), x + u) - (A(x - u), x - u)],$$

for all $x, u \in H$, provided A is self-adjoint (real case); and

$$(A(x), u) = \frac{1}{4}[(A(x + u), x + u) - (A(x - u), x - u)$$

$$+ i(A(x + iu), x + iu) - i(A(x - iu), x - iu)]$$

(complex case).

As a consequence of these identities, we have:

Proposition 3.191 *(a) If H is a real Hilbert space, $A \in \mathcal{L}(H)$ is self-adjoint and $(A(x), x) = 0$ for all $x \in H$, then $A = 0$. (b) If it is a complex Hilbert space, $A \in \mathcal{L}(H)$ and $(A(x), x) = 0$ for all $x \in H$, then $A = 0$.*

Remark 3.192 In the real case, the above result fails if A is not self-adjoint. For example, let $A\binom{x_1}{x_2} = \binom{-x_2}{x_1}$ for all $\hat{x} = \binom{x_1}{x_2}$ (this operator performs a counterclockwise rotation by $\pi/2$). Then $(A(\hat{x}), \hat{x}) = 0$ for all $\hat{x} \in \mathbb{R}^2$, but $A \neq 0$. In this case the matrix $A = \begin{bmatrix} 0 & -1 \\ 1 & 0 \end{bmatrix}$ is not symmetric.

Proposition 3.193 *If H is a Hilbert space and $A \in \mathcal{L}(H)$ is self-adjoint, then* $\|A\|_{\mathcal{L}} = \sup[|(A(x), x)| : \|x\| \leq 1] = M_A$.

Proof Let $\|x\| \leq 1$. Then $|(A(x), x)| \leq \|A(x)\| \|x\| \leq \|A\|_{\mathcal{L}} \|x\|^2 \leq \|A\|_{\mathcal{L}}$. Hence

$$M_A \leq \|A\|_{\mathcal{L}}. \tag{3.22}$$

Next let $u \in H$, $u \neq 0$ and set $\lambda = (\frac{\|A(u)\|}{\|u\|})^{1/2}$, $v = \frac{1}{\lambda}A(u)$. We have

$$\|A(u)\|^2 = (A(\lambda u), \frac{1}{\lambda}A(u)) = (A(\lambda u), v)$$

$$= \frac{1}{4}[(A(\lambda u + v), \lambda u + v) - (A(\lambda u - v), \lambda u - v)]$$

(see Proposition 3.190)

$$\leq \frac{1}{4}M_A[\|\lambda u + v\|^2 + \|\lambda u - v\|^2]$$

$$= \frac{1}{2}M_A[\|\lambda u\|^2 + \|v\|^2] = \frac{1}{2}M_A[\lambda^2\|u\|^2 + \frac{1}{\lambda^2}\|A(u)\|^2]$$

$$= M_A\|u\|\|A(u)\|,$$

$\Rightarrow \|A(u)\| \leq M_A\|u\|$ for all $u \in H$, thus

$$\|A\|_{\mathcal{L}} \leq M_A. \tag{3.23}$$

From (3.22) and (3.23) it follows that $\|A\|_{\mathcal{L}} = M_A$.   □

Proposition 3.194 *If H is a Hilbert space and $A \in \mathcal{L}(H)$ is self-adjoint, then $\|A^n\|_{\mathcal{L}} = \|A\|_{\mathcal{L}}^n$ for all $n \in \mathbb{N}$.*

Proof We have

$$\|A\|_{\mathcal{L}}^2 = \sup[(A(x), A(x)) : \|x\| \leq 1]$$

$$= \sup[(A^*A(x), x) : \|x\| \leq 1] \leq \|A^*A\|_{\mathcal{L}}$$

$$\leq \|A^*\|_{\mathcal{L}}\|A\|_{\mathcal{L}} = \|A\|_{\mathcal{L}}^2,$$

thus $\|A\|_{\mathcal{L}}^2 = \|A^*A\|_{\mathcal{L}} = \|A^2\|_{\mathcal{L}}$ (since A is self-adjoint). For every $k \in \mathbb{N}$, A^k is self-adjoint and so from the previous part we have $\|A^{2^k}\|_{\mathcal{L}} = \|A\|_{\mathcal{L}}^{2^k}$ for all $k \in \mathbb{N}$. Let $n \in \mathbb{N}$ such that $1 \leq n \leq 2^k$. Then

$$\|A\|_{\mathcal{L}}^{2^k} = \|A^{2^k}\|_{\mathcal{L}} = \|A^n A^{2^k-n}\|_{\mathcal{L}}$$

$$\leq \|A^n\|_{\mathcal{L}}\|A\|_{\mathcal{L}}^{2^k-n}$$

$$\leq \|A\|_{\mathcal{L}}^n\|A\|_{\mathcal{L}}^{2^k-n} = \|A\|_{\mathcal{L}}^{2^k},$$

thus $\|A^n\|_{\mathcal{L}}\|A\|_{\mathcal{L}}^{2^k-n} = \|A\|_{\mathcal{L}}^{2^k}$ and so $\|A^n\|_{\mathcal{L}} = \|A\|_{\mathcal{L}}^n$ for all $n \in \mathbb{N}$.   □

The notion of orthogonality in Hilbert spaces leads to a special kind of projection operators (see Definition 3.76(a)).

Definition 3.195 Let H be an inner product space and $P : H \to H$ is a projection. We say that P is an "orthogonal projection" if $R(P) \perp \ker P$.

For such projections, we have the following characterization.

Proposition 3.196 *If H is an inner space and $P : H \to H$ is a projection, then P is an orthogonal if and only if $(A(x), u) = (x, A(u))$ for all $x, u \in H$.*

Proof $\Rightarrow$: For $x, u \in H$ we have $x = y + v$, $u = \widehat{y} + \widehat{v}$ with $y, \widehat{y} \in R(P)$, $v, \widehat{v} \in \ker P$. Then

$$(P(x), u) = (y, \widehat{y} + \widehat{v}) = (y, \widehat{y}) \text{ and } (x, P(u)) = (y + v, \widehat{y}) = (y, \widehat{y}).$$

$\Leftarrow$: Let $y \in R(P)$ and $v \in \ker P$. Then $P(y) = y$ and $P(v) = 0$ and so $(y, v) = (P(y) + P(v), v) = 0 \Rightarrow R(P) \perp \ker P$. $\square$

In fact an orthogonal projection is automatically continuous.

Proposition 3.197 *If H is an inner product space and $P : H \to H$ is an orthogonal projection, then $P \in \mathcal{L}(H)$ and $\|P\|_{\mathcal{L}} = 1$.*

Proof For $x \in H$ we have $x = y + v$ with $y \in R(P)$ and $v \in \ker P$, $y \perp v$. Then $\|x\|^2 = \|P(x)\|^2 + \|v\|^2$ (see Theorem 3.177), thus

$$\|P(x)\| \leq \|x\| \text{ and so } P \in \mathcal{L}(H), \|P\|_{\mathcal{L}} \leq 1. \tag{3.24}$$

Since P is a projection, $P^2 = P$ (see Definition 3.76(a)). Hence

$$\|P\|_{\mathcal{L}} = \|P^2\|_{\mathcal{L}} \leq \|P\|_{\mathcal{L}}^2 \text{ and } 1 \leq \|P\|_{\mathcal{L}}. \tag{3.25}$$

From (3.24) and (3.25) it follows that $\|P\|_{\mathcal{L}} = 1$. $\square$

Combing Propositions 3.196 and 3.197, we obtain.

Corollary 3.198 *If H is Hilbert space and $P : H \to H$ is a projection, then P is an orthogonal projection if and only if P is self-adjoint.*

Definition 3.199 Let H be a Hilbert space and $A \in \mathcal{L}(H)$. We say that A is "positive" ("monotone") if $(A(x), x) \geq 0$ for all $x \in H$. We write $A \geq 0$.

Remark 3.200 In a complex Hilbert space, a positive operator is self-adjoint. This is no longer true for real Hilbert space. (see Proposition 3.190). For any $A \in \mathcal{L}(H)$, we have $A^*A \geq 0$. If $A, K \in \mathcal{L}(H)$, then we say that $K \leq A$ if and only if $A - K \geq 0$.

Proposition 3.201 *If H is Hilbert space and $P \in \mathcal{L}(H)$ is an orthogonal projection, then $P \geq 0$.*

Proof For every $x \in H$, we have $(P(x), x) = (P^2(x), x) = (P(x), P(x)) = \|P(x)\|^2 \geq 0$ (see Corollary 3.198). $\square$

Now we turn our attention to the spectrum of a bounded linear operator. Spectral theory examines the properties of the set

$$\{\lambda \in F : \lambda I - A \text{ is not invertible }\},$$

with $F = \mathbb{R}$ or $\mathbb{C}$. Such a study has important applications in many parts of mathematics.

Definition 3.202 Let X be a Banach space over $F(= \mathbb{R} \text{ or } \mathbb{C})$ and $A \in \mathcal{L}(X)$. The "spectrum" $\sigma(A)$ of A is the set

$$\sigma(A) = \{\lambda \in F : \lambda I - A \text{ is not an isomorphism }\}.$$

The "resolvent set" $\varrho(A)$ is defined by $\varrho(A) = F \setminus \sigma(A)$. The points of $\varrho(A)$ are called "regular values" of A and if $\lambda \in \varrho(A)$, the operator $R(\lambda) = (\lambda I - A)^{-1} \in \mathcal{L}(X)$ is the "resolvent operator."

In general the operator $\lambda I - A$ is not invertible if and only if

(a) $\lambda I - A$ is not injective (that is, $\ker(\lambda I - A) \neq \{0\}$.
(b) $\lambda I - A$ is injective but $R(\lambda I - A)$ is not dense in X.
(c) $\lambda I - A$ is injective, $R(\lambda I - A)$ is dense in X but $(\lambda I - A)^{-1}$ is not continuous.

If $\lambda \in \sigma(A)$ corresponds to (a), then $\lambda \in \sigma_p(A) = $ point spectrum of A.
If $\lambda \in \sigma(A)$ corresponds to (b), then $\lambda \in \sigma_r(A) = $ residual spectrum of A.
If $\lambda \in \sigma(A)$ corresponds to (c), then $\lambda \in \sigma_c(A) = $ continuous spectrum of A.
    We have

$$\sigma(A) = \sigma_\varrho(A) \cup \sigma_r(A) \cup \sigma_c(A).$$

For complex Banach spaces, we have $\sigma(A) \neq \emptyset$. This is no longer true for real Banach spaces. From basic Linear Algebra we have the following result.

Proposition 3.203 *If X is a finite dimensional vector space and $A : X \to X$ is a linear operator, then A is injective if and only if A is surjective.*

So, in finite dimensional spaces the spectral theory simplifies considerably.

Corollary 3.204 *If X is a finite dimensional vector space and $A : X \to X$ is a linear operator, then $\sigma(A) = \sigma_p(A)$.*

For the elements of the point spectrum, we have the classical terminology from Linear Algebra.

Definition 3.205 Let X be a Banach space and $A \in \mathcal{L}(X)$. The elements of $\sigma_p(A)$ are called "eigenvalues" of A. If λ is an eigenvalue of A, then the elements $u \in X \setminus \{0\}$ such that $A(u) = \lambda u$ are called "eigenvectors" of A corresponding to λ. The subspace $E_\lambda = \{u \in X : A(u) = \lambda u\}$ is the "eigenspace" corresponding to the eigenvalue $\lambda \in \sigma_p(A)$.

The next proposition gives the basic relation between eigenvalues and the corresponding eigenvectors.

Proposition 3.206 *If X is a Banach space, $A \in \mathcal{L}(X)$, $\{\lambda_k\}_{k=1}^{n}$ are distinct eigenvalues of A and $\{u_k\}_{k=1}^{n} \subseteq X$ are corresponding eigenvectors, then $\{u_k\}_{k=1}^{n}$ are linearly independent.*

Proof We argue by induction. Evidently $\{u_1\}$ is linearly independent. Suppose that $\{u_k\}_{k=1}^{m-1} \subseteq X$ are linearly independent and set $u_m = \sum_{k=1}^{m-1} \vartheta_k u_k \ \vartheta_k \in F$. We have

$$\lambda_m u_m = A(u_m) = \sum_{k=1}^{m-1} \vartheta A(u_k) = \sum_{k=1}^{m-1} \vartheta_k \lambda_k u_k,$$

hence $\sum_{k=1}^{m-1} (\lambda_m - \lambda_k)\vartheta_k u_k = 0$. But by hypothesis $\{u_k\}_{k=1}^{m-1}$ are linearly independent and $\lambda_m - \lambda_k \neq 0$ for all $k = 1, \ldots, m-1$. Hence $\vartheta_k = 0$ for all $k = 1, \ldots, m-1$. Therefore $\{u_k\}_{k=1}^{m} \subseteq X$ are linearly independent. $\square$

Using Proposition 3.143 and Definition 3.184, we have the following result.

Proposition 3.207 *(a) If X is a Banach space and $A \in \mathcal{L}(X)$, then $\sigma(A) = \sigma(A^*)$. (b) If X is a complex Hilbert space and $A \in \mathcal{L}(H)$, then $\lambda \in \sigma(A)$ if and only if $\bar{\lambda} \in \sigma(A^*)$.*

Proposition 3.208 *If H is a complex Hilbert space and $A \in \mathcal{L}(H)$ is self-adjoint, then $(A(u), u) \in \mathbb{R}$ for all $u \in X$, and $\sigma_p(A) \subseteq \mathbb{R}$.*

Proof For $u \in X$, we have $(A(u), u) = (u, A(u)) = \overline{(A(u), u)}$, hence $(A(u), u) \in \mathbb{R}$. Suppose $\lambda \in \sigma_p(A)$ and $u \in H$ a corresponding eigenvector. Then we have

$$(A(u), u) = (\lambda u, u) = \lambda \|u\|^2,$$

hence $\lambda = \frac{(A(u), u)}{\|u\|^2} \in \mathbb{R}$. $\square$

Proposition 3.209

(a) *If X is a Banach space and $A \in \mathcal{L}(X)$, then $\sigma(A)$ is compact and*

$$|\sigma(A)| = \sup\{|\lambda| : \lambda \in \sigma(A)\} \leq \|A\|_{\mathcal{L}}.$$

(b) *If H is a Hilbert space and $A \in \mathcal{L}(H)$ is self-adjoint, then $\lambda \in \sigma(A)$ if and only if*

$$\inf[\|(\lambda I - A)(x)\| : \|x\| = 1] = 0.$$

Proof

(a) By Remark 3.28, $\sigma(A)$ is closed. If $|\lambda| > \|A\|_{\mathcal{L}}$, then

$$(\lambda I - A)^{-1} = \frac{1}{\lambda}(I - \frac{1}{\lambda}A)^{-1} = \sum_{n \geq 0} \frac{1}{\lambda^{n+1}} A^n$$

and so $\lambda \notin \sigma(A)$ (see Remark 3.28). Therefore $\sigma(A)$ is bounded, hence compact.

(b) $\Rightarrow$ Suppose that $\inf[\|(\lambda I - A)(x)\| : \|x\| = 1] = m > 0$. Then

$$\|(\lambda I - A)(x)\| \geq m\|x\| \text{ for all } x \in H,$$

thus $\lambda I - A$ is an isomorphism (see Proposition 3.26), $\Rightarrow \lambda \in \varrho(A)$.
$\Leftarrow$ If $\lambda \in \varrho(A)$, then $\lambda I - A$ is an isomorphism. For $x \in H$ with $\|x\| = 1$ we have

$$1 = \|x\| = \|(\lambda I - A)^{-1}(\lambda I - A)(x)\| \leq \|(\lambda I - A)^{-1}\|_{\mathcal{L}}\|(\lambda I - A)(x)\|,$$

thus $0 < \|(\lambda I - A)^{-1}\|_{\mathcal{L}}^{-1} \leq \inf[\|(\lambda I - A)(x)\| : \|x\| = 1]$. $\qquad\square$

Proposition 3.210 *If H is Hilbert space, $A \in \mathcal{L}(H)$ is self-adjoint and*

$$m_A = \inf[(A(x), x) : \|x\| = 1] \text{ and } M_A = \sup[(A(x), x) : \|x\| = 1],$$

then $\sigma(A) \subseteq [m_A, M_A] \subseteq \mathbb{R}$ and $m_A, M_A \in \sigma(A)$.

Proof First we show that the spectrum of A is real (in fact we already have that $\sigma_p(A) \subseteq \mathbb{R}$, see Proposition 3.208). So, we assume that H is complex and consider $\lambda = \vartheta + i\mu$. For $x \in H$ with $\|x\| = 1$, we have since A is self-adjoint

$$(\lambda x - A(x), x) - (x, \lambda x - A(x)) = [\lambda - \bar{\lambda}]\|x\|^2 = 2i\mu,$$

thus

$$\begin{aligned} 2|\mu| &= |(\lambda x - A(x), x) - (x, \lambda x - A(x))| \\ &\leq |(\lambda x - A(x), x)| + |(x, \lambda x - A(x))| \\ &\leq 2\|(\lambda I - A)(x)\| \text{ (recall } \|x\| = 1), \end{aligned}$$

thus $|\mu| \leq \inf[\|(\lambda I - A)(x)\| : \|x\| = 1]$. By Proposition 3.209, $\lambda \in \sigma(A)$ implies that $\mu = 0$ and so $\lambda \in \mathbb{R}$.
If $\widehat{A} = A + \eta I$, then

$$\sigma(\widehat{A}) = \sigma(A) + \eta, \ m_{\widehat{A}} = m_A + \eta, \ M_{\widehat{A}} = M_A + \eta.$$

So, by doing such a translation if necessary, we may assume that $0 \le m_A \le M_A$. From Proposition 3.193 we have that $\|A\|_{\mathcal{L}} = M_A$. Therefore

$$\sigma(A) \subseteq [-M_A, M_A]$$

(recall $\sigma(A) \subseteq \mathbb{R}$ and see Proposition 3.209(a)).

Let $\tau > 0$ and $\lambda = m_A - \tau$. For $x \in H$ with $\|x\| = 1$, we have

$$(A(x) - \lambda x, x) = (A(x), x) - \lambda\|x\|^2 \ge m_A - \lambda = \tau,$$

thus $0 < \tau < \|(\lambda I - A)(x)\|$ for all $x \in H$ with $\|x\| = 1$. Then Proposition 3.209(b) implies that $\lambda \notin \sigma(A)$. Therefore we conclude that $\sigma(A) \subseteq [m_A, M_A]$.

Consider a sequence $\{x_n\}_{n \in \mathbb{N}} \subseteq H$, $\|x_n\| = 1$ for all $n \in \mathbb{N}$ such that $(A(x), x_n) \uparrow M_A$. Since $0 \le m_A \le M_A$, we have $M_A = \|A\|_{\mathcal{L}}$ and so $\|A(x_n)\| \le M_A$. Then

$$0 \le \|(M_A I - A)(x_n)\|^2$$
$$= M_A^2 - 2M_A(A(x_n), x_n) + \|A(x_n)\|^2$$
$$\le 2M_A^2 - 2M_A(A(x_n), x_n) \to 0,$$

thus $\|(M_A I - A)(x_n)\| \to 0$ and so $M_A \in \sigma(A)$ (see Proposition 3.209(b)). Let $K = A - M_A I$, then $m_K \le M_K = 0$ and so $|m_k| = \|A\|_{\mathcal{L}}$. Then reasoning as above, we obtain $m_K \in \sigma(A)$. $\qquad\square$

Corollary 3.211 *If H is a Hilbert space and $A \in \mathcal{L}(H)$ is self-adjoint, then $\|A\|_{\mathcal{L}} \in \sigma(A)$.*

The next proposition improves Proposition 3.206 in the case of self-adjoint operators on a Hilbert space.

Proposition 3.212 *If H is a Hilbert space and $A \in \mathcal{L}(H)$ is self-adjoint, then eigenvectors corresponding to distinct eigenvalues are orthogonal.*

Proof Let $\lambda, \mu \in \sigma(A) \subseteq \mathbb{R}$, $\lambda \ne \mu$, and $u, v \in H$ be the corresponding eigenvectors. We have

$$(A(u), v) = \lambda(u, v) \text{ and } (A(u), v) = (u, A(v)) = \mu(u, v),$$

thus $(\lambda - \mu)(u, v) = 0$. Since $\lambda \ne \mu$, we must have $(u, v) = 0$. $\qquad\square$

So, the eigenspaces corresponding to distinct eigenvalues are orthogonal. Moreover, they are invariant for the operator (that is, if E is an eigenspace, then $A(E) \subseteq E$).

Proposition 3.213 *If H is a Hilbert space. $A \in \mathcal{L}(H)$ is self-adjoint and $V \subseteq H$ is a closed subspace which is A-invariant, then $V^\perp$ is A-invariant too and if furthermore $K_1 = A|_V$, $K_2 = A|_{V^\perp}$, then*

$$K_1 \in \mathcal{L}(V) \text{ and } K_2 \in \mathcal{L}(V^\perp) \text{ are self-adjoint,}$$

and $R(A) = R(K_1) \oplus R(K_2)$ and $\sigma(A) = \sigma(K_1) \cup \sigma(K_2)$.

Proof Let $y \in V^\perp$. We have $A(v) \in V$ for all $v \in V$ hence $0 = (A(v), y) = (v, A(y))$ and so $A(y) \in V^\perp$, that is, $V^\perp$ is A-invariant. Then $K_1 = A|_V \in \mathcal{L}(V)$ and $K_2 = A|_{V^\perp} \in \mathcal{L}(V^\perp)$ are self-adjoint (since A is). Also

$$R(A) = R(K_1) \oplus R(K_2).$$

Let $\lambda \in \sigma(K_1)$. Then according to Proposition 3.209(b) we can find $x_n \in V$ $\|x_n\| = 1$ $n \in \mathbb{N}$ such that $\|\lambda x_n - K_1(x_n)\| \to 0$, hence $\|\lambda x_n - A(x_n)\| \to 0$ and so $\lambda \in \sigma(A)$ (again by Proposition 3.209(b)). Therefore $\sigma(K_1) \subseteq \sigma(A)$. Similarly we show that $\sigma(K_2) \subseteq \sigma(A)$. Therefore

$$\sigma(K_1) \cup \sigma(K_2) \subseteq \sigma(A).$$

Now suppose that $\lambda \notin \sigma(K_1) \cup \sigma(K_2)$. Then we can find $c > 0$ such that

$$\|\lambda v - A(v)\| \geq c\|v\| \text{ for } v \in V \text{ and } \|\lambda y - A(y)\| \geq c\|y\| \text{ for } y \in V^\perp \qquad (3.26)$$

(see Proposition 3.26). Let $x \in H$. We write $x = v + y$ with $v \in V$, $y \in V^\perp$. Then $\lambda v - A(v) \in V$ and $\lambda y - A(y) \in V^\perp$, thus

$$\|\lambda x - A(x)\|^2 = \|\lambda v - A(v)\|^2 + \|\lambda y - A(y)\|^2$$

$$\text{(see Theorem 3.177)}$$

$$\geq c^2[\|v\|^2 + \|y\|^2] \text{ (see (3.26))}$$

$$= c^2\|x\|^2 \text{ (see Theorem 3.177),}$$

thus $\|\lambda x - A(x)\| \geq c\|x\|$ for all $x \in H$, $\Rightarrow \lambda \in \varrho(A)$ (see Proposition 3.26), hence $\sigma(A) = \sigma(K_1) \cup \sigma(K_2)$. $\qquad \square$

Corollary 3.214 *If H is a Hilbert space and $A \in \mathcal{L}_c(H)$ is self-adjoint, then* $\sigma_p(A) \neq \emptyset$.

Proof If $A = 0$, then $\lambda = 0 \in \sigma_p(A)$. If $A \neq 0$, then by Corollary 3.211 $\|A\|_\mathcal{L} \in \sigma(A)$ and $\lambda = \|A\|_\mathcal{L} \neq 0$. We claim $\lambda \in \sigma_p(A)$. If this is not the case, then $\ker(\lambda I - A) = \{0\}$ and so by Theorem 3.159(c), we have $R(\lambda I - A) = H$. Therefore $\lambda I - A$ is an isomorphism and so $\lambda \in \varrho(A)$ a contradiction. $\qquad \square$

An interesting byproduct of the above proof is the following corollary.

Corollary 3.215 *If X is Banach space and $A \in \mathcal{L}_c(X)$, then*

$$\sigma(A) = \sigma_p(A) \cup \{0\}.$$

In fact we can have the following improved version of this corollary (see Rudin [232], p.103).

Proposition 3.216 *If X is a Banach space and $A \in \mathcal{L}_c(X)$, then $\sigma(A) = D \cup \{0\}$ with D (possible empty) either a finite set or a sequence $\lambda_k \to 0$ and $D \subseteq \sigma_p(A)$.*

We will apply this proposition in the context of compact self-adjoint operators in a Hilbert space. First let us recall a basic notion for Hilbert spaces.

Definition 3.217 Let H be a Hilbert space and $E \subseteq H$. We say that E is an "orthonormal set" if $\|u\| = 1$ for all $u \in E$, and

$$(u_\alpha, u_\beta) = 0 \text{ for all } u_\alpha \neq u_\beta \text{ from } E.$$

A maximal orthonormal set in the sense of inclusion is an "orthonormal basis" for H.

Remark 3.218 Starting with a sequence $\{x_n\}_{n \in \mathbb{N}}$ of linearly independent vectors in H and using the Gram-Schmidt orthonormalization process, we can generate an orthonormal set. It is an easy consequence of Zorn's lemma that every Hilbert space has an orthonormal basis. Note that such a basis is a Schauder basis for H (see Definition 3.37).

The next theorem can be found in Kubrusly [165], p.357.

Theorem 3.219 *A Hilbert space H is separable if and only if it has a countable orthonormal basis.*

Now we can have the first theorem concerning the spectrum of a compact self-adjoint operator in a Hilbert space.

Theorem 3.220 *If H is an infinite dimensional Hilbert space and $A \in \mathcal{L}_c(H)$ is nonzero and self-adjoint, then*

$$\sigma(A) = D \cup \{0\} \text{ with } \|A\|_{\mathcal{L}} \in D \subseteq \sigma_p(A),$$

where the set D either is finite or $D = \{\lambda_k\}_{k \in \mathbb{N}}$, with $\lambda_k \to 0$; moreover, H has an orthonormal basis consisting of eigenvectors of A.

Proof Since H is infinite dimensional and $A \in \mathcal{L}_c(H)$, we have $0 \in \sigma(A)$. Also from Corollary 3.211, we know that $\|A\|_{\mathcal{L}} \in \sigma(A) \setminus \{0\}$. Then by Proposition 3.216 we have $\|A\|_{\mathcal{L}} \in D \subseteq \sigma_p(A)$, with D as postulated by the theorem.

Let $\lambda \in D \subseteq \sigma_p(A)$ and let $E_\lambda = \ker(\lambda I - A)$. Let B_λ be an orthonormal basis for E_λ. We set $B = \bigcup_{\lambda \in D} B_\lambda$. Then $B \subseteq H$ is orthonormal. Suppose $\overline{\text{span}}B \neq H$. Let $V = \overline{\text{span}}B$. Since each eigenspace is A-invariant, so is V. Hence $V^\perp$ is A-invariant and if $K_1 = A|_V$, $K_2 = A|_{V^\perp}$, then $\sigma(A) = \sigma(K_1) \cup \sigma(K_2)$ (see Proposition 3.213). Since K_2 is self-adjoint, $\sigma_p(K_2) \neq \emptyset$ (see Corollary 3.214) and so we can find $\lambda \in \sigma_p(K_2)$ and a corresponding eigenvector y. Since $\lambda \in \sigma_p(A)$ we have $y \in V \cap V^\perp$, hence $y = 0$, a contradiction. So $V = H$. $\qquad\square$

Corollary 3.221 *If H is a Hilbert space and $A \in \mathcal{L}_c(H)$ is self-adjoint, then $\sigma(A) = \overline{\sigma_\varrho(A)}$.*

Next we state the second theorem from the spectral theory of compact self-adjoint operators on a Hilbert space. The result is known as the "Spectral Decomposition Theorem."

Theorem 3.222 *If H is an infinite dimensional separable Hilbert space and $A \in \mathcal{L}_c(H)$ is self-adjoint. then there is an orthonormal basis $\{u_n\}_{n\in\mathbb{N}}$ consisting of eigenvectors corresponding to the real eigenvalues $\{\lambda_n\}_{n\in\mathbb{N}}$ of A and for every $x \in H$ we have $A(x) = \sum_{n\in\mathbb{N}} \lambda_n(x, v_n)u_n$.*

Proof From Theorems 3.219 and 3.220, we have an orthonormal basis $\{u_n\}_{n\in\mathbb{N}}$ consisting of eigenvectors of A. Let $\widehat{A}(x) = \sum_{n\in\mathbb{N}} \lambda_n(x, u_n)u_n$. Then for every $x \in H$ with $\|x\| \leq 1$, we have

$$\|\sum_{n=1}^{m} \lambda_n(x, u_n)u_n\|^2 = \sum_{n=1}^{m} \lambda_n^2(x, u_n)^2 \text{ (see Theorem 3.177)}$$

$$\leq \|A\|_{\mathcal{L}}^2 \|x\|^2 \text{ for all } m \in \mathbb{N},$$

thus $\widehat{A} \in \mathcal{L}(H)$. Note that $A(u_n) = \widehat{A}(u_n)$ for all $n \in \mathbb{N}$ and since $\{u_n\}_{n\in\mathbb{N}}$ is an orthonormal basis, from the continuity and linearity of A, $\widehat{A}$ it follows that $A = \widehat{A}$. $\qquad\square$

Remark 3.223 We can also show that if $\lambda \in \varrho(A)$, then for all $x \in H$

$$R(\lambda)(x) = \sum_{n\in\mathbb{N}} \frac{(x, u_n)}{\lambda - \lambda_n} u_n.$$

We conclude this section with a result known in the literature as "Stampacchia's Theorem," which is important in the study of variational inequalities.

Theorem 3.224 *Assume that H is a Hilbert space, $a : H \times H \rightarrow \mathbb{R}$ is a bilinear form such that (i) $|a(x, u)| \leq M\|x\|\|u\|$ for all $x, u \in H$, some $M > 0$ (continuity), (ii) $a(x, x) \geq c\|x\|^2$ for all $x \in H$, some $c > 0$ (coercivity). Let $C \subseteq H$ be closed, convex, $h \in H$. Then there exists unique $\widehat{u} \in C$ such that*

$$a(\widehat{u}, x - \widehat{u}) \geq (h, x - \widehat{u}) \text{ for all } x \in C. \tag{3.27}$$

Proof We fix $v \in H$. The map $y \rightarrow a(v, y)$ belongs in H^* (see hypothesis (i)). Theorem 3.181 (the Riesz-Frechet Representation Theorem), says that there exists unique $A(v) \in H$ such that

$$(A(v), y) = a(v, y) \text{ for all } y \in H.$$

Clearly $A(\cdot)$ is linear and $\|A(v)\| \leq M\|v\|$, $(A(v), v) \geq c\|v\|^2$ for all $v \in H$. Therefore $A \in \mathcal{L}(H)$ and (3.27) is equivalent to finding $\widehat{u} \in C$ such that

$$(A(\widehat{u}), x - \widehat{u}) \geq (h, x - \widehat{u}) \text{ for all } x \in C. \tag{3.28}$$

Let $\vartheta > 0$. Then (3.28) is equivalent to

$$(\vartheta h - \vartheta A(\widehat{u}) + \widehat{u} - \widehat{u}, x - \widehat{u}) \leq 0 \text{ for all } x \in C,$$

thus $\widehat{u} = p_C(\widehat{u} - \vartheta A(\widehat{u}) + \vartheta h)$ (see Proposition 3.176(a)).

Consider the map $f : C \to C$ defined by

$$f(x) = p_C(x - \vartheta A(x) + \vartheta h).$$

By Proposition 3.176(b), we have for all $x, w \in C$

$$\|f(x) - f(w)\| \leq \|x - w - \vartheta(A(x) - A(w))\|,$$

thus

$$\|f(x) - f(w)\|^2 \leq \|x - w\|^2 - 2\vartheta(x - w, A(x - w)) + \vartheta^2\|A(x) - A(w)\|^2$$

$$\leq [1 - 2\vartheta c + \vartheta^2 M^2]\|x - w\|^2 \text{(see hypotheses (i),(ii)).} \tag{3.29}$$

Choose $\vartheta \in (0, \frac{2c}{M^2})$, then (3.29) implies that f is a contraction and so by the Banach fixed point theorem, there exist unique $\widehat{u} \in C$ such at $f(\widehat{u}) = \widehat{u}$. $\quad\square$

A consequence of this theorem is the so-called Lax-Milgram Theorem, which a useful tool in the study of semilinear elliptic equations.

Corollary 3.225 *If H is a Hilbert space, $a : H \times H \to \mathbb{R}$ is a bilinear form which is continuous and coercive (see hypotheses (i), (ii) of Theorem 3.124) and $h \in H$, then there exists unique $\widehat{u} \in H$ such that $a(\widehat{u}, x) = (h, x)$ for all $x \in H$.*

Proof Apply Theorem 3.224 for the particular case when $C = H$. $\quad\square$

3.9 Unbounded Linear Operators

In this section, we have a brief encounter with unbounded linear operators.

Definition 3.226 Let X, Y be Banach spaces. An "unbounded linear operator" from X to Y is a pair $(A, D(A))$, where $D(A) \subseteq X$ is a subspace called the "domain of A" and $A : D(A) \to Y$ is a linear operator. We say that $A(\cdot)$ is "densely defined" if $\overline{D(A)} = X$; and we say that $A(\cdot)$ is "closed" if $\operatorname{Gr} A = \{(x, A(x)) : x \in D(A)\} \subseteq X \times Y$ is closed. As always by $R(A)$ we denote the range of A and by $\ker A$ the kernel of A.

Remark 3.227 If $A : D(A) \subseteq X \to Y$ is closed, then $\ker A \subseteq X$ is closed, but $R(A)$ need not be closed. The domain $D(A)$ of an unbounded operator $A : D(A) \subseteq X \to Y$ is a normed space with the "graph norm"

$$|x| = \|x\|_X + \|A(x)\|_Y \text{ for all } x \in D(A)$$

(see the proof of Theorem 3.86). So, an unbounded linear operator can be viewed as a bounded linear operator from $D(A)$ furnished with the graph norm into Y. Note that $A(\cdot)$ is closed if and only if $(D(A), | \cdot |)$ is a Banach space.

We can extend the notion of dual operator (see Definition 3.137) to unbounded linear operators.

Definition 3.228 Let X, Y be Banach spaces and $A : D(A) \subseteq X \to Y$ is an unbounded linear operator which is densely defined. The "dual operator" of A, is the unbounded linear operator $A^* : D(A^*) \subseteq Y^* \to X^*$ defined as follows:

$$D(A^*) = \{y^* \in Y^* : |\langle y^*, A(x) \rangle| \leq c\|x\|_X \text{ for all } x \in D(A) \text{ and some } c > 0\}$$
$$(3.30)$$

and for any $y^* \in D(A^*)$, $A^*(y^*) \in X^*$ is the unique element such that

$$\langle A^*(y^*), x \rangle_X = \langle y^*, A(x) \rangle_Y \text{ for all } x \in D(A).$$

Remark 3.229 For $y^* \in D(A^*)$, consider the map $f : D(A) \to \mathbb{R}$ defined by

$$f(x) = \langle y^*, A(x) \rangle \text{ for all } x \in D(A),$$

thus $|f(x)| \leq c\|x\|_X$ (see (3.30)), that is $f \in D(A)^*$. By Proposition 3.54, there exists $\widehat{f} \in X^*$ such that $\widehat{f}|_{D(A)} = f$. In fact $\widehat{f}$ is unique since $\overline{D(A)} = X$. Therefore we can define $A^*(y^*) = \widehat{f}$. So, the basic relation, linking A and A^* is the following one

$$\langle y^*, A(x) \rangle_Y = \langle A^*(y^*), x \rangle_X \text{ for all } x \in D(A) \text{ and } y^* \in D(A^*). \quad (3.31)$$

Note that in general A^* is not densely defined.

Proposition 3.230 *If X, Y are Banach spaces and $A : D(A) \subseteq X \to Y$ is a densely defined unbounded linear operator, then the dual operator $A^* : D(A^*) \subseteq Y^* \to X^*$ is closed.*

Proof Let $\{y_n^*\}_{n \geq 1} \subseteq D(A^*)$ be such that $y_n^* \to y^*$ in Y^* and $A^*(y_n^*) \to x^*$ in X^*. We have

$$\langle y_n^*, A(x) \rangle_Y = \langle A^*(y_n^*), x \rangle_X \text{ for all } x \in D(A) \text{ and } n \in \mathbb{N},$$

thus $\langle y^*, A(x) \rangle_Y = \langle x^*, x \rangle_X$ for all $x \in D(A)$. Hence,

$$|\langle y^*, A(x) \rangle_Y| \leq \|x^*\|_{X^*} \|x\|_X$$

and so $y^* \in D(A^*)$ and $A^*(y^*) = x^*$, therefore A^* is closed. $\qquad\square$

Remark 3.231 There is a simple relation between $\text{Gr}\, A$ and $\text{Gr}\, A^*$. Let

$$j : Y^* \times X^* \to X^* \times Y^*$$

be defined by $j(y^*, x^*) = (-x^*, y^*)$. Then given $A : D(A) \subseteq X \to Y$ a densely defined unbounded linear operator, we have $j(\text{Gr}\, A^*) = \text{Gr}\, A^{\perp}$.

Corollary 3.232 *If X, Y are Banach spaces and $A : D(A) \subseteq X \to Y$ is an unbounded linear operator which is densely defined and closed, then*

(a) $\ker A = {}^{\perp} R(A^*)$.
(b) $\ker A^* = R(A)^{\perp}$.
(c) $\overline{R(A^*)} \subseteq (\ker A)^{\perp}$.
(d) $\overline{R(A)} = (\ker A^*)^{\perp}$.

The next result characterizes those linear operators with a closed range. For a proof we refer to Brezis [49], p.46.

Proposition 3.233 *If X, Y are Banach spaces and $A : D(A) \subseteq X \to Y$ is an unbounded linear operator which is density defined and closed, then the following statements are equivalent:*

(a) $R(A) \subseteq Y$ *is closed.*
(b) $R(A^*) \subseteq X^*$ *is closed.*
(c) $R(A) = {}^{\perp} \ker A^*$.
(d) $R(A^*) = (\ker A)^{\perp}$.

Finally we characterize surjective operators. The result generalizes Proposition 3.29.

Proposition 3.234 *If X, Y are Banach spaces and $A : D(A) \subseteq X \to Y$ is an unbounded linear operator which is densely defined and closed, then the following statements are equivalent:*

(a) $A(\cdot)$ *is surjective (that is, $R(A) = Y$).*

(b) $\|y^*\|_{Y^*} \leq c \|A^*(y^*)\|_{X^*}$ *for all* $y^* \in D(A^*)$, *some* $c > 0$.
(c) $\ker A^* = \{0\}$ *and* $R(A^*) \subseteq X^*$ *is closed*.

Proof *(a)* $\Rightarrow$ *(b)* Let $E_* = \{y^* \in D(A^*) : \|A^*(y^*)\|_{X^*} \leq 1\}$ and $y_0 \in Y$. By hypothesis we can find $x_0 \in D(A)$ such that $A(x_0) = y_0$. Then for every $y^* \in E_*$ we have

$$\langle y^*, y_0 \rangle_Y = \langle y^*, A(x_0) \rangle_Y = \langle A^*(y^*), x_0 \rangle_X (\text{see } (3.31)),$$

thus $|\langle y^*, y_0 \rangle_Y| \leq \|x_0\|_X$, $\Rightarrow E_* \subseteq Y^*$ is bounded (since $y_0 \in Y$ is arbitrary). Finally by homogeneity, we have (b).
(b) $\Rightarrow$ *(a)* Let $A^*(y_n^*) \to x^*$ in X^*. We have

$$\|y_n^* - y_m^*\|_{Y^*} \leq c \|A^*(y_n^* - y_m^*)\|_{X^*} \text{ for all } n, m \in \mathbb{N}.$$

So, $\{y_n^*\}_{n \in \mathbb{N}} \subseteq Y^*$ is Cauchy and we have $y_n^* \to y^*$ in Y^*. But by Proposition 3.230, $A(\cdot)$ is closed. Hence $A^*(y^*) = x^*$. Therefore $R(A^*) \subseteq X^*$ is closed. It is clear from (b) that $\ker A^* = \{0\}$.
(c) $\Rightarrow$ *(a)* From Proposition 3.233, we have $R(A) = {}^\perp \ker A^* = Y$ $\square$

There is a "dual" version of this result.

Proposition 3.235 *If* X, Y *are Banach spaces and* $A : D(A) \subseteq X \to Y$ *is an unbounded linear operator which is densely defined and closed, then the following statements are equivalent:*

(a) $A^*(\cdot)$ *is surjective (that is,* $R(A^*) = X^*$).
(b) $\|x\|_X \leq c \|A(x)\|_Y$ *for all* $x \in D(A)$, *some* $c > 0$.
(c) $\ker A = \{0\}$ *and* $R(A) \subseteq Y$ *is closed*.

So, summarizing we can say the following:

(a) If X, Y are Banach spaces with either X or Y finite dimensional and $A : D(A) \subseteq X \to Y$ is a linear operator, then

- A is surjective if and only if A^* is injective.
- A^* surjective if and only if A is injective.

(b) In the general case we have

- A is surjective $\Rightarrow A^*$ is injective.
- A^* is surjective $\Rightarrow A$ is injective.

3.10 Remarks

(3.1) The theory of locally convex spaces and of normed and Banach spaces is the outgrowth of a tendency toward abstraction which characterizes mathematical analysis in the first half of the twentieth century. First came the study of normed spaces, with the first abstract axiomatic treatment being the celebrated thesis of the Polish mathematician Banach [21]. The book of Banach [24] published ten years later (1932), was extremely influential and so justifiably these spaces were named after him. Important contributions were made also by Hahn [130] and Wiener [282]. Up until the mid-forties, functional analysts focused almost exclusively on normed spaces. The first major work on locally convex spaces, is the paper of Dieudonne and Schwartz [88]. The rapid development of the theory owes to the theory of distributions developed by Schwartz in the fifties. Schauder bases (see Definition 3.37) were introduced by Schauder [247, 248] who provided basic sequences for the spaces

$$c_0, l^p (1 \leq p < \infty), C[0, 1] \text{ and } L^p[0, 1](1 \leq p < \infty).$$

For further details of topological vector spaces and locally convex spaces, the reader can consult the books of Kelley and Namioka [158], Rudin [242], Schaefer [249], Treves [268] and Wilansky [284].

(3.2) The Hahn-Banach Theorem dates back to the work of Helly [138]. The present version of the theorem is due to Hahn [131] and Banach [22]. The work of Hahn [131] is the starting point of the duality theory for general normed spaces. Theorem 3.67 is of course due to Caratheodory. Some extensions of it can be found in Rockafellar [233]. The first formulation of the Hahn-Banach theorem in terms of separating convex sets, is due to Mazur [190]. The modern form (see Theorem 3.71) is due to Edelheit [97].

(3.3) The Open Mapping Theorem (see Theorem 3.82) was proved by Schauder [247] for Banach spaces and it was extended to Frechet spaces by Banach [23, 24]. The Open Mapping Theorem and the Hahn-Banach Theorem, found applications in the theory at partial differential equations, see Hörmander [143]. The Closed Graph Theorem (see Theorem 3.86) is due to Banach [24]. The Uniform Boundedness Theorem (see Theorem 3.90) was first proved by Hahn [131] using a different method. The use of the Baire category theory in proving the result is due to Banach-Steinhaus [25]. For this reason the result appears also under the name "Banach-Steinhaus Theorem."

(3.4) Explicit description of weak neighborhoods in a Hilbert space, were first given by von Neumann [206]. Banach [23] deals only with weakly convergent sequences and as we already mentioned sequences do not suffice to describe the weak and weak* topologies. The weak continuity of bounded linear operators (see Proposition 3.94(d)) was first established by Banach [22]. The converse is due to Dunford [92] (p.317), see also Dunford–Schwartz [95] (p.422). Theorem 3.105 is due to Mazur [190]. Gillespie-Hurwitz [125] and Zalcwasser [289] proved results similar to Proposition 3.107 for weakly null sequences in the Banach space $C[0, 1]$. Theorem 3.108, was proved by Alaoglu [3], while working on the differentiation of vector measures. In fact the result was already known to Banach [23] for separable Banach spaces.

For this reason, sometimes the result is called the "Banach-Alaoglu" theorem. Theorem 3.110 is due to Goldstine [126] (p.128), with a different proof.

(3.5) Banach [22] was the first to relate metrizability of the unit ball with the weak topology and separability (see Theorems 3.117 and 3.118). The fundamental Theorem 3.121 is due to Smulian [258] and Eberlein [96]. On weakly compact sets, fundamental results were proved by James [151, 152].

Definition 3.236 Let p be a properly of normed spaces and X is a normed space, $V \subseteq X$ a closed subspace. We say that p is a "three space property," if when two of the spaces $X, V, X/V$ have this property, then so does the third space.

Remark 3.237 In Proposition 3.22, we saw that completeness is a three space property.

Proposition 3.238 *Separability and reflexivity are three space properties.*

In the next theorem, we have put together the main criteria for reflexivity.

Theorem 3.239 *For a Banach space X, the following are equivalent*

(a) X is reflexive.

(b) X^* is reflexive.

(c) $\overline{B}_1 = \{x \in X : \|x\| \leq 1\}$ is w-compact.

(d) Every bounded sequence in X has a weakly convergent subsequence.

(e) Every continuous linear functional is norm attaining (James [152]).

(f) If $\{C_n\}_{n \in \mathbb{N}}$ is a decreasing sequence of nonempty, bounded, closed, convex sets, then $\bigcap_{n \in \mathbb{N}} C_n \neq \emptyset$ (Smulian [257]).

(g) Every separable closed subspace of X is reflexive (that is, reflexivity is a separably determined property).

Definition 3.240

(a) A Banach space X is said "strictly convex" (or "rotund") if $\|x\| = \|u\| = 1$ and $\|x + u\| = 2$ imply $x = u$.

(b) A Banach space X is said "locally uniformly convex" if $\{x_n, x\}_{n \in \mathbb{N}} \subseteq X$,

$$2\|x_n\|^2 + 2\|x\|^2 - \|x_n + x\|^2 \to 0$$

imply $\|x_n - x\| \to 0$.

(c) A Banach space X is said "uniformly convex" if for

$$\{x_n, u_n\}_{n \in \mathbb{N}} \subseteq \overline{B}_1 = \{x \in X : \|x\| \leq 1\},$$

$\|x_n + u_n\| \to 2$ implies $\|x_n - u_n\| \to 0$.

Uniformly convex spaces were introduced by Clarkson [70] who also proved that all L^p-spaces $(1 < p < \infty)$, are uniformly convex. More generally, there is the following result known in the literature as the "Milman-Pettis Theorem."

Theorem 3.241 *Every uniformly convex Banach space X is reflexive.*

Definition 3.242 Let X be a Banach space. We say that X has the "Kadec–Klee" property if

$$x_n \xrightarrow{w} x \text{ in } X \text{ and } \lim_{n \to \infty} \sup \|x_n\| \leq \|x\|,$$

then $x_n \to x$ in X.

Remark 3.243 Another way to define the Kadec–Klee property is to say that the weak and norm topologies coincide on $\partial B_1 = \{x \in X : \|x\| = 1\}$.

Proposition 3.244 *A locally uniformly convex space has the Kadec–Klee property.*

In Dunford and Schwartz [95](p.434) and Diestel and Uhl [85](p.51), Theorems 3.129 and 3.130 are proved using Bochner integration.

Definition 3.245 Let X be a Banach space and $C \subseteq X$. We say that $x \in C$ is an "extreme point" of C, if there are no $u, y \in C$ such that $x = (1 - t)u + ty$ for some $0 < t < 1$. The set of all extreme points of C is denoted by $ext\,C$.

The next result is known as the "Krein-Milman Theorem," see Dunford and Schwartz [95], p.440.

Theorem 3.246 *If X is a Banach space, then*

(a) for $K \subseteq X$ w-compact and convex, we have $K = \overline{conv}\,ext\,K$.
(b) For $K \subseteq X^*$ w^*-compact and convex, we have $K = \overline{conv}^{w^*} ext\,K$.

More on Banach space theory, the interested reader can find in the books of Brezis [50], Bollobas [37], Day [76], Diestel [84], Dunford and Schwartz [95] (a classical reference), Fabian et al. [106], Megginson [192], Rudin [242], Yosida [287].

(3.6) The notion of dual operator (see Definition 3.137) goes back to the works of Riesz [228] (for L^p-spaces $1 < p < \infty$) and Banach [23] (general case). The notion of compact operator (see Definition 3.146), is essentially due to Hilbert from his work on integral equations. The present definition is due to Riesz [229] who did a detailed study of such operators. Theorem 3.158 is due to Schauder [248]. Projection operators were first considered by Schmidt [245]. See also Riesz [231, 232]

(3.7) The axiomatic approach to Hilbert spaces is due to von Neumann [206], [207] who developed the spectral theory of operators acting on such spaces. Using the parallelogram law (see Proposition 3.169(a)), we see that a Hilbert space is uniformly convex, hence reflexive (see Theorem 2.341). This is an alternative way to prove Corollary 3.183. Theorem 3.181, is usually called "Riesz Representation Theorem" and it was stated by Riesz [229] and Frechet [117] in separate notes in the same issue of the Comptes Rendus. According to this theorem we can identify a Hilbert space with its dual. However, we should be careful in doing so. We cannot do it in every occasion. Here is such a situation. Let X, H be Hilbert spaces with $X \hookrightarrow H$ continuously and densely. Using Theorem 3.181 we identify H^* with H. Let $h \in H$. We consider the map $A(h) : X \to \mathbb{R}$ defined by $A(h)(x) = (x, h)_H$ for all $x \in X$. We have

$$|(x, h)| \leq \|x\|_H \|h\|_H \text{ (see Proposition 3.168)}$$

$$\leq c\|x\|_V \|h\|_H \text{ for some } c > 0 \text{ (since } X \hookrightarrow H \text{ continuously),}$$

thus $A(h) \in X^*$ for all $h \in H$. Clearly $A(\cdot)$ is linear and so $A \in \mathcal{L}(H, X^*)$ and we have $\|A(h)\|_{X^*} \leq c\|h\|_H$

If $A(h) = 0$, then $(x, h)_H = 0$ for all $x \in X$ and since $X \hookrightarrow H$ densely we infer that $h = 0$. This proves that $A(\cdot)$ is injective. We claim that $\overline{R(A)} = X^*$. To this end let $\widehat{f} \in X^{**}$ such that $\widehat{f}|_{R(A)} = 0$. Since X is reflexive, we can find $\widehat{x} \in X$ such that $A(h)(\widehat{x}) = 0$ for all $h \in H$, that is $(\widehat{x}, h)_H = 0$ for all $h \in H$. Let $h = \widehat{x} \in X \subseteq H$. Then $\|\widehat{x}\|_H = 0$ and so $\widehat{x} = 0$. This proves that $\overline{R(A)} = X^*$. Therefore we have

$$X \hookrightarrow H = H^* \hookrightarrow X^*$$

with both inclusions continuous and dense. So, it is clear that we cannot identify X and X^* although X is a Hilbert space. This situation arises in many evolution equations with X a Sobolev space and $H = L^2$. Theorem 3.222 is essentially due to Riesz [232] and Schauder [248].

More on Hilbert spaces can be found in Akhiezer-Glazman [2], and Halmos [134].

(3.8) For more on unbounded linear operators, we refer to Brezis [50], Kato [155] and Reed and Simon [225].

3.11 Problems

Problem 3.1 Let X be a normed space and $C, D \subseteq X$ two convex sets. Show that conv$[C \cup D] = \{\lambda x + (1 - \lambda)u : x \in C, u \in D, 0 \leq \lambda \leq 1\}$.

Problem 3.2 On $C[0, 1]$ we consider the following two norms,

$$\|u\|_\infty = \max_{0 \leq t \leq 1} |u(t)| \text{ and } \|u\| = \vartheta \|u\|_1 = \vartheta \int_0^1 |u(t)|dt, \vartheta > 0.$$

We set $|u| = \min\{\|u\|_\infty, \|u\|\}$ for all $u \in C[0, 1]$. Show that $|\cdot|$ is a norm on $C[0, 1]$ if and only if $0 < \vartheta \leq 1$.

Problem 3.3 Let $V \subseteq l^1$ be an infinite dimensional subspace. Show that V cannot be reflexive.

Problem 3.4 Let X be a Banach space. Show that X is reflexive if and only if for any $C \subseteq X$ closed, convex and $x \notin V$, we can find $\widehat{u} \in C$ such that

$$\|x - \widehat{u}\| = \inf[\|x - u\| : u \in C].$$

Problem 3.5 Let X be a vector space and $\|\cdot\|, |\cdot|$ two equivalent norms on X. Let

$$\overline{B}_1^{\|\cdot\|} = \{x \in X : \|x\| \leq 1\} \text{ and } \overline{B}_1^{|\cdot|} = \{x \in X : |x| \leq 1\}.$$

Show that $\overline{B}_1^{\|\cdot\|}$ and $\overline{B}_1^{|\cdot|}$ are homeomorphic metric spaces.

Problem 3.6 Let $p(\cdot)$ be a polynomial on $[0, 1]$ of degree at most $k \in \mathbb{N}$. Show that we can find $c_k > 0$ such that $|p(0)| \leq c_k \int_0^1 |p(t)|dt$.

Problem 3.7 Let X, Y be Banach spaces and $A \in \mathcal{L}(X, Y)$. Show that A is an isometric isomorphism if and only if A^* is an isometric isomorphism.

Problem 3.8 Let X be a Banach space with a separable dual. Show that $B(X^*) = B(X_{w*}^*)$ (recall that for any topological space Y, by $B(Y)$ we denote the Borel σ-field of Y).

Problem 3.9 Let X be a Banach space, Y a normed space and $A \in \mathcal{L}(X, Y)$ an open map. Show that Y is a Banach space too.

Problem 3.10 Let X, Y be nontrivial normed spaces and assume that $\mathcal{L}(X, Y)$ with the operator norm is a Banach space. Show that Y is a Banach space.

Problem 3.11 Show that a Banach space X is separable if and only if it is isometrically isomorphic to a closed subspace of $C(K)$ with K a compact metric space.

Problem 3.12 Let X be a Banach space and $A \in \mathcal{L}_c(X) \setminus \mathcal{L}_f(X)$ (that is, $R(A)$ is infinite dimensional). Show that $0 \in \overline{A(\partial B_1)}$, where $\partial B_1 = \{x \in X : \|x\| = 1\}$.

Problem 3.13 Let X, Y be Banach spaces with X reflexive and $A \in \mathcal{L}(X, Y)$. Show that $A(\overline{B}_1^X) \subseteq Y$ is closed (recall $\overline{B}_1^X = \{x \in X : \|x\|_X \leq 1\}$.

Problem 3.14 Let X be a normed space, $V \subseteq X$ a subspace and $i : V \to X$ the inclusion (embedding) map. Show that $i^* : X^* \to V^*$ is the restriction map (that is, $i^*(x^*) = x^*|_V$).

Problem 3.15 Let X be a reflexive Banach space, Y a Banach space and $A \in \mathcal{L}(X, Y)$ is surjective. Show that Y is reflexive too.

Problem 3.16 Let H be a Hilbert space and $A, L : H \to H$ two linear operators such that $(A(u), x) = (u, L(x))$ for all $u, x \in H$. Show that $A \in \mathcal{L}(H)$ and $L = A^*$.

Problem 3.17 Let X, Y be Banach spaces and $A \in \mathcal{L}_c(X, Y)$. Show that $R(A) \subseteq Y$ is separable.

Problem 3.18 Let X, Y be Banach spaces, with X reflexive, $A \in \mathcal{L}_c(X, Y)$, $\|\cdot\|_X$ is the norm on X and $|\cdot|$ is another norm on X weaker than $\|\cdot\|_X$. Show that given $\epsilon > 0$, we can find $c_\epsilon > 0$ such that

$$\|A(x)\|_Y \le \epsilon \|x\|_X + c_\epsilon |x| \text{ for all } x \in X.$$

Problem 3.19 Let H be a Hilbert space, $\{A_n\}_{n \in \mathbb{N}} \subseteq \mathcal{L}(H)$ and suppose

$$\sup_{n \ge 1} |(A_n(x), u)| < \infty \text{ for all } x, u \in H.$$

Show that $\sup_{n \in \mathbb{N}} \|A_n\|_\mathcal{L} < \infty$.

Problem 3.20 Let X be a Banach space, $Y \subseteq X$ a closed subspace and $x_0 \notin Y$. Show that $span\{Y, x_0\}$ is a closed subspace of X.

Problem 3.21 Let X, Y be Banach spaces and $A : X \to Y$ a linear operator. Assume that if $x_n \to 0$ in X, then $\{A(x_n)\}_{n \in \mathbb{N}} \subseteq Y$ is bounded. Show that $A \in \mathcal{L}(X, Y)$.

Problem 3.22 Let X be a Banach space, $C \subseteq X$ and assume that for every $V \in X$ closed, separable subspace, we have that $C \cap V$ is w-compact. Show that C is w-compact.

Problem 3.23 Let X be a Banach space and $C \subseteq X$ a nonempty, closed, convex set. Show that C is w-compact if and only if for every $\{C_n\}_{n \in \mathbb{N}}$ decreasing sequence of nonempty, closed, convex subsets of C we have $\bigcap_{n \ge 1} C_n \ne \emptyset$.

Problem 3.24 Let H be a Hilbert space and $A \in \mathcal{L}(H)$ with $\|A\|_\mathcal{L} \le 1$. Show that $I - A^*A \ge 0$.

Problem 3.25 Let X, Y be Banach spaces, $A : X \to Y$ a linear operator and suppose that $y^* \circ A \in X^*$ for every $y^* \in Y^*$. Show that $A \in \mathcal{L}(X, Y)$.

Problem 3.26 Let X, Y be Banach spaces with X nonreflexive and Y reflexive and $A \in \mathcal{L}(X, Y)$ is injective. Show that $A(X) \subseteq Y$ is not closed.

Problem 3.27 Let X, Y be Banach spaces. Show that X is reflexive if and only if for every $A \in \mathcal{L}_c(X, Y)$ we can find $\widehat{u} \in X$ with $\|\widehat{u}\|_X \leq 1$ such that $\|A\|_{\mathcal{L}} = \|A(\widehat{u})\|_X$.

Chapter 4
Function Spaces

In this chapter we present the main function spaces which we encounter in most applications. We start with the classical L^p-spaces (the Lebesgue spaces). We determine when these spaces are separable and reflexive. Also we study the weak convergence in such spaces and provide necessary and sufficient condition for weak compactness in $L^1(\Omega)$. In connection with this, we also present the "Bitting Theorem." In Sect. 4.2,we have a brief presentation of Lebesgue spaces with variable exponents. These are generalizations of the classical Banach spaces, in which the constant exponent p is replaces by variable exponent function $p(z)$.The resulting function spaces $L^{p(z)}(\Omega)$ have many properties similar to the L^p-spaces but also differ in interesting ways. In Sect. 4.3, we deal with Sobolev spaces. These spaces are the cornerstone of the modern treatment of boundary value problems. For this reason, we have a rather detailed presentation of their properties with special emphasis on those that are needed in the study of partial differential equations. In Sect. 4.4 we present the basic items concerning the integration theory of Banach spaces-valued functions and the resulting L^p and Sobolev like spaces. These spaces are essential tools in the study of evolution equations. Finally in Sect. 4.5 we consider spaces of measures. We conduct this study initially in a purely measure theoretic framework and eventually in a topological one. This chapter provides the concepts and results for applications in many different areas.

4.1 Lebesgue Spaces

In Sect. 2.6, we introduced and studied L^p-spaces (Lebesgue spaces) for $1 \leq p \leq \infty$. In this section we continue this investigation, using also the new tools that we have from Chap. 3.

So, let us begin with a brief review of these spaces. Let (X, Σ, μ) be a measure space. If $f, h : X \to \mathbb{R}^* = [-\infty, +\infty]$ are two Σ-measurable functions, then we

© The Author(s), under exclusive license to Springer Nature Switzerland AG 2022
S. Hu, N. S. Papageorgiou, *Research Topics in Analysis, Volume I*, Birkhäuser
Advanced Texts Basler Lehrbücher, https://doi.org/10.1007/978-3-031-17837-5_4

say that they are "equivalent" if and only if $f(x) = h(x)$ for μ-a.a $x \in X$. We write $f \sim h$. This is an equivalence relation. If $1 \leq p < \infty$, then

$$\mathscr{L}^p(X) = \{f : X \to \mathbb{R}^* \text{ is } \Sigma - measurable, \int_X |f|^p d\mu < \infty\},$$

$$L^p(X) = \mathscr{L}^p(X)/N.$$

If $p = \infty$, then

$$\mathscr{L}^\infty(X) = \{f : x \to \mathbb{R}^* \text{ is } \Sigma - measurable, \inf\{M > 0 : |f(X)| \leq M \mu - a.e\} < \infty\}$$

$$L^\infty(X) = \mathscr{L}^\infty(X)/\sim.$$

We furnish $L^p(X)$ $1 \leq p < \infty$ with the norm

$$\|f\|_p = \left[\int_X |f|^p d\mu\right]^{1/p}$$

and $L^\infty(X)$ with the norm

$$\|f\|_\infty = \inf\{M > 0 : |f(x)| \leq M \mu - a.e\}.$$

We know that the spaces $L^p(X)$ $1 \leq p \leq \infty$ equipped with the above norms are Banach spaces (see Theorem 2.126).

Remark 4.1 If $X = \mathbb{N}$ and $\mu(\cdot)$ is the counting measure, then the spaces $L^p(\mathbb{N})$ are the sequence spaces denoted by ℓ^p and we have

$$\ell^p = \{\widehat{x} = (x_n)_{n\in\mathbb{N}} : \|\widehat{x}\|_p = (\sum_{n\geq 1} |x_n|^p)^{1/p} < \infty\} \ 1 \leq p < \infty,$$

$$\ell^\infty = \{\widehat{x} = (x_n)_{n\in\mathbb{N}} : \|\widehat{x}\|_\infty = \sup_{n\in\mathbb{N}} |x_n| < \infty\}.$$

We have seen in Proposition 2.124 that if $\mu(\cdot)$ is finite, then the spaces $L^p(X)$, $1 \leq p \leq \infty$ decrease with $p \geq 1$, that is

$$\text{"}1 \leq p \leq q \leq \infty \Rightarrow L^q(X) \subseteq L^p(X).\text{"}$$

The next result shows that this inclusion fails if $\mu(\cdot)$ is not finite.

Proposition 4.2 *If (X, Σ, μ) is a measure space and $1 \le p < q < \infty$, then $L^q(X)$ is not contained in $L^p(X)$ if and only if there are Σ-sets of arbitrarily large finite μ−measure.*

Proof $\Rightarrow$ We can find $f \in L^q(X)$ such that $\int_X |f|^p d\mu = +\infty$. We introduce the sets

$$A_n = \{x \in X : \frac{1}{n+1} < |f(x)| \le \frac{1}{n}\} \in \Sigma, \ n \in \mathbb{N},$$

$$A_\infty = \{x \in X : 0 < |f(x)| \le 1\} = \bigcup_{n \in \mathbb{N}} A_n \in \Sigma.$$

Assume $\mu(A_\infty) < \infty$. Then $\int_X |f|^p d\mu = \int_{\{|f| \le 1\}} |f|^p d\mu + \int_{\{|f| > 1\}} |f|^p d\mu \le \mu(A_\infty) + \int_{\{|f| > 1\}} |f|^q d\mu < +\infty$, a contradiction. Therefore $\mu(A_\infty) = +\infty$. On the other hand, we have

$$\frac{1}{(n+1)^q} \mu(A_n) \le \int_{A_n} |f|^q d\mu \le \int_X |f|^q d\mu < \infty \text{ for all } n \in \mathbb{N}. \tag{4.1}$$

Let $C_m = \bigcup_{n=1}^{m} A_n$, $m \in \mathbb{N}$. From (4.1) we see that $\mu(C_m) < \infty$ and from Proposition 2.17(c) we have $\mu(C_m) \uparrow \mu(A_\infty) = +\infty$ as $m \to \infty$. So, X contains Σ-sets of arbitrarily large finite measure.

$\Leftarrow$ Let $\{A_n\}_{n \in \mathbb{N}} \subseteq \Sigma$ be pairwise disjoint such that $\mu(A_n) \uparrow \infty$ as $n \to \infty$. Choose $\{\xi_n\}_{n \in \mathbb{N}} \subseteq (0, \infty)$ such that $\xi_n \downarrow 0$ and

$$\sum_{n \ge 1} \xi_n^q \mu(A_n) < \infty, \quad \sum_{n \ge 1} \xi_n^p \mu(A_n) = \infty.$$

Then $f = \sum_{n \ge 1} \xi_n \chi_{A_n} \in L^q(X) \setminus L^p(X)$. $\qquad\square$

In Propositions 2.128 and 2.131 we produced two dense subspaces of $L^p(X)$ $1 \le p < \infty$. In the next proposition, we provide another such density result.

Proposition 4.3 *If X is a metric space, $\Sigma = B(X)$ and μ finite, then $L^p(X) \cap C_b(X)$ is dense in $L^p(X)$ for all $1 \le p < l$.*

Proof On account of Proposition 2.128, it suffices to approximate χ_A, $A \in \Sigma = B(X)$ by functions in $L^p(X) \cap C_b(X)$. Given $\epsilon > 0$ and using Theorem 2.154, we can find $U \supseteq A$ open and $C \subseteq A$ closed such that $\mu(U \setminus C) \le \varepsilon^p$. Also Theorem 1.42 (Urysohn's Lemma) says that there exists a continuous $f : X \to [0, 1]$ such that $f|_C \equiv 1$ and $f|_{X \setminus U} \equiv 0$. Evidently $f \in L^p(X) \cap C_b(X)$ and we have

$$\int_X |\chi_A - f|^p d\mu = \int_{U \setminus C} |\chi_A - f|^p d\mu \leq \mu(U \setminus C) \leq \epsilon^p.$$

$\square$

Proposition 4.4 *If (X, Σ, μ) is a σ-finite measure space, $p_0 \geq 1$ and $f \in L^p(X)$ for all $p_0 \leq p < \infty$, then $\|f\|_\infty = \lim\limits_{p \to +\infty} \|f\|_p$.*

Proof If $\|f\|_\infty = 0$, then the conclusion of the proposition is obvious. So, we assume that $\|f\|_\infty > 0$. For $\eta \in (0, \|f\|_\infty)$, the set $A_\eta = \{x \in X : |f(x)| > \eta\}$ has positive μ-measure. For $p_0 \leq p < \infty$, we have

$$\eta \mu(A_\eta)^{1/p} \leq \|f\|_p,$$

$$\Rightarrow \eta \leq \liminf_{p \to \infty} \|f\|_p.$$

Since $\eta \in (0, \|f\|_\infty)$ is arbitrary, it follows that $\|f\|_\infty \leq \liminf\limits_{p \to \infty} \|f\|_p$.

Suppose that $f \in L^\infty(\Omega)$. We will show that $\limsup\limits_{p \in \infty} \|f\|_p \leq \|f\|_\infty$.

We assume that $\|f\|_\infty > 0$, otherwise $\|f\|_p = 0$ for all $p \geq 1$ and so we are done. Invoking Proposition 2.123 (the Interpolation Inequality), we have

$$\|f\|_p \leq \|f\|_{p_0}^{\frac{p_0}{p}} \|f\|_\infty^{1 - \frac{p_0}{p}},$$

$$\Rightarrow \limsup_{p \to \infty} \|f\|_p \leq \|f\|_\infty.$$

So, when $f \in L^\infty(\Omega)$, we have that

$$\|f\|_\infty = \lim_{p \to \infty} \|f\|_p.$$

If $f \notin L^\infty(\Omega)$, then $\|f\|_\infty = +\infty$ and so from the first part of the proof we have $\infty = \|f\|_\infty = \lim\limits_{p \to \infty} \|f\|_p$.

$\square$

We recall the so-called Clarkson's Inequalities. For a proof of these inequalities we refer to Hewitt and Stromberg [139] (pp.225, 227).

Theorem 4.5 *If (X, Σ, μ) is a measure space and $f, h \in L^p(X)$ $1 < p < \infty$ then*

(a) $\|\frac{f+h}{2}\|_p^\varrho + \|\frac{f-h}{2}\|_p^\varrho \leq \frac{1}{2}\|f\|_p^\varrho + \frac{1}{2}\|h\|_p^\varrho$ *if $p \geq 2$.*

(b) $\|\frac{f+h}{2}\|_p^{p'} + \|\frac{f-h}{2}\|_p^{p'} \leq [\frac{1}{2}\|f\|_p^p + \frac{1}{2}\|h\|_p^p]^{p'-1}$ *if $1 < p < 2$*
    (recall $p' = \frac{p}{p-1}$, the conjugate exponent to p).

Using these inequalities, we can have the uniform convexity of the Lebesgue spaces $L^p(X)$ $(1 < p < \infty)$; see Definition 3.240(c).3

Theorem 4.6 *If (X, Σ, μ) is a measure space and $1 < p < \infty$, then $L^p(X)$ is uniformly convex.*

Proof Let $\{f\}_{n\in\mathbb{N}}, \{h_n\}_{n\in\mathbb{N}} \subseteq \overline{B}_1^{L^p}$ and assume that $\|f_n + h_n\|_p \to 2$. Then from the inequalities of Theorem 4.5, we obtain $\|f_n - h_n\|_p \to 0$, hence $L^p(X)$ is uniformly convex, see Definition 2.340(c). □

Using Theorem 3.241(the Milman-Pettis Theorem), we have:

Corollary 4.7 *If (X, Σ, μ) is a measure space and $1 < p < \infty$, then the Banach space $L^p(X)$ is reflexive.*

Next we examine the separability of the Lebesgue spaces.

Theorem 4.8 *If (X, Σ, μ) is a separable σ-finite measure space (see Definition 2.168) and $1 \leq p < \infty$, then $L^p(X)$ is separable.*

Proof Let $\Sigma = \sigma(\{A_n\}_{n\in\mathbb{N}})$ (see Definition 2.168(c)), let $\mathscr{S}$ be the algebra generated by $\{A_n\}_{n\in\mathbb{N}}$. We consider the set g^* of simple functions of the form $\sum_{k=1}^{n} \vartheta_k \chi_{C_k}$, where $n \in \mathbb{N}$, $\vartheta_k \in Q$ and $C_k \in \mathscr{S}$ with $\mu(C_k) < \infty$, $k = 1, ..., n$. We will show that g^* is countable and dense in $L^p(X), l \leq p < \infty$. Since simple functions are dense in $L^p(X)$ (see Proposition 2.128), it suffices to approximate χ_A with $A \in \Sigma, \mu(A) < \infty$, by $\{\chi_{C_n}\}_{n\in\mathbb{N}}$ with $C_n \in \mathscr{S}, \mu(C_n) < \infty$ for all $n \in \mathbb{N}$.

First we assume that $\mu(\cdot)$ is finite.

Let $\mathscr{D} = \{A \in \Sigma : \text{there exists } \{C_n\}_{n\in\mathbb{N}} \subseteq \mathscr{S} \text{ such that } \chi_{C_n} \to \chi_A \text{ in } L^p(X)\}$. We claim that $\mathscr{D}$ is a σ-algebra. Since $\mathscr{S} \subseteq \mathscr{D}$, we see that $\emptyset, X \in \mathscr{D}$. Next we show that $\mathscr{D}$ is closed under complementation. So, let $A \in \mathscr{D}$. Then by definition, we can find $\{C_n\}_{n\in\mathbb{N}} \subseteq \mathscr{S}$ such that $\chi_{C_n} \to \chi_A$ in $L^p(X)$. We have $\{X \setminus C_n = C_n^1\}_{n\in\mathbb{N}} \subseteq \mathscr{S}$ and $\chi_{C_n^2} = 1 - \chi_{C_n} \to 1 - \chi_A = \chi_{A^c}$ in $L^p(X)$.

Let $A_1, A_2 \in \mathscr{D}$ and let $\{C_n^1\}_{n\in\mathbb{N}}, \{C_n^2\}_{n\in\mathbb{N}} \subseteq \mathscr{S}$ such that $\chi_{C_n^1} \in \chi_{A_1}$ and $\chi_{C_n^2} \to \chi_{A_2}$ in $L^p(X)$ as $n \in \infty$.

By passing to a subsequence if necessary, we can also have

$$\chi_{C_n^1}(x) \to \chi_{A_1}(x) \text{ and } \chi_{C_n^2} \to \chi_{A_2}(x) \ \mu \cdot a.e \text{ on } X$$

(see Proposition 2.127). Then by the Lebesgue, Dominated Convergence Theorem (see Theorem 2.86), we have that

$$\chi_{C_n^1 \cup C_n^2} \in \chi_{A_1 \cup A_2} \text{ in } L^p(X) \text{ as } n \to \infty,$$

$$\Rightarrow \mathscr{D} \text{ is an algebra} .$$

Finally to show that $\mathscr{D}$ is a σ-algebra, we use Proposition 2.6. So, let $\{A_n\}_{n\in\mathbb{N}} \subseteq \mathscr{D}$ and assume that $A_n \uparrow A$. For every $n \in \mathbb{N}$, we can find $\{C_n^k\}_{k\in\mathbb{N}} \subseteq \mathscr{S}$ such that $\chi_{C_n^k} \to \chi_{A_n}$ in $L^p(X)$ as $k \in \infty$. From the Monotone Convergence Theorem (see Theorem 2.80), we have $\chi_{A_n} \to \chi_A$ in $L^p(X)$ as $n \to \infty$. Invoking the double limit lemma (see Lemma 1.193), we can find a sequence $\{k_n\}_{n\in\mathbb{N}}$ with $k_n \to +\infty$ such that $\chi_{C_n^{k_n}} \to \chi_A$ in $L^p(\Omega)$. Hence by Proposition 2.6, $\mathscr{D}$ is a σ-algebra. Since $\mathscr{S} \subseteq \mathscr{D}$, and $\Sigma = \sigma(\{A_n\}_{n\in\mathbb{N}})$, it follows that $\Sigma = \mathscr{D}$.

Next we remove the restriction that $\mu(\cdot)$ is finite.

Since $\mu(\cdot)$ is σ-finite. So, $X = \bigcup_{n\in\mathbb{N}} B_n$, $B_n \in \Sigma$, $\mu(B_n) < \infty$. Then from the first part of the proof for every $n \in \mathbb{N}$, we can find $\{C_n^k\}_{k\in\mathbb{N}} \subseteq \mathscr{S}$ such that

$$\chi_{C_n^k \cap B_n} \to \chi_{A \cap B_n} \ in \ L^p(X) \ as \ k \to \infty,$$

with $A \in \Sigma$, $\mu(A) < \infty$. Since by the Lebesgue Dominated Convergence Theorem (see Theorem 2.86), we have

$$\chi_{A \cap B_n} \to \chi_A \ in \ L^p(X) \ as \ n \to \infty,$$

as above using the double limit lemma, we obtain that $A \in \mathscr{D}$, hence $\Sigma = \mathscr{D}$.

Finally note that $\mathscr{S}$ is countable and so g^* is countable and we conclude that $L^p(X)$ $1 \le p < \infty$ is separable. $\square$

Remark 4.9 A nice byproduct of the above proof is that if $A \in \Sigma$, then we can find $\{C_n\}_{n\in\mathbb{N}} \subseteq \mathscr{S}$ such that $\chi_{C_n} \to \chi_A$ in $L^p(X)$.

The following particular case of Theorem 4.8 arises in applications.

Corollary 4.10 *If X is a second countable, locally compact topological space, $\Sigma = B(X)$, $\mu(\cdot)$ is a σ-finite Borel measure and $1 \le p < \infty$, then $L^p(X)$ is separable.*

Remark 4.11 Since a locally compact space is regular, then from the Urysohn Metrization Theorem (see Theorem 1.116), we have that X is metrizable.

Let (X, Σ, μ) be a finite measure space and let Σ_μ be the μ-completion of Σ.

Definition 4.12 We say that $A, B \in \Sigma_\mu$ are "μ-equivalent," if $\mu(A \Delta B) = 0$. We write $A \sim B$. This is an equivalence relation on Σ_μ. We set

$$\widehat{\Sigma}_\mu = \Sigma_\mu / \sim$$

that is, we identify μ-equivalent sets (so every μ-null set is identified with the empty set).

Consider the map $\xi : \widehat{\Sigma}_\mu \to L^1(X)$ defined by $\xi(A) = \chi_A$. We have

$$\min\{\chi_A, \chi_B\} = \chi_{A\cap B}, \ \max\{\chi_A, \chi_B\} = \chi_{A\cup B}$$

$$\chi_{A\backslash B} = \chi_A - \chi_{A\cap B}.$$

So, we have an embedding into $L^1(X)$ and we can view $\widehat{\Sigma}_\mu$ as a subset of $L^1(X)$. Then we define the metric

$$\widehat{d}(A, B) = \|\chi_A - \chi_B\|_1 = \mu(A \Delta B).$$

We see that $(\widehat{\Sigma}_\mu, \widehat{d})$ is a metric space. In fact we can say more.

Proposition 4.13 $\widehat{\Sigma}_\mu$ *is a closed subset of* $L^1(X)$ *and so* $(\widehat{\Sigma}_\mu, \widehat{d})$ *is a complete metric space.*

Proof Let $\chi_{A_n} \to f$ is $L^1(X)$. According to Proposition 2.127, at least for a subsequence, we also have $\chi_{A_n}(x) \to f(x)$ $\mu \cdot a.e.$ Therefore $f = \chi_A$ for some $A \in \widehat{\Sigma}_\mu$ and this proves the proposition. $\qquad\square$

Remark 4.14 The maps $(A, B) \to A \cup B$, $(A, B) \to A \cap B$, $(A, B) \to A \setminus B$ from $\widehat{\Sigma}_\mu \times \widehat{\Sigma}_\mu$ into $\widehat{\Sigma}_\mu$ are uniformly continuous. Let us check this for the union the proof for the other two maps being similar. So, we have

$$\widehat{d}(A \cup B, C \cup D) = \int_X |\chi_{A \cup B} - \chi_{C \cup D}| d\mu$$

$$= \int_X |\max\{\chi_A, \chi_B\} - \max\{\chi_C, \chi_D\}| d\mu$$

$$\leq \int_X |\max\{\chi_A, \chi_B\} - \max\{\chi_C, \chi_B\}| d\mu$$

$$+ \int_X |\max\{\chi_C, \chi_B\} - \max\{\chi_C, \chi_D\}| d\mu$$

$$\leq \int_X |\chi_A - \chi_C| d\mu + \int_X |\chi_B - \chi_D| d\mu$$

$$= \widehat{d}(A, C) + \widehat{d}(B, D).$$

Proposition 4.15 $(\widehat{\Sigma}_\mu, \widehat{d})$ *is a separable metric space if and only if* $L^1(X)$ *is separable.*

Proof $\Rightarrow$ Let $\{A_n\}_{n \in \mathbb{N}} \subseteq \widehat{\Sigma}_\mu$ be dense. Then $span_{\mathbb{Q}}\{\chi_{A_n}\}_{n \in \mathbb{N}}$ is a countable dense subset of $L^1(X)$, hence $L^1(X)$ is separable.

$\Leftarrow$ Since $\widehat{\Sigma}_\mu \subseteq L^1(X)$, the separability of $L^1(X)$ implies that of $(\widehat{\Sigma}_\mu, \widehat{d})$. $\qquad\square$

From Theorem 2.133 we know that:

Theorem 4.16 *If* (X, Σ, μ) *is a* σ-*finite measure space and* $1 \leq p < \infty$, *then* $L^p(X)^* = L^{p'}(X)$ *with* $\frac{1}{p} + \frac{1}{p'} = 1$ (*if* $p = 1$, *then* $p' = \infty$) *and the duality brackets are defined by*

$$\langle f, h \rangle = \int_X f h d\mu \ for \ all \ f \in L^p(X), \ all \ h \in L^{p'}(X).$$

So, we can define weak convergence in the Lebesgue spaces.

Definition 4.17 Let (X, Σ, μ) be a σ-finite measure space, $1 \leq p < \infty$ and $\{f_n\}_{n \in \mathbb{N}} \subseteq L^p(X)$. We say that the sequence $\{f_n\}_{n \in \mathbb{N}}$ converges weakly to $f \in L^p(X)$ if and only if

$$\langle f_n, h \rangle = \int_\Omega f_n h d\mu \to \int_\Omega f h d\mu = \langle f, h \rangle \ for \ all \ h \in L^{p'}(X).$$

We denote this convergence by $f_n \overset{w}{\to} f$ in $L^p(X)$.
For $p = +\infty$, we define the w^*-convergence. So, if $\{f_n\}_{n \in \mathbb{N}} \subseteq L^\infty(\Omega)$, then we say that the sequence $\{f_n\}_{n \in \mathbb{N}}$ w^*-converges to $f \in L^\infty(X)$ if and only if

$$\langle f_n, h \rangle = \int_\Omega f_n h d\mu \to \int_X f h d\mu = \langle f, h \rangle \ for \ all \ h \in L^1(X).$$

We denote this convergence by $f_n \overset{w^*}{\to} f$ in $L^\infty(X)$.

The results of Chaps. 2 and 3 yield the following proposition.

Proposition 4.18 *If* (X, Σ, μ) *is a σ-finite measure space*, $1 < p < \infty$, $\{f_n\}_{n \in \mathbb{N}} \subseteq L^p(X)$ *is bounded and* $f_n \to f$ $\mu \cdot a.e$ *or in measure, then* $f \in L^p(X)$ *and* $f_n \overset{w}{\to} f$ *in* $L^p(X)$.

Proof We know that $L^p(X)$ is reflexive (see Corollary 4.7). So, by Proposition 3.124 $\{f_n\}_{n \in \mathbb{N}} \subseteq L^p(X)$ is relatively w-compact and by the Eberlein–Smulian Theorem (see Theorem 3.121), it is relatively sequentially weakly compact. So, by passing to a suitable subsequence if necessary, we may assume that $f_n \overset{w}{\to} \widehat{f} \in L^p(X)$. For every $h \in L^{p'}(X)$ we have $(f_n h)(x) \to (f h)(x)$ for $\mu \cdot a.a \ x \in X$. Evidently $\{f_n h\}_{n \in \mathbb{N}} \subseteq L^1(X)$ satisfies the conditions in Vitali's Convergence Theorem (see Theorem 2.147) and so we have $\int_X f_n h d\mu \to \int_X \widehat{f} h d\mu$ for all $h \in L^{p'}(X)$. Therefore $\widehat{f} = f \in L^p(X)$ and for the initial sequence we have $f_n \overset{w}{\to} f$ in $L^p(X)$.   □

Proposition 4.19 *If* (X, Σ, μ) *is a σ-finite measure space*, $1 \leq p \leq \infty$, $\{f_n\}_{n \in \mathbb{N}} \subseteq L^p(X)$ *and* $f_n \overset{w}{\to} f$ *in* $L^p(X)$ *if* $1 \leq p < \infty$ *and* $f_n \overset{w}{\to} f$ *in* $L^\infty(X)$, *then* $\|f\|_p \leq \liminf_{n \to \infty} \|f_n\|_p \leq \sup_{n \in \mathbb{N}} \|f_n\|_p < \infty$.

Proof This is a consequence of Propositions 3.104 and 3.101.   □

Proposition 4.20 *If (X, Σ, μ) is a σ-finite measure space, $1 < p < \infty$, $\{f_n\}_{n\in\mathbb{N}} \subseteq L^p(X)$, $f_n \overset{w}{\to} f$ in $L^p(X)$ and $\|f_n\|_p \to \|f\|_p$, then $f_n \to f$ in $L^p(X)$.*

Proof From Theorem 4.6 we know that $L^p(X)$ is uniformly convex. So $L^p(X)$ has the Kadec–Klee property (see Proposition 3.244 and Definition 3.242). It follows that $f_n \to f$ in $L^p(X)$. $\qquad\square$

Proposition 4.21 *If (X, Σ, μ) is a σ-finite measure space, $1 < p < \infty$ and $\{f_n\}_{n\in\mathbb{N}} \subseteq$ is bounded then there exists a subsequence $\{f_{n_k}\}_{k\in\mathbb{N}}$ of $\{f_n\}_{n\in\mathbb{N}}$ such that $f_{n_k} \overset{w}{\to} f$ in $L^p(X)$ for some $f \in L^p(X)$.*

Proof Since $L^p(X)$ is reflexive, bounded sets are relatively sequentially weakly compact. $\qquad\square$

The next result is an easy consequence of the definition of weak convergence.

Proposition 4.22 *If (X, Σ, μ) is a σ-finite measure space, $1 \leq p < \infty$ and $\{f_n, f\}_{n\in\mathbb{N}} \subseteq L^p(X)$, then $f_n \overset{w}{\to} f$ in $L^p(X)$ as $n \to \infty$ if and only if $\{f_n\}_{n\in\mathbb{N}} \subseteq L^p(X)$ is bounded and $\int_A f_n d\mu \to \int_A f d\mu$ for all $A \in \Sigma$ with $\mu(A) < \infty$.*

Proof This is a consequence of Definition 4.17 and of the fact that simple functions are dense in $L^p(X)$ (see Proposition 2.128). $\qquad\square$

Remark 4.23 Proposition 4.22 is also true for $p = \infty$, provided that the weak convergence is replaced by the w^*-convergence. On the other hand, Proposition 4.21 fails for $p = 1$. To see this let $X = \mathbb{R}$, $\Sigma = B(\mathbb{R})$ and $\mu = dx$ the Lebesgue measure. Consider the sequence $\{f_n = n\chi_{[0,\frac{1}{n}]}\}_{n\in\mathbb{N}} \subseteq L^1(X)$. Then $f_n \overset{\mu}{\to} 0$, $\|f_n\|_{L^1(\mathbb{R})} = 1$ for all $n \in \mathbb{N}$, but by Proposition 4.22 we see that we do not have weak convergence.

When $1 < p < \infty$, the relatively weakly compact sets in $L^p(X)$ and the relatively w^*-compact subsets of $L^\infty(X)$, are the bounded sets (see Proposition 3.124 and Theorem 3.108 respectively). So, it remains to characterize the relatively weakly compact subsets of $L^1(X)$. For this we will need following theorem on setwise convergent sequences of signed measures. The result is known as the "Vitali-Hahn-Saks Theorem."

Theorem 4.24 *If (X, Σ) is a measurable space, $\{\lambda_n\}_{n\in\mathbb{N}}$ are finite signed measures μ is a finite measure, $\lambda_n \ll \mu$ for all $n \in \mathbb{N}$ and for all $A \in \Sigma$*

$$\lambda(A) = \lim_{n\to\infty} \lambda_n(A) \; exists,$$

then $\lambda : \Sigma \to \mathbb{R}$ is a signed measure and $\lambda \ll \mu$.

Proof Using the Jordan Decomposition Theorem (see Theorem 2.103), we see that we may assume that each λ_n is a measure.

First we show that $\{\lambda_n\}_{n\in\mathbb{N}}$ is uniformly absolutely continuous with respect to μ, that is, given $\varepsilon > 0$, we can find $\delta > 0$ such that

$$\text{"}\mu(A) \leq \delta \Rightarrow \lambda_n(A) \leq \varepsilon \text{ for all } n \in \mathbb{N} \text{ (see Proposition 2.110)."}$$

We have

$$|\lambda_n(A) - \lambda_n(B)| \leq \lambda_n(A\Delta B) \text{ for all } A, B \in \Sigma, \lambda_n \ll \mu \text{ for all } n \in \mathbb{N}.$$

The map $\lambda_n : (\widehat{\Sigma}_\mu, \widehat{d}) \to [0, +\infty), n \in \mathbb{N}$ is well defined and continuous. Let

$$E_k = \{A \in \widehat{\Sigma}_\mu : |\lambda_n(A) - \lambda_m(A)| \leq \varepsilon \text{ for all } n, m \geq k\}, k \in \mathbb{N}.$$

These are closed sets in $\widehat{\Sigma}_\mu$ and $\widehat{\Sigma}_\mu = \bigcup_{k\in\mathbb{N}} E_k$. Using Proposition 4.13, Proposition 1.173 (the Baire Category Theorem) and Proposition 1.174, we can find $k \in \mathbb{N}$ such that $\text{int} E_k \neq \emptyset$. So, there exist $\widehat{A} \in E_k$ and $\delta_1 > 0$ such that

$$\text{"}A \in \widehat{\Sigma}_\mu \text{ and } \mu(A\Delta\widehat{A}) \leq \delta_1 \Rightarrow A \in E_k''.$$

By hypothesis $\lambda_m \ll \mu$ for $m \in \{1, ..., k\}$. Hence according to Proposition 2.110, we can find $\delta \in (0, \delta_1)$ such that

$$\text{"}A \in \Sigma \text{ and } \mu(A) \leq \delta \Rightarrow \lambda_m(A) \leq \varepsilon \text{ for } m \in \{1, ..., k\}."$$

If $A \in \Sigma$ and $\mu(A) \leq \delta$, then $\mu\big((A \cup \widehat{A})\Delta\widehat{A}\big) \leq \mu(A) \leq \delta \leq \delta_1$ and so

$$|\lambda_n(A) - \lambda_k(A)| = \big|(\lambda_n - \lambda_k)(A \cup \widehat{A}) - (\lambda_n - \lambda_k)(\widehat{A} \setminus A)\big|$$

$$\leq \big|(\lambda_n - \lambda_k)(A \cup \widehat{A})\big| + |(\lambda_n - \lambda_k)(\widehat{A} \setminus A)\big| \leq 2\varepsilon \text{ for all } n \geq k.$$

Therefore we have

$$\text{"}A \in \Sigma \text{ and } \mu(A) \leq \delta \Rightarrow \lambda_n(A) \leq 2\varepsilon + \lambda_k(A) \leq 3\varepsilon \text{ for all } n \in \mathbb{N}."$$

So, we have the uniform absolute continuity with respect to μ.

Suppose $\{A_n\}_{n\in\mathbb{N}} \subseteq \Sigma$ be pairwise disjoint sets and $\varepsilon > 0$. We set $A = \bigcup_{n\in\mathbb{N}} A_n \in \Sigma$. Let $\delta > 0$ be as postulated by the uniform absolute continuity with respect to μ established above. Let $k \in \mathbb{N}$ be such that $\mu(A \setminus \bigcup_{i=1}^{k} A_i) \leq \delta$. Then we have

$$\left|\lambda_n(A) - \sum_{i=1}^{m}\lambda_n(A_i)\right| = |\lambda_n(A \setminus \bigcup_{i=1}^{m} A_i)| \leq \varepsilon \text{ for all } n, m \geq k,$$

$$\Rightarrow |\lambda(A) - \sum_{i=1}^{m} \lambda(A_i)| \leq \varepsilon \text{ for all } m \geq k.$$

Since $\varepsilon > 0$ is arbitrary, we infer that $\lambda(A) = \sum_{n \in \mathbb{N}} \lambda(A_n)$ which means that $\lambda(\cdot)$ is a measure. Finally it is clear from the first part of the proof and Proposition 2.110 that $\lambda \ll \mu$. □

A consequence of this theorem is the so-called Nikodym's Theorem.

Theorem 4.25 *If (X, Σ) is a measurable space, $\{\lambda_n\}_{n \in \mathbb{N}}$ a sequence of nonzero finite measures defined on Σ and for all $A \in \Sigma$, $\lambda(A) = \lim_{n \to \infty} \lambda_n(A)$ exists, then $\lambda(\cdot)$ is a finite measure.*

Proof Let $\mu : \Sigma \to [0, +\infty)$ be defined by

$$\mu(A) = \sum_{n \in \mathbb{N}} \frac{1}{2^n} \frac{\lambda_n(A)}{\lambda_n(X)} \text{ for all } A \in \Sigma.$$

This is a finite measure on Σ and $\lambda_n \ll \mu$ for all $n \in \mathbb{N}$. So, we can apply Theorem 4.24 and conclude that $\lambda(\cdot)$ is a measure on Σ. □

Now we are ready to characterize the relatively weakly compact sets in $L^1(X)$. The result is known as the "Dunford-Pettis Theorem."

Theorem 4.26 *If (X, Σ, μ) is a finite measure space and $K \subseteq L^1(X)$, then K is relatively w-compact if and only if it is bounded and uniformly integrable.*

Proof $\Rightarrow$ Evidently $K \subseteq L^1(X)$ is bounded (see Proposition 3.101(a)). Suppose that K is not uniformly integrable. Then we can find $\{f_n\}_{n \in \mathbb{N}} \subseteq K$, $\{A_n\}_{n \in \mathbb{N}} \subseteq \Sigma$ and $\varepsilon_0 > 0$ such that

$$\mu(A_n) \to 0 \text{ and } \int_{A_n} |f_n| d\mu \geq \varepsilon_0 \text{ for all } n \in \mathbb{N}. \tag{4.2}$$

By the Eberlein–Smulian Theorem (see Theorem 3.121) we may assume that $f_n \xrightarrow{w} f$ in $L^1(X)$, $f \in L^1(X)$. For every $A \in \Sigma$ we have

$$\lambda_n(A) = \int_A f_n d\mu \to \int_A f d\mu = \lambda(A) \text{ as } n \to \infty.$$

From the proof of Theorem 4.24, we know that we can find $\delta > 0$ such that

$$|\int_A f_n d\mu| \leq \frac{\varepsilon_0}{3} \text{ for all } n \in \mathbb{N}, \text{ all } A \in \Sigma \text{ with } \mu(A) \leq \delta. \tag{4.3}$$

Let $n_0 \in \mathbb{N}$ so that $\mu(A_n) \leq \delta$ for all $n \geq n_0$. In (4.3) we use $A = A_n \cap \{f_n \geq 0\}$ and $A = A_n \cap \{f < 0\}$. We obtain

$$\int_{A_n} |f_n| d\mu \leq \frac{2\epsilon_0}{3} \text{ for all } n \geq n_0. \tag{4.4}$$

Comparing (4.2) and (4.4) we have a contradiction.

$\Leftarrow$ First we assume that Σ is countably generated. Hence $L^1(X)$ is separable. Since $L^1(X) \hookrightarrow L^\infty(X)^*$, we see that K is bounded in $L^\infty(X)^*$. We view K as a subset of $L^\infty(X)^*$ and so we can consider $\overline{K}^{w^*}$. By the Alaoglu Theorem (see Theorem 3.108), $\overline{K}^{w^*}$ is w^*-compact in $L^\infty(X)^*$. Recall that

$$w(L^1(X), L^\infty(X)) = w(L^\infty(X)^*, L^\infty(X))|_{L^1(X)}.$$

So, we will prove the implication if we show that $\overline{K}^{w^*} \subseteq L^1(X)$. So, let $u \in \overline{K}^{w^*}$. Then we can find a net $\{f_\vartheta\}_{\vartheta \in D} \subseteq K$ such that $f_\vartheta \xrightarrow{w^*} u$ in $L^\infty(X)^*$. We have

$$\int_X f_\vartheta h d\mu \longrightarrow u(h) \text{ for all } h \in L^\infty(X). \tag{4.5}$$

The uniform integrability of K implies that given $\varepsilon > 0$, we can find $\delta > 0$ such that

$$\text{"} A \in \Sigma \text{ and } \mu(A) \leq \delta \Rightarrow \int_A |f_\vartheta| d\mu \leq \varepsilon \text{ for all } \vartheta \in \mathbb{D}.\text{"}$$

From (4.5) it follows that

$$\lambda(A) = u(\chi_A) \leq \varepsilon \text{ for all } A \in \Sigma.$$

Also, if $\lambda_\vartheta(A) = \int_A f_\vartheta d\mu$, then $\lambda_\vartheta(A) \to \lambda(A)$ for all $A \in \Sigma$ and so by Theorem 4.24 $\lambda(\cdot)$ is a finite signed measure on Σ and $\lambda \ll \mu$. Then the Radon–Nikodym Theorem (see Theorem 2.113) implies that we can find $f \in L^1(X)$ such that $\lambda(A) = \int_A f d\mu$ for all $A \in \Sigma$. Recall that simple functions are dense in $L^\infty(X)$ (see Proposition 2.129). So, we infer that

$$u(h) = \int_X f h d\mu \text{ for all } h \in L^\infty(X),$$

$$\Rightarrow f \in L^1(X) \text{ and so } \overline{K}^{w^*} = \overline{K}^w \text{ is w-compact.}$$

Finally, we remove the assumption that Σ is countably generated. Then, we may replace Σ by Σ', which is generated by

$$\{x \in X : f_n(x) > \eta\} \text{ and } \{x \in X : f_n(x) < -\eta\}$$

for $n \in \mathbb{N}$ and $\eta \in \mathbb{Q}$, and replace X by

$$V = \bigcup_{n \in \mathbb{N}} \{x \in X : f_n(x) \neq 0\}.$$

Note that for any $h \in L^\infty(V, \Sigma)$ there exists $h' \in L^\infty(V, \Sigma')$ such that $\int_V f h d\mu = \int_V f h' d\mu$ for all $f \in L^1(V, \Sigma')$. □

In fact we can remove the restriction that $\mu(\cdot)$ is finite and have the following more general version of the Dunford-Pettis Theorem(see Dunford and Schwartz [95], Theorem 8.9, p. 292).

Theorem 4.27 *If (X, Σ, μ) is a σ-finite measure space and $K \subseteq L^1(X)$, then K is relatively weakly compact in $L^1(X)$ if and only if K is bounded, uniformly integrable and for every $\varepsilon > 0$, there exists $A \in \Sigma$ with $\mu(A) < \infty$ and $\sup_{f \in K} \int_{X \setminus A} |f| d\mu \leq \varepsilon$.*

The next proposition allows us to identify the limit of a weakly convergent sequence in $L^1(X)$.

Proposition 4.28 *If (X, Σ, μ) is a finite measure space and $\{f_n\}_{n \in \mathbb{N}} \subseteq L^1(X)\}$ is relatively w-compact, then $f_n \overset{w}{\to} f$ in $L^1(X)$ if and only if for all $h \in L^1(X)$*

$$\int_X |f + h| d\mu \leq \liminf_{n \to \infty} \int_X |f_n + h| d\mu.$$

Proof $\Rightarrow$ We have $f_n + h \overset{w}{\to} f + h$ in $L^1(X)$ and so from Proposition 3.104 we have

$$\int_X |f + h| d\mu \leq \liminf_{n \to \infty} \int_X |f_n + h| d\mu.$$

$\Leftarrow$ From the Eberlein–Smulian theorem (see Theorem 3.121) and by passing to a subsequence if necessary, we may assume that $f_n \overset{w}{\to} \widehat{f}$ in $L^1(X)$. Let $A = \{\widehat{f} > f\} \in \Sigma$. The sequence $\{f_n - f\}_{n \in \mathbb{N}}$ is uniformly integrable (see Theorem 4.26). So, given $\varepsilon > 0$ we can find $c > 0$ such that

$$\int_{\{|f_n - f| > c\}} |f_n - f| d\mu \leq \varepsilon \text{ for all } n \in \mathbb{N}.$$

Let $h = c\chi_{A^c} - c\chi_A - f \in L^1(X)$ (recall $\mu(\cdot)$ is finite). We have

$$|f_n + h| \leq c + 2|f_n - f|\chi_{\{|f_n - f| > c\}} + (f_n - f)\chi_{A^c} - (f_n - f)\chi_A$$

$$\Rightarrow c\mu(X) = \int_X |f + h|d\mu \le \liminf_{n\to\infty} \int_X |f_n + h|d\mu$$

$$\le c\mu(X) - \int_A (\widehat{f} - f)d\mu + \int_{A^c} (\widehat{f} - f)d\mu + 2\varepsilon$$

$$= c\mu(X) - \int_X |\widehat{f} - f|d\mu + 2\varepsilon,$$

$$\Rightarrow \int_X |\widehat{f} - f|d\mu \le 2\varepsilon.$$

Let $\varepsilon \to 0^+$ to conclude that $\widehat{f} = f$ and so $f_n \xrightarrow{w} f$ in $L^1(X)$.   □

A bounded sequence in $L^1(X)$ need not have a weakly convergent subsequence (see the example in Remark 4.23). However, if we exclude a decreasing sequence of measurable sets whose measure goes to zero, then we can extract a weakly convergent subsequence. The result is known as the "Biting Theorem."

Theorem 4.29 *If (X, Σ, μ) is a finite measure space and $\{f_n\}_{n\in\mathbb{N}} \subseteq L^1(X)$ is bounded, then there exists a subsequence $\{f_{n_k}\}_{k\in\mathbb{N}}$ of $\{f_n\}_{n\in\mathbb{N}}$ and a nondecreasing sequence $\{A_m\}_{m\in\mathbb{N}} \subseteq \Sigma$ such that $\mu(A_m) \uparrow \mu(\Omega)$ and $f_{n_k} \xrightarrow{w} f$ in $L^1(A_m)$ for all $m \in \mathbb{N}$ as $k \to \infty$.*

Proof Let $m \in \mathbb{N}$ and define

$$\xi_m(c) = \sup_{n\ge m} \int_{\{|f_n|>c\}} |f_n|d\mu \text{ for all } c \ge 0.$$

We see that for each $m \in \mathbb{N}$, $\xi_n : \mathbb{R}_+ \to \mathbb{R}_+$ and it is decreasing. So $\lim_{c\to\infty} \xi_m(c)$ exists for all $m \in \mathbb{N}$. Let

$$\vartheta_0 = \lim_{c\to+\infty} \xi_1(c).$$

Since for every $m \in \mathbb{N}$, $\{f_k\}_{k=1}^{m-1}$ is uniformly integrable, we have

$$\vartheta_0 = \lim_{c\to+\infty} \xi_m(c).$$

Let $\{c_k\}_{k\in\mathbb{N}}$ an increasing sequence such that $c_k \uparrow +\infty$ and

$$\vartheta_0 \le \xi_1(c_k) \le \vartheta_0 + \frac{1}{k} \text{ for all } k \in \mathbb{N}.$$

Since $\vartheta_0 \le \xi_m(c)$ for all $m \in \mathbb{N}$, $c \ge 0$, we can find a strictly increasing sequence $\{n_k\}_{k\in\mathbb{N}}$ such that

$$\vartheta_0 - \frac{1}{k} \leq \int_{\{|f_{n_k}| > c_k\}} |f_{n_k}| d\mu.$$

We set $\widehat{E}_k = \{|f_{n_k}| > c_k\}$. We have

$$c_k \mu(\widehat{E}_k) \leq \sup_{n \in \mathbb{N}} \|f_n\|_1,$$

$$\Rightarrow \mu(\widehat{E}_k) \to 0 \text{ as } k \to +\infty.$$

Let $\widehat{A}_k = X \setminus \widehat{E}_k$. We claim that $\{\chi_{\widehat{A}_k} f_{n_k}\}_{k \in \mathbb{N}}$ is uniformly integrable. To this and let

$$\xi(c) = \sup_{k \in \mathbb{N}} \int_{\widehat{A}_K \cap \{|f_{n_k}| > c\}} |f_{n_k}| d\mu.$$

We need to show that $\xi(c) \to 0$ as $c \to +\infty$. We have

$$\xi(c_l) = \sup_{k \geq l} \int_{\{c_l < |f_{n_k}| \leq c_k\}} |f_{n_k}| d\mu$$

$$= \sup_{k \geq l} \left[\int_{\{|f_{n_k}| > c_k\}} |f_{n_k}| d\mu - \int_{\{|f_{n_k}| > c_l\}} |f_{n_k}| d\mu \right]$$

$$\leq \sup_{k \geq l} \left[\vartheta_0 + \frac{1}{l} - (\xi - \frac{1}{k}) \right] \leq \frac{2}{l},$$

$$\Rightarrow \xi(c) \to 0^+ \text{ as } c \to +\infty.$$

Finally replace $\{\widehat{E}_k\}_{k \in \mathbb{N}}$ by a sequence $\{E_k\}_{k \in \mathbb{N}}$ decreasing to $\emptyset$ (or to a μ-null set). Then we can find a strictly increasing sequence $\{i_j\}_{j \in \mathbb{N}}$ such that $\mu(E_{i_j}) \leq \frac{1}{2^j}$ for all $j \in \mathbb{N}$. Set $B_m = \bigcup_{j \geq m} E_{i_j}$. Then we see that $\{B_m\}_{m \in \mathbb{N}}$ decreases to a μ-null set. Then if $A_m = X \setminus B_m$ $m \in \mathbb{N}$, we have that $\{\chi_{A_m} f_{n_k}\}_{k \in \mathbb{N}}$ is uniformly integrable and so an application of the Dunford-Pettis theorem (see Theorem 4.26), gives us the desired result. $\qquad \square$

4.2 Variable Exponent Lebesgue Spaces

In this section, we present the so-called Variable Exponent Lebesgue Spaces, which emerged from the needs of mathematical modeling of various physical processes.

Our presentation is concise and we do not aim to the greatest generality, instead we want to give the basic ideas and notions concerning this class of spaces.

So, let $\Omega \subseteq \mathbb{R}^N$ be a measurable set (usually an open set) furnished with the Lebesgue measure. By $L^0(\Omega)$ we denote the set of all measurable functions p : $\Omega \to \mathbb{R}$. We identity two such functions which differ only on a Lebesgue-null set. We set

$$p_- = \underset{\Omega}{\text{ess inf}}\, p \text{ and } p_+ = \underset{\Omega}{\text{ess sup}}\, p_+$$

and $E_1 = \{p \in L^0(\Omega) : 1 < p_- \leq p_+ < \infty\}$. Given $p \in E_1$, the "conjugate exponent" $p'(\cdot)$ is defined by

$$\frac{1}{p(z)} + \frac{1}{p'(z)} = 1 \text{ for a.a } z \in \Omega,$$

$$\Rightarrow p'(z) = \frac{p(z)}{p(z) - 1} \text{ for a.a } z \in \Omega.$$

Evidently $p' \in E$, too. We have

$$p'_+ = (p_-)' \text{ and } p'_- = (p_+)'.$$

Definition 4.30 Given $p \in E_1$, the "Variable Exponent Lebesgue Space" $L^{p(z)}(\Omega)$ is defined by

$$L^{p(z)}(\Omega) = \{f \in L^0(\Omega) : \int_\Omega |f|^{p(z)} d(z) < \infty\}.$$

The function $p \in E_1$ is the "variable exponent" and the function

$$\varrho_p(f) = \int_\Omega |f(z)^{p(z)} dz$$

is known as the "modular function."

Remark 4.31 A simple function $s(z) = \sum_{k=1}^{m} a_k \chi_{A_k}(z)$, with $a_k \in \mathbb{R}$ and $A_k \subseteq \Omega$ measurable and pairwise disjoint with $\lambda^N(A_k) < \infty$ ($\lambda^N(\cdot)$ being the Lebesgue measure on $\mathbb{R}^N$), belongs in $L^{p(z)}(\Omega)$. Indeed we have

$$\varrho_p(f) = \sum_{k=1}^{m} \int_{A_k} |a_k|^{p(z)} dz \leq \sum_{k=1}^{m} \max\{1, |a_k|^{p_1}\}\lambda^N(A_k) < +\infty.$$

We want to provide a norm for these spaces. The next lemma will be helpful in this respect.

Lemma 4.32 *If $p \in E_1$ and $f \in L^{p(z)}(\Omega)$, then $\varphi(t) = \varrho_p(\frac{f}{t})\, t > 0$, has finite values and is nonincreasing, continuous and $\lim_{t \to +\infty} \varphi(t) = 0$.*

Proof Since $f \in L^{p(z)}(\Omega)$, we have $\varphi(1) < \infty$ and since $\varphi(\cdot)$ is clearly nonincreasing, we have $\varphi(t) < +\infty$ for all $t \geq 1$. For $t \in (0, 1)$ we see that $\varphi(t) \leq \frac{1}{t^{p+}}\varphi(1) < \infty$. Therefore $\varphi(\cdot)$ has finite values. Also, we have

$$\lim_{t \to t_0} |\varphi(t) - \varphi(t_0)| = \lim_{t \to t_0} \int_\Omega |f(z)|^{p(z)} |\frac{1}{t^{p(z)}} - \frac{1}{t_0^{p(z)}}| dz.$$

If $t \geq 1$, then $\frac{1}{t^{p(z)}} \leq 1$ and we can apply the Lebesgue Dominated Convergence Theorem.

If $0 < t < 1$, then $\frac{1}{t^{p(z)}} \leq \frac{c}{t^{p+}}$ for some $c > 0$ and t close to t_0. So, again the Lebesgue Dominated Convergence Theorem applies and we have

$$\varphi(t) \to \varphi(t_0) \text{ as } t \to t_0,$$

$$\Rightarrow \varphi(\cdot) \text{ is continuous.}$$

Finally, a new application of the Lebesgue Dominated Convergence Theorem gives that $\varphi(t) \to 0$ as $t \to +\infty$. $\square$

Proposition 4.33 *If $p \in E_1$ and for $f \in L^{p(z)}(\Omega)$ we set*

$$\|f\|_{p(z)} = inf\left\{t > 0 : \int_\Omega \left(\frac{|f(z)|}{t}\right)^{p(z)} dz \leq 1\right\}, \tag{4.6}$$

then $\| \cdot \|_{p(z)}$ is a norm on $L^{p(z)}(\Omega)$.

Proof From Lemma 4.32, we see that $\| \cdot \|_{p(z)}$ has finite values on $L^{p(z)}(\Omega)$. Also for $\|f\|_{p(z)} \neq 0$, we have $\varrho_p(\frac{f}{\|f\|_{p(z)}}) = 1$. So, to show that this is a norm, we need to verify the triangle inequality. But this is an immediate consequence of the convexity of the function $t \to t^r \ (r \geq 1)$. $\square$

There is a close relation between this norm $\| \cdot \|_{p(z)}$ and the modular function $\varrho_p(\cdot)$.

Proposition 4.34 *If $p \in E_1$, then for every $f \in L^{p(z)}(\Omega)$ we have*

$$\left(\frac{\|f\|_{p(z)}}{t}\right)^{p+} \leq \varrho_p\left(\frac{f}{t}\right) \leq \left(\frac{\|f\|_{p(z)}}{t}\right)^{p-} \quad \text{if } t \geq \|f\|_{p(z)}, \tag{4.7}$$

$$\left(\frac{\|f\|_{p(z)}}{t}\right)^{p_-} \leq \varrho_p\left(\frac{f}{t}\right) \leq \left(\frac{\|f\|_{p(z)}}{t}\right)^{p_+} \quad if\ 0 < t < \|f\|_{p(z)}. \tag{4.8}$$

Proof We rewrite (4.7) and (4.8) as

$$t^{p_+} \leq \varrho_p\left(\frac{t}{\|f\|_{p(z)}}f\right) \leq t^{p_-} \quad if\ 0 < t \leq 1, \tag{4.9}$$

$$t^{p_-} \leq \varrho_p\left(\frac{t}{\|f\|_{p(z)}}f\right) \leq t^{p_+} \quad if\ 1 \leq t. \tag{4.10}$$

Then (4.9) and (4.10) follow from the fact that

$$\varrho_p\left(\frac{f}{\|f\|_{p(z)}}\right) = 1 \quad for\ f \neq 0. \tag{4.11}$$

$\square$

Corollary 4.35 *If $p \in E_1$ and $f \in L^{p(z)}(\Omega)$, then*

(a) $\|f\|_{p(z)}^{p_+} \leq \varrho_p(f) \leq \|f\|_{p(z)}^{p_-} \ if\ \|f\|_{p(z)} \leq 1.$
(b) $\|f\|_{p(z)}^{p_-} \leq \varrho_p(f) \leq \|f\|_{p(z)}^{p_+} \ if\ 1 \leq \|f\|_{p(z)}.$

Corollary 4.36 *If $p \in E_1$ and $A \subseteq \Omega$ is a measurable subset, then*

(a) $\lambda^N(A)^{\frac{1}{p_-}} \leq \|\chi_A\|_{p(z)} \leq \lambda^N(A)^{\frac{1}{p_+}} \ if\ \lambda^N(A) \leq 1.$
(b) $\lambda^N(A)^{\frac{1}{p_+}} \leq \|\chi_A\|_{p(z)} \leq \lambda^N(A)^{\frac{1}{p_-}} \ if\ 1 \leq \lambda^N(A).$

Remark 4.37 It follows that $\|\chi_A\|_{p(z)} = 1$ if and only if $\lambda^N(A) = 1$.

We can also have a version of Holder's inequality.

Theorem 4.38 *If $p \in E_1$, $p' \in E_1$ is the conjugate exponent and $f \in L^{p(z)}(\Omega)$, $h \in L^{p'(z)}(\Omega)$, then $\int_\Omega |fh|dz \leq \eta\|f\|_{p(z)}\|h\|_{p'(z)}$ with $\eta = \frac{1}{p_-} + \frac{1}{p'_-} = ess\sup_{\Omega} \frac{1}{p(\cdot)} + ess\sup_{\Omega} \frac{1}{p'(\cdot)}$.*

Proof Using Young's inequality

$$\left|\frac{f(z)h(z)}{\|f\|_{p(z)}\|h\|_{p(z)}}\right| \leq \frac{1}{p(z)}\left|\frac{f(z)}{\|f\|_{p(z)}}\right|^{p(z)} + \frac{1}{p'(z)}\left|\frac{h(z)}{\|h\|_{p'(z)}}\right|^{p'(z)}$$

$$\Rightarrow \int_\Omega \frac{|f(z)h(z)|}{\|f\|_{p(z)}\|h\|_{p(z)}}dz \leq \frac{1}{p_-} + \frac{1}{p'_-} \quad (see (4.11)),$$

$$\Rightarrow \int_{\Omega} |f(z)h(z)|dz \leq \eta \|f\|_{p(z)} \|h\|_{p'(z)}.$$

□

We can also have a version of the generalized Hölder inequality (see Corollary 2.121). First note that if $a, b > 0$ and $1 \leq p, q, r$ satisfy $\frac{1}{p} + \frac{1}{q} = \frac{1}{r}$, then we have

$$(ab)^r \leq \frac{r}{p} a^p + \frac{r}{q} b^q. \tag{4.12}$$

To see this note that $\frac{r}{p} + \frac{r}{q} = 1$ and so by Young's inequality we have

$$(ab)^r \leq \frac{r}{q} (a^r)^{\frac{p}{r}} + \frac{r}{q} (b^r)^{\frac{q}{r}} = \frac{r}{p} a^p + \frac{r}{q} b^q.$$

Proposition 4.39 If $p, q, r \in E_1$, $\frac{1}{p(z)} + \frac{1}{q(z)} = \frac{1}{r(z)}$ for a.a $z \in \Omega$ and $f \in L^{p(z)}(\Omega)$, $h \in L^{q(z)}(\Omega)$, then $fh \in L^{r(z)}(\Omega)$ and $\|fh\|_{r(z)} \leq c\|f\|_{p(z)}\|h\|_{q(z)}$ with $c = ess \sup_{\Omega} \frac{r}{p} + ess \sup_{\Omega} \frac{r}{q}$.

Proof Using (4.12) we have

$$|f(z)h(z)|^{r(z)} \leq \frac{r(z)}{p(z)}|f(z)|^{p(z)} + \frac{r(z)}{q(z)}|h(z)|^{q(z)} \text{ for a.a } z \in \Omega,$$

$$\Rightarrow \varrho_r(fh) \leq c_1 \varrho_p(f) + c_2 \varrho_q(h) \text{ with } c_1 = ess \sup_{\Omega} \frac{r}{p}, \ c_2 = ess \sup_{\Omega} \frac{r}{q},$$

$$\Rightarrow \int_{\Omega} \left(\frac{|fh|}{\|f\|_{p(z)}\|h\|_{p'(z)}}\right)^{r(z)} dz \leq c_1 \int_{\Omega} \left(\frac{|f|}{\|f\|_{p(z)}}\right)^{p(z)} dz + c_2 \int_{\Omega} \left(\frac{|h|}{\|h\|_{q(z)}}\right)^{q(z)} dz$$

$$= c_1 + c_2,$$

$$\Rightarrow \|fh\|_{r(z)} \leq c\|f\|_{p(z)}\|h\|_{q(z)}.$$

□

Now we can show that $L^{p(z)}(\Omega)$ furnished with the norm given by (4.6) is in fact a Banach space.

Theorem 4.40 If $p \in E_1$, then $L^{p(z)}(\Omega)$ is a Banach space.

Proof Let $\{f_n\}_{n \in \mathbb{N}}$ be a Cauchy sequence. We can extract a subsequence $\{f_{n_k}\}_{k \in \mathbb{N}}$ ($n_1 < n_2 < \dots < n_k < \dots$) such that

$$\|f_{n_{k+1}} - f_{n_k}\|_{p(z)} \leq \frac{1}{2^k} \text{ for all } k \in \mathbb{N},$$

$$\Rightarrow \sum_{k \geq 1} \|f_{n_{k+1}} - f_{n_k}\|_{p(z)} < \infty.$$

For $\vartheta > 0$, let $\Omega_\vartheta = \{z \in \Omega : |z| < \vartheta\}$. Using Theorem 4.38 (Hölder's inequality), we have

$$\sum_{k \geq 1} \int_{\Omega_\vartheta} |f_{n_{k+1}} - f_{n_k}| dz \leq c_\vartheta \sum_{k \geq 1} \|f_{n_{k+1}} - f_{n_k}\|_{L^{p(z)}(\Omega_\vartheta)} < \infty \qquad (4.13)$$

with $c_\vartheta = [\frac{1}{p_-} + \frac{1}{p'_-}]\|\chi_{\Omega_\vartheta}\|_{L^{p'(z)}(\Omega_\vartheta)} \in \mathbb{R}_+$.

From (4.13) it follows that $\{f_{n_k}\}_{k \in \mathbb{N}} \subseteq L^1(\Omega_\vartheta)$ is Cauchy. So, $f(z) = \lim_{k \in \infty} f_{n_k}(z)$ for a.a $z \in \Omega_\vartheta$ for every $\vartheta > 0$, hence the convergence is true for a.a $z \in \Omega$.

Since $\{f_n\}_{n \in \mathbb{N}} \subseteq L^{p(z)}(\Omega)$ is Cauchy, we have

$$\|f_n - f_{n_k}\|_{p(z)} < \epsilon \text{ for all big } n, k \in \mathbb{N}, \ (0 < \epsilon \leq 1)$$

$$\Rightarrow \int_\Omega |f_n(z) - f_{n_k}(z)|^{p(z)} dz \leq \epsilon^{p_-} \leq \epsilon,$$

$$\Rightarrow \int_\Omega |f_n(z) - f(z)|^{p(z)} dz \leq \epsilon \ (\text{ by Fatou's lemma}),$$

$$\Rightarrow \|f_n - f\|_{p(z)} \to 0 \text{ as } n \to \infty. \qquad \qquad \square$$

In the constant exponent case, we know that if $1 \leq r, p < \infty$, then

$$\||f|^r\|_p = \|f\|_{pr}^r \text{ for all } f \in L^{pr}(\Omega). \qquad (4.14)$$

This is an easy consequence of the Hölder inequality. In the variable exponent case, equality (4.14) takes the following form.

Proposition 4.41 *If $p, q \in E_1$, $q(z) \leq p(z)$ for a.a $z \in \Omega$ and $f \in L^{p(z)}(\Omega)$, then*

$$\|f\|_{p(z)}^{q+} \leq \||f|^q\|_{(\frac{p}{q})(z)} \leq \|f\|_{p(z)}^{q-} \ if \ \|f\|_{p(z)} \leq 1, \qquad (4.15)$$

$$\|f\|_{p(z)}^{q-} \leq \||f|^q\|_{(\frac{p}{q})(z)} \leq \|f\|_{p(z)}^{q+} \ if \ 1 \leq \|f\|_{p(z)}. \qquad (4.16)$$

Proof Let $\eta = \|f\|_{p(z)}$ and $\vartheta = \||f|^q\|_{(\frac{p}{q})(z)}$. We have

$$\int_\Omega |\frac{|f|^{q(z)}}{\vartheta}|^{(\frac{p}{q})(z)} dz = \int_\Omega |\frac{f}{\eta}|^{p(z)} dz = 1,$$

$$\Rightarrow \int_\Omega |f|^{p(z)} \frac{\eta^{p(z)} - \vartheta^{(\frac{p}{q})(z)}}{\eta^{p(z)} \vartheta^{(\frac{p}{q})(z)}} dz = 0. \qquad (4.17)$$

Suppose that $0 < \eta \leq 1$. If $\vartheta > \eta^{q-}$, then $\vartheta^{(\frac{p}{q})(z)} > \eta^{(\frac{p}{q})(z)q-}$ and so in (4.17) the integrand is negative, a contradiction. If $\eta < \vartheta^{q-}$, then the integrand is positive, again a contradiction. So $\eta = \vartheta^{q-}$ and (4.15) follows. Similarly if $\eta > 1$, then $\eta = \vartheta^{q+}$ and then (4.16) follows. $\qquad\square$

Remark 4.42 If $p, q \in C(\Omega)$, then from (4.17) it follows that there exists $z_0 \in \Omega$ such that $\eta^{p(z_0)} = \vartheta^{(\frac{p}{q})(z_0)}$ and so

$$\||f|^q\|_{(\frac{p}{q})(z)} = \|f\|_{p(z)}^{q(z_0)}.$$

Moreover, taking $q = p$ and assuming that $p \in C(\Omega)$, we see that there exists $z_0 \in \Omega$ such that $\|f\|_{p(z)} = (\int_\Omega |f|^{p(z)} dz)^{1/p(z_0)}$.

The next inclusion property is an easy consequence of the Hölder inequality.

Proposition 4.43 *If $\Omega \subseteq \mathbb{R}^N$ has finite Lebesgue measure, $p, q \in E_1$, and $q(z) \leq p(z)$ for a.a $z \in \Omega$ then $L^{p(z)}(\Omega) \hookrightarrow L^{q(z)}(\Omega)$ continuously.*

The next lemma collects analogues of the well-known convergence results for the Lebesgue integral.

Proposition 4.44 *If $p \in E_1$ and $\{f_n, f, h\}_{n \in \mathbb{N}} \subseteq M(\Omega)$ then*

(a) $f_n(z) \to f(z)$ *a.e. in $\Omega \Rightarrow \varrho_p(f) \leq \liminf_{n \to \infty} \varrho_p(f_n)$.*
(b) $|f_n| \uparrow |f| \Rightarrow \varrho_p(f_n) \uparrow \varrho_p(f)$.
(c) *If $f_n \to f$ a.e. in Ω, $|f_n| \leq h$ for a.a $z \in \Omega$, all $n \in \mathbb{N}$ and $h \in L^{p(z)}(\Omega)$, then $f_n \to f$ in $L^{p(z)}(\Omega)$.*

Proof

(a) Follows from Fatou's lemma.
(b) Follows from the Monotone Convergence Theorem.
(c) We have $|(f_n - f)(z)| \to 0$ a.e. and $|f| \leq h$. Therefore

$$|f_n - f| \leq 2h.$$

So, we can use the dominated convergence theorem and have

$$\lim_{n \to \infty} \varrho_p(|f_n - f|) = 0, \Rightarrow f_n \to f \text{ in } L^{p(z)}(\Omega). \qquad\square$$

Let $S_{p(z)}$ denote the set of simple functions in $L^{p(z)}(\Omega)$. Then from Propositions 2.70 and 4.44(c), we have

Proposition 4.45 *If $p \in E_1$, then $\overline{S_{p(z)}}^{\|\cdot\|_{p(z)}} = L^{p(z)}(\Omega)$.*

Proposition 4.46 *If $p \in E_1$, then $C_c^\infty(\Omega)$ is dense in $L^{p(z)}(\Omega)$.*

Proof From Proposition 4.45 we know that $S_{p(z)}$ is dense in $L^{p(z)}(\Omega)$. A simple function belongs in $L^{p-}(\Omega) \cap L^{p+}(\Omega)$ and so by Proposition 2.132 it can be approximated by a sequence of $C_c^{\infty}(\Omega)$ functions in the space $L^{p-}(\Omega) \cap L^{p+}(\Omega)$. But $L^{p-}(\Omega) \cap L^{p+}(\Omega) \hookrightarrow L^{p(z)}(\Omega)$ continuously. This proves the Proposition.
$\square$

A useful consequence of this density result, is the following proposition.

Proposition 4.47 *If $p \in E_1$, then $L^{p(z)}(\Omega)$ is separable.*

Proof If $f \in L^{p(z)}(\Omega)$, it can be approximated by a continuous function with compact support which in turn is approximated by a polynomial with rational coefficients. The latter set is countable. We conclude that $L^{p(z)}(\Omega)$ is separable.
$\square$

Next we identify the dual of the Banach space $L^{p(z)}(\Omega)$. Recall that if $p \in E_1$, then $p' \in E_1$ denotes the conjugate exponent (that is, $p'(z) = \frac{p(z)}{p(z)-1}$ for a.a $z \in \Omega$).

Theorem 4.48 *If $\Omega \subseteq \mathbb{R}^N$ has finite Lebesgue measure and $p \in E_1$, then $L^{p(z)}(\Omega)^* = L^{p'(z)}(\Omega)$.*

Proof From Hölder's inequality (see Proposition 4.39), we have

$$L^{p'(z)}(\Omega) \subseteq L^{p(z)}(\Omega)^*. \tag{4.18}$$

Let $\varphi \in L^{p(z)}(\Omega)^*$ and let $\mu(A) = \varphi(\chi_A)$ for all $A \subseteq \Omega$ measurable. Since $\chi_{A \cup B} = \chi_A + \chi_B - \chi_{A \cap B}$, we see that $\mu(\cdot)$ is an additive set function. We will show σ-additivity. So, let $\{A_n\}_{n \in \mathbb{N}}$ pairwise disjoint measurable subsets of Ω and let $A = \bigcup_{n \geq 1} A_n$. We set $B_m = \bigcup_{n=1}^{m} A_n$ and we have

$$\|\chi_A - \chi_{B_m}\|_{p(z)} \leq c\|\chi_A - \chi_{B_m}\|_{p+} = c\lambda^N(A \setminus B_m)^{1/p+}.$$

Note that $\lambda^N(A \setminus B_m) \to 0$ as $m \to \infty$ (see Proposition 2.17(d) and recall that by hypothesis $\lambda^N(\Omega) < +\infty$. So, we have

$$\chi_{B_m} \to \chi_A \text{ in } L^{p(z)}(\Omega) \text{ as } m \to \infty,$$

$$\Rightarrow \varphi(\chi_{B_m}) \to \varphi(\chi_A),$$

$$\Rightarrow \mu(\cdot) \text{ is countably additive.}$$

So, we have that $\mu(\cdot)$ is a signed measure. Moreover, if $\lambda^N(A) = 0$, then $|\mu(A)| = |\varphi(\chi_A)| \leq \|\varphi\|_* \|\chi_A\|_{p(z)} = 0$, which means that $\mu << \lambda^N$ so by the Radon–Nikodym Theorem (see Theorem 2.113), we can find $h \in L^1(\Omega)$ such

$$\varphi(\chi_A) = \mu(A) = \int_{\Omega} \chi_A h \, dz.$$

Then by linearity, we have

$$\varphi(s) = \int_{\Omega} sh\,dz \text{ for all s=simple.}$$

Since simple functions are dense in $L^{p(z)}(\Omega)$ (see Proposition 4.45), it follows that $h \in L^{p'(z)}(\Omega)$ and

$$\varphi(f) = \int_{\Omega} fh\,dz \text{ for all } f \in L^{p(z)}(\Omega), \|\varphi\|_{x} = \|h\|_{p'(z)},$$

$$\Rightarrow L^{p(z)}(\Omega)^{*} \subseteq L^{p'(z)}(\Omega),$$

$$\Rightarrow L^{p(z)}(\Omega)^{*} = L^{p'(z)}(\Omega) \text{ (see (4.18))}.$$

(that is $L^{p(z)}(\Omega)^{*}$ and $L^{p'(z)}(\Omega)$ are isometrically isomorphic Banach spaces). □

Remark 4.49 The above representation theorem remains valid if we remove the restriction that $\Omega \subseteq \mathbb{R}^{N}$ has finite Lebesgue measure (see Cruz Uribe and Fiorenza [73], p.65).

Corollary 4.50 *If $p \in E_{1}$, then $L^{p(z)}(\Omega)$ is reflexive.*

Remark 4.51 In fact it is uniformly convex. (see Diening et al. [83] Theorem 5.4.9, p.87)

4.3 Sobolev Spaces

Sobolev spaces are special subspaces of the Lebesgue L^{p}-spaces, whose elements exhibit some differentiability properties.

Let $\Omega \subseteq \mathbb{R}^{N}$ be open. Let $C_{c}^{\infty}(\Omega)$ denote the space of all C^{∞}-functions with compact support, that is, if $f \in C_{c}^{\infty}(\Omega)$, then $\text{supp } f = \overline{\{z \in \Omega : f(z) \neq 0\}}$ is a compact subset of Ω.

Let $\alpha = (\alpha_{k})_{k=1}^{N} \in \mathbb{N}^{N}$ be a "multi-index" we write $|\alpha| = \sum_{k=1}^{N} \alpha_{k}$ (the "length" of the multi-index). If $f \in C^{|\alpha|}(\Omega)$, then we define $D^{\alpha} f = \frac{\partial^{|\alpha|} f}{\partial z_{1}^{\alpha_{1}} \ldots \partial z_{N}^{\alpha_{N}}}$.

We can define a mode of convergence in $C_{c}^{\infty}(\Omega)$ as follows. We say that $\{f_{n}\}_{n \in \mathbb{N}} \subseteq C_{c}^{\infty}(\Omega)$ converges to 0 in $C_{c}^{\infty}(\Omega)$ if we can find a compact set $K \subseteq \Omega$ such that $\bigcup_{n \in \mathbb{N}} \text{supp } f_{n} \subseteq K$ and $D^{\alpha} f_{n} \to 0$ uniformly for all $\alpha \in \mathbb{N}^{N}$.

Definition 4.52 A "distribution" is a linear functional $L : C_c^\infty(\mathbb{R}) \to \mathbb{R}$ such that "$f_n \to 0$ in $C_c^\infty(\Omega) \Rightarrow L(f_n) \to 0$." By $\mathcal{D}'(\Omega)$ we denote the space of distributions.

If $f, h \in C^m(\Omega)$ and $\alpha \in \mathbb{N}^N$ satisfies $|\alpha| \leq m$, then we can easily verify that $\int_\Omega (D^\alpha f)h\,dz = (-1)^{|\alpha|} \int_\Omega f(D^\alpha h)\,dz$ (integration by parts).
This leads to the following definition.

Definition 4.53 Given a distribution $L \in \mathcal{D}'(\Omega)$ and $\alpha \in \mathbb{N}^N$, we define the α-derivative of L, to be the distribution $D^\alpha L$ defined by

$$(D^\alpha L)(f) = (-1)^{|\alpha|} L(D^\alpha f) \text{ for all } f \in C_c^\infty(\Omega).$$

So, every distribution has derivatives of arbitrary order.

Example 4.54 Let $h \in L_{loc}^1(\Omega)$ (that is, $f|_K \in L^1(K)$ for all $K \subseteq \Omega$ compact). Then we can define the distribution $L_h \in \mathcal{D}'(\Omega)$ by

$$L_h(f) = \int_\Omega hf\,dz \text{ for all } f \in C_c^\infty(\Omega).$$

We see that $L_h = L_g$ if and only if $h(z) = g(z)$ for a.a $z \in \Omega$.
Moreover, if $h, g \in L_{loc}^1(\Omega)$ and $\alpha \in \mathbb{N}^N$, then we write that $g = D^\alpha h$ if and only if $L_g = D^\alpha L_h$. Hence we have

$$\int_\Omega fg\,dz = (-1)^{|\alpha|} \int_\Omega h(D^\alpha f)\,dz \text{ for all } f \in C_c^\infty(\Omega).$$

We will say that $D^\alpha h \in L_{loc}^1(\Omega)$, if there exists $g \in L_{loc}^1(\Omega)$ such that $D^\alpha h = g$. Similarly, we say that $D^\alpha h \in L^p(\Omega)$ $1 \leq p \leq \infty$, if there exists $g \in L^p(\Omega)$ such that $D^\alpha h = g$. This way we see that we extended the notion of derivative to all L^p-functions ("weak derivatives").

Using the notion of weak derivative, we can now introduce the Sobolev spaces.

Definition 4.55 Given $m \in \mathbb{N}$ and $1 \leq p \leq \infty$, we define the Sobolev space $W^{m,p}(\Omega)$ as follows

$$W^{m,p}(\Omega) = \{u \in L^p(\Omega) : D^\alpha u \in L^p(\Omega) \text{ for all } \alpha \in \mathbb{N}^N \text{ with } |\alpha| \leq m\}.$$

We furnish this space with following norm

$$\|u\|_{m,p} = \Big[\sum_{|a|\leq m} \|D^\alpha u\|_p^\varrho \Big]^{1/p} \text{ if } 1 \leq p < \infty,$$

$$\|u\|_{m,\infty} = \sum_{|\alpha|\leq m} \|D^\alpha u\|_\infty.$$

Also we set $W_0^{m,p}(\Omega) = \overline{C_c^\infty(\Omega)}^{\|\cdot\|_{m,p}}$.

Proposition 4.56 *If $m \in \mathbb{N}$ and $1 \leq p \leq \infty$,*
then $W^{m,p}(\Omega)$ is a Banach space which is separable if $1 \leq p < \infty$.

Proof Let $\{u_n\}_{n \in \mathbb{N}} \subseteq W^{m,p}(\Omega)$ be a Cauchy sequence. Then for all $\alpha \in \mathbb{N}^N$ with $|\alpha| \leq m$, we see that $\{D^\alpha u_n\}_{n \in \mathbb{N}} \subseteq L^p(\Omega)$ is Cauchy. So, we can find $u_\alpha \in L^p(\Omega)$ such that $D^\alpha u_n \to u_\alpha$ in $L^p(\Omega)$. Also let $u = \lim_{n \in \infty} u_n$ in $L^p(\Omega)$. We have

$$\int_\Omega f(D^\alpha u_n)dz = (-1)^{|\alpha|} \int_\Omega (D^\alpha f)u_n dz,$$

$$\Rightarrow \int_\Omega f u_\alpha dz = (-1)^{|\alpha|} \int_\Omega (D^\alpha f)u \, dz,$$

$$\Rightarrow u_\alpha = D^\alpha u.$$

Therefore $W^{m,p}(\Omega)$ is a Banach space. Consider the linear map $S : W^{m,p}(\Omega) \to L^p(\Omega)^d$ with

$$d = card\{\alpha \in \mathbb{N}^N : |\alpha| \leq m\}$$

defined by $S(u) = (D^\alpha u)_{|\alpha| \leq m}$. This map is an isometry and $S(W^{m,p}(\Omega))$ is a closed subspace of $L^p(\Omega)^d$ which is separable, if $1 \leq p < \infty$. (see Theorem 4.8). So, $W^{m,p}(\Omega)$ is separable if $1 \leq p < \infty$. $\qquad\square$

Also using the isometry S from the above proof together with Theorem 4.6 and Corollary 4.7, we obtain:

Proposition 4.57 *If $m \in \mathbb{N}$ and $1 < p < \infty$,*
then $W^{m,p}(\Omega)$ is uniformly convex, hence reflexive too.

Note that $W_o^{m,p}(\Omega)$ is a closed subspace of $W^{m,p}(\Omega)$.

Proposition 4.58 *If $m \in \mathbb{N}$ and $1 \leq p \leq \infty$,*
then $W_o^{m,p}(\Omega)$ is a Banach space which is separable if $1 \leq p < \infty$ and uniformly convex (thus reflexive) if $1 < p < \infty$.

Remark 4.59 The space $W^{m,2}(\Omega)$ (that is, p=2) deserves special mention. This space is a Hilbert space with inner product given by

$$(u, v) = \sum_{|\alpha| \leq m} (D^\alpha u, D^\alpha v)_{L^2} = \sum_{|\alpha| \leq m} \int_\Omega (D^\alpha u)(D^\alpha v)dz \quad \text{for all } u, v \in W^{m,2}(\Omega).$$

To emphasize the Hilbert space structure we usually write

$$H^m(\Omega) = W^{m,2}(\Omega).$$

In what follows we focus primarily on the first order Sobolev spaces (m=1). First we examine approximations of the elements of $W^{1,p}(\Omega)$ $1 \leq p < +\infty$ by regular (smooth) functions. To this end we introduce "mollifiers."

Definition 4.60 A "mollifier" is a function $\xi_\epsilon \in C_c^\infty(\mathbb{R}^N)$ such that

$$\operatorname{supp}\xi_\epsilon = B_\epsilon = \{z \in \mathbb{R}^N : |z| < \epsilon\}, \xi_\epsilon \geq 0 \text{ and } \int_{\mathbb{R}^N} \xi_\epsilon(z)dz = 1.$$

Remark 4.61 The following family $\{\xi_\epsilon\}_{\epsilon>0}$ are known as the "standard mollifiers."

$$\xi(z) = \begin{cases} c\exp[\frac{1}{|z|^2-1}], & \text{if } |z| < 1, \\ 0, & \text{if } 1 \leq |z|, \end{cases}$$

with $c > 0$ such that $\int_{\mathbb{R}^N} \xi(z)dz = 1$. We define

$$\xi_\epsilon(z) = \frac{1}{\epsilon^N}\xi(\frac{z}{\epsilon}) \text{ for all } z \in \Omega, \text{ all } \epsilon > 0.$$

As their name suggests, we will use mollifiers to regularize "rough" functions. So, let $f \in L_{loc}^1(\Omega)$ and let $\{\xi_\epsilon\}_{\epsilon>0}$ be a family of mollifiers. We extend f to all $\mathbb{R}^N$, by $f|_{\mathbb{R}^N \setminus \Omega} = 0$ and then define

$$f_\epsilon(x) = (\xi_\epsilon * f)(x) = \int_{\mathbb{R}^N} \xi_\epsilon(x-z)f(z)dz = \int_{\mathbb{R}^N} \xi_\epsilon(z)f(x-z)dz.$$

Note that $f_\epsilon \in C^\infty(\mathbb{R}^N)$ and $\operatorname{supp} f_\epsilon \subseteq \operatorname{supp} f + \overline{B}_\epsilon$.

Proposition 4.62

(a) If $f \in C_c(\Omega)$, then $f_\epsilon \to f$ uniformly on Ω as $\epsilon \to 0^+$.
(b) If $f \in L^p(\Omega)$ $1 \leq p < \infty$, then $\|f_\epsilon\|_p \leq \|f\|_p$ for all $\epsilon > 0$ and $f_\epsilon \to f$ in $L^p(\Omega)$ as $\epsilon \to 0^+$.
(c) If $f \in W^{1,p}(\Omega)$ $1 \leq p < \infty$, then $f_\epsilon \to f$ in $W^{1,p}(\Omega)$ as $\epsilon \to 0^+$.

Proof

(a) We have

$$|f_\epsilon(x) - f(x)| \leq \int_{\mathbb{R}^N} |f(x-z) - f(x)|\xi_\epsilon(z)dz$$

$$\leq \sup[|f(x-z) - f(x)| : x \in \operatorname{supp} f, |z| < \epsilon]. \tag{4.19}$$

The uniform continuity of $f(\cdot)$ and (4.19) imply that

$$\sup[|f_\epsilon(x) - f(x)| : x \in \Omega] \to 0 \text{ as } \epsilon \to 0^+.$$

(b) First let $p = 1$ using the Fubini-Tonelli Theorem (see Theorem 2.150), we have

$$\|f_\epsilon\|_1 \leq \int_\Omega \int_{\mathbb{R}^N} |f(x-z)|\xi_\epsilon(z)dzdx$$

$$= \int_{\mathbb{R}^N} \int_\Omega |f(x-z)|dx\xi_\epsilon(z)dz$$

$$= \int_\Omega |f(x)|dx \int_{\mathbb{R}^N} \xi_\epsilon(z)dz = \|f\|_1.$$

Next let $1 < p < \infty$ and let $\vartheta \in C_c(\Omega)$. Using Hölder's inequality, we have

$$|\int_{\mathbb{R}^N} f_\epsilon(x)\vartheta(x)dx| \leq \int_{\mathbb{R}^N} \int_{\mathbb{R}^N} |f(x-z)\vartheta(x)|dx\xi_\epsilon(z)dz$$

$$\leq \int_{\mathbb{R}^N} \|f\|_p\|\vartheta\|_{p'}\xi_\epsilon(z)dz \quad (\frac{1}{p} + \frac{1}{p'} = 1)$$

$$= \|f\|_p\|\vartheta\|_{p'}. \tag{4.20}$$

Recall that $C_c(\Omega)$ is dense in $L^{p'}(\Omega)$ (see Proposition 2.132). Then from (4.20) we infer that

$$\|f_\epsilon\|_p \leq \|f\|_p.$$

We prove the convergence part of (b). Given $\delta > 0$, we can find $h \in C_c(\Omega)$ such that $\|f - h\|_p < \delta$ (see Proposition 2.132). It follows that $\|f_\epsilon - h_\epsilon\|_p < \delta$ and so we have

$$\|f - f_\epsilon\|_p \leq \|f - g\|_p + \|g - g_\epsilon\|_p + \|g_\epsilon - f_\epsilon\|_p < 2\delta + \|g - g_\epsilon\|_p. \tag{4.21}$$

From (a) we have $\|g - g_\epsilon\|_p \to 0$ as $\epsilon \to 0^+$. So, if in (4.21) we pass to the limit as $\epsilon \to 0^+$ we obtain $\limsup_{\epsilon \to 0^+} \|f - f_\epsilon\|_p \leq 2\delta$. But $\delta > 0$ is arbitrary. We conclude that $f_\epsilon \to f$ in $L^p(\Omega)$ as $\epsilon \to 0^+$.

(c) By the dominated convergence theorem, we have

$$D_k f_\epsilon(x) = \int_{\mathbb{R}^N} \frac{\partial}{\partial x_k}\xi_\epsilon(x - z)f(z)dz \quad (D_k = \frac{\partial}{\partial x_k}, k = 1, ..., N)$$

$$= -\int_{\mathbb{R}^N} \frac{\partial}{\partial z_k}\xi_\epsilon(x - z)f(z)dz$$

$$= \int_{\mathbb{R}^N} \xi_\epsilon(x - z) \frac{\partial f}{\partial z_k}(z) dz$$

$$= \xi_\epsilon * D_k f.$$

Then the result of this part of the proposition follows from part (b).

□

A useful byproduct of the above proof, is the following corollary.

Corollary 4.63 *If* $m \in \mathbb{N}$, $1 \leq p < \infty$ *and* $f \in W^{m,p}(\Omega)$, *then* $D^\alpha f_\varepsilon = \xi_\varepsilon * D^\alpha f = (D^\alpha f)_\varepsilon$ *for every* $\varepsilon > 0$ *and* $\alpha \in \mathbb{N}^N$ *with* $|\alpha| \leq m$.

Remark 4.64 Also from Proposition 4.62, we infer that $C_c^\infty(\Omega)$ is dense in $L^p(\Omega)$ for all $1 \leq p < \infty$ (see Proposition 2.132).

For our first approximation result, we will need the notion of partition of unity introduced in Definition 1.190. In Proposition 1.191, we characterized paracompact spaces through the existence of a partition of unity. We will need a "smooth" version of that proposition. It can be found in Adams and Fournier [1], Theorem 3.15, p.65 and it asserts the following.

Proposition 4.65 *If* $D \subseteq \mathbb{R}^N$ *is nonempty,* $\mathscr{S} = \{U_\alpha\}_{\alpha \in J}$ *is an open cover of* D, *then we can find a collection* $\mathcal{P}$ *of functions* $\varphi \in C_c^\infty(\mathbb{R}^N)$ *such that*

 (i) $0 \leq \varphi(z) \leq 1$ *for all* $z \in \mathbb{R}^N$ *and all* $\varphi \in \mathcal{P}$.
 (ii) *If* $K \subseteq D$ *is compact, then all but finitely many* $\varphi \in \mathcal{P}$ *vanish on* K.
 (iii) *For every* $\varphi \in \mathcal{P}$, *there exist* $U_\alpha \in \mathscr{S}$ *such that* supp $\varphi \subseteq U_\alpha$.
 (iv) $\sum_{\varphi \in \mathcal{P}} \varphi(z) = 1$ *for all* $z \in D$,
    (that is, $\mathcal{P}$ *is a smooth partition of unity subordinate to the cover* $\mathscr{S}$*)*.

Using this fact, we can have the first approximation result.

Theorem 4.66 *If* $\Omega \subseteq \mathbb{R}^N$ *is open,* $m \in \mathbb{N}$, $1 \leq p < \infty$ *and* $u \in W^{m,p}(\Omega)$, *then we can find* $u_n \in W^{m,p}(\Omega) \cap C^\infty(\Omega)$, $n \in \mathbb{N}$ *such that*

$$u_n \to u \text{ in } W^{m,p}(\Omega)$$

(that is $W^{m,p}(\Omega) = \overline{W^{m,p}(\Omega) \cap C^\infty(\Omega)}^{\|\cdot\|_{m,p}}$*).*

Proof We can find $\{\Omega_k\}_{k \in \mathbb{N}}$ open subsets of Ω such that, $\Omega_k \subseteq \Omega_{k+1}$, $\overline{\Omega}_k \subseteq \Omega$ is compact for all $k \in \mathbb{N}$ and $\Omega = \bigcup_{k \in \mathbb{N}} \Omega_k$ (see Proposition 1.115). According to Proposition 4.65, we can have a partition of unity $\mathcal{P} = \{\varphi_k\}_{k \in \mathbb{N}} \subseteq C^\infty(\Omega)$ subordinate to the open cover $\mathscr{S} = \{U_k = \Omega_{k+1} \setminus \overline{\Omega}_{k-1}\}_{k \in \mathbb{N}}$ with $\Omega_0 = \emptyset$. Given $\delta > 0$, we can find $0 < \varepsilon_k \leq d(\Omega_k, \partial\Omega_{k+1})$, $k \in \mathbb{N}$ such that

$$\|(\varphi_k u)_{\varepsilon_k} - \varphi_k u\|_{m,p} \leq \frac{\delta}{2^k}.(\text{ see Proposition 4.62}). \tag{4.22}$$

We set $u_k = (\varphi_k u)_{\varepsilon_k}$, $k \in \mathbb{N}$. At any $\Omega' \subseteq \Omega$ open with $\overline{\Omega'} \subseteq \Omega$ compact only a finite number of the functions u_k do not vanish identically (see Proposition 4.65). So, if $u_\delta = \sum_{k \in \mathbb{N}} u_k$, then u_δ is well defined and $u_\delta \in C^\infty(\Omega)$ and we have

$$\|u - u_\delta\|_{m,p} \leq \sum_{k \in \mathbb{N}} \|u_k - \varphi_k u\|_{m,p} \leq \delta \text{ (see (4.22))}.$$

Therefore, $u_\delta \in W^{m,p}(\Omega) \cap C^\infty(\Omega)$ and $u_\delta \to u$ as $\delta \to 0^+$. □

Remark 4.67 We stress that the approximating smooth functions belong in $C^\infty(\Omega)$ and not in $C^\infty(\overline{\Omega})$. For this reason the result is usually called "local approximation theorem." This theorem suggests that we can have a definition of the Sobolev space $W^{m,p}(\Omega)$, $m \in \mathbb{N}$, $1 \leq p < \infty$, without any mention of weak derivatives. Indeed let $V = \{u \in C^\infty(\Omega) : D^\alpha u \in L^p(\Omega) \text{ for all } |\alpha| \leq m\}$. Then $W^{m,p}(\Omega) = \overline{V}^{\|\cdot\|_{m,p}}$. We mention that Theorem 4.66 fails if $p = \infty$. To see this let $\Omega = (-1, 1)$ and consider the function $u(z) = \begin{cases} 0, & \text{if } z \leq 0 \\ z, & \text{if } 0 \leq z. \end{cases}$ Evidently, $u'(z) = \begin{cases} 0, & \text{if } z < 0 \\ 1, & \text{if } 0 < z. \end{cases}$ Let $\varepsilon > 0$ and $\varphi \in W^{1,\infty}(\Omega) \cap C^\infty(\Omega)$ such that $\|u - \varphi\|_{1,\infty} \leq \varepsilon$. Then we will have $\|u' - \varphi'\|_\alpha < \varepsilon$. If $z < 0$, then $|\varphi'(z)| < \varepsilon$ and if $z > 0$, then $|1 - \varphi'(z)| < \varepsilon \Rightarrow 1 - \varepsilon < \varphi'(z)$. Let $z \to 0$. Then $\varphi'(0) \leq \varepsilon$, $\varphi'(0) \geq 1 - \varepsilon$, a contradiction (take $\varepsilon \in (0, 1/2)$). This shows that functions in $W^{1,\infty}(\Omega)$ cannot be approximated by smooth ones.

We can have a "global approximation theorem" (that is, approximation by functions which are smooth up to the boundary), provided we impose some restrictions on the set Ω. We will say that $\Omega \subseteq \mathbb{R}^N$ open has Lipschitz boundary $\partial\Omega$, if $\partial\Omega$ locally is the graph of a Lipschitz function. For a precise definition using local charts, see Adams and Fournier [1], Definition 4.9, p.83. Note that for such a boundary $\partial\Omega$, by Rademacher's theorem (see Evans and Gariepy [105], Theorem 2, p.81), the outward unit normal $n(\cdot)$ on $\partial\Omega$ exists at almost all points of $\partial\Omega$.

Using the notion of Lipschitz boundary, we can have a global approximation result. For its proof, we refer to Evans and Gariepy [105], Theorem 3, p.127.

Theorem 4.68 *If $\Omega \subseteq \mathbb{R}^N$ is bounded open with Lipschitz boundary $\partial\Omega$, $m \in \mathbb{N}$, $1 \leq p < \infty$ and $u \in W^{m,p}(\Omega)$, then we can find $\{u_n\}_{n \in \mathbb{N}} \subseteq C^\infty(\overline{\Omega})$ such that*

$$u_n \to u \text{ in } W^{m,p}(\Omega) \text{ as } n \to \infty.$$

Remark 4.69 Theorem 4.68 implies that for any bounded open $\Omega \subseteq \mathbb{R}^N$ with Lipschitz boundary $\partial\Omega$, for any $u \in W^{1,p}(\Omega)$ ($1 \leq p < \infty$), we can find $\{\widehat{u}_n\}_{n \in \mathbb{N}} \subseteq C_c^\infty(\mathbb{R}^N)$ such that $\widehat{u}_n|_\Omega \to u$ in $W^{1,p}(\Omega)$. In general given an $\Omega \subseteq \mathbb{R}^N$

open and $u \in W^{1,p}(\Omega)$ $(1 \le p < \infty)$, we can find $\{\widehat{u}_n\}_{n \in \mathbb{N}} \subseteq C_c^\infty(\mathbb{R}^N)$ such that

$$\widehat{u}_n \to u \text{ in } L^p(\Omega) \text{ and } \frac{\partial \widehat{u}_n}{\partial z_k}\Big|_{\Omega'} \to \frac{\partial u}{\partial z_k}\Big|_{\Omega'} \text{ for all } k \in \{1, \cdots, N\}$$

with $\Omega' \subseteq \Omega$ open, $\overline{\Omega'} \subseteq \Omega$ compact. This approximation result is often known as "Friedrich's Theorem." We point out that Theorem 4.68 fails for $p = \infty$.

Next we provide a characterization of the elements of $W^{1,p}(\Omega)$.

Proposition 4.70 *If* $\Omega \subseteq \mathbb{R}^N$ *is open,* $1 < p \le \infty$ *and* $u \in L^p(\Omega)$, *then the following statements are equivalent:*

(a) $u \in W^{1,p}(\Omega)$.
(b) $\left| \int_\Omega u \frac{\partial h}{\partial z_k} dz \right| \le c\|h\|_{L^{p'}(\Omega)}$ *for some* $c > 0$, *all* $h \in C_c^\infty(\Omega)$, *all* $k = 1, \cdots, N$, $(\frac{1}{p} + \frac{1}{p'} = 1)$.

Proof $(a) \Rightarrow (b)$ Clear using integration by parts and Holder's inequality.
$(b) \Rightarrow (a)$ Consider the linear functional $\xi_k : C_c^\infty(\Omega) \to \mathbb{R}$ defined by

$$\xi_k(h) = \int_\Omega u \frac{\partial h}{\partial z_k} dz, k = 1, \cdots, N.$$

This is defined on a dense subset of $L^{p'}(\Omega)$ (see Proposition 2.132 and note that $1 \le p' < \infty$) and it is uniformly continuous on that subset. So it can be extended continuously on $L^{p'}(\Omega)$. We denote the extension by $\widehat{\xi}_k$. From the Riesz Representation Theorem (see Theorem 2.133), we can find $g_k \in L^p(\Omega)$ such that

$$\langle \widehat{\xi}_k, h \rangle = \int_\Omega g_k h \, dz \text{ for all } h \in L^{p'}(\Omega),$$

$$\Rightarrow \int_\Omega u \frac{\partial h}{\partial z_k} dz = \int_\Omega g_k h \, dz \text{ for all } h \in C_c^\infty(\Omega), k = 1, \cdots, N,$$

$$\Rightarrow u \in W^{1,p}(\Omega) \text{ (see Definition 4.55)}. \qquad \square$$

Remark 4.71 For $W^{1,\infty}(\Omega)$ with Ω connected, we can say that every $u \in W^{1,\infty}(\Omega)$ has a continuous representative which satisfies

$$|u(z) - u(x)| \le \|Du\|_\infty d_g(z, x) \text{ for all } z, x \in \Omega,$$

with $D = (\frac{\partial}{\partial z_k})_{k=1}^N$ (the gradient) and $d_g(z, x)$ the geodesic distance between z and x. If Ω is convex, then $d_g(z, x) = |z - x|$. Therefore we can think of the functions in $W^{1,\infty}(\Omega)$ as Lipschitz continuous functions.

Definition 4.72 For $1 \leq p < \infty$, let $W^{1,p}_{loc}(\Omega) = \{u \in L^p(\Omega) : u \in W^{1,p}(\Omega')$ for every $\Omega' \subseteq \Omega$ open with $\overline{\Omega'} \subseteq \Omega$ compact $\}$. Also, if $A \subseteq \mathbb{R}^N$ and $u : A \to \mathbb{R}$, we say that $u(\cdot)$ is locally Lipschitz, it for every $z \in A$ we can find $U \in \mathcal{N}(z)$ and a constant $k_U > 0$ (depends on U) such that

$$|u(x) - u(y)| \leq k_U |x - y| \text{ for all } x, y \in U.$$

Remark 4.73 Since we consider functions on $\mathbb{R}^N$, the above definition of locally Lipschitz function is equivalent to saying that the function restricted on every compact subset of A is Lipschitz continuous. As we already mentioned, by Rademacher's theorem such a function is almost everywhere differentiable.

Proposition 4.74 *If $\Omega \subseteq \mathbb{R}^N$ is open and $u : \Omega \to \mathbb{R}$ is locally Lipschitz, then $u \in W^{1,p}_{loc}(\Omega)$, $1 \leq p < \infty$.*

Proof Let $U \subseteq \Omega$ open with $\overline{U} \subseteq \Omega$ compact. Multiplying u with a cut off function $\varphi \in C^\infty_c(\Omega)$, $\varphi|_{\overline{U}} = 1$, $0 \leq \varphi \leq 1$, we see that we may assume that u is Lipschitz and bounded in $\mathbb{R}^N$. Let $\xi(\cdot)$ be the standard mollifier (see Remark 4.61) and for $\varepsilon_n = \frac{1}{n}$, $n \in \mathbb{N}$, let $u_n = \xi_{\varepsilon_n} * u \in C^\infty(\mathbb{R}^N)$. We have

$$\|u_n\|_\infty \leq \|u\|_\infty \text{ and } u_n \to u \text{ uniformly on } U \text{ (Proposition 4.62).} \tag{4.23}$$

Also we have

$$\frac{\partial u_n}{\partial z_k}(z) \to \frac{\partial u}{\partial z_k}(z) \text{ for a.a } z \in \Omega, \text{ all } k = 1, \cdots, N, \tag{4.24}$$

and

$$\|\frac{\partial u_n}{\partial z_k}\|_\infty \leq \|\frac{\partial u}{\partial z_k}\|_\infty \text{ for all } n \in \mathbb{N}, \text{ all } k = 1, \cdots, N. \tag{4.25}$$

From (4.23), (4.24), and the Lebesgue Dominated Convergence Theorem, it follows that

$$\|u_n - u\|_p \to 0 \text{ and } \|Du_n - Du\|_p \to 0.$$

(recall that $D = \left(\frac{\partial}{\partial z_k}\right)^N_{k=1}$). Hence $u_n \to u$ in $W^{1,p}(\Omega)$ and so $u \in W^{1,p}(\Omega)$. $\square$

Proposition 4.75 *If $\Omega \subseteq \mathbb{R}^N$ is open, $1 \leq p < \infty$ and $u \in W^{1,p}_{loc}(\Omega) \cap L^p(\Omega)$ with $Du \in L^p(\Omega, \mathbb{R}^N)$, then $u \in W^{1,p}(\Omega)$.*

Proof Let $\{\Omega_k\}_{k \in \mathbb{N}}$ be an increasing sequence of open subsets of Ω, with $\overline{\Omega_k} \subseteq \Omega$ compact for all $k \in \mathbb{N}$ and $\Omega = \bigcup_{k \in \mathbb{N}} \Omega_k$. According to Proposition 4.65, we can find a C^∞ partition of unity $\{\varphi_k\}_{k\in\mathbb{N}}$ subordinate to the open cover $\{\Omega_{k+1} \setminus \overline{\Omega}_{k-1}\}_{k\in\mathbb{N}}$, $\Omega_0 = \emptyset$. Given $\varepsilon > 0$, we can find $\psi_k \in C_c^\infty(\Omega_{k+1} \setminus \overline{\Omega}_{k-1})$ such that

$$\|\psi_k - \varphi_k u\|_{1,p} \leq \frac{\varepsilon}{2^k}. \tag{4.26}$$

Let $\psi = \sum_{k\in\mathbb{N}} \psi_k \in C^\infty(\Omega)$. We have

$$\|\psi - u\|_{1,p} = \left\| \sum_{k\in\mathbb{N}} \psi_k - \sum_{k\in\mathbb{N}} \varphi_k u \right\|_{1,p}$$

$$\leq \sum_{k\in\mathbb{N}} \|\psi_k - \varphi_k u\|_{1,p} \leq \varepsilon \text{ (see (4.26))},$$

$$\Rightarrow u \in W^{1,p}(\Omega).$$

$\square$

Next we present a chain rule for Sobolev functions. As we will see the result has some important consequences.

Theorem 4.76 *If $\Omega \subseteq \mathbb{R}^N$ is open, $\xi : \mathbb{R} \to \mathbb{R}$ is Lipschitz continuous, $\xi(0) = 0$, $1 \leq p \leq \infty$ and $u \in W^{1,p}(\Omega)$, then $\xi \circ u \in W^{1,p}(\Omega)$ and $D(\xi \circ u) = \xi^*(u)Du$ with $\xi^* : \mathbb{R} \to \mathbb{R}$ being any bounded Borel function such that $\xi^x(z) = \xi'(z)$ for a.a $z \in \Omega$; the result remains true if we replace $W^{1,p}(\Omega)$ with $W_0^{1,p}(\Omega)$.*

Proof We do the proof for the case $\xi \in C^1(\mathbb{R})$ with bounded derivative. For the general case we refer to Marcus and Mizel [186].

By the mean value theorem we have $|\xi(t)| \leq c|t|$ for all $t \in \mathbb{R}$ and so $|\xi \circ u| \leq c|u|$. This means that $\xi \circ u \in L^p(\Omega)$ and $(\xi' \circ u)\frac{\partial u}{\partial z_k} \in L^p(\Omega)$ for all $k = 1, \cdots, N$.

If $1 \leq p < \infty$, then using Theorem 4.66, we can find $\{u_n\}_{n\in\mathbb{N}} \subseteq W^{1,p}(\Omega) \cap C^\infty(\Omega)$, such that $u_n \to u$ in $W^{1,p}(\Omega)$. We have $\xi \circ u_n$ is locally Lipschitz and

$$D(\xi \circ u_n) = \xi'(u_n)Du_n \in L^p(\Omega, \mathbb{R}^N) \text{ for all } n \in \mathbb{N}.$$

Also $|\xi(u_n)| \leq c|u_n|$ for all $n \in \mathbb{N}$. Hence $\xi \circ u_n \in L^p(\Omega)$ for all $n \in \mathbb{N}$. Then from Proposition 4.75, it follows that $\xi \circ u_n \in W^{1,p}(\Omega)$.

We have

$$\int_\Omega \left| \xi \circ u_n - \xi \circ u \right|^p dz \leq C^p \left\| u_n - u \right\|_p^\varrho \to 0, \tag{4.27}$$

and

$$\int_\Omega |\xi'(u_n)Du_n - \xi'(u)Du|^p dz \leq \widehat{c} \left[\|u_n - u\|_p^\varrho + \int_\Omega |Du|^p |\xi'(u_n) - \xi'(u)|^p dz \right]$$

for some $\widehat{c} > 0$. Hence we have

$$\int_\Omega |\xi'(u_n)Du_n - \xi'(u)Du|^p dz \to 0. \tag{4.28}$$

We know that

$$\int_\Omega \big(\xi'(u_n)Du_n\big)h dz = -\int_\Omega \xi(u_n)Dh dz \text{ for all } n \in \mathbb{N}, \text{ all } h \in C_c^\infty(\Omega),$$

$$\Rightarrow \int_\Omega \big(\xi'(u)Du\big)h dz = -\int_\Omega \xi(u)Dh dz \text{ for all } h \in C_c^\infty(\Omega) \text{ (see (4.27-28))},$$

$$\Rightarrow \xi \circ u \in W^{1,p}(\Omega).$$

Similarly if $W^{1,p}(\Omega)$ is replaced by $W_0^{1,p}(\Omega)$.
When $p = \infty$, the result is a consequence of the Rademacher theorem. □

Remark 4.77 A careful inspection of the above proof, reveals that if $\Omega \subseteq \mathbb{R}^N$ is bounded, then we can drop the requirement that $\xi(0) = 0$.

A first consequence of this theorem is the following proposition.

Proposition 4.78 *If $\Omega \subseteq \mathbb{R}^N$ is open, $1 \le p \le \infty$ and $u \in W^{1,p}(\Omega)$ (respectively, $u \in W_0^{1,p}(\Omega)$), then $u^+, u^-, |u| \in W^{1,p}(\Omega)$ (respectively, $u^+, u^-, |u| \in W_0^{1,p}(\Omega)$) and we have*

$$Du^+ = \begin{cases} Du & \text{a.e on } \{u > 0\} \\ 0 & \text{a.e. on } \{u \le 0\}, \end{cases}$$

$$Du^- = \begin{cases} -Du & \text{a.e on } \{u < 0\} \\ 0 & \text{a.e. on } \{u \ge 0\}, \end{cases}$$

$$D(u) = \begin{cases} -Du & \text{a.e. on } \{u < 0\} \\ 0 & \text{a.e. on } \{u = 0\} \\ Du & \text{a.e. on } \{u > 0\}. \end{cases}$$

Proof Just consider the Lipschitz maps

$$\xi_+(x) = x^+ = \max\{x, o\}, \ \xi_-(x) = -x^- = \min\{x, 0\}, \ \xi_0(x) = |x| \text{ for all } x \in \mathbb{R},$$

and apply the chain rule (Theorem 4.76). □

Corollary 4.79 *If $\Omega \subseteq \mathbb{R}^N$ is open, $1 \le p \le \infty$ and $u \in W^{1,p}(\Omega)$, then for all $\lambda \in \mathbb{R}$, $Du(z) = 0$ for a.a $z \in \{u = \lambda\}$.*

Proof By replacing u, with $u - \lambda$ if necessary, we may assume that $\lambda = 0$. Then the result follows from Proposition 4.78 since $Du = Du^+ - Du^-$. □

Remark 4.80 In fact we have a more general version of this corollary. Namely, if $D \subseteq \mathbb{R}$ is Lebesgue-null, then $Du = 0$ a.e. on $u^{-1}(D)$. The result is known as "Stampacchia's theorem."

Proposition 4.81 *If $\Omega \subseteq \mathbb{R}^N$ is open, $1 \leq p < \infty$ and $u, v \in W^{1,p}(\Omega)$ or $W_0^{1,p}(\Omega)$, then $\max\{u, v\}, \min\{u, v\} \in W^{1,p}(\Omega)$ or $W_0^{1,p}(\Omega)$ and we have*

$$D \max\{u, v\} = \begin{cases} Du & a.e \ on \ \{u \geq v\} \\ Dv & a.e. \ on \ \{u \leq v\} \end{cases}$$

$$D \min\{u, v\} = \begin{cases} Du & a.e \ on \ \{u \leq v\} \\ Dv & a.e. \ on \ \{v \leq u\} \end{cases}$$

Proof Simply observe that

$$\max\{u, v\} = (u - v)^+ + v \text{ and } \min\{u, v\} = u - (u - v)^+$$

and then apply Proposition 4.78.   □

Proposition 4.82 *If $\Omega \subseteq \mathbb{R}^N$ is open, $1 \leq p < \infty$, $\{u_n, u\}_{n \in \mathbb{N}} \subseteq W^{1,p}(\Omega)$, $u \geq 0$ and $u_n \to u$ in $W^{1,p}(\Omega)$, then $u_n^+ \to u$ in $W^{1,p}(\Omega)$.*

Proof We have $|u_n^+ - u| \leq |u_n - u|$ and so $u_n^+ \to u$ in $L^p(\Omega)$.
  Also using Proposition 4.78 and Corollary 4.79, we have

$$\int_\Omega |Du_n^+ - Du|^p dz = \int_{\{u_n > 0\}} |Du_n - Du|^p dz = \|Du_n - Du\|_p^\varrho - \int_{\{u_n < 0\}} |Du|^p dz. \tag{4.29}$$

If $u = 0$, then $Du = 0$ and so from (4.29) we are done. So, we assume that $u \neq 0$. Then for $\Omega_n = \{z \in \Omega : u_n(z) < 0\}$, we have $\chi_{\Omega_n} \xrightarrow{\lambda^N} 0$ with λ^N being the Lebesgue measure on $\mathbb{R}^N$. We have $\chi_{\Omega_n}|Du|^p \xrightarrow{\lambda^N} 0$ (see Proposition 2.138) and then by Vitali's theorem (see Theorem 2.147), we have

$$\int_{\{u_n < 0\}} |Du|^p dz = \int_\Omega \chi_{\Omega_n} |Du|^p dz \to 0 \text{ as } n \to \infty,$$

$$\Rightarrow Du_n^+ \to Du \text{ in } L^p(\Omega, \mathbb{R}^N) \text{ (see (4.29)),}$$

$$\Rightarrow u_n^+ \to u \text{ in } W^{1,p}(\Omega) \text{ as } n \to \infty.$$ □

Proposition 4.83 *If $\Omega \subseteq \mathbb{R}^N$ is open, $1 \leq p < \infty$, $u \in W_0^{1,p}(\Omega)$ and $u \geq 0$, then we can find a sequence $\{\vartheta_n\}_{n \in \mathbb{N}} \subseteq C_c^\infty(\Omega)$, $\vartheta_n \geq 0$ for all $n \in \mathbb{N}$ such that $\vartheta_n \to u$ in $W_0^{1,p}(\Omega)$.*

Proof Since $u \in W_0^{1,p}(\Omega)$, we can find $\{\widehat{\vartheta}_n\}_{n\in\mathbb{N}} \subseteq C_c^\infty(\Omega)$ such that $\vartheta_n \to u$ in $W_0^{1,p}(\Omega)$. By Proposition 4.82, we have $\widehat{\vartheta}_n^+ \to u$ in $W_0^{1,p}(\Omega)$. For every $n \in \mathbb{N}$, via mollification and using Proposition 4.62, we can find $\{\widehat{\vartheta}_m^n\}_{m\in\mathbb{N}} \subseteq C_c^\infty(\Omega)$ such that $\widehat{\vartheta}_m^n \geq 0$ and $\widehat{\vartheta}_m^n \to \widehat{\vartheta}_n^+$ in $W_0^{1,p}(\Omega)$. Then we can find a subsequence $\{m_n\}_{n\in\mathbb{N}}$ such that $\widehat{\vartheta}_{m_n}^n \to u$ in $W_0^{1,p}(\Omega)$. $\qquad\square$

Next we prove a product rule for Sobolev functions.

Proposition 4.84 *If $\Omega \subseteq \mathbb{R}^N$ is open, $1 \leq p < \infty$ and $u, v \in W^{1,p}(\Omega) \cap L^\infty(\Omega)$, then $uv \in W^{1,p}(\Omega) \cap L^\infty(\Omega)$ and $D(uv) = u\,Dv + (Du)v$.*

Proof Since $u, v \in L^\infty(\Omega)$, without any loss of generality, we may assume that $|u(z)|, |v(z)| \leq 1$ for a.a $z \in \Omega$. By Theorem 4.66, we can find $\{\widehat{u}_n\}_{n\in\mathbb{N}}, \{\widehat{v}_n\}_{n\in\mathbb{N}} \subseteq W^{1,p}(\Omega) \cap C^\infty(\Omega)$ such that $\widehat{u}_n \to u, \widehat{v}_n \to v$ in $W^{1,p}(\Omega)$. We define

$$u_n(z) = \begin{cases} -1, & \text{if } \widehat{u}_n(z) < -1 \\ \widehat{u}_n(z), & \text{if } -1 \leq \widehat{u}_n(z) \leq 1, \\ 1, & \text{if } 1 < \widehat{u}_n(z) \end{cases} \quad v_n(z) = \begin{cases} -1, & \text{if } \widehat{v}_n < -1 \\ \widehat{v}_n(z), & \text{if } -1 \leq \widehat{v}_n(z) \leq 1 \\ 1, & \text{if } 1 < \widehat{v}_n(z). \end{cases}$$

Then these are locally Lipschitz functions and then so is $u_n v_n$ and by Rademacher's theorem we have

$$D(u_n v_n) = u_n(Dv_n) + (Du_n)v_n \text{ for all } n \in \mathbb{N}. \tag{4.30}$$

From Proposition 4.74 we know that $u_n v_n \in W_{loc}^{1,p}(\Omega)$. Then (4.30) and Proposition 4.75 imply that $u_n v_n \in W^{1,p}(\Omega)$ for all $n \in \mathbb{N}$. We have

$$u_n \to u \text{ and } v_n \to v \text{ in } W^{1,p}(\Omega),$$

$$\Rightarrow u_n v_n \to uv \text{ in } L^p(\Omega).$$

Moreover, we have

$$\int_\Omega \left| (u_n(Dv_n) + (Du_n)v_n) - (u(Dv) + (Du)v) \right|^p dz$$

$$\leq 2^{p-1}\left[\int_\Omega \left| u_n(Dv_n) - u(Dv) \right|^p dz + \int_\Omega \left| (Du_n)v_n - (Du)v \right|^p dz \right]$$

$$\leq 2^{2(p-1)}\left[\int_\Omega |u_n|^p \left| Dv_n - Dv \right|^p dz + \int_\Omega |Dv|^p |u_n - u|^p dz \right.$$

$$\left. + \int_\Omega |v_n|^p \left| Du_n - Du \right|^p dz + \int_\Omega |Du|^p |v_n - v|^p dz \right] \to 0,$$

$$\Rightarrow uv \in W^{1,p}(\Omega) \cap L^\infty(\Omega) \text{ and } D(uv) = u(Dv) + (Du)v \text{ (see (4.30))}. \quad\square$$

Remark 4.85 The result remains true if $W^{1,p}(\Omega)$ is replaced by $W_0^{1,p}(\Omega)$.

Proposition 4.86 *If* $\Omega \subseteq \mathbb{R}^N$ *is open,* $1 \leq p < \infty$, $\{u_n\}_{n\in\mathbb{N}} \subseteq W^{1,p}(\Omega)$ *and*

$$g = \sup_{n\in\mathbb{N}} |Du_n| \in L^p(\Omega), \quad v = \sup_{n\in\mathbb{N}} u_n,$$

then $v \in W^{1,p}(\Omega)$ *and* $|Dv(z)| \leq g(z)$ *for a.a* $z \in \Omega$.

Proof Let $v_k = \max\limits_{1 \leq n \leq k} u_n$, $k \in \mathbb{N}$. From Proposition 4.81, we know that $v_k \in W^{1,p}(\Omega)$ and $|Dv_k(z)| \leq \max\limits_{1 \leq n \leq k} |Du_n(z)| \leq g(z)$ for a.a $z \in \Omega$. Note that $\{v_k\}_{k\in\mathbb{N}}$ is increasing and $v_k(z) \uparrow v(z)$ for a.a $z \in \Omega$. Also $\{Dv_k\}_{k\in\mathbb{N}} \subseteq L^p(\Omega, \mathbb{R}^N)$ is bounded. So we can say that

$$v_k \to v \text{ in } L^p(\Omega) \text{ (monotone convergence theorem)}, \tag{4.31}$$

$$Dv_k \xrightarrow{w} f \text{ in } L^p(\Omega, \mathbb{R}^N) \text{ (Eberlein–Smulian theorem)} \tag{4.32}$$

(for the second convergence we may have to pass to a subsequence). For every $\psi \in C_c^\infty(\Omega)$ and $m = 1, \cdots, N$, we have (see (4.31-32)

$$\int_\Omega \left(\frac{\partial v_k}{\partial z_m}\right) \psi dz = -\int_\Omega v_k \left(\frac{\partial \psi}{\partial z_m}\right) dz$$

$$\Rightarrow \int_\Omega f_k \psi dz = -\int_\Omega v \left(\frac{\partial \psi}{\partial z_k}\right) dz \text{ with } f = (f_k)_{k=1}^N.$$

Therefore, $f = Dv$ and $|Dv(z)| \leq g(z)$ for a.a $z \in \Omega$ (see (4.31) and Proposition 3.107). $\qquad\square$

As an immediate consequence of the previous proofs, we have:

Proposition 4.87 *If* $\Omega \subseteq \mathbb{R}^N$ *is open,* $1 \leq p < \infty$, $\{u_n\}_{n\in\mathbb{N}} \subseteq W^{1,p}(\Omega)$ *and we have*

$$u_n \xrightarrow{w} u \text{ in } L^p(\Omega), \ Du_n \xrightarrow{w} f \text{ in } L^p(\Omega, \mathbb{R}^N),$$

then $u \in W^{1,p}(\Omega)$, $Du = f$.

The next proposition is an easy consequence of Proposition 4.81.

Proposition 4.88 *If* $\Omega \subseteq \mathbb{R}^N$ *is open,* $1 \leq p < \infty$ *and* $\{u_n\}_{n\in\mathbb{N}}$, $\{v_n\}_{n\in\mathbb{N}} \subseteq W^{1,p}(\Omega)$ *satisfy* $u_n \to u$, $v_n \to v$ *in* $W^{1,p}(\Omega)$, *then* $\min\{u_n, v_n\} \to \min\{u, v\}$, $\max\{u_n, v_n\} \to \max\{u, v\}$ *in* $W^{1,p}(\Omega)$.

Proof Note that $\max\{u, v\} = (u - v)^+ - v$ and $\min\{u, v\} = v - (v - u)^+$. So, it suffices to show that "$u_n \to u$ in $W^{1,p}(\Omega) \Rightarrow u_n^+ \to u^+$ in $W^{1,p}(\Omega)$." But this was already proved in Proposition 4.82. □

Next we prove some results which help us identify those functions in $W^{1,p}(\Omega)$ which belong in $W_0^{1,p}(\Omega)$.

Proposition 4.89 *If $\Omega \subseteq \mathbb{R}^N$ is open, $1 \leq p < \infty$ and $u \in W^{1,p}(\Omega)$ vanishes outside a compact set $K \subseteq \Omega$, then $u \in W_0^{1,p}(\Omega)$.*

Proof Let $K \subseteq \Omega' \subseteq \Omega$ be bounded open and $\varphi \in C_c^\infty(\mathbb{R}^N)$ with $\varphi|_K = 1$. Evidently $\varphi u = u$. also, let $\vartheta_n \in C_c^\infty(\Omega)$ such that

$$\vartheta_n \to u \text{ in } L^p(\Omega) \text{ and } D\vartheta_n \to Du \text{ in } L^p(\Omega', \mathbb{R}^N) \text{ (see Remark 4.69)}.$$

We have $\varphi\vartheta_n \to \varphi u$ in $W^{1,p}(\Omega)$ and $\varphi\vartheta_n \in C_c^\infty(\Omega)$. Therefore $\varphi u = u \in W_0^{1,p}(\Omega)$. □

Proposition 4.90 *If $\Omega \subseteq \mathbb{R}^N$ is open. $1 \leq p < \infty$, $u \in W_0^{1,p}(\Omega)$, $v \in W^{1,p}(\Omega)$ and $0 \leq v(z) \leq u(z)$ for a.a $z \in \Omega$, then $v \in W_0^{1,p}(\Omega)$.*

Proof We can find $\{\vartheta_n\}_{n\in\mathbb{N}} \subseteq C_c^\infty(\Omega)$, $\vartheta_n \geq 0$ for all $n \in \mathbb{N}$ such that $\vartheta_n \to u$ in $W^{1,p}(\Omega)$(see Proposition 4.83). Let $g_n = \min\{v, \vartheta_n\}$. Evidently for each $n \in \mathbb{N}$ $g_n(\cdot)$ vanishes outside a compact subset of Ω. Therefore by Proposition 4.89 we have $g_n \in W_0^{1,p}(\Omega)$. Moreover, by Proposition 4.88, we have $g_n \to v$ in $W^{1,p}(\Omega)$. Therefore $v \in W_0^{1,p}(\Omega)$. □

Remark 4.91 An alternative proof of this result can be based on the monotonicity of the trace map. The trace theory will be discussed below.

Proposition 4.92 *If $\Omega \subseteq \mathbb{R}^N$ is open, $1 \leq p < \infty$, $u \in W_0^{1,p}(\Omega)$, $v \in W^{1,p}(\Omega)$ and there is compact set $K \subseteq \Omega$ such that*

$$|v(z)| \leq |u(z)| \text{ for a.a } z \in \Omega \setminus K,$$

then $v \in W_0^{1,p}(\Omega)$.

Proof Consider $\varphi \in C_c^\infty(\Omega)$ with $\varphi|_K \equiv 1$, $0 \leq \varphi \leq 1$. We set

$$\widehat{u} = (1 - \varphi)|u| + \varphi v^+.$$

Then $\widehat{u} \in W_0^{1,p}(\Omega)$ and $0 \leq v^+ \leq \widehat{u}$. So, Proposition 4.90 implies $v^+ \in W_0^{1,p}(\Omega)$. In a similar fashion we show that $v^- \in W_0^{1,p}(\Omega)$ and so we conclude that $v \in W_0^{1,p}(\Omega)$. □

The last result in this direction motivates also the trace theory which will follow.

Proposition 4.93 *If $\Omega \subseteq \mathbb{R}^N$ is bounded open, $1 \leq p < \infty$ and $u \in W^{1,p}(\Omega)$ such that*

$$\lim_{z \to x} u(z) = 0 \text{ for all } x \in \partial\Omega,$$

then $u \in W_0^{1,p}(\Omega)$.

Proof Recall that $u = u^+ - u^-$. So, without any loss of generality, we may assume that $u \geq 0$. For $\epsilon > 0$, let $u_\epsilon \in W^{1,p}(\Omega)$ be the mollification of u. We see that u_ϵ vanishes outside a compact set in Ω and so by Proposition 4.89 we have $u_\epsilon \in W_0^{1,p}(\Omega)$. Since $u_\epsilon \to u$ in $W^{1,p}(\Omega)$ as $\epsilon \to 0^+$ (see Proposition 4.62) we conclude that $u \in W_0^{1,p}(\Omega)$.   □

Corollary 4.94 *If $\Omega \subseteq \mathbb{R}^N$ is bounded open, $1 \leq p < \infty$, $u \in W^{1,p}(\Omega) \cap C(\overline{\Omega})$ and $u|_{\partial\Omega} = 0$, then $u \in W_0^{1,p}(\Omega)$.*

Remark 4.95 If $\Omega \subseteq \mathbb{R}^N$ is bounded open with C^1-boundary $\partial\Omega$, then the converse is also true, that is, $u \in W_0^{1,p}(\Omega) \cap C(\overline{\Omega})$ implies $u|_{\partial\Omega} = 0$ (see Brezis [48], Theorem 9.17, p.288).

Definition 4.96 Let $\Omega \subseteq \mathbb{R}^N$ a bounded, open, connected set (a bounded domain). Given $u \in C(\Omega)$, let

$$Z_u = \{z \in \Omega : u(z) = 0\} \text{(the zero set of u)}.$$

A connected component of $\Omega \setminus Z_u$ is said to be a "nodal domain" of u.

The next result can be found on Le [172], Theorem 3.

Proposition 4.97 *If $\Omega \subseteq \mathbb{R}^N$ is a bounded domain, $1 \leq p < \infty$, $X = W^{1,p}(\Omega)$ or $W_0^{1,p}(\Omega)$ and $u \in X \cap C(\Omega)$, then for any nodal domain D of u we have*

$$\chi_D u \in X \text{ and } D(\chi_D u) = \chi_D u.$$

Now we are ready to discuss the trace theory for Sobolev spaces. This theory provides a way to assign boundary values to a Sobolev function. Proposition 4.93 and Corollary 4.94 suggest that a function $u \in W^{1,p}(\Omega)$ belongs in $W_0^{1,p}(\Omega)$ if it vanishes on the boundary. But since $u \in L^p(\Omega)$, it is not meaningful to refer to the boundary values of u, since $\partial\Omega$ is Lebesgue-null Trace theory serves this purpose and provides an effective way to assign boundary values to a Sobolev function.

We start with a lemma, whose proof is essentially based on the notion of Lipschitz boundary.

Lemma 4.98 *If $\Omega \subseteq \mathbb{R}^N$ is bounded open with Lipschitz boundary $\partial\Omega$ and $1 \leq p < \infty$, then there exists $c_0 > 0$ such that*

$$\int_{\partial\Omega} |u|^p d\sigma^{N-1} \leq c_0 \|u\|_{1,p}^p \text{ for all } u \in C^1(\overline{\Omega})$$

with σ^{N-1} being the $(N-1)$-dimensional Hausdorff (surface) measure on $\partial\Omega$.

Proof Since $\partial\Omega$ is Lipschitz, for any $z \in \partial\Omega$ we can find $\varrho > 0$ and a Lipschitz function $\xi : \mathbb{R}^{N-1} \to \mathbb{R}$ such that $\Omega \cap B_\varrho(z) = \{x = (x_k)_{k=1}^N \in \Omega : \xi(x_1, ..., x_{N-1}) < x_N\} \cap B_\varrho(x)$ (we may have to rotate and relabel the coordinate axes). To simplify the notation, we write $B = B_\varrho(z)$ and assume for the moment that $u|_{\Omega \cap B^c} \equiv 0$. We have $0 < \frac{1}{[1+(Lip\xi)^2]^{1/2}} \leq -(e_N, n)_{\mathbb{R}^N} \sigma^{N-1}$-a.e on $\partial\Omega \cap B$, with $Lip\xi$ being the Lipschitz constant of ξ, $e_N = (0, 0, ..., 0, 1) \in \mathbb{R}^N$ and $n(\cdot)$ is the outward unit normal on $\partial\Omega$.

For $\epsilon > 0$, we set $\mu_\epsilon(t) = [t^2 + \epsilon^2]^{1/2} - \epsilon$ for all $t \in \mathbb{R}$. We have

$$\int_{\partial\Omega} \mu_\epsilon(u) d\sigma^{N-1} = \int_{\partial\Omega \cap B} \mu_\epsilon(u) d\sigma^{N-1}$$

$$\leq c_0 \int_{\partial\Omega \cap B} \mu_\epsilon(u)(-e_N, n)_{\mathbb{R}^N} d\sigma^{N-1} \text{ with } c_0 > 0$$

$$= -c_0 \int_{\Omega \cap B} \frac{\partial}{\partial x_N} \mu_\epsilon(u) dx \text{ (by the divergence theorem)}$$

$$\leq c_0 \int_{\Omega \cap B} |\mu_\epsilon'(u)| \|Du| dx$$

$$\leq c_0 \int_\Omega |Du| dx.$$

Passing to the limit as $\epsilon \to 0^+$, we obtain

$$\int_{\partial\Omega} |u| d\sigma^{N-1} \leq c_0 \int_\Omega |Du| dx. \tag{4.33}$$

We remove the extra condition that $u|_{\Omega \cap B^c} = 0$. To this end, by compactness we cover $\partial\Omega$ by a finite number of balls and we choose a partition of unity subordinate to this cover. Using (4.33) we obtain

$$\int_{\partial\Omega} |u| d\sigma^{N-1} \leq \widehat{c} \int_\Omega [|Du| + |u|] dx = c_0 \|u\|_{1,p} \tag{4.34}$$

for all $u \in C^1(\overline{\Omega})$, some $\widehat{c} > 0$.

For $1 < p < \infty$, in (4.34) we replace $|u|$ with $|u|^p$ and obtain

$$\int_{\partial\Omega} |u|^p d\sigma^{N-1} \leq \widehat{c} \int_{\Omega} [|Du| |u|^{p-1} + |u|^p] dx$$

$$\leq c \int_{\Omega} [|Du|^p + |u|^p] dx$$

some $c > 0$ (using Young's inequality),

$$\Rightarrow \int_{\partial\Omega} |u|^p d\sigma^{N-1} \leq c_0 \|u\|_{1,p}^p \text{ for all } u \in C^1(\overline{\Omega}).$$ □

Using this lemma, we have the following theorem.

Theorem 4.99 *If $\Omega \subseteq \mathbb{R}^N$ is bounded open with Lipschitz boundary $\partial\Omega$ and $1 \leq p < \infty$, then there exists $\gamma_0 \in \mathcal{L}(W^{1,p}(\Omega), L^p(\partial\Omega))$(we furnish $\partial\Omega$ with the surface measure σ^{N-1}) unique such that*

$$\gamma_0(u) = u|_{\partial\Omega} \text{ for all } u \in W^{1,p}(\Omega) \cap C(\overline{\Omega})$$

and for all $u \in W^{1,p}(\Omega)$ and $\psi \in C^1(\mathbb{R}^N, \mathbb{R}^N)$, we have

$$\int_{\Omega} u(div\psi) dz + \int_{\Omega} (Du, \psi)_{\mathbb{R}^N} dz = \int_{\partial\Omega} (\psi, n)_{\mathbb{R}^N} \gamma_0(u) d\sigma^{N-1}. \tag{4.35}$$

Proof Let $\gamma_0(u) = u|_{0\Omega}$ for all $u \in C^1(\overline{\Omega})$. From Lemma 4.98 we see that $\gamma_0(\cdot)$ is continuous, linear for the $W^{1,p}(\Omega)$ norm topology. Exploiting the density of $C^1(\overline{\Omega})$ in $W^{1,p}(\Omega)$(see Theorem 4.68), we can obtain $\gamma_0 \in \mathcal{L}(W^{1,p}(\Omega), L^p(\partial\Omega))$ satisfying $\gamma_0(u) = u|_{\partial\Omega}$ for all $u \in W^{1,p}(\Omega) \cap C(\overline{\Omega})$. Clearly $\gamma_0(\cdot)$ is uniquely defined.

Finally (4.35) follows using a standard approximation argument and the divergence theorem. □

Definition 4.100 The unique map $\gamma_0 \in \mathcal{L}(W^{1,p}(\Omega), L^p(\partial\Omega))$ from Theorem 4.99, is known as the "trace map" and we interpret $\gamma_0(u)$ as the boundary values of the Sobolev function $u \in W^{1,p}(\Omega)$.

Remark 4.101 We have $ker \gamma_0 = W_0^{1,p}(\Omega)$ and $im \gamma_0 = W^{\frac{1}{p'},p}(\partial\Omega)$ ($p' = \frac{p}{p-1}$). Exploiting the density of $C^1(\overline{\Omega})$ in $H^1(\Omega)$(see Theorem 4.68), from (34), we obtain the so-called Green's identity.

Theorem 4.102 *If $\Omega \subseteq \mathbb{R}^N$ is bounded open with Lipschitz boundary $\partial\Omega$, then for all $u, v \in H^1(\Omega)$ and all $1 \leq k \leq N$, we have*

$$\int_{\Omega} u \frac{\partial v}{\partial z_k} dz + \int_{\Omega} \frac{\partial u}{\partial z_k} v dz = \int_{\partial\Omega} \gamma_0(u) \gamma_0(v) n_k d\sigma^{N-1}$$

with $n = (n_k)_{k=1}^N$ being the outward unit normal on $\partial\Omega$.

Corollary 4.103 *If $\Omega \subseteq \mathbb{R}^N$ is bounded open with Lipschitz boundary $\partial\Omega$, then for all $u \in H^2(\Omega)$ and all $v \in H^1(\Omega)$, we have*

$$\int_\Omega (\Delta u)v \, dz + \int_\Omega (Du, Dv)_{\mathbb{R}^N} dz = \int_{\partial\Omega} \left(\frac{\partial u}{\partial n}\right)v \, d\sigma^{N-1}.$$

The result can be extended to more general operators than the Laplacian. So, let $\Omega \subseteq \mathbb{R}^N$ be bounded open with Lipschitz boundary $\partial\Omega$ and let $1 < q < \infty$. We introduce the following space

$$V^q(\Omega, div) = \{h \in L^q(\Omega, \mathbb{R}^N) : div \, h \in L^q(\Omega)\}.$$

Recall that $div \, h = \sum_{k=1}^N \frac{\partial h}{\partial z_k}$. We equip $V^q(\Omega, div)$ with the norm

$$\|h\|_{V^q} = [\|h\|_q^q + \|div \, h\|_q^q]^{1/q}.$$

The $V^q(\Omega, div)$ becomes a separable and uniformly convex (thus reflexive) Banach space and $C^\infty(\overline{\Omega}, \mathbb{R}^N)$ is dense in $V^q(\Omega, div)$. Then we have the following generalization of Theorem 4.99 due to Casas and Fernandez [60], Lemma 4 and Kenmochi [160], Proposition 1.4.

Theorem 4.104 *If $\Omega \subseteq \mathbb{R}^N$ is bounded open with Lipschitz boundary $\partial\Omega$, $1 < p < \infty$ and $p' = \frac{p}{p-1}$ (the conjugate exponent to p), then there exists a unique bounded linear operator*

$$\widehat{\gamma} : V^{p'}(\Omega, div) \to W^{-\frac{1}{p'}, p'}(\partial\Omega)$$

such that $\gamma_0(u) = (u, n)_{\mathbb{R}^N}$ for all $u \in C^1(\overline{\Omega}, \mathbb{R}^N)$ and

$$\int_\Omega u(Dv) \, dz + \int_\Omega v(div \, u) \, dz = (\widehat{\gamma}(u), \gamma_0(v))_{\partial\Omega}$$

for all $u \in V^{p'}(\Omega, div)$, all $v \in W^{1,p}(\Omega)$ (by $(\cdot, \cdot)_{\partial\Omega}$ we denote the duality brackets for the pair $(W^{\frac{1}{p'}, p}(\partial\Omega), W^{-\frac{1}{p'}, p'}(\partial\Omega))$).

Remark 4.105 The space $W^{\frac{1}{p'}, p}(\partial\Omega)$ is a fractional Sobolev space defined by

$$W^{\frac{1}{p'}, p}(\partial\Omega) = \{u \in L^p(\partial\Omega) : \frac{|u(x) - u(z)|}{|x - z|^{\frac{1}{p} + \frac{N}{p'}}} \in L^{p'}(\partial\Omega \times \partial\Omega)\}.$$

(see Adams and Fournier [1], p.250).

Consider a nonlinear differential operator in divergence form defined by

$$u \to -div(a(z, u, Du))$$

with $a : \Omega \times \mathbb{R} \times \mathbb{R}^N \longrightarrow \mathbb{R}^N a$ Caratheodory function such that

$$|a(z, x, y)| \le \widehat{c}[\eta(z) + |x|^{p-1} + |y|^{p-1}] \text{ for a.a } z \in \Omega, \text{ all } x \in \mathbb{R}, \text{ all } y \in \mathbb{R}^N,$$

with $\widehat{c} > 0$ and $\eta \in L^{p'}(\Omega)$.

Then we have the following generalized Green's identity (see Casas and Fernandez [60]. Corollary 2, p.71).

Theorem 4.106 *If $u \in W^{1,p}(\Omega)$ with div $a(\cdot, u(\cdot), Du(\cdot)) \in L^{p'}(\Omega)$ $(p' = \frac{p}{p-1})$,*
then there exists unique element of $W^{-\frac{1}{p'}, p'}(\partial\Omega)$ which we denote by $\frac{\partial u}{\partial n_a}$ such that

$$\int_\Omega (a(z, u, Du), Dh)_{(\mathbb{R}^N)}dz + \int_\Omega (diva(z, u, Du))hdz = (\frac{\partial u}{\partial n_a}, \gamma_0(h))_{\partial\Omega}$$

for all $h \in W^{1,p}(\Omega)$.

Specializing this nonlinear Green's identity to the case of the p-Laplace differential operator $\Delta_p u = div(|Du|^{p-2}Du)$ $1 < p < \infty$, $u \in W^{1,p}(\Omega)$, which appears often in the literature, we have:

Theorem 4.107 *If $u \in W^{1,p}(\Omega)$ with $\Delta_p u \in L^{p'}(\Omega)$, then there exists a unique element*

$$\frac{\partial u}{\partial n_p} = \widehat{\gamma}(|Du|^{p-2}Du) \in W^{-\frac{1}{p'}, p'}(\partial\Omega)$$

such that

$$\int_\Omega (\Delta_p u)hdz + \int_\Omega |Du|^{p-2}(Du, Dh)_{\mathbb{R}^N}dz = (\frac{\partial u}{\partial n_p}, \gamma_0(h))_{\partial\Omega}$$

for all $h \in W^{1,p}(\Omega)$.

To establish properties of the Sobolev functions $u \in W^{1,p}(\Omega)$, it is often convenient to start with the case $\Omega = \mathbb{R}^N$. So, it is important to know if we can extend a Sobolev function $u \in W^{1,p}(\Omega)$ to a function $\widehat{u} \in W^{1,p}(\mathbb{R}^N)$. Note that extension by zero on $\mathbb{R}^N \setminus \Omega$ works for functions in $W_0^{1,p}(\Omega)$, but not for functions in $W^{1,p}(\Omega) \setminus W_0^{1,p}(\Omega)$, since it may create a bad discontinuity along $\partial\Omega$ and then the extended function need not have weak derivatives. However, for Ω sufficiently smooth the extension is possible. More precisely, we have the following theorem whose proof can be found in Brezis [50], Theorem 9.7, p.272 and in Evans-Gariepy [105], Theorem 1, p.135.

Theorem 4.108 *If $\Omega \subseteq \mathbb{R}^N$ is bounded open with Lipschitz boundary $\partial\Omega$, $1 \leq p < \infty$, $U \subseteq \mathbb{R}^N$ is open with $\overline{\Omega} \subseteq U$, then there exists a bounded linear operator*

$$E : W^{1,p}(\Omega) \to W^{1,p}(\mathbb{R}^N)$$

such that $E(u)|_\Omega = u$ and $E(u)$ vanishes outside U, namely supp $E(u) \subseteq U$).

Remark 4.109 We call $E(u)$ the "extension" of u.

Now we want to describe the dual of $W^{1,p}(\Omega)$, $1 \leq p < \infty$.

Proposition 4.110 *If $\Omega \subseteq \mathbb{R}^N$ is an open set, $1 \leq p < \infty$ and $\eta \in W^{1,p}(\Omega)^*$, then $\eta(u) = \int_\Omega uh_0 dz + \int_\Omega (Du, h_1)_{\mathbb{R}^N} dz$ for all $u \in W^{1,p}(\Omega)$ with $h_0 \in L^{p'}(\Omega)$, $h_1 \in L^{p'}(\Omega, \mathbb{R}^N)$ ($p' = \frac{p}{p-1}$).*

Proof Clearly $u \to \int_\Omega uh_0 dz + \int_\Omega (Du, h_1)_{\mathbb{R}^N} dz$ defines an element $\eta \in W^{1,p}(\Omega)^*$ and we have

$$\|\eta\|_* \leq c\|\widehat{h}\|_{L^p(\Omega,\mathbb{R}^{N+1})}, \widehat{h} = (h_0, h_1) \in L^p(\Omega, \mathbb{R}^{N+1}).$$

Conversely, let $\eta \in W^{1,p}(\Omega)^*$. We identify $W^{1,p}(\Omega)$ with a closed subspace or $L^p(\Omega, \mathbb{R}^{N+1})$ via the operator $T : W^{1,p}(\Omega) \to L^p(\Omega, \mathbb{R}^{N+1})$ defined by $T(u) = (u, Du)$. We define $\widehat{\eta} : R(T) \to \mathbb{R}$ by $\widehat{\eta}(T(u)) = \eta(u)$ for all $u \in W^{1,p}(\Omega)$. Then using Proposition 3.34 we can find a norm preserving extension $\widehat{\eta}_0$ on all of $L^p(\Omega, \mathbb{R}^{N+1})$. Using the Riesz Representation Theorem (see Theorem 2.133), we can find $h_0 \in L^{p'}(\Omega)$, $h_1 \in L^{p'}(\Omega, \mathbb{R}^N)$ such that

$$\widehat{\eta}_0((y, \sim y)) = \int_\Omega yh_0 dz + \int_\Omega (\widetilde{y}, h_1)_{\mathbb{R}^N} dz \tag{4.36}$$

for all $(y, \widetilde{y}) \in L^p(\Omega, \mathbb{R}^{N+1}) = L^p(\Omega) \times L^p(\Omega, \mathbb{R}^N)$.

Then if $u \in W^{1,p}(\Omega)$, we can identify it with $(u, Du) \in L^p(\Omega, \mathbb{R}^{N+1})$ and therefore

$$\eta(u) = \widehat{\eta}(T(u)) = \widehat{\eta}_0(T(u)) = \int_\Omega uh_0 dz + \int_\Omega (Du, h_1)_{\mathbb{R}^N} dz$$

for all $u \in W^{1,p}(\Omega)$ (see (4.36)).   □

Using the language of distributions we see that η restricted on $C_c^\infty(\Omega)$ coincides with the distribution

$$L = h_0 - div Dh_1. \tag{4.37}$$

However, not every distribution of the form (4.37) is necessarily in $W^{1,p}(\Omega)^*$. But if we replace $W^{1,p}(\Omega)$ by $W_0^{1,p}(\Omega)$, then (4.37) can completely describe the dual space $W_0^{1,p}(\Omega)^*$ since $L(\cdot)$ admits a unique extension on $W_0^{1,p}(\Omega)$ (see Definition 4.55). In the case of $W^{1,p}(\Omega)$ this extension is not unique. So, we have:

Theorem 4.111 *If $\Omega \subseteq \mathbb{R}^N$ is open and $1 \le p < \infty$, then $W_0^{1,p}(\Omega)^*$ consists of all distributions L of the form*

$$L = h_0 - div Dh_1 \text{ with } h_0 \in L^{p'}(\Omega), h_1 \in L^{p'}(\Omega, \mathbb{R}^N).$$

Moreover, if $\Omega \subseteq \mathbb{R}^N$ is bounded, then we can take $h_0 = 0$.
Finally if $\eta \in W_0^{1,p}(\Omega)^*$ is the corresponding functional, then

$$< \eta, u >= \int_\Omega u h_0 dz + \int_\Omega (Du, Dh)_{\mathbb{R}^N} dz$$

for all $u \in W_0^{1,p}(\Omega)$,
and

$$\|\eta\|_* = \max\{\|h_0\|_{p'}, \|h_1\|_{L^{p'}(\Omega, \mathbb{R}^N)}.$$

The next result is known in the literature as the "Poincare Inequality."

Theorem 4.112 *If $\Omega \subseteq \mathbb{R}^N$ is bounded open and $1 \le p < \infty$, then there exists $c > 0$ such that*

$$\|u\|_p \le c\|Du\|_p$$

for all $u \in W_0^{1,p}(\Omega)$.

Proof Let $\vartheta > 0$ such that $\Omega \subseteq [-\vartheta, \vartheta]^N$. Let $\psi \in C_c^\infty(\Omega)$. Then for all $x = (x_k)_{k=1}^N \in \Omega$, we have

$$\psi(x) = \int_{-\vartheta}^{x_N} \frac{\partial}{\partial s} \psi(x_1, ..., x_{N-1}, s) ds.$$

Using Holder's inequality, we have

$$|\psi(x_1, \ldots, x_{N-1}, x_N)|^p \le (2\vartheta)^{p-1} \int_{-\vartheta}^{\vartheta} |\frac{\partial}{\partial s} \psi(x_1, \ldots, x_{N-1}, s)|^p ds,$$

$$\Rightarrow \int_{[-\vartheta, \vartheta]^{N-1}]} |\psi(x_1, \ldots, x_{N-1}, s)|^p ds] dx_1 \cdots dx_{N-1},$$

$$\le (2\vartheta)^{p-1} \int_{[-\vartheta, \vartheta]^{N-1}} [\int_{-\vartheta}^{\vartheta} |\frac{\partial}{\partial s} \psi(x_1, \ldots, x_{N-1}, s|^p ds] dx_1 \cdots dx_{N-1},$$

$$\Rightarrow \|\psi\|_p^\varrho \le [2\vartheta]^p \int_\Omega |\frac{\partial \psi}{\partial x_N}|^p dx = [2\vartheta]^p \|D\psi\|_p^\varrho,$$

$$\Rightarrow \|\psi\|_p \le 2\vartheta \|D\psi\|_p.$$

Exploiting the density of $C_c^\infty(\Omega)$ in $W_0^{1,p}(\Omega)$ we conclude that $\|u\|_p \leq c\|Du\|_p$ for all $u \in W_0^{1,p}(\Omega)$, $c = 2\vartheta > 0$. $\qquad\square$

Corollary 4.113 *If $\Omega \subseteq \mathbb{R}^N$ is bounded open and $1 \leq p < \infty$, then $\|Du\|_p$ for all $u \in W_0^{1,p}(\Omega)$ is an equivalent norm on $W_0^{1,p}(\Omega)$.*

We can have extensions of the Poincare inequality (Theorem 4.112). To prove them, we will need the following fundamental theorem for Sobolev spaces known as the Sobolev Embedding Theorem (see Adams and Fournier [1]. Theorem 4.12, p.85).

Theorem 4.114 *Assume that $\Omega \subseteq \mathbb{R}^N$ is bounded open with Lipschitz boundary $\partial\Omega$, $k \in \mathbb{N}$, $1 \leq p \leq \infty$. (a) If $kp < N$, then $W^{k,p}(\Omega) \to L^q(\Omega)$ continuously for all $1 \leq q \leq \frac{Np}{N-kp}$ and the embedding is compact if $1 \leq q < \frac{Np}{N-kp}$; (b) If $0 \leq m < k - \frac{N}{p} < m+1$, then we have $W^{k,p}(\Omega) \hookrightarrow C^{m,\vartheta}(\overline{\Omega})$ continuously for all $0 \leq \vartheta \leq k - m - \frac{N}{p}$ and the embedding is compact if $0 \leq \vartheta < k - m - \frac{N}{p}$.*

In particular, we have the following embedding theorem known in the literature as the "Rellich-Kondrachov Embedding Theorem."

Theorem 4.115 *If $\Omega \subseteq \mathbb{R}^N$ is a bounded open set with Lipschitz bounding $\partial\Omega$, then (a) If $1 \leq p < N$ and $1 \leq q < p^* = \frac{Np}{N-p}$, then $W^{1,p}(\Omega) \hookrightarrow L^q(\Omega)$ compactly; (b) if $p = N$ and $1 \leq q < \infty$, then $W^{1,p}(\Omega) \hookrightarrow L^q(\Omega)$ compactly; (c) if $N < p < \infty$, then $W^{1,p}(\Omega) \hookrightarrow C(\overline{\Omega})$ compactly.*

Now we have the tools to state and prove the extensions at the Poincare inequality (see Theorem 4.112).

Theorem 4.116 *If $\Omega \subseteq \mathbb{R}^N$ is a bounded, open, connected set with Lipschitz boundary $\partial\Omega$, $1 \leq p < \infty$, $V \subseteq W^{1,p}(\Omega)$ is a closed subspace with the only constant function belonging in V being the zero function, then there exists a constant $c > 0$ such that*

$$\|u\|_p \leq c\|Du\|_p$$

for all $u \in V$.

Proof We argue indirectly. So, suppose that the assertion of the theorem is not true. Then we can find $\{v_n\}_{n \in \mathbb{N}} \subseteq V$, $\|v_n\|_p = 1$ such that $\|Dv_n\|_p \to 0$. Then $\{v_n\}_{n \in \mathbb{N}} \subseteq W^{1,p}(\Omega)$ is bounded and so by passing to a suitable subsequence if necessary we may assume that $v_n \xrightarrow{w} v$ in $W^{1,p}(\Omega)$ and $v_n \xrightarrow{u} v$ in $L^p(\Omega)$ (see Theorem 4.115). Then $\|v\|_p = 1$ and $\|Dv\|_p \leq \liminf \|Dv_n\|_p = 0$. Since Ω is connected it follows that v is constant, hence zero (since $v \in V$), a contradiction. $\qquad\square$

Remark 4.117 In the sequel an open, connected set in $\mathbb{R}^N$ will be called a "domain."

Corollary 4.118 *If $\Omega \subseteq \mathbb{R}^N$ is a bounded domain with Lipschitz boundary $\partial\Omega$, $\Gamma_0 \subseteq \partial\Omega$ has positive surface (Hausdorff) measure, $1 \leq p < \infty$ and $V = \{u \in W^{1,p}(\Omega) : \gamma_0(u) = 0 \text{ on } \Gamma_0\}$ with γ_0 being the trace map, then there exists a constant $c > 0$ such that*

$$\|u\|_p \leq c\|Du\|_p \text{ for all } u \in W^{1,p}(\Omega).$$

Proof On account of the linearity and continuity of the trace map $\gamma_0(\cdot)$, we see that V is a closed subspace of $W^{1,p}(\Omega)$. If $u = \vartheta \in \mathbb{R}$ (constant function), then since Γ_0 has positive surface measure, we have $\gamma_0(u) = 0$ on Γ_0 and so if $u \in V$ then $u \equiv 0$. $\qquad\square$

The next result is known in the literature as the "Poincare-Wirtinger' Inequality".

Theorem 4.119 *If $\Omega \subseteq \mathbb{R}^N$ is a bounded domain with a Lipschitz boundary $\partial\Omega$ and $1 \leq p < \infty$ then there exists $c > 0$ such that*

$$\|u - \overline{u}\|_p \leq c\|Du\|_p \text{ for all } u \in W^{1,p}(\Omega)$$

with $\overline{u} = \frac{1}{\lambda^N(\Omega)} \int_\Omega u dz$, $\lambda^N(\cdot)$ being the Lebesgue measure on $\mathbb{R}^N$.

Proof Let $V = \{v \in W^{1,p}(\Omega) : \int_\Omega v dz = 0\}$. This is a closed subspace of $W^{1,p}(\Omega)$. Evidently $u \equiv 0$ is the only constant function in V. Also, if $u \in W^{1,p}(\Omega)$, then $u - \overline{u} \in V$. So, we can apply Theorem 4.116 and find $c > 0$ such that

$$\|u - \overline{u}\|_p \leq c\|Du\|_p \text{ for all } u \in W^{1,p}(\Omega).$$

$\qquad\square$

When $N = 1$ (functions of one variable), the situation is simpler. As we will see the reason is that in this case every Sobolev function is absolutely continuous, hence it has a classical derivative almost everywhere. In what follows by $\frac{du}{dt}$ we will denote the classical derivative of u and by u' the weak derivative of u.

Definition 4.120 Let $T \subseteq \mathbb{R}$ be an interval. We say that $u : T \to \mathbb{R}$ is "absolutely continuous," if there exists $g \in L^1_{loc}(T)$ such that for all $t_1, t_2 \in T$

$$u(t_2) - u(t_1) = \int_{t_1}^{t_2} g(s)ds.$$

Hence, an absolutely continuous function is continuous on T and differentiable (in the classical sense) almost everywhere and $\frac{du}{dt} = g$ a.e. If $g \in L^1(T)$, then $u \in C(\overline{T})$. An equivalent definition of absolute continuity is to say that for every $\varepsilon > 0$, there exists $\delta = \delta(\epsilon) > 0$ such that for every family $\{(a_k, b_k)\}_{k=1}^n$ of mutually disjoint intervals satisfying $\sum_{k=1}^n (b_k - a_k) < \delta$, then we have $\sum_{k=1}^n |f(b_k) - f(a_k)| < \varepsilon$.

For such functions, we have the following integration by parts formula.

Proposition 4.121 *If $T \subseteq \mathbb{R}$ is an interval and $u, v : T \to \mathbb{R}$ are absolutely continuous functions with derivatives a.e. g and h respectively, then $\int_a^b u(t)h(t)dt + \int_a^b v(t)g(t)dt = u(b)v(b) - u(a)v(a)$ for all $a, b \in T$.*

Proof Let $\{g_n, h_n\}_{n \in \mathbb{N}} \subseteq C_c(T)$ such that

$$g_n \to g \text{ and } h_n \to h \text{ in } L^1[a, b] \text{ (see Proposition 2.131)}. \tag{4.38}$$

Let $u_n(t) = u(a) + \int_a^t g_n(s)ds$, $v_n(t) = v(a) + \int_a^t h_n(s)ds$ for all $t \in [a, b]$, all $n \in \mathbb{N}$. We have

$$\int_a^b u_n(t)h_n(t)dt + \int_a^b v_n(t)g_n(t)dt = u_n(b)v_n(b) - u_n(a)v_n(a) \text{ for all } n \in \mathbb{N}. \tag{4.39}$$

Note that

$$|u_n(t) - u(t)| \le \|g_n - g\|_{L^1[a,b]}, \ |v_n(t) - v(t)| \le \|h_n - h\|_{L^1[a,b]}$$

(for all $t \in [a, b]$, all $n \in \mathbb{N}$). Thus, $u_n \to u$ and $v_n \to v$ uniformly on T (see (4.38)). Therefore, $\int_a^b u(t)h(t)dt + \int_a^b v(t)g(t)dt = u(b)v(b) - u(a)v(a)$ (see (4.39)). $\qquad\square$

Proposition 4.122 *If $T \subseteq \mathbb{R}$ is an interval, $u \in L_{loc}^1(T)$ and $u' \in L_{loc}^1(T)$, then (a) for every $t_0 \in T$, there exists $c \in \mathbb{R}$ such that*

$$u(t) = c + \int_{t_0}^t u'(s)ds \qquad \text{for a.a } t \in T.$$

(b) When $u \in C(T)$, then $u(t) = u(t_0) + \int_{t_0}^t u'(s)ds$ for all $t \in T$ and so $u(\cdot)$ is absolutely continuous with $\frac{du}{dt} = u'$ a.e.
(c) When $u' \in L^1(T)$, then u is absolutely continuous on $\overline{T}$.
(d) when $u, u' \in C(T)$, then $u \in C^1(T)$ and if $u' \in L^1(T)$ we have $u \in C^1(\overline{T})$.

Proof

(a) Let $v(t) = u(t_0) + \int_{t_0}^t u'(s)ds$. Let $f \in C_C^\infty(T)$ and $a, b \in \mathbb{R}$ such that supp $f \subseteq [a, b]$. Consider the distribution L_Y corresponding to v. Using Definition 4.53 and Proposition 4.121, we have

$$(L_v)'(f) = -\int_T vf'dt = -v(b)f(b) + v(a)f(a) + \int_T v'f dt$$

$$= \int_T u'f dt = L_{u'}(f) = (L_u)'(f)$$

$$\Rightarrow v = u + c \text{ with } c \in \mathbb{R}.$$

(b) If $u \in C(T)$, then $v(t) = u(t) + c$ for all $t \in T$.

(c) If $u' \in L^1(T)$, then $v \in C(\overline{T})$ and so $u \in C(\overline{T})$.

(d) Follows from (c). $\square$

Corollary 4.123 *If* $T \subseteq \mathbb{R}$ *is an interval,* $u \in L^1_{loc}(T)$ *and* $u' \in L^1_{loc}(T)$ *then* $u(\cdot)$ *is almost everywhere equal to a function* $v \in C(T)$ *such that* $u' = v'$ *a.e.*

Remark 4.124 In the sequel we will always use this continuous representative of the function u.

Using these facts, we can characterize the Sobolev functions of one variable.

Theorem 4.125 *If* $T \subseteq \mathbb{R}$ *is an open interval and* $1 \leq p \leq \infty$, *then* $u \in W^{1,p}(T)$ *if and only if* $u \in L^p(T)$, *u is absolutely continuous on T and* $\frac{du}{dt} \in L^p(T)$.

Proof $\Rightarrow$ According to Proposition 4.122 u equals almost everywhere an absolutely continuous function and $\frac{du}{dt} = u'$ a.e.

$\Leftarrow$ Let $f \in C_c^\infty C(T)$ and $a, b \in \mathbb{R}$ such that $supp f \subseteq [a, b]$. We have

$$(L_u)'(f) = -\int_T u f' dt = -\int_a^b u f' dt$$

$$= -u(b) f(b) + u(a) f(a) + \int_a^b \frac{du}{dt} f dt$$

(see Proposition 4.121)

$$= \int_a^b \frac{du}{dt} f dt = (L_{\frac{du}{dt}})(f),$$

$$\Rightarrow u' = \frac{du}{dt} \text{ a.e. and so } u \in W^{1,p}(\Omega).$$ $\square$

Corollary 4.126 *As a consequence of Theorem 4.125,* $W^{1,p}(T)$ *is embedded continuously in* $L^\infty(T)$.

Remark 4.127 According to this Theorem, for all $1 \leq p \leq \infty$, $W^{1,p}(T) \subseteq C(\overline{T})$.

Theorem 4.128 *If* $T \subseteq \mathbb{R}$ *is an open interval and* $1 \leq p \leq \infty$ *then there exists* $c > 0$ *such that*

$$\|u\|_\infty \leq c\|u\|_{1,p} \qquad \text{for all } u \in W^{1,p}(T).$$

Proof The result is clear for $p = +\infty$. So, suppose $1 \leq p < \infty$. According to Theorem 4.108 it suffices to consider the case $T = \mathbb{R}$. Let $f \in C_C^1(\mathbb{R})$ and $E(t) = |t|^{p-1}t$. Then the function $y = \xi(u) \in W^{1,p}(\mathbb{R})$ and $y' = \xi'(u)u' = p|u|^{p-1}u'$ (see Theorem 4.76). For $t \in \mathbb{R}$ we have

$$\xi(u(t)) = \int_{-\infty}^{t} p|u(s)|^{p-1}u'(s)ds,$$

$$\Rightarrow |u(t)|^p \leq p\|u\|_p^{p-1}\|u'\|_p \text{ (using Holder's inequality)},$$

$$\Rightarrow \|u\|_\infty \leq c\|u\|_{1,p}, \ c > 0, u \in C_c^1(\mathbb{R}).$$

If $u \in W^{1,p}(\mathbb{R})$, then we can find $\{u_n\}_{n\in\mathbb{N}} \subseteq C_c^1(\mathbb{R})$ such that $u_n \to u$ in $W^{1,p}(\mathbb{R})$. From (4.38), it follows that $\{u_n\}_{n\in\mathbb{N}}$ is Cauchy in $L^\infty(\mathbb{R})$. Hence

$$u_n \to u \text{ in } L^\infty(\mathbb{R}), \ \Rightarrow \|u\|_\infty \leq c\|u\|_{1,p}.$$

$\square$

Remark 4.129 The constant $c > 0$ is universal (that is, independent of $p \geq 1$). This is seen if we recall that $\sqrt[p]{p} \leq \sqrt[e]{e}$ for all $p \geq 1$.

When T is bounded, we have compact embedding into $C(\overline{T})$.

Theorem 4.130 *If $T \subseteq \mathbb{R}$ is a bounded open interval, then for $1 < p \leq \infty$ $W^{1,p}(T) \hookrightarrow C(\overline{T})$ compactly.*

Proof Let $\overline{B}_1 \subseteq W^{1,p}(T)$ be the unit ball. For any $u \in \overline{B}_1$, we have

$$|u(t) - u(s)| = |\int_s^t u'(t)dt| \leq \|u'\|_p|t-s|^{1/p'} \leq |t-s|^{1/p'}$$

$$\text{for all } t, s \in T \text{ (recall } p' = \frac{p}{p-1}).$$

So by the Arzela–Ascoli theorem we infer that in $C(\overline{T})$, $\overline{B}_1$ is relatively compact. Therefore $W^{1,p}(T) \hookrightarrow C(\overline{T})$ compactly. $\square$

Remark 4.131 For $p = 1$, we have $W^{1,1}(T) \hookrightarrow L^q(T)$ compactly for all $1 \leq q < \infty$. The embedding $W^{1,1}(T) \hookrightarrow C(\overline{T})$ is continuous but never compact. Also from Theorem 4.128, we know that if T is unbounded and $1 < p \leq \infty$, then $W^{1,p}(T) \hookrightarrow L^\infty(T)$ continuously but never compactly.

As a consequence of Theorem 4.130, we have the following simple definition of the space $W_0^{1,p}(T)$.

Proposition 4.132 *If $T = (a, b) \subseteq \mathbb{R}$ is a bounded open interval and $1 \leq p \leq \infty$, then $W_0^{1,p}(a, b) = \{u \in W^{1,p}(a, b) : u(a) = u(b) = 0\}$.*

Another consequence of Theorem 4.130 is the following proposition.

Proposition 4.133 *If $T \subseteq \mathbb{R}$ is a bounded open interval, $1 < p < \infty$, $\{u_n\}_{n\in\mathbb{N}} \subseteq W^{1,p}(a, b)$ and $u_n \overset{w}{\to} u$ in $W^{1,p}(a, b)$, then $u_n \to u$ uniformly on $[a, b]$.*

The Poincare Inequality (see Theorem 4.112) takes the following form.

Theorem 4.134 *If $T = (a, b)$ is a bounded open interval, $1 \leq p \leq \infty$, $t_0 \in [a, b]$ and*

$$V = \{u \in W^{1,p}(a, b) : u(t_0) = 0\},$$

then there exists $c > 0$ such that

$$\|u\|_p \leq c\|u'\|_p \qquad for\ all\ u \in V.$$

Proof For every $t \in T = (a, b)$ we have

$$u(t) = u(t_0) + \int_{t_0}^t u'(s)ds = \int_{t_0}^t u'(s)ds \qquad \text{(see Proposition 4.122)}.$$

If $p = 1$ or if $p = +\infty$, then the assertion of the theorem is obvious. So, suppose that $1 < p < \infty$. Then using Holder's inequality, we have

$$|u(t)| \leq |b - a|^{1/p'} \|u'\|_p \ (p' = \frac{p}{p-1}),$$

$$\Rightarrow \|u\|_p \leq c\|u'\|_p \text{ for some } c > 0, \text{ all } u \in W^{1,p}(a, b). \qquad \square$$

Remark 4.135 So, according to Theorem 4.134, on $W_0^{1,p}(a, b)$, we can use the equivalent norm $\|u\|_{1,p} = \|u'\|_p$ for all $u \in W_0^{1,p}(a, b)$.

We conclude this section by mentioning some useful equivalent norms on the Sobolev spaces. For details we refer to Zeidler [292], p.1032.

Theorem 4.136

(a) *If $\Omega \subseteq \mathbb{R}^N$ is bounded with Lipschitz boundary $\partial\Omega$ and $1 \leq p < \infty$, then $|u| = \|u\|_q + \|Du\|_p$ for all $u \in W^{1,p}(\Omega)$ is an equivalent norm provided that $1 \leq q \leq p^* = \frac{Np}{N-p}$ if $1 \leq p < N$, $1 \leq q < \infty$ if $p = N$, $1 \leq q \leq \infty$ if $p > N$.*

(b) *If $T \subseteq \mathbb{R}$ is a bounded open interval, $1 \leq p \leq \infty$ and $1 \leq q \leq \infty$, then $|u| = \|u\|_q + \|u'\|_p$ for all $u \in W^{1,p}(T)$ is an equivalent norm.*

4.4 Lebesgue-Bochner Spaces

Throughout this section (Ω, Σ, μ) is a complete finite measure space and X is a Banach space. Recall that if (Ω, Σ, μ) is a measure space and $(\Omega, \Sigma, \widehat{\mu})$ is its completion (see Proposition 2.19), then the corresponding L^p spaces coincide.

Since we will be dealing primarily with the spaces and with integrals, there is no real loss of generality in assuming that μ is complete.

First we need to clarify what we mean by a measurable X-valued function. Indeed the natural definition of Borel measurability (that is, for $u : \Omega \to X$, we have $u^{-1}(A) \in \Sigma$ for all $A \in B(X) = $ the Borel Σ-algebra of X) may not be so useful since $B(X)$ is "too big." In general we can have that $\sigma(X^*) \subsetneq B(X)$, with $\sigma(X^*)$ being the smallest σ-algebra on X for which every $x^* \in X^*$ is measurable. When this is true, then we encounter serious difficulties in using standard tools from functional analysis. For this reason we are led to the following definition.

Definition 4.137

(a) A function $s(z) = \sum_{k=1}^{m} \chi_{A_k}(z) x_k$ with $A_k \in \Sigma$, $x_k \in X$, $m \in \mathbb{N}$ and $z \in \Omega$, is said to be a "simple function." (see also Definition 2.68(b))

(b) A function $u : \Omega \to X$ is said to be "strongly μ-measurable" if there exists a sequence $\{s_n\}_{n \in \mathbb{N}}$ of simple functions such that $s_n(z) \to u(z)$ for μ a.a $z \in \Omega$.

(c) A function $u : \Omega \to X$ is said to be "weakly μ-measurable" if for every $x^* \in X^*$ the $\mathbb{R}$-valued function $z \to \langle x^*, u(z) \rangle$ is Σ-measurable (by $\langle \cdot, \cdot \rangle$ we denote the duality brackets for the pair (X^*, X)).

(d) A function $u : \Omega \to X$ is said to be "μ-essentially separably valued," if there exists $N \in \Sigma$ with $\mu(N) = 0$ such that $f(\Omega \setminus N)$ is norm separable in X.

The relation between strong and weak measurability is given by the so-called Pettis Measurability Theorem.

Theorem 4.138 *A function $u : \Omega \to X$ is strongly μ-measurable if and only if u is weakly μ-measurable and μ-essentially separably valued.*

Proof $\Downarrow$ Let $\{s_n\}_{n \in \mathbb{N}}$ be a sequence of simple functions converging μ-a.e to u. Let V be the closed subspace spanned by the union of the ranges of these simple functions. Evidently this union is countable and so $V \subseteq X$ is separable. Therefore $u(\cdot)$ is μ-essentially separably valued. Also for every $x^* \in X^*$ $\langle x^*, s_n(\cdot) \rangle$ is Σ-measurable and $\langle x^*, s_n(z) \rangle \to \langle x^*, u(z) \rangle$ for μ- a.a $z \in \Omega$ and so $\langle x^*, u(\cdot) \rangle$ is Σ-measurable. Therefore $u(\cdot)$ is μ-weakly measurable.

$\Uparrow$ Let $V \subseteq X$ be the closed, separable space which essentially contains the range of u. Let $\{x_n^*\}_{n \geq 1} \subseteq \partial \overline{B}_1^* = \{x^* \in X^* : \|x^*\|_* = 1\}$ be a norming sequence for V (that is, $\|v\| = \sup_{n \in \mathbb{N}} \langle x_n^*, v \rangle$ for all $v \in V$). Since by hypothesis $u(\cdot)$ is μ-weakly measurable, we have that for every $v \in V$, the function

$$z \to \|u(z) - v\| = \sup_{n \in \mathbb{N}} \langle x_n^*, u(z) - v \rangle$$

is Σ-measurable. Let $\{x_n\}_{n \in \mathbb{N}} \subseteq V$ be dense and let $\vartheta_n(z) = \|f(z) - x_n\|$ for every $n \in \mathbb{N}$. We have just seen that for every $n \in \mathbb{N}$, $\vartheta_n(\cdot)$ is Σ-measurable. Let $\epsilon > 0$ and $A_n = \{z \in \Omega : \vartheta_n(z) < \epsilon\} \in \Sigma$. We define $\widehat{s}_\epsilon : \Omega \to X$ by

$$\widehat{s_\epsilon}(z) = \begin{cases} x_n & \text{if } z \in A_n \setminus \bigcup_{k=1}^{n-1} A_k \\ 0 & \text{otherwise.} \end{cases}$$

Then $\|u(z) - s_\varepsilon(z)\| \le \epsilon$ for μ-a.a $z \in \Omega$. So, $u(\cdot)$ is μ−essentially uniformly approximated by the countably valued function $\widehat{s_\varepsilon}(\cdot)$. Let $\epsilon_n = \frac{1}{n}$, $n \in \mathbb{N}$ and $\widehat{s}_n = \widehat{s_{\varepsilon_n}}$. We have

$$\|u(z) - \widehat{s}_n(z)\| \le \frac{1}{n} \text{ for } \mu - \text{a.a } z \in \Omega, \text{ all } n \in \mathbb{N}.$$

Since $\mu(\cdot)$ is finite, terminating the summation in the definition of $\widehat{s}_n(\cdot)$ at an appropriately big $m \in \mathbb{N}$, we can produce a sequence $\{s_n\}_{n \in \mathbb{N}}$ of simple functions such that

$$s_n(z) \to u(z) \text{ for } \mu - a.a \ z \in \Omega,$$

$$\Rightarrow u(\cdot) \text{ is } \mu - \text{strongly measurable.} \qquad \square$$

An interesting byproduct of the above proof is the following

Corollary 4.139 $u : \Omega \to X$ *is μ−strongly measurable if and only if $u(\cdot)$ is μ−almost everywhere uniform limit of a sequence of countably valued Σ−measurable functions.*

We list a few more simple consequences of Definition 4.137 and of Theorem 4.138 and of its proof.

Corollary 4.140 *If $u : \Omega \to X$ is strongly μ−measurable, then there exists a sequence $\{s_n\}_{n \in \mathbb{N}}$ of simple functions such that $\|s_n(z)\| \le \|f(z)\|$ for a.a $z \in \Omega$, all $n \in \mathbb{N}$ and $s_n(z) \to f(z)$ for μ a.a $z \in \Omega$.*

Corollary 4.141 *If $u_n : \Omega \to X \ n \in \mathbb{N}$ is a sequence of strongly μ−measurable functions and $u_n(z) \to u(z) \ \mu$−a.e, then $u : \Omega \to X$ is strongly μ−measurable.*

Proposition 4.142 *If X is separable, then $\sigma(X^*) = B(X)$ (recall that $\sigma(X^*)$ is the smallest σ−algebra on X making all $x^* \in X^*$ measurable and $B(X)$ is the Borel σ−algebra of X).*

Proof Evidently $\sigma(X^*) \subseteq B(X)$. We need to show that the opposite inclusion also holds. We know that every open set in X is the countable union of open balls (recall X is separable). Moreover, every open ball is the countable union of closed balls. So, it suffices to show that for every $x \in X$ and $r > 0$, $\overline{B}_r(x) = \{u \in X : \|u - x\| \le r\} \in \sigma(X^*)$. Let $\{x_n^*\}_{n \in \mathbb{N}} \subseteq X^*$ be a norming sequence of unit vectors (see the proof of Theorem 4.138). We have

$$\overline{B}_r(x) = \bigcap_{n\in\mathbb{N}} \{u \in X : |\langle x_n^*, u - x\rangle| \le r\} \in \sigma(X^*),$$
$$\Rightarrow \sigma(X^*) = B(X).$$
□

An immediate consequence of this proposition is the following result.

Proposition 4.143 *If X is separable and $u : \Omega \to X$, then $u(\cdot)$ is Borel measurable if and only if $u(\cdot)$ is weakly μ-measurable.*

Proof $\Rightarrow$ Clear.

$\Leftarrow$ Using the notation from the proof of Proposition 4.142, we have

$$u^{-1}(\overline{B}_r(x)) = \bigcap_{n\in\mathbb{N}} \{z \in \Omega : |\langle x_n^*, u(z) - x\rangle| \le r\} \in \Sigma.$$

But the balls generate $B(X)$. So $u(\cdot)$ is Borel measurable. □

Combining this proposition with Theorem 4.138, we have the following theorem which shows that in the context of separable Banach spaces the question of measurability of vector valued function simplifies.

Theorem 4.144 *If X is separable and $u : \Omega \to X$, then the following statements are equivalent*

(a) $u(\cdot)$ is strongly μ-measurable.
(b) $u(\cdot)$ is Borel measurable.
(c) $u(\cdot)$ is weakly μ-measurable.

Example 4.145 If we drop the separability hypothesis, the result fails. To see this let $X = l^2[0, 1]$ (a nonseparable Hilbert space) and let $u : [0, 1] \to l^2[0, 1]$ be defined by $u(t) = e_t$ (as always $\{e_t\}_{t\in[0,1]}$ is the orthonormal basis of $l^2[0, 1]$). For every $x^* \in l^2[0, 1]^* = l^2[0, 1]$, we have $\langle x^*, u\rangle = 0$ a.e on $[0, 1]$ and so $u(\cdot)$ is weakly λ-measurable (on $[0, 1]$ we consider the Lebesgue measure $\lambda(\cdot)$). If $D \subseteq [0, 1]$, then $u([0, 1] \setminus D)$ is separable if and only if $[0, 1] \setminus D$ is countable and so $u(\cdot)$ is not λ-essentially separably valued. Hence it is not Borel measurable or strongly λ-measurable.

Proposition 4.146 *If $u : \Omega \to X$ is strongly μ-measurable, Y is a Banach space and $\xi : X \to Y$ is Borel measurable, then $\xi \circ u : \Omega \to Y$ is strongly μ-measurable.*

Proof Evidently $\xi \circ u$ is Borel measurable. By Theorem 4.138 u is μ-essentially separably valued. So, there exists $N \in \Sigma$ with $\mu(N) = 0$ such that $V = u(\Omega \setminus N) \subseteq X$ is norm separable. Hence $\xi(V) \subseteq Y$ is norm separable and this by Proposition 4.143 and Theorem 4.138 implies that $\xi \circ u$ is strongly μ-measurable. □

Proposition 4.147 *If $u, v : \Omega \to X$ are strongly μ-measurable functions such that $\langle x^*, u(z)\rangle = \langle x^*, v(z)\rangle$ for $z \in \Omega \setminus N_{x^*}$, $\mu(N_{x^*}) = 0$, then $u = v$ μ-a.e.*

Proof According to Theorem 4.138, we can find $V \subseteq X$ closed separable subspace and $N_0 \in \Sigma$, $\mu(N_0) = 0$ such that $u(\Omega \setminus N_0)$, $v(\Omega \setminus N_0) \in V$. The separability of V implies that we can find a sequence $\{x_n^*\}_{n \in \mathbb{N}}$ which separates points in V. We have

$$\langle x_n^*, u(z) \rangle = \langle x_n^*, v(z) \rangle \text{ for all } z \in \Omega \setminus N_n, N_n = N_{x_n^*}, \mu(N_n) = 0.$$

So, if $N = \left[\bigcup_{n \in \mathbb{N}} N_n \right] \cup N_0$, then we have

$$\langle x_n^*, u(z) \rangle = \langle x_n^*, v(z) \rangle \text{ for } n \in \mathbb{N}, z \in \Omega \setminus N, \mu(N) = 0.$$
$$\Rightarrow u = v \ \mu - a.e.$$
$\square$

Remark 4.148 If Y is another Banach and we consider the Banach space $\mathcal{L}(X, Y)$ (with the operator norm), then $\mathcal{L}(X, Y)$ is rarely separable and for this reason we modify the definition of strong μ-measurability as follows. A function $u : \Omega \rightarrow \mathcal{L}(X, Y)$ is said to be strongly μ-measurable, if for all $x \in X$ the function $\vartheta(z) = u(z)(x) \in Y$ is strongly μ-measurable. So, what we have done is to replace the uniform operator topology on $\mathcal{L}(X, Y)$ by the so-called strong operator topology.

Now we are ready to introduce and study the vector valued extension of the Lebesgue integral. This extension is known as "Bochner integral."

Definition 4.149 A strongly μ-measurable function $u : \Omega \rightarrow X$ is said to be "Bochner integrable with respect to μ," if there exists a sequence $\{s_n\}_{n \in \mathbb{N}}$ of simple functions such that

$$\int_\Omega \|u(z) - s_n(z)\| d\mu \rightarrow 0 \text{ as } n \rightarrow \infty.$$

In this case, $\int_\Omega u d\mu = \lim_{n \in \infty} \int_\Omega s_n d\mu$ is called the "Bochner integral of u."

Remark 4.150 Note that for all $n, m \in \mathbb{N}$

$$\left\| \int_\Omega s_n d\mu - \int_\Omega s_m d\mu \right\| \leq \int_\Omega \|s_n - s_m\| d\mu \leq \int_\Omega \|s_n - u\| d\mu + \int_\Omega \|u - s_m\| d\mu,$$

$\Rightarrow \left\{ \int_\Omega s_n d\mu \right\}_{n \in \mathbb{N}} \subseteq X$ is Cauchy, hence it converges to a limit in X.

Therefore the Bochner integral of u is well defined. If the measure μ is well understood, we drop it and simply say that $u(\cdot)$ is Bochner integrable. Of course if u is Bochner integrable and $u = v$, $\mu - a.e.$, then v is Bochner integrable too and $\int_\Omega u d\mu = \int_\Omega v d\mu$. Definition 4.149 is independent of the sequence $\{s_n\}_{n \in \mathbb{N}}$.

Proposition 4.151 *A strongly μ-measurable function $u : \Omega \rightarrow X$ is Bochner integrable with respect to μ if and only if $\|u(\cdot)\| \in L^1(\Omega)$ and $\|\int_\Omega u d\mu\| \leq \int_\Omega \|u\| d\mu$.*

Proof $\Rightarrow$ According to Definition 4.149 we can find $\{s_n\}_{n \in \mathbb{N}}$ simple functions such that $\int_\Omega \|u - s_n\| d\mu \rightarrow 0$. We have

$$\int_\Omega \|u\| d\mu \leq \int_\Omega \|u - s_n\| d\mu + \int_\Omega \|s_n\| d\mu \leq 1 + \int_\Omega \|s_n\| d\mu$$

for $n \in \mathbb{N}$ big and fixed, $\Rightarrow \|u\| \in L^1(\Omega)$.

Note that for every simple function $s(\cdot)$ we have $\|\int_\Omega s d\mu\| \leq \int_\Omega \|s\| d\mu$. Finally use simple functions $\{s_n\}_{n \in \mathbb{N}}$ as in Corollary 4.140 to approximate $u(\cdot)$ and then pass to the limit as $n \rightarrow \infty$. We obtain $\|\int_\Omega u d\mu\| \leq \int_\Omega \|u\| d\mu$.

$\Leftarrow$ Use Corollary 4.140 and the Lebesgue Dominated Convergence Theorem in order to obtain a sequence $\{s_n\}_{n \in \mathbb{N}}$ of simple functions such that $\int_\Omega \|u - s_n\| d\mu \rightarrow 0$ as $n \in \infty$. $\qquad\square$

Remark 4.152 If $u : \Omega \rightarrow X$ is Bochner integrable and $A \in \Sigma$, then from Proposition 4.151 we infer that $\chi_A u$ is Bochner integrable, and $\int_A u d\mu = \int_\Omega \chi_A u d\mu$.

In general many of the results from the theory of the Lebesgue integral pass to the Bochner integral unchanged. We state one such result which is the "Dominated Convergence Theorem."

Theorem 4.153 *If $u_n, u : \Omega \rightarrow X$ $n \in \mathbb{N}$, each $u_n(\cdot)$ is Bochner integrable, $u_n(z) \rightarrow u(z)$ for $\mu - a.a$ $z \in \Omega$ and $\|u_n(z)\| \leq \vartheta(z)$ for $\mu - a.a$ $z \in \Omega$, all $n \in \mathbb{N}$ with $\vartheta \in L^1(\Omega)$, then $u(\cdot)$ is Bochner integrable and $\int_\Omega \|u_n - u\| d\mu \rightarrow 0$; in particular, $\lim\limits_{n \in \infty} \int_\Omega u_n d\mu = \int_\Omega u d\mu$.*

Proof We have $u(\cdot)$ is strongly μ-measurable (see Corollary 4.141). Then since $\|u_n(z) - u(z)\| \leq 2\vartheta(z)$ for $\mu - a.a$ $z \in \Omega$, all $n \in \mathbb{N}$, we apply the Lebesgue Dominated Convergence Theorem and conclude that

$$u(\cdot) \text{ is Bochner integrable, } \int_\Omega \|u_n - u\| d\mu \rightarrow 0 \text{ as } n \rightarrow \infty,$$

$$\int_\Omega u_n d\mu \rightarrow \int_\Omega u d\mu \text{ in } X \text{ as } n \rightarrow \infty.$$

$\qquad\square$

On the other hand, the next theorem has no counterpart in the Lebesgue integration theory.

Theorem 4.154 *If Y is another Banach space, $A : D \subseteq X \rightarrow Y$ is a closed, linear operator and $u : \Omega \rightarrow X$, $A \circ u : \Omega \rightarrow Y$ are both Bochner integrable, then $A\left(\int_C u d\mu\right) = \int_C (A \circ u) d\mu$ for all $C \in \Sigma$.*

Proof Directly from the definition of the Bochner integral (see Definition 4.149), we see that if $A \in \mathcal{L}(X, Y)$, then

$$A \left(\int_C u d\mu \right) = \int_C (A \circ u) d\mu \text{ for all } C \in \Sigma.$$

For the general case of $A : D \subseteq X \to Y$ linear, closed, consider the map $g : \Omega \to X \times Y$ defined by

$$g(z) = (u(z), A(u(z))) \text{ for all } z \in \Omega.$$

Note that $R(g) \subseteq \text{Gr } A$ and the latter furnished with the graph norm

$$|x| = \|x\|_X + \|A(x)\|_Y \text{ for all } x \in X$$

is a Banach space. (see the proof of Theorem 3.86). We have

$$\int_C g d\mu = \left(\int_C \mu d\mu, \int_C A(u) d\mu \right) \in \text{Gr } A \text{ for all } C \in \Sigma,$$
$$\Rightarrow A \left(\int_C u d\mu \right) = \int_C A(u) d\mu \text{ for all } C \in \Sigma. \qquad \square$$

Corollary 4.155 *If $u : \Omega \to X$ is Bochner integrable and $x^* \in X^*$, then $\langle x^*, \int_\Omega u d\mu \rangle = \int_\Omega \langle x^*, u \rangle d\mu$.*

Using this corollary, we can have the following version of the Fubini-Tonelli Theorem for vector valued functions.

Theorem 4.156 *If $(S, \mathscr{S}, \vartheta)$ is another finite measure space and $g : \Omega \times S \to X$ is Bochner integrable for the product measure $\mu \otimes \vartheta$, then*

(a) for $\mu - a.a \ z \in \Omega$, the map $s \to g(z, s)$ is Bochner integrable.
(b) For $\vartheta - a.a \ s \in S$, the map $z \to g(z, x)$ is Bochner integrable.
(c) The functions $z \to \int_S \|g(z, s)\| d\vartheta$, $s \to \int_\Omega \|g(z, s)\| d\mu$ are Bochner integrable and

$$\int_{\Omega \times S} g d(\mu \otimes \vartheta) = \int_S \int_\Omega g d\mu d\vartheta = \int_\Omega \int_S g d\vartheta d\mu.$$

We prove some more properties of the Bochner integral.

Proposition 4.157 *If $u : \Omega \to X$ is Bochner integrable and $A \in \Sigma$ with $\mu(A) > 0$, then $\frac{1}{\mu(A)} \int_A u d\mu \in \overline{\text{conv}} u(A)$.*

Proof Let $C = \overline{\text{conv}} u(A)$ and arguing by contradiction suppose that

$$\frac{1}{\mu(A)} \int_A u d\mu \notin C.$$

By the strong separation theorem (see Theorem 3.71(b)), we can find $x^* \in X^* \setminus \{0\}$ and $\epsilon > 0$ such that

$$\sigma(x^*, C) = sup[\langle x^*, c \rangle : c \in C] \leq \left\langle x^*, \frac{1}{\mu(A)} \int_A u d\mu \right\rangle - \epsilon. \tag{4.40}$$

On the other hand, $u(z) \in C$ for μ-a.a $z \in A$, hence $\langle x^*, u(z) \rangle \leq \sigma(x^*, C)$,

$$\left\langle x^*, \int_A u d\mu \right\rangle = \int_A \langle x^*, u \rangle d\mu \leq \mu(A)\sigma(x^*, C)$$

(see Corollary 4.155). Therefore,

$$\left\langle x^*, \frac{1}{\mu(A)} \int_A u d\mu \right\rangle \leq \sigma(x^*, C). \tag{4.41}$$

Comparing (4.40) and (4.41), we have a contradiction. Therefore

$$\frac{1}{\mu(A)} \int_A u d\mu \in \overline{conv}u(A) \text{ for all } A \in \Sigma.$$

$\square$

There is a converse to the above proposition.

Proposition 4.158 *If $u : \Omega \to X$ is Bochner integrable, $C \subseteq X$ is closed, convex and for all $A \in \Sigma$ with $\mu(A) > 0$ we have $\frac{1}{\mu(A)} \int_A u d\mu \in C$, then $u(z) \in C$ for $\mu - a.a \ z \in \Omega$.*

Proof On account of the Pettis Measurability Theorem (see Theorem 4.136) we may assume that X is separable. Using Corollary 4.155, for every $x^* \in X^*$, we have for all $A \in \Sigma$

$$\int_A \langle x^*, u \rangle d\mu = \langle x^*, \int_A u d\mu \rangle \leq \mu(A)\sigma(x^*, C) = \int_A \sigma(x^*, C)d\mu,$$
$$\Rightarrow \langle x^*, u(z) \rangle \leq \sigma(x^*, C) \text{ for all } z \in \Omega \setminus N_{x^*}, \text{ with } \mu(N_{x^*}) = 0.$$

Since X is separable, as in the proof of Proposition 4.147, we conclude that $\langle x^*, u(z) \rangle \leq \sigma(x^*, C)$ for all $x^* \in X^*$, all $z \in \Omega \setminus N$, with $\mu(N) = 0$, hence $u(z) \in C$ for $\mu - a.a \ z \in \Omega$. $\square$

Proposition 4.159 *If $u, v : \Sigma \to X$ are Bochner integrable, $\tau \subseteq \Sigma$ is closed under finite intersections, $\Omega \in \tau$, $\sigma(\tau) = \Sigma$ and $\int_C u d\mu = \int_C v d\mu$ for all $C \in \tau$, then $u(z) = v(z)$ for $\mu - a.a \ z \in \Omega$.*

Proof For every $x^* \in X^*$, we consider the $\mathbb{R}$-valued measure $k_{x^*}(A) = \int_A \langle x^*, u - v \rangle d\mu$ for all $A \in \Sigma$. By hypothesis $k_{x^*}|_\tau = 0$ and since $\sigma(\tau) = \Sigma$, we infer that

$$k_{x^*}(A) = \int_A \langle x^*, u - v \rangle d\mu = 0 \text{ for all } A \in \Sigma.$$

Using Proposition 4.158 with $C = \{0\}$, we infer that

$$\langle x^*, u(z) \rangle = \langle x^*, v(z) \rangle \text{ for a.a } z \in \Omega \setminus N_{x^*}, \mu(N_{x^*}) = 0,$$

$$\Rightarrow u(z) = v(z) \text{ for } \mu - a.a \ z \in \Omega \text{ (see Proposition 4.147).} \qquad \square$$

Definition 4.160

(a) A set function $m : \Sigma \rightarrow X$ is a "vector measure," if for all $\{A_n\}_{n \in \mathbb{N}} \subseteq \Sigma$ mutually disjoint we have

$$m(\bigcup_{n \in \mathbb{N}} A_n) = \sum_{n \in \mathbb{N}} m(A_n) \text{ (convergence in the norm of X).}$$

Since we may permute the $A'_n s$, the convergence of the sum is unconditional.

(b) The "variation" of the vector measure $m(\cdot)$, $|m| : \Sigma \rightarrow [0, +\infty]$ is defined by

$$|m|(A) = \sup_{\pi_A} \sum_{C \in \pi_A} \|m(C)\|,$$

where π_A is the family of all finite Σ-partitions of A. We say that $m(\cdot)$ is of "bounded variation," if $|m|(\Omega) < \infty$.

(c) m is "absolutely continuous" with respect to μ, denoted by $m << \mu$, if $\mu(A) = 0$ for $A \in \Sigma \Rightarrow m(A) = 0$.

Remark 4.161 It is routine to check that if $m : \Sigma \rightarrow X$ is a vector measure of bounded variation, then $|m|(\cdot)$ is a finite measure. Then $m << \mu$ if and only if $|m| << \mu$ (see Definition 2.108). Therefore using Proposition 2.110, we conclude that

$$m << \mu \Leftrightarrow |m| << \mu$$
$\Leftrightarrow$ for every $\epsilon > 0$, there exists $\delta > 0$ such that $\mu(A) < \delta \Rightarrow \|m(A)\| < \epsilon$
$\Leftrightarrow$ for every $\epsilon > 0$, there exists $\delta > 0$ such that $\mu(A) < \delta \Rightarrow |m|(A) < \epsilon$.

The next proposition extends to Bochner integral Corollary 2.111.

Proposition 4.162 *If $u : \Omega \rightarrow X$ is Bochner integrable and for all $A \in \Sigma$ $m(A) = \int_A u d\mu$, then $m(\cdot)$ is a vector measure of bounded variation, $|m|(A) = \int_A \|u\| d\mu$ for all $A \in \Sigma$ and $m << \mu$.*

Proof Note that $\|m(A)\| \leq \int_A \|u\| d\mu$ for all $A \in \Sigma$. This fact and Corollary 2.111 imply that $m(\cdot)$ is a vector measure. Moreover, it follows that

$$|m|(A) \leq \int_A \|u\| d\mu \text{ for all } A \in \Sigma. \qquad (4.42)$$

We need to show that opposite inequality is also true. On account of Theorem 4.138, we may assume that X is separable and then we can find a norming sequence $\{x_n^*\}_{n\in\mathbb{N}} \subseteq X^*$, $\|x_n^*\|_* = 1$ for all $n \in \mathbb{N}$. Given $\epsilon > 0$, let

$$A_n^\epsilon = \{z \in \Omega : \|u(z)\| - \epsilon \leq \langle x_n^*, u(z)\rangle\} \in \Sigma. \tag{4.43}$$

We have $\Omega = \bigcup_{n\in\mathbb{N}} A_n^\epsilon$ (we ignore a μ-null set since its integral will be zero). Let $C_1^\epsilon = A_1^\epsilon$ and $C_{n+1}^\epsilon = A_{n+1}^\epsilon \setminus \bigcup_{k=1}^{n} C_k^\epsilon$, $n \in \mathbb{N}$. These are mutually disjoint Σ-sets and $\Omega = \bigcup_{n\in\mathbb{N}} C_n^\epsilon$. Then for all $A \in \Sigma$, we have

$$\int_A \|u\|d\mu = \sum_{n\in\mathbb{N}} \int_{A\cap C_n^\epsilon} \|u\|d\mu \leq \sum_{n\in\mathbb{N}} \int_{A\cap C_n^\epsilon} \langle x_n^*, u\rangle d\mu + \epsilon\mu(A) \text{ (see (4.43))}$$

$$= \sum_{n\in\mathbb{N}} \langle x_n^*, m(A \cap C_n^\epsilon)\rangle + \epsilon\mu(A) \text{ (see Corollary 4.155)}$$

$$\leq \sum_{n\in\mathbb{N}} \|m(A \cap C_n^\epsilon)\| + \epsilon\mu(A)$$

$$\leq |m|(A) + 2\epsilon\mu(A).$$

Let $\epsilon \to 0^+$ to conclude that $\int_A \|u\|d\mu = |m|(A)$ (see (4.41)). Finally it is clear that $m << \mu$. $\qquad\square$

Now we will introduce generalizations of the Lebesgue spaces L^p, $1 \leq p \leq \infty$. If $u, v : \Omega \to X$ are strongly μ-measurable, then $u \sim v$ if and only if they differ only on a μ-null set. This is an equivalence relation.

Definition 4.163

(a) Let $1 \leq p < \infty$. Then $L^p(\Omega, X)$ is the vector space of all equivalence classes for $\sim$ of strongly μ-measurable functions $u : \Omega \to X$ such that $\int_\Omega \|u\|^p d\mu < \infty$.

(b) $L^\infty(\Omega, X)$ is the vector space of all equivalence classes for $\sim$ of strongly μ-measurable functions such that $\mu\{\|u\| > M\} = 0$ for some $M > 0$.

The spaces $L^p(\Omega, X)$ $1 \leq p \leq \infty$ are known as "Lebesgue-Bochner spaces." These spaces become normed spaces when endowed with the following norms

$$\|u\|_p = (\int_\Omega \|u\|^p d\mu)^{1/p} \text{ and } \|u\|_\infty = \inf[M \geq 0 : \mu\{\|u\| > M\} = 0].$$

As we did for the scalar case (see Theorem 2.126), we can show that $L^p(\Omega, X)$ $1 \leq p \leq \infty$ are all Banach spaces.

Remark 4.164 In fact as in the scalar case(see Proposition 2.127), we can show that if $u_n \to u$ in $L^p(\Omega, X)$, then we can extract a subsequence $\{u_{n_k}\}_{k \in \mathbb{N}}$ such that $u_{n_k}(z) \to u(z)$ for $\mu - a.a \ z \in \Omega$. Moreover, note that $u \in L^p(\Omega, X)$ if and if $\|u\| \in L^p(\Omega)$ ($1 \leq p \leq \infty$). Also simple functions are dense in $L^p(\Omega, X)$ $1 \leq p < \infty$. For $L^\infty(\Omega, X)$ we can say that if $u \in L^\infty(\Omega, X)$ and $\epsilon > 0$, then we can find a simple function $s : \Omega \to X$ and $A \in \Sigma$ with $\mu(\Omega \setminus A) < \epsilon$ such that $\|s\|_\infty \leq \|u\|_\infty$ and $\|u - s\|_{L^\infty(A)} < \epsilon$.

Proposition 4.165 *If* $u : \Omega \to X$ *is a strongly* μ-*measurable function then* $u \in L^\infty(\Omega, X)$ *if and only if* $\langle x^*, u(\cdot) \rangle \in L^\infty(\Omega)$ *for all* $x^* \in X^*$; *moreover, we have* $\|u\|_\infty = sup[\|\langle x^*, u \rangle\|_\infty : \|x^*\|_* \leq 1]$.

Proof $\Rightarrow$ Obvious.

$\Leftarrow$ As before, on account of Theorem 4.136, we may assume that X is separable. We consider the map $L_u : X^* \to L^\infty(\Omega)$ defined by $L_u(x^*)(\cdot) = \langle x^*, u(\cdot) \rangle$. It is easy to check that $L_u \in \mathcal{L}(X^*, L^\infty(\Omega))$. Let $\eta = \|L_u\|_{\mathcal{L}} = sup[\|L_u(x^*)\|_\infty : \|x^*\|_* \leq 1]$. If $\{x_n^*\}_{n \in \mathbb{N}} \subseteq X^*$ is a norming sequence with $\|x_n^*\|_* = 1$, then we have

$$\|u(z)\| = \sup_{n \in \mathbb{N}} |\langle x_n^*, u(z) \rangle|.$$

Suppose $\mu\{\|u\| > \eta\} > 0$. Then $\mu(\bigcup_{n \in \mathbb{N}}\{|\langle x_n^*, u \rangle| > \eta\}) > 0$. So, for some $n_0 > 0$, we have $\mu\{|\langle x_{n_0}^*, u \rangle| > \eta\} > 0$, a contradiction to the definition of $\eta > 0$. Therefore $u \in L^\infty(\Omega, X)$ and $\eta = \|u\|_\infty$. $\qquad\square$

As for the scalar case (see Theorem 4.8), we can have the following result.

Proposition 4.166 *If* $1 \leq p < \infty$, Σ *is a separable (see Definition 2.168(c)) and* X *is separable, then* $L^p(\Omega, X)$ *is separable.*

Remark 4.167 Also for $1 < p < \infty$, $L^p(\Omega, X)$ is uniformly convex if and only if X is uniformly convex. If X is a Hilbert space, then $L^2(\Omega, X)$ is a Hilbert space with inner product $(u, v) = \int_\Omega (u(z), v(z))_X dz$ (with $(\cdot, \cdot)_X$ denoting the inner inner product on X). If Y is another Banach space with $X \hookrightarrow Y$ continuously and $1 \leq p \leq r \leq \infty$, then using Hölder's inequality we can easily check that $L^r(\Omega, X) \hookrightarrow L^p(\Omega, Y)$ continuously.

What about the dual of $L^p(\Omega, X)$? Can we have a result similar to the Riesz Representation Theorem (see Theorem 2.133) for the Lebesgue-Bochner spaces? It turns out that this depends on the structure of X. We want to identify those Banach spaces for which the converse of Proposition 4.162 holds. This leads to the following notion which in turn will provide us the Riesz Representation Theorem for vector valued functions.

Definition 4.168 W say that the Banach space X has the "Radon–Nikodym Property" (the RNP for short) with respect to the measure space (Ω, Σ, μ), if every vector measure $m : \Sigma \to X$ of bounded variation such that $m << \mu$ admits a Radon–Nikodym derivative, that is, there exists $u \in L^1(\Omega, X)$ such that

$$m(A) = \int_A u(z)d\mu \text{ for all } A \in \Sigma.$$

Remark 4.169 This definition is in fact unsatisfactory since it depends also on the measure space (Ω, Σ, μ). However, it turns out that it is enough to consider the unit interval $[0, 1]$, namely " X has the RNP with respect to $[0, 1]$ if and only if it has the RNP with respect to every finite measure space (Ω, Σ, μ)" (universality of $[0, 1]$). Not every Banach space has this property.

Example 4.170 The Banach spaces $c_0, L^\infty, C[0, 1], L^\infty[0, 1]$ all fail the RNP (with respect to $([0, 1], \mathcal{L}[0, 1], \lambda)$, where $\mathcal{L}[0, 1]$ =the Lebesgue σ-algebra of $[0, 1]$ and λ is the Lebesgue measure). Consider the vector measure $m : \mathcal{L}[0, 1] \to c_0$ defined by

$$m(A) = \int_A sin(2^n nt)d\lambda)_{n\in\mathbb{N}}.$$

The Riemann-Lebesgue lemma guarantees that m is c_0-valued, and $\|m(A)\| \leq \lambda(A)$ for all $A \in \mathcal{L}[0, 1]$ and so $m(\cdot)$ is of bounded variation and $m << \lambda$. Suppose we could find $u : [0, 1] \to c_0$ Bochner integrable such that $m(A) = \int_A u \, d\lambda$ for all $A \in \mathcal{L}[0, 1]$. Then $u = (u_n)_{n\in\mathbb{N}} = (sin 2^n \pi t)_{n\in\mathbb{N}}$. But for irrational $t \in (0, 1)$ we see that $u(t) \in l^\infty \setminus c_0$, a contradiction. So, c_0 does not have the RNP. Note that c_0 is a closed subspace of $l^\infty, C[0, 1], L^\infty[0, 1]$ and the RNP is a hereditary property of closed subspaces (see Diestel and Uhl [85], Theorem 2, p.81). So, none of these spaces has the RNP. It can be shown that $L^1[0, 1]$ too fails the RNP (see Diestel and Uhl [85], Example 2, p.61).

However, there are two important classes of Banach spaces with the RNP (see Diestel and Uhl [85], Corollary 13,p.76 and Theorem 1, p.79).

Theorem 4.171 *The following classes of Banach spaces have the RNP.*
(a) Reflexive Banach spaces; (b) Separable dual spaces.

Now let $1 \leq p \leq \infty$ and let $p' = \frac{p}{p-1}$ if $p > 1$, $p' = \infty$ if $p = 1$, $p' = 1$ if $p = \infty$. Let $v^* \in L^{p'}(\Omega, X^*)$. Such a function defines a continuous linear functional $\xi_{v^*} \in L^p(\Omega, X)^*$ by the formula

$$\langle \xi_{v^*}, u \rangle = \int_\Omega \langle v^*(z), u(z) \rangle_X d\mu$$

with $\langle \cdot, \cdot \rangle_X$ denoting the duality brackets for the pair (X^*, X).

Lemma 4.172 $\|\xi_{v^*}\|_* = \|v^*\|_{p'}$.

Proof It is clear using Hölder's inequality that $\|\xi_{v^*}\|_* \leq \|v^*\|_{p'}$. We will show that the opposite inequality also holds. We may assume that $\|v^*\|_{p'} = 1$.

First let $1 < p < \infty$. If $v_n^* \to v^*$ in $L^{p'}(\Omega, X^*)$, then given $\epsilon > 0$, we can find $n_0 \in \mathbb{N}$ such that

$$\|\xi_{v^*}\|_* \geq \|\xi_{v_n^*}\|_* - \|\xi_{v^*-v_n^*}\|_*$$
$$\geq \|\xi_{v_n^*}\|_{x^*} - \|v^* - v_n^*\|_{p'}$$
$$\geq \|\xi_{v_n^*}\|_* - \epsilon \text{ for all } n \geq n_0.$$

Therefore it suffices to show that $\|\xi_{v^*}\| \geq 1$ for every v^* simple X^*-valued function such that $\|v^*\|_{p'} = 1$.

So, let $v^* = \sum_{n=1}^{m} \chi_{A_n} x_n^*$ with $A_n \in \Sigma$, $x_n^* \in X^* \setminus \{0\}$. Given $\epsilon > 0$, we can find $x_n \in X$, $\|x_n\| = 1$ for all $n \in \mathbb{N}$ such that

$$(1 - \epsilon)\|x_n^*\|_{X^*} \leq \langle x_n^*, x^n \rangle.$$

Let $u = \sum_{n=1}^{m} \chi_{A_n} \|x^*\|_{X^*}^{p'-1} x_n$. Then

$$\|u\|_p^\varrho = \sum_{n=1}^{m} \mu(A_n)\|x_n^*\|_{X^*}^{p(p'-1)} = \sum_{n=1}^{m} \mu(A_n)\|x_n^*\|_{X^*}^{p'} = \|v^*\|_{p'}^{p'} = 1$$

and

$$\langle \xi_{v^*}, u \rangle = \sum_{n=1}^{m} \mu(A_n)\|x_n^*\|_{X^*}^{p'-1} \langle x_n^*, x_n \rangle$$
$$\geq (1 - \epsilon) \sum_{n=1}^{m} \mu(A_n)\|x_n^*\|_{X^*}^{p'}$$
$$= 1 - \epsilon.$$

Let $\epsilon \to 0^+$ to conclude that $\|\xi_{v^*}\|_* \geq 1$.

Now let $p = \infty$. In this case we let $u = \sum_{n=1}^{m} \chi_{A_n} x_n$. Then $\|u\|_\infty = 1$ and

$$\langle \xi_{v^*}, u \rangle = \sum_{n=1}^{m} \mu(A_n)\langle x_n^*, x_n \rangle_X \geq (1 - \epsilon) \sum_{n=1}^{m} \mu(A_n)\|x_n^*\|_{X^*} = 1 - \epsilon.$$

Again let $\epsilon \to 0^+$ to conclude that $\|\xi_{v^*}\|_* \geq 1$.

Finally let $p = 1$. Let $\epsilon > 0$ and set $A_\epsilon = \{z \in \Omega : \|v^*(z)\|_{X^*} > 1 - \epsilon\}$. Then $\mu(A_\epsilon) > 0$. We may assume that the range of $v^*|_{A_\epsilon}$ is separable. We can find $x^* \in X^*$ such that

$$A_\epsilon^{x^*} = A_\epsilon \cap \{z \in \Omega : \|v^*(z) - x^*\|_{X^*} < \epsilon\}$$

has positive $\mu-$measure. Then $\|x^*\|_{X^*} \geq 1 - 2\epsilon$, we choose $x \in X$, $\|x\| = 1$ such that $\langle x^*, x \rangle \geq \|x^*\|_{X^*} - \epsilon$. With $u = \frac{1}{\mu(A_\epsilon^{x^*})} \chi_{A_\epsilon^{x^*}} x$, we have $\|u\|_1 = 1$ and

$$\begin{aligned}
\langle \xi_{v^*}, u \rangle &= \frac{1}{\mu(A_\epsilon^{x^*})} | \int_{A_\epsilon^{x^*}} \langle v^*, x \rangle d\mu | \\
&\geq \frac{1}{\mu(A_\epsilon^{x^*})} \int_{A_\epsilon^{x^*}} \langle x^*, x \rangle d\mu - \epsilon \\
&\geq \|x^*\|_{X^*} - 2\epsilon \\
&\geq 1 - 4\epsilon.
\end{aligned}$$

Then letting $\epsilon \to 0^+$ again we have $\|\xi_{v^*}\|_* \geq 1$.   □

Now we are ready for the Riesz Representation Theorem for Lebesgue-Bochner spaces.

Theorem 4.173 *If* $1 \leq p < \infty$, $p' = \frac{p}{p-1}$ ($p' = +\infty$, *if* $p = 1$) *and* X^* *has the RNP with respect to* (Ω, Σ, μ), *then* $L^p(\Omega, X)^* = L^{p'}(\Omega, X^*)$ *(that is, the map* $v^* \to \xi_{v^*}$ *is on isometric isomorphism).*

Proof Let $\xi^* \in L^p(\Omega, X)^*$. We will show that $\xi^* = \xi_{v^*}$ for some $v^* \in L^{p'}(\Omega, X^*)$ and then invoke Lemma 4.172. Consider the vector valued set function $m : \Sigma \to X^*$ defined by

$$\langle m(A), x \rangle = \xi^*(\chi_A x) \quad \text{for all } A \in \Sigma, \ x \in X.$$

Clearly this is on X^*-valued vector measure. If $\{A_k\}_{k=1}^n \subseteq \Sigma$ is a partition of Ω and $\{x_k\}_{k=1}^n \subseteq X$ with $\|x_k\| \leq 1$ for all $k \in \{1, ..., n\}$. We have

$$\begin{aligned}
| \sum_{k=1}^n \langle m(A_n), x_k \rangle | &= | \sum_{k=1}^n \xi^*(\chi_{A_k} x_k) | \\
&\leq \|\xi^*\|_* \| \sum_{k=1}^n \chi_{A_k} x_k \|_p \\
&\leq \|\xi^*\|_* \mu(\Omega)^{1/P},
\end{aligned}$$

$$\Rightarrow |m|(\Omega) \leq \|\xi^*\|_* \mu(\Omega)^{1/p},$$
$$\Rightarrow m(\cdot) \text{ is of bounded variation.}$$

Also $m << \mu$. Since by hypothesis X^* has the RNP we can find $v^* \in L^1(\Omega, X^*)$ such that

$$m(A) = \int_A v^* d\mu \text{ for all } A \in \Sigma.$$
$$\Rightarrow \xi^*(\chi_A x) = \langle m(A), x \rangle = \int_\Omega \langle v^*(z), \chi_A(z)x \rangle d\mu.$$

So, for every simple function $s \in L^p(\Omega, X)$ we have

$$\xi^*(s) = \int_\Omega \langle v^*(z), s(z) \rangle d\mu.$$

Let $\{A_n\}_{n \in \mathbb{N}} \subseteq \Sigma$ be an increasing sequence such that $\Omega = \bigcup_{n \in \mathbb{N}} A_n$ and $v^*|_{A_n}$ is bounded for all $n \in \mathbb{N}$.

Fix $n_0 \in \mathbb{N}$ and note that $u \to \int_{A_{n_0}} \langle v^*, u \rangle d\mu$ is a bounded linear functional on $L^p(\Omega, X)$ and agrees with ξ^* on the simple functions which vanish outside $A_{n_0} \in \Sigma$. Therefore

$$\xi^*(\chi_{A_{n_0}} u) = \int_\Omega \langle \chi_{A_{n_0}} v^*, u \rangle d\mu.$$

The function $\chi_{A_{n_0}}^{v^*} \in L^{p'}(\Omega, X^*)$ (being bounded). Hence

$$\|\chi_{A_{n_0}} v^*\|_{p'} \leq \|\xi^*\|_*. \tag{4.44}$$

Since $n_0 \in \mathbb{N}$ is arbitrary, from (42) and the Monotone Convergence Theorem, we infer that $v^* \in L^{p'}(\Omega, X^*)$ and we have

$$\xi^*(u) = \int_\Omega \langle v^*, u \rangle d\mu \text{ for all } u \in L^p(\Omega, X),$$
$$\Rightarrow \xi^* = \xi_{v^*}$$
$$\Rightarrow \|\xi^*\|_* = \|v^*\|_{p'}.(see \text{ } Lemma \text{ } 4.172)$$

We conclude that $L^p(\Omega, X)^* = L^{p'}(\Omega, X^*)$. $\qquad\qquad\qquad\qquad\qquad\square$

Remark 4.174 In fact the converse is also true. Namely, $L^p(\Omega, X)^* = L^{p'}(\Omega, X^*)$ implies X^* has the RNP (see Diestel and Uhl [85], Theorem 1, p.98). Moreover, the result is also true if the measure space (Ω, Σ, μ) is σ-finite and in fact if $1 < p < \infty$, then even the σ-finiteness condition can be dropped (as in the scalar case).

Next we determine $L^1(\Omega, X)^*$ when X^* fails to satisfy the RNP.

Definition 4.175 Suppose $u^*, v^* : \Omega \to X^*$ two w^*-measurable functions (that is, for all $x \in X$, the $\mathbb{R}$-valued functions $z \to \langle u^*(z), x \rangle$ and $z \to \langle v^*(z), x \rangle$ are both

Borel measurable). We say that $u^* \sim v^*$ if

$$\langle u^*(z), x \rangle = \langle v^*(z), x \rangle$$

for all $z \in \Omega \setminus N_x$ with $\mu(N_x) = 0$ and all $x \in X$. This is an equivalence relation. Then by $L^\infty(\Omega, X^*_{w*})$ we denote the linear space of the equivalence classes for the relation $\sim$ of the functions $u^* : \Omega \to X^*$ which are w^*-measurable and there exists $c > 0$ such that

$$|\langle u^*(z), x \rangle| \leq c\|x\| \text{ for all } z \in \Omega \setminus N_x \text{ with } \mu(N_x) = 0, \text{ all } x \in X.$$

The infimum of all such $c > 0$ is denoted by $\|u^*\|_{L^\infty(\Omega, X^*_{w*})}$ and it is a norm on $L^\infty(\Omega, X^*_{w*})$.

Remark 4.176 If X is separable and $u^* \in L^\infty(\Omega, X^*_{w*})$, then $z \to \|u^*(z)\|_x$ belongs in $L^\infty(\Omega)$ and we have $\|u^*\|_{L^\infty(\Omega, X^*_{w*})} = \text{esssup}_\Omega \|u^*(\cdot)\|_*$.

The next theorem identifies $L^1(\Omega, X)^*$ for a general Banach space X. It is known as the "Dinculeanu-Foias Theorem" and can be found in Ionescu Tulcea [147], p.95.

Theorem 4.177 *If X is any Banach space, then $L^1(\Omega, X)^* = L^\infty(\Omega, X^*_{w*})$ and the duality brackets for this pair are given by*

$$\langle u^*, h \rangle = \int_\Omega \langle u^*(z), h(z) \rangle_X d\mu$$

for all $h \in L^1(\Omega, X)$, all $u^* \in L^\infty(\Omega, X^*_{w*})$.

Remark 4.178 Simple functions are dense in $L^\infty(\Omega, X)$ if and only if $dim\, X < \infty$.

A consequence of Theorem 4.173 (see also Remark 4.174) is the following result.

Proposition 4.179 *If X is reflexive and $1 < p < \infty$, then $L^p(\Omega, X)$ is reflexive too.*

Next we move beyond the Lebesgue-Bochner spaces and consider vector valued functions with integrable "weak derivative." So, let $I = (a, b)$ with $a, b \in \mathbb{R} \cup \{\pm\infty\}$ and X be a Banach space. We introduce vector valued distributions (see Definition 4.52). Recall that, if $\{u_n\}_{n \in \mathbb{N}} \subseteq C_c^\infty(I)$, then we say that $u_n \to u$ in $C_c^\infty(I)$ if $\{u_n, u\}_{n \in \mathbb{N}}$ are all supported in the same compact set and $u_n^{(k)} \to u^{(k)}$ uniformly for all $k \in \mathbb{N}_0$.

Definition 4.180 The space $\mathcal{D}'(1, X)$ of vector valued distributions is defined by $\mathcal{D}'(I, X) = \mathcal{L}(C_c^\infty(I), X)$.

Remark 4.181 For any function $u \in L^1_{loc}(I, X)$, we define a vector valued distribution $L_u(\vartheta) = \int_I u\vartheta dz$ for all $\vartheta \in C_c^\infty(\Omega)$. In fact for our analysis, this is the most important case.

Definition 4.182 Given $L \in \mathcal{D}'(I, X)$, the derivative L^1 of L is the distribution

$$L^1(\vartheta) = -L(\vartheta') \text{ for all } \vartheta \in C_c^\infty(I).$$

To be able to work with distributions generated by L^p-functions, we will need the following proposition which follows from the Lebesgue Differentiation Theorem for vector valued functions, which we recall.

Theorem 4.183 *If* $1 \le p < \infty$, $u \in L^p(\mathbb{R}, X)$ *and* $M_h u(t) = \dfrac{1}{h} \displaystyle\int_t^{t+h} u(s)ds$
$(h > 0)$, *then* $M_h u \in L^p(\mathbb{R}, X) \cap C(\mathbb{R}, X)$ *and* $M_h u \to u$ *in* $L^p(\mathbb{R}, X)$ *and a.e.*

Two important consequences of this Theorem.

Proposition 4.184 *If* $u \in L^1_{loc}(I, X)$ *and* $L_u = 0$, *then* $u(t) = 0$ *for a.a* $t \in I$.

Proof Let $S \subseteq I$ be bounded open subinterval. Then $\chi_S u \in L^1(S, X)$. Consider a sequence $\{\vartheta_n\}_{n\in\mathbb{N}} \subseteq C_c^\infty(S)$ such that $|\vartheta_n(t)| \le 1$ for all $t \in S$ and $\vartheta_n \to \chi_S$ a.e. Using the Dominated Convergence Theorem (see Theorem 4.153), we have

$$\int_S u dt = \lim_{n\to\infty} \int_I u\vartheta_n dt = \lim_{n\in\infty} L_u(\vartheta_n) = 0,$$
$$\Rightarrow M_h \chi_S u(t) = 0 \text{ for all } t \in S, \text{ all } h > 0 \text{ small,}$$
$$\Rightarrow \chi_S u = 0 \ a.e. \text{ (see Theorem 4.183),}$$
$$\Rightarrow u = 0. \qquad\qquad\qquad\qquad\qquad\qquad\qquad\qquad\qquad \square$$

Proposition 4.185 *If* $h \in L^1_{loc}(I, X)$, $t_0 \in I$ *and* $u(t) = \int_{t_0}^t h(s)ds$ *for all* $t \in I$, *then* $u \in C(I, X)$, $L'_u = L_h$ *and* $u(\cdot)$ *is differentiable a.e and* $u' = h$.

Proof Extending by zero h on all of $\mathbb{R}$, we may assume that $I = \mathbb{R}$.
Let $\vartheta \in C_c^\infty(I)$. We have

$$\langle L'_u, \vartheta \rangle = -\langle L_u, \vartheta' \rangle$$
$$= -\int_\mathbb{R} u(t) \lim_{\tau\to 0} \frac{\vartheta(t+\tau) - \vartheta(t)}{\tau} dt$$
$$= -\lim_{\tau\to 0} \int_\mathbb{R} u(t) \left[\frac{\vartheta(t+\tau) - \vartheta(t)}{\tau} \right] dt$$
$$= -\lim_{\tau\to 0} \left[\int_\mathbb{R} u(t) \frac{\vartheta(t+\tau)}{\tau} dt - \int_\mathbb{R} u(t) \frac{\vartheta(t)}{\tau} dt \right]$$
$$= -\lim_{\tau\to 0} \left[\int_\mathbb{R} \left(\frac{1}{\tau} \int_t^{t-\tau} h(s)ds \right) \vartheta(t)dt \right]$$
$$= -\lim_{\tau\to 0} \int_\mathbb{R} M_{-\tau} h(t) \vartheta(t) dt$$
$$= \langle L_h, \vartheta \rangle \text{(see Theorem 4.183)},$$

$$\Rightarrow L'_u = L_h.$$

Note that $M_\tau h(t) = \dfrac{u(t+\tau) - u(\tau)}{\tau}$. So the last assertions of the proposition follow from Theorem 4.183. □

Proposition 4.186 *If* $L \in \mathcal{D}'(I, X)$ *and* $L' = 0$, *then* L *is constant (that is,* $L(\psi) = (\int_I \psi\, dt)\widehat{x}$ *for all* $\psi \in C_c^\infty(I)$ *and some* $\widehat{x} \in X$.

Proof We consider $\vartheta \in C_c^\infty(I)$ such that $\int_I \vartheta(t)dt = 1$ and let $\widehat{x} = L(\vartheta)$. We have $supp\,\vartheta \subseteq [a, b]$ for some $a, b \in \mathbb{R}$ and let $t_0 < a$. Then for any $\psi \in C_c^\infty(I)$ we set

$$g(t) = \int_{t_0}^t \left[\psi(s) - \vartheta(s) \int_I \psi(\tau)d\tau \right] ds.$$

Evidently $\psi \in C_c^\infty(I)$ and we have

$$0 = -L'(g) = L(g') = L\left(\psi - \vartheta \int_I \psi \right) = L(\psi) - \widehat{x}\int_I \psi,$$

$$\Rightarrow L(\psi) = \left(\int_I \psi \right)\widehat{x} \text{ for all } \psi \in C_c^\infty(I).$$

□

Now we can define Sobolev spaces as in the scalar case.

Definition 4.187 Let $1 \leq p \leq \infty$. The vectorial Sobolev space $W^{1,p}(I, X)$ is defined by

$$W^{1,p}(I, X) = \{u \in L^p(I, X) : \text{ there exists } v \in L^p(I, X) \text{ such that } L'_u = L_v\}.$$

Then we write $v = u'$ and we endow $W^{1,p}(I, X)$ with the norm $\|u\|_{1,p} = \|u\|_p + \|u'\|_p$.

It is routine to check that:

Proposition 4.188 *If* $1 \leq p \leq \infty$, *then* $(W^{1,p}(I, X), \|\cdot\|_{1,p})$ *is a Banach space. If* X *is a Hilbert space, then* $W^{1,2}(I, X)$ *is a Hilbert space with inner product* $(u, h) = \int_I uh\,dz + \int_I u'h'\,dz$ *for all* $u, h \in W^{1,2}(I, X)$.

Proposition 4.189 *If* $1 < p < \infty$ *and* X *is reflexive, then so is* $W^{1,p}(I, X)$.

Proof From Proposition 4.179 we know that $L^p(I, X)$ is reflexive. Hence so is $L^p(I, X) \times L^p(I, X)$. Consider the map $\xi : W^{1,p}(I, X) \to L^p(I, X) \times L^p(I, X)$ defined by $\xi(u) = (u, u')$. This map is an isometry and $\xi(W^{1,p}(I, X))$ is closed in $L^p(I, X) \times L^p(I, X)$ and so it is reflexive (see Proposition 3.126). Since $\xi(\cdot)$ is an isometry, we conclude that $W^{1,p}(I, X)$ is reflexive. □

We can extend the notion of absolute continuity to vector valued functions by simply replacing absolute values with norms. (see Definition 4.120)

Definition 4.190 Let $I \subseteq \mathbb{R}$ be an interval and $u : I \to X$. We say that $u(\cdot)$ is "absolutely continuous" on I, if for every $\epsilon > 0$, there exists $\delta = \delta(\epsilon) > 0$ such that for every family $\{(a_k, b_k)\}_{k=1}^{n}$, $n \in \mathbb{N}$, of mutually disjoint intervals with $[a_k, b_k] \subseteq I$ for all $k \in \{1, .., n\}$ we have

$$\sum_{k=1}^{n} (b_k - a_k) < \delta \Rightarrow \sum_{k=1}^{n} \|u(b_k) - u(a_k)\| < \epsilon.$$

We say that $u(\cdot)$ is "locally absolutely continuous," if it is absolutely continuous on every compact subinterval $[a, b]$ of I.

Remark 4.191 We know that X has the RNP if and only if every locally absolutely continuous function is differentiable almost everywhere. In fact we can replace locally absolutely continuous functions by locally Lipschitz ones. Note that every Lipschitz function is absolutely continuous (see Diestel and Uhl [85], pp.217–218).

Proposition 4.192 *If $1 \le p \le \infty$ and $u \in W^{1,p}(I, X)$, then there exists $t_0 \in I$ such that*

$$u(t) = u(t_0) + \int_{t_0}^{t} u'(s)ds \ for \ a.a \ t \in I.$$

Proof Let $\widehat{t} \in I$ and set $v(t) = \int_{\widehat{t}}^{t} u'(s)ds$ and let $y(t) = u(t) - v(t)$ for all $t \in I$. By Proposition 4.185, we have $T_y' = 0$ and so $T_y = \widehat{x} \in X$ (see Proposition 4.186). Then we have $u(t) = \widehat{x} + \int_{\widehat{t}}^{t} u'(s)ds$ for $a.a \ t \in T$. Let $t_0 \in T$ be a point at which this last equality is true. Then $u(t) = u(t_0) + \int_{t_0}^{t} u'(s)ds$ for $a.a \ t \in I$. $\square$

Remark 4.193 So, according to this proposition $u \in W^{1,p}(I, X)$ is $a.e$ equal to a locally absolutely continuous function and it is almost everywhere differentiable in the classical sense with the derivative being $u' \in L^p(I, X)$.

Definition 4.194 For $1 \le p \le \infty$, let $AC^{1,p}([a, b], X)$ be the space of absolutely continuous functions $u : [a, b] \to X$ which are differentiable (in the classical sense) almost everywhere and the derivative belongs in $L^p(I, X)$.

According to Propositions 4.185 and 4.192, we have

Theorem 4.195 *If $1 \le p \le \infty$, then $W^{1,p}((a, b), X) = AC^{1,p}([a, b], X)$.*

Remark 4.196 In fact if $u \in L^p([a, b], X)$ and there exists $u' \in L^p([a, b], X)$ such that for any $x^* \in X^*$ the function $\vartheta_{x^*}(t) = \langle x^*, u(t) \rangle$ is absolutely continuous on $[a, b]$ and $\vartheta_{x^*}'(\cdot) = \langle x^*, u'(\cdot) \rangle$, then $u \in W^{1,p}((a, b), X) = AC^{1,p}([a, b], X)$. Also if X has the RNP, $1 \le p \le \infty$ and $u \in L^p([a, b], X)$, then $u \in W^{1,p}((a, b), X)$

if and only if $\|u(t) - u(s)\| \leq |\int_s^t h(t) dt|$ for all $t, s \in [a, b]$, some $h \in L^p[a, b]$. This and Remark 4.191 lead to the following result.

Proposition 4.197 *If X, Y are Banach spaces, with Y having the RNP, $1 \leq p \leq \infty$, $u \in W^{1,p}((a, b), X)$ and $\xi : X \to Y$ is Lipschitz continuous, then $\xi \circ u \in W^{1,p}((a, b), Y)$.*

Corollary 4.198 *If $1 \leq p \leq \infty$ and $u \in W^{1,p}((a, b), X)$, then $\|u(\cdot)\| \in W^{1,p}(a, b)$.*

Finally using Theorem 4.195, we see that we have the following embedding theorem.

Proposition 4.199 *If $1 \leq p \leq \infty$, then $W^{1,p}((a, b), X) \hookrightarrow L^\infty([a, b], X)$ continuously and every element of $W^{1,p}((a, b), X)$ has an absolutely continuous representative.*

The modern strategy in the study of parabolic equations is to introduce and make use of many function spaces. For this reason in the last part of this section we deal with more than one spaces.

We start with a lemma which will help us deal with the different spaces.

Lemma 4.200 *If X, Y are Banach spaces and $X \hookrightarrow Y$ continuously and densely, then*

(a) $Y^* \hookrightarrow X^*$ continuously.
(b) If X is reflexive, then $Y^* \hookrightarrow X^*$ densely.

Proof

(a) Since $X \hookrightarrow Y$ continuously, we can find $c_1 > 0$ such that

$$\|x\|_Y \leq c_1 \|x\|_X \quad for \ all \ x \in X. \tag{4.45}$$

If $y^* \in Y^*$, then

$$|\langle y^*, x \rangle_Y| \leq \|y^*\|_{Y^*} \|x\|_Y \leq c_1 \|y^*\|_{Y^*} \|x\|_X \ (\text{see } (4.45)). \tag{4.46}$$

Let $\widehat{y}^* = y^*|_X$. Then $\langle \widehat{y}^*, x \rangle_X = \langle y^*, x \rangle_Y$ for all $x \in X$ and so from (4.46) we have

$$\|\widehat{y}^*\|_{X^*} \leq c_1 \|y^*\|_{Y^*}. \tag{4.47}$$

Note that since X is embedded densely in Y, we have that

$$\widehat{y}^* = 0 \Rightarrow y^* = 0.$$

Therefore the map $j : Y^* \to X^*$ defined by $j(y^*) = \widehat{y}^*$ is injective and continuous (see (4.47)). We conclude that $Y^* \hookrightarrow X^*$ continuously.

(b) Suppose that the embedding $Y^* \hookrightarrow X^*$ (see (a)) is not dense. Then Y^* is a proper closed subspace of X^*. Then by Proposition 3.60 and since X is reflexive, we can find $u \in X \setminus \{0\}$ such that

$$\langle x^*, u \rangle = 0 \quad for\ all\ x^* \in Y^*,$$

$$\Rightarrow u = 0 \quad (\text{since } u \in X \subseteq Y)$$

a contradiction.   □

Now we can introduce the notion of "evolution triple" which is a major tool in the study of evolution equations.

Definition 4.201 By an "evolution triple" (or "Gelfand triple") we mean a triple of spaces (X, H, X^*) such that

(a) X is a separable reflexive Banach space with X^* its dual.
(b) H is a separable Hilbert space.
(c) $X \hookrightarrow H$ continuously and densely.

Remark 4.202 For an evolution triple (X, H, X^*), using Theorem 3.181 we identify H with its dual. Then as we explained in Sect. 3.10 (see the remarks for Sect. 3.8), when X is Hilbert too, then we cannot identify X with its dual. If $(\cdot, \cdot)_H$ denotes the inner product of H and $\langle \cdot, \cdot \rangle_X$ denotes the duality brackets for the pair (X, X^*), then we have $\langle \cdot, \cdot \rangle_X|_{H \times X} = (\cdot, \cdot)_H|_{H \times X}$. Note that Lemma 4.200 implies that $H \hookrightarrow X^*$ continuously and densely. In the sequel by $| \cdot |$ we denote the norm of the Hilbert space H, by $\| \cdot \|$ the norm of X and by $\| \cdot \|_*$ the norm of X^*. On account of Definition 4.201, we can find $\widehat{c_1}, \widehat{c_2} > 0$ such that

$$|u| \leq \widehat{c_1} \|u\| \quad for\ all\ u \in X,$$

$$\|v\|_* \leq \widehat{c_2} |v| \quad for\ all\ v \in H.$$

Using evolution triples, we can define the following Sobolev-like space which is important in the study of evolution equations.

Definition 4.203 Let (X, H, X^*) be an evolution triple and $1 < p < \infty$. We define

$$W_p(0, b) = \{u \in L^p(T, X) : u' \in L^{p'}(T, X^*)\}$$

with $T = [0, b]$ and $p' = \frac{p}{p-1}$. In this definition, u' is understood as a weak derivative (see Definition 4.182) and by Theorem 4.195, if we view $u \in W_p(0, b)$ as an X^*-valued function, the derivative is at almost all $t \in (0, b)$ a classical derivative and the function $u : [0, b] \to X^*$ is absolutely continuous. We furnish $W_p(0, b)$

with the norm

$$\|u\|_{W_p(0,b)} = \|u\|_{L^p([0,b],X)} + \|u'\|_{L^{p'}([0,b],X^*)}.$$

Clearly normed this way $W_p(0,b)$ is a separable, reflexive Banach space.

Theorem 4.204 *If (X, H, X^*) is an evolution triple and $1 < p < \infty$, then $W_p(0,b) \hookrightarrow C([0,b], H)$ continuously and densely.*

Proof The space of X-valued polynomials is dense in $W_p(0,b)$ and so it follows that $C^1([0,b], X) \hookrightarrow W_p(0,b)$ continuously and densely.

Let $u, v \in C^1([0,b], X)$. We have

$$\frac{d}{dt}(u(t), v(t))_H = (u'(t), v(t))_H + (u(t), v'(t))_H$$

$$= \langle u'(t), v(t) \rangle_X + \langle u(t), v'(t) \rangle_X \quad \text{(see Remark 4.202)}.$$

Thus, for all $0 \le s \le t \le b$ we have

$$\Rightarrow (u(t), v(t))_H - (u(s), v(s))_H = \int_s^t [\langle u'(\tau), v(\tau) \rangle_X + \langle u(\tau), v'(\tau) \rangle_X] d\tau.$$
(4.48)

We choose $\vartheta \in C^1(\mathbb{R})$, with $\vartheta(0) = 0$, $\vartheta(1) = 1$, $|\vartheta| + |\vartheta'| \le 1$ on $\mathbb{R}$ and set $v = \vartheta u$. Then $v' = \vartheta' u + \vartheta u'$ and using Hölder's inequality from (4.48) we have

$$|u(t)|^2 \le c_1 \|u\|^2_{W_p(0,b)} \quad \text{for all } t \in [0,b], \text{ and some } c_1 > 0,$$

$$\Rightarrow \|u\|_{C(T,X)} \le c_1^{\frac{1}{2}} \|u\|_{W_p(0,b)},$$

$$\Rightarrow (C^1([0,b], X), \|\cdot\|_{W_p(0,b)}) \hookrightarrow C(T, H) \text{ continuously and densely},$$

$$\Rightarrow W_p(0,b) \hookrightarrow C(T, H) \text{ continuously and densely}. \qquad \square$$

From (4.48) and the density of $C^1([0,b], X)$ in $W_p(0,b)$, we infer the following "integration by parts" formula for the space $W_p(0,b)$.

Theorem 4.205 *If $1 < p < \infty$ and $u, v \in W_p(0,b)$, then*

$$(u(t), v(t))_H - (u(s), v(s))_H = \int_s^t [\langle u'(\tau), v(\tau) \rangle_X + \langle u(\tau), v'(\tau) \rangle_X] d\tau$$

$$for \ all \ 0 \le s \le t \le b.$$
(46)

The embedding of $W_p(0,b)$ in $C([0,b], H)$ is continuous but in general it is not compact. However, as we will see below in a more general setting $W_p(0,b) \hookrightarrow L^p([0,b], H)$ compactly.

First we prove an interpolation lemma, known in the literature as "Ehrling's inequality."

Lemma 4.206 *If X, Y, V are Banach spaces, $X \hookrightarrow Y$ compactly and $Y \hookrightarrow V$ continuously, then for every $\varepsilon > 0$, there exists $c(\varepsilon) > 0$ such that*

$$\|x\|_Y \leq \varepsilon \|x\|_X + c(\varepsilon) \|x\|_V \quad \text{for all } x \in X.$$

Proof We argue by contradiction. So, suppose that the assertion of the lemma is not true. Then there exists $\varepsilon > 0$ and a sequence $\{x_n\}_{n \in \mathbb{N}} \subseteq X$,

$$\|x_n\|_Y > \varepsilon \|x_n\|_X + n \|x_n\|_V \quad \text{for all } n \in \mathbb{N}. \tag{4.49}$$

Let $y_n = \frac{x_n}{\|x_n\|_X}$, $n \in \mathbb{N}$. Then from (4.49) we have

$$\|y_n\|_Y > \varepsilon + n \|y_n\|_V. \tag{4.50}$$

Since $\|y_n\|_X = 1$ for all $n \in \mathbb{N}$ and $X \hookrightarrow Y$ compactly, it follows that $\{y_n\}_{n \in \mathbb{N}} \subseteq Y$ is relatively compact. So, by passing to a subsequence if necessary and using the fact that $Y \hookrightarrow V$ continuously, we can say that

$$y_n \to y \text{ in } Y \text{ and in } V \text{ as } n \to \infty.$$

From (4.50) it follows that $\|y\|_V = 0$ and $\|y\|_Y \geq \varepsilon$, a contradiction. $\qquad \square$

Now let X, Y, V be Banach spaces with X, V reflexive and assume that $X \hookrightarrow Y$ compactly, $Y \hookrightarrow V$ continuously. For $1 < p, q < \infty$ we introduce the following extension of the space $W_p(0, b)$,

$$W_{pq}(0, b) = \{u \in L^p([0, b], X) : u' \in L^q([0, b], V)\}.$$

We equip $W_{pq}(0, b)$ with the norm

$$\|u\|_{W_{pq}(0,b)} = \|u\|_{L^p([0,b],X)} + \|u'\|_{L^q([0,b],V)} \text{ for all } u \in W_{pq}(0, b).$$

Then $W_{pq}(0, b)$ becomes a separable, reflexive Banach space.

Theorem 4.207 $W_{pq}(0, b) \hookrightarrow L^p([0, b], Y)$ *compactly and densely.*

Proof Clearly the embedding is dense. So, we need to show that it is also compact. Let $\{u_n\}_{n \in \mathbb{N}} \subseteq W_{pq}(0, b)$ be bounded. The reflexivity of $W_{pq}(0, b)$ implies that at least for a subsequence we have

$$u_n \overset{w}{\to} u \text{ in } W_{pq}(0, b),$$

$$\Rightarrow u_n \xrightarrow{w} u \text{ in } L^p([0, b], X) \text{ and } u_n' \xrightarrow{w} u' \text{ in } L^q([0, b], V).$$

(note that $W_{pq}(0, b) \hookrightarrow C([0, b], V)$ continuously, see Theorem 4.195).

We may assume that $u = 0$. We show that $u_n(t) \to 0$ in V for all $t \in [0, b]$. We do this for $t = 0$ and the proof is similar for the other t. We have

$$u_n(0) = u_n(t) - \int_0^t u'(\tau) d\tau \text{ for all } t \in [0, b], \text{ and all } n \in \mathbb{N},$$

$$\Rightarrow u_n(0) = \frac{1}{s} \int_0^s u_n(t) dt - \frac{1}{s} \int_0^s \int_0^t u_n'(\tau) d\tau dt = a_n + c_n \text{ for all } n \in \mathbb{N}$$

(4.51)

with $a_n = \frac{1}{s} \int_0^s u_n(t) dt$ and $c_n = -\frac{1}{s} \int_0^s \int_0^t u_n'(\tau) d\tau dt = -\frac{1}{s} \int_0^s (s - t) u_n'(t) dt$. Given $\varepsilon > 0$, we can find $s > 0$ small such that

$$\|c_n\|_V \le \int_0^s \|u_n'(t)\|_V dt \le \frac{\varepsilon}{2}.$$

(4.52)

We have $a_n \xrightarrow{w} 0$ in X, , thus $a_n \to 0$ in V (since $X \hookrightarrow V$ is compactly). Therefore,

$$\|a_n\|_V \le \frac{\varepsilon}{2} \text{ for all } n \ge n_0.$$

(4.53)

From (4.51)–(4.53) it follows that

$$\|u_n(0)\|_V \le \varepsilon \text{ for all } n \in n_0,$$

$$\Rightarrow u_n(0) \to 0 \text{ in } V,$$

$$\Rightarrow u_n(t) \to 0 \text{ in } V \text{ for all } t \in [0, b].$$

(4.54)

Since $W_{pq}(0, b) \hookrightarrow C([0, b], V)$ continuously, we have that

$$\{u_n\}_{n \in \mathbb{N}} \subseteq C([0, b], V) \text{ is bounded.}$$

(4.55)

Then from (4.54)–(4.55) and Theorem 4.153 (the Dominated Convergence Theorem), we have

$$u_n \to 0 \quad \text{in } L^p([0, b], V) \text{ as } n \to \infty.$$

Using Lemma 4.206, for any $\varepsilon > 0$, we can find $c(\varepsilon) > 0$ such that

$$\|u_n\|_{L^p([0,b],Y)} \le \varepsilon \|u_n\|_{L^p([0,b],X)} + c(\varepsilon) \|u_n\|_{L^p([0,b],V)}$$

$$\leq \varepsilon \widehat{c} + c(\varepsilon)\|u_n\|_{L^p([0,b],V)} \text{ for some } \widehat{c} > 0, \text{ all } n \in \mathbb{N}$$

$$\Rightarrow u_n \to 0 \text{ in } L^p([0,b],Y) \text{ as } n \to \infty,$$

which proves that $W_{pq}(0,b) \hookrightarrow L^p([0,b],Y)$ compactly.   □

A final result on the subject is the following proposition.

Proposition 4.208 *If X, Y are Banach spaces with X reflexive, $X \hookrightarrow Y$ continuously and $u \in L^\infty([0,b],X) \cap C([0,b],Y_w)$, then $u \in C([0,b],X_w)$ (by X_w (respectively, Y_w), we denote the Banach space X (respectively, Y) furnished with the weak topology).*

Proof Replacing Y with $\overline{X}^{\|\cdot\|_X}$ if necessary, we may assume that the embedding $X \hookrightarrow Y$ is also dense. Then by Lemma 4.200 $Y^* \hookrightarrow X^*$ continuously and densely. Let $t_n \to t$ in $[0,b]$. Since $u \in C([0,b],Y_w)$, for every $y^* \in Y^*$ we have

$$\langle y^*, u(t_n)\rangle_Y \to \langle y^*, u(t)\rangle_Y$$

(by $\langle \cdot, \cdot \rangle_Y$ we denote the duality brackets for the pair (Y, Y^*)).

Extend u by zero outside $[0,b]$ and denote this extension by $\widehat{u}$. We regularize $\widehat{u}$ (by mollification) and so we can have a sequence $\{u_n\}_{n\in\mathbb{N}} \subseteq C^1([0,b],X)$ such that

$$\|u_n(t)\|_X \leq \|u\|_{L^\infty([0,b],X)} \text{ for all } t \in T, \text{ and all } n \in \mathbb{N},$$

$$\langle y^*, u_n(t)\rangle_Y \to \langle y^*, u(t)\rangle_Y \text{ as } n \to \infty, \text{ for all } y^* \in Y^*.$$

Denoting by $\langle \cdot, \cdot \rangle_X$ the duality brackets for (X, X^*), we have

$$|\langle y^*, u_n(t)\rangle_Y| = |\langle y^*, u_n(t)\rangle_X| \quad (\text{recall } Y^* \hookrightarrow X^*),$$

$$\leq \|y^*\|_{X^*}\|u_n(t)\|_X$$

$$\leq \|y^*\|_{X^*}\|u\|_{L^\infty([0,b],X)} \text{ for all } n \in \mathbb{N}.$$

Passing to the limit as $n \to \infty$, we obtain

$$|\langle y^*, u(t)\rangle_Y| = |\langle y^*, u(t)\rangle_X| \leq \|y^*\|_{X^*}\|u\|_{L^\infty([0,b],X)}.$$

The density of Y^* in X^* implies that

$$u(t) \in X \text{ for all } t \in T, \|u(t)\|_X \leq \|u\|_{L^\infty([0,b],X)}. \tag{4.56}$$

Now let $x^* \in X^*$. Then we can find $\{y_k^*\}_{k\in\mathbb{N}} \subseteq Y^*$ such that $y_k^* \to x^*$ in X^* as $k \to \infty$. We have

$$\langle y_k^*, u(t_n) \rangle_X \to \langle y_k^*, u(t) \rangle_X \text{ as } n \to \infty, \text{ for all } k \in \mathbb{N},$$

$$\langle y_k^*, u(t) \rangle_X \to \langle x^*, u(t) \rangle_X \text{ as } k \to \infty.$$

Then we can find a sequence $n \to k(n)$ with $k(n) \to \infty$ as $n \to \infty$ such that

$$\langle y_{k(n)}^*, u(t_n) \rangle_X \to \langle x^*, u(t) \rangle_X \text{ as } n \to \infty. \tag{4.57}$$

We have

$$|\langle x^*, u(t_n) \rangle_X - \langle x^*, u(t) \rangle_X|$$

$$\leq |\langle x^*, u(t_n) \rangle_X - \langle y_{k(n)}^*, u(t_n) \rangle_X| + |\langle y_{k(n)}^*, u(t_n) \rangle_X - \langle x^*, u(t) \rangle_X|$$

$$\leq \|x^* - y_{k(n)}^*\|_{X^*} \|u(t_n)\|_X + |\langle y_{k(n)}^*, u(t_n) \rangle_X - \langle x^*, u(t) \rangle_X|$$

$$\Rightarrow \langle x^*, u(t_n) \rangle_X \to \langle x^*, u(t) \rangle_X \text{ as } n \to \infty \quad (\text{see}(4.56 - 57)),$$

$$\Rightarrow u \in C([0, b], X_w).$$

$\square$

4.5 Spaces of Measures

In this section we present the main properties of the spaces of measures.

So, we consider a measurable space (Ω, Σ).

Definition 4.209

(a) By $ba(\Sigma)$ we denote the space of all $\mathbb{R}$-valued, bounded, additive set functions.
(b) By $ca(\Sigma)$ we denote the space of all $\mathbb{R}$-valued signed measures on Σ. Both spaces are furnished with the total variation norm.
(c) By $ba(\Sigma)_+$ (respectively, $ca(\Sigma)_+$) we denote the subset of all nonnegative set functions $\mu \in ba(\Sigma)$ (respectively, of all measures on Σ).

Remark 4.210 On account of Proposition 2.104, we know that every $\mu \in ca(\Sigma)$ is bounded, hence $ca(\Sigma) \subseteq ba(\Sigma)$. On the other hand, an additive set function $\mu : \Sigma \to \mathbb{R}$ need not be bounded. Recall that for every additive set function $\mu : \Sigma \to \mathbb{R}$, the total variation norm is defined by

$$\|\mu\| = |\mu|(\Omega) = \sup\{\sum_{k=1}^n |\mu(A_k)| : \{A_k\}_{k=1}^n \text{ is a } \Sigma - \text{partition of } \Omega\}.$$

We can define another norm on these spaces, namely the norm of uniform convergence defined by

$$\|\mu\|_\infty = \sup\{|\mu(A)| : A \in \Sigma\}.$$

Next we show that these are equivalent norms.

Proposition 4.211 $\|\cdot\|$ *and* $\|\cdot\|_\infty$ *are norms on* $ba(\Sigma)$ *(hence on* $ca(\Sigma)$ *too) which are equivalent.*

Proof On account of Proposition 2.115 (an inspection of the proof reveals that the result is also true for additive set functions), we have

$$\|\mu\|_\infty \leq \|\mu\| \leq 2\|\mu\|_\infty \quad \text{for all } \mu \in ba(\Sigma). \tag{4.58}$$

So, if we show that $\|\cdot\|$, $\|\cdot\|_\infty$ are norms on $ba(\Sigma)$, then from (4.58) we infer that they are equivalent norms.

We have $\|\mu\| = 0 \iff \|\mu\|_\infty = 0 \iff |\mu(A)| = 0$ for all $A \in \Sigma \iff \mu = 0$. Also for every $t \in \mathbb{R}$, we have

$$\|t\mu\| = |t\mu|(\Omega) = |t|\,|\mu|(\Omega) = |t|\|\mu\|,$$

$$\|t\mu\|_\infty = \sup\{|t\mu(A)| : A \in \Sigma\} = |t|\,\|\mu\|_\infty.$$

Therefore it remains to prove the triangle inequality.
Given $\varepsilon > 0$, we can find $\{A_k\}_{k=1}^n \subseteq \Sigma$ a partition of Ω such that

$$\|\mu_1 + \mu_2\| - \varepsilon \leq \sum_{k=1}^n |\mu_1(A_k) + \mu_2(A_k)| \leq \sum_{k=1}^n |\mu_1(A_k)| + \sum_{k=1}^n |\mu_2(A_k)|$$

$$\leq \|\mu_1\| + \|\mu_2\|.$$

Since $\varepsilon > 0$ is arbitrary, we let $\varepsilon \to 0^+$ to conclude that

$$\|\mu_1 + \mu_2\| \leq \|\mu_1\| + \|\mu_2\| \text{ for all } \mu_1, \mu_2 \in ba(\Sigma).$$

Similarly for $\|\cdot\|_\infty$. Given $\varepsilon > 0$, we can find $A_\varepsilon \in \Sigma$ such that

$$\|\mu_1 + \mu_2\|_\infty - \varepsilon \leq |(\mu_1 + \mu_2)(A)| \leq |\mu_1(A)| + |\mu_2(A)| \leq \|\mu_1\|_\infty + \|\mu_2\|_\infty.$$

As before let $\varepsilon \to 0^+$ to conclude that

$$\|\mu_1 + \mu_2\|_\infty \leq \|\mu_1\|_\infty + \|\mu_2\|_\infty.$$

Therefore we conclude that $\|\cdot\|$, $\|\cdot\|_\infty$ are norms on $ba(\Sigma)$ and they are equivalent. $\square$

Proposition 4.212 *The space* $ba(\Sigma)$ *furnished with one of the equivalent norms* $\|\cdot\|$ *or* $\|\cdot\|_\infty$ *is a Banach space and* $ca(\Sigma)$ *is a closed subspace (hence a Banach space too).*

Proof From (4.58) it is clear that

$$\mu_n \xrightarrow{\|\cdot\|} \mu \iff \mu_n \xrightarrow{\|\cdot\|_\infty} \mu \iff \sup_{A\in\Sigma} |\mu_n(A) - \mu(A)| \to 0.$$

So, if $\{\mu_n\}_{n\in\mathbb{N}}$ is $\|\cdot\|$-Cauchy (equivalently $\|\cdot\|_\infty$-Cauchy), then $\{\mu_n(A)\}_{n\in\mathbb{N}}$ is Cauchy in X uniformly in $A \in \Sigma$. So, there exists $\mu : \Sigma \to \mathbb{R}$ such that

$$\sup_{A\in\Sigma} |\mu_n(A) - \mu(A)| \to 0 \quad \text{as } n \to \infty,$$

$$\Rightarrow \mu(\cdot) \text{ is additive and bounded, that is, } \mu \in ba(\Sigma).$$

Therefore $(ba(\Sigma), \|\cdot\|)$ and $(ba(\Sigma), \|\cdot\|_\infty)$ are Banach spaces.

Next we show that $ca(\Sigma)$ is a closed subspace of $ba(\Sigma)$. To this end, we use the $\|\cdot\|$-norm and assume $\mu_n \xrightarrow{\|\cdot\|} \mu$ with $\{\mu_n\}_{n\in\mathbb{N}} \subseteq ca(\Sigma)$. We have $\mu \in ba(\Sigma)$. We need to show that $\mu \in ca(\Sigma)$. So, let $\{A_n\}_{n\in\mathbb{N}} \subseteq \Sigma$ be pairwise disjoint. Let $A = \bigcup_{n\in\mathbb{N}} A_n \in \Sigma$. We have that $\sup_{C\in\Sigma} |\mu_n(C) - \mu(C)| \to 0$ and so given $\varepsilon > 0$ we can find $n_0 \in \mathbb{N}$ such that

$$|\mu_n(C) - \mu(C)| \leq \frac{\varepsilon}{2} \quad \text{for all } n \geq n_0, \ \text{all } C \in \Sigma. \tag{4.59}$$

Since $\mu_{n_0} \in ca(\Sigma)$, we can find $n_1 > n_0$ such that

$$\Big|\mu_{n_0}(A) - \sum_{k=1}^{n} \mu_{n_0}(A_k)\Big| \leq \frac{\varepsilon}{2} \quad \text{for all } n \geq n_1. \tag{4.60}$$

Then for all $n \in \mathbb{N}, n \geq n_1$ we have

$$\Big|\mu(A) - \sum_{k=1}^{n} \mu(A_k)\Big| = \Big|\mu\Big(A \setminus \bigcup_{k=1}^{n} A_k\Big)\Big|$$

$$\leq \Big|\mu\Big(A \setminus \bigcup_{k=1}^{n} A_k\Big) - \mu_{n_0}\Big(A \setminus \bigcup_{k=1}^{n} A_k\Big)\Big| + \Big|\mu_{n_0}\Big(A \setminus \bigcup_{k=1}^{n} A_k\Big)\Big|$$

$$= \Big|\mu\Big(A \setminus \bigcup_{k=1}^{n} A_k\Big) - \mu_{n_0}\Big(A \setminus \bigcup_{k=1}^{n} A_k\Big)\Big| + \Big|\mu_{n_0}(A) - \sum_{k=1}^{n} \mu_{n_0}(A_k)\Big|$$

$$\leq \frac{\varepsilon}{2} + \frac{\varepsilon}{2} = \varepsilon \quad \text{for all } n \geq n_1 \text{ (see (4.59-60))},$$

$$\Rightarrow \mu \in ca(\Sigma) \text{ (since } \varepsilon > 0 \text{ is arbitrary).} \qquad \square$$

Proposition 4.213 *If* $\lambda \in ca(\Sigma)_+$, $f \in L^1(\Omega, \lambda)$ *and* $\mu(A) = \int_A f d\lambda$ *for* $A \in \Sigma$, *then* $\mu \in ca(\Sigma)$ *and* $\|\mu\| = |\mu|(\Omega) = \int_\Omega |f| d\lambda$.

Proof From Proposition 2.90 applied to $f^+, f^- \in L^1(\Omega, \lambda)$, we have that $\mu \in ca(\Sigma)$ and $\|\mu\| = |\mu|(\Omega) = \mu^+(\Omega) + \mu^-(\Omega) = \int_\Omega f^+ d\lambda + \int_\Omega f^- d\lambda = \int_\Omega |f| d\lambda$. $\qquad \square$

If $\lambda \in ca(\Sigma)$ (that is, it is in general a signed measure (and not a measure)), then $L^1(\Omega, \lambda) = L^1(\Omega, \lambda^+) \cap L^1(\Omega, \lambda^-))$ and

$$\int_A f d\lambda = \int_A f d\lambda^+ - \int_A f d\lambda^- \quad \text{for all } A \in \Sigma, \text{ all } f \in L^1(\Omega, \lambda).$$

Proposition 4.214 *If* $\lambda \in ca(\Sigma)$, *then*

(a) $L^1(\Omega, \lambda) = L^1(\Omega, |\lambda|)$ *and* $|\int_A f d\lambda| \leq \int_A |f| d|\lambda|$ *for all* $A \in \sum$, *all* $f \in L^1(\Omega, \lambda)$.

(b) $|\lambda|(A) = \sup \left\{ |\int_A f d\lambda| : f \in L^1(\Omega, \lambda), |f(x)| \leq 1 \ \lambda - a.e. \right\}$.

(c) $\|f\|_1 = \int_\Omega |f| d|\lambda|$ *is a norm on* $L^1(\Omega, \lambda)$ *and* $L^1(\Omega, \lambda)$ *furnished with this norm is a Banach space.*

Recall the metric $\widehat{d}$ on $\widehat{\Sigma}_\mu$ from Sect. 4.1. From Proposition 4.13 we know that $(\widehat{\Sigma}_\mu, \widehat{d})$ is a complete metric space.

Proposition 4.215 *If* $\mu \in ca(\Sigma)_+$ *and* $\lambda \in ca(\Sigma)$, *then* $\lambda \ll \mu$ *if and only if* $\lambda : (\widehat{\Sigma}_\mu, \widehat{d}) \to \mathbb{R}$ *is continuous.*

Proof $\Rightarrow$ According to Proposition 2.110, given $\varepsilon > 0$, we can find $\delta > 0$ such that

$$A \in \Sigma_\mu, \ \mu(A) \leq \delta \Rightarrow |\lambda|(A) \leq \varepsilon. \tag{4.61}$$

Then let $A, B \in \widehat{\Sigma}_\mu$ with $\widehat{d}(A, B) \leq \delta$. We have $\mu(A \triangle B) \leq \delta$ and so by (4.61) we have $|\lambda|(A \triangle B) \leq \varepsilon$. Hence we have

$$|\lambda(A) - \lambda(B)| = |\lambda(A) - \lambda(A \cap B) + \lambda(A \cap B) - \lambda(B)|$$
$$\leq |\lambda(A \setminus B) - \lambda(B \setminus A)|$$
$$\leq |\lambda(A \setminus B)| + |\lambda(B \setminus A)|$$
$$\leq |\lambda|(A \triangle B) \leq \varepsilon,$$
$$\Rightarrow \lambda \text{ is } \widehat{d}\text{-continuous.}$$

$\Leftarrow$ The continuity of λ implies that given $\varepsilon > 0$, we can find $\delta \in (0, 1)$ such that for every $A \in \widehat{\Sigma}_\mu$ with $\widehat{d}(A, \varnothing) = \mu(A) \leq \delta$ implies $|\lambda(A)| \leq \varepsilon$. Let $A \in \Sigma$ with $\mu(A) = 0$. Then $\widehat{d}(A, \varnothing) \leq \delta$ and so $|\lambda(A)| \leq \varepsilon$. But $\varepsilon > 0$ is arbitrary. Let $\varepsilon \to 0^+$ to conclude $|\lambda(A)| = 0$. Therefore $\lambda \ll \mu$. $\square$

Definition 4.216 Let $\mu \in ca(\Sigma)_+$ and $\lambda \in ca(\Sigma)$. We say that λ is "μ-continuous" if $\lambda : (\widehat{\Sigma}_\mu, \widehat{d}) \to \mathbb{R}$ is continuous. By $ca_\mu(\Sigma)$ we denote the subset of all μ-continuous signed measures $\lambda \in ca(\Sigma)$. A subset $C \subseteq ca(\Sigma)$ $\widehat{d}$-equicontinuous at $\varnothing$ (hence on Σ, see Theorem 4.217 below), is called "μ-equicontinuous."

The next theorem describes equicontinuous subsets of $ca(\Sigma)$. It can be found in Diestel [84], Theorem 9, p.87.

Theorem 4.217 If $C \subseteq ca(\Sigma)$, then the following properties are equivalent

(a) C is $\widehat{d}$-equicontinuous at some $A \in \Sigma$.
(b) C is $\widehat{d}$-equicontinuous at $\varnothing$.
(c) C is uniformly $\widehat{d}$-equicontinuous on Σ.
   Moreover, each one of these statements implies:
(d) C is uniformly countably additive.

Using this Theorem, we can restate the Vitali-Hahn-Saks Theorem (see Theorem 4.24) and the Nikodym Theorem (see Theorem 4.25) as follows.

Theorem 4.218 If $\mu \in ca(\Sigma)_+, \{\lambda_n\}_{n \in \mathbb{N}} \subseteq ca_\mu(\Sigma)$ and $\lim_{n \to \infty} \lambda_n(A) = \lambda(A)$ exists for all $A \in \Sigma$, then

(a) $\{\lambda_n\}_{n \in \mathbb{N}}$ is μ-equicontinuous.
(b) $\lambda \in ca(\Sigma)$.
(c) $\lambda \in ca_\mu(\Sigma)$.

Theorem 4.219 If $\{\lambda_n\}_{n \in \mathbb{N}} \in ca(\Sigma)$ and for every $A \in \Sigma$, $\lim_{n \to \infty} \lambda_n(A) = \lambda(A) \in \mathbb{R}$ exists, then

(a) $\lambda \in ca(\Sigma)$.
(b) $\{\lambda_n\}_{n \in \mathbb{N}}$ is bounded in the Banach space $ca(\Sigma)$.
(c) $\{\lambda_n\}_{n \in \mathbb{N}}$ is uniformly countably additive.

Continuing with the Banach space $ca(\Sigma)$ we have the following characterization of weak convergence in $ca(\Sigma)$.

Proposition 4.220 If $\{\lambda_n, \lambda\}_{n \in \mathbb{N}} \subseteq ca(\Sigma)$, then $\lambda_n \xrightarrow{w} \lambda$ in $ca(\Sigma)$ if and only if for all $A \in \Sigma$, we have $\lambda_n(A) \to \lambda(A)$.

Proof $\Rightarrow$ For $A \in \Sigma$, let $x_A^* : ca(\Sigma) \to \mathbb{R}$ be defined by $x_A^*(\lambda) = \lambda(A)$. Evidently x_A^* is linear and $|x_A^*(\lambda)| = |\lambda(A)| \leq |\lambda|(A) \leq \|\lambda\|$. Therefore $x_A^* \in ca(\Sigma)^*$. Hence $\langle x_A^*, \lambda_n \rangle \to \langle x_A^*, \lambda \rangle \Rightarrow \lambda_n(A) \to \lambda(A)$.

$\Leftarrow$ By Theorem 4.219, $\{\lambda_n\}_{n \in \mathbb{N}} \subseteq ca(\Sigma)$ is bounded. Let $\mu : \Sigma \to \mathbb{R}_+$ be defined by

$$\mu(A) = \sum_{n \in \mathbb{N}} \frac{1}{2^n} |\lambda_n|(A) \text{ for all } A \in \Sigma.$$

It is routine to check that $\lambda \in ca(\Sigma)_+$. Also $\lambda_n \ll \mu$ for every $n \in \mathbb{N}$. So, by the Radon–Nikodym Theorem (see Theorem 2.113), we can find $u_n \in L^1(\Omega, \mu)$ such that $\lambda_n(A) = \int_A u_n d\mu$ for all $A \in \Sigma$, all $n \in \mathbb{N}$.

Also $\lambda \ll \mu$ (see Theorem 4.218) and so we can find $u \in L^1(\Omega, \lambda)$ such that $\lambda(A) = \int_A u d\mu$ for all $A \in \Sigma$.

We have

$$\int_A u_n d\mu \to \int_A u d\mu \text{ for all } A \in \Sigma,$$

$$\Rightarrow u_n \xrightarrow{w} u \quad L^1(\Omega, \mu) \text{ (see Proposition 4.22).}$$

But $L^1(\Omega, \mu)$ is isometrically isomorphic to $ca_\mu(\Sigma)$. Therefore

$$\lambda_n \xrightarrow{w} \lambda \text{ in } ca(\Sigma). \qquad \square$$

The next theorem characterizes w-compact subsets in $ca(\Sigma)$ and can be found in Diestel [84], Theorem 13, p.92.

Theorem 4.221 *If $C \subseteq ca(\Sigma)$, then the following statements are equivalent:*

(a) C is relatively w-compact.
(b) C is bounded and uniformly countably additive.
(c) C is bounded and there exists $\mu \in ca(\Sigma)_+$ such that C is μ-equicontinuous.

Now we introduce topological structure on the set Ω. So, we assume that Ω is a locally compact space. We introduce the following vector spaces of continuous functions.

- $C_c(\Omega)$ =space of continuous functions on Ω with compact support.
- $C_0(\Omega)$ =space of continuous functions which vanish at infinity, that is, $u \in C_0(X)$ if and only if given $\varepsilon > 0$, we can find $K_\varepsilon \subseteq \Omega$ compact such that $|u(z)| < \varepsilon$ for all $z \in \Omega \setminus K_\varepsilon$.
- $C_b(\Omega)$ =space of bounded continuous functions on Ω. (see Definition 1.64(b)).

Evidently we have

$$C_c(\Omega) \subseteq C_0(\Omega) \subseteq C_b(\Omega).$$

If Ω is compact, then these spaces coincide. If Ω is not compact, then the above inclusions are strict.

On $C_b(\Omega)$ we introduce the supremum norm defined by

$$\|u\|_{C_b(\Omega)} = \sup[|u(z)| : z \in \Omega]. \tag{4.62}$$

By restriction this norm is also inherited by $C_0(\Omega), C_c(\Omega)$. The restriction of the norm on $C_0(\Omega)$ is denoted by $\|\cdot\|_{C_0(\Omega)}$ and we have

$$\|u\|_{C_0(\Omega)} = \max[|u(z)| : z \in \Omega],$$

that is, in the case of $C_0(\Omega)$ the supremum in (4.62) is actually attained.

Proposition 4.222 $(C_b(\Omega), \|\cdot\|_{C_b(\Omega)})$ *is a Banach space.*

Proof Suppose $\{u_n\} \subseteq C_b(\Omega)$ is Cauchy. Then for every $z \in \Omega$ $\{u_n(z)\}_{n \in \mathbb{N}} \subseteq \mathbb{R}\}$ is Cauchy and so $u(z) = \lim\limits_{n \to \infty} u_n(z)$ exists. We claim that $u \in C_b(\Omega)$ and

$$\|u_n - u\|_{C_b(\Omega)} \to 0. \tag{4.63}$$

Recalling that the uniform limit of continuous functions is continuous, it suffices to show (4.63). Given $\varepsilon > 0$, we can find $n_0 \in \mathbb{N}$ such that

$$\|u_n - u_m\|_{C_b(\Omega)} < \frac{\varepsilon}{2} \text{ for all } m, n \geq n_0. \tag{4.64}$$

Fix $z \in \Omega$ and choose $m \in \mathbb{N}$ (depending in general on z and ε) such that

$$m \geq n_0 \text{ and } |u_m(z) - u(z)| < \frac{\varepsilon}{2}. \tag{4.65}$$

We have

$$|u_n(z) - u(z)| \leq |u_n(z) - u_m(z)| + |u_m(z) - u(z)|$$

$$< \frac{\varepsilon}{2} + \frac{\varepsilon}{2} = \varepsilon \text{ for all } n \geq n_0 \ (see \ (4.62), (4.63)).$$

Recall that $n_0 \in \mathbb{N}$ is independent of z and $z \in \Omega$ is arbitrary. So, it follows that

$$\|u_n - u\|_{C_b(\Omega)} \leq \varepsilon \text{ for all } n \geq n_0,$$

which proves that (4.63) is true. $\square$

Proposition 4.223 $C_0(X)$ *is a closed subspace of* $C_b(\Omega)$.

Proof Let $\{u_n\}_{n \in \mathbb{N}} \subseteq C_0(\Omega)$ and assume that $u_n \to u$ in $C_b(\Omega)$. Then $u \in C_b(\Omega)$ (see Proposition 4.222). Given $\varepsilon > 0$, we can find $n \in \mathbb{N}$ such that

$$\|u_n - u\|_{C_b(\Omega)} < \frac{\varepsilon}{2}$$

$$\Rightarrow |u(z)| \leq |u_n(z)| + \frac{\varepsilon}{2} \text{ for all } z \in \Omega,$$

$$\Rightarrow |u(z)| < \varepsilon \text{ for all } z \in \Omega \setminus K_\varepsilon, \ K_\varepsilon \subseteq \Omega \text{ compact (since } u_n \in C_0(\Omega))$$

$\Rightarrow u \in C_0(\Omega)$ and so $C_0(\Omega)$ is closed in $C_b(\Omega)$. $\square$

Corollary 4.224 $(C_0(\Omega), \|\cdot\|_{C_0(\Omega)})$ *is a Banach space.*

Proposition 4.225 $C_0(\Omega)$ *is the closure of* $C_c(\Omega)$ *in the* $\|\cdot\|_{C_b(\Omega)}$ *norm.*

Proof Let $\{u_n\}_{n\in\mathbb{N}} \subseteq C_c(\Omega)$ and suppose that $u_n \to u$ in $C_b(\Omega)$. Then $u \in C_b(\Omega)$ (see Proposition 4.222). Given $\varepsilon > 0$, we can find $n \in \mathbb{N}$ such that

$$\|u_n - u\|_{C_b(\Omega)} < \varepsilon,$$

$$\Rightarrow |u(z)| < \varepsilon \text{ for all } z \notin \operatorname{supp} u_n$$

$$\Rightarrow u \in C_0(\Omega).$$

Conversely, let $u \in C_0(\Omega)$. For every $n \in \mathbb{N}$, let $K_n = \{z \in \Omega : |u(z)| \geq \frac{1}{n}\}$. Then $K_n \subseteq \Omega$ is compact. Let $h_n \in C_c(\Omega)$ with $0 \leq h_n \leq 1, h_n|_{K_n} = 1$ (see Theorem 1.47). We set $u_n = h_n u$. Then $u_n \in C_c(\Omega)$ and $u_n \to u$ in $C_0(\Omega)$. $\square$

Since Ω is locally compact it has an Alexandrov one-point compactification $Y = \Omega \cup \{+\infty\}$ (see Theorem 1.109 and Definition 1.108). We set

$$Z = \{u \in C(Y) : u(\infty) = 0\}.$$

Evidently Z is a closed subspace of $C(Y)$.

Proposition 4.226 $C_0(\Omega)$ *and* Z *are isometrically isomorphic.*

Proof Consider the map $L : Z \to C_0(\Omega)$ defined by $L(u) = u|_\Omega$. This map is well defined since if $u \in Z$, $L(u) \in C_b(\Omega)$. Moreover, the continuity of u on Y implies that given $\varepsilon > 0$, we can find $K_\varepsilon \subseteq \Omega$ compact such that

$$|u(z) - u(\infty)| = |u(z)| < \varepsilon \text{ for all } z \in \Omega \setminus K_\varepsilon \text{ (see Definition 4.111)}$$

$$\Rightarrow u \in C_0(\Omega).$$

Evidently $L(\cdot)$ is linear and $\|L(u)\|_{C_0(\Omega)} = \|u\|_{C_Y}$. Moreover, if $u \in C_0(\Omega)$, then we can extend $u(\cdot)$ to Y by setting $u(\infty) = 0$. Since $u \in C_0(\Omega)$, we see that $u(\cdot)$ is continuous at ∞, hence $u \in Z$. $\square$

Let $ca(B(Y))_R$ be the Radon signed measures λ on Ω such that $|\lambda|$ is finite. We furnish $ca(B(\Omega))_R$ with the following norm

$$\|\lambda\|_{ca(B(\Omega))_R} = \int_\Omega d|\lambda| = |\lambda|(\Omega).$$

Recall that Y is the Alexandrov one-point compactification of Ω. We define

$$V = \{\lambda \in ca(B(Y)) : \lambda(\{\infty\}) = 0\}.$$

Evidently V is a closed subspace of $ca(B(Y))$ (for the total variation norm).

Proposition 4.227 $ca(B(Y))_R$ *is isometrically isomorphic to* V.

Proof We consider the linear map $S : V \to ca(B(\Omega))_R$ defined by $S(\lambda) = \lambda|_{B(\Omega)}$ (recall $B(\Omega) \subseteq B(Y)$). Note that every $B \in B(Y)$ can be written as

$$B = A \cup \{\infty\} \quad A \in B(\Omega) \quad \text{or} \quad B = A \quad A \in B(\Omega).$$

In the first case the additivity of $\lambda(\cdot)$ implies $\lambda(B) = \lambda(A)$. □

Now we are ready to identify the dual of $C_0(\Omega)$.

Theorem 4.228 *If* Ω *is locally compact, then* $C_0(\Omega)^*$ *is isometrically isomorphic to* $ca(B(\Omega))_R$.

Proof Let $M = \{u^* \in C(Y)^* : u^*|_Z = 0\}$. Then $Z^* = C(Y)^*/M$. The space Z is of codimension 1. Therefore $\dim M = 1$. Let δ_∞ denote the Dirac measure concentrated on ∞, that is

$$\delta_\infty(A) = \begin{cases} 1 & \text{if } \infty \in A \\ 0 & \text{if } \infty \notin A \end{cases} \quad \text{with } A \in B(Y).$$

Then $M = \mathbb{R}\delta_\infty$ and so we have $V = ca(B(Y))_R/M$, hence $C_0(\Omega)^* = ca(B(Y))_R/M = V = ca(B(\Omega))_R$ (see Proposition 4.227)(all are isometries). □

Using the spaces $C_c(\Omega), C_0(\Omega), C_b(\Omega)$ we can define three different modes of convergence in $ca(B(\Omega))$.

Definition 4.229

(a) Let $\{\lambda_n, \lambda\}_{n \in \mathbb{N}} \subseteq ca(B(\Omega))$. $\{\lambda_n\}_{n \in \mathbb{N}}$ is said to "converge vaguely" to λ, if for all $u \in C_c(\Omega)$ we have

$$\int_\Omega u d\lambda_n \to \int_\Omega u d\lambda.$$

We denote this convergence by $\lambda_n \overset{v}{\to} \lambda$ in $ca(B(\Omega))$.

(b) Let $\{\lambda_n, \lambda\}_{n \in \mathbb{N}} \subseteq ca(B(\Omega))$. We say that $\{\lambda_n\}_{n \in \mathbb{N}}$"converges weakly" to λ,if for all $u \in C_0(\Omega)$ we have

$$\int_\Omega u d\lambda_n \to \int_\Omega u d\lambda.$$

We denote this convergence by $\lambda_n \overset{w}{\to} \lambda$ in $ca(B(\Omega))$.

(c) Let $\{\lambda_n, \lambda\}_{n \in \mathbb{N}} \subseteq ca(B(\Omega))$. We say that $\{\lambda_n\}_{n \in \mathbb{N}}$ "converges narrowly" to λ, if for all $u \in C_b(\Omega)$ we have

$$\int_\Omega u d\lambda_n \to \int_\Omega u d\lambda.$$

We denote this convergence by $\lambda_n \xrightarrow{n} \lambda$ in $ca(B(\Omega))$.

Remark 4.230 These convergences also hold on $ca(B(\Omega))_R \subseteq ca(B(\Omega))$. Then on account of Theorem 4.228, the weak convergence is in fact the *weak** convergence on $ca(B(\Omega))_R = C_0(\Omega)^*$. These modes of convergence are topological(that is, correspond to topologies on $ca(B(\Omega))$) and we have

$$norm\ convergence \Rightarrow narrow\ convergence$$

$$\Rightarrow weak\ convergence$$

$$\Rightarrow vague\ convergence.$$

Proposition 4.231 *If $\{\lambda_n, \lambda\}_{n \in \mathbb{N}} \subseteq ca(B(\Omega))_R$, $\lambda_n \xrightarrow{w} \lambda$ and for every $\varepsilon > 0$ there exists $K_\varepsilon \subseteq \Omega$ compact such that $|\lambda_n|(\Omega \setminus K_\varepsilon) < \varepsilon$ for all $n \geq n_0$, then $\lambda_n \xrightarrow{n} \lambda$ in $ca(B(\Omega))_R$.*

Proof Let $\varepsilon > 0$ be given and let $K_\varepsilon \subseteq \Omega$ be the compact set postulated by the hypothesis. Let $\vartheta \in C_c(\Omega)$ such that $\vartheta|_{K_\varepsilon} \equiv 1$. Then given $u \in C_b(\Omega)$, we have

$$u = \vartheta u + \eta \quad \text{with} \quad \text{supp}\,\eta \subseteq \Omega \setminus K_\varepsilon.$$

$$\Rightarrow \int_\Omega u d\lambda_n = \int_\Omega \vartheta u d\lambda_n + \int_\Omega \eta d\lambda_n.$$

Since $\vartheta u \in C_c(\Omega)$, we have

$$\int_\Omega \vartheta u d\lambda_n \to \int_\Omega \vartheta u d\lambda.$$

Also we have

$$|\int_\Omega \eta d\lambda_n| \leq \|\eta\|_{C_b(\Omega)} |\lambda_n|(\Omega \setminus K_\varepsilon) \leq \varepsilon \|u\|_{C_b(\Omega)} \quad \text{for all } n \geq n_0.$$

We conclude that $\lambda_n \xrightarrow{n} \lambda$ in $ca(B(Y))_R$.   □

Motivated by this proposition, we make the following definition.

Definition 4.232 Let $C \subseteq ca(B(\Omega))_R$. We say that C is "uniformly tight," if for every $\varepsilon > 0$, we can find $K_\varepsilon \subseteq \Omega$ compact such that

$$|\lambda|(\Omega \setminus K_\varepsilon) < \varepsilon \quad \text{for all } \lambda \in C.$$

This notion is closely related to compactness in the narrow topology. The result is known as "Prohorov Theorem" and can be found in Bogachev [36], Theorem 4.5.3, p.161 (recall that a locally compact space is completely regular).

Theorem 4.233 *If $C \subseteq ca(B(\Omega))_\mathbb{R}$ is norm bounded and uniformly tight, then C is relatively compact for the narrow topology.*

Proposition 4.234 *If $\{\lambda_n\}_{n\in\mathbb{N}} \subseteq ca(B(Y))_\mathbb{R}$, $\|\lambda_n\| \le c$ for all $n \in \mathbb{N}$ and there exists $D \subseteq C_0(\Omega)$ dense such that*

$$\lim_{n\to\infty} \int_\Omega u d\lambda_n \ \text{exists for all } u \in D,$$

then there exists $\lambda \in ca(B(Y))_R$ such that $\lambda_n \xrightarrow{w} \lambda$ in $ca(B(Y)_R$.

Proof Consider $\xi_n \in C_0(\Omega)^*$ defined by $\xi_n(u) = \int_\Omega u d\lambda_n$ for all $n \in \mathbb{N}$. Then

$$|\xi_n(u - v)| \le c\|u - v\|_{C_0(\Omega)}$$

$$\Rightarrow \{\xi_n\}_{n\in\mathbb{N}} \text{ is equicontinuous.}$$

By hypothesis it is uniformly bounded and converges on a dense subset of $C_b(\Omega)$. The Arzela–Ascoli Theorem implies that

$$\xi_n(u) \to \xi(u) \ \text{for all } u \in C_0(\Omega).$$

Using Theorem 4.228 we can find $\lambda \in ca(B(\Omega))_R$ such that

$$\xi(u) = \int_\Omega u d\lambda,$$

$$\Rightarrow \lambda_n \xrightarrow{w} \lambda \ \text{in } ca(B(Y))_R.$$

$\square$

Proposition 4.235 *If $\{\lambda_n, \lambda\}_{n\in\mathbb{N}} \subseteq ca(B(Y))_R$, $\lambda_n \ge 0$, $\lambda_n \xrightarrow{w} \lambda$ and $\lambda_n(\Omega) \to \lambda(\Omega) \in \mathbb{R}_+$ as $n \to \infty$, then $\lambda_n \xrightarrow{n} \lambda$ in $ca(B(\Omega))_R$.*

Proof According to Proposition 4.231, it suffices to show that $\{\lambda_n\}_{n\ge n_0}$ is uniformly tight. Since $\lambda \in ca(B(Y))_R$, given $\varepsilon > 0$, we can find $K_\varepsilon \subseteq \Omega$ compact such that $|\lambda|(\Omega \setminus K_\varepsilon) < \varepsilon$. Let $u \in C_c(\Omega)$ with $\text{supp}\, u \subseteq K_\varepsilon$, $0 \le u \le 1$ and

$$\|\lambda\| - \varepsilon \le \int_\Omega u d\lambda.$$

We can find $n_0 \in \mathbb{N}$ such that

$$|\lambda_n(\Omega) - \lambda(\Omega)| < \frac{\varepsilon}{3} \quad \text{and} \quad |\int_\Omega u d\lambda_n - \int_\Omega u d\lambda| < \frac{\varepsilon}{3} \quad \text{for all } n \geq n_0.$$

Then we have

$$\lambda_n(\Omega \setminus K_\varepsilon) \leq \|\lambda_n\| - \int_\Omega u d\lambda_n \quad for \ n \geq n_0,$$

$$\Rightarrow \lambda_n(\Omega \setminus K_\varepsilon) \leq \varepsilon \text{ for all } n \geq n_0.$$

□

Let δ_z be the Dirac measure concentrated at $z \in \Omega$, that is,

$$\delta_z(A) = \begin{cases} 1 & \text{if } z \in A \\ 0 & \text{if } z \notin A \end{cases} \quad \text{for all } A \in B(\Omega).$$

Evidently a Dirac measure is Radon. Consider the vector space $ca(B(\Omega))_R^d$ of finite linear combinations of Dirac measures.

Proposition 4.236 *If Ω is a metric space and $\lambda \in ca(B(\Omega))_R$, then we can find $\{\lambda_n\}_{n \in \mathbb{N}} \subseteq ca(B(\Omega))_R^d$ such that $\lambda_n \xrightarrow{n} \lambda$.*

Proof Consider $\{\vartheta_n\}_{n \in \mathbb{N}} \subseteq C_c(\Omega)$ increasing such that

$$0 \leq \vartheta_n \leq 1 \text{ for all } n \in \mathbb{N}, \ \vartheta \uparrow 1 \text{ as } n \to \infty.$$

We set $\widehat{\lambda}_n(A) = \int_A \vartheta_n d\lambda$ for all $A \in B(\Omega)$. Then $\widehat{\lambda}_n \in ca(B(Y))_R$, it has compact support and $\|\widehat{\lambda}_n - \lambda\| \to 0$. So it suffices to prove the result for λ of compact support K. Since K is totally bounded we can find balls $\{B_{\frac{1}{n}}(z_k)\}_{k=1}^{m_n}$ of radius $\frac{1}{n}$ which cover K. Let $A_{n,1} = B_{\frac{1}{n}}(z_1) \cap K$, $A_{n,2} = B_{\frac{1}{n}}(z_2) \cap B_{\frac{1}{n}}(z_1)^c \cap K$ and so on. Then each $A_{n,k}$ has diameter less than $\frac{1}{n}$ and these sets form a partition of K. We consider only these sets $A_{n,k} \neq \emptyset$ and choose $z_{n,k} \in A_{n,k}$. We set

$$\lambda_n = \sum_{k=1}^{m_n} \lambda(A_{n,k}) \delta_{z_{n,k}}.$$

For $u \in C_b(\Omega)$ we have

$$\int_\Omega u d\lambda = \sum_{k=1}^{m_n} \int_{A_{n,k}} u d\lambda.$$

The function $u(\cdot)$ is uniformly continuous on K. So, we can find $\varepsilon_n \to 0^+$ such that

$$d_\Omega(z, z') < \frac{1}{n} \Rightarrow |u(z) - u(z')| < \varepsilon_n.$$

Therefore

$$\int_\Omega \chi_{A_{n,k}} u d\lambda = u(z_{n,k}) \int \chi_{A_{n,k}} d\lambda + \eta_k \varepsilon_n |\lambda|(A_{n,k}) \quad |\eta_k| \le 1$$

Summing up in $k \in \{1, \cdots, m_n\}$, we have

$$\left| \int_\Omega u d\lambda_n - \int_\Omega u d\lambda \right| \le \varepsilon_n |\lambda|(K) \to 0.$$

$\square$

Now suppose that Ω is a Polish space. Recall that every finite Borel measure is Radon (see Theorem 2.157). Let $ca(B(\Omega))^1_+$ be the set of probability measures. In this setting we have several equivalent conditions for narrow convergence. The result is often called "Portmanteau Theorem" and can be found in Parthasarathy [217], Theorem 6.1, p.40.

Theorem 4.237 *If Ω is a Polish space and $\{\lambda_n, \lambda\}_{n \in \mathbb{N}} \subseteq ca(B(Y))^1_+$, then the following statements are equivalent:*

(a) $\lambda_n \xrightarrow{n} \lambda$ in $ca(B(Y))$.
(b) If $u \in U(\Omega)$=space of bounded uniformly continuous functions on Ω, then

$$\int_\Omega u d\lambda_{a_n} \to \int_\Omega u d\lambda.$$

(c) For every $C \subseteq \Omega$ closed,

$$\limsup_{n \to \infty} \lambda_n(C) \le \lambda(C).$$

(d) For every $U \subseteq \Omega$ open,

$$\lambda(U) \le \liminf_{n \to \infty} \lambda_n(U).$$

(e) For every $A \in B(\Omega)$ with $\lambda(A) = 0$, we have

$$\lambda_n(A) \to \lambda(A).$$

4.6   Remarks

(4.1) The L^2-metric was introduced by Riesz [226] and Fischer [112] proved that
the resulting metric space is complete. Riesz [228] introduced L^p-spaces for
$p \neq 2$ and proved their completeness (see also Riesz [229]). The inequalities
of Theorem 4.5 are due to Clarkson [70]. Theorem 4.16 for $p = 2$ (Hilbert
space case) was proved simultaneously by Frechet [117] and Riesz [227] for
$\Omega = [0, 1]$. For $p \neq 2, 1 < p < \infty$ and $\Omega = [0, 1]$ was proved by Riesz
[230]. For a finite measure space (Ω, Σ, μ) is due to Dunford [93] and for σ-
finite measure spaces (in fact arbitrary measure spaces if $1 < p < \infty$) is due
to McShane [191]. The fact that $(L^1)^* = L^\infty$ was first proved by Steinhaus
[262]. Schwartz [251] found a characterization of $(L^1)^*$ which is valid for
arbitrary measure spaces.

Brezis and Lieb [51] proved a refinement of Fatou's lemma by finding the
"missing term" in the inequality in the lemma of Fatou. More precisely they
proved the following theorem.

Theorem 4.238 *If (Ω, Σ, μ) is a measure space, $\{u_n\}_{n \in \mathbb{N}} \subseteq L^p(\Omega)$ $(1 \leq p < \infty)$ is bounded and $u_n(z) \to u(z)$ $\mu - a.e$, then*

$$\lim_{n \to \infty} [\|u_n\|_p^p - \|u_n - u\|_p^p] = \|u\|_p^p.$$

Corollary 4.239 *If (Ω, Σ, μ) is a measure space, $\{u_n\}_{n \in \mathbb{N}} \subseteq L^p(\Omega)$ $(1 \leq p < \infty)$ is bounded, $u_n(z) \to u(z)$ $\mu - a.e$ and $\|u_n\|_p \to \|u\|_p$, then $u_n \to u$ in $L^p(\Omega)$.*

Although Theorem 4.24 usually goes under the name "Vitali-Hahn-Saks
Theorem" other people contributed to its formulation, like Lebesgue and
Nikodym. The general form presented here, first appears in Saks [245].
Theorem 4.24 and its offspring Theorem 4.25, are very useful in Measure
Theory and in Functional Analysis.

Theorem 4.26 is due to Dunford and Pettis [94]. It illustrates the impor-
tance of the notion of "uniform integrability" (see Definition 2.144) in
Analysis. The "Bitting Theorem" (Theorem 4.29), is due to Brooks and
Chacon [52].

Additional results on L^p-spaces can be found in Aliprantis and Border
[10], Castillo and Rafeiro [63], Denkowski et al. [81], Hewitt and Stromberg
[139], Kufner et al. [166], Papageorgiou and Winkert [216], Pick et al. [222].

(4.2) Variable exponent Lebesgue spaces were first studied by Orlicz [213].
Later Nakano [205] introduced modular spaces and used variable exponent
Lebesgue spaces as a particular example of modular spaces. The revival of the
interest on the variable exponent Lebesgue spaces, occurred with the paper of
Kovacik and Rakosnik [163]. A comprehensive treatment of these spaces can
be found in the books of Cruz Uribe and Fiorenza [73] and of Diening et al.

[83]. Applications of these spaces to problems in the Calculus of Variations, can be found in Zhikov [293].

(4.3) Sobolev spaces are an essential tool in the treatment of boundary value problems. Presentations of the theory of Sobolev spaces can be found in Adams and Fournier [1], Brezis [50], Evans and Gariepy [105], Gasinski and Papageorgiou [121], Kufner et al. [166], Leoni [179], Papageorgiou and Winkert [216], Ziemer [295]. See also the paper of Le [172].

(4.4) The Pettis Measurability Theorem (see Theorem 4.138) goes back to Pettis [220], while the notion of strong measurability (see Definition 4.137(b)), is due to Bochner [34]. There is an even more general form of Theorem 4.138 which can be found in Vakhania et al. [275] (see Propositions 1.1.9 and 1.1.10).

Proposition 4.240 *If (Ω, Σ) is a measure space, X is a complete metric space $\mathcal{D} \subseteq C(X, \mathbb{R})$ which separates points of X and $u : \Omega \to X$ is a function with separable range, then the following statements are equivalent:*

(a) u is Σ-measurable.

(b) u is strongly measurable (in Definition 4.137(b) replace the norm by the metric $d_X(\cdot, \cdot)$ on X).

(c) $\vartheta \circ u$ is Σ-measurable for all $\vartheta \in \mathcal{D}$.

The equality in Proposition 4.142 fails if X is not separable. For example, if I is an uncountable set, then $\sigma(l^2(I)^*) \subsetneq B(l^2(I))$. Recall that $l^2(I)$ is a nonseparable Hilbert space.

The Bochner integral (see Definition 4.149) was introduced by Bochner [34] (see also Dunford [93]). Expositions of Bochner integration, can be found in Dunford and Schwartz [95], Hille and Phillips [141] and the more recent and modern one in Diestel and Uhl [85].

The standard reference for the Radon–Nikodym Property (RNP, see Definition 4.168), is the book of Diestel and Uhl [85]. The RNP plays a central role in the theory of vector measures and in infinite dimensional convexity (see Bourgin [44]). As we mentioned in Remark 4.174, Theorem 4.173 can be found in Diestel and Uhl [84] and in fact the RNP of X^* is also a necessary condition for the duality to hold.

Note that if X^* is separable, the $L^\infty(\Omega, X^*_{w^*}) = L^\infty(\Omega, X^*)$. (see Definition 4.175).

For vector valued Sobolev spaces and evolution spaces, we refer to Barbu [27] and Zeidler [291, 292].

(4.5) For the information on the Banach space $ca(\Sigma)$, our main source is Diestel [84]. Theorem 4.228, is the outgrowth of the contributions of many mathematicians. The story starts with the work of Riesz [229] where $\Omega = [0, 1]$. It was extended to $\Omega \subseteq \mathbb{R}^N$ compact subset, by Radon [223], to Ω a compact metric space by Saks [245] (he attributes the result to Banach) and to Ω compact topological space by Kakutani [154]. The noncompact case was examined by Markov [188]. The result is also true for vector valued functions. So, let K be a compact topological space, X a Banach space, $cabv(B(K), X^*)$ the X^*-valued vector measures. By $rcabv(B(K), X^*)$ we denote the space of $m \in cabv(B(K), X^*)$ which are regular. The following theorem can be found in Singer [256] (Lemma 2.1.6).

Theorem 4.241 $C(K, X)^* = rcabv(B(K), X^*)$.

For the three types of convergence of measures, we refer to Billingsley [32], Bogachev [36], Bourbaki [43], Constantinescu [72], Papageorgiou and Winkert [216], Parthasarathy [217].

4.7 Problems

Problem 4.1 Let (Ω, Σ, μ) be a σ-finite measure space and $\mu : \Omega \rightarrow \mathbb{R}$ a nontrivial Σ-measurable function. Set $\tau_u = \{p \in [1, +\infty] : \|u\|_p < +\infty\}$. Show that τ_u is an interval (possibly empty).

Problem 4.2 Let $\Omega \subseteq \mathbb{R}^N$ be an open set, $u \in L^1_{loc}(\Omega)$ and assume that $\int_\Omega u\vartheta \, dz = 0$ for all $\vartheta \in C_c(\Omega)$. Show that $u(z) = 0$ for $a.a \ z \in \Omega$.

Problem 4.3 Let $C \in L^p[0, 1](1 < p < \infty)$ be bounded. Show that C is uniformly integrable.

Problem 4.4 Let (Ω, Σ, μ) be a finite measure space and $u \in L^p(\Omega)$ for all $1 \leq p < \infty$. Show that $u \in L^\infty(\Omega)$ if and only if $\lim_{p \to \infty} \|u\|_p$ is finite (then $\|u\|_\infty = \lim_{p \to \infty} \|u\|_p$).

Problem 4.5 Let (Ω, Σ, μ) be a measure space, $1 < p < \infty, C \subseteq L^p(\Omega)$ nonempty, closed convex. If $u \notin C$, show that there exists $v_0 \in C$ such that $\|u - v_0\|_p = d(u, C)$.

Problem 4.6 Let $\Omega \subseteq \mathbb{R}^N$ be an open set and $1 \leq p < \infty$. Show that $L^1(\Omega) \cap L^\infty(\Omega) \subseteq L^p(\Omega)$ and it is dense in $L^p(\Omega)$.

Problem 4.7 Let $v : \mathbb{R} \rightarrow \mathbb{R}$ be a measurable function such that $\int_\mathbb{R} |uv| dt < \infty$ for all $u \in L^p(\mathbb{R})$ ($1 \leq p \leq \infty$ fixed). Show that there exists $c > 0$ such that $\int_\mathbb{R} |uv| dt \leq c\|u\|_p$ for all $u \in L^p(\mathbb{R})$.

Problem 4.8 Let (Ω, Σ, μ) be a measure space and let $C = \{u \in L^1(\Omega) : u(z) \geq 1 \ for \ \mu - a.a \ z \in \Omega\}$. Is C w-closed in $L^1(\Omega)$? Justify your answer.

Problem 4.9 Suppose $u \in L^{p_0}[0, 1]$ for some $p_0 > 1$. Show that $\lim_{p \to 1} \|u\|_p = \|u\|_1$.

Problem 4.10 Let (Ω, Σ, μ) be a finite measure space, $1 < p < \infty$, $\{u_n, u\}_{n \in \mathbb{N}} \subseteq L^p(\Omega)$ and assume that

$$\|u_n\|_p \leq c \text{ for all } n \in \mathbb{N}, \, u_n(z) \to u(z) \, \mu - a.e.$$

Show that for every $q \in [1, p)$, we have $u_n \to u$ in $L^q(\Omega)$.

Problem 4.11 Let (Ω, Σ, μ) be a finite measure space, $\{u_n, h_n, v_n\}_{n \in \mathbb{N}} \subseteq L^1(\Omega)$ such that

$$u_n(z) \leq h_n(z) \leq v_n(z) \quad \mu - a.e \, for \, all \, n \in \mathbb{N}$$

$$u_n(z) \to u(z), h_n(z) \to h(z), v_n(z) \to v(z) \quad \mu - a.e.$$

with $u, v \in L^1(\Omega)$. Suppose that

$$\int_\Omega u_n d\mu \to \int_\Omega u d\mu, \int_\Omega v_n d\mu \to \int_\Omega v d\mu.$$

Show that $h \in L^1(\Omega)$ and $\int_\Omega h_n d\mu \to \int_\Omega h d\mu$.

Problem 4.12 Find a sequence of Σ-measurable functions $u_n : \Omega \to \mathbb{R} \quad n \in \mathbb{N}$ which converges in μ-measure and μ-a.e, but does not converge weakly in $L^p(\Omega)$ for any $1 \leq p \leq \infty$.

Problem 4.13 Let (Ω, Σ, μ) be a finite measure space, X a Banach space and $\Sigma_0 \subseteq \Sigma$ a sub-Σ-algebra. Show that there exists a unique operator $E^{\Sigma_0} \in \mathcal{L}(L^1(\Omega, \Sigma; X), L^1(\Omega, \Sigma_0; X))$ such that $\int_A u d\mu = \int_A E^{\Sigma_0} u d\mu$ for all $A \in \Sigma_0$, all $u \in L^1(\Omega, \Sigma; X)$.
(Remark: E^{Σ_0} is called the conditional expectation with respect to Σ_0 operator).

Problem 4.14 Let (Ω, Σ, μ) be a finite measure space, X a Banach space and $u^* : \Omega \to X^*$ a function such that

$$\langle u^*(\cdot), x \rangle \in L^1(\Omega) \text{ for all } x \in X.$$

Show that for every $A \in \Sigma$, there exists $x_A^* \in X^*$ such that

$$\langle x_A^*, x \rangle = \int_A \langle u^*(z), x \rangle d\mu \text{ for all } x \in X.$$

Problem 4.15 Let (Ω, Σ, μ) be a finite measure space, X a reflexive Banach space, $1 < p < \infty$ and $\{u_n\}_{n \in \mathbb{N}} \subseteq L^p(\Omega, X)$ a bounded sequence such that

$$u_n(z) \to u(z) \, \mu - a.e \text{ in } X.$$

Show that

$$u \in L^p(\Omega, X) \text{ and } u_n \xrightarrow{w} u \text{ in } L^p(\Omega, X).$$

Problem 4.16 Let (Ω, Σ, μ) be a measure space, $1 \leq p \leq \infty$, $p' = \frac{p}{p-1}$, $u :$ $\Omega \to [-\infty, \infty]$ is Σ-measurable, $uh \in L^1(\Omega)$ for every $h \in L^{p'}(\Omega)$ and $\{u \neq 0\}$ has σ-finite measure. Show that $u \in L^p(\Omega)$.

Problem 4.17 Let $\Omega \subseteq \mathbb{R}^N$ be a bounded open set, $1 < p < \infty$, $p' = \frac{p}{p-1}$, $u \in$ $W^{1,p}(\Omega)$, $v \in W^{1,p'}(\Omega)$. Show that $uv \in W^{1,1}(\Omega)$ and

$$\frac{\partial}{\partial z_k}(uv) = u\frac{\partial v}{\partial z_k} + v\frac{\partial u}{\partial z_k} \quad for \text{ all } k \in \{1, \cdots, N\}.$$

Problem 4.18 Let $1 < p < \infty$, $\{u_n\}_{n \in \mathbb{N}} \subseteq L^p(0, b)$ such that $u_n \xrightarrow{w} u$ in $L^p(0, b)$ and $u'_n \to u'$ in $W^{-1,p}(0, b)$. Show that $u_n \to n$ in $L^p(0, b)$.

Problem 4.19 Suppose that Ω is a Polish space, $C \subseteq \Omega$ is closed, $U \in \Omega$ is open and $\eta \in \mathbb{R}$. Show that

(a) the set $\{\lambda \in ca(B(\Omega)^1_+ : \lambda(C) \geq \eta\}$ is narrowly closed.
(b) The set $\{\lambda \in ca(B(\Omega)^1_+ : \lambda(U) > \eta\}$ is narrowly open.

Problem 4.20 Let $\Omega \subseteq \mathbb{R}^N$ be a bounded open set, $1 < p < \infty$, $C \subseteq W^{1,p}(\Omega)$ is closed convex, $\{u_n\}_{n \in \mathbb{N}} \subseteq C$, $u \in L^p(\Omega)$, $y \in L^p(\Omega, \mathbb{R}^N)$. We assume that

$$u_n \xrightarrow{w} u \text{ in } L^p(\Omega), \, Du_n \xrightarrow{w} y \text{ in } L^p(\Omega, \mathbb{R}^N).$$

Show that $u \in C$ and $y = Du$.

Problem 4.21 Let $1 \leq p \leq \infty$ and set $\overline{B}_1(p) = \{u \in W^{1,p}(0, 1) : \|u\| \leq 1\}$. Show that for $1 < p \leq \infty$, $\overline{B}_1(p)$ is compact in $L^p(0, 1)$, but $\overline{B}_1(1)$ is not closed in $L^1(0, 1)$.

Problem 4.22 Let $u \in W^{1,2}(a, b)$. Show that $\|u\|_\infty \leq c\|u\|$ $(\|u\| = \|u\|_2 + \|u'\|_2)$ with $c = \max\{(b-a)^{-\frac{1}{2}}, (b-a)^{\frac{1}{2}}\}$.

Problem 4.23 Let $\Omega \subseteq \mathbb{R}^N$ be a bounded, open, connected set (a bounded domain) with Lipschitz boundary $\partial \Omega$ and $1 < p < \infty$. Show that there exists $c = c(\Omega) > 0$ such that

$$\|u\|_p \leq c \left[\|\mathcal{D}u\|_p + \left| \int_\Omega u dz \right| \right] \quad for\ all\ u \in W^{1,p}(\Omega).$$

Problem 4.24 Let $1 < p < \infty$ and define

$$\widehat{\lambda}_1 = \inf \left[\frac{\|u'\|_p^\varrho}{\|u\|_p^\varrho} : u \in W_0^{1,p}(0,1), u \neq 0 \right].$$

Show that $\widehat{\lambda}_1 > 0$ and there exists $\widehat{u}_1 \in W_0^{1,p}(0,1)$ such that

$$\widehat{\lambda}_1 = \frac{\|\widehat{u'}_1\|_p^\varrho}{\|\widehat{u}_1\|_p^\varrho}.$$

Problem 4.25 Let (Ω, Σ, μ) be a finite measure space and $u : \Omega \in \mathbb{R}$ a Σ-measurable function. Show that $u \in L^p(\Omega)$ if and only if $\sum_{k \geq 0} k^{p-1} \mu(A_k)$ with $A_k = \{|f| \geq k\}$.

Problem 4.26 Let (Ω, Σ, μ) be a measure space, $1 < p < \infty$, $\{u_n\}_{n \in \mathbb{N}} \subseteq L^p(\Omega)$ and assume that

$$u_n \xrightarrow{w} u \text{ in } L^p(\Omega) \text{ and } \limsup_{n \to \infty} \|u_n\|_p \leq \|u\|_p.$$

Show that $u_n \to u$ in $L^p(\Omega)$.

Problem 4.27 Let $1 \leq p, q \leq \infty$. Show that there exists $c > 0$ such that

$$\|u\|_q \leq c \|u'\|_p \text{ for all } u \in W_0^{1,p}(0,b)$$

and $c > 0$ is realized if $1 \leq q \leq \infty$, $1 < p \leq \infty$.

Problem 4.28 Let Ω be a Polish space and consider the map $\xi : C_b(\Omega) \times ca(B(\Omega))_+^1 \to \mathbb{R}$ defined by

$$\xi(u, \lambda) = \int_\Omega u d\lambda.$$

Show that $\xi(\cdot, \cdot)$ is sequentially continuous for the narrow convergence.

Problem 4.29 Show that $L^1[-1, 1]$ is a strict subspace of $L^\infty[-1, 1]^*$.

Problem 4.30 Let (Ω, Σ, μ) be a finite measure space and $\{u_n\}_{n \geq 1} \subseteq L^2(\Omega)$ which satisfies

$$u_n \xrightarrow{w} u \text{ in } L^2(\Omega), u_n^2 \xrightarrow{w} u^2 \text{ in } L^1(\Omega).$$

Show that

$$u_n \to u \text{ in } L^2(\Omega).$$

Problem 4.31 If (Ω, Σ, μ) is a measure space, show that $L^\infty(\Omega)$ is not separable; similarly for $ca(\Sigma)$.

Chapter 5
Multivalued Analysis

Multivalued Analysis deals with the study of maps whose values are sets. Such maps are usually called multifunctions or set-valued functions/correspondences, mostly used already by mathematical economists. Multivalued Analysis grew from the needs of applied fields such as optimal control, optimization, game theory, and mathematical economics. Also, it had a parallel development and symbiotic relationship with "Nonsmooth Analysis," which is part of the next chapter.

In Sect. 5.1 we deal with the topological properties of multifunctions. We introduce various continuity notions and study their properties and relations among them. In Sect. 5.2 we examine the measure theoretic properties of multifunctions. We introduce various notions of measurability of a multifunction and study their relations, which highlight the importance of separability in the range space. Section 5.3 examines the fundamental question of existence of continuous and measurable selections for a multifunction. We state and prove the main results in this direction, namely the Michael theorem, the Kyratowski–Ryll Nardzewski theorem, and the Yankov–von Neumann–Aumann theorem. In connection with the third result, we also present some projection theorems, which are of independent interest. In Sect. 5.4, we introduce and investigate the properties of decomposable sets. The results of this section reveal the close relation between decomposable sets and the set of measurable selections of a multifunction. Moreover, we show that in many occasions decomposability is a good substitute of convexity. In Sect. 5.5 we introduce an integral for multifunctions, the so-called Aumann integral, and establish its properties. Finally, in Sect. 5.6 we consider Caratheodory functions and multifunctions and prove parametric versions of the classical Lusin theorem.

© The Author(s), under exclusive license to Springer Nature Switzerland AG 2022
S. Hu, N. S. Papageorgiou, *Research Topics in Analysis, Volume I*, Birkhäuser
Advanced Texts Basler Lehrbücher, https://doi.org/10.1007/978-3-031-17837-5_5

5.1 Continuity of Multifunctions

First let us fix our notation. So, suppose that X is a Hausdorff topological space. We define

$$P_f(X) = \{A \subseteq X : A \text{ is nonempty and closed }\},$$

$$P_k(X) = \{A \subseteq X : A \text{ is nonempty and compact }\}.$$

If X is also a normed space, then we define

$$P_{f_c}(X) = \{A \subseteq X : A \text{ is nonempty, closed, convex}\},$$

$$P_{(w)k(c)}(X) = \{A \subseteq X : A \text{ is nonempty, (weakly-) compact (and convex)}\},$$

$$P_{bf(c)} = \{A \subseteq X : A \text{ is nonempty, bounded, closed (and convex)}\}.$$

Suppose that X and Y are two sets and $F : X \to 2^Y$ a set-valued map (a multifunction). Given $C \subseteq Y$, we can define two kinds of inverse images of C under $F(\cdot)$

$$F^+(C) = \{x \in X : F(x) \subseteq C\} \text{ (the "strong inverse image" of } C),$$

$$F^-(C) = \{x \in X : F(x) \cap C \neq \emptyset\} \text{ (the "weak inverse image" of } C).$$

We present some straightforward facts concerning these notions.

Proposition 5.1 *If X and Y are sets and $F : X \to 2^Y$ a multifunction, then*

(a) $A \subseteq F^+(F(A)), F^-(F(A))$ for all $A \subseteq X$.
(b) $F^+(Y \setminus C) \subseteq X \setminus F^+(C), F^-(Y \setminus C) = X \setminus F^-(C)$ for all $C \subseteq Y$.
(c) If $\{C_i\}_{i \in I}$ are subsets of Y and I an arbitrary index set, then

$$\bigcup_{i \in I} F^+(C_i) \subseteq F^+(\bigcup_{i \in I} C_i), \quad \bigcup_{i \in I} F^-(C_i) = F^-(\bigcup_{i \in I} C_i),$$

$$\bigcap_{i \in I} F^+(C_i) = F^+(\bigcap_{i \in I} C_i), \quad F^-(\bigcap_{i \in I} C_i) \subseteq \bigcap_{i \in I} F^-(C_i).$$

If X and Y are sets and $F, G : X \to 2^Y$ two multifunctions, then we define $(F \cup G)(x) = F(x) \cup G(x)$ and $(F \cap G)(x) = F(x) \cap G(x)$ for all $x \in X$.

Proposition 5.2 *If X and Y are sets, $F, G : X \to 2^Y$ are two multifunctions, and $C \subseteq Y$, then*

(a) $(F \cup G)^+(C) = F^+(C) \cap G^+(C)$
(b) $F^+(C) \cap G^+(C) \subseteq (F \cap G)^+(C)$

(c) $(F \cup G)^-(C) = F^-(C) \cup G^-(C)$
(d) $(F \cap G)^-(C) \subseteq F^-(C) \cap G^-(C)$

If X, Y, and V are sets and $F : X \to 2^Y$, $G : X \to 2^V$, and $H : Y \to 2^V$ three multifunctions, then we define

$$(F \times G)(x) = F(x) \times G(x), \ (H \circ F)(x) = H(F(x)) = \bigcup_{y \in F(x)} H(y), \ x \in X.$$

Proposition 5.3 *If X, Y, and V are sets and $F : X \to 2^Y$, $G : X \to 2^V$, and $H : Y \to 2^V$ three multifunctions, then*

(a) $(F \times G)^+(C \times D) = F^+(C) \cap G^+(D)$, $(F \times G)^-(C \times D) = F^-(C) \cap G^-(D)$
    for all $C \subseteq Y$ and all $D \subseteq V$ (it also true for arbitrary products).
(b) $(H \circ F)^+(D) = F^+(H^+(U))$, $(H \circ F)^-(D) = F^-(H^-(D))$ *for all $D \subseteq V$.*

Now we will introduce the first continuity concepts for multifunctions. In what follows, X and Y are Hausdorff topological spaces. Additional hypotheses will be introduced as needed.

Definition 5.4 Consider a multifunction $F : X \to 2^Y$.

(a) We say that $F(\cdot)$ is "upper semicontinuous at $x_0 \in X$" ("usc at x_0" for short), if for all $V \subseteq Y$ open with $F(x_0) \subseteq V$, we can find $U \in \mathcal{N}(x_0) =$ filter of neighborhoods of x_0 such that

$$F(x) \subseteq V \ for \ all \ x \in U;$$

if $F(\cdot)$ is usc at every $x \in X$, then we simply say that $F(\cdot)$ is usc.
(b) We say that $F(\cdot)$ is "lower semicontinuous at $x_0 \in X$" ("lsc at x_0" for short), if for all $V \subseteq Y$ open with $F(x_0) \cap V \neq \emptyset$, we can find $U \in \mathcal{N}(x_0)$ such that

$$F(x) \cap V \neq \emptyset \ for \ all \ x \in U;$$

if $F(\cdot)$ is lsc at every $x \in X$, then we simply say that $F(\cdot)$ is lsc.
(c) We say that $F(\cdot)$ is "continuous" or "Victoria continuous," at $x_0 \in X$, if it is both usc and lsc at x_0; if $F(\cdot)$ is continuous at every $x \in X$, then we simply say that $F(\cdot)$ is continuous.

From this definition and Proposition 5.1, we infer the following propositions that provide a complete characterization of these continuity concepts.

Proposition 5.5 *If $F : X \to 2^Y$ is multifunction, then the following statements are equivalent:*

(a) *$F(\cdot)$ is usc.*
(b) *For every $C \subseteq Y$ closed, $F^-(C) \subseteq X$ is closed.*
(c) *If $x \in X$, $F(x) \subseteq V =$ open set in Y and $\{x_\alpha\}_{\alpha \in J} \subseteq X$ a net such that $x_\alpha \to x$,*

then $F(x_\alpha) \subseteq V$ for all $\alpha \in J$, $\alpha \succeq \alpha_0$.

Proposition 5.6 *If $F : X \to 2^Y$ is a multifunction, then the following statements are equivalent:*

(a) *$F(\cdot)$ is lse.*
(b) *For every $C \subseteq Y$ closed, $F^+(C) \subseteq X$ is closed.*
(c) *If $x \in X$, $V =$ open set in Y, $F(x) \cap V \neq \emptyset$ and $\{x_\alpha\}_{\alpha \in J} \subseteq X$ a net such that $x_\alpha \to x$, then $F(x_\alpha) \cap V \neq \emptyset$ for all $\alpha \in J$, $\alpha \succeq \alpha_0$.*
(d) *If $y \in F(x)$ and $\{x_\alpha\}_{\alpha \in J}$ is a net such that $x_\alpha \to x$, then we can find $y_\alpha \in F(x_\alpha)$ $\alpha \in J$ such that $y_\alpha \to y$ in Y.*

Remark 5.7 Both notions of usc and lsc in the case of a single-valued function reduce to the notion of continuity. The concepts of usc and lsc are distinct. Roughly speaking, usc allows upward jumps in terms of inclusion, while lsc allows downward jumps. This is easily seen in the following example.

Example 5.8 Let $F_1, F_2 : \mathbb{R} \to 2^{\mathbb{R}} \setminus \{\emptyset\}$ be two multifunctions defined by

$$F_1(x) = \begin{cases} \{0\} & \text{if } x \neq 0 \\ [0, 1] & \text{if } x = 0 \end{cases}, \quad F_2(x) = \begin{cases} [0, 1] & \text{if } x \neq 0 \\ \{0\} & \text{if } x = 0 \end{cases}, \quad x \in \mathbb{R}.$$

Then $F_1(\cdot)$ is usc and $F_2(\cdot)$ is lsc.

Proposition 5.9 *If $F : X \to 2^Y$ is a multifunction, then the following statements are equivalent:*

(a) *$F(\cdot)$ is continuous.*
(b) *For every $C \subseteq Y$ closed, $F^-(C), F^+(C) \subseteq X$ are both closed.*
(c) *If $x \in X$, $V \subseteq Y$ is open, $F(x) \subseteq V$ or $F(x) \cap V \neq \emptyset$ and $\{x_\alpha\}_{\alpha \in J} \subseteq X$ is a net such that $x_\alpha \to x$, then $F(x_\alpha) \subseteq V$ or $F(x_\alpha) \cap V \neq \emptyset$ for all $\alpha \in J$, $\alpha \succeq \alpha_0$.*

For a multifunction $F : X \to 2^Y$, its "graph" is defined in the obvious way by

$$\text{Gr } F = \{(x, y) \in X \times Y : y \in F(x)\}.$$

Definition 5.10 We say that a multifunction $F : X \to 2^Y$ is "closed" if $\text{Gr } F \subseteq X \times Y$ is closed.

There is a close relation between usc and closed multifunctions.

Proposition 5.11 *If Y is regular and $F : X \to P_f(Y)$ is usc, then $F(\cdot)$ is closed.*

Proof Arguing by contradiction, we suppose that the assertion of the proposition is not true. So, we can find a net $\{(x_\alpha, y_\alpha)\}_{\alpha \in J} \subseteq \text{Gr } F$ such that $(x_\alpha, y_\alpha) \to (x, y)$ in $X \times Y$ and $y \notin F(x)$.

Then, on account of the regularity of Y, we can find $V_1, V_2 \subseteq Y$ open such that

$$y \in V_1, F(x) \subseteq V_2 \text{ and } V_1 \cap V_2 = \emptyset. \tag{5.1}$$

Using Proposition 5.5, we can find $\alpha_0 \in J$ such that

$$y_\alpha \in V_1, F(x_\alpha) \subseteq V_2 \text{ for all } \alpha \in J, \ \alpha \succeq \alpha_0,$$

$$\Rightarrow (x_\alpha, y_\alpha) \notin \text{Gr } F \text{ for all } \alpha \in J, \alpha \succeq \alpha_0 \quad (\text{see } (5.1)),$$

a contradiction. This proves that $F(\cdot)$ is closed. $\qquad \square$

Remark 5.12 The converse is not in general true. To see this, let $X = Y = \mathbb{R}_+$ and consider the multifunction $F : X \to P_k(Y)$ defined by

$$F(x) = \begin{cases} \{0\} & \text{if } x = 0 \\ [0, 1] \cup \{1/x\} & \text{if } 0 < x. \end{cases}$$

Evidently $\text{Gr } F \subseteq X \times Y$ is closed, but $F(\cdot)$ is not usc at $x = 0$, since $F(0) \subseteq V = [0, 1)$, but $F^+(V) = \{0\}$, which is not open (see Proposition 5.5). For $P_k(Y)$-valued multifunctions, we can drop the regularity requirement on Y and show that $F(\cdot)$ is usc if and only if for every net $\{(x_\alpha, y_\alpha)\}_{\alpha \in J} \subseteq \text{Gr } F$ such that $x_\alpha \to x$ in X, $\{y_\alpha\}_{\alpha \in J}$ has a cluster point in $F(x)$ (see Hu-Papageorgious [145], p.41).

For the converse to be true, we need additional conditions on the multifunction $F(\cdot)$.

Proposition 5.13 *If $F : X \to P_k(Y)$ is closed and locally compact (that is, for every $x \in X$, we can find $U \in \mathcal{N}(x)$ such that $\overline{F(U)} \subseteq Y$ is compact), then $F(\cdot)$ is usc.*

Proof Let $C \subseteq Y$ be closed. We need to show that $F^-(C) \subseteq X$ is closed (see Proposition 5.5). So, let $\{x_\alpha\}_{\alpha \in J} \subseteq F^-(C)$ be a net and assume that $x_\alpha \to x$ in X. By hypothesis, we can find $U \in \mathcal{N}(x)$ such that $\overline{F(U)} \subseteq Y$ is compact. We have $x_\alpha \in U$ for all $\alpha \in J, \alpha \succeq \alpha_0$. Hence if $y_\alpha \in F(x_\alpha) \ \alpha \in J, \alpha \succeq \alpha_0$, then $\{y_\alpha\}_{\alpha \succeq \alpha_0} \subseteq Y$ is relatively compact and so we can find a subnet $\{y_\beta\}_{\beta \in I}$ such that $y_\beta \to y \in \overline{F(U)}$. Since $F(\cdot)$ is closed, we have $(x, y) \in \text{Gr } F$. Also $y \in C$. Hence $y \in F(x) \cap C$ and so $F^-(C) \subseteq X$ is closed. This proves that $F(\cdot)$ is usc. $\qquad \square$

Proposition 5.14 *If $F : X \to P_k(Y)$ is usc and $K \subseteq X$ is compact, then $F(K) = \bigcup_{x \in K} F(x) \subseteq Y$ is compact.*

Proof Consider a net $\{y_\alpha\}_{\alpha \in J} \subseteq F(K)$. We have $y_\alpha \in F(x_\alpha)$ for some $x_\alpha \in K$. Since K is compact, we can find a subnet $\{x_\beta\}_{\beta \in I}$ of $\{x_\alpha\}_{\alpha \in J}$ such that $x_\beta \to x \in K$ (see Proposition 1.81). We claim that $\{y_\beta\}_{\beta \in I}$ has a cluster point in $F(x)$ (see Definition 1.28). Arguing by contradiction, suppose that for all $y \in F(x)$, we can find $\beta_0(y) \in I$ and $V(y) \in \mathcal{N}(y)$ such that $y_\beta \notin V(y)$ for all $\beta \geq \beta_0(y)$. The family $\{V(y)\}_{y \in K}$ is an open cover of K and so we can find $\{y_k\}_{k=1}^m \subseteq K$ such that $K \subseteq \bigcup_{k=1}^m V(y_k)$. Let $\beta_0 \in T$ such that $\beta_0 \geq \beta_0(y_k)$ $k = 1, \ldots, m$ and $y_\beta \notin V(y_k)$ for all $\beta \geq \beta_0$, all $k = 1, \ldots, m$. Hence $y_\beta \notin V \supseteq F(x)$ for all $\beta \geq \beta_0$ contradicting Proposition 5.5. Therefore $\{y_\beta\}_{\beta \in I} \subseteq Y$ has a cluster point in $F(x)$, and by Proposition 1.31 we can find a subnet $\{y_\gamma\}_{\gamma \in y}$ of $\{y_\beta\}_{\beta \in I}$ such that $y_\gamma \to y \in F(x) \subseteq F(K)$ and so $F(K)$ is compact.   □

Given a function $f : X \to Y$, we can associate with it the multifunction $F(y) = f^{-1}(y) = \{x \in X : f(x) = y\}$ for all $y \in Y$.

Proposition 5.15

(a) $f(\cdot)$ is closed (see Definition 1.75) if and only if $F(\cdot)$ is usc.
(b) $f(\cdot)$ is open (see Definition 1.75) if and only if $F(\cdot)$ is lsc.

Proof

(a) $\Rightarrow$ Let $y \in Y$ and suppose $F(y) \subseteq V =$ open set in X. Let $W = Y \setminus f(X \setminus V)$. Since $f(\cdot)$ is closed, $f(X \setminus V) \subseteq Y$ is closed and so $W \subseteq Y$ is open. Moreover, $W \in \mathcal{N}(y)$ and $F(y') \subseteq V$ for all $y' \in W$. By Proposition 5.5, this implies that $F(\cdot)$ is usc.
$\Leftarrow$ Let $C \subseteq X$ be closed and let $y \in Y \setminus f(C)$. Then $F(y) \subseteq X \setminus C$. Since $F(\cdot)$ is usc, we can find $V \in \mathcal{N}(y)$ such that $F(y') \subseteq X \setminus C$ for all $y' \in V$. Then $f(C) \cap V = \emptyset$ and so $V \subseteq Y \setminus f(C)$. Since $y \in Y \setminus f(C)$ is arbitrary, we infer that $Y \setminus f(C)$ is open and so $f(C) \subseteq Y$ is closed, that is, $f(\cdot)$ is a closed function.
(b) $\Rightarrow$ Let $V \subseteq X$ be open and suppose $F(y) \cap V \neq \emptyset$. Then $f(V) \subseteq Y$ is open, and for all $y' \in f(V)$ we have $y' = f(x')$ with $x' \in V$ and so $F(y') \cap V \neq \emptyset$ for all $y' \in f(V)$, which shows that $F(\cdot)$ is lsc (see Proposition 5.6).
$\Leftarrow$ Let $U \subseteq X$ be open and $y \in f(U)$. Then $F(y) \cap U \neq \emptyset$. Since $F(\cdot)$ is lsc, we can find $V \in \mathcal{N}(y)$ such that $F(y') \cap U \neq \emptyset$ for all $y' \in V$ (see Proposition 5.6) and so $y' \in f(U)$. This proves that $f(U)$ is open, and hence $f(\cdot)$ is an open function.   □

Directly from Definition 5.4(b), we see that the following result holds.

Proposition 5.16 *If $F : X \to 2^Y$ is a multifunction with $\text{Gr } F \subseteq X \times Y$ open, then $F(\cdot)$ is lsc.*

Next we introduce more structure on the range space Y and derive new characterizations for lsc and usc multifunctions. For, this we will need the following two notions.

Definition 5.17

(a) If (Y, d) is a metric space and $C \subseteq Y$, then we define $d(y, C) = \inf[d(y, c) : c \in C]$ for all $y \in Y$ (here we use the convention that $\inf_{\emptyset} = +\infty$). This is the "distance function" from the set C.

(b) If Y is a normed space and Y^* its dual and $\langle \cdot, \cdot \rangle$ are the duality brackets for the pair (Y, Y^*) and $C \subseteq Y$, then we define $\sigma(y^*, C) = \sup\left[\langle y^*, c \rangle : c \in C\right]$ for all $y^* \in Y^*$ (here we use the convention that $\sup_{\emptyset} = -\infty$). This is the "support function" of the set C.

Remark 5.18 The distance function from a set $C \neq \emptyset$ is nonexpansive, that is,

$$|d(y, C) - d(y', C)| \leq d(y, y') \text{ for all } y, y' \in Y.$$

The support function of a set $C \neq \emptyset$ takes values in $\overline{\mathbb{R}} = \mathbb{R} \cup \{+\infty\}$ and it is sublinear, that is, positively homogeneous and subadditive and $\sigma(\cdot, C) = \sigma(\cdot, \overline{\text{conv}}C)$. On account of Proposition 1.182, $\sigma(\cdot, C)$ is w^*-lower semicontinuous. Also using Theorem 3.71(b), we see that for all $C \in P_{f_c}(Y)$, we have

$$C = \{y \in Y : \langle y^*, y \rangle \leq \sigma(y^*, C) \text{ for all } y^* \in Y^*\}.$$

The support function is very useful when working with closed, convex sets because it allows us to derive properties of such sets by working on a functional analytic level instead of a set theoretic one.

Proposition 5.19 *If X is a Hausdorff topological space, (Y, d) is a metric space, and $F : X \to 2^Y \setminus \{\emptyset\}$ is a multifunction, then $F(\cdot)$ is lsc if and only if for every $y \in Y$, $d(y, F(\cdot))$ is an upper semicontinuous, $\mathbb{R}_+$-valued function.*

Proof $\Rightarrow$ According to Definition 1.176, we need to show that for every $y \in Y$ and every $\lambda \geq 0$, the set $L_\lambda^u = \{x \in X : d(y, F(x)) \geq \lambda\}$ is closed. So, let $\{x_\alpha\}_{\alpha \in J} \subseteq L_\lambda^u$ be a net such that $x_\alpha \to x$ in X. Let $v \in F(x)$. Since $F(\cdot)$ is lsc, by Proposition 5.6, we can find $v_\alpha \in F(x_\alpha)$ such that $v_\alpha \to v$ in Y. Then

$$d(y, v_\alpha) \geq d(y, F(x_\alpha)) \geq \lambda \text{ for all } \alpha \in J,$$

$$\Rightarrow d(y, v) \geq \lambda \text{ for all } v \in F(x),$$

$$\Rightarrow d(y, F(x) \geq \lambda \text{ and so } x \in L_\lambda^u, \text{ that is, } L_\lambda^u \subseteq X \text{ is closed.}$$

This proves that $d(y, F(\cdot))$ is upper semicontinuous for all $y \in Y$.
$\Leftarrow$ Let $V \subseteq Y$ be open. We need to show that $F^-(V) \subseteq X$ is open. Let $x \in F^-(V)$ and choose $y \in F(x) \cap V$. We can find $\varepsilon > 0$ such that $B_\varepsilon(y) = \{y' \in Y : d(y', y) <$

$\varepsilon\} \subseteq V$. By hypothesis $d(y, F(\cdot))$ is upper semicontinuous. So, by Definition 1.176, we can find $U \in \mathcal{N}(x)$ such that

$$d(y, F(x')) < d(y, F(x)) + \varepsilon = \varepsilon \text{ for all } x' \in U,$$

$$\Rightarrow F(x') \cap B_\varepsilon(y) \neq \emptyset \text{ for all } x' \in U,$$

$$\Rightarrow F(x') \cap V \neq \emptyset \text{ for all } x' \in U \text{ and so } F(\cdot) \text{ is lsc }. \qquad \square$$

For upper semicontinuous multifunctions, the situation is more complicated.

Proposition 5.20 *If X is a Hausdorff topological space, (Y, d) is a metric space, and $F : X \to 2^Y \setminus \{\emptyset\}$ is a multifunction, then*

(a) *$F(\cdot)$ is usc $\Rightarrow$ for every $y \in Y$ $x \to d(y, F(x))$ is lower semicontinuous.*
(b) *If $F(\cdot)$ is locally compact (see Proposition 5.13), then $F(\cdot)$ is usc $\Leftrightarrow$ for every $y \in X$, $x \to d(y, F(x))$ is lower semicontinuous.*

Proof

(a) We need to show that for every $\lambda \geq 0$ the set

$$L_\lambda^l = \{x \in X : d(y, F(x)) \leq \lambda\} \text{ is closed.}$$

So, let $\{x_\alpha\}_{\alpha \in J} \subseteq L_\lambda^l$ be a net such that $x_\alpha \to x$. By hypothesis, $F(\cdot)$ is usc. Therefore given $\varepsilon > 0$, we have

$$F(x_\alpha) \subseteq F(x)_\varepsilon = \{y \in Y : d(y, F(x)) < \varepsilon\} \text{ for all } \alpha \in J, \alpha \succeq \alpha_0,$$

$$\Rightarrow d(y, F(x)) < d(y, F(x_\alpha)) + \varepsilon \leq \lambda + \varepsilon.$$

Since $\varepsilon > 0$ is arbitrary, we let $\varepsilon \to 0^+$ and obtain $d(y, F(x)) \leq \lambda$, that is, $x \in L_\lambda^l$. Hence $d(y, F(\cdot))$ is lower semicontinuous for all $y \in Y$.
(b) $\Rightarrow$ Follows from (a).
$\Leftarrow$ According to Proposition 5.13, it suffices to show that $F(\cdot)$ is closed. So, let $\{x_\alpha, v_\alpha)\}_{\alpha \in J} \subseteq \text{Gr } F$ be a net such that

$$(x_\alpha, v_\alpha) \to (x, v) \text{ in } X \times Y.$$

Then, for all $\alpha \in J$, we have

$$d(v, F(x_\alpha)) \leq d(v, v_\alpha),$$

$$\Rightarrow d(v, F(x)) \leq \liminf_{\alpha \in J} d(v, F(x_\alpha)) = 0,$$

$$\Rightarrow v \in F(x) \text{ (since } F(\cdot) \text{ is } P_k(Y)\text{-valued)},$$

$$\Rightarrow F(\cdot) \text{ is closed and so usc.}$$

$\square$

Now we see how the support function is related to the upper semi-continuity of a multifunction. We will need the following lemma which can be found in Hu and Papageorgious [145] (p. 47).

Lemma 5.21 *If Y is a normed space, $K \in P_{wk}(Y)$, $V \subseteq Y$ is w-open, and $K \subseteq V$, then we can find $\{y_k^*\}_{k=1}^m \subseteq Y^*$ and $\varepsilon > 0$ such that*

$$V_0 = \{y \in Y : \langle y_k^*, y \rangle < \sigma(y_k^*, K) + \varepsilon \text{ for all } k = 1, \ldots m\} \subseteq V.$$

Proposition 5.22

(a) *If X is a Hausdorff topological space, Y is a normed space equipped with the weak topology (denoted by Y_w), and $F : X \to 2^Y \setminus \{\emptyset\}$ is usc, then for all $y^* \in Y^*$, the $\overline{\mathbb{R}} = \mathbb{R} \cup \{+\infty\}$-valued function $x \to \sigma(y^*, F(x))$ is upper semicontinuous.*

(b) *If X is a Hausdorff topological space, Y is a normed space, and $F : X \to P_{wkc}(Y)$ is a multifunction such that for all $y^* \in Y^*$ $x \to \sigma(y^*, F(x))$ is an $\mathbb{R}$-valued upper semicontinuous function, then $F(\cdot)$ is usc from X into Y_w.*

Proof

(a) We fix $y^* \in Y^*$ and $\varepsilon > 0$ and consider the set

$$U(y^*, \varepsilon) = \{y \in Y : \langle y^*, y \rangle < \varepsilon\}.$$

This set is weakly open in Y and $0 \in U(y^*, \varepsilon)$. For $x \in X$, since $F(\cdot)$ is usc, we can find $V \in \mathcal{N}(x)$ such that

$$F(v) \subseteq F(x) + U(y^*, \varepsilon) \text{ for all } v \in V,$$

$$\Rightarrow \sigma(y^*, F(v)) \leq \sigma(y^*, F(x)) + \varepsilon \text{ for all } v \in V,$$

$$\Rightarrow x \to \sigma(y^*, F(x)) \text{ is upper semicontinuous for all } y^* \in Y^*.$$

(b) Let $V \subseteq Y$ be w-open. According to Lemma 5.21, we can find $\varepsilon > 0$ and $\{y_k^*\}_{k=1}^m \subseteq Y^*$ such that

$$V_0 = \{y \in Y : \langle y_k^*, y \rangle < \sigma(y_k^*, F(x)) + \varepsilon, k = 1, \ldots m\} \subseteq V.$$

By hypothesis for every $k = 1, \ldots, m$, $\sigma(y_k^*, F(\cdot))$ is upper semicontinuous. Therefore, we can find $U \in \mathcal{N}(x)$ such that

$$\sigma(y_k^*, F(x')) < \sigma(y_k^*, F(x)) + \varepsilon \text{ for all } x' \in U,$$

$$\Rightarrow F(x') \subseteq V_0 \subseteq V \text{ for all } x' \in U,$$

$$\Rightarrow F(\cdot) \text{ is usc from } X \text{ into } Y_w. \qquad \square$$

In the results that follow, we will see how the continuity concepts behave with respect to certain basic operations. We start with the closure operation. So, we consider the multifunction $\overline{F} : X \rightarrow 2^Y \setminus \{\emptyset\}$ defined by $\overline{F}(x) = \overline{F(x)}$ for all $x \in X$.

Proposition 5.23 $F(\cdot)$ is lsc at x_0 if and only if $\overline{F}(\cdot)$ is lsc at x_0.

Proof Let $V \subseteq Y$ be open. Then $F(x_0) \cap V \neq \emptyset$ if and only if $\overline{F}(x_0) \cap V = \overline{F(x_0)} \cap V \neq \emptyset$. So, the result follows from this (see Definition 5.4(b)). $\qquad \square$

Proposition 5.24 If Y is normal and $F(\cdot)$ is usc at x_0, then $\overline{F}(\cdot)$ is usc at x_0.

Proof Let $V \subseteq Y$ be open and $\overline{F}(x_0) \subseteq V$. Since Y is normal by Proposition 1.41, we can find $W \subseteq Y$ open such that $\overline{F}(x_0) \subseteq W \subseteq \overline{W} \subseteq V$. Because $F(\cdot)$ is usc at $x_0 \in X$, we can find $U \in N(x_0)$ such that $F(x) \subseteq W$ for all $x \in U$, and hence $\overline{F}(x) \subseteq \overline{W} \subseteq V$ for all $x \in U$. This shows that $\overline{F}(\cdot)$ is usc at x_0. $\qquad \square$

Corollary 5.25 If Y is normal and $F : X \rightarrow 2^Y$ is continuous at x_0, then $\overline{F}(\cdot)$ is continuous at x_0.

Next we deal with the composition operation. Using Proposition 5.3(b), we obtain the following result. As always, X, Y, and V are Hausdorff topological spaces and $F : X \rightarrow 2^Y$, and $G : Y \rightarrow 2^Y$ two multifunctions.

Proposition 5.26 If F and G are usc (resp., lsc), then $G \circ F$ is usc (resp., lsc).

Remark 5.27 However the composition of two closed multifunctions need not be closed.

Now we deal with the operations of union and intersection. For the union, the situation is very pleasant.

Proposition 5.28 If $F, G : X \rightarrow 2^Y$ are two multifunctions, then

(a) F and G are usc $\Rightarrow F \cup G$ is usc.
(b) F and G are lsc $\Rightarrow F \cup G$ is lsc.

(c) F and G are closed $\Rightarrow$ F $\cup$ G is closed.

Proof

(a) Use Proposition 5.2(a).
(b) Use Proposition 5.2(c).
(c) Just observe that $\text{Gr}(F \cup G) = \text{Gr}\,F \cup \text{Gr}\,G$. □

Remark 5.29 Proposition 5.28(b) is in fact valid for arbitrary unions since Proposition 5.2(c) remains valid if instead of a finite union we have an arbitrary one.

For the intersection, the situation is more involved.

Proposition 5.30 *If $F, G : X \to 2^Y \setminus \{\emptyset\}$ are two multifunctions such that $F(x) \cap G(x) \neq \emptyset$ for all $x \in X$, then*

(a) If Y is normal and F, G are $P_f(Y)$-valued and usc, then $F \cap G$ is usc.
(b) If F is closed and G is $P_k(Y)$-valued and usc, then $F \cap G$ is usc.

Proof

(a) Let $x \in X$ and $V \subseteq Y$ open such that $(F \cap G)(x) = F(x) \cap G(x) \subseteq V$. Consider the sets $F(x)$ and $G(x) \setminus V$. These are disjoint closed sets in Y. Since Y is normal, we can find open sets $V_1, V_2 \subseteq Y$ such that

$$F(x) \subseteq V_1, G(x) \setminus V \subseteq V_2, V_1 \cap V_2 = \emptyset. \tag{5.2}$$

Let $V_3 = V_2 \cup V \subseteq Y$ open. Then $G(x) \subseteq V_3$. Since by hypothesis F and G are usc, we can find $U \in \mathcal{N}(x)$ such that

$$F(x') \subseteq V_1 \text{ and } G(x') \subseteq V_3 \text{ for all } x' \in U. \tag{5.3}$$

We have

$$(F \cap G)(x') = F(x') \cap G(x') \subseteq V_1 \cap V_2 \text{ (see (5.3))}$$
$$\subseteq (Y \setminus V_2) \cap (V_2 \cup V) \text{ (see (5.2))}$$
$$\subseteq V \text{ for all } x' \in U,$$

$$\Rightarrow F \cap G \text{ is usc.}$$

(b) Let $x \in V$ and $V \subseteq Y$ open such that $(F \cap G)(x) = F(x) \cap G(x) \subseteq V$. In the present setting, $F(x)$ and $G(x) \setminus V$ are disjoint sets with $F(x)$ closed and $G(x) \setminus V$ is compact. If $G(x) \setminus V = \emptyset$, then for $U = G^+(V) \in \mathcal{N}(x)$ we have $(F \cap G)(x') \subseteq V$ for all $x' \in U$ and so we are done. Therefore we assume that $G(x) \setminus V \neq \emptyset$ and take $y \in G(x) \setminus V$. Note that $(x, y) \notin \text{Gr}\,F$ which is closed in $X \times Y$. So we can find $U_y \in \mathcal{N}(x)$ and $V_y \in \mathcal{N}(y)$ such that

$$(U_y \times V_y) \cap \mathrm{Gr}\, F = \emptyset.$$

The compactness of $G(x) \setminus V$ implies that we can find $\{y_k\}_{k=1}^m \subseteq G(x) \setminus V$ such that $G(x) \setminus V \subseteq \bigcup_{k=1}^m V_{y_k} = V_0$. Let $W = G^+(V \cup V_0)$ and $U = W \cap \left[\bigcap_{k=1}^m U_{y_k}\right] \in \mathcal{N}(x)$. Then we have $(F \cap G)(x') \subseteq V$ for all $x' \in U$ and this proves that $F \cap G$ is usc.   □

In general the intersection of two lsc multifunctions need not be lsc. However, we can prove the following result.

Proposition 5.31 *If $F, G : X \to 2^Y$ are two multifunctions, $F(x) \cap G(x) \neq \emptyset$, $F(\cdot)$ is lsc, and $\mathrm{Gr}\, G \subseteq X \times Y$ is open, then $F \cap G$ is lsc.*

Proof Let $\emptyset \neq V \subseteq Y$ be open, $x \in (F \cap G)^-(V)$ and $y \in F(x) \cap G(x) \cap V$. Then $(x, y) \in \mathrm{Gr}\, G \cap (X \times V)$. By hypothesis, $\mathrm{Gr}\, G \subseteq X \times Y$ is open. Therefore $\mathrm{Gr}\, G \cap (X \times Y) \subseteq X \times Y$ is open and so we can find $U_1(x) \in \mathcal{N}(x)$ and $V_1(y) \in \mathcal{N}(y)$ such that $U_1(x) \times V_1(y) \subseteq \mathrm{Gr}\, G \cap (X \times V)$. We have $F(x) \cap V_1(y) \neq \emptyset$, and since F is lsc, we can find $U_2(x) \in \mathcal{N}(x)$ such that $F(x') \cap V_1(y) \neq \emptyset$ for all $x' \in U_2(x)$. We set $U(x) = U_1(x) \cap U_2(x) \in \mathcal{N}(x)$. We have

$$F(x') \cap V_1(y) \neq \emptyset \text{ for all } x' \in U_2(x) \text{ and } U(x) \cap V_1(y) \subseteq \mathrm{Gr}\, G \cap (X \times V). \quad (5.4)$$

From (5.4), we see that

$$F(x') \cap G(x') \neq \emptyset \text{ for all } x' \in U(x),$$

$$\Rightarrow (F \cap G)^-(V) \subseteq X \text{ is open},$$

$$\Rightarrow F \cap G \text{ is lsc (see Proposition 5.6).}$$

□

For products of multifunctions, the situation is reversed, namely lsc multifunctions behave well, while the situation for usc multifunctions is more delicate. Again X, Y_1, and Y_2 are Hausdorff topological spaces.

Proposition 5.32 *If $F : X \to 2^{Y_1}$ and $G : X \to 2^{Y_2}$ are lsc multifunctions, then so is $x \to (F \times G)(x) = F(x) \times G(x)$.*

Proof Let $W \subseteq Y_1 \times Y_2$ be nonempty open. On account of Proposition 5.3(a), we may assume that W is basic open. So $W = V_1 \times V_2$ with $V_1 \subseteq Y_1, V_2 \subseteq Y_2$ open. Then $(F \times G)^-(W) = F^-(V_1) \cap G^-(V_2)$ (see Proposition 5.3(a)). Since F, G are lsc we see that $F^-(V_1), G^-(V_2) \subseteq X$ are open and so $(F \times G)^-(W) \subseteq X$ is open, which means that $F \times G$ is lsc.   □

For usc multifunctions, we have the following result.

Proposition 5.33 *If $F : X \to P_k(Y_1)$ and $G : X \to P_k(Y_2)$ are usc multifunctions, then $F \times G : X \to P_k(X \times Y)$ is usc too.*

Proof Let $V \subseteq Y_1 \times Y_2$ be nonempty open and $x \in (F \times G)^+(V)$. Since $(F \times G)(x) = F(x) \times G(x) \in P_k(Y_1 \times Y_2)$ (see Theorem 1.102, Tychonov's Theorem), we can find $W \subseteq Y_1 \times Y_2$ basic open such that $(F \times G)(x) \subseteq W \subseteq V$. Then $W = V_1 \times V_2$ with $V_1 \subseteq Y_1, V_2 \subseteq Y_2$ open and we have

$$(F \times G)^+(W) = F^+(V_1) \cap G^+(V_2) \text{ (see Proposition 5.3(a))},$$

$$\Rightarrow (F \times G)^+(W) \subseteq X \text{ is open (since F,G are both usc)},$$

$$\Rightarrow F \times G \text{ is usc}.$$

$\square$

Now we introduce linear structure on Y and consider the operations of vector addition and of convexification. So X is a Hausdorff topological space and Y is a normed space.

Proposition 5.34 *If $F, G : X \to 2^Y \setminus \{\emptyset\}$ are lsc multifunctions, then so is $x \to (F + G)(x) = F(x) + G(x)$.*

Proof From Proposition 5.32, we know that $x \to (F \times G)(x)$ is lsc. Also the vector addition map $k : Y \times Y \to Y$ defined by $k((y, v)) = y + v$ is continuous, hence lsc viewed as a multifunction. Then we have

$$(F + G)(x) = k \circ (F \times G)(x) \text{ for all } x \in X,$$

$$\Rightarrow x \to (F + G)(x) \text{ is lsc } (see \ Proposition \ 5.26).$$

$\square$

In a similar way, using this time Proposition 5.33, we prove the following result.

Proposition 5.35 *If $F, G : X \to P_k(Y)$ are usc multifunctions, then so is $x \to (F + G)(x) = F(x) + G(x)$.*

Proposition 5.36 *If $F : X \to 2^Y \setminus \{\emptyset\}$ is lsc, then so is $x \to \text{conv } F(x)$.*

Proof Let $x \in X$ and $V \subseteq Y$ be open such that $\text{conv } F(x) \cap V \neq \emptyset$. Let $y \in \text{conv } F(x) \cap V$. Then we can find $\{v_k\}_{k=1}^m \subseteq F(x)$ and $\{\lambda_k\}_{k=1}^m \subseteq [0, 1]$ such that

$$y = \sum_{k=1}^m \lambda_k v_k \in V, \text{ with } \sum_{k=1}^m \lambda_k = 1.$$

Since $V \subseteq Y$ is open, we can find $\epsilon > 0$ such that

$$B_\epsilon(y) = \{y' \in Y : \|y' - y\| < \epsilon\} \subseteq V.$$

Consider the open bulls $\{B_\epsilon(v_k)\}_{k=1}^m$. Since by hypothesis F is lsc, for every $k = 1, \ldots m$, we can find $u_k \in \mathcal{N}(x)$ such that

$$F(x') \cap B_\epsilon(v_k) \neq \emptyset \text{ for all } x' \in U_k.$$

We set $U = \cap_{k=1}^m U_k$ and have

$$F(x') \cap B_\epsilon(v_k) \neq \emptyset \text{ for all } x' \in U, \text{ all } k = 1, \dots, m.$$

Consider $v'_k \in F(x') \cap B_\epsilon(v_k)$. Then

$$\| \sum_{k=1}^{m} \lambda_k v'_k - \sum_{k=1}^{m} \lambda_k v_k \| \leq \sum_{k=1}^{m} \lambda_k \|v'_k - v_k\| < \epsilon,$$

$$\Rightarrow \text{conv } F(x') \cap V \neq \emptyset \quad \text{for all } x' \in U.$$

Therefore $x \to \text{conv } F(x)$ is lsc. □

Proposition 5.37 *If* $F : X \to P_{wkc}(Y)$ *is usc into* Y_w,
then $x \to \overline{\text{conv}} F(x)$ *is usc into* Y_w.

Proof For every $y^* \in Y^*$, the function $x \to \sigma(y^*, F(x))$ is upper semicontinuous (see Proposition 5.22(a)). Note that $\sigma(y^*, F(x)) = \sigma(y^*, \overline{\text{conv}} F(x))$. Hence $x \to \sigma(y^*, \overline{\text{conv}} F(x))$ is upper semicontinuous and this by Proposition 5.22(b) implies that $x \to \overline{\text{conv}} F(x)$ is usc into Y_w. □

Proposition 5.38 *If* Y *is a Banach space and* $F : X \to P_k(Y)$ *is usc,*
then $x \to \overline{\text{conv}} F(x)$ *is* $P_k(Y)$*-valued and usc.*

Proof By Theorem 3.130 (Mazur's Theorem), the multifunction $x \to \overline{\text{conv}} F(x)$ is $P_k(Y)$-valued.

Let $x_0 \in X$ and $0 < \epsilon' < \epsilon$. Since $F(\cdot)$ is by hypothesis usc, we can find $\delta > 0$ such that

$$F(B_\sigma(x_0)) \subseteq F(x_0)_{\epsilon'} = \{y \in Y : d(y, F(x_0)) < \epsilon'\},$$

$$\Rightarrow F(B_\sigma(x_0)) \subseteq [\overline{\text{conv}} F(x_0)]_{\epsilon'},$$

$$\Rightarrow \text{conv } F(B_\sigma(x_0)) \subseteq [\overline{\text{conv}} F(x_0)]_{\epsilon'} \text{ (since } [\overline{\text{conv}} F(x_0)]_{\epsilon'} \text{ is convex)}$$

$$\Rightarrow \overline{\text{conv}} F(B_\sigma(x)) \subseteq \overline{[\overline{\text{conv}} F(x_0)]_{\epsilon'}} \subseteq [\overline{\text{conv}} F(x_0)]_{\epsilon'}$$

$$\Rightarrow x \to \overline{\text{conv}} F(x) \text{ is lcs and } P_k(Y) - \text{valued.}$$

□

Now we assume that X is a metric space and introduce an extended (generalized) metric on $P_f(X)$. Using this metric, we will be able to introduce new continuity concepts for multifunctions.

Definition 5.39 Let (X, d) be a metric space and $A, C \subseteq X$. We set

$$h^*(A, C) = \sup[d(a, C) : a \in A] = \inf[\epsilon > 0 : A \subseteq C_\epsilon],$$

where $C_\varepsilon = \{x \in X : d(x, C) < \epsilon\}$ (recall that $\inf \emptyset = +\infty$). Then the "Hausdorff distance" between A and C is defined by

$$h(A, C) = \max\{h^*(A, C), h^*(C, A)\}$$
$$= \inf[\epsilon > 0 : A \subseteq C_\epsilon, C \subseteq A_\epsilon].$$

Remark 5.40 Evidently $h(A, C) = 0$ if and only if $\overline{A} = \overline{C}$. So $(P_f(X) \cup \emptyset, h)$ is an extended (generalized) metric space (that is, a metric space in which the distance function takes values in $\overline{\mathbb{R}}_+ = \mathbb{R}_+ \cup \{+\infty\}$) and $\emptyset$ is an isolated point. Of course $(P_{bf}(X), h)$ is a standard metric space. We call $h(\cdot, \cdot)$ the "Hausdorff metric."

In the next two propositions, we state some straightforward consequences of Definition 5.39.

Proposition 5.41 *Suppose X is a normed space. Then we have*

(a) $h(\lambda A, \lambda C) = |\lambda| h(A, C)$ *for all* $\lambda \in \mathbb{R}$.
(b) $h(A_1 + C_1, A_2 + C_2) \leq h(A_1, A_2) + h(C_1, C_2)$.
(c) $h(\lambda A_1 + (1 - \lambda)C_1, \lambda A_2 + (1 - \lambda)C_2) \leq \lambda h(A_1, C_1) + (1 - \lambda)h(A_2, C_2)$ *for all* $\lambda \in [0, 1]$.
(d) $h(\overline{\text{conv}}A, \overline{\text{conv}}C) \leq h(A, C)$.
(e) $h^*(A, C) = \sup[d(x, C) - d(x, A) : x \in X]$.
(f) *For* $A, C \in P_{bfc}(X)$, *we have*

$$h^*(A, C) = \sup[\sigma(x^*, A) - \sigma(x^*, C) : \|x^*\|_* \leq 1].$$

Proposition 5.42

(a) *If (X,d) is a metric space and $A, C \subseteq X$ nonempty, then $h(A, C) = \sup[|d(x, A) - d(x, C)| : x \in X]$.*
(b) *If X is a normed space and $A, C \in P_{bfc}(X)$, then $h(A, C) = \sup[|\sigma(x^*, A) - \sigma(x^*, C)| : \|x^*\|_* \leq 1]$. (This expression of the Hausdorff distance between $A, C \in P_{bfc}(X)$ is known as "Hormander's formula.")*

The following facts about the metric space $(P_f(X), h)$ can be found in Hu and Papageorgiou [145] (p.8).

Proposition 5.43

(a) *If (X, d) is a complete metric space, so is $(P_f(X), h)$.*
(b) *$P_{bf}(X)$ and $P_k(X)$ are closed subsets of $(P_f(X).h)$.*
(c) *If X is separable, then so is $(P_k(X), h)$.*
(d) *If X is a normed space, then $P_{kc}(X), P_{bfc}(X)$, and $P_{fc}(X)$ are closed subsets of $(P_f(X), h)$.*

Now we can introduce continuity concepts on multifunctions based on the Hausdorff metric.

Definition 5.44 Let X be a Hausdorff topological space, (Y, d) a metric space, and $F : X \to 2^Y \setminus \{\emptyset\}$ a multifunction.

(a) We say that $F(\cdot)$ is "h-upper semicontinuous at $x_0 \in X$" ("h-usc at x_0" for short) if the function $x \to h^*(F(x), F(x_0))$ is continuous at x_0. If $F(\cdot)$ is h-usc at every $x \in X$, then we simply say that $F(\cdot)$ is "h-usc."
(b) We say that $F(\cdot)$ is "h-lower semicontinuous at $x_0 \in X$" (h-lsc at x_0 for short) if the function $x \to h^*(F(x_0), F(x))$ is continuous at x_0. If $F(\cdot)$ is h-lsc at every $x \in X$, then we simply say that $F(\cdot)$ is "h-lsc."
(c) We say that $F(\cdot)$ is "h-continuous at $x_0 \in X$" if it is both h-usc and h-lsc at x_0. If $F(\cdot)$ is h-continuous at every $x \in X$, then we simply say that $F(\cdot)$ is "h-continuous" (hence h-continuity of $F(\cdot)$ is continuity from X into the pseudometric space $(2^Y \setminus \{\emptyset\}, h)$.

Of course we would like to know how these new continuity notions are related to the ones introduced in Definition 5.4. In the next propositions, we investigate this question. In what follows, X is a Hausdorff topological space and (Y, d) is a metric space.

Proposition 5.45 *If $F : X \to 2^Y \setminus \{\emptyset\}$ is usc, then $F(\cdot)$ is h-usc.*

Proof Since $F(\cdot)$ is usc, given $\epsilon > 0$, we have $F^+(F(x)_\epsilon) = U_\epsilon \in \mathcal{N}(x)$ (recall that $F(x)_\epsilon = \{y \in V : d(y, F(x)) < \epsilon\}$, the ϵ-enlargement of $F(x)$). If $x' \in U_\epsilon$, we have $F(x') \subseteq F(x)_\epsilon$, which implies that $h^*(F(x'), F(x)) < \epsilon$ and so we conclude that $F(\cdot)$ is h-usc (see Definition 5.44(a)). $\qquad\square$

Remark 5.46 The converse is not in general true. To see this, let $X = [0, 1]$, $Y = \mathbb{R}$ and consider the multifunction $F : X \to 2^Y \setminus \{\emptyset\}$ defined by

$$F(x) = \begin{cases} [0, 1] & \text{if } 0 \leq x < 1 \\ [0, 1) & \text{if } x = 1. \end{cases}$$

Evidently $F(\cdot)$ is h-usc, but not usc since $F^+((-1, 1)) = \{1\}$.

Proposition 5.47 *If $F : X \to 2^Y \setminus \{\emptyset\}$ is h-lsc, then $F(\cdot)$ is lsc.*

Proof Let $C \subseteq Y$ be nonempty and closed. We show that $F^+(C) \subseteq X$ is closed. So, let $\{x_\alpha\}_{\alpha \in J} \subseteq F^+(C)$ be a net such that $x_\alpha \to x$. We have $F(x_\alpha) \subseteq C$ for all $\alpha \in J$. Since by hypothesis $F(\cdot)$ is h-lsc, given $\epsilon > 0$, we can find $\alpha_0 \in J$ such that

$$h^*(F(x), F(x_\alpha)) \leq \epsilon \text{ for all } \alpha \in J, \alpha \geq \alpha_0,$$

$$\Rightarrow F(x) \subseteq F(x_\alpha)_\epsilon \subseteq C_\epsilon \text{ for all } \alpha \in, \alpha \geq \alpha_0.$$

Since $\epsilon > 0$ is arbitrary, we let $\epsilon \to 0^+$ and obtain

$$F(x) \subseteq C \text{ (recall that } C \text{ is closed)},$$

$$\Rightarrow x \in F^+(C) \text{ and so } F^+(C) \subseteq X \text{ is closed}.$$

This by Proposition 5.6 implies that $F(\cdot)$ is lsc. □

Remark 5.48 Again the converse is not in general true. To see this, let $X = [0, 1]$, $Y = \mathbb{R}_+^2$ and let $F : X \to 2^Y \setminus \{\emptyset\}$ be the multifunction defined by

$$F(x) = \{(t, tx) : t > 0\}.$$

This multifunction is lsc but not h-lsc.

For $P_k(Y)$-valued multifunctions, these continuity concepts coincide.

Proposition 5.49 *If $F : X \to P_k(Y)$ is multifunction, then*

(a) $F(\cdot)$ is usc if and only if h-usc.
(b) $F(\cdot)$ is lsc if and only if h-lsc.

Proof

(a) $\Rightarrow$: Follows from Proposition 5.45.

$\Leftarrow$: Let $C \subseteq Y$ be nonempty and closed. We show that $F^-(C) \subseteq X$ is closed. So, let $\{x_\alpha\}_{\alpha \in J} \subseteq F^-(C)$ be a net such that $x_\alpha \to x$ in X. Let $y_\alpha \in F(x_\alpha) \cap C$ for all $\alpha \in J$. We have

$$d\big(y_\alpha, F(x)\big) \le h^*\big(F(x_\alpha), F(x)\big) \to 0 \text{ (since } F(\cdot) \text{ is } h\text{-usc)}. \tag{5.5}$$

On account of the compactness of $F(x)$, we can find $v_\alpha \in F(x)$ such that $d(y_\alpha, v_\alpha) = d(y_\alpha, F(x))$. Since $\{v_\alpha\}_{\alpha \in J} \subseteq F(x)$ and the latter is compact, we can find a subnet $\{v_\beta\}_{\beta \in I}$ of $\{v_\alpha\}_{\alpha \in J}$ such that $v_\beta \to v \in F(x)$. Hence $y_\beta \to v$ (see (5.5)). Then $d(y, F(x)) = 0$ and so $y \in F(x)$. Therefore, $y \in F(x) \cap C$ and so $x \in F^-(C)$. This proves that $F^-(C) \subseteq X$ is closed and this by Proposition 5.5 implies that $F(\cdot)$ is usc.

(b) $\Rightarrow$: Consider a net $\{x_\alpha\}_{\alpha \in J} \subseteq X$ such that $x_\alpha \to x$ in X. Since $F(x) \subseteq Y$ is compact, we can find $v_\alpha \in F(x)$ such that $h^*(F(x), F(x_\alpha)) = d(v_\alpha, F(x_\alpha))$. Also, we can find $\{v_\beta\}_{\beta \in I}$ a subnet of $\{v_\alpha\}_{\alpha \in J}$ such that $v_\beta \to v \in F(x)$. Exploiting the fact that $F(\cdot)$ is lsc, given $\varepsilon > 0$, we can find $\beta_0 \in I$ such that

$$F(x_\beta) \cap B_{\varepsilon/2}(v) \ne \emptyset, v_\beta \in B_{\varepsilon/2}(v) \text{ for all } \beta \in I, \beta \succeq \beta_0 \tag{5.6}$$

(see Proposition 5.6). Then for $\beta \ge \beta_0$, we have

$$h^*(F(x), F(x_\beta)) \leq d(v_\beta, v) + d(v, F(x_\beta)) < \varepsilon \text{ (see (5.6))},$$

$$\Rightarrow h^*(F(x), F(x_\beta)) \to 0.$$

By Urysohn's criterion for the convergence of nets, we conclude that $F(\cdot)$ is h-lsc.

$\Leftarrow$: Follows from Proposition 5.47. $\square$

Corollary 5.50 *If $F : X \to P_k(Y)$ is a multifunction, then $F(\cdot)$ is continuous if and only if it is h-continuous.*

Next we present some results that are useful for those working in constrained optimization. In what follows, X and Y are Hausdorff topological spaces.

Proposition 5.51 *If $f : X \times Y \to \mathbb{R}$ is a lower semicontinuous function, $F : X \to 2^Y \setminus \{\emptyset\}$ is an lsc multifunction, and $m : X \to \overline{\mathbb{R}} = \mathbb{R} \cup \{+\infty\}$ is defined by*

$$m(x) = \sup \big[f(x, y) : y \in F(x) \big],$$

then $m(\cdot)$ is lower semicontinuous.

Proof Given any $\lambda \in \mathbb{R}$, we need to show that the set $L_\lambda^l = \{x \in X : m(x) \leq \lambda\}$ is closed. So, let $\{x_\alpha\}_{\alpha \in J} \subseteq L_\lambda^l$ be a net such that $x_\alpha \to x$ in X. Given $\varepsilon > 0$, we can find $y \in F(x)$ such that

$$f(x, y) \geq \min \Big\{ m(x), \frac{1}{\varepsilon} \Big\}. \tag{5.7}$$

Since $F(\cdot)$ is lsc, using Proposition 5.6, we can find $y_\alpha \in F(x_\alpha)$, $\alpha \in J$, such that $y_\alpha \to y$ in Y. We have

$$\lambda \geq \liminf_{\alpha \in J} m(x_\alpha) \geq \liminf_{\alpha \in J} f(x_\alpha, y_\alpha) \geq f(x, y) \geq \min \Big\{ (m(x), \frac{1}{\varepsilon} \Big\} \text{ (see (5.7))}.$$

Recall that $\varepsilon > 0$ is arbitrary. So, we let $\varepsilon \to 0^+$ to conclude that

$$m(x) \leq \lambda, \text{ that is, } x \in L_\lambda^l.$$

Therefore, $L_\lambda^l \subseteq X$ is closed and this proves the lower semicontinuity of $m(\cdot)$.
$\square$

Remark 5.52 If $f(\cdot, \cdot)$ is upper semicontinuous, $F(\cdot)$ is an lsc multifunction and $m(x) = \inf \big[f(x, y) : y \in F(x) \big]$, then $m(\cdot)$ is upper semicontinuous. Just multiply with -1 Proposition 5.51.

Proposition 5.53 *If $f : X \times Y \to \mathbb{R}$ is an upper semicontinuous function, $F : X \to P_k(Y)$ is an usc multifunction, and $m : X \to \overline{\mathbb{R}} = \mathbb{R} \cup \{+\infty\}$ is defined by*

$$m(x) = \max \left[f(x, y) : y \in F(x) \right],$$

then $m(\cdot)$ is upper semicontinuous.

Proof Let $\lambda \in \mathbb{R}$. We need to show that $L_\lambda^l = \{x \in X : m(x) < \lambda\}$ is open. To this end, let $x_0 \in L_\lambda^l$ and consider the set $\widehat{U} = \{(x, y) \in X \times Y : f(x, y) < \lambda\}$. The upper semicontinuity of $f(\cdot, \cdot)$ implies that $\widehat{U} \subseteq X \times Y$ is open. Moreover, if $(x_0, y) \in \operatorname{Gr} F$, then $(x_0, y) \in \widehat{U}$ (that is, $\{x_0\} \times F(x_0) \subseteq \widehat{U}$).

Let $y \in F(x_0)$. On account of Proposition 5.33, we can find $U_y \in \mathcal{N}(x_0)$ and $V_y \in \mathcal{N}(y)$ such that $U_y \times V_y \subseteq \widehat{U}$. The collection $\{V_y\}_{y \in F(x_0)}$ is an open cover of $F(x_0)$. So, we can find a finite subcover $\{V_{y_k}\}_{k=1}^m$. Let

$$U = \bigcap_{k=1}^m U_{y_k} \text{ and } V = \bigcup_{k=1}^m V_{y_k}.$$

Then $U \in \mathcal{N}(x_0)$ and $V \subseteq Y$ is open. So, $U_0 = U \cap F^+(V) \in \mathcal{N}(x_0)$. Let $x \in U_0$. If $y \in F(x)$, then $(x, y) \in U \times V \subseteq W$ and so $f(x, y) < \lambda$, hence $m(x) < \lambda$. This proves that L_λ^l is open and so $m(\cdot)$ is upper semicontinuous. $\qquad \square$

Remark 5.54 If $f(\cdot, \cdot)$ is lower semicontinuous, $F : X \to P_k(Y)$ is usc, and $m(x) = \min[f(x, y) : y \in F(x)]$, then $m(\cdot)$ is lower semicontinuous.

The next result combines and completes the above propositions. The result is sometimes called the "Maximum Theorem."

Theorem 5.55 *If $f : X \times Y \to \mathbb{R}$ is a continuous function, $F : X \to P_k(Y)$ is a continuous multifunction, and $m : X \to \mathbb{R}$ and $S : X \to 2^Y \setminus \{\emptyset\}$ are defined by*

$$m(x) = \max \left[f(x, y) : y \in F(x) \right],$$
$$S(x) = \left\{ y \in F(x) : m(x) = f(x, y) \right\} \text{ for all } x \in X,$$

then

(a) $m(\cdot)$ is continuous.
(b) $S : X \to P_k(Y)$ is usc.

Proof

(a) Follows from Propositions 5.51 and 5.53.
(b) Let $C \subseteq Y$ be closed. We need to show that $S^-(C) \subseteq X$ is closed (see Proposition 5.5). Let $\{x_\alpha\}_{\alpha \in J} \subseteq S^-(C)$ be a net such that $x_\alpha \to x$ in X. Then for $y_\alpha \in S(x_\alpha) \cap C, \alpha \in J$, we have

$$y_\alpha \in F(x_\alpha) \text{ and } m(x_\alpha) = f(x_\alpha, y_\alpha) \text{ for all } \alpha \in J. \tag{5.8}$$

Since $F(\cdot)$ is $P_k(Y)$-valued and usc, according to Remark 5.12, we know that $\{y_\alpha\}_{\alpha \in J}$ has a cluster point $y \in F(x)$. Proposition 1.31 says that we can find a subnet $\{y_\beta\}_{\beta \in I}$ of $\{y_\alpha\}_{\alpha \in J}$ such that $y_\beta \to y$ in Y. From part (a) and the continuity of f, we have

$$m(x_\beta) \to m(x) \text{ and } f(x_\beta, y_\beta) \to f(x, y),$$

$$\Rightarrow m(x) = f(x, y) \text{ (see (5.8)) and } y \in C,$$

$$\Rightarrow x \in S^-(C) \text{ and so } S^-(C) \subseteq X \text{ is closed.}$$

This proves that $S(\cdot)$ is usc. $\square$

Remark 5.56 Of course the result remains true if instead of constrained maximization, we have constrained minimization.

For another continuity notion for multifunctions see Bressan [46]

5.2 Measurability of Multifunctions

In this section we switch to the study of the measure theoretic properties of multifunctions.

The standing hypotheses are that (Ω, Σ) is a measurable space and (X, d) is separable metric space. Additional conditions will be introduced as needed.

Definition 5.57 Let $F : \Omega \to 2^X$ be a multifunction.

(a) We say that $F(\cdot)$ is "measurable" if for all $U \subseteq X$ open $F^-(U) \in \Sigma$.
(b) We say that $F(\cdot)$ is "graph measurable" if

$$\text{Gr } F = \big\{(w, x) \in \Omega \times X : x \in F(w)\big\} \in \Sigma \otimes B(X)$$

(recall $B(X)$ is the Borel σ-field of X).
(c) If X is a normed space, then we say that $F(\cdot)$ is "scalarly measurable" if for all $x^* \in X^*$, the function

$$w \to \sigma(x^*, F(w))$$

is Σ-measurable.

Remark 5.58 Let dom $F = \{w \in \Omega : F(w) \neq \emptyset\}$. Evidently if $F(\cdot)$ is measurable, then dom $F \in \Sigma$. So, in the sequel, we consider measurable multifunctions with dom $F = \Omega$.

Proposition 5.59 *If $F^-(C) \in \Sigma$ for all $C \subseteq X$ closed, then $F(\cdot)$ is measurable.*

Proof From Proposition 1.157, we know that every open set $U \subseteq X$ is F_σ (see Definition 1.153). So, $U = \bigcup_{n \in \mathbb{N}} C_n$ with $C_n \subseteq X$ closed. We have

$$F^-(U) = F^-\left(\bigcup_{n \in \mathbb{N}} C_n\right) = \bigcup_{n \in \mathbb{N}} F^-(C_n) \text{ (see Proposition 5.1(c))},$$

$$\Rightarrow F^-(U) \in \Sigma, \text{ that is, } F(\cdot) \text{ is measurable.} \qquad \square$$

For the converse to hold, we need to impose additional conditions on $F(\cdot)$.

Proposition 5.60 *If $F : \Omega \to P_k(X)$ is measurable, then $F^-(C) \in \Sigma$ for all $C \subseteq X$ closed.*

Proof Evidently we may assume that $C \neq \emptyset$. For every $n \in \mathbb{N}$, let $U_n = \{x \in X : d(x, C_n) > \frac{1}{n}\}$. Then $\{U_n\}_{n \in \mathbb{N}}$ is an increasing sequence of open subsets of X. We have $X \backslash C = \bigcup_{n \in \mathbb{N}} U_n = \bigcup_{n \in \mathbb{N}} \overline{U}_n$. Suppose $F(x) \subseteq X \backslash C$. The compactness of $F(x)$ implies that we can find $n \in \mathbb{N}$ such that $F(x) \subseteq U_n \subseteq \overline{U}_n$. Hence we have

$$F^-(C) = X \setminus F^+(X \setminus C) = X \setminus \bigcup_{n \in \mathbb{N}} F^+(\overline{U}_n) = \bigcap_{n \in \mathbb{N}} F^-(X \setminus \overline{U}_n) \in \Sigma.$$

$\square$

Combining Propositions 5.59 and 5.60, we have the following corollary:

Corollary 5.61 *If $F : \Omega \to P_k(X)$, then $F(\cdot)$ is measurable if and only if $F^-(C) \in \Sigma$ for all $C \subseteq X$ closed.*

The next proposition provides a very convenient characterization of measurability of multifunctions.

Proposition 5.62 *The multifunction $F : \Omega \to 2^X \setminus \{\emptyset\}$ is measurable if and only if for all $x \in X$, $w \to d(x, F(w))$ is Σ-measurable.*

Proof $\Rightarrow$: According to Proposition 2.33, we need to show that for every $\lambda \geq 0$, $\mathring{L}^l_\lambda(x) = \{w \in \Omega : d(x, F(w)) < \lambda\} \in \Sigma$. Note that $\mathring{L}^l_\lambda(x) = F^-(B_\lambda(x))$ (recall $B_\lambda(x) = \{u \in X : d(u, x) < \lambda\}$). Hence $\mathring{L}^l_\lambda(x) \in \Sigma$ and so $w \to d(x, F(w))$ is Σ-measurable.

$\Leftarrow$: For every $x \in X$ and every $\lambda > 0$, $F^-(B_\lambda(x)) \in \Sigma$. The separability of X implies that it is second countable (see Proposition 1.54). So, if $U \subseteq X$ is open, $U = \bigcup_{n \in \mathbb{N}} B_{\lambda_n}(x_n)$. Then $F^-(U) = F^-\left(\bigcup_{n \in \mathbb{N}} B_{\lambda_n}(x_n)\right) = \bigcup_{n \in \mathbb{N}} F^-\left(B_{\lambda_n}(x_n)\right) \in \Sigma$ (see Proposition 5.1(c)). $\square$

Corollary 5.63 *$F : \Omega \to 2^X \setminus \{\emptyset\}$ is measurable if and only if $\overline{F}(\cdot)$ is.*

Next we see how measurability and graph measurability are related.

Proposition 5.64 *If $F : \Omega \to P_f(X)$ is measurable, then $F(\cdot)$ is graph measurable.*

Proof Note that $\text{Gr}\, F = \{(w, x) \in \Omega \times X : d(x, F(w)) = 0\}$ (since $F(\cdot)$ is $P_f(X)$-valued). The function $(w, x) \to d(x, F(w))$ is the Caratheodory (see Definition 2.46). Since X is separable from Theorem 2.47, we have that $(w, x) \to d(x, F(w))$ is $\Sigma \otimes B(X)$-measurable. Therefore, $\text{Gr}\, F \in \Sigma \otimes B(X)$. $\qquad\square$

Proposition 5.65 *If* $F : \Omega \to P_f(X)$ *is measurable, then* $F^-(K) \in \Sigma$ *for all* $K \subseteq X$ *compact.*

Proof On account of Theorem 1.116, we can view X as a dense subset of a compact of metric space Y. Consider the multifunction $G : \Omega \to P_k(Y)$ defined by $G(w) = \overline{F(w)}$. By Corollary 5.63, $G(\cdot)$ is measurable. For $K \subseteq X$ compact, we have

$$F^-(K) = \{w \in \Omega : G(w) \cap X \cap K \neq \emptyset\} = G^-(K),$$

$$\Rightarrow F^-(K) \in \Sigma \text{ (see Proposition 5.60)}. \qquad\square$$

Proposition 5.66 *If* X *is* σ*-compact and* $F : \Omega \to P_f(X)$*, then the following properties of* $F(\cdot)$ *are equivalent:*

(a) $F^-(C) \in \Sigma$ *for all* $C \subseteq X$ *closed.*
(b) $F(\cdot)$ *is measurable.*
(c) $F^-(K) \in \Sigma$ *for all* $K \subseteq X$ *compact.*

Proof $(a) \Rightarrow (b)$ This implication follows from Proposition 5.59.
$\quad (b) \Rightarrow (c)$ This implication follows from Proposition 5.65.
$\quad (c) \Rightarrow (a)$ Since X is σ-compact, we have $X = \bigcup_{n \in \mathbb{N}} K_n$ with $K_n \subseteq X$ compact for all $n \in \mathbb{N}$. If $C \subseteq X$ is nonempty closed, then $C = \bigcup_{n \in \mathbb{N}} C \cap K_n$ with $C \cap K_n \in P_k(X)$. From Proposition 5.1(c), we have

$$F^-(C) = F^-\left(\bigcup_{n \in \mathbb{N}} (C \cap K_n)\right) = \bigcup_{n \in \mathbb{N}} F^-(C \cap K_n) \in \Sigma. \qquad\square$$

To discuss further the measurability of multifunctions, we need to stop for a moment and study the projection on Ω of a Borel set in $\Omega \times X$. It is well known that if $B \subseteq \mathbb{R}^2$ is a Borel set, then $proj_{\mathbb{R}} B$ need not be Borel. This observation was the starting point of the theory of analytic sets by Souslin (1917).

Recall that $\overline{\Sigma}$ is the universal σ-algebra corresponding to Σ (see Definition 2.168(a)). We know that if (Ω, Σ, μ) is a complete, σ finite measure space, then $\Sigma = \overline{\Sigma}$.

Definition 5.67 A measurable space (Ω, Σ) is said to be complete if $\Sigma = \overline{\Sigma}$.

Remark 5.68 If $X = \mathbb{R}^N$ and $\Sigma = B(\mathbb{R}^N)$, then $\overline{\Sigma} = \mathcal{L}_{\mathbb{R}^N}$: the Lebesgue σ algebra of $\mathbb{R}^N$.

Lemma 5.69 *For every* $n \in \mathbb{N}$*, let*

$$D_n = \{\sum_{k \geq 1} \frac{\xi(k)}{4^k} : \xi : \mathbb{N} \to \{0, 1\}, \xi(n) = 1\}, \ n \in \mathbb{N},$$

and then $D_n \subseteq \mathbb{R}$ is compact. Moreover, for every $\xi : \mathbb{N} \to \{0, 1\}$,

$$\sum_{k \geq 1} \frac{\xi(k)}{4^k} \in D_n \ \text{if and only if} \ \xi(n) = 1, n \in \mathbb{N}.$$

Proof Note that $X = \{0, 1\}^{\mathbb{N}}$ is compact and $A_n = \{\xi : \mathbb{N} \to \{0, 1\} : \xi(n) = 1\}$ is a closed subset of X. Hence A_n is compact. The function

$$\psi(\varepsilon) = \sum_{k \geq 1} \frac{\xi(k)}{4^k}$$

is the uniform limit of continuous functions, and therefore $\psi(\cdot)$ is continuous. It follows that $D_n = \psi(A_n) \ n \in \mathbb{N}$ is compact. Moreover, $\psi(\cdot)$ is strictly increasing for the lexicographic order on $\{0, 1\}^{\mathbb{N}}$ and so it is injective. It follows that $\psi(\xi) \in D_n = \psi(A_n)$ if and only if $\xi \in A_n$. □

Lemma 5.70 *If (Ω, Σ) is a measurable space, Y is a Souslin space and $A \in \Sigma \otimes B(Y)$, then there exist $C \in B(\mathbb{R} \times Y)$ and $\psi : \Omega \to \mathbb{R}$ a Σ-measurable function such that*

$$A = \{(w, y) \in \Omega \times Y : (\psi(w), y) \in C\}.$$

Proof On account of Problem 2.27 and since Y is Souslin, we can find $\{E_n\}_{n \in \mathbb{N}} \subseteq \Sigma$ and $\{H_n\}_{n \in \mathbb{N}} \subseteq B(Y)$ such that

$$A \in D = \sigma(\{E_n \times H_n\}_{n \in \mathbb{N}}). \tag{5.9}$$

Let $\psi : \Omega \to \mathbb{R}$ be defined by

$$\psi(w) = \sum_{k \geq 1} \frac{1}{4^k} \chi_{E_k}(w).$$

Using Lemma 5.69, we see that

$$\psi(w) \in D_n \Leftrightarrow \chi(w) = 1 \Leftrightarrow w \in E_n$$

(here D_n is as defined in Lemma 5.69). Evidently $\psi(\cdot)$ being the limit of Σ-measurable functions is Σ-measurable. We set

$$k(w, y) = (\psi(w), y) \ \text{for all} \ (w, y) \in \Omega \times Y$$

and introduce the σ-algebra

$$\mathscr{S} = \{k^{-1}(s) : s \in B(\mathbb{R} \times Y)\}.$$

Note that $\psi^{-1}(D_n) = E_n$ for every $n \in \mathbb{N}$ and then

$$k^{-1}(D_n \times H_n) = \psi^{-1}(D_n) \times H_n = E_n \times H_n, n \in \mathbb{N},$$

$$\Rightarrow E_n \times H_n \in \mathscr{S}, n \in \mathbb{N}.$$

Note that $D \subseteq \mathscr{S}$. Hence $A \in \mathscr{S}$ (see (5.9)) and so from the definition of ψ, we can find $C \in B(\mathbb{R} \times Y)$ such that

$$A = k^{-1}(C),$$

$$\Rightarrow A = \{(w, y) \in \Omega \times Y : (\psi(w), y) \in C\}. \qquad \square$$

The next proposition concerns the inverse images of Souslin sets under the action of measurable maps.

Proposition 5.71 *If (Ω, Σ) is a complete measurable space, Y is a metric space and $f : \Omega \to Y$ is Σ-measurable, then for every $A \in \alpha(Y)$, $f^{-1}(A) \in \Sigma$.*

Proof Arguing by contradiction, suppose that for some $A \in \alpha(Y)$, $f^{-1}(A) \notin \Sigma = \overline{\Sigma}$. So, we can find a finite measure $\mu : \Sigma \to \mathbb{R}_+$ such that $f^{-1}(A) \notin \Sigma_\mu$. Let $\vartheta(A) = \mu(f^{-1}(A))$. Then $(Y, B(Y), \vartheta)$ is a finite measure space and so

$$A = A_0 \cup N \text{ with } A_0 \in B(Y), N = \vartheta - \text{null},$$

$$\Rightarrow f^{-1}(A) = f^{-1}(A_0) \cup f^{-1}(N) \in \Sigma_\mu = \Sigma \text{ (since } \mu(f^{-1}(N)) = 0),$$

a contradiction. Therefore $f^{-1}(A) \in \Sigma$ for all $A \in \alpha(Y)$. $\qquad \square$

Now we are ready for the measurable projection theorem, usually called the "Yankov–von Neumann–Aumann Projection Theorem."

Theorem 5.72 *If (Ω, Σ) is a complete measurable space, Y is a Souslin space, and $A \in \Sigma \bigotimes B(Y)$, then $proj_\Omega A \in \Sigma$.*

Proof On account of Lemma 5.70, there is a $C \in B(\mathbb{R} \times Y) = B(\mathbb{R}) \otimes B(Y)$ and a Σ-measurable function $\psi : \Omega \to \mathbb{R}$ such that

$$A = \{(w, y) \in \Omega \times Y : (\psi(w), y) \in C\},$$
$$\Rightarrow proj_\Omega A = \psi^{-1}(proj_\mathbb{R} C). \tag{5.10}$$

Since $\mathbb{R} \times Y$ is Souslin (see Proposition 1.166), from Proposition 2.54, it follows that C is Souslin. Then, using Proposition 5.71, from (5.10), we infer that $proj_\Omega A \in \Sigma$. $\qquad \square$

Corollary 5.73 *If* (Ω, Σ, μ) *is a complete,* σ-*finite measure space,* Y *is a Souslin space, and* $A \in \Sigma \otimes B(Y)$, *then* $proj_\Omega A \in \Sigma$.

There is another measurable projection theorem due to Levin [182].

Definition 5.74 Let Y be a Hausdorff topological space. We say that Y is of "class σ-MK" if $Y = \bigcup_{n \in \mathbb{N}} K_n$ with each K_n compact, metrizable.

Example 5.75 Let X be a separable Banach space. Then $X^*_{w^*} = X^*$ equipped with the w^*-topology is of class σ-MK. Just note that $X^* = \bigcup_{n \in \mathbb{N}} n \overline{B}^*_l$ and usc, Theorem 3.117.

The projection theorem of Levin [182] is the following:

Theorem 5.76 *If* V *is a Borel space* (*see Definition 2.58*), Y *is a* σ-*MK space* (*see Definition 5.74*), *and* $A \in B(V \times Y) = B(V) \otimes B(Y)$ *with* $A_v = \{y \in Y : (v, y) \in A\} \in P_f(Y)$ *for all* $v \in V$, *then* $proj_V A \in B(V)$.

Using Theorem 5.72, we can enrich further our analysis of measurable multifunctions.

Proposition 5.77 *If* (Ω, Σ) *is a complete measurable space,* Y *is a Souslin space, and* $F : \Omega \to 2^Y \setminus \{\emptyset\}$ *is a multifunction such that*

$$\text{Gr } F \in \Sigma \otimes B(Y),$$

then for all $D \in B(Y)$, $F^-(D) \in \Sigma$.

Proof We have

$$F^{-1}(D) = \{w \in \Omega : F(w) \cap D \neq \emptyset\}$$
$$= proj_\Omega[\text{Gr } F \cap (\Omega \times D)] \in \Sigma \text{ (by Theorem 5.72).} \qquad \square$$

At this point, we can have the first theorem characterizing measurable multifunctions. In the next section, where we will discuss the existence of measurable selections, we will enrich this theorem with additional characterizations of measurability.

Theorem 5.78 *If* (Ω, Σ) *is a measurable space,* (X, d) *is a separable metric space,* $F : \Omega \to P_f(X)$ *is a multifunction, and we consider the following properties:*

(a) $F^-(D) \in \Sigma$ *for all* $D \in B(X)$
(b) $F^-(C) \in \Sigma$ *for all* $C \subseteq X$ *closed*
(c) $F^-(U) \in \Sigma$ *for all* $U \subseteq X$ *open* (*that is,* $F(\cdot)$ *is measurable*)
(d) *For every* $x \in X$, $w \to d(x, F(w))$ *is* Σ-*measurable*
(e) $\text{Gr } F \in \Sigma \otimes B(X)$(*that is,* $F(\cdot)$ *is graph measurable*)

then the following implications hold:

[1] (a) $\Rightarrow$ (b) $\Rightarrow$ (c) $\Leftrightarrow$ (d) $\Rightarrow$ (e).

[2] If X is σ-compact, then (b) $\Leftrightarrow$ (c).
[3] If $\Sigma = \overline{\Sigma}$ (that is (Ω, Σ) is complete) and X is Polish, then all the above properties (a) $\to$ (e) are equivalent.

Remark 5.79 Sometimes in the literature, multifunctions satisfying (a) (resp., (b)) are called Borel measurable (resp., strongly measurable).

Next let us examine the behavior of the measurability properties of $F(\cdot)$ under some basic operations. Recall our standing hypotheses that (Ω, Σ) is a measurable space and (X, d) a separable metric space.

Proposition 5.80 *If $F_n : \Omega \to 2^X \setminus \{\emptyset\}$ $n \in \mathbb{N}$ is a sequence of measurable (resp., strongly measurable, Borel measurable, graph measurable) multifunctions, then so is $w \to \cup_{n \in \mathbb{N}} F_n(w)$.*

Proof Follows from Proposition 5.1(c). $\square$

Proposition 5.81 *If $F_n : \Omega \to P_f(X)$ $n \in \mathbb{N}$ is a sequence of measurable multifunctions and for some $n_0 \in \mathbb{N}$ $F_{n_0}(\cdot)$ is $P_k(X)$-valued, then $w \to F(w) = \bigcap_{n \in \mathbb{N}} F_n(w)$ is measurable.*

Proof First assume that all the multifunctions are $P_k(X)$-valued and consider the multifunction $w \to \Gamma(w) = \prod_{n \in \mathbb{N}} F_n(w)$. On account of Proposition 5.3(a), we have that $F : \Omega \to P_k(X^{\mathbb{N}})$ is measurable. Then, by Corollary 5.61, we have for all $C \subseteq X$ closed $F^-(C) = \{w \in \Omega : \Gamma(w) \cap \Delta \cap C^{\mathbb{N}} \neq \emptyset\} \in \Sigma$, where $\Delta \subseteq X^{\mathbb{N}}$ is the diagonal set in the Cartesian product $X^{\mathbb{N}}$. Hence $F(\cdot)$ is measurable.

Now we drop the requirement that all the F_n are $P_k(X)$-valued and assume that only $F_{n_0}(\cdot)$ is. Using Theorem 1.116, we view X as a dense subset of a compact metric space Y (a closed subset of the Hilbert cube $I^{\mathbb{N}}$). Then, for each $n \in \mathbb{N}$, let $\widehat{F}_n(w) = \overline{F}_n^Y(w)$. From Theorem 5.78, we know that $\widehat{F}_n(\cdot)$ is measurable for each $n \in \mathbb{N}$ and then from the first part of the proof so is $w \to \widehat{F}(w) = \bigcap_{n \in \mathbb{N}} \widehat{F}_n(w)$. Since $F_{n_0} = \widehat{F}_{n_0}$, it follows that $F = \widehat{F}$ and so $w \to F(w)$ is measurable. $\square$

Proposition 5.82 *If X is a normed space, then $F(\cdot)$ is scalarly measurable if and only if $\overline{\mathrm{conv}} F(\cdot)$ is.*

Proof Follows from the fact that for all $x^* \in X^*$, $\sigma(x^*, F(\cdot)) = \sigma(x^*, \overline{\mathrm{conv}} F(\cdot))$.
                                                                                                                   $\square$

We will have a more detailed discussion of scalar measurability in Sect. 5.3 after developing the tool of measurable selections.

Let Ω be a Hausdorff topological space with a Radon measure μ and X a Polish space. If we view a $P_k(X)$-valued multifunction as a map from Ω into the Polish space $(P_k(X), h)$, then using Theorem 2.158 we have the following result.

Theorem 5.83 *$F : \Omega \to P_k(X)$ is measurable if and only if for every $\epsilon > 0$, we can find $K_\epsilon \subseteq \Omega$ compact such that $\mu(\Omega \setminus K_\epsilon) < \epsilon$ and $F|_{K_\epsilon}$ is continuous and h-continuous.*

5.3 Continuous and Measurable Selections

Given X and Y two sets and a multifunction $F : X \to 2^Y \setminus \{\emptyset\}$, a "selection" of $F(\cdot)$ is a function $f : X \to Y$ such that $f(x) \in F(x)$ for all $x \in X$. For multifunctions with topological properties, we look for continuous selections, while for multifunctions with measurability properties, we look for measurable selections.

For continuous selections, we have to choose between usc and lsc multifunctions. The next example illustrates that usc multifunctions should be excluded.

Example 5.84 Let $X = Y = \mathbb{R}$ and consider the multifunction $F : X \to P_k(Y)$ defined by

$$F(x) = \begin{cases} \{-1\} & \text{if } x \le 0 \\ [-1, 1] & \text{if } x = 0 \\ \{1\} & \text{if } 0 < x \end{cases}.$$

This multifunction allows upward jumps in terms of inclusion (at $x = 0$) and so it is usc (see Remark 5.7). However, we cannot have a continuous selection for $F(\cdot)$.

So, we need to focus on lsc multifunctions. For such multifunctions, we have the celebrated "Michael Selection Theorem."

Theorem 5.85 *If X is a paracompact space, Y is a Banach space, and $F : X \to P_{f_c}(Y)$ is an lsc multifunction, then $F(\cdot)$ admits a continuous selection, that is, there exists a continuous function $f : X \to Y$ such that $f(x) \in F(x)$ for all $x \in X$.*

Proof We produce an approximate continuous selection, namely given $\varepsilon > 0$ we produce a continuous function $f_\varepsilon : X \to Y$ such that

$$f_\varepsilon(x) \in F(x) + \varepsilon B_1 \quad \text{for all } x \in X$$

$$\text{with} \quad B_1 = \{y \in Y : \|y\| < 1\}. \tag{5.11}$$

So, let $x \in X$ and let $y_x \in F(x)$. Since $F(\cdot)$ is lsc, the set $F^-(y_x + \varepsilon B_1) \subseteq X$ is an open neighborhood of x. The family $\{F^-(y_x + \varepsilon B_1)\}_{x \in X}$ is an open cover of X. The paracompactness of X implies that we can find a locally finite refinement $\{F^-(y_l + \varepsilon B_1)\}_{i \in I}$ and a corresponding continuous partition of unity $\{\psi_i\}_{i \in I}$ subordinate to it. We set

$$f_\epsilon(x) = \sum_{i \in J} \psi_i(x) u_i, \tag{5.12}$$

with $u_i \in (y_i + \varepsilon B_i) \subseteq (y_{x_i} + \varepsilon B_i)$. The local finiteness of the cover implies that $f_\varepsilon(\cdot)$ is well defined and the convexity of the values of $F(\cdot)$ and (5.12) imply that (5.11) holds.

Having established such an approximate continuous selection inductively, we will generate $\{f_n\}_{n\in\mathbb{N}}$ a sequence of continuous functions from X into Y such that

$$f_n(x) \in F(x) + \frac{1}{2^n}B \quad \text{for all } x \in X, \text{ all } n \in \mathbb{N}, \tag{5.13}$$

$$\|f_{n+1}(x) - f_n(x)\| < \frac{1}{2^{n-1}} \quad \text{for all } x \in X, \text{ all } n \in \mathbb{N}. \tag{5.14}$$

For $n = 1$, the first part of the proof with $\varepsilon = \frac{1}{2}$ gives a continuous function $f_1 : X \to Y$ such that $f_1(x) \in F(x) + \frac{1}{2}B_1$ for all $x \in X$.

Now suppose we have produced functions $\{f_k\}_{k=1}^n$ satisfying (5.13) and (5.14). Consider the multifunction

$$G_n(x) = F(x) \cap [f_n(x) + \frac{1}{2^n}B_1] \quad \text{for all } x \in X, \text{ all } n \in \mathbb{N}.$$

Note that $G_n(\cdot)$ has nonempty, convex values and by Proposition 5.31 $G_n(\cdot)$ is lsc. Then from the first part of the proof, we know that we can find a continuous map $f_{n+1} : X \to Y$ such that

$$f_{n+1}(x) \in G_n(x) + \frac{1}{2^{n+1}}B_1 \quad \text{for all } x \in X,$$

$$\Rightarrow \|f_{n+1}(x) - f_n(x)\| < \frac{1}{2^{n-1}} \quad \text{for all } x \in X.$$

Therefore, by induction, we have the sequence $\{f_n\}_{n\in\mathbb{N}} \subseteq C(X, Y)$ such that (5.13) and (5.14) hold. From (5.14), it follows that $\{f_n\}_{n\in\mathbb{N}} \subseteq C(X, Y)$ is Cauchy and so we have $f_n \xrightarrow{u} f \in C(X, Y)$. Then from (5.13) and since $F(\cdot)$ is $P_f(X)$-valued, we conclude that $f(x) \in F(x)$ for all $x \in X$. $\square$

Remark 5.86 In fact the existence of a continuous selection for every such $F(\cdot)$ implies the paracompactness of X (see Michael [193]).

Corollary 5.87 *If X is a paracompact space, Y is a Banach space, and $F : X \to P_{fc}(Y)$ is lsc, then for every $(\widehat{x}, \widehat{y}) \in \operatorname{Gr} F$, there exists continuous function $\widehat{f} : X \to Y$ such that $\widehat{f}(\widehat{x}) = \widehat{y}$ and $\widehat{f})(x) \in F(x)$ for all $x \in X$.*

Proof Consider the multifunction $\widehat{F} : X \to P_{fc}(Y)$ defined by

$$\widehat{F}(x) = \begin{cases} F(x), & x \neq \widehat{x} \\ \{\widehat{y}\}, & x = \widehat{x}. \end{cases}$$

This multifunction is lsc (it allows downward jumps with respect to inclusion) and $\widehat{F}(x) \subseteq F(x)$. Using Theorem 5.84, we produce $\widehat{f} : X \to Y$ continuous such that $\widehat{f}(x) \in \widehat{F}(x)$ for all $x \in X$. Hence

$$\widehat{f}(x) \in F(x) \quad \text{for all } x \in X, \widehat{f}(\widehat{x}) = \widehat{y}.$$

$\square$

An interesting application of the Michael Selection Theorem (Theorem 5.84) is the following result from Operator Theory.

Proposition 5.88 *If X and Y are Banach spaces and $A \in \mathcal{L}(X, Y)$ is surjective, then there exists $f : Y \to X$ continuous (not necessarily linear) such that $A \circ f = id_X$.*

Proof Let $F(y) = A^{-1}(y)$. From Theorem 3.62, we know that A is an open map. Hence Proposition 5.15(b) implies that $F(\cdot)$ is lsc. Also the continuity and linearity of A imply that $F(y) \in P_{f_c}(X)$ for all $y \in Y$. So, we can apply Theorem 5.84 and find $f : Y \to X$ a continuous map such that $f(y) \in F(y)$ for all $y \in Y$. Evidently we have $A \circ f = id_X$. $\square$

If X is a metric space and Y a separable Banach space, then we can improve Theorem 5.84 and generate a whole sequence of continuous selections dense in $F(x)$. Recall that a metric space is paracompact (see Theorem 1.189(a)).

Theorem 5.89 *If X is a metric space, Y is a separable Banach space, and $F : X \to P_{f_c}(Y)$ is an lsc multifunction, then there is a sequence $\{f_n\}_{n \in \mathbb{N}} \subseteq C(X, Y)$ such that*

$$f_n(x) \in F(x) \quad \text{for all } x \in X, \text{ all } n \in \mathbb{N},$$

$$F(x) = \overline{\{f_n(x)\}}_{n \in \mathbb{N}} \quad \text{for all } x \in X.$$

Proof Let $\{y_n\}_{n \in \mathbb{N}} \subseteq Y$ be dense and let $U_{nm} = F^-(B_{\frac{1}{2^m}}(y_n))$ $n, m \in \mathbb{N}$. Each U_{nm} is open and so by Proposition 1.157 we have $U_{nm} = \bigcup_{k \in \mathbb{N}} C_{nmk}$ with $C_{nmk} \subseteq Y$ closed for all $k \in \mathbb{N}$. We set

$$F_{nmk}(x) = \begin{cases} F(x), & x \notin C_{nmk}, x \in X \\ \overline{F(x) \cap B_{\frac{1}{2^m}}(y_n)}, & x \in C_{nmk}. \end{cases}$$

This multifunction is $P_{f_c}(Y)$-valued and lsc. So we can find $f_{nmk} \in C(X, Y)$ such that $f_{nmk}(x) \in F_{nmk}(x)$ for all $x \in X$ (see Theorem 5.84). Consider the sequence $\{f_{nmk}\}_{n,m,k \in \mathbb{N}} \subseteq C(X, Y)$. Evidently all these functions are continuous selections of $F(\cdot)$ and satisfy $F(x) = \overline{\{f_{nmk}(x)\}}_{n,m,k \in \mathbb{N}}$ for all $x \in X$. $\square$

There is one more continuous selection theorem, also due to Michael [193] (Theorem 3.1), where the interested reader can find its proof (see also Hu and Popageogiou [145], Theorem 4.19, p.97).

Theorem 5.90 *If X is a metric space, Y is a separable Banach space, and $F : X \to 2^Y \setminus \{\emptyset\}$ is convex valued and either its images are finite dimensional or have nonempty interior, then there exists $f : X \to Y$ continuous map such that*

$$f(x) \in F(x) \quad \text{for all } x \in X.$$

What about usc multifunction? We have already seen in Example 5.83 that an usc multifunction need not have a continuous selection. However, a moment's reflection on that example reveals that the multifunction has an approximate continuous selection. More generally we can state the following theorem.

Theorem 5.91 *If X is a metric space, Y is a Banach space, and $F : X \to 2^Y \setminus \{\emptyset\}$ is an usc multifunction with convex values, then given any $\varepsilon > 0$, we can find a locally Lipschitz function $f_\varepsilon : X \to Y$ such that*

$$f_\varepsilon(X) \subseteq \text{conv } F(X) \text{ and } \text{Gr } f_\varepsilon \subseteq \text{Gr } F + \varepsilon B_1, \ (B_1 = \{y \in Y : \|y\| < 1\}).$$

Proof We fix $\varepsilon > 0$. Since by hypothesis $F(\cdot)$ is usc, it is also h-usc (see Proposition 5.45) and so given $x \in X$, we can find $\delta = \delta(x) > 0$ such that

$$F(x') \subseteq F(x) + \frac{\varepsilon}{2} B_1 \quad \text{for all} x' \in B_\delta(x).$$

We can always have $\delta = \delta(x) < \frac{\varepsilon}{2}$. The family $\{B_{\frac{\delta}{4}}(x)\}_{x \in X}$ is a cover of X which is paracompact (being a metric space; see Theorem 1.189). So, we can find a locally finite refinement $\{U_j\}_{j \in J}$. Let $\{\psi_j(\cdot)\}_{j \in J}$ be a locally Lipschitz partition of unity subordinate to it. For each $j \in J$, let $\{(x_j, y_j)\}_{j \in J} \in \text{Gr } F \bigcap (U_j \times Y)$ and define

$$f_\varepsilon(x) = \sum_{j \in J} \psi_j(x) y_j.$$

This is a well-defined locally Lipschitz function and

$$f_\varepsilon(X) \subseteq \text{conv } F(X).$$

Fix $x \in X$ and let $J(x) = \{j \in J : \psi_j(x) > 0\}$. We know that $J(x)$ is finite. For $j \in J(x)$, let $x_j \in X$ such that $U_j \subseteq B_{\sigma(x_j)}(x_j)$. Set $\delta_i = \max\{\delta_{x_j}\}_{j \in J(x)}$. Then $x_j \in B_{\frac{\delta_i}{4}}(x_i)$ and so $U_j \subseteq B_{\delta_i}(x_i)$. Hence, for any $j \in J(x)$, we have

$$y_j \in F(U_j) \subseteq F(x_i) + \frac{\varepsilon}{2} B_i.$$

So, there exists $y_i \in F(x_i)$ such that $\|f_\varepsilon(x) - y_i\| < \frac{\varepsilon}{2}$ and

$$d((x, f_\varepsilon(x)), (x_i, y_i)) = d(x, x_i) + \|f_\varepsilon(x) - y_i\| < \varepsilon,$$

$$\Rightarrow (x, f_\varepsilon(x)) \subseteq \text{Gr } F + \varepsilon B_1,$$

$$\Rightarrow h^*(\text{Gr } F_\varepsilon, \text{Gr } F) < \varepsilon. \qquad \square$$

This leads to the following definition.

Definition 5.92 Let X and Y be Hausdorff topological spaces and $F : X \to 2^Y \setminus \{\emptyset\}$ a multifunction. We say that $F(\cdot)$ is "σ-selectionable" if there exists a sequence $\{F_n\}_{n \in \mathbb{N}}$ of $P_k(Y)$-valued closed multifunctions such that

(a) Each $F_n(\cdot)$ admits a continuous selection.
(b) $F(x) = \bigcap_{n \in \mathbb{N}} F_n(x)$ for all $x \in X$.

The following result shows that the class of σ-selectionable maps is big. The proof of this result can be found in Aubin and Cellina [15] (Theorem 1, p.86).

Theorem 5.93 *If X is a metric space, Y is a Banach space, and $F : X \to P_{f_c}(Y)$ is usc, then $f(\cdot)$ is σ-selectionable.*

When the range space has special structure, then we can have continuous selections of special nature produced in a constructive way.

Proposition 5.94 *If X is a metric space, Y is a reflexive, strictly convex Banach space with the Kadec–Klee property (see Definition 3.242), and $F : X \to P_{f_c}(Y)$ is an lsc multifunction with Gr F sequentially closed in $X \times Y_w$, then the map $x \to m(x) = proj_{F(x)}(0)$ is a continuous selection of $F(\cdot)$ (known as the "minimal selection" of $F(\cdot)$).*

Proof Since X is reflexive and strictly convex and $F(\cdot)$ is $P_{f_c}(X)$-valued, the minimal selection map is well defined.

Let $x_n \to x$. We claim that $\{m(x_n)\}_{n \in \mathbb{N}} \subseteq Y$ is bounded. To see this, let $y \in F(x)$ and consider $y_n \in F(x_n)$ $n \in \mathbb{N}$ such that $y_n \to y$. Such a sequence exists since $F(\cdot)$ is lsc (see Proposition 5.6(d)). We have

$$\|m(x_n)\| \leq \|y_n\| \leq M \quad \text{for some } M > 0, \text{ all } n \in \mathbb{N}. \tag{5.15}$$

Since Y is reflexive, then (5.15) and the Eberlein–Smulian Theorem (see Theorem 3.121) imply that by passing to a suitable subsequence, we can have $m(x_n) \xrightarrow{w} y$ in Y. By hypothesis, $(x, y) \in \text{Gr } F$. If $u \in F(x)$, then since $F(\cdot)$ is lsc, we can find $u_n \in F(x_n)$ $n \in \mathbb{N}$ such that $u_n \to u$ in Y (see Proposition 5.6(d)). Hence

$$\|y\| \leq \liminf \|m(x_n)\| \leq \lim \|u_n\| = \|u\|,$$

$$\Rightarrow y = m(x).$$

Therefore $m(x_n) \xrightarrow{w} m(x)$ in Y. Once again let $v_n \in F(x_n)$ $n \in \mathbb{N}$ such that $v_n \to m(x)$ in Y. Then $\|v_n\| \to \|m(x)\|$. Since $\|m(x_n)\| \leq \|v_n\|$, it follows that $\limsup_{n \to \infty} \|m(x_n)\| \leq \|m(x)\|$. Since we already have

$$\|m(x)\| \leq \liminf_{n \to \infty} \|m(x_n)\|,$$

we infer that $\|m(x_n)\| \to \|m(x)\|$. Then the Kadec–Klee property implies that $m(x_n) \to m(x)$ in X and so $m(\cdot)$ is a continuous selection of $F(\cdot)$. $\qquad\square$

Suppose that the domain is $\mathbb{R}^N$ ordered by a closed convex cone K (that is, $K \in P_{f_c}(\mathbb{R}^N)$, $K \cap (-K) = \{0\}$ and $\lambda K \subseteq K$ for all $\lambda \geq 0$). Then we can speak of K-continuous maps and of a K-continuous selection of a multifunction $F(\cdot)$.

Definition 5.95 Let $K \subseteq \mathbb{R}^N$ be a closed, convex cone, and (Y, d) is a metric space. We say that a function $f : \mathbb{R}^N \to Y$ is "K-continuous at $x_0 \in \mathbb{R}^N$" if and only if for every $\varepsilon > 0$, we can find $\delta = \delta(x_0) > 0$ such that

$$d(f(x), f(x_0)) < \varepsilon \text{ for all } x \in B_\delta(x_0) \cap (x_0 + K).$$

If $f(\cdot)$ is K-continuous at every $x \in X$, then we simply say that $f(\cdot)$ is "K-continuous."

From the above definition, we easily infer the following characterizations of K-continuity.

Proposition 5.96

(a) *F is K continuous at $x_0 \in \mathbb{R}^N$ if and only if $f(x_n) \to f(x_0)$ for every sequence $\{x_n\}_{n \in \mathbb{N}} \subseteq \mathbb{R}^N$ such that*

$$x_n \to x_0 \text{ in } \mathbb{R}^N, x_n - x_0 \in K \text{ for all } n \in \mathbb{N}.$$

(b) *If $K, \widehat{K} \subseteq \mathbb{R}^N$ are closed convex cones, $\widehat{K} \subseteq K$, and $f : \mathbb{R}^N \to Y$ K-continuous, then $f(\cdot)$ is $\widehat{K}$-continuous too.*

(c) *If $\{f_n\}_{n \in \mathbb{N}}$ is a sequence of K-continuous functions and $f_n \overset{u}{\to} f$, then F is K-continuous too.*

Using the notion of K-continuity, we can have a version of Theorem 5.85 (Michael's Selection Theorem), for multifunctions with nonconvex values. The result is due to Bressan [46] (Theorem 1, p.462) where the interested reader can find its proof.

Theorem 5.97 *If $K \subseteq \mathbb{R}^N$ is a closed convex cone, Y is a metric space, and $F : \mathbb{R}^N \to P_f(Y)$ is an lsc multifunction, then $F(\cdot)$ admits a K-continuous selection.*

Before proceeding to measurable selections, we mention an approximation result which is very useful in many occasions including the continuous selection problem.

Proposition 5.98 *If (X, d) is a metric space, Y is a normed space, and $f : X \to Y$ is a continuous function, then given $\varepsilon > 0$, we can find a locally Lipschitz function $f_\varepsilon : X \to Y$ such that*

$$\|f(x) - f_\varepsilon(x)\| < \varepsilon \text{ for all } x \in X.$$

Proof Let $\varepsilon > 0$. For every $x \in X$, we introduce the open set

$$U_\varepsilon(x) = \{u \in X : \|f(u) - f(x)\| < \frac{\varepsilon}{2}\}.$$

Then $\mathscr{S} = \{U_\varepsilon(x)\}_{x \in X}$ is an open cover of X. Since X is paracompact (see Theorem 1.189), we can find a locally finite refinement $D = \{V_j\}_{j \in J}$. Then we can find $\{\varphi_j\}_{j \in J}$ a locally finite partition of unity subordinate to D. We set

$$f_\varepsilon(x) = \sum_{j \in J} \varphi_j(x) f(u_j) \text{ with } u_j \in V_j, j \in J.$$

This function is well defined and locally Lipschitz. Moreover, we have

$$\|f(x) - f_\varepsilon(x)\| = \|\sum_{j \in J} \varphi_j(x) f(x) - \sum_{j \in J} \varphi_j(x) u_j\|$$

$$\leq \sum_{j \in J} \varphi_j(x) \|f(x) - f(u_j)\| < \varepsilon.$$

$\square$

Now we turn our attention to the existence of "measurable selections." We start with a result known in the literature as the "Kuratowski–Ryll Nardzewski Theorem."

Theorem 5.99 *If (Ω, Σ) is a measurable space, X is a Polish space, and $F : \Omega \to P_f(X)$ is a measurable multifunction, then $F(\cdot)$ admits a Σ-measurable selection (that is, there exists a function $f : \Omega \to X$, which is $(\Sigma, B(X))$-measurable and $f(w) \in F(w)$ for all $w \in \Omega$).*

Proof The proof is similar to the one of Theorem 5.85 (Michael's Selection Theorem). Let $D = \{x_m\}_{m \in \mathbb{N}}$ be dense in X and $d(\cdot, \cdot)$ a bounded metric generating the topology on X. We may assume that for this metric we have $diam X < 1$.

For every $m, n \in \mathbb{N}$, we define $B_{m,n} = \{x \in X : d(x_m, x) < \frac{1}{2^n}\}$. Inductively we will generate a sequence $\{f_n\}_{n \in \mathbb{N}_0}$ of Σ-measurable functions from Ω into X such that

$$d(f_n(w), F(w)) < \frac{1}{2^n} \text{ for all } w \in \Omega \text{ and } n \in \mathbb{N}_0; \tag{5.16}$$

$$d(f_n(w), f_{n+1}(w)) < \frac{1}{2^n} \text{ for all } w \in \Omega \text{ and } n \in \mathbb{N}_0. \tag{5.17}$$

Starting, we set $f_0(w) = x$, for all $w \in \Omega$. Evidently $f_0(\cdot)$ satisfies (5.16). (Recall that we assume that $diam X < 1$). Now suppose that $f_n(\cdot)$ satisfies (5.16). Then, for every $w \in \Omega$, we can find $u \in F(w)$ such that $d(u, f_n(w)) < \frac{1}{2^n}$. The density of D in X implies that we can find $x_m \in D$ such that

$$d(x_m, u) + d(u, f_n(w)) < \frac{1}{2^n}, \quad d(x_m, F(w)) < \frac{1}{2^{n+1}},$$

$\Rightarrow w \in C_m = F^-(B_{m,n+1}) \cap f_n^{-1}(B_{m,n})$.

For each $w \in \Omega$, let $m_n(w)$ denote the smallest $m \in \mathbb{N}$ such that $w \in C_m$. We define $f_{n+1}(w) = x_{m_n(w)}$. Then

$$d(f_{n+1}(w), F(w)) < \frac{1}{2^{n+1}}, \ d(f_n(w), f_{n+1}(w)) < \frac{1}{2^n} \text{ for all } w \in \Omega, \text{ all } n \in \mathbb{N}_0.$$

If we show that each $f_n(\cdot)$ is Σ-measurable, then we will complete the induction process. Let $A \in B(X)$. Then

$$f_{n+1}^{-1}(A) = \{w \in \Omega : x_{m_n(w)} \in A\}$$
$$= \bigcup_{x_m \in A} f_{n+1}^{-1}(\{x_m\})$$
$$= \bigcup_{x_m \in A} \{w \in \Omega : m_n(w) = m\}$$
$$= C_m \setminus \bigcup_{k=1}^{m-1} C_k \in \Sigma,$$

$\Rightarrow f_n(\cdot)$ is $\sum$-measurable for every $n \in \mathbb{N}$.

This completes the induction.

From (5.17), we see that $f_n \xrightarrow{u} f$, with $f : \Omega \to X$. So, $f(\cdot)$ is Σ-measurable and $f(w) \in F(w)$ for all $w \in \Sigma$ (see (5.16)).   □

We can have a representation analogous to Theorem 5.89.

Theorem 5.100 *If (Ω, Σ) is a measurable space, X is a Polish space, and $F : \Omega \to P_f(X)$ is a multifunction, then $F(\cdot)$ is measurable if and only if there exists a sequence $\{f_n\}_{n \in \mathbb{N}}$ of Σ-measurable selections of $F(\cdot)$ such that $F(w) = \overline{\{f_n(w)\}_{n \in \mathbb{N}}}$ for all $w \in \Omega$.*

Proof $\Rightarrow$ Let $\{U_n\}_{n \in \mathbb{N}}$ be a countable basis for the metric topology of X, and for every $n \in \mathbb{N}$ consider the multifunction $G_n : \Omega \to 2^X \setminus \{\emptyset\}$ defined by

$$G_n(w) = \begin{cases} F(w) \cap U_n & \text{if } F(w) \cap U_n \neq \emptyset, \\ F(w) & \text{otherwise.} \end{cases}$$

For every $V \subseteq X$ open, we have

$$G_n^-(V) = F^-(V \cap U_n) \cup [F^-(V) \cap (\Omega \setminus F^-(U_n))] \in \Sigma,$$
$$\Rightarrow F_n = \overline{G}_n \text{ is measurable (see Corollary 5.63).}$$

Apply Theorem 5.99 (the Kuratowski–Ryll Nardzewski Selection Theorem) to find a Σ-measurable function $f_n : \Omega \to X$ such that $f_n(w) \in F_n(w) \subseteq F(w)$ for all $w \in \Omega$, all $n \in \mathbb{N}$. Then

$$\overline{\{f_n(w)\}_{n \in \mathbb{N}}} = F(w) \text{ for all } w \in \Omega.$$

$\Leftarrow$ For every $x \in X$, we have

$$d(x, F(w)) = \inf_{n \in \mathbb{N}} d(x, f_n(w)) \text{ for all } w \in \Omega,$$
$$\Rightarrow w \to d(x, F(w)) \text{ is } \Sigma\text{-measurable},$$
$$\Rightarrow F(\cdot) \text{ is measurable (see Theorem 5.78[1])}. \qquad \square$$

We can complete Theorem 5.78 as follows.

Theorem 5.101 *Let (Ω, Σ) be a measurable space, X is a separable metrizable space, and $F : \Omega \to P_f(X)$ is a multifunction. We consider the following statements:*

(a) $F^-(D) \in \Sigma$ for all $D \in B(X)$.
(b) $F^-(C) \in \Sigma$ for all $C \subseteq X$ closed.
(c) $F^-(U) \in \Sigma$ for all $U \subseteq X$ open.
(d) $w \to d(x, F(w))$ is Σ measurable for every $x \in X$ and every compatible metric $d(\cdot, \cdot)$ on X.
(e) There exists a sequence $\{f_n\}_{n \in \mathbb{N}}$ of Σ-measurable selections of $F(\cdot)$ such that $F(w) = \overline{\{f_n(w)\}_{n \in \mathbb{N}}}$ for all $w \in \Omega$.
(f) Gr $F \in \Sigma \otimes B(X)$.

Then the following implications are true:

[1] $(a) \Rightarrow (b) \Rightarrow (c) \Leftrightarrow (d) \Rightarrow (f)$.
[2] If X is Polish, then $(c) \Leftrightarrow (d) \Leftrightarrow (e)$.
[3] If X is σ-compact, then $(b) \Leftrightarrow (c)$.
[4] If $\Sigma = \overline{\Sigma}$ and X is Polish, then $(a) \to (f)$ are equivalent.

There is a version of this theorem for nonmetrizable spaces (see Klei [161]).

Theorem 5.102 *Let (Ω, Σ) be a complete measurable space, X is a regular Souslin space, and $F : \Omega \to P_f(X)$ is a multifunction. We consider the following statements:*

(a) $F^-(D) \in \Sigma$ for all $D \in B(X)$.
(b) $F^-(C) \in \Sigma$ for all $C \subseteq X$ closed.
(c) $F^-(U) \in \Sigma$ for all $U \subseteq X$ open.
(d) There exists a sequence $\{f_n\}_{n \in \mathbb{N}}$ of Σ-measurable selections of $F(\cdot)$ such that $F(w) = \overline{\{f_n(w)\}_{n \in \mathbb{N}}}$ for all $w \in \Omega$.
(e) Gr $F \in \Sigma \otimes B(X)$.
(f) For every continuous function $\beta : X \to \mathbb{R}$, the function $w \to \widehat{m}(w) = \sup[\beta(x) : x \in F(w)]$ is Σ-measurable. Then the following implications are true:

[1] $(a) \Leftrightarrow (d) \Leftrightarrow (e) \Leftrightarrow (f) \Rightarrow (b) \Rightarrow (c)$.

[2] If X is second countable, then (a) → (f) are equivalent.

We know that if X is a separable Banach space, then $X^*_{w^*}$ (= the dual X^* equipped with the w^*-topology) is a regular and second countable Souslin space. So, we have the following result.

Proposition 5.103 *If (Ω, Σ) is a complete measurable space, X is a separable Banach space, and $F : \Omega \to 2^{X^*} \setminus \{\emptyset\}$ is a multifunction with w^*-compact convex values, then $F^-(U) \in \Sigma$ for all $U \subseteq X^*$ w^*-open if and only if for every $x \in X$, the $\mathbb{R}$-valued function $w \to \sigma(x; F(w))$ is Σ-measurable.*

Given a multifunction $F : \Omega \to 2^X \setminus \{\emptyset\}$ (X is a Banach space) by $ext\,F(\cdot)$, we denote the multifunction which to each $w \in \Omega$ associates the extreme points of the set $F(w)$ (see Definition 3.245). In general we can have $ext\,F(w) = \emptyset$. However if $F(w) \in P_{wkc}(X)$, then $ext\,F(w) \neq \emptyset$ (see Theorem 3.246).

The next result is useful in the study of differential inclusions, as well as in applied areas such as control theory.

Proposition 5.104 *If (Ω, Σ) is a measurable space, X is a separable Banach space, and $F : \Omega \to P_{wkc}(X)$ is a measurable multifunction, then $w \to ext\,F(w)$ is graph measurable.*

Proof Consider X^* with the Mackey topology $m(X^*, X)$ (see Definition 3.114). Then X^* is m-separable. Let $\{x^*_n\}_{n\in\mathbb{N}} \subseteq \overline{B}^*_1 = \{x^* \in X^* : \|x^*\|_x \leq 1\}$ be m-dense in $\overline{B}^*_1$. Consider the function $\vartheta_F : \Omega \times X \to \overline{\mathbb{R}}_+ = \mathbb{R}_+ \cup \{+\infty\}$ defined by

$$\vartheta_F(w, x) = \begin{cases} \sum_{n\in\mathbb{N}} \frac{\langle x^*_n, x\rangle^2}{2^n}, & \text{if } x \in F(w), \\ +\infty, & otherwise. \end{cases}$$

Clearly $\vartheta_F(\cdot, \cdot)$ is $\Sigma \otimes B(X)$-measurable, and for every $w \in \Omega$ we see that $\vartheta_F(w, \cdot)|_{F(w)}$ is continuous. Let $\mathcal{A}$ be the collection of all continuous affine functions $a : X \to \mathbb{R}$. Let

$$\widehat{\vartheta}_F(w, x) = \inf[a(x) : a \in \mathcal{A}, a(u) > \vartheta_F(w, u) \text{ for all } u \in F(w)].$$

Using Theorem 5.100, we can find a sequence $\{f_n\}_{n\in\mathbb{N}}$ of Σ-measurable selections of $F(\cdot)$ such that

$$F(w) = \overline{\{f_n(w)\}}_{n\in\mathbb{N}} \text{ for all } w \in \Omega. \tag{5.18}$$

For every $(w, x^*) \in \Omega \times X^*$, we define

$$\gamma_{x^*}(w) = \sup[\vartheta_F(w, x) - \langle x^*, x\rangle : x \in F(w)]$$

Note that for every $w \in \Omega$, $x^* \to \gamma_{x^*}(w)$ is m-continuous, while $(w, x^*) \to \gamma_{x^*}(w)$ is $\Sigma \otimes B(X^*_m) = \Sigma \otimes B(X^*)$ measurable (see (5.18)). If $\{u^*_n\}_{n\in\mathbb{N}} \subseteq X^*$ is m-dense, then

$$\widehat{\vartheta}_F(w, x) = \inf_{n \in \mathbb{N}} [\langle u_n^*, x \rangle + \gamma_{u_n^*}(w)] \in \Sigma \otimes B(X^*).$$

But we know that

$$ext F(w) = \{x \in X : \widehat{\vartheta}_F(w, x) = \vartheta_F(w, x)\} \text{ (see Choquet [66], Chap. 6),}$$
$$\Rightarrow w \to ext f(w) \text{ is graph measurable.} \qquad \square$$

Remark 5.105 We set $\xi_F(\omega, x) = \widehat{\vartheta}_F(\omega, x) - \vartheta_F(\omega, x)$. This is known as the "Choquet function" for the multifunction $F(\cdot)$. For every $\omega \in \Omega$, $\xi_F(\omega, \cdot)$ is strictly concave and upper semicontinuous on X.

In Theorem 5.99 (the Kuratowski–Ryll Nardzewski Selection Theorem), we required that the multifunction $F(\cdot)$ is closed valued. The next measurable selection theorem drops this requirement. The result known as the "Yankov–von Neumann–Aumann Selection Theorem."

Theorem 5.106 *If* (Ω, Σ) *is a complete measurable space,* X *is a Souslin space and* $F : \Omega \to 2^X \setminus \{\emptyset\}$ *is a graph measurable multifunction, then* $F(\cdot)$ *admits a* Σ-*measurable selection (that is, there exists a* Σ-*measurable function* $f : \Omega \to X$ *such that* $f(\omega) \in F(\omega)$ *for all* $\omega \in \Omega$ *).*

Proof From Lemma 5.70, we know that there exist $C \in B(\mathbb{R} \times Y)$ and $\psi : \Omega \to \mathbb{R}$ a Σ-measurable function such that

$$\operatorname{Gr} F = \{(\omega, x) \in \Omega \times X : (\psi(\omega), x) \in C\}.$$

The set $C \in B(\mathbb{R} \times X)$ is Souslin in $\mathbb{R} \times X$ (see Proposition 2.54). Hence we can find a Polish space and a continuous surjection $\eta : Y \to C$. Let

$$K(\omega) = \eta^{-1}(\{\psi(\omega)\} \times X).$$

Since $F(\omega) \neq \emptyset$ for all $\omega \in \Omega$ and η is continuous surjective, it follows that

$$K(\omega) \in P_f(Y) \text{ for all } \omega \in \Omega.$$

If $U \subseteq Y$ is nonempty open, then U is Polish (see Proposition 1.147). Then on account of the continuity of η, we have that

$$proj_{\mathbb{R}}(\eta(U)) \in \alpha(\mathbb{R}) \text{ (see Corollary 2.55).}$$

The function ψ is Σ-measurable. So

$$K^-(U) = \psi^{-1}(proj_{\mathbb{R}}(\eta(U))) \in \Sigma,$$

and thus $K(\cdot)$ is a measurable, $P_f(Y)$-valued multifunction. We apply Theorem 5.99 and obtain $g : \Omega \to Y$ a Σ-measurable selection of $K(\cdot)$. For every $\omega \in \Omega$,

$$\eta(g(\omega)) \in (\{\psi(\omega)\} \times X) \cap C.$$

Therefore $proj_{\mathbb{R}}(\eta(g(\omega))) = \psi(\omega)$ for all $\omega \in \Omega$. Let

$$f(\omega) = proj_X(\eta(g(\omega))) \text{ for all } \omega \in \Omega.$$

Then $f(\cdot)$ is Σ-measurable and for every $\omega \in \Omega$

$$(\psi(\omega), f(\omega)) = \eta(g(\omega)) \in C \text{ for all } \omega \in \Omega,$$
$$\Rightarrow (\omega, f(\omega)) \in \text{Gr } F \text{ for all } \omega \in \Omega$$
$$\Rightarrow f(\cdot) \text{ is a } \Sigma\text{-measurable selection of } F(\cdot). \qquad \square$$

In fact in analogy to Theorem 5.100, we can also have the following representation theorem (see Theorem 5.102[1]).

Theorem 5.107 *If (Ω, Σ) is a complete measurable space, X is a regular Souslin space, and $F : \Omega \to 2^X \setminus \{\emptyset\}$ is a graph measurable multifunction, then there is a sequence $\{f_n\}_{n \in N}$ of Σ-measurable selections of $F(\cdot)$ such that $F(\omega) \subseteq \overline{\{f_n(\omega)\}_{n \in N}}$ for all $\omega \in \Omega$.*

The next theorem is useful in many occasions (for example, in the existence of extremal constant sign solutions for elliptic equations). It establishes the existence of the essential supremum function of a family of measurable functions. Recall that, if $\mathscr{F} \subseteq M(\Omega) =$ set of all measurable functions $f : \Omega \to \mathbb{R}$, then we say that $\mathscr{F}$ is upward (resp., downward) directed, if for all $u, v \in \mathscr{F}$, we can find $y \in \mathscr{F}$ such that

$$u \le y, v \le y(\text{ resp., } y \le u, y \le v).$$

Also, we recall the definition of the notion of the supremum of a family of measurable functions.

Definition 5.108 Let (Ω, Σ, μ) be a measure space and $\mathcal{F} = \{f_\beta\}_{\beta \in I}$ is a family of Σ-measurable functions. We say that a Σ-measurable function $\widehat{f} : \Omega \to \mathbb{R}^* = \mathbb{R} \cup \{\pm\infty\}$ is the "essential supremum function" of $\mathcal{F}$ if the following properties are satisfied:
(a) $f_\beta(\omega) \le \widehat{f}(\omega)$ for $\mu - a.a \ \omega \in \Omega$ and all $\beta \in I$.
(b) If $f : \Omega \to \mathbb{R}^* = \mathbb{R} \cup \{\pm\infty\}$ is a Σ-measurable function such that $f_\beta(\omega) \le f(\omega)$ for $\mu - a.a \ \omega \in \Omega$ and all $\beta \in I$, then $\widehat{f}(\omega) \le f(\omega)$ for all $\mu - a.a \ \omega \in \Omega$.

Theorem 5.109 *If (Ω, Σ, μ) is a σ-finite measure space and $\mathcal{F} = \{f_\beta\}_{\beta \in I}$ is a family of Σ-measurable functions $f_\beta : \Omega \to \mathbb{R}^* = \mathbb{R} \cup \{\pm\infty\}$, then there exists a countable set $I_c \subseteq I$ such that*

$$\widehat{f}(\omega) = \sup_{\beta \in I_c} f_\beta(\omega), \tag{5.19}$$

which is Σ-measurable, is the essentially supremum of the family $\mathcal{F}$(that is, $\widehat{f}(\omega) \le g(\omega)$ for all $g : \Omega \to \mathbb{R}$ Σ-measurable such that $f \le g$ $\mu - a.e$ for all $f \in \mathcal{F}$); if $\mathcal{F}$ is upward directed, then we can assume that $\{f_\beta\}_{\beta \in I_c}$ is increasing.

Proof If I is itself countable, then $I_c = I$ and $\widehat{f}(\cdot)$ given by (5.19) satisfies the requirements of Definition 5.108.

So, suppose that I is uncountable. We may assume without any loss of generality that $\mu(\cdot)$ is finite. Moreover, since only the order structure of $\mathbb{R}^*$ is involved, we may assume that the functions take their values in $[-1, 1]$ (just consider an increasing bijection mapping $\mathbb{R}^*$ onto $[-1, 1]$, for example, we can consider $\frac{2}{\pi}tan^{-1}x$).

Let $\mathcal{D}$=the family of all countable subsets of I and let

$$\widehat{m} = \sup \left[\int_\Omega (\sup_J f_\beta) d\mu : J \in \mathcal{D} \right]. \tag{5.20}$$

Then $\widehat{m} \le \mu(X)$. For every $n \in \mathbb{N}$, we can find $J_n \in \mathcal{D}$ such that

$$\widehat{m} \le \int_\Omega (\sup_{J_n} f_\beta) d\mu + \frac{1}{n} \text{ for all } n \in \mathbb{N}. \tag{5.21}$$

Set $I_c = \bigcup_{n \in \mathbb{N}} J_n$ and let $\widehat{f}(\cdot)$ be defined by (5.19). Clearly Definition 5.108(b) is satisfied. We need to show that Definition 5.108(a) is satisfied. From (5.20) and (5.21), $\widehat{m} - \frac{1}{n} \le \int_\Omega (\sup_{J_n} f_\beta) d\mu \le \int_\Omega \widehat{f} d\mu \le \widehat{m}$ for all $n \in \mathbb{N}$, thus

$$\int_\Omega \widehat{f} d\mu = \widehat{m} \text{ (just let } n \to \infty). \tag{5.22}$$

Also using (5.22), for every $\vartheta \in I$, we have

$$\widehat{m} = \int_\Omega \widehat{f} d\mu \le \int_\Omega max\{\widehat{f}, f_\vartheta\} d\mu$$
$$\le \int_\Omega (\sup_{I_c \cup \{\vartheta\}} f_\beta) d\mu \le \widehat{m} \text{ (see (5.20))}.$$

Then all the inequalities are equalities and we conclude that

$$f_\vartheta(\omega) \le \widehat{f}(\omega) \text{ for } \mu\text{-a.a } \omega \in \Omega.$$

Finally if $\mathcal{F}$ is upward directed, then it is clear that we can choose $\{f_\beta\}_{\beta \in I_c}$ to be increasing. $\square$

An analogous result can be stated for closed-valued measurable multifunctions. First the definition of the essential supremum multifunction.

Definition 5.110 Let (Ω, Σ, μ) be a measure space, X is a separable metric space, and $\{F_\beta\}_{\beta \in I}$ is a family of $P_f(X)$-valued, measurable multifunctions. A measurable multifunction $\widehat{F} : \Omega \to P_f(X)$ is the "essential supremum multifunction" of the family if the following properties are satisfied:

(a) $F_\beta(\omega) \subseteq \widehat{F}(\omega)$ for $\mu - a.a\ \omega \in \Omega$ and all $\beta \in I$.
(b) If $F : \Omega \to P_f(X)$ is a measurable multifunction such that

$$F_\beta(\omega) \subseteq F(\omega) \text{ for } \mu - a.a\ \omega \in \Omega \text{ and all } \beta \in I,$$

then $\widehat{F}(\omega) \subseteq F(\omega)$ for $\mu - a.a\ \omega \in \Omega$.

Proposition 5.111 *If (Ω, Σ, μ) is a σ-finite measure space, X is a separable metric space, and $\mathscr{S} = \{F_\beta\}_{\beta \in I}$ is a family of $P_f(X)$-valued measurable multifunctions, then there exists a countable set $I_c \subseteq I$ such that*

$$\widehat{F}(\omega) = \overline{\bigcup_{\beta \in I_c} F_\beta(\omega)} \text{ for all } \omega \in \Omega \tag{5.23}$$

is the essential supremum multifunction of the family $\mathscr{S}$.

Proof Let $\{x_m\}_{m \in \mathbb{N}} \subseteq X$ be dense and consider the countable family of open balls $B_{\frac{1}{k}}(x_m)$, $m, k \in \mathbb{N}$. Using Theorem 5.109, we can find $J_{m,k} \subseteq I$ countable such that

$$\bigcup_{\beta \in J_{m_1 k}} F_\beta^- \left(B_{\frac{1}{k}}(x_m) \right)$$

is the essential union of the family of sets $\left\{ F_\beta^- \left(B_{\frac{1}{k}}(x_m) \right) \right\}_{m,k \in \mathbb{N}}$. Then for every $\beta \in I$, we have

$$F_\beta^- \left(B_{\frac{1}{k}}(x_m) \right) \subseteq \left[\bigcup_{\beta \in J_{m_1 k}} F_\beta^- \left(B_{\frac{1}{k}}(x_m) \right) \right] \cup N, \quad N = \mu - \text{null}. \tag{5.24}$$

Set $I_c = \bigcup_{m,k \in \mathbb{N}} J_{m,k}$. This is countable and we define $\widehat{F}(\cdot)$ as in (5.23). First we show that $\widehat{F} : \Omega \to P_f(X)$ is measurable. So, let $U \subseteq X$ be open and let $w \in \widehat{F}^-(U)$. Then

$$U \cap \bigcup_{\beta \in I_c} F_\beta(w) \neq \emptyset,$$

thus $\widehat{F}^-(U) = \bigcup_{\beta \in I_c} F_\beta^-(U) \in \Sigma$.

Next we show that $\widehat{F}(\cdot)$ satisfies (a) and (b) of Definition 5.110. Property (b) is clear. We show (a). So, let $\beta \in I$. Then from (5.24), we see that up to μ-null set, we have

$$F_\beta^- \left(B_{\frac{1}{k}} (x_m) \right) \subseteq \widehat{F}^- \left(B_{\frac{1}{k}} (x_m) \right) \text{ for all } m, k \in \mathbb{N}, \tag{5.25}$$

Let $\omega \in \Omega \setminus N_\beta$ and $u \in F_\beta(\omega)$. Let $\{x_{m_{l_\omega}}\}_{l \in \mathbb{N}} \subseteq \{x_m\}_{m \in \mathbb{N}}$ and $\{k_{l_w}\} \subseteq \mathbb{N}$ subsequences such that

$$k_{l_w} \to \infty \text{ and } u \in B_{\frac{1}{k_{l_\omega}}} (x_{m_l}).$$

It follows that

$$\omega \in F_\beta^- \left(B_{\frac{1}{k_{l_\omega}}} (x_{m_{l_\omega}}) \right) \subseteq \widehat{F}^- \left(B_{\frac{1}{k}} (x_m) \right). \text{ (see (5.25))}$$

Hence we can find $u_{l_\omega} \in \widehat{F}(\omega) \cap B_{\frac{1}{k_{l_\omega}}} (x_{m_{l_\omega}})$ and $u_{l_\omega} \to u \in \widehat{F}(\omega)$ (since $\widehat{F}(\cdot)$ is closed valued). This proves that

$$F_\beta(\omega) \subseteq \widehat{F}(\omega) \text{ for } \mu - a.a. \; \omega \in \Omega, \text{ all } \beta \in I,$$

$$\Rightarrow \widehat{F}(\cdot) \text{ is the essential supremum multifunction.} \qquad \square$$

5.4 Decomposable Sets

The notion of decomposability is an effective substitute of convexity. Also the notion is closely related to multifunctions, since closed decomposable sets in $L^p(\Omega, X)$ are the sets of L^p-selections of a measurable, closed valued multifunction.

The standing hypotheses in this section are the following: (Ω, Σ, μ) is a σ-finite measure space and X is a separable Banach space. Additional hypotheses will be introduced as needed. By $L^0(\Omega, X)$, we denote the space of the equivalence classes in the space $M(\Omega, X) = \{f : \Omega \to X : \Sigma - \text{measurable}\}$ with the equivalence relation $f \sim g$ if and only if $f(\omega) = g(\omega)$ μ-a.e.

Definition 5.112 Let $D \subseteq L^0(\Omega, X)$. We say that D is "decomposable" if for every triple $(A, f_1, f_2) \in \Sigma \times D \times D$, we have

$$\chi_A f_1 + \chi_{\Omega \setminus A} f_2 \in D.$$

Remark 5.113 Since $\chi_{\Omega \setminus A} = 1 - \chi_A$, the notion of decomposability formally looks like that convexity. Only now the coefficients are not in $[0, 1]$ but are functions with values $\{0, 1\}$. Nevertheless we will see that decomposable sets often behave like convex ones.

Given a multifunction $F : \Omega \to 2^X \setminus \{\emptyset\}$ and $1 \leqslant p \leqslant \infty$, let

$$S_F^p = \left\{ f \in L^p(\Omega, X) : f(w) \in F(\omega) \ \mu - \text{a.e.} \right\}.$$

and $S_F = \left\{ f \in L^0(\Omega, X) : f(\omega) \in F(w) \ \mu - \text{a.e.} \right\}.$

Clearly both these sets are decomposable.

Proposition 5.114 *If $F : \Omega \to 2^X \setminus \{\emptyset\}$ is graph measurable, $1 \leq p \leq \infty$, and $S_F^p \neq \emptyset$, then we can find a sequence $\{f_n\}_{n \in \mathbb{N}} \subseteq S_F^p$ such that*

$$F(w) \subseteq \overline{\{f_n(w)\}}_{n \in \mathbb{N}} \ \mu - a.e.$$

Proof On account of Theorem 5.107, we can find $\{g_n\}_{n \in \mathbb{N}} \subseteq S_F$ such that

$$F(w) \subseteq \overline{\{g_i(w)\}}_{i \in \mathbb{N}} \ \mu - a.e.$$

Also since $\mu(\cdot)$ is σ-finite, we can find $\{A_m\}_{m \in \mathbb{N}} \subseteq \Sigma$, a partition of Ω such that $\mu(A_m) < \infty$ for all $m \in \mathbb{N}$. Let $f \in S_F^p$ and define

$$E_{imk} = \left\{ w \in \Omega : k - 1 \leq \|g_i(w)\| < k \right\} \cap A_m \in \Sigma$$

$$\widehat{f}_{imk} = \chi_{E_{imk}} g_i + \chi_{\Omega \setminus E_{imk}} f \in S_F^p, \ i, m, k \in \mathbb{N}.$$

We see that $f_n = \widehat{f}_{imk} \in S_F^p$ is the desired dense sequence. $\qquad \square$

Corollary 5.115 *If $F, G : \Omega \to 2^X \setminus \{\emptyset\}$ are graph measurable multifunctions and for some $1 \leq p \leq \infty$ we have $S_F^p = S_G^p \neq \emptyset$, then $F(w) = G(w) \ \mu - a.e.$*

Some more properties of decomposable sets which are straightforward consequences of Definition 5.112.

Proposition 5.116

(a) *If $\{D_\alpha\}_{\alpha \in I} \subseteq L^p(\Omega, X) \ 1 \leq p \leq \infty$ are (closed) decomposable sets, then so is $\bigcap_{\alpha \in I} D_\alpha \subseteq L^p(\Omega, X)$.*

(b) *If $\{D_n\}_{n \in \mathbb{N}} \subseteq L^p(\Omega, X) \ 1 \leq p \leq \infty$ is an increasing sequence of decomposable sets, then so is $\bigcup_{n \in \mathbb{N}} D_n \subseteq L^p(\Omega, X)$.*

(c) *If $D \subseteq L^p(\Omega, X) \ 1 \leq p \leq \infty$ is decomposable, then so is $\overline{D} \subseteq L^p(\Omega, X)$.*

(d) If $D \subseteq L^p(\Omega, X)$ $1 \leq p \leq \infty$ is decomposable and $g \in L^p(\Omega, X)$, then $D - g \subseteq L^p(\Omega, X)$ is decomposable.

From this proposition, it follows that given $E \subseteq L^p(\Omega, X)$ $1 \leq p \leq \infty$, there is a smallest (closed) decomposable set containing E. We call this minimal decomposable set the "(closed) decomposable hull" of E and is denoted by $dec_p E$ (resp., $cldec_p E$).

Proposition 5.117 If $E = \{f, g\} \subseteq L^p(\Omega, X), 1 \leq p \leq \infty$, then $dec_p E = cldec_p E = \{\chi_A f + \chi_{\Omega \backslash A} g : A \in \Sigma\}$.

Proof The set $K = \{\chi_A f + \chi_{\Omega \backslash A} g : A \in \Sigma\}$ is closed and $f, g \in K$. Then

$$dec_p E \subseteq cldec_p E \subseteq K.$$

On the other hand, if $u, v \in K$ and $C \in \Sigma$, then

$$\chi_C u + \chi_{\Omega \backslash C} v = \chi_C [\chi_A f + \chi_{\Omega \backslash A} g] + \chi_{\Omega \backslash C} [\chi_B f + \chi_{\Omega \backslash B} g]$$
$$= \chi_M f + \chi_{\Omega \backslash M} g,$$

for some $A, B \in \Sigma$, with $M = C \cap A \cup [(\Omega \backslash C) \cap B] \in \Sigma$.

Therefore, K is decomposable.

We conclude that

$$dec_p E = cldec_p E = K.$$

$\square$

Let $\overline{B_1} = \{f \in L^p(\Omega, X) :\| f \|_p \leq 1\}, 1 \leq p \leq \infty$. We have the following proposition:

Proposition 5.118 $dec_p \overline{B_1} = L^p(\Omega, X), 1 \leq p \leq \infty$.

Proof Let $f \in L^p(\Omega, X)$ and $n \in \mathbb{N}$ such that $\| f \|_p^p \leq n$. We consider the measure $m(A) = \int_A | f |^p d\mu$ for all $A \in \Sigma$. We can find $\{A_t\}_{t \in I = [0,1]} \subseteq \Sigma$ such that $m(A_t) = t m(A)$ for all $t \in I$. We set $C_m = A_{\frac{m}{n}} \backslash A_{\frac{m-1}{n}} \in \Sigma, m \in \{1, \ldots, n\}$. Evidently $\{C_m\}_{m=1}^n$ is a Σ-partition of Ω and $m(C_m) = \frac{1}{n} \int_\Omega | f |^p dy \leq 1$. Consider the functions $g_m = \chi_{C_m} f \in \overline{B_1}$. We have

$$f = \sum_{m=1}^n \chi_{C_m} f = \sum_{m=1}^n \chi_{C_m} g_m,$$

$$\Rightarrow f \in dec_p \overline{B_1},$$

$$\Rightarrow dec_p \overline{B_1} = L^p(\Omega, X).$$

$\square$

Proposition 5.119 If $K \subseteq L^p(\Omega, X), 1 \leq p \leq \infty$, is decomposable and relatively compact, then K is a singleton.

Proof On account of Proposition 5.116(d) without any loss of generality, we may assume that $0 \in K$. Arguing by contradiction, suppose that K is not a singleton. Then we can find $f \in K$, $f \neq 0$, and so for $\varepsilon > 0$ small, we will have that

$$\mu(C_\varepsilon) > 0 \quad \text{with} \ C_\varepsilon = \{w \in \Omega : | f(w) | \geq \varepsilon\}.$$

Since $0 \in K$, we have that $\chi_{C_\varepsilon} f \in K$. Then

$$\widehat{K} = \{\chi_A f : A \in \Sigma_{C_\varepsilon} = \Sigma \cap C_\varepsilon\} = cldec_p\{\chi_{C_\varepsilon} f, 0\} \subseteq \overline{K},$$

$$\Rightarrow \widehat{K} \subseteq L^p(\Omega, X) \ \text{is compact.}$$

For $A, B \in \Sigma \cap C_\varepsilon$, we have

$$\| \chi_A - \chi_B \| \leq \frac{1}{\varepsilon^p} \| \chi_A f - \chi_B f \|_p^p,$$

$$\Rightarrow \chi_A f \to \chi_A \ \text{is a continuous map,}$$

$$\Rightarrow \{\chi_A : A \in \Sigma_{C_\varepsilon}\} \subseteq L^p(\Omega) \ \text{is compact, a contradiction.} \qquad \square$$

As we already mentioned, decomposable sets are closely related to measurable closed valued multifunctions. In fact, we will show next that a closed decomposable set in $L^p(\Omega, X)$, $1 \leq p \leq \infty$, equals the set S_F^p for some measurable multifunction $F : \Omega \to P_f(X)$. To this end, we will need the following lemma.

Lemma 5.120 *If $F : \Omega \to 2^X \setminus \{\varnothing\}$ is graph measurable, $1 \leq p \leq \infty$, $\{f_n\}_{n \in \mathbb{N}} \subseteq S_F^p$ is such that $F(w) \subseteq \overline{\{f_n(w)\}}_{n \in \mathbb{N}}$ μ-a.e., $f \in S_F^p$, and $\varepsilon > 0$, then we can find a finite Σ-partition $\{D_n\}_{n=1}^N$ of Ω such that*

$$\| f - \sum_{m=1}^N \chi_{D_n} f_n \|_p < \varepsilon.$$

Proof Without any loss of generality, we may assume that $f(w) \in F(w)$ for all $w \in \Omega$. We choose a function $\eta \in L^1(\Omega)$ such that $\eta(w) > 0$ for all $w \in \Omega$ and $\int_\Omega \eta d\mu < \varepsilon^p/3$. Let $\{A_n\}_{n \in \mathbb{N}}$ be a Σ-partition of Ω such that

$$\| f(w) - f_n(w) \|^p < \eta(w) \ \text{for all} \ w \in A_n, \ \text{all} \ n \in \mathbb{N}.$$

We can find $N \geq 1$ big such that

$$\sum_{n \geq N+1} \int_{A_n} \| f \|^p \, dy < \frac{\varepsilon^p}{3.2^p} \quad \text{and} \quad \sum_{n \geq N+1} \int_{A_n} \| f_1 \|^p \, dy < \frac{\varepsilon^p}{3.2^p}.$$

Let $\{D_n\}_{n=1}^N \subseteq \Sigma$ be defined by

$$D_n = A_1 \cup [\cup_{n \geq N+1} A_n] \quad \text{and} \quad D_n = A_n \quad \text{for} \ n = 2, \ldots, N.$$

This too is a finite Σ-partition of Ω. We have

$$\left\| f - \sum_{n=1}^{N} \chi_{D_n} f_n \right\|_p^p = \sum_{n=1}^{N} \int_{A_n} \| f - f_n \|^p d\mu + \sum_{n \geq N+1} \int_{A_n} \| f - f_1 \|^p d\mu$$

$$\leq \int_{\Omega} \eta d\mu + \sum_{n \geq N+1} 2^{p-1} \int_{A_n} [\| f \|^p + \| f_1 \|^p] d\mu < \varepsilon^p.$$

$\square$

Using this lemma, we can have the following complete characterization of closed decomposable sets in $L^p(\Omega, X)$, $1 \leq p < \infty$.

Theorem 5.121 *If $K \subseteq L^p(\Omega, X)$ $1 \leq p < \infty$ is nonempty and closed, then K is decomposable if and only if $K = S_F^p$ for some $F : \Omega \to P_f(X)$ measurable.*

Proof $\Rightarrow$ Proposition 5.114 implies that we can find a sequence $\{f_n\}_{n \in N} \subseteq L^p(\Omega, X)$ such that $X = \overline{\{f_n(\omega)\}}_{n \in N}$ $\mu - a.e.$ We define

$$\vartheta_n = inf[\| f_n - k \|_p : k \in K] \quad \text{for all} n \in N.$$

Consider $\{k_{nm}\}_{m \in N} \subseteq K$ such that $\| f_n - k_{nm} \|_p \downarrow \vartheta_n$ as $m \to \infty$. We set $F(w) = \overline{\{k_{nm}(w)\}}_{n,m \in N}$ for all $w \in \Omega$. Clearly $F(\cdot)$ is a measurable, $P_f(X)$-valued multifunction (see Theorem 5.100). We will show that $K = S_F^p$. To this end, let $f \in S_F^p$ and $\varepsilon > 0$. Using Lemma 5.120, we can find a finite Σ-partition $\{D_i\}_{i=1}^{N}$ of Ω and $\{g_i\}_{i=1}^{N} \subseteq \{k_{nm}\}_{n,m \in N}$ such that

$$\left\| f - \sum_{i=1}^{N} \chi_{D_i} g_i \right\|_p < \varepsilon.$$

The decomposability of K implies that $\Sigma_{i=1}^{N} \chi_{D_i} g_i \in K$. Hence $f \in K$ (recall that K is closed) and so we have

$$S_F^p \subseteq K. \tag{5.26}$$

Suppose that the inclusion in (5.26) is strict. Then we can find $f \in K$, $A \in \Sigma$ with $\mu(A) > 0$ and $\delta > 0$ such that

$$\| f(w) - k_{nm}(w) \| \geq \delta \quad \text{for all} w \in A, \text{ all } n, m \in N. \tag{5.27}$$

Let $n \in N$ be such that $\mu(C) > 0$, with $C = A \cap \{w \in \Omega : \| f(w) - f_n(w) \| < \frac{\delta}{3}\}$. We set

$$h_m = \chi_C f + \chi_{\Omega \setminus C} k_{nm}, \quad m \in N. \tag{5.28}$$

Then $h_m \in K$ and we have

$$\|k_{nm}(w) - f_n(w)\| \geq \|k_{nm}(w) - f(w)\| - \|f(w) - f_n(\omega)\|$$

$$\geq \delta - \frac{\delta}{3} = \frac{2\delta}{3} \quad \text{for all } \omega \in C \text{ (see (5.27))}. \tag{5.29}$$

It follows that

$$\|f_n - k_{nm}\|_p^p - \vartheta_n^p$$

$$\geq \|f_n - k_{nm}\|_p^p - \|f_n - h_m\|_p^p$$

$$\geq \int_C [\|f_n - k_{nm}\|^p - \|f_n - f\|^p]d\mu \quad \text{(see (5.28))} \tag{5.30}$$

$$\geq [(\frac{2\delta}{3})^p - (\frac{\delta}{3})^p]\mu(C) > 0 \quad \text{(see (5.29))}.$$

In (5.30), we let $m \to \infty$ and have a contradiction (recall the choice of the sequence $\{k_{nm}\}_{m\in N}$). Hence the inclusion in (5.26) cannot be strict and we have $K = S_F^p$.

$\Leftarrow$ Obvious. $\qquad\qquad\qquad\qquad\qquad\qquad\qquad\qquad\qquad\qquad\qquad\qquad\qquad$ $\square$

An interesting consequence of this theorem is the following proposition.

Proposition 5.122 *If $F : \Omega \to 2^X \setminus \{\emptyset\}$ is a measurable multifunction, $1 \leq p < \infty$ and $S_F^p \neq \emptyset$, then $\overline{S_F^p} = S_{\overline{F}}^p$.*

Proof From Proposition 5.116(c), we know that $\overline{S_F^p}$ is decomposable. So, according to Theorem 5.121, we can find $G : \Omega \to P_f(X)$ measurable multifunction such that $\overline{S_F^p} = S_G^p$. Evidently $\overline{F}(w) \subseteq G(w)$ μ-a.e. (see Theorem 5.100). On the other hand, $S_{\overline{F}}^p$ is clearly closed and decomposable and contains S_F^p. Hence $\overline{S_F^p} = S_G^p \subseteq S_{\overline{F}}^p$ and so $G(w) \subseteq \overline{F}(w)$ μ-a.e. We conclude that $G = \overline{F}$. $\qquad\qquad\qquad$ $\square$

Next we provide necessary and sufficient conditions for $S_F^p \neq \emptyset$.

Proposition 5.123 *If $F : \Omega \to 2^X \setminus \{\emptyset\}$ is graph measurable and $1 \leq p < \infty$, then $S_F^p \neq \emptyset$ if and only if $\inf[\|u\| : u \in F(w)] \leq \vartheta(w)$ μ-a.e. with $\vartheta \in L^p(\Omega)$.*

Proof Without any loss of generality, we may assume that $\mu(\cdot)$ is finite. From Theorem 5.107, we have

$$\vartheta_0(w) = \inf[\|u\| : u \in F(w)] = \inf_{n\geq 1} \|f_n(w)\| \text{ with } \{f_n\}_{n\in\mathbb{N}} \subseteq S_F,$$
$$\Rightarrow \vartheta_0(\cdot) \text{ is } \Sigma\text{-measurable, hence } \vartheta_0 \in L^p(\Omega).$$

Let $\widehat{\vartheta}_0 = \vartheta_0 + 1 \in L^p(\Omega)$ and consider the multifunction

$$G(w) = F(w) \cap \overline{B}_{\widehat{\vartheta}_0(w)}(0).$$

Evidently $G(\cdot)$ is graph measurable and so we can apply Theorem 5.106 and infer that $S_G \neq \emptyset$. If $f \in S_G$, then $f \in S_F^p \neq \emptyset$. $\square$

Remark 5.124 In fact if $m = \mathrm{essinf}\{\|f\| : f \in \overline{S_F^p}\}$ and $u \in L^p(\Omega)$ satisfies

$$m(w) < u(w) \ \mu - a.e.,$$

then we can find $f \in \overline{S_F^p} = S_{\overline{F}}^p$ (see Proposition 5.121) such that $\|f(w)\| < u(w)\ \mu$-a.e. Moreover, $\|m\|_p = \inf[\|f\|_p : f \in \overline{S_F^p}]$.

Proposition 5.125 *If $\mu(\cdot)$ is finite, Y is a Banach space and $K \subseteq L^1(\Omega, Y)$ is bounded and decomposable, then K is uniformly integrable.*

Proof Let $|K| = \{\|f(\cdot)\| : f \in K\} \subseteq L^1(\Omega)$ and let $g = \mathrm{esssup}\,|K|$. Theorem 5.109 implies that we can find $\{f_n\}_{n \in \mathbb{N}} \subseteq K$ such that

$$g(w) = \sup_{n \in \mathbb{N}} \|f_n(w)\| \ \mu - a.e.$$

Note that the decomposability of K implies that $|K|$ is upward directed and so we may assume that the sequence $\{\|f_n(\cdot)\|\}_{n \in \mathbb{N}}$ is increasing. The Monotone Convergence Theorem (see Theorem 2.80) and the boundedness of $K \subseteq L^1(\Omega, X)$ imply that $g \in L^1(\Omega)$. Since $\|f(w)\| \leq g(w)\ \mu - a.e.$ for all $f \in K$, we conclude that K is uniformly integrable. $\square$

The next theorem reveals the power of the notion of decomposability and it is a useful tool in many applications.

Theorem 5.126 *If $\varphi : \Omega \times X \to \mathbb{R}^* = \mathbb{R} \cup \{\pm\infty\}$ is a $\Sigma \otimes B(X)-$measurable function, $F : \Omega \to 2^X \setminus \{\emptyset\}$ is a graph measurable multifunction, the integral functional*

$$I_\varphi(u) = \int_\Omega \varphi(w, u(w))d\mu \ \text{for } u \in L^p(\Omega, X), \ 1 \leq P < \infty$$

is defined for all $u \in S_F^p$, and there exists $u_0 \in S_F^p$ such that

$$I_\varphi(u_0) > -\infty,$$

then $\sup[I_\varphi(u) : u \in S_F^p] = \int_\Omega \sup[\varphi(w, x) : x \in F(w)]d\mu$.

Proof Let $\widehat{m}(w) = \sup[\varphi(w, x) : x \in F(w)]$. From Problem 5.9, we know that $\widehat{m}(\circ)$ is $\Sigma_\mu-$measurable. For $u \in S_F^p$, we have $\varphi(w, u(w)) \leq \widehat{m}(w)$, hence

$$\varphi(w, u_0(w)) \leq \widehat{m}(w) \ \mu - a.e.,$$
$$\Rightarrow \int_\Omega \widehat{m} d\mu \in \overline{\mathbb{R}} = \mathbb{R} \cup \{+\infty\}.$$

We have

$$\sup[I_\varphi(u) : u \in S_F^p] \leq \int_\Omega \widehat{m} d\mu.$$

If $I_\varphi(u_0) = +\infty$, then we are done. So, we assume that $I_\varphi(u_0) \in \mathbb{R}$. Let $\vartheta < \int_\Omega \widehat{m} d\mu$ and consider a sequence $\{A_n\}_{n \in \mathbb{N}} \subseteq \Sigma$ such that $\bigcup_{n \in \mathbb{N}} A_n = \Omega$, $\mu(A_n) < \infty$ for all $n \in \mathbb{N}$. Let $k \in L^1(\Omega)$ be such that $0 < k(w) \ \mu - a.e.$ We set

$$C_n = A_n \cap \{w \in \Omega : \varphi(w, u_0(w)) \leq n\}$$

and

$$\widehat{m}_n(w) = \begin{cases} m(w) - \frac{k(w)}{n}, & if \ w \in C_n, k(w) \leq n \\ n - \frac{k(w)}{n}, & if \ w \in C_n, k(w) > n, \ n \in \mathbb{N} \\ \varphi(w, u_0(w)) - \frac{k(w)}{n}, & if \ w \in \Omega \setminus C_n. \end{cases}$$

Evidently $\widehat{m}_n \in L^1(\Omega)$ for all $n \in \mathbb{N}$ and $\widehat{m}_n \uparrow m$. Then the Monotone Convergence Theorem (see Theorem 2.80) implies that we can find $n_0 \in \mathbb{N}$ such that $\vartheta < \int_\Omega \widehat{m}_{n_0} d\mu$. Let $\beta = \widehat{m}_{n_0}$. Then

$$\vartheta < \int_\Omega \beta d\mu \ and \ \beta(w) < \widehat{m}(w) \ \mu - a.e.$$

We introduce the multifunction

$$G(w) = F(w) \cap \{x \in X : \beta(w) \leq \varphi(w, x)\} \neq \emptyset \text{ for all } w \in \Omega. \tag{5.31}$$

Evidently $G(\cdot)$ is graph measurable and so using Theorem 5.106 (the Yankov–von Neumann–Aumann Selection Theorem), we can find $u \in S_G$. We define

$$D_n = C_n \cap \{w \in \Omega : \|u(w)\| \leq n\}, v_n = \chi_{D_n} u + \chi_{\Omega \setminus D_n} u_0, n \in \mathbb{N}.$$

We see that $\{v_n\}_{n \in \mathbb{N}} \subseteq S_F^p$ and we have

$$I_\varphi(v_n) = \int_{D_n} \varphi(w, u) d\mu + \int_{\Omega \setminus D_n} \varphi(w, u_0) d\mu$$

$$\geq \int_{D_n} \beta d\mu + \int_{\Omega \setminus D_n} \varphi(w, u_0) d\mu \ (see\ (5.31))$$

$$= \int_\Omega \beta d\mu + \int_{\Omega \setminus D_n} [\varphi(w, u_0) - \beta] d\mu.$$

Recall that $\int_\Omega \beta d\mu > \vartheta$ and $D_n \uparrow \Omega$. Therefore, for $n \in \mathbb{N}$ big, we will have

$$I_\varphi(u_n) > \vartheta.$$

Since $\vartheta < \int_\Omega \widehat{m} d\mu$ is arbitrary, we conclude that

$$\sup[I_\varphi(u) : u \in S_F^p] = \int_\Omega \sup[\varphi(w, x) : x \in F(w)] d\mu.$$

$\square$

A couple of useful consequences of this theorem.

Proposition 5.127 *If $F : \Omega \to 2^X \setminus \{\emptyset\}$ is graph measurable and $S_F^p \neq \emptyset$, $1 \leq p < \infty$, then $\overline{\mathrm{conv}} S_F^p = S_{\overline{\mathrm{conv}} F}^p$.*

Proof Evidently $S_{\overline{\mathrm{conv}} F}^p \in P_{f_c}(L^p(\Omega, X))$ and contains S_F^p. Hence

$$\overline{\mathrm{conv}} S_F^p \subseteq S_{\overline{\mathrm{conv}} F}^p. \tag{5.32}$$

Suppose that the inclusion in (5.32) is strict. This means that we can find $g \in S_{\overline{\mathrm{conv}} F}^p$ such that $g \notin \overline{\mathrm{conv}} S_F^p$. Using Theorem 3.71, we can find $u^* \in L^{p'}(\Omega, X_{w^*}^*) = L^p(\Omega, X)^*$ ($\frac{1}{p} + \frac{1}{p'} = 1$, see Remark 4.178) such that

$$\sigma(u^*, \overline{\mathrm{conv}} S_F^p) < \langle u^*, g \rangle,$$

$$\Rightarrow \sigma(u^*, S_F^p) = \sup[\langle u^*, f \rangle : f \in S_F^p]$$

$$= \int_\Omega \sup[\langle u^*(w), x \rangle : x \in F(w)] d\mu \ (see\ Proposition\ 5.126)$$

$$= \int_\Omega \sigma(u^*(w), F(w)) d\mu$$

$$= \int_\Omega \sigma(u^*(w), \overline{\mathrm{conv}} F(w)) d\mu$$

$$< \int_\Omega \langle u^*(w), g(w) \rangle d\mu$$

$$\leq \int_\Omega \sigma(u^*(w), \overline{\mathrm{conv}} F(w)) d\mu,$$

a contradiction. So, we conclude that $\overline{\mathrm{conv}} S_F^p = S_{\overline{\mathrm{conv}} F}^p$. $\square$

From the above proof, we have the following corollary.

Corollary 5.128 *If $F : \Omega \to 2^X \setminus \{\emptyset\}$ is graph measurable, $S_F^p \neq \emptyset$, $1 \leq p < \infty$, and*

$$u^* \in L^{p'}(\Omega, X_{w*}^*) = L^p(\Omega, X)^* \; (\frac{1}{p} + \frac{1}{p'} = 1),$$

then $\sigma(u^*, S_F^p) = \int_\Omega \sigma(u^*(w), F(w))d\mu$.

The following theorem is another illustration of the close relation between decomposability and convexity, and its proof can be found in Hu-Papageorgiou [145] (Theorem 3.17, p.101).

Theorem 5.129 *If (Ω, Σ, μ) is σ-finite and nonatomic, Y is a Banach space, and $K \subseteq L^p(\Omega, Y), 1 \leq p \leq \infty$ is nonempty, decomposable, and w-closed, then, K is convex.*

Using this theorem, we can have the following proposition.

Proposition 5.130 *If (Ω, Σ, μ) is σ-finite and nonatomic, $F : \Omega \to 2^X \setminus \{\emptyset\}$ is graph measurable, and $S_F^p \neq \emptyset, 1 \leq p \leq \infty$, then, $\overline{S_F^p}^w = S_{\overline{conv}F}^p$.*

Proof The set $\overline{S_F^p}^w$ is nonempty, decomposable, and w-closed. So, Theorem 5.129 implies that $\overline{S_F^p}^w$ is convex. Therefore

$$\overline{conv}S_F^p = \overline{S_F^p}^w,$$

$$\Rightarrow S_{\overline{conv}F}^p = \overline{S_F^p}^w \; \text{(See Theorem 5.126).} \qquad \square$$

A consequence of this proposition is the following result.

Proposition 5.131 *If $F : \Omega \to 2^X \setminus \{\emptyset\}$ is graph measurable and S_F^p $1 \leqslant p < \infty$ is*

$$(i) Nonempty \ and \ closed$$
$$resp., \ (ii) Nonempty, \ closed, \ and \ convex \ and \ \mu(\cdot) \ is \ nonatomic,$$
$$then \ (a) F(w) \in P_f(X) \ for \ \mu - a.a \ w \in \Omega$$
$$resp., \ (b) F(w) \in P_{f_c}(X) \ for \ \mu - a.a \ w \in \Omega.$$

Proof

(a) From hypothesis (i) and Proposition 5.122, we have $S_F^p = S_{\overline{F}}^p$ and so $F(w) = \overline{F}(w) \in P_f(X)$ for $\mu - a.a \ w \in \Omega$ (see Corollary 5.115).

(b) From hypothesis (ii) and Proposition 5.130, we have

$$\overline{S_F^p}^w = S_{\overline{conv}F}^p, \text{ hence } \overline{S_F^p} = S_F^p = S_{\overline{conv}F}^p \text{ (see Theorem 3.105)}$$

and so $F(w) = \overline{conv}F(w) \in P_{fc}(X)$ for $\mu - a.a\ w \in \Omega$. $\qquad\square$

Definition 5.132 A multifunction $F : \Omega \to 2^X \setminus \{\emptyset\}$ is said to be "p-integrably bounded" $(1 \leqslant p \leqslant \infty)$ if there exists $\vartheta \in L^p(\Omega)$ such that

$$|F(w)| = \sup[\|x\| : x \in F(\omega)] \leq \vartheta(w)\ \mu - a.e.$$

If $p = 1$, then we say that $F(\cdot)$ is "integrably bounded."

The next result is very useful in many applications.

Theorem 5.133 *If $F : \Omega \to P_{wkc}X$ is graph measurable and integrably bounded, then $S_F^1 \in P_{wkc}(L^1(\Omega, X))$.*

Proof It is clear that S_F^1 is nonempty, closed, and convex. Let $u^* \in L^1(\Omega, X)^* = L^\infty(\Omega, X_{w*}^*)$ (see Theorem 4.177). We have

$$\sup[\langle u^*(\omega), f\rangle : f \in S_F^1] = \sup[\int_\Omega \langle u^*(\omega), f(\omega)\rangle d\mu : f \in S_F']$$

$$\int_\Omega \sup[\langle u^*(\omega), x\rangle : x \in F(\omega)]d\mu \text{ (see Theorem 5.126)}.$$

We introduce the multifunction $L : \Omega \to 2^X$ defined by

$$L(\omega) = \{u \in F(\omega) : \langle u^*(\omega), u\rangle = \sup[\langle u^*(\omega), x\rangle : x \in F(\omega)]\}$$
$$= \{u \in F(\omega) : \langle u^*(\omega), u\rangle = \sigma(u^*(\omega), F(\omega))\}.$$

Evidently $L(\omega) \in P_{wkc}(X)$ for all $\omega \in \Omega$, and using Theorem 5.100, we see that $\omega \to \sigma(u^*(\omega), F(\omega))$ is Σ-measurable. Therefore

$$\text{Gr } L \in \Sigma \otimes B(X).$$

Invoking Theorem 5.106 (the Yankov–von Neumann–Aumann Selection Theorem), we can find $\widehat{u} : \Omega \to X$ Σ-measurable such that

$$\widehat{u}(\omega) \in L(\omega)\ \mu - a.e.,$$

$$\Rightarrow \widehat{u} \in S_F^1 \text{ and } \sigma(u^*(\omega), S_F^1) = \langle u^*, \widehat{u}\rangle.$$

Invoking Theorem 3.128 (James' Theorem), we conclude that

$$S_F^1 \in P_{wkc}(L^1(\Omega, X)).$$

□

In fact the converse is also true (see Klei [161], p.313).

Theorem 5.134 *If* $F : \Omega \to 2^X \setminus \{\emptyset\}$ *is graph measurable, integrably bounded and* $S_F^1 \subseteq L^1(\Omega, X)$ *is w-compact, convex, then* $F(\omega) \in P_{wkc}(X)$ $\mu - a.e.$

Proposition 5.135 *If* $F : \Omega \to P_{wkc}(X)$ *is scalarly measurable, then* $ext S_F = S_{ext F} \neq \emptyset$ *in* $L^0(\Omega, X)$.

Proof On account of Proposition 5.104 and Theorem 5.106 (the Yankov–von Neumann–Aumann Selection Theorem), we can say that $S_{ext F} \neq \emptyset$. Moreover, directly from Definition 3.245, we see that $S_{ext F} \subseteq ext S_F$. Suppose that the inclusion is strict. This means that we can find $f \in ext S_F$ and a set $A \in \Sigma$ with $\mu(A) > 0$ such that $f(\omega) \notin ext F(\omega)$ for all $\omega \in A$. We consider the multifunction $L : A \to 2^{X \times X}$ defined

$$L(\omega) = \{(x, u) \in F(\omega) \times F(\omega) : f(\omega) = \frac{1}{2}(x + u), x \neq u\}.$$

Evidently $L(\cdot)$ is graph measurable and so we can find $\widehat{f}_1, \widehat{f}_2 : A \to X$ $\Sigma_A = \Sigma \cap A$-measurable functions such that

$$(\widehat{f}_1(\omega), \widehat{f}_2(\omega)) \in L(\omega) \ for \ \mu - a.e. \ \omega \in A.$$

We define

$$f_1 = \chi_A \widehat{f}_1 + \chi_{\Omega \setminus A} f, \ f_2 = \chi_A \widehat{f}_2 + \chi_{\Omega \setminus A} f.$$

Evidently $f_1, f_2 \in S_F$ and $f = \frac{1}{2}[f_1 + f_2]$, and hence $f \notin ext S_F$ a contradiction.   □

From this, it follows easily the following result.

Proposition 5.136 *If* $F : \Omega \to P_{wkc}(X)$ *is scalar measurable and* $S_F^p \neq 0$, $1 \leq p \leq \infty$, *then* $ext S_F^p = S_{ext F}^p$.

We conclude this section with a "decomposable" version of the Michael Selection Theorem. This is another illustration that decomposability can serve as a substitute of convexity. The result is due to Bressan and Colombo [47] and Fryszkowski [119, 120].

Theorem 5.137 *If* V *is a separable metric space,* X *is a separable Banach space, and* $S : V \to P_f(X)$ *is an lsc multifunction with decomposable values, then,* $S(\cdot)$ *admits a sequence* $\{s_n\}_{n \in \mathbb{N}}$ *of continuous selections such that* $S(v) = \overline{\{s_n(v)\}}_{n \in \mathbb{N}}$ *for all* $v \in V$.

5.5   Set-Valued Integral

In this section we introduce an integral for multifunctions and study its properties.

The definition of the set-valued integral (often called the "Aumann integral") is a straightforward generalization of the Minkowski sum of sets.

As before the standing hypotheses in this section are that (Ω, Σ, μ) is a σ-finite measure space and X is a separable Banach space.

Definition 5.138   Given a multifunction $F : \Omega \to 2^X \backslash \{\emptyset\}$, its "(Aumann)-integral" is defined by

$$\int_\Omega F d\mu = \{\int_\Omega f d\mu : f \in S_F^1\}.$$

Using Theorem 5.126, we have the following proposition:

Proposition 5.139   If $F : 2^X \setminus \{\emptyset\}$ is graph measurable and $S_F^1 \neq \emptyset$, then we have for all $x^* \in X^*$, we have

$$\sigma(x^*, \int_\Omega F d\mu) = \int_\Omega \sigma(x^*, F) d\mu.$$

Also, from Proposition 5.126, we infer the following result.

Proposition 5.140   If $F : \Omega \to 2^X \setminus \{\emptyset\}$ is graph measurable and $S_F^1 \neq \emptyset$, then

$$cl \int_\Omega \operatorname{conv} F d\mu = \overline{\operatorname{conv}} \int_\Omega F d\mu = cl \int_\Omega \overline{\operatorname{conv}} F d\mu.$$

The set-valued integral has some remarkable intrinsic convexity properties, which are a consequence of the so-called Lyapunov Convexity Theorem, which can be found in Diestel and Uhl [85] (pp. 264, 266). Let us recall this result that has interesting uses in many applied areas such as Control Theory, Game Theory, Mathematical Economics, and others.

Theorem 5.141   If (Ω, Σ) is a measurable space, then:

(a) Given a nonatomic vector measure $m : \Sigma \to \mathbb{R}^N$, we have $m(\Sigma) \subseteq \mathbb{R}^N$ is compact and convex.

(b) If Y is a Banach space with the RNP and $m : \sum \to Y$ is a nonatomic vector measure of bounded variation, then $\overline{m(\Sigma)} \subseteq Y$ is compact and convex.

This theorem leads to the following remarkable convexity property of the set-valued integral.

Theorem 5.142   If μ is nonatomic and $F : \Omega \to 2^X \setminus \{\emptyset\}$ is graph measurable with $S_F^1 \neq \emptyset$, then $cl \int_\Omega F d\mu$ is convex. Moreover, if X is finite dimensional, then $\int_\Omega F d\mu$ is convex.

Proof Let $u_1, u_2 \in \int_\Omega F d\mu$. We have

$$u_1 = \int_\Omega f_1 d\mu \text{ and } u_2 = \int_\Omega f_2 d\mu \text{ with } f_1, f_2 \in S_F^1 \text{ (see Definition 5.138)}.$$

We consider the vector measure $m : \Sigma \to X \times X$ defined by

$$m(A) = \left(\int_A f_1 d\mu, \int_A f_2 d\mu \right) \text{ for all } A \in \Sigma.$$

Then $m(\cdot)$ is nonatomic, and by Theorem 5.141(b), we have $\overline{m(\Sigma)} \subseteq X \times X$ convex. We have

$$m(\emptyset) = 0 \text{ and } m(\Omega) = (u_1, u_2).$$

Given $\epsilon > 0$ and $t \in [0, 1]$, we can find $A \in \Sigma$ such that

$$\left\| t \int_\Omega f_i d\mu - \int_A f_i d\mu \right\| < \frac{\epsilon}{2} \text{ for } i = 1, 2.$$

Let $f = \chi_A f_1 + \chi_{\Omega \setminus A} f_2$. Then $f \in S_F^1$ and

$$\left\| t u_1 + (1 - t)u_2 - \int_\Omega f d\mu \right\| < \epsilon,$$

$$\Rightarrow cl \int_\Omega F d\mu \text{ is convex}.$$

If X is finite dimensional, then Theorem 5.141(a) implies that $\int_\Omega F d\mu$ is convex. □

Corollary 5.143 *If μ is nonatomic and $F : \Omega \to 2^X \setminus \{\emptyset\}$ is graph measurable with $S_F^1 \neq \emptyset$, then*

$$cl \int_\Omega \overline{conv} F d\mu = cl \int_\Omega F d\mu.$$

Proof On account of Proposition 5.139, for every $x^* \in X^*$, we have

$$\sigma\left(x^*, \int_\Omega \overline{conv} F d\mu\right) = \int_\Omega \sigma(x^*, \overline{conv} F) d\mu = \int_\Omega \sigma(x^*, F) d\mu = \sigma\left(x^*, \int_\Omega F d\mu\right),$$

$$\Rightarrow cl \int_\Omega \overline{conv} F d\mu = cl \int_\Omega F d\mu \text{ (see Theorem 5.142)}.$$ □

Using Theorem 5.133, we have the following result.

Proposition 5.144 *If* $F : \Omega \to P_{wkc}(X)$ *is graph measurable and integrably bounded, then* $\int_A F d\mu \in P_{wkc}(X)$ *for all* $A \in \Sigma$.

Next we look for conditions that guarantee openness of the set-valued integral.

Proposition 5.145 *If* $F : \Omega \to 2^X \setminus \{\emptyset\}$ *is graph measurable with open values and* $S_F^1 \neq \emptyset$, *then* $\int_A F d\mu \in X$ *is open for all* $A \subseteq \Sigma$.

Proof Without any loss of generality, we may assume that $A = \Omega$. Moreover, we may also assume that $0 \in F(\omega)$ for all $\omega \in \Omega$. Indeed let $f \in S_F^1$ and consider the multifunction $G(\omega) = F(\omega) - f(\omega)$, $\omega \in \Omega$. Evidently $G(\cdot)$ has the same properties as $F(\cdot)$, namely it is graph measurable and open valued. So, we see that there is no loss of generality in assuming that $0 \in F(\omega)$ for all $\omega \in \Omega$.

Let $\vartheta_F(\omega) = d(0, X \setminus F(\omega))$, and for $\lambda > 0$, let $L_\lambda = \{\omega \in \Omega : \vartheta_F(\omega) < \lambda\}$. Note that

$$L_\lambda = proj_\Omega K \text{ where } K = (\Omega \times B_\lambda) \cap \mathrm{Gr}(X \setminus F) \in \Sigma \otimes B(X).$$

Invoking Theorem 5.72 (the Yankov–von Neumann–Aumann Projection Theorem), we have that

$$L_\lambda \in \Sigma_\mu,$$

$$\Rightarrow \vartheta_F \text{ is } \Sigma_\mu\text{-measurable.}$$

Since $F(\cdot)$ is open valued, we have $\vartheta_F(\omega) > 0$ for all $\omega \in \Omega$. It follows that we can find $A \in \Sigma$ with $\mu(A) > 0$ and $\epsilon > 0$ such that

$$\vartheta_F(\omega) \geq \epsilon \text{ for all } \omega \in A,$$

$$\Rightarrow B_\epsilon \subseteq F(\omega) \text{ for all } \omega \in A,$$

$$\Rightarrow \mu(A) B_\epsilon \subseteq \int_A F d\mu \subseteq \int_\Omega F d\mu \text{ (the last inclusion since } 0 \in F(\omega)),$$

$$\Rightarrow \int_\Omega F d\mu \subseteq X \text{ is open.}$$

$\square$

We can also show that the interior operator (int) and the integral commute. To do this, we will need two simple lemmata that we prove next.

Lemma 5.146 *If* Y *is a Banach space and* $U_1 \subseteq U_2 \subseteq Y$ *are nonempty open sets with* U_1 *convex and dense in* U_2, *then* $U_1 = U_2$.

Proof We have $U_2 \subseteq \inf \overline{U_1}$. Also the convexity of U_1 implies that $U_1 = int \overline{U_1}$ (see Proposition 3.64). Hence $U_1 = U_2$.

$\square$

Lemma 5.147 *If $F : \Omega \to 2^X \setminus \{\emptyset\}$ is graph measurable and $\operatorname{int} F(\omega) \neq 0$ for all $\omega \in \Omega$, then $\operatorname{Gr}(\operatorname{int} F) \in \Sigma_\mu \otimes B(X)$.*

Proof We have $\operatorname{int} F(\omega) = \overline{F(\omega)} \setminus \operatorname{bd} F(\omega)$ for all $\omega \in \Omega$ (see Remark 1.10). Also, we know that $\operatorname{bd} F(\omega) = \overline{F(\omega)} \cap \overline{(X \setminus F(\omega))}$ for all $\omega \in \Omega$ (see Definition 1.9(c)). Hence $\operatorname{Gr} \operatorname{bd} F \in \Sigma_\mu \otimes B(X)$ (see the proof of Proposition 5.144) and so it follows that $\operatorname{Gr}(\operatorname{int} F) = \operatorname{Gr} F \cap (\Omega \times X \setminus \operatorname{Gr} \operatorname{bd} F) \in \Sigma_\mu \otimes B(X)$.   □

Now we are ready to show that the operations of taking interior and of integration commute.

Theorem 5.148 *If μ is finite, $F : \Omega \to 2^X \setminus \{\emptyset\}$ is graph measurable and convex valued, $\operatorname{int} F(\omega) \neq \emptyset$ for $\mu - a.a$ $\omega \in \Omega$ and $S_F^1 \neq \emptyset$, then $\int_A F d\mu = \int_A \operatorname{int} F d\mu$ for all $A \in \Sigma$.*

Proof Let $f \in S_F^1$ and $\epsilon > 0$ and consider the multifunction $G_\epsilon : \Omega \to 2^X \setminus \{\emptyset\}$ defined by

$$G_\epsilon(\omega) = \{x \in \operatorname{int} F(\omega) : \|x - f(\omega)\| < \frac{\epsilon}{\mu(A)}\}$$

with $A \in \Sigma$, $\mu(A) > 0$. Evidently, $G_\epsilon(\omega) \neq \emptyset$ $\mu - a.e.$, and without any loss of generality we may assume that $G_\epsilon(\omega) \neq \emptyset$ for all $\omega \in \Omega$ (since we integrate, we can ignore μ-null sets). Also, on account of Lemma 5.147, we have

$$\operatorname{Gr} G_\epsilon \in \Sigma_\mu \otimes B(X).$$

So, using Theorem 5.106 (the Yankov–von Neumann–Aumann Selection Theorem), we can find $g_\epsilon : \Omega \to X$ Σ-measurable map such that

$$g_\epsilon(\omega) \in G_\epsilon(\omega) \quad \mu - a.e.,$$

$$\Rightarrow g_\epsilon \in S_{\operatorname{int} F}^1 \text{ and } \|g_\epsilon(\omega) - f(\omega)\| < \frac{\epsilon}{\mu(A)} \quad \mu - a.e.,$$

$$\Rightarrow \| \int_A (g_\epsilon - f) d\mu \| < \epsilon,$$

$$\Rightarrow \int_A \operatorname{int} F d\mu \text{ is dense in } \int_A F d\mu,$$

$$\Rightarrow \int_A \operatorname{int} F d\mu = \int_A F d\mu \quad \text{(see Proposition 5.145 and Lemma 5.146)}.$$

□

When X is finite dimensional, we can improve the above theorem.

Theorem 5.149 *If μ is finite and nonatomic, X is finite dimensional, $F : \Omega \to 2^X \setminus \{\emptyset\}$ is graph measurable, int $F(\omega)$ is dense in $F(\omega)$ for all $\omega \in \Omega$ and $S_F^1 \neq \emptyset$, then int $\int_A F d\mu = \int_A$ int $F d\mu$ for all $A \in \Sigma$.*

Finally we use the set-valued integral to decide whether $f \in L^1(\Omega, X)$ is an integrable selection of $F(\cdot)$.

Proposition 5.150

(a) *If $F : \Omega \to P_{wkc}(X)$ is graph measurable and integrably bounded, then $f \in S_F^1$ if and only if $\int_A f d\mu \in cl \int_A F d\mu = \int_A F d\mu$ for all $A \in \Sigma$.*
(b) *If X^* is separable and $F : \Omega \to P_{fc}(X)$ is graph measurable and integrably bounded, then $f \in S_F^1$ if and only if $\int_A f d\mu \in cl \int_A F d\mu$ for all $A \in \Sigma$.*

Proof $\Rightarrow$ Obvious (see Proposition 5.144).
$\Leftarrow$ For every $x^* \in X^*$ and $A \in \Sigma$, we have

$$\langle x^*, \int_A f d\mu \rangle = \int_A \langle x^*, f \rangle d\mu \leq \int_A \sigma(x^*, F) d\mu = \sigma(x^*, \int_A F d\mu),$$

$$\Rightarrow \langle x^*, f(\omega) \rangle \leq \sigma(x^*, F(\omega)) \; for \; all \; \omega \in \Omega \setminus N(x^*) \; with \; \mu(N(x^*)) = 0.$$

Let $\{x_n^*\}_{n \in \mathbb{N}} \subseteq X^*$ be m-dense. Since $F(\cdot)$ is $P_{wkc}(X)$-valued, we have that $\sigma(\cdot, F(\omega))$ is m-continuous. Then

$$\langle x^*, f(\omega) \rangle \leq \sigma(x^*, F(\omega)) \; for \; all \; \omega \in N = \bigcup_{n \in \mathbb{N}} N(x_n^*), \; all \; x^* \in X^*,$$

$$\Rightarrow f(\omega) \in F(\omega) \; \mu - a.e.,$$

$$\Rightarrow f \in S_F^1.$$

(b)$\Rightarrow$ Obvious.
$\Leftarrow$ In this case use the fact that X^* is separable and $\sigma(\cdot, F(\omega))$ is strongly continuous. $\square$

5.6 Caratheodory Multifunctions

In this section we deal with multifunctions $F(\omega, x)$ of two variables and examine their measurability and continuity properties.

We start with a parametric version of Lusin's Theorem (see Theorem 2.158), known in the literature as the "Scorza-Dragoni Theorem."

Theorem 5.151 *If T and X are Polish spaces, Y is a separable metric space, $\mu(\cdot)$ is a finite, tight Borel measure on T, and $f : T \times X \to Y$ is a Caratheodory function,* *then for every* $\epsilon > 0$ *there exists a compact set* $K_\epsilon \subseteq T$ *such that* $\mu(T \setminus K_\epsilon) < \epsilon$ *and* $f|_{K_\epsilon \times X}$ *is continuous.*

Proof From Theorem 1.116, we know that Y is homeomorphic to a subset of the Hilbert cube $\mathbb{H} = [0, 1]^{\mathbb{N}}$. Let $h = (h_n)_{n \in \mathbb{N}} : Y \to \mathbb{H}$ be this homeomorphism. We have that $f(\cdot, \cdot)$ is Caratheodory if and only if $h_n \circ f : T \times X \to [0, 1]$ is Caratheodory for every $n \in \mathbb{N}$. Hence, without any loss of generality, we may assume that $Y = [0, 1]$.

We consider $\{U_n\}_{n \in \mathbb{N}}$ a basis for the topology of X and $\{x_m\}_{m \in \mathbb{N}} \subseteq X$ a dense set. Given $q \in [0, 1] \cap \mathbb{Q}$, let $\vartheta_{nq} : X \to [0, 1]$ be defined by

$$\vartheta_{nq}(x) = q \chi_{U_n}(x).$$

Since U_n is open, $\vartheta_{nq}(\cdot)$ is lower semicontinuous (see Definition 1.176(a)), and if $\psi : X \to Y = [0, 1]$ is lower semicontinuous, then $\psi(x) = \sup[\vartheta_{nq}(x) : \vartheta_{nq} \leq \psi]$, $x \in X$. So, we define

$$A_{nqm} = \{t \in T : \vartheta_{nq}(x_m) \leq f(t, x_m)\} \in B(T)$$

and

$$A_{nq} = \bigcap_{m \in \mathbb{N}} A_{nqm} \in B(T).$$

We see that $A_{nq} = \{t \in T : \vartheta_{nq}(t) \leq f(t, x) \ for \ all \ x \in X\}$. We set

$$\xi_{nq}(t, x) = \chi_{A_{nq}}(t) \vartheta_{nq}(x),$$

$$\Rightarrow \xi_{nq} \leq f \ and \ f(t, x) = \sup_{n, q} \xi_{nq}(t, x) \ for \ all \ (t, x) \in T \times X,$$

$$\Rightarrow f = \sup_{k \in \mathbb{N}} \chi_{C_k} g_k.$$

with $C_k \in B(T)$ and $g_k : X \to Y$ lower semicontinuous.

Let $V_k \subseteq T$ be open and $K_k \subseteq T$ compact such that

$$K_k \subseteq C_k \subseteq V_k \ and \ \mu(V_k \setminus K_k) \leq \frac{\epsilon}{2^{k+2}} \ for \ all \ k \in \mathbb{N}. \tag{5.33}$$

Let $D_k = K_k \cup (X \setminus V_k)$, $k \in \mathbb{N}$. Then $\chi_{C_k}|_{D_k}$ is continuous (see (5.33)) and so $\chi_{C_k} g_k$ is lower semicontinuous. Let $D = \bigcap_{k \in \mathbb{N}} D_k \subseteq T$ be compact. We have $\mu(T \setminus D) \leq \frac{\varepsilon}{2}$ and $f|_{D \times X}$ is lower semicontinuous (see Proposition 1.182). The same argument applied on $1 - f$ produces another compact set $\widehat{D} \subseteq T$ such that $\mu(T \setminus \widehat{D}) \leq \frac{\varepsilon}{2}$ and $(1 - f)|_{\widehat{D} \times x}$ is lower semicontinuous. We set $K_\varepsilon = D \cap \widehat{D} \subseteq T$ compact. Then

$$\mu(T \setminus K_\varepsilon) \leq \varepsilon \text{ and } f|_{K_\varepsilon \times X} \text{ is continuous.}$$

□

Now we introduce a notion that is important in many applications and extends the notion of Caratheodory function (see Definition 2.46).

Definition 5.152 Let (Ω, Σ) be a measurable space, Y be a Hausdorff topological space, and $f : \Omega \times Y \to \overline{\mathbb{R}} = \mathbb{R} \cup \{+\infty\}$. We say that $f(\cdot, \cdot)$ is a normal integrand if it has the following properties:

(a) f is $\Sigma \otimes B(Y)$-measurable (that is, it is jointly measurable).
(b) For all $\omega \in \Omega, y \to f(\omega, y)$ is lower semicontinuous.

Remark 5.153 In contrast to Caratheodory functions, here we require that $f(\cdot, \cdot)$ is jointly measurable. The reason is that a function $f(w, x)$ measurable in $\omega \in \Omega$ and lower semicontinuous in $x \in X$ need not be jointly measurable (see Hu-Papageorgious [145], Example 7.1, p.226 and Sierpinski [255]).

Proposition 5.154 *If* (Ω, Σ, μ) *is a complete measure space,* Y *is a Polish space,* $f : \Omega \times Y \to \overline{\mathbb{R}} = \mathbb{R} \cup \{+\infty\}$ *is a normal integrand, and there exists a Caratheodory function* $h(\omega, x)$ *such that* $h \leq f$, *then there exists a sequence* $\{f_n\}_{n \in \mathbb{N}}$ *of Caratheodory functions on* $\Omega \times X$ *such that* $f_n \uparrow f$ *as* $n \to \infty$.

Proof We reason as in the proof of Proposition 1.183. We set

$$f_n(\omega, y) = \inf[f(\omega, u) + nd(y, u) : u \in Y].$$

If $\{u_m\}_{m \in \mathbb{N}} \subseteq Y$ is dense, then

$$f_n(\omega, y) = \inf_{m \in \mathbb{N}} [f(\omega, u_m) + nd(y, u_m)]$$

$$\Rightarrow f_n(\cdot, \cdot) \text{ is jointly measurable.}$$

Also for all $\omega \in \Omega$, $f_n(\omega, \cdot)$ is continuous and $h \leq f_n$ for all $n \in \mathbb{N}$. Finally we have $f_n \uparrow f$ (see the proof of Proposition 1.183). □

This proposition leads to the following extension of Theorem 5.151 (the Scorza-Dragoni Theorem).

Theorem 5.155 *If* T *and* X *are Polish spaces,* μ *is a finite, tight Borel measure on* T, *and* $f : T \times X \to \overline{\mathbb{R}} = \mathbb{R} \cup \{+\infty\}$ *is a normal integrand bounded below by a Caratheodory function* h, *then given* $\varepsilon > 0$, *we can find* $K_\varepsilon \subseteq T$ *compact such that* $\mu(T \setminus K_\varepsilon) < \varepsilon$ *and* $f|_{K_\varepsilon \times X}$ *is lower semicontinuous.*

Proof On account of Proposition 5.154, we can find a sequence $\{f_n\}_{n \in \mathbb{N}}$ of Caratheodory functions on $\Omega \times X$ such that

$$h \leq f_n \text{ for all } n \in \mathbb{N}, \quad f_n \uparrow f \text{ as } n \to \infty. \tag{5.34}$$

Invoking Theorem 5.151, for every $n \in \mathbb{N}$, we can find $K_n \subseteq T$ compact such that

$$\mu(T \setminus K_n) \leq \frac{\varepsilon}{2^n} \text{ and } f_n|_{K_n \times X} \text{ is continuous for all } n \in \mathbb{N}. \tag{5.35}$$

We set $K_\varepsilon = \bigcap_{n \in \mathbb{N}} K_n \subseteq T$ compact. Then

$$\mu(T \setminus K_\varepsilon) \leq \varepsilon \text{ and } f|_{K_\varepsilon \times X} \text{ is lower semicontinuous (see (5.34) and (5.35)).} \quad \square$$

Proposition 5.156 *If (Ω, Σ) is a measurable space, X is separable metrizable, Y is metrizable, and $F : \Omega \times X \to P_k(Y)$ is a multifunction such that:*

(a) For every $x \in X$, $\omega \to F(\omega, x)$ is measurable
(b) For every $x \in X$, $x \to F(\omega, x)$ is continuous

then $F(\cdot, \cdot)$ is $\Sigma \otimes B(X)$-measurable.

Proof From Corollary 5.50, we know that $F(\omega, \cdot)$ is h-continuous. Hence for every $y \in Y$, the function $(\omega, x) \to d(y, F(\omega, x))$ is Caratheodory, thus jointly measurable (see Theorem 2.47). From Theorem 5.101, we conclude that $F(\cdot, \cdot)$ is $\Sigma \otimes B(X)$-measurable. $\quad \square$

Proposition 5.157 *If T and X are Polish spaces, Y is a separable metrizable space, $\mu(\cdot)$ is a finite, tight Borel measure on T, and $F : T \times X \to P_f(X)$ is a multifunction such that:*

(a) $F(\cdot, \cdot)$ is $B(T \times X) = B(T) \otimes B(X)$ measurable
(b) For all $t \in T$, $F(t, \cdot)$ is lsc

then given $\varepsilon > 0$, we can find $K_\varepsilon \subseteq T$ compact such that $\mu(T \setminus K_\varepsilon) < \varepsilon$ and $F|_{K_\varepsilon \times X}$ is lsc.

Proof Let $\{u_m\}_{m \in \mathbb{N}} \subseteq Y$ be dense in Y and $d(\cdot, \cdot)$ a compatible metric on Y. Let $g_m(t, x) = d(u_m, F(t, x))$ for all $(t, x) \in T \times X$, all $m \in \mathbb{N}$. According to Proposition 5.19, for every $m \in \mathbb{N}$, $(t, x) \to g_m(t, x)$ is upper semicontinuous. By Theorem 5.151 (the Scorza-Dragoni Theorem), given $\varepsilon > 0$, we can find $K_m \subseteq T$ such that

$$\mu(T \setminus K_m) < \frac{\varepsilon}{2^m} \text{ and } g_m|_{K_m \times X} \text{ is upper semicontinuous.}$$

Set $K_\varepsilon = \bigcap_{m \in \mathbb{N}} K_m \subseteq T$ compact, $\mu(T \setminus K_\varepsilon) < \varepsilon$ and

$$g_m|_{K_\varepsilon \times X} \text{ is upper semicontinuous for all } m \in \mathbb{N}. \tag{5.36}$$

Suppose $(t_n, x_n) \to (t, x) \in K_\varepsilon \times X$. Given $u \in Y$, let $\{u_k\}_{k \in \mathbb{N}} \subseteq \{u_m\}_{m \in \mathbb{N}}$ such that $u_k \to u$ in Y. We have

$$d(u, F(t_n, x_n)) = d(u, F(t_n, x_n)) - d(u_k, F(t_n, u_n)) + d(u_k, F(t_n, x_n))$$

$$\leq d(u, u_k) + d(u_k, F(t_n, x_n)) \ for \ all \ k, n \in \mathbb{N},$$

$$\Rightarrow \limsup_{n \to \infty} d(u, F(t_n, x_n)) \leq d(u, u_k) + d(u_k, F(t, x)) \ \text{for} k \in \mathbb{N}(\text{see } (5.36)),$$

$$\Rightarrow \limsup_{n \to \infty} d(u, F(t_n, x_n)) \leq d(u, F(t, x)).$$

So, for every $u \in Y$, $(t, x) \to d(u, F(t, x))$ is upper semicontinuous on $K_\varepsilon \times X$, and this by Proposition 5.19 implies that $F|_{K_\varepsilon \times X}$ is lsc. $\qquad\square$

Finally we mention the following "measurable" version of Theorem 5.88. Its proof can be found in Fryszkowski [120] (Theorem 33, p.103).

Theorem 5.158 *If T and X are Polish spaces, Y is a separable Banach space, $\mu(\cdot)$ is a finite, tight Borel measure on $B(T)$, and $F : T \times X \to P_{f_c}(Y)$ is a multifunction such that:*

(a) $F(\cdot, \cdot)$ is $B(T \times X) = B(T) \otimes B(X)$-measurable
(b) For $\mu - a.a \ t \in T$, $F(t, \cdot)$ is lsc

then we can find a sequence $\{u_n\}_{n \in \mathbb{N}}$ of Caratheodory functions defined on $T \times X$ with values in Y such that

$$F(t, x) = \overline{\{u_n(t, x)\}}_{n \in \mathbb{N}} \ for \ \mu - a.a \ t \in T, \ all \ x \in X.$$

5.7 Remarks

5.1 We can describe the notions of upper and lower semicontinuity and of (Vietoris) continuity of multifunctions, using hyperspace topologies. So, let (X, τ) be a Hausdorff topological space and consider the family $2^X \setminus \{\emptyset\}$ (the nonempty subsets of X).

Definition 5.159

(a) Let $U \in \tau$ and let $U^+ = \{A \in 2^X \setminus \{\emptyset\} : A \subseteq U\}$. Then $\mathcal{B} = \{U^+ : U \in \tau\}$ is basis for a topology on $2^X \setminus \{\emptyset\}$ known as the "upper Vietoris topology" on $2^X \setminus \{\emptyset\}$, denoted by $\widehat{\tau}_{uv}$.
(b) Let $U \in \tau$ and let $U^- = \{A \in 2^X \setminus \{\emptyset\} : A \cap U \neq \emptyset\}$. Then $\mathcal{S} = \{U^- : U \in \tau\}$ is a subbasis for a topology on $2^X \setminus \{\emptyset\}$ known as the "lower Vietoris topology" on $2^X \setminus \{\emptyset\}$, denoted by $\widehat{\tau}_{LV}$.
(c) The "Vietoris topology" on $2^X \setminus \{\emptyset\}$ is generated by $\widehat{\tau}_{UV} \cup \widehat{\tau}_{LV}$ and is denoted by $\widehat{\tau}_V$.

Remark 5.160 An alternate characterization of $\widehat{\tau}_V$ is the following. Let $\{U_k\}_{k=1}^n \subseteq \tau$ and let $\mathcal{D}(\{U_k\}_{k=1}^n) = \{A \in 2^X \setminus \{\emptyset\} : A \subseteq \bigcup_{k=1}^n U_k, A \cap U_k \neq \emptyset \text{ for all } k = 1, \cdots, n\}$. Then the family of all such collections $\mathcal{D}(\{U_k\}_{k=1}^n)$ is a subbasis for $\widehat{\tau}_V$. Of course we can consider restrictions of these topologies on $P_f(X)$ and on $P_k(X)$.

Proposition 5.161 *If X and Y are Hausdorff topological space and $F : X \to 2^Y \setminus \{\emptyset\}$ a multifunction, then*

(a) *$F(\cdot)$ is usc if and only if it is continuous from X into $(2^Y \setminus \{\emptyset\}, \widehat{\tau}_{UV})$.*
(b) *$F(\cdot)$ is lsc if and only if it is continuous from X into $(2^Y \setminus \{\emptyset\}, \widehat{\tau}_{LV})$.*
(c) *$F(\cdot)$ is continuous if and only if it is continuous from X into $(2^X \setminus \{\emptyset\}, \widehat{\tau}_V)$.*

Remark 5.162 One can show that if $\alpha_1 \subseteq (2^X \setminus \{\emptyset\}, \widehat{\tau}_{UV})$ is connected and each element in α_1 is connected, then so is $C = \bigcup_{A \in \alpha_1} A$. Similarly if $\alpha_2 \subseteq (2^X \setminus \{\emptyset\}, \widehat{\tau}_{LV})$ is connected and each element of α_2 is connected, then so is $C = \bigcup_{A \in \alpha_2} A$. This observation leads to the following result.

Proposition 5.163 *If X and Y are Hausdorff topological spaces and $F : X \to 2^Y \setminus \{\emptyset\}$ is an usc or lsc multifunction with connected values, then for all $A \subseteq X$ connected, $F(A) \subseteq Y$ is connected too.*

For the Vietoris topology, we can improve this result.

Proposition 5.164

(a) *If $\alpha_3 \subseteq (2^X \setminus \{\emptyset\}, \widehat{\tau}_v)$ is connected and there is at least one connected element in α_3, then $\cup_{A \in \alpha_3}$ is connected too.*
(b) *If $F : X \to 2^Y \setminus \{\emptyset\}$ is continuous and for some $x \in X$, $F(x)$ is connected, then for all $A \subseteq X$ connected, $F(A) \subseteq Y$ is connected too.*

Hyperspace topologies and continuity of multifunctions are discussed in the books of Aliprantis and Border [8], Aubin and Cellina [15], Aubin and Frankowska [16], Beer [29], Berge [30], Castaing and Valadier [62], Denkowski et al. [81], Gorniewicz [127], Hu and Papageorgiou [145, 146], Klein and Thompson [162], and Papageorgiou and Kyritsi [215] (see also Flachsmeyer [113]).

5.2 The systematic study of the measurability properties of multifunctions started with the works of Castaing [61] and Jacobs [149], who adopted a "topological" point of view, since the domain of the multifunctions is a locally compact space equipped with a Radon measure. Castaing [61] deals with compact valued multifunctions, while Jacobs drops this restriction. Rockafellar [234, 238] considered as domain a measure space (Ω, Σ, μ) and as range space $\mathbb{R}^N$. Subsequently more general settings were adopted by Himmelberg [142], Leese [178], Parthasarathy [218], Wagner [279, 280], and the books mentioned earlier.

Let (Ω, Σ) be a measurable space and X a separable metrizable space. As we did in the topological setting above, we can interpret the measurability of a multifunction $F : \Omega \to P_k(X)$ using a σ-algebra on the hyperspace $P_k(X)$.

Definition 5.165 The "Effros σ-algebra" on $P_k(X)$ is defined as the σ-algebra generated by $U^- = \{C \in P_k(X) : C \cap U \neq \emptyset\}$, $U \subseteq X$ open. We denote the Effros σ-algebra by $\mathscr{S}_\varepsilon$.

Remark 5.166 The Effros σ-algebra is also generated by $U^+ = \{C \in P_k(X) : C \subseteq U\}$, $U \subseteq X$ open.

On $P_k(X)$, we consider the Hausdorff metric topology. It is separable metric.

Proposition 5.167 $\mathcal{B}(P_k(X)) = \mathscr{S}_\varepsilon$.

Then we can interpret the measurability of $F : \Omega \to P_k(X)$ as follows.

Proposition 5.168 $F : \Omega \to P_k(X)$ *is measurable if and only if it is $\mathscr{S}_\varepsilon$-measurable.*

5.3 Theorem 5.85 is due to Michael [193]. For Lipschitz selections, there is the following result due to Yost [288].

Proposition 5.169 *Let X be a metric space and Y a Banach space. Every h-Lipschitz multifunction $F : X \to P_{bfc}(Y)$ admits a Lipschitz continuous selection if and only if $\dim Y < \infty$.*

Approximate continuous selections for multifunctions (see Theorem 5.91) were first obtained by Cellina [64]. Related is the notion of σ-selectionable multifunctions (see Definition 5.92), which was introduced by Haddad and Lasry [129] who also proved Theorem 5.93.

Theorem 5.99 was proved by Kuratowski and Ryll Nardzewski [169]. The other major measurable selection theorem, Theorem 5.106, was proved in its present general form (with X=Souslin) by Saint-Beuve [244]. However, there were earlier versions of it with more restrictions on X by Yankov [286], von Neumann [208], and Aumann [19] (X=Polish). For the essential infimum and supremum of a functions, we refer also to Neveu [210].

We mention the following result due to Gasinski and Papaeorgiou [123] which can be useful in applications.

Proposition 5.170 *If (Ω, Σ, μ) is a finite measure space, X is a separable Banach space, and $G : \Omega \to P_{wkc}(X)$ is a multifunction with $S_G \neq \emptyset$, then there exists a measurable multifunction $\widehat{G} : \Omega \to P_{wkc}(X)$ such that*

$$\widehat{G}(\omega) \subseteq G(\omega)) \quad \mu - a.e. \text{ and } S_{\widehat{G}} = S_G.$$

5.4 The notion of decomposability has its origin from applied areas (see Boltyanski et al. [38] and Neustadt [209]). We saw that it looks like the notion of convexity and it serves as a substitute of it. Decomposability is also called "convexity with respect to switching." Decomposability is used in many different contexts, see the works of Boltyanski, Gamkrelidze, Pontryagin [38], Bressan and Colombo [47], Fryszkowski [119], Hiai and Umehaki [140], Olech [212], and Rockafellar [238]. A detailed presentation of the properties of decomposable sets can be found in the books of Fryszkowski [120] and Hu and Papageorgiou [145] (Sect. 2.3).

5.5 The integral of a multifunction was introduced by Aumann [18] in order to treat problems in Mathematical Economics (equilibrium theory). It is a natural extension of the classical Minkowski sum of sets. Aumann [18] also proved the following result.

Theorem 5.171 *If* $F : \Omega \to 2^{\mathbb{R}^N} \setminus \{\emptyset\}$ *is graph measurable,* $S_F^1 \neq \emptyset$, *and* $F(\omega) \subseteq \mathbb{R}_+^N$ $\mu - a.e.$, *then* conv $\int_\Omega F d\mu = \int_S$ conv $F d\mu$.

Corollary 5.172 *If* $\mu(\cdot)$ *is nonatomic,* $F : \Omega \to 2^{\mathbb{R}^N} \setminus \{\emptyset\}$ *is graph measurable,* $S_F^1 \neq \emptyset$ *and* $F(\omega) \subseteq \mathbb{R}_+^N$ $\mu - a.e.$, *then* $\int_\Omega F d\mu = \int_\Omega$ conv $F d\mu$

There is an alternative approach to the integration of $P_{kc}(X)$-valued multi-functions (X=Separable Banach space). It was introduced by Debreu [77] and it is based on the following embedding theorem due to Rädström [224].

Theorem 5.173 *If* X *is a Banach space, then there exists a normed space* V *such that* $(P_{kc}(X), h)$ *is embedded as a convex cone with vertex at the origin in* V *and the embedding is isometric.*

Then we can view a multifunction $F : \Omega \to P_{kc}(X)$ as a V-valued function. Then the Aumann and the Bochner integral of $F(\cdot)$ (as a V-valued map) coincide.

5.6 More on Caratheodory multifunctions can be found in Fryszkowski [120] and Hu and Papageorgiou [145].

5.8 Problems

Problem 5.1 Let X and Y be Hausdorff topological spaces and $f_k : X \to Y, k \in \{1, \cdots, N\}$ functions which are continuous at x_0. Let $F : X \to f_k(Y)$ be defined by $F(X) = \{f_k(x)\}_{k=1}^N$. Show that $F(\cdot)$ is continuous at $x_0 \in X$.

Problem 5.2 Let X, V, and Y be Hausdorff topological spaces and $f : X \times V \to Y$. (a) If $f(\cdot, u)$ is continuous on X for every $u \in V$ and $F(x) = f(x, V)$ for all $x \in X$, then show that $F(\cdot)$ is lsc on X.

Problem 5.3 Let X be a Hausdorff topological space, Y a Banach space, and $F : X \to P_{wkc}(X)$ a multifunction. Show that $F(\cdot)$ is usc from X into Y_w (=Y with the weak topology) if and only if for every $y^* \in Y^*$ the function $x \to \sigma(y^*, F(x))$ is upper semicontinuous.

Problem 5.4 Let X be a Hausdorff topological space, Y a Banach space, and $F : X \to P_{(w)k}(Y)$ a multifunction such that $F(X) \subseteq K \in P_k(Y)$ (resp.,$\in P_{wk}(Y)$). Show that F is lsc into Y (resp., Y_w) if and only if for every $y^* \in Y^*$ the function $x \to \sigma(y^*, F(x))$ is lower semicontinuous.

Problem 5.5 Let X be a Hausdorff topological space and $F : X \to P_{kc}(\mathbb{R}^N)$ a multifunction such that for all $y^* \in \mathbb{R}^*$, $x \to \sigma(y^*, F(x))$ is continuous. Show that $F(\cdot)$ is continuous.

Problem 5.6 Let X be a metric space, Y a Banach space, $C \in P_f(X)$, and $F : C \to P_f(Y)$ a continuous multifunction. Show that there exists a continuous multifunction $\widehat{F} : X \to P_f(Y)$ such that

$$\widehat{F}|_C = F \text{ and } \widehat{F}(X) \subseteq \text{conv } F(C).$$

Problem 5.7 Let (Ω, Σ) be a measurable space, X is a separable metrizable space, Y is a Hausdorff topological space, $f : \Omega \times X \to Y$ is a Caratheodory function, and $U \subseteq Y$ is nonempty open. Show that the multifunction $\omega \to F(\omega) = \{x \in X : f(\omega, x) \in U\}$ is measurable.

Problem 5.8 Let (Ω, Σ) be a measurable space, X a separable metrizable space, and $f : \Omega \times X \to \mathbb{R}$ a Caratheodory function. Consider the multifunction

$$F(\omega) = \{x \in X : f(\omega, x) = 0\},$$

and assume that $F(\omega) \neq 0$ for all $\omega \in \Omega$. Show that

(a) $F(\cdot)$ is graph measurable.
(b) If X is compact, then $F(\cdot)$ is measurable.

Problem 5.9 Let (Ω, Σ) be a complete measurable space, X a Souslin space, $f : \Omega \times X \to \mathbb{R}^* = \mathbb{R} \cup \{\pm\infty\}$ a $\Sigma \otimes B(X)$-measurable function, and $F : \Omega \to 2^X \setminus \{\emptyset\}$ a graph measurable multifunction. We set

$$m(\omega) = \inf[f(\omega, x) : x \in F(\omega)].$$

Show that $\omega \to m(\omega)$ is Σ-measurable.

Problem 5.10 Let (Ω, Σ) be a measurable space, X a separable metrizable space, $f : \Omega \times X \to \mathbb{R}$ a Caratheodory function, and $F : \Omega \to P_k(X)$ a measurable multifunction. We set

$$m(\omega) = \min[f(\omega, x) : x \in F(\omega)]$$

$$and \ s(\omega) = \{x \in F(\omega) : m(\omega) = f(\omega, x)\}.$$

Show that both $m(\cdot)$ and $s(\cdot)$ are Σ-measurable.

Problem 5.11 Let (Ω, Σ) be a measurable space, X is a Polish space, Y is a metric space, $f : \Omega \times X \rightarrow Y$ is a Caratheodory function, $F : \Omega \rightarrow P_k(X)$ is a measurable multifunction, and $h : \Omega \rightarrow Y$ is a Σ-measurable function such that $h(\omega) \in f(\omega, F(\omega))$ for all $\omega \in \Omega$. Show that there exists $u : \Omega \rightarrow X$ Σ-measurable function such that

$$h(\omega) = f(\omega, u(\omega)) \quad for \ all \ \omega \in \Omega.$$

Problem 5.12 Let (Ω, Σ) be a measurable space, (X, d) is a separable, complete metric space, and $F : \Omega \rightarrow P_k(X), h : \Omega \rightarrow X$ are both Σ-measurable. Show that there exists $f : \Omega \rightarrow X$ Σ-measurable such that $f(\omega) \in F(\omega)$ and $d(h(\omega), f(\omega)) = d(h(\omega), F(\omega))$ for all $\omega \in \Omega$.

Problem 5.13 Let X be a metric space and $\varphi, \psi : X \rightarrow \mathbb{R}$ two functions such that $\varphi(\cdot), -\psi(\cdot)$ are upper semicontinuous and

$$\varphi(x) < \psi(x) \quad for \ all \ x \in X.$$

Show that there exists a continuous function $f : \Omega \rightarrow X$ such that

$$\varphi(x) < f(x) < \psi(x) \quad for \ all \ x \in X.$$

Problem 5.14 Let X and Y be two Hausdorff topological space, $D \subseteq X$ nonempty, and $F : X \rightarrow 2^Y \setminus \{\emptyset\}$ a multifunction which is lsc on $\overline{D} \setminus D$. Show that $F(\overline{D}) \subseteq \overline{F(D)}$.

Problem 5.15 Let (Ω, Σ) be a measurable space, X is a Polish space, Y is a metric space, $f : \Omega \times X \rightarrow Y$ is a Caratheodory function, and $L : \Omega \rightarrow P_f(X)$ is a measurable multifunction. We set

$$H(\omega) = f(\omega, L(\omega)) \quad \omega \in \Omega.$$

Show that $H : \Omega \rightarrow 2^Y \setminus \{\emptyset\}$ is measurable.

Problem 5.16 Suppose (Ω, Σ) be a measurable space, X is a separable Banach space, and $u : \Omega \rightarrow X, r : \Omega \rightarrow (0, +\infty)$ are two Σ-measurable functions. We consider the multifunction

$$\omega \to B_{r(\omega)}(u(\omega)) = \{x \in X : \|x - u(\omega)\| < r(\omega)\} = F(\omega)$$

Show that $F(\cdot)$ is measurable.

Problem 5.17 Suppose (Ω, Σ) is a measurable space, X a Polish space, and $u : X \to \Omega$ a map which satisfies

(a) $u^{-1}(\{\omega\}) \in P_f(X)$ for all $\omega \in \Omega$.
(b) $u(V) \in \Sigma$ for all $V \subseteq X$ open.

Show that we can find a Σ-measurable function $f : \Omega \to X$ such that $u(f(\omega)) = \omega$ for all $\omega \in \Omega$.

Problem 5.18 Suppose (Ω, Σ) is a complete measurable space, X, Y are Polish spaces with Y being σ-compact, $F : \Omega \to P_f(X \times Y)$ and $G : \Omega \times X \to P_f(Y)$ satisfies

$$G(\omega, x) = \{y \in Y : (x, y) \in F(\omega)\}.$$

Show that $F(\cdot)$ is measurable if and only if $G(\cdot, \cdot)$ is.

Problem 5.19 Let (Ω, Σ, μ) be a σ-finite measure space, X a separable Banach space, and $F : \Omega \to 2^X \setminus \{\emptyset\}$ a graph measurable multifunction. Show that $S_F^p \subseteq L^p(\Omega, X)$ $(1 \le p \le \infty)$ is bounded if and only if $F(\cdot)$ is p-integrably bounded.

Problem 5.20 Let $C \in P_{fc}(\mathbb{R}^N)$ and $t_1, t_2 \in [0, b], t_1 < t_2$. Show that

$$\int_{t_1}^{t_2} C dt = (t_2 - t_1)C.$$

Problem 5.21 Suppose $F : [0, b] \to P_k(\mathbb{R}^N)$ is measurable and integrably bounded. Show that

$$\lim_{n \to 0} h\left(\frac{1}{h} \int_t^{t+h} F(s)ds, \text{conv } F(t)\right) = 0 \quad for \ a.a \ t \in [0, b].$$

Problem 5.22 Suppose X is a compact space, Y is a topological vector space, and $F : X \to 2^Y \setminus \{\emptyset\}$ is a multifunction with convex values such that $F^-(\{y\}) \subseteq X$ is open for all $y \in Y$. Show that $F(\cdot)$ admits a continuous selection.

Problem 5.23 Let X be a Hausdorff topological space and $\varphi : X \to \mathbb{R}$ a function. Show that

(a) $\varphi(\cdot)$ is lower semicontinuous if and only if $x \to U_\varphi(x) = \{\lambda \in \mathbb{R} : \varphi(x) \le \lambda\}$ is usc.

(b) $\varphi(\cdot)$ is upper semicontinuous if and only if $x \to U_\varphi(x)$ is lsc.

Problem 5.24 Suppose X is a paracompact space, Y is a Banach space, $F : X \to 2^Y \setminus \{\emptyset\}$ is an lsc multifunction, and $\varepsilon > 0$. Show that there exists a continuous map $f : X \to Y$ such that

$$d(f(x), F(x)) < \varepsilon \quad for\ all\ x \in X.$$

Problem 5.25 Let (Ω, Σ, μ) be a σ-finite measure space, X a separable Banach, and $F, G : \Omega \to P_f(X)$ two measurable multifunctions such that $S_F^p, S_G^p \neq \emptyset, 1 \leq p < \infty$. Let $H(\omega) = \overline{F(\omega) + G(\omega)}$ for all $\omega \in \Omega$. Show that $S_H^p = S_F^p + S_G^p$.

Chapter 6
Smooth and Nonsmooth Calculus

In this chapter, we provide an introduction to the classical smooth calculus and the nonsmooth calculus by developing the subdifferential theory of convex and of locally Lipschitz functions.

Section 6.1 deals with the smooth calculus of functions between Banach spaces. We focus on the notions of Gateaux and Frechet derivatives and develop some calculus rules for them.

In Sect. 6.2, we enter the realm of nonsmooth analysis and examine convex functions and their subdifferential theory. First we establish the remarkable continuity properties that convex functions exhibit and then introduce the notion of "convex subdifferential" that extends the concept of derivative to nonsmooth convex functions, and develop the corresponding calculus.

Duality is in the core of convex analysis. Section 6.3 deals with this aspect of the theory of convex functions. We introduce the notion of "conjugate function" and develop its properties and how it relates to the notion of subdifferential.

In Sect. 6.4, we introduce the operation of infimal convolution, which in some sense is dual to the operation of addition. Also, we consider regular approximations of a convex function (Moreau–Yosida regularization) and also examine coercive functions that are a basic tool in variational analysis.

Finally, in Sect. 6.5, we deal with Lipschitz and locally Lipschitz functions. Lipschitz functions are central in modern geometric measure theory, while locally Lipschitz functions provide the right framework for the development of a subdifferential theory that extends the convex one.

© The Author(s), under exclusive license to Springer Nature Switzerland AG 2022   339
S. Hu, N. S. Papageorgiou, *Research Topics in Analysis, Volume I*, Birkhäuser
Advanced Texts Basler Lehrbücher, https://doi.org/10.1007/978-3-031-17837-5_6

6.1 Differential Calculus in Normed Spaces

In this section, we introduce the generalizations of the derivative of functions from $\mathbb{R}$ to $\mathbb{R}$. We develop the relevant calculus and also discuss extrema, namely, maxima and minima of $\mathbb{R}$-valued functions defined on a normed space.

Definition 6.1

(a) Let X be a topological vector space, Y a normed space, and $U \subseteq X$ open. $f : U \to Y$ is called "Gateaux differentiable" at $x \in U$ if there exists $f'_G(x) \in \mathcal{L}(X, Y)$ such that for all $h \in X$

$$f'_G(x)(h) = \lim_{t \to 0} \frac{f(x + th) - f(x)}{t}.$$

(b) Let X, Y be normed spaces, and $U \subseteq X$ open. $f : U \to Y$ is called "Frechet differentiable" at $x \in U$ if there exists $f'_F(x) \in \mathcal{L}(X, Y)$ such that

$$\lim_{\|h\|_X \to 0} \frac{\|f(x + h) - f(x) - f'_F(x)(h)\|_Y}{\|h\|_X} = 0.$$

Obviously, we may have the following more convenient formulation of the Frechet differentiability.

Proposition 6.2 *Let X, Y be normed space, $U \subseteq Y$ open, and $f : U \to Y$ Frechet differentiable at $x \in U$. Then $f(x + h) = f(x) + f'_F(x)(h) + e(x, h)$, where*

$$\lim_{\|h\|_X \to 0} \frac{\|e(x, h)\|_Y}{\|h\|_X} = 0.$$

It is clear from the definitions that Frechet differentiability implies Gateaux differentiability, though the converse is not true.

Example 6.3 Let $X = \mathbb{R}^2$, $Y = \mathbb{R}$ and consider $f : X \to Y$, defined by

$$f(x, y) = \begin{cases} \dfrac{x^5}{(y - x^2)^2 + x^4} & \text{if } (x, y) \neq (0, 0) \\ 0 & \text{if } (x, y) = (0, 0). \end{cases}$$

This function is Gateaux differentiable at the origin and $f'_G(0) = 0$, but if it were Frechet differentiable at $x = 0$, we would have $f'_F(0) = 0$. However, in Definition 6.1(b), if we pass to the limit as h moves to the origin along the parabola $y = x^2$, then the limit is equal to 1. Therefore, f is not Frechet differentiable at $x = 0$.

To be able to claim that Gateaux differentiability implies Frechet differentiability, we need some additional hypotheses.

Proposition 6.4 *Let X, Y be normed spaces, $U \subseteq X$ open, $f : U \to X$ Gateaux differentiable, and $u \to f'_G(u)$ is continuous at $x \in U$. Then f is Frechet differentiable at x and $f'_F(x) = f'_G(x)$.*

Proof Let $e(x, h) = f(x + h) - f(x) - f'_G(x)(h)$, and $y^* \in Y^*$, with $\|y^*\|_{Y^*} = 1$ such that $\|e(x, h)\|_Y = |\langle y^*, e(x, h) \rangle_Y|$, where $\langle \cdot, \cdot \rangle_Y$ denotes the duality brackets for the pair (Y, Y^*). We have

$$\|e(x, h)\|_Y = |\langle y^*, f(x + h) - f(x) \rangle_Y - \langle y^*, f'_G(x)(h) \rangle_Y|$$
$$= |\langle y^*, (f'_G(x + th) - f'_G(x))h \rangle \text{ for some } t \in (0, 1)$$
$$\leq \|f'_G(x + th) - f'_G(x)\|_{\mathcal{L}} \|h\|_X.$$

Thus,

$$\frac{\|e(x, h)\|_Y}{\|h\|_X} \leq \|f'_G(x + th) - f'_G(x)\|_{\mathcal{L}},$$

and therefore, $\lim_{\|h\|_X \to 0} \frac{\|e(x,h)\|_Y}{\|h\|_X} \to 0$ since $x \to f'_G(x)$ is continuous. Hence, f is Frechet differentiable at x and $f'_F(x) = f'_G(x)$. $\square$

Remark 6.5 The above proposition is a generalization of the well-known result from calculus, which says that the existence of all partial derivatives at an open set and the continuity of these partial derivatives at a point imply differentiability of the function at the point. Simple existence of the partial derivatives does not suffice.

Proposition 6.6 *Let X, Y be normed spaces, $U \subseteq X$ open, and $f : X \to Y$ Frechet differentiable at $x \in U$. Then f is continuous at x.*

Proof From Definition 6.1(b), we can find $\delta > 0$ such that $\|f(x + h) - f(x) - f'_F(x)(h)\|_Y \leq \|h\|_Y$ for all $\|h\|_X \leq \delta$. Thus,

$$\|f(x + h) - f(x)\| \leq \left[1 + \|f'_F(x)\|_{\mathcal{L}}\right] \|h\|_X$$

for $\|h\|_X \leq \delta$. Hence, f is continuous at x. $\square$

Remark 6.7 In contrast, Gateaux differentiability does not imply continuity (see Example 6.3).

Definition 6.8 Let X, Y be normed spaces, $U \subseteq X$ open, and $f : U \to Y$. f is said to be "strongly (resp., weakly) hemicontinuous at $x \in U$" if $t_n \to 0$ implies $f(x + t_n h) \to f(x)$ in Y (resp., $f(x + t_n h) \xrightarrow{w} f(x)$ in Y) as $n \to \infty$ for all $h \in X$.

Proposition 6.9 *Let X, Y be normed spaces, $U \subseteq X$ open and $f : U \to Y$ Gateaux differentiable at $x \in U$. Then f is strongly hemicontinuous at $x \in U$.*

Proof Let $\varphi(t) = f(x + th)$ for $t \in \mathbb{R}, h \in X$. Then φ is differentiable at $t = 0$ and $\varphi'(0) = f'_G(x)(h)$. φ is then continuous at $t = 0$. Hence,

$$\varphi(t_n) \to \varphi(0) \text{ as } t_n \to 0,$$

and thus $f(x + t_n h) \to f(x)$ as $n \to \infty$. Therefore, f is strongly hemicontinuous at $x \in U$. $\square$

Gateaux and Frechet derivatives are linear operations. Composition of two Gateaux differentiable functions need not be Gateaux differentiable. However, Frechet differentiability yields the following "chain rule."

Proposition 6.10 *Let X, Y, V be normed spaces, $U \subseteq X$ and $W \subseteq Y$ open sets, $f : U \to Y$ Gateaux differentiable at $x \in U$ with $f(x) \in W$, and $g : W \to V$ Frechet differentiable at $f(x)$. Then $k = g \circ f : U \to V$ is Gateaux differentiable at $x \in U$ and*

$$k'(x) = (g \circ f)'_G(x) = g'_F(f(x)f'_G(x). \tag{6.1}$$

Moreover, if f is Frechet differentiable at $x \in U$, then so is $g \circ f$ and (6.1) holds.

Proof Let $t \neq 0$. We have

$$\frac{1}{|t|}\|k(x + th) - k(x) - tg'_F(f(x))f'_G(x)\|_V$$

$$\leq \frac{1}{|t|}\|g(f(x + th)) - g(f(x)) - tg'_F(f(x))[f(x + th) - f(x)]\|_V \tag{6.2}$$

$$+ \|g'_F(f(x))[f(x + th) - f(x) - tf'_G(x)(h)]\|_V.$$

The Gateaux differentiability of f implies that

$$\|g'_F(f(x))[f(x + th) - f(x) - tf'_G(x)(h)]\|_V \to 0 \text{ as } t \to 0. \tag{6.3}$$

Also, since g is Frechet differentiable and $\|f(x + th) - f(x)\|_Y \to 0$ (see Proposition 6.9), we have

$$\frac{1}{|t|}\|g(f(x + th)) - g(f(x)) - g'_F(f(x))[f(x + th) - f(x)]\|_Y \to 0 \text{ as } t \to 0. \tag{6.4}$$

Returning to (6.2) and using (6.3) and (6.4), we obtain

$$\frac{1}{|t|}\|k(x + th) - k(x) - tg'_F((f(x))f'_G(x)\|_V \to 0 \text{ as } t \to 0.$$

Thus, k is Gateaux differentiable. In a similar way, we can show the Frechet differentiability of k if both f and g are Frechet differentiable. □

Now we present some characteristic examples often seen in applications.

Example 6.11 Let X, Y be normed spaces and $A \in \mathcal{L}(X, Y)$. Then for $v_0 \in Y$, the function $f(x) = A(x) + v_0$ is Frechet differentiable and $f'_F(x)(h) = A(h)$ for all $x, h \in X$.

Remark 6.12 The example illustrates the interpretation of the derivative as the best local linear approximation of the function at a point.

The next example is in the same vein as the previous one.

Example 6.13 Let H be a Hilbert space and $\alpha : H \times H \to R$ be a symmetric bilinear continuous form. Continuity means that $|\alpha(x, u)| \leq c\|x\|\|u\|$ for all $x, u \in H$, some $c > 0$. For $v_0 \in H$, we introduce the function $f : H \to R$ defined by

$$f(x) = \frac{1}{2}\alpha(x, x) + (v_0, x)$$

for all $x \in H$, with $(\cdot, \cdot)$ being the inner product of H. We have

$$f(x + h) - f(x) = \alpha(x, h) + \frac{1}{2}\alpha(h, h) + (v_0, h) \leq \alpha(x, h) + \frac{c}{2}\|h\|^2 + (v_0, h),$$

$$\Rightarrow \frac{\|f(x + h) - f(x) - (\alpha(x, h) + (v_0, h))\|}{\|h\|} \leq \frac{c}{2}\|h\|,$$

$\Rightarrow f$ is Frechet differentiable and $(f'_F(x), h) = \alpha(x, h) + (v_0, h)$ for all $x, h \in H$.

The Nemytskii (or superposition) map appears in many occasions.

Definition 6.14 Let (Ω, Σ, μ) be a measure space and $f : \Omega \times \mathbb{R} \to \mathbb{R}$ a measurable function (e.g., a Caratheodory function, see Theorem 2.47). The Nemytskii (superposition) map $N_f : M(\Omega) \to M(\Omega)$ ($M(\Omega)$ is the vector space of an $\mathbb{R}$-valued Σ-measurable function on Ω), is defined by

$$N_f(u)(\cdot) = f(\cdot, u(\cdot))$$

for all $u \in M(\Omega)$.

We want to know when $N_f(\cdot)$ maps a Lebesgue space to another Lebesgue space. A straightforward application of the Lebesgue dominated convergence theorem (see Theorem 2.86) gives the following result.

Proposition 6.15 *If $f : \Omega \times \mathbb{R} \to \mathbb{R}$ is a Caratheodory function such that*

$$|f(\omega, x)| \le \alpha(\omega) + b|x|^q, \text{ for } \mu - a.a \ \omega \in \Omega, \text{ all } x \in \mathbb{R}, \tag{6.5}$$

with $\alpha \in L^p(\Omega)$, $1 \le p \le \infty$, and $q \ge 1$, then N_f maps $L^{pq}(\Omega)$ into $L^p(\Omega)$, and it is continuous and bounded (i.e., maps bounded sets to bounded sets).

A remarkable result of Krasnoselskii shows that the sufficient condition (5) is also necessary, and we have continuity of the Nemytskii map (see Krasnoselskii [164], Theorems 2.1, 2.2, 2.3, pp. 22–27).

Theorem 6.16 *If N_f maps $L^p(\Omega)$ into $L^q(\Omega)$, $1 \le p, q \le \infty$, then N_f is continuous and bounded, and there exist $a \in L^q(\Omega)$ and $b > 0$ such that $|f(\omega, x)| \le \alpha(\omega) + b|x|^{p/q}$ for all $\omega \in \Omega$, all $x \in \mathbb{R}$.*

We want to investigate the Frechet differentiability of the Nemytskii map. So, suppose (Ω, Σ, μ) is a finite measure space, and $f : \Omega \times \mathbb{R} \to \mathbb{R}$ is a Caratheodory function such that for all $\omega \in \Omega$, $f(\omega, \cdot) \in C^1(\mathbb{R})$. Then $f_x'(w, x)$ is a Caratheodory function, and we assume that

$$|f_x'(\omega, x)| \le a(\omega) + b|x|^r \text{ for } \mu - a.a \ \omega \in \Omega, \text{ all } x \in \mathbb{R}, \tag{6.6}$$

with $a \in L^\tau(\Omega)$, $1 \le \tau < \infty$, $r > 0$. Integrating (6.6), we obtain

$$|f(\omega, x)| \le c(\omega) + \alpha(\omega)|x| + \frac{b}{r+1}|x|^{r+1} \text{ for } \mu - a.a \ \omega \in \Omega, \text{ all } x \in \mathbb{R}. \tag{6.7}$$

Here $c(\cdot)$ is an arbitrary function. We use Young's inequality on the second term in (6.7) and obtain

$$|f(\omega, x)| \le c(\omega) + \frac{r}{r+1}\alpha(\omega)^{\frac{r+1}{r}} + \frac{b+1}{r+1}|x|^{r+1} \text{ for } \mu - a.a \ \omega \in \Omega, \text{ all } x \in \mathbb{R}.$$

Note that $\alpha(\cdot)^{\frac{r+1}{r}} \in L^q(\Omega)$ with $q = \frac{r\tau}{r+1}$. So, if we have $c \in L^q(\Omega)$, then using Proposition 6.15, we have

$$N_f : L^p(\Omega) \to L^q(\Omega) \text{ with } p = r\tau, q = \frac{r\tau}{r+1}, \tag{6.8}$$

$$N_{f_x'} : L^p(\Omega) \to L^\tau(\Omega). \tag{6.9}$$

Theorem 6.17 *If $f, f_x' : \Omega \times \mathbb{R} \to \mathbb{R}$ are as above and satisfy (6.6), (6.7), then $N_f : L^p(\Omega) \to L^q(\Omega)$ (see (6.8)) is continuously Frechet differentiable and $(N_f)'(u) = N_{f_x'}(u)$, that is*

$$(N_f)'(u)(h) = N_{f_x'}(u)(h) \text{ for all } u, h \in L^p(\Omega).$$

Proof Using the generalized Hölder inequality (see Corollary 2.181), we see that

$$N_{f'_x}u)(h) \in L^q(\Omega) \text{ for all } u, h \in L^p(\Omega) \text{ (see (6.9))}.$$

We fix $u \in L^p(\Omega)$ and set

$$w(h) = N_f(u + h) - N_f(u) - N_{f'_x}(u)(h).$$

We have

$$f(u(\omega) + h(\omega)) - f(u(\omega)) = \int_0^1 \frac{d}{dt} f(\omega, u(\omega) + th(\omega)) dt$$

$$= \int_0^1 f'_x(\omega, u(\omega) + th(\omega)) h(\omega) dt.$$

Thus,

$$\int_\Omega |w(h)(\omega)|^q d\mu = \int_\Omega |[\int_0^1 |f'_x(\omega, u(\omega) + th(\omega)) - f'_x(w, u(w)h(\omega))| dt|]^q d\mu$$

and

$$\int_\Omega |w(h)(\omega)|^q d\mu \leq (\int_0^1 \int_\Omega |f'_x(\omega, u(\omega) + th(\omega)) - f'_x(w, u(w))|^\tau d\mu dt)^{q/\tau} \|h\|_p^q$$

(using Hölder's inequality and Fubini's theorem). Hence, $\|w(h)\|_q \to 0$. We conclude that N_f is continuously Frechet differentiable and

$$(N_f)'(u)(h) = N_{f'_x}(u)(h) \text{ for all } u, h \in L^p(\Omega).$$

$\square$

Remark 6.18 Note that $q < p$ (since $r > 0$). If $r = 0$, then from (6.6), we see that $|f'_x(\omega, x)| \leq \alpha(\omega)$ for μ-a.a $\omega \in \Omega$, all $x \in \mathbb{R}$, with $\alpha \in L^\tau(\Omega)$, $1 \leq \tau \leq \infty$. Then

$$N_{f'_x} : L^p(\Omega) \to L^\tau(\Omega) \text{ for all } p \geq 1,$$

and arguing as above, we have

$$N_f : L^p(\Omega) \to L^q(\Omega) \text{ for all } p \geq 1, q = \frac{\tau p}{p + \tau}$$

and so the previous result applies.

Now assume that $\tau = +\infty$ and $\Omega \subseteq \mathbb{R}^N$ is a bounded open set. We have

$$|f'_x(\omega, x)| \leq M \text{ for } \mu - a.a \, \omega \in \Omega, \text{ all } x \in \mathbb{R}, \text{ some } M > 0. \tag{6.10}$$

Integrating (6.10), we obtain

$$|f(\omega, x)| \leq \alpha(\omega) + c|x| \text{ for } \mu - a.a \ \omega \in \Omega, \text{ all } x \in \mathbb{R}, \tag{6.11}$$

with $c > 0$, $\alpha(\cdot)$ an arbitrary function. From (6.10) and (6.11), it follows that

$$N_{f_x'} : L^p(\Omega) \to L^\infty(\Omega) \text{ for all } 1 \leq p \leq \infty,$$

$$N_f : L^p(\Omega) \to L^p(\Omega) \text{ for all } 1 \leq p \leq \infty \ (taking \ \alpha \in L^p(\Omega)).$$

In this case, if N_f is Frechet differentiable, then $f(\emptyset, \cdot)$ is affine.

Proposition 6.19 *If $\Omega \subseteq \mathbb{R}^N$ is bounded open, $f_x' : \Omega \times \mathbb{R} \to \mathbb{R}$ satisfies (6.10), and $N_f : L^p(\Omega) \to L^p(\Omega)$ is Frechet differentiable, then $f(\emptyset, x) = \alpha(\emptyset) + c(\emptyset)x$ with $\emptyset \in L^p(\Omega), c \in L^\infty(\Omega)$.*

Proof Let

$$w_t(\emptyset) = \frac{1}{t}[f(\emptyset, u(\emptyset) + th(\emptyset)) - f(\emptyset, u(\emptyset))] - f_x'(\emptyset, u(\emptyset))h(\emptyset));$$

we have

$$w_t(\emptyset) = \int_0^1 [f_x'(\emptyset, u(\emptyset) + th(\emptyset)) - f_x'(\emptyset, u(\emptyset))]h(\emptyset)dt;$$

hence,

$$\int_\Omega |w_t(\emptyset)|^p d\mu \leq \int_0^1 \int_\Omega |f_x'(\emptyset), u(\emptyset) + sth(\emptyset) - f_x'(\emptyset, u(\emptyset))|^p |h(\emptyset)|^p d\mu ds.$$

By the Lebesgue dominated convergence theorem, we have

$$(N_f)_G'(u)(h) = N_{f_x'}(u)h \text{ for all } u, h \in L^p(\Omega). \tag{6.12}$$

Suppose that N_f is in fact Frechet differentiable. Then

$$(N_f)_F'(u) = (N_f)_G'(u) = N_{f_x'}(u) \ (\text{see } (6.12)).$$

Assuming that $f(\emptyset, 0) = 0 \ \mu - a.e$, we have

$$\frac{1}{\|u\|_p} \|N_f(u) - N_{f_x'}(0)u\|_p \to 0 \text{ as } \|u\|_p \to 0. \tag{6.13}$$

We fix $\vartheta \in \mathbb{R}$ and $\emptyset_0 \in \Omega$ and consider the functions

$$u(\emptyset) = \vartheta \chi_{B_\delta(\emptyset_0)}, \ \delta > 0.$$

We have that

$$\frac{1}{\|u\|_p} \|N_f(u) - N_{f'_x}(0)u\|_p^p = \frac{1}{\vartheta^2 \lambda^N(B_\delta(\emptyset_0))} \int_{B_\delta(\emptyset_0)} |f(\emptyset, \vartheta) - f'_x(\emptyset, 0)\vartheta^p| d\lambda^N,$$

(6.14)

with $\lambda^N(\cdot)$ being the Lebesgue measure on $\mathbb{R}^N$. From (6.13) and (6.14), it follows that

$$\frac{1}{\vartheta^p} |f(\emptyset, \vartheta) - f'_x(\emptyset, 0)\vartheta| \to 0 \ a.e. \ in \ \Omega, \ as \ \delta \to 0^+,$$

$$\Rightarrow f(\emptyset, \vartheta) = f'_x(\emptyset, 0)\vartheta,$$

$$\Rightarrow f(\emptyset, x) = \alpha(\emptyset)x \ with \ a \in L^\infty(\Omega). \qquad \square$$

Next we will determine the derivative of a functional that we encounter in the use of variational methods to solve elliptic boundary value problems.

So, let $\Omega \subseteq \mathbb{R}^N$ be a bounded domain and $f : \Omega \times \mathbb{R} \to \mathbb{R}$ a Caratheodory function such that

$$|f(\emptyset, x)| \le a(\emptyset) + c|x|^{r-1} \ for \ a.a \ \emptyset \in \Omega, \ all \ x \in \mathbb{R},$$

with $a \in L^p(\Omega), 1 \le p \le \infty, c > 0, r > 1$. We consider the primitive of $f(\emptyset, \cdot)$, namely the function $F(\emptyset, x) = \int_0^x f(\emptyset, s)ds$. We see that

$$|F(\emptyset, x)| \le \widehat{a}(\emptyset) + \widehat{c}|x|^r \ for \ a.a \ \emptyset \in \Omega, \ all \ x \in \mathbb{R},$$

with $\widehat{a} \in L^{p/r}(\Omega), \widehat{c} > 0$. Then from Proposition 6.10, we have $N_f : L^p(\Omega) \to L^{p/r-1}(\Omega)$ and $N_F : L^p(\Omega) \to L^{p/r}(\Omega)$.

If $p = r > 1$, then we have

$$|f(\emptyset, x)| \le a(\emptyset) + c|x|^{r-1} \ for \ a.a \ \emptyset \in \Omega, \ all \ x \in \mathbb{R}, \qquad (6.15)$$

with $a \in L^{p'}(\Omega), (\frac{1}{p} + \frac{1}{p'} = 1), c > 0$ and

$$|F(\emptyset, x)| \le \widehat{a}(\emptyset) + \widehat{c}|x|^r \ for \ a.a \ \emptyset \in \Omega, \ all \ x \in \mathbb{R},$$

with $\widehat{a} \in L^1(\Omega), \widehat{c} > 0$. Hence

$$N_f : L^p(\Omega) \to L^{p'}(\Omega) \ and \ N_F : L^p(\Omega) \to L^1(\Omega).$$

Proposition 6.20 *If $f : \Omega \times \mathbb{R} \to \mathbb{R}$ is a Caratheodory function satisfying (6.15) and $\sigma : L^p(\Omega) \to \mathbb{R}(1 \leq p \leq \infty)$ is defined by*

$$\sigma(u) = \int_{\Omega} F(\emptyset, u(\emptyset))d\lambda^N \text{ for all } u \in L^p(\Omega),$$

then $\sigma(\cdot)$ is continuously Frechet differentiable and

$$\sigma'(u) = N_f(u) \text{ for all } u \in L^p(\Omega).$$

Proof Let

$$w(h) = \int_{\Omega} F(\emptyset, u + h)d\lambda^N - \int_{\Omega} F(\emptyset, u)d\lambda^N - \int_{\Omega} f(\emptyset, u)hd\lambda^N.$$

We have (see the proof of Theorem 6.17)

$$w(h) = \int_{\Omega} \int_0^1 [f(\emptyset, u + th) - f(\emptyset, u)]hdtd\lambda^N.$$

Thus, using Fubini's Theorem and Hölder's Inequality,

$$|w(h)| \leq \int_0^1 \|N_f(u + th) - N_f(u)\|_{p'}dt\|h\|_p.$$

$$\Rightarrow \int_{\Omega} |w(h)|d\lambda^N \to 0,$$

$$\Rightarrow \sigma'(u) = N_f(u) \text{ for all } u \in L^p(\Omega).$$

Since $N_f(\cdot)$ is continuous (see Proposition 6.15), we conclude that $\sigma(\cdot)$ is continuously Frechet differentiable.   □

Corollary 6.21 *If $\Omega \subseteq \mathbb{R}^N$ is a bounded domain, $1 < p < \infty$, and $\sigma_p : L^p(\Omega) \to \mathbb{R}$ is defined by $\sigma_p(u) = \frac{1}{p}\|u\|_p^p$ for all $u \in L^p(\Omega)$, then $\sigma_p(\cdot)$ is everywhere Frechet differentiable and*

$$\sigma_p'(u) = p|u|^{p-2}u \in L^{p'}(\Omega) \text{ for all } u \in L^p(\Omega).$$

Now let $\{X_k, Y\}_{k=1}^m$ be normed spaces, $X = \prod_{k=1}^m X_k$, $U \subseteq X$ be open, and $f : U \to Y$ a map. Given $(x_0^k)_{k=1}^m \in X$, let $E_k : X_k \to X$, $k = 1, \ldots, m$, be defined by

$$\xi_k(x_k) = (x_0^1, \ldots, x_0^{k-1}, x_k, x_o^{k+1}, \ldots, x_0^m).$$

Proposition 6.22 *If* $f : U \to Y$ *is Frechet differentiable at* $x_0 \in U$, *then for every* $k \in \{1, \ldots, m\}$, $f \circ \xi_k$ *is differentiable at* $x_0^k \in X_k$ *and*

$$f_F'(x_0)(h) = \sum_{k=1}^m (f \circ \xi_k)'(x_0^k)h_k, \text{ for all } h = (h_k)_{k=1}^m \in X.$$

Proof Let $i_k : X_k \to X$ be the injection map. We have

$$\xi_k(x_k) = x_0 + i_k(x_k \cdot x_0^k),$$

$$\Rightarrow \xi_k'(x_k) = i_k \text{ for all } x_k \in X_k \text{ (see Example 6.11)}.$$

Hence, $(f \circ \xi_k)(\cdot)$ is Frechet differentiable at x_k and

$$(f \circ \xi_k)(\cdot)'(x_0^k) = f'(x_0) \circ i_k.$$

Let $p_k : X \to X_k$ be the projection map. Then

$$\sum_{k=1}^m i_k \circ p_k = id_X,$$

$$\Rightarrow \sum_{k=1}^m (f'(x_0) \circ i_k) \circ p_k = f'(x_0).$$

$\square$

Definition 6.23 We call $(f \circ \xi_k)'(x_0^k)$ the "$k = $-partial derivative of $f(\cdot)$ at $x_0 \in U$"." We denote the partial derivative by $f_{x_k}'(x_0)$.

Example 6.24

(a) Let $U \subseteq \mathbb{R}^N$ be open, and $f : U \to \mathbb{R}$ is differentiable at $x_0 \in U$. Then we have

$$f'(x_0) = (f_{x_k}'x_0))_{k=1}^N.$$

Usually, this vector is denoted by $\nabla f(x_0)$ (the gradient of $f(\cdot)$ at $x_0 \in U$).

(b) Let $U \subseteq \mathbb{R}^N$ be open, and $f = (f_i)_{i=1}^m : U \to \mathbb{R}^m$ is differentiable at $x_0 \in U$. Then we have the $m \times N$ matrix, known as the Jacobian matrix of f at x_0:

$$f'(x_0) = ((f_i)_{x_k}'(x_0))_{i=1,k=1}^{m,N}.$$

Remark 6.25 We already mentioned (see Remark 6.5) that mere existence of all partial derivatives does not imply differentiability of the function. Additional conditions are needed.

Next we will examine to what extent the mean value theorem holds in the present setting.

Definition 6.26 Let V be a vector space and $u, v \in V$. We define

$$[u, v] = \{x \in V : x = (1 - t)u + tv, 0 \le t \le 1\} \text{ (the closed interval)},$$

$$(u, v) = \{x \in V : x = (1 - t)u + tv, 0 < t < 1\} \text{ (the open interval)}.$$

We have the following versions of the classical mean value theorem.

Theorem 6.27 *If X is a normed space, $U \subseteq X$ an open set, $u, v \in X$ with $[u, v] \subseteq U$, and $f : U \to \mathbb{R}$ a Frechet differentiable function, then there exists $x_0 \in (u, v)$ such that*

$$f(v) - f(u) = f'(x_0)(v - u).$$

Proof Let $\xi : [0, 1] \to X$ be defined by $\xi(t) = (1 - t)u + tv = u + t(v - u)$. This function is continuous on $[0, 1]$ and differentiable on $(0, 1)$. Then $g : [0, 1] \to \mathbb{R}$ defined by $g = f \circ \xi$ is continuous on $[0, 1]$ and differentiable on $(0, 1)$. By the chain rule, we have

$$g'(t) = f'(\xi(t))\xi'(t) = f'(\xi(t))(v - u) \text{ for all } t \in (0, 1). \tag{6.16}$$

The classical mean value theorem implies the existence of $t_0 \in (0, 1)$ such that

$$g(1) - g(0) = g'(t_0),$$

$$\Rightarrow f(v) - f(u) = f'(\xi(t_0))(u - v), \text{ (see (6.16))}.$$

Setting $x_0 = \xi(t_0)$, we have the mean value theorem. $\square$

Now we check the mean value theorem for vector valued functions. In this case, the result takes an inequality form.

Theorem 6.28 *If X, Y are normed spaces, $U \subseteq X$ an open set, $[u, v] \subseteq U$, and $f : U \to Y$ a Frechet differentiable function, then*

$$\|f(v) - f(u)\|_Y \le \sup_{x \in [u,v]} \|f'(x)\|_{\mathscr{L}} \|v - u\|_X.$$

Proof Let $M = \sup\limits_{x \in [u,v]} \|f'(x)\|_\zeta$. Consider $y^* \in Y^*$, and let $g_{y^*} : X \to \mathbb{R}$ be defined by $g_{y^*}(x) = \langle y^*, f(x) \rangle$. Let $x, h \in X$ such that $x, x + h \in [u, v]$. Then from Theorem 6.27, we have

$$g_{y^*}(x + h) - g_{y^*}(x) = g'_{y^*}(x + th)h \text{ with } 0 < t < 1,$$

$$\Rightarrow \langle y^*, f(x + h) - f(x) \rangle = \langle y^*, f'(x + th)h \rangle. \tag{6.17}$$

Let $y^* \in Y^*$, $\|y^*\|_{Y^*} = 1$ such that

$$|\langle y^*, f(x + h) - f(x) \rangle| = \|f(x + h) - f(x)\|_Y. \tag{6.18}$$

Then from (17) and (18), we infer that

$$\|f(x + h) - f(x)\|_Y \leq M\|h\| \ (\text{ recall } \|y^*\|_{Y^*} = 1).\qquad \square$$

Corollary 6.29 *If X, Y are normed spaces, $f : X \to Y$ is Frechet differentiable, and*

$$\|f'(x)\|_\zeta \leq M \text{ for all } x \in X,$$

then $f(\cdot)$ is M-Lipschitz.

Corollary 6.30 *If X, Y are normed spaces, $U \subseteq X$ an open connected set, and $f : U \to Y$ a Frechet differentiable map such that $f'(x) = 0$, then f is a constant function.*

Using Theorem 6.28, we can easily prove the following result.

Proposition 6.31 *If $\{X_k, Y\}_{k=1}^m$ are normed spaces, $X = \prod\limits_{k=1}^m X_k$, $U \subseteq X$ open, and $f : U \to Y$ has partial derivatives at every $x \in U$ and the maps $x \to f'_{x_k}(x)\ k = 1, \ldots, m$ are continuous at $x_0 \in U$, then $f(\cdot)$ is Frechet differentiable at x_0.*

When $f : U \to X$ is Frechet differentiable at every $x \in U$ and the map $x \to f'(x)$ from U into $\mathcal{L}(X, Y)$ is continuous (on $\mathcal{L}(X, Y)$ we consider the operator norm topology), then we say that $f \in C^1(U))$.

Now let X, Y be normed spaces, $U \subseteq X$ an open set, and $f : U \to X$ a Frechet differentiable map. Then $x \to f'(x)$ is a map from U into the normed space $\mathcal{L}(X, Y)$, and so we can speak about the differentiability of this map. If this derivative exists, then it belongs in $\mathcal{L}(X, \mathcal{L}(X, Y))$ and is the second derivative of $f(\cdot)$ and it is denoted by $f''(x)$.

Definition 6.32 A map $f : U \to Y$ is said to be "twice differentiable," if $f''(x) \in \mathcal{L}(X, \mathcal{L}(X, Y))$ exists at every $x \in U$. Moreover, if the map $x \to f''(x)$ from U into $\mathcal{L}(X, \mathcal{L}(X, Y))$ is continuous, then we say that $f \in C^2(U)$.

Remark 6.33 The space $\mathcal{L}(X, \mathcal{L}(X, Y))$ is isomorphic to $\mathcal{L}_2(X, Y) =$ the space of all continuous bilinear forms from $X \times X$ into Y. Therefore, we can think of $f''(x)$ as a bilinear form on $X \times X$, that is,

$$f''(x)(u, v) = (f''(x)u)v \text{ for all } u, v \in X. \tag{6.19}$$

Note that $f''(x)u \in \mathcal{L}(X, Y)$ and so (6.19) makes sense. Moreover, it can be shown that the continuous bilinear form $f''(x)(\cdot, \cdot)$ is symmetric, that is, $f''(x)(u, v) = f''(x)(v, u)$ for all $u, v \in X$ (see Cartan [59], Théoréme 5.11, p. 65).

Suppose that $X = \prod_{k=1}^{m} X_k$, $U \subseteq X$ an open set and $f : U \to V$ is twice differentiable at $x_0 \in U$. Then

$$f''(x_0)(u, v) = (f''(x_0)u)v = \sum_{i,j=1}^{m} (\frac{\partial^2 f}{\partial x_i \partial x_j}(x_0)u_i)v_j,$$

$$= \sum_{i,j=1}^{m} (\frac{\partial^2 f}{\partial x_j \partial x_i}(x_0)v_j)u_i,$$

$$\Rightarrow \frac{\partial^2 f}{\partial x_i \partial x_j}(x_0) = \frac{\partial^2 f}{\partial x_j \partial x_i}(x_0).$$

As for the first order partial derivatives, the existence of the second order partial derivatives implies that the function $f(\cdot)$ is twice differentiable at x_0, provided that the second order partial derivatives are also continuous of x_0.

Example 6.34 Let H be a Hilbert space, and let $f : H \to \mathbb{R}$ be as in Example 6.13. Then $f''(x)(u, v) = a(u, v)$ for all $x, u, v \in H$.

Proposition 6.35 *If X is a Banach space, $\overline{B}_r = \{x \in X : \|x\| \leq r\}$, $f : X \to \mathbb{R}$ is C^1, and $f''(x)(u, u) \geq 0$ for all $x \in \overline{B}_r$, all $u \in X$, then f is weakly lower semicontinuous on $\overline{B}_r$.*

Proof Using Theorem 6.27, we have

$$f(u) - f(x) = f'(x + t(u - x))(u - x) \text{ with } t \in (0, 1)$$

$$= f'(x)(u - x) + [f'(x + t(u - x)) - f'(x)](u - x)$$

$$= f'(x)(u - x) + tf''(x + t'(u - x))(u - x, u - x) \text{ with } t' \in (0, 1),$$

$$\Rightarrow f(u) - f(x) \geq f'(x)(u - x) \text{ (by the hypothesis)},$$

$$\Rightarrow f(\cdot) \text{ is weakly lower semicontinuous on } \overline{B}_r. \qquad \square$$

Proposition 6.36 *If X is a Banach space, $f : X \to \mathbb{R}$ is C^1,*

$$\langle f'(x), x \rangle \geq \vartheta(\|x\|) \tag{6.20}$$

with $\vartheta : (0, +\infty) \to \mathbb{R}_+$ continuous such that $\int_{t_0}^{+\infty} \frac{\vartheta(s)}{s} ds = +\infty$ for some $t_0 > 0$ and $f|_{\partial B_{t_0}}$ is bounded below, then $f(\cdot)$ is coercive.

Proof From (20), we see that for all $h \in \partial B_{t_0}$ and all $t > t_0$, we have

$$f(th) - f(t_0 h) = \int_{t_0}^{t} \frac{d}{ds} f(sh) ds$$

$$= \int_{t_0}^{t} \langle f'(sh), h \rangle ds \geq \int_{t_0}^{t} \frac{\vartheta(s)}{s} ds \text{ (see (6.20))}.$$

So, if $\|h\| \to \infty$, then

$$f(h) \geq f(\frac{t_0}{\|h\|} h) + \int_{t_0}^{\|h\|} \frac{\vartheta(s)}{s} ds \to +\infty. \qquad \square$$

The next result can be viewed as a kind of Rolle's Theorem for weakly lower semicontinuous functionals.

Proposition 6.37 *If X is a reflexive Banach space, $D \subseteq X$ is a bounded open set. $\partial_w D = \overline{D}^w \setminus D$, $f : \overline{D}^w \to \mathbb{R}$ is weakly lower semicontinuous, $f(x) \geq f(x_0)$ for some $x_0 \in D$, all $x \in \partial_w D$, and then there exists $\widehat{x} \in D$ such that $f(\widehat{x}) = \min_{\overline{D}^w} f$, and if $f(\cdot)$ is Frechet differentiable on D, then $f'(\widehat{x}) = 0$.*

Proof By Proposition 1.180, we know that there exists $\widehat{x} \in \overline{D}^w$ such that

$$f(\widehat{x}) = \min[f(x) : x \in \overline{D}^w].$$

On account of the hypothesis that $f(x_0) \leq f(x)$ for all $x \in \partial_w D$, we see that we may assume that $\widehat{x} \in D$. Hence, if $f(\cdot)$ is Frechet differentiable on D, then $f'(\widehat{x}) = 0$. $\qquad \square$

Now we will introduce some notions that are important in variational calculus and in its applications.

Definition 6.38 Let X, Y be Banach spaces and $U \subseteq X$, $W \subseteq Y$ are open sets.

(a) Let $f : U \to \mathbb{R}$ be a function. We say that $x_0 \in U$ is a "local minimizer (resp., local maximizer)," if we can find $\varrho > 0$ such that $B_\varrho(x_0) = \{u \in X : \|u - x_0\| < \varrho\} \subseteq U$ and

$$f(x_0) \leq f(u) \text{ for all } u \in B_\varrho(x_0)$$

$$(resp., \ f(u) \leq f(x_0) \text{ for all } u \in B_\varrho(x_0)).$$

We also call them "local extrema."

(b) Let $g : U \to W$ be a Frechet differentiable map. We say that $x_0 \in U$ is a "critical point" of $g(\cdot)$, if $g'(x_0) \in \mathcal{L}(X, Y)$ is not surjective.

(c) Let $g : U \to W$ be a Frechet differentiable map. We say that $x_0 \in U$ is a "regular point" of $g(\cdot)$, if $g'(x_0) \in \mathcal{L}(X, Y)$ is surjective.

Remark 6.39 In (a), if $f(x_0) \leq f(x)$ (resp., $f(x) \leq f(x_0)$) for all $x \in U$, then we have a global minimum (resp., a global maximum) of $f(\cdot)$ on U. We also call them global extrema. In (b) and (c), if $Y = \mathbb{R}$, then $x_0 \in U$ is a critical point of $g(\cdot)$ (resp., a regular point of $g(\cdot)$), if $g'(x_0) = 0$ (resp., $g'(x_0) \neq 0$).

The next proposition is a direct generalization of the "Fermat Rule" from classical calculus.

Proposition 6.40 *If X is a Banach space, $f : X \to \overline{\mathbb{R}} = \mathbb{R} \cup \{+\infty\}$ has a local extremum at $x_0 \in X$, and it is Gateaux differentiable at x_0, then $f'_G(x_0) = 0$ in X^*.*

Proof To fix things, we assume that x_0 is a local minimizer. The proof is similar if x_0 is a local maximizer. Then for all $h \in X$ and for $t > 0$ small, we have

$$\frac{f(x_0 + th) - f(x_0)}{t} \geq 0,$$

thus $\langle f'_G(x_0), h \rangle \geq 0$ for all $h \in X$, and $f'_G(x_0) = 0$ in X^*.   □

When f is C^2, we can characterize local extrema using the second order derivative. For this purpose, we will need the following generalization of the Taylor expansion of a C^2 functional (see Cartan [59],Theorem 5.6.3, p.78).

Proposition 6.41 *If X is a Banach space, $U \subseteq X$ is an open set, and $f \in C^2(U)$, then for every $x \in U$, there exists $\varrho = \varrho(x) > 0$ such that for all $h \in B_\varrho$ we have*

$$f(x_0 + h) = f(x_0) + f'(x_0)(h) + \frac{1}{2}f''(x_0)(h, h) + R_2(x_0, h)(h, h)$$

with $R_2(x_0, h) \to 0$ as $\|h\| \to 0$ in $\mathcal{L}_2(X, \mathbb{R})$.

Using this Taylor expansion, we can characterize local extrema using the second derivative.

Proposition 6.42 *If X is a Banach space, $U \subseteq X$ is an open set, $f \in C^2(U)$, and $x_0 \in U$ is a local minimizer (resp., local maximizer) of $f(\cdot)$, then $f''(x_0) \in \mathcal{L}_2(X, \mathbb{R})$ is nonnegative (resp., nonpositive), that is, $f''(x_0)(h, h) \geq 0$ (resp., $f''(x_0)(h, h) \leq 0$) for all $h \in X$.*

Proof Again we do the proof for a local minimizer and the proof for a local maximizer being similar. So, we can find $\varrho > 0$ such that

$$B_\varrho(x_0) \subseteq U \quad \text{and} \quad f(x_0) \leq f(x) \text{ for all } x \in B_\varrho(x_0). \tag{6.21}$$

Using Proposition 6.41 and (6.21), we have for all $h \in B_\varrho(x_0)$

$$f(x_0) \leq f(x_0 + h) = f(x_0) + f'(x_0)(h) + \frac{1}{2}f''(x_0)(h, h) + R_2(x_0, h)(h, h). \tag{6.22}$$

From Proposition 6.40, we know that $f'(x_0) = 0$. Then from (6.22), we obtain that

$$0 \leq \frac{1}{2}f''(x_0)(h, h) + R_2(x_0, h)(h, h) \text{ for all } h \in B_\varrho,$$

$$\Rightarrow 0 \leq t^2[\frac{1}{2}f''(x_0)(h, h) + R_2(x_0, th)(h, h)]$$

$$\text{for all } 0 < t \leq 1, \text{ all } h \in B_\varrho,$$

$$\Rightarrow 0 \leq f''(x_0)(h, h) + 2R_2(x_0, th)(h, h) \text{ for all } h \in B_\varrho.$$

Passing to the limit as $t \to 0^+$, we obtain

$$0 \leq f''(x_0)(h, h) \text{ for all } h \in B_\varrho.$$

For a general $h \in X$, we can find $s > 0$ such that $sh \in B_\varrho$. Then

$$0 \leq f''(x_0)(sh, sh) = s^2 f''(x_0)(h, h);$$

hence, $0 \leq f''(x_0)(h, h)$ for all $h \in X$. $\qquad \square$

We have an "almost" converse of the above result.

Proposition 6.43 *If X is a Banach space, $U \subseteq X$ is an open set, $f \in C^2(U)$, and for $x_0 \in U$, we have:*

(a) $f'(x_0) = 0$ *in* X^*.
(b) $f''(x_0)$ *is positive (resp., negative), that is,*

$$\inf[f''(x_0)(h, h) : \|h\| = 1] > 0$$

$$(resp. \inf[f'(x_0)(h, h) : \|h\| = 1] < 0),$$

and then x_0 is a strict local minimizer (resp., strict local maximizer) of $f(\cdot)$, that is, there exists $\varrho > 0$ such that

$$f(x_0) < f(x_0 + h) \text{ for all } h \in B_\varrho \setminus \{0\}$$

$$(resp., f(x + h) < f(x_0) \text{ for all } h \in B_\varrho \setminus \{0\}).$$

Proof We do the proof when $f''(x_0)$ is positive. The reasoning is similar if $f''(x_0)$ is negative. Let $\varrho > 0$ be as postulated by Proposition 6.41. On account of hypothesis (a), we have

$$f(x_0 + h) = f(x_0) + \frac{1}{2} f''(x_0)(h, h) + R_2(x_0, h)(h, h) \text{ for all } h \in B_\varrho,$$

thus $f(x_0 + th) - f(x_0) = t^2[f''(x_0)(h, h) + R_2(x_0, th)(h, h)]$ for all $0 < t \leq 1$, and all $h \in B_\varrho$. Since for $t \in (0, 1]$ small we have $f''(x_0)(h, h) + R_2(x_0, th)(h, h) > 0$(see(b)), it follows that x_0 is a strict local minimizer of $f(\cdot)$. $\square$

Next we will consider constrained minimization problems. To deal with such problems, we will need one of the fundamental results of differential calculus, known as the "Implicit Function Theorem." The proof of this result uses the so-called Banach Fixed Point Theorem. Here for the convenience of the reader, we recall the statement of the Banach Fixed Point Theorem that we will use in the proof of Theorem 6.45 below.

Theorem 6.44 *If X is a Banach space, $C \in X$ is a closed set, and $\varphi : C \to C$ satisfies*

$$\|\varphi(x) - \varphi(u)\| \leq k\|x - u\| \text{ for all } x, u \in C \text{ with } k \in (0, 1), \tag{6.23}$$

then there exists a unique $v \in C$ such that $\varphi(v) = v$; moreover, if we have a family $\{\varphi_y\}_{y \in W}$ (with $W \subseteq Y$ open, where Y is a Banach space) that satisfies (6.23) with the same $k \in (0, 1)$, then the unique fixed point $V(y) \in C$ depends continuously on y.

Now we are ready to deal with the "Implicit Function Theorem."

Theorem 6.45 *If X, Y, V are Banach spaces, $U \subseteq X \times Y$ is an open set, $(x_0, y_0) \in U$, $f \in C^1(U, V)$, $f(x_0, y_0) = 0$, and $f'_y(x_0, y_0) \in \mathcal{L}(Y, V)$ is an isomorphism, then there exist $V_1 \in \mathcal{N}(x_0), V_2 \in \mathcal{N}(y_0), V_1 \times V_2 \subseteq U$, and a differentiable function $\xi : V_1 \to V_2$ such that*

$$f(x, \xi(x)) = 0 \ for \ all \ x \in V_1, \tag{6.24}$$

and $\xi'(x) = -[f_y'(x, \xi(x))]^{-1} \circ f_x'(x, \xi(x))$ for all $x \in V_1$. Finally, $\xi(x)$ is the unique solution of $f(x, \xi(x)) = 0$ for every $x \in V_1$.

Proof Let $A_0 = f_y'(x_0, y_0) \in \mathcal{L}(Y, V)$. The equation $f(x, y) = 0$ is equivalent to the fixed point problem $y = g(x, y)$ with $g(x, y) = y - A_0^{-1} f(x, y)$. We have

$$g(x, y_1) - g(x, y_0) = A_0^{-1}[A_0(y_1 - y_2) - (f(x, y_1) - f(x, y_0))].$$

The differentiabilities of f and the continuity of A_0^{-1} imply that we can find $\delta_1 > 0$ and $\vartheta > 0$ such that

$$\|x - x_0\|_X \le \delta_1, \|y_1 - y_0\|_Y, \|y_2 - y_0\|_Y \le \vartheta \ (\text{hence } \|y_1 - y_2\|_Y \le 2\vartheta)$$

imply that

$$\|g(x, y_1) - g(x, y_2)\|_V \le \frac{1}{2}\|y_1 - y_2\|_Y. \tag{6.25}$$

Also let $\delta_2 > 0$ such that if $\|x - x_0\|_X \le \delta_2$, then

$$\|g(x, y_0) - g(x_0, y_0)\|_V = \|g(x, y_0) - y_0\|_V \le \frac{\vartheta}{2}. \tag{6.26}$$

So, if $\|y - y_0\|_Y \le \vartheta,$, then from (6.25) and (6.26), we have

$$\|g(x, y_0) - y_0\|_V \le \frac{1}{2}\|y - y_0\|_V + \frac{\vartheta}{2} \le \vartheta$$

for every $x \in \overline{B}_\delta(x_0)$ with $\delta = \min\{\delta_1, \delta_2\}$.

We have proved that for every $x \in \overline{B}_\delta(x_0), g(x, \cdot)$ maps $\overline{B}_\vartheta(y_0)$ onto itself (similarly for the open ball $B_\vartheta(y_0)$). So, by Theorem 6.44 for every $x \in \overline{B}_\delta(x_0)$, we can find a unique $y(x) \in \overline{B}_\vartheta(y_0)$ such that

$$g(x, y(x)) = y(x),$$

and thus $f(x, y(x)) = 0$ and $y(x)$ depends continuously on x.

Let $V_1 = B_\delta(x_0), and V_2 = B_\vartheta(y_0)$. Then $V_1 \times V_2 \subseteq U$, and we denote the unique $y(x)$ fixed point of $g(x, \cdot)$ by $\xi(x)$. We have $\xi : V_1 \to V_2$, and it remains to show the differentiability of $\xi(\cdot)$. Let $(x_1, y_1) \in V_1 \times V_2, y_1 = \xi(x_1)$. We set $B = f_x'(x_1, y_1) and E = f_y'(x_1, y_1)$. Then on account of Proposition 6.22, we have

$$f(x, y) = B(x - x_1) + E(y - y_1) + e(x, y) \ for \ all \ (x, y) \in U \tag{6.27}$$

with

$$\frac{e(x, y)}{\|(x - x_1, y - y_1)\|_{X \times Y}} \to 0 \text{ as } (x, y) \to (x_0, y_0). \tag{6.28}$$

Since $f(x, \xi(x)) = 0$ for all $x \in V_1$, we have

$$\xi(x) = -E^{-1}B(x - x_1) + y_1 - E^{-1}e(x, y) \text{ (see (6.27)).} \tag{6.29}$$

Then (6.28) implies that there exist $\varrho_1 > 0, \varrho_2 > 0$ such that

$$\|x - x_1\|_X \le \varrho_1, \|y - y_1\|_Y \le \varrho_2, \|e(x, y)\|_V \le \frac{1}{2\|E^{-1}\|_{\mathcal{L}}}[\|x - x_1\|_X + \|y - y_1\|_Y];$$

hence,

$$\|e(x, \xi(x))\|_V \le \frac{1}{2\|E^{-1}\|_{\mathcal{L}}}[\|x - x_1\|_X + \|\xi(x) - \xi(x_1)\|_Y]. \tag{6.30}$$

From (6.29) and (6.30), we have

$$\|\xi(x) - \xi(x_1)\|_Y \le \|E^{-1}B\|_{\mathcal{L}}\|x - x_1\|_X + \frac{1}{2}\|x - x_1\|_X + \frac{1}{2}\|\xi(x) - \xi(x_1)\|_Y;$$

thus,

$$\|\xi(x) - \xi(x_1)\|_Y \le k\|x - x_1\|_X \text{ with } k = 2\|E^{-1}B\|_{\mathcal{L}} + 1. \tag{6.31}$$

We let $\Upsilon(x) = -E^{-1}e(x, \xi(x))$, and from (6.29), we have

$$\xi(x) - \xi(x_1) = -E^{-1}B(x - x_1) + \Upsilon(x). \tag{6.32}$$

Also since $\|\Upsilon(x)\|_Y \le \|E^{-1}\|_{\mathcal{L}}\|e(x, \xi(x))\|_V$ and using (6.28), we infer that

$$\lim_{x \to x_1} \frac{\Upsilon(x)}{\|x - x_1\|_X} = 0. \tag{6.33}$$

Then (6.32) and (6.33) imply that $\xi(\cdot)$ is differentiable at x_1 and

$$\xi'(x_1) = -E^{-1}B = -(f_y'(x_1, y_1))^{-1}f_x'(x_1, y_1). \qquad \square$$

Using this theorem, we can easily prove the other fundamental result of differential calculus, namely the "Inverse Function Theorem."

Theorem 6.46 *If X, Y are Banach spaces, $U \subseteq X$ is an open set, $f \in C^1(U, Y)x_0 \in U$, and $f'(x_0)$ is an isomorphism, then there exist $V \subseteq U$ open,*

$x_0 \in V$, and $W \in \mathcal{N}(y_0)$, $y_0 = f(x_0)$ such that $f : V \to W$ is a diffeomorphism and

$$(f^{-1})'(y_0) = f'(x_0)^{-1}.$$

Proof Let $\eta(y, x) = f(x) - y$. We have $\eta'_x(y_0, x_0) = f'(x_0)$, which by hypothesis is an isomorphism. So, by Theorem 6.45, there exist $V_1 \in \mathcal{N}(y_0)$, $V_2 \in \mathcal{N}(x_0)$, and a differentiable map $\xi : V_1 \to X$ such that $\xi(V_1) \subseteq V_2$. $\eta(y, \xi(y)) = 0$. Hence, $f(\xi(y)) = y$ for all $y \in V_1, \xi(y_0) = x_0$. Clearly, $\xi(\cdot)$ is injective on V_1, hence a bijection between V_1 and $\xi(V_1)$. Moreover, $\xi(V_1) = f^{-1}(V_1)$ (recall $f(\cdot)$ is continuous). Let $W = \xi(V_1)$. Then f is a bijection between V_1 and W. Setting $V = V_1$, we conclude the proof of the theorem.   □

Now we can focus on minimization problems with constraints. Our aim is to derive a general "Lagrange Multiplier Rule."

So, let X be a Banach space, $\varphi, f \in C^1(X)$, and we consider the following constrained minimization problem:

$$\inf[\varphi(x) : x \in M] \text{ with } M = \{x \in X : f(x) = 0\}. \tag{6.34}$$

Theorem 6.47 *If X, φ, f are as above, $f'(x) \neq 0$ for all $x \in M$, and $x_0 \in M$ is a solution of problem (6.34), then there exists $\widehat{\lambda} \in \mathbb{R}$ (known as the Lagrange multiplier for x_0) such that $\varphi'(x_0) = \widehat{\lambda} f'(x_0)$.*

Proof Let $h \in X$ such that $f'(x_0)(h) = \langle f'(x_0), h \rangle \neq 0$. Then for any $x \in X$, we see that $x - \frac{f'(x_0)(x)}{f'(x_0)(h)} h \in \ker f'(x_0)$. It follows that

$$X = \ker f'(x_0) \oplus \mathbb{R}h.$$

Consider the function $\vartheta : \ker f'(x_0) \times \mathbb{R} \to \mathbb{R}$ defined by

$$\vartheta(x, \lambda) = f(x_0 + x + \lambda h) \text{ for all } (x, \lambda) \in \ker f'(x_0) \times \mathbb{R}.$$

We see that $\vartheta(\cdot, \cdot)$ is C^1 and

$$\vartheta(0, 0) = f(x_0) = 0 \text{ and } \vartheta'_\lambda(0, 0) = f'(x_0)(h) = \langle f'(x_0), h \rangle \neq 0.$$

So, we can apply Theorem 6.45 and find $V_1 \in \mathcal{N}(0)$ in $\ker f'(x_0)$, $V_2 \in \mathcal{N}(0)$ in $\mathbb{R}$, and a unique $C^1 - function$ $\xi : V_1 \to V_2$ such that

$$\xi(0) = 0, \tag{6.35}$$

$$\vartheta(x, \lambda) = 0 \Leftrightarrow \lambda = \xi(x) \text{ for all } (x, \lambda) \in V_1 \times V_2 \tag{6.36}$$

$$\xi'(0) = -\vartheta'_\lambda(0,0)^{-1}\vartheta_x(0,0) = -f'(x_0)(h)^{-1}f'(x_0)|_{\ker f'(x_0)}. \tag{6.37}$$

Let $\widehat{\varphi} : V_1 \to \mathbb{R}$ be defined by

$$\widehat{\varphi}(x) = \varphi(x_0 + x + \xi(x)h) \ for \ all \ x \in V_1.$$

Note $\widehat{\varphi} \in C^1(V_1)$ and 0 is a minimizer of $\widehat{\varphi}$ (see (6.36), (6.37)). Hence,

$$\widehat{\varphi}'(0) = 0. \tag{6.38}$$

But $\widehat{\varphi}'(0) = \varphi'(x_0)|_{\ker f'(x_0)}$ and so from (6.38), we infer that

$$\ker f'(x_0) \subseteq \ker \varphi'(x_0). \tag{6.39}$$

For $x \in X$, recall that $x - \frac{f'(x_0)(x)}{f'(x_0)(h)}h \in \ker f'(x_0)$; hence,

$$x - \frac{f'(x_0)(x)}{f'(x_0)(h)}h \in \ker \varphi'(x_0) \ (\text{see (6.39)}),$$

$$\Rightarrow \varphi'(x_0)(x) = \frac{f'(x_0)(x)}{f'(x_0)(h)}\varphi'(x_0)(h) \ \text{for all } x \in X.$$

So, choosing $\widehat{\lambda} = \frac{\varphi'(x_0)(h)}{f'(x_0)(h)} \in \mathbb{R}$, we have

$$\varphi'(x_0) = \lambda f'(x_0).$$

$\square$

We conclude with another result about constrained minimization problems. The result is known as "Lyusternik's Theorem," and its proof can be found in Zeidler [291] (Theorem 43C, p.286). First a definition.

Definition 6.48 Let X be a Banach space and $C \subseteq X$ a nonempty set.

(a) A vector $h \in X$ is said to be "tangent to C at $x_0 \in \overline{C}$" if there exist $\varepsilon > 0$ and a path $s : [0, \varepsilon] \to X$ such that

$$x_0 + \lambda h + s(\lambda) \in C \ for \ all \ \lambda \in [0, \varepsilon],$$

$$\frac{\|s(\lambda)\|_X}{\lambda} \to 0 \ as \ \lambda \to 0^+.$$

(b) The set of all vectors $h \in X$ that are tangent to C at $x_0 \in \overline{C}$ forms a closed cone denoted by $T_C(x_0)$. If the cone is in fact a subspace, then we say that $T_C(x_0)$ is the "tangent space" to C at $x_0 \in \overline{C}$.

(c) Let $Y \subseteq X$ be a closed subspace. We say that a subspace $V \subseteq X$ is a "topological complement" of Y if [1] V is closed and [2] $Y \cap V = 0$ and $X = Y + V$. We write $X = Y \oplus V$.

Remark 6.49 Evidently, Y admits a topological complement, if there is a projection operator $P_Y \in \mathcal{L}(X)$ onto Y. If Y is finite dimensional or if Y is finite codimensional, then Y admits a topological complement.

Let X, Y be Banach spaces and $f : X \to Y$ be a map. We consider the constraint set

$$C = \{x \in X : f(x) = 0\}.$$

The Lyusternik theorem identifies the tangent space $T_C(x_0)$.

Theorem 6.50 *If $U \subseteq X$ is an open set, $f \in C^1(U, Y)$, $C = \{x \in X : f(x) = 0\}$, $x_0 \in C$ is a regular point of $f(\cdot)$, and $\ker f'(x_0)$ admits a topological complement, then $T_C(x_0) = \ker f'(x_0)$.*

6.2 Convex Functions–Subdifferential Theory

Convex functions are used in many applied fields such as optimization, control theory, game theory, and mathematical economics. Convex functions have many useful properties, which together with the corresponding theory of convex sets form what we call the theory of "Convex Analysis." Our aim is in the rest of this chapter to present some aspects of this theory.

In this section, we focus primarily on the subdifferential theory of convex functions. The subgradients and the subdifferential set they form are a very fruitful substitute of the usual gradient, for functions that are not differentiable in the classical sense. The notion of subdifferential lends to a calculus that parallels in many respects the classical one.

Let X be a Banach space and $\varphi : X \to \overline{\mathbb{R}} = \mathbb{R} \cup \{+\infty\}$ a function. We always assume that $\varphi \not\equiv +\infty$ (i.e., there exists $x \in X$ such that $\varphi(x) < +\infty$).

Definition 6.51 We say that $\varphi(\cdot)$ is "*convex*" if

$$\varphi((1-t)x + tu) \leq (1-t)\varphi(x) + t\varphi(u)$$

$$for \ all \ x, u \in X, \ all \ 0 \leq t \leq 1.$$

If the inequality is strict for $x \neq u$ and $0 < t < 1$, then we say that $\varphi(\cdot)$ is "strictly convex."

An alternative equivalent definition of a convex function is in terms of its "epigraph"

$$\text{epi}\,\varphi = \{(x, \lambda) \in X \times \mathbb{R} : \varphi(x) \leq \lambda\}.$$

Then $\varphi(\cdot)$ is convex if and only if epi $\varphi \subseteq X \times \mathbb{R}$ is a convex set.

Remark 6.52 If $\varphi(\cdot)$ is convex and $\{x_k\}_{k=1}^m \subseteq X$, $\{\lambda_k\}_{k=1}^m \subseteq [0, 1]$ with $\sum_{k=1}^m \lambda_k = 1$, then $\varphi(\sum_{k=1}^m \lambda_k x_k) \leq \sum_{k=1}^m \lambda_k \varphi(x_k)$. This follows from the above definition by induction. We set dom $\varphi = \{x \in X : \varphi(x) < +\infty\}$ (the "effective domain" of $\varphi(\cdot)$).

The next proposition collects some immediate consequences of the above definition.

Proposition 6.53 (a) *If φ_1, φ_2 are convex functions and $\lambda \geq 0$, then $\lambda\varphi_1 + \varphi_2$ is convex too.*
(b) *If $\{\varphi_i\}_{i \in I}$ is a family of convex functions, then $\varphi = \sup_i \varphi_i$ is convex and*

$$\text{epi}\,\varphi = \bigcap_{i \in I} \text{epi}\,\varphi_i.$$

(c) *$\varphi(\cdot)$ is convex and lower semicontinuous if and only if epi $\varphi \subseteq X \times \mathbb{R}$ is convex and closed.*
(d) *If $\varphi(\cdot)$ is convex, then for every $\lambda \in \mathbb{R}$, $L_\lambda = \{x \in X : \varphi(x) \leq \lambda\}$ is convex.*

Remark 6.54 Convexity of the sublevel sets $L_\lambda (\lambda \in \mathbb{R})$ does not characterize convex functions. For example, the function $f(x) = \ln x, x > 0$. Convexity of the sublevel sets L_λ leads to a broader class of functions, known as "quasiconvex function." If φ is convex (and closed), $A \in \mathcal{L}(X), u_0 \in X$, and $\psi(x) = \varphi(A(x) + u_0)$, then $\psi(\cdot)$ is convex (and closed).

Convex functions exhibit remarkable continuity properties.

Proposition 6.55 *If $\varphi : X \to \overline{\mathbb{R}} = \mathbb{R} \cup \{+\infty\}$ is convex, $x_0 \in \text{dom}\,\varphi$ (see Remark 6.52) and $\varphi(u) \leq \widehat{c}$ for all $u \in U \in \mathcal{N}(x_0)$, then $\varphi(\cdot)$ is continuous at x_0.*

Proof By translating things if necessary, we may assume that $x_0 = 0$ and

$$\varphi(u) \leq 1 \; for \; all \; u \in B_\varrho = \{x \in X : \|x\| < \varrho\}, \varphi(0) = 0. \tag{6.40}$$

Let $\lambda > 0$ and $u \in B_{\lambda\varrho} = \lambda B_\varrho$. We have

$$0 = \varphi(0) = \varphi\left(\frac{1}{1+\lambda}u + \left(1 - \frac{1}{1+\lambda}\right)\left(-\frac{u}{\lambda}\right)\right)$$

$$\leq \frac{1}{1+\lambda}\varphi(u) + \left(1 - \frac{1}{1+\lambda}\right)\varphi\left(-\frac{u}{\lambda}\right)$$

$$\Rightarrow -\lambda \leq \varphi(u) \text{ (since } -\frac{u}{\lambda} \in B_\varrho).$$

Moreover, we have

$$\varphi(u) = \varphi\left(\lambda\frac{u}{\lambda} + (1-\lambda)\,0\right) \leq \lambda\varphi\left(\frac{u}{\lambda}\right) + (1-\lambda)\,\varphi(0) = \lambda\varphi\left(\frac{u}{\lambda}\right)$$

$$\Rightarrow \varphi(u) \leq \lambda \text{ (since } \frac{u}{\lambda} \in B_\varrho).$$

Therefore, we conclude that $|\varphi(u)| \leq \lambda$ for all $u \in \lambda B_\varrho = B_{\lambda\varrho}$; hence, $\varphi(\cdot)$ is continuous at $x_0 = 0$. □

Theorem 6.56 *If $\varphi : X \to \overline{\mathbb{R}} = \mathbb{R} \cup \{+\infty\}$ is convex and bounded above in a neighborhood of a point, then $\varphi\mid_{\text{intdom}\,\varphi}$ is locally Lipschitz (i.e., if $x \in \text{intdom}\,\varphi$, we can find $U \in \mathcal{N}(x)$ and $k_U > 0$ such that $|\varphi(u) - \varphi(v)| \leq k_U\|u - v\|$ for all $u, v \in U$).*

Proof Let $x \in \text{intdom}\,\varphi$. Arguing as in the proof of Proposition 6.55, we show that $\varphi(\cdot)$ is bounded in a neighborhood of x. We may always assume that $\varphi(\cdot)$ is bounded above by $c > 0$ on $\lambda B \subseteq \text{intdom}\,\varphi$ with $\lambda > 0$, $B \subseteq X$ an open ball. Let $r > 1$ such that $u = rx \in \text{intdom}\,\varphi$ and let $t = \frac{1}{r}$. We set

$$U = \{y \in X : y = (1-t)x' + tu : x' \in \lambda B\}$$

and observe that $U \in \mathcal{N}(tu)$ and in fact is a ball with radius $(1-t)\lambda$. From the convexity of $\varphi(\cdot)$ for all $y \in U$, we have

$$\varphi(y) \leq (1-t)\varphi(x') + t\varphi(u) \leq c + t\varphi(u)$$

$$\Rightarrow \varphi \text{ is bounded above in a neighborhood of } x \in \text{intdom}\varphi.$$

From the proof of Proposition 6.55, we have that $\varphi(\cdot)$ is bounded from below too in that neighborhood.

Therefore, we can say that

$$|\varphi(v)| \leq \widehat{c} \text{ for all } v \in B_{2\delta}(x),\, \delta > 0. \tag{6.41}$$

Consider $x_1, x_2 \in B_\delta(x)$ and consider $x_3 = x_2 + \frac{\delta}{\mu}(x_2 - x_1)$, $\mu = \|x_2 - x_1\|$; we have $x_3 \in B(x)_{2\delta}$. Also

$$x_2 = \frac{\delta}{\mu + \delta} x_1 + \frac{\mu}{\mu + \delta} x_3.$$

From the convexity of $\varphi(\cdot)$, we have

$$\varphi(x_2) \leq \frac{\delta}{\mu + \delta} \varphi(x_1) + \frac{\mu}{\mu + \delta} \varphi(x_3),$$

$$\Rightarrow \varphi(x_2) - \varphi(x_1) \leq \frac{\mu}{\mu + \delta} [\varphi(x_3) - \varphi(x_1)] \leq \frac{\mu}{\delta} [\varphi(x_3) - \varphi(x_1)].$$

Since $x_1, x_3 \in B(x)_{2\delta}$, from (6.41), we infer that

$$\varphi(x_2) - \varphi(x_1) \leq \frac{2\widehat{c}}{\delta} \|x_2 - x_1\| \text{ (recall } \mu = \|x_2 - x_1\|).$$

Interchanging the roles of x_1 and x_2 in the above argument, we conclude that

$$|\varphi(x_2) - \varphi(x_1)| \leq \frac{2\widehat{c}}{\delta} \|x_2 - x_1\|,$$

$$\Rightarrow \varphi \mid_{\text{intdom } \varphi} \text{ is locally Lipschitz.} \qquad \square$$

Corollary 6.57 *If $\varphi : X \to \overline{\mathbb{R}} = \mathbb{R} \cup \{+\infty\}$ is convex, $|\varphi(x)| \leq c$ for all $x \in U$ open convex set, and $C_\delta \subseteq U$ with $C_\delta = \{u \in X : d(u, C) < \delta\}$ with $\delta > 0$, then $\varphi \mid_C$ is Lipschitz continuous with Lipschitz constant $\frac{2c}{\delta}$.*

Corollary 6.58 *If X is finite dimensional and $\varphi : X \to \overline{\mathbb{R}} = \mathbb{R} \cup \{+\infty\}$ is convex, then $\varphi \mid_{\text{intdom } \varphi}$ is locally Lipschitz.*

Let $\varphi : X \to \overline{\mathbb{R}} = \mathbb{R} \cup \{+\infty\}$ be a convex function and $x \in \text{dom} \varphi$. Then for $h \in X$ and $0 < s \leq t$, we have

$$\varphi(x + sh) = \varphi\left(\frac{s}{t}(x + th) + \left(1 - \frac{s}{t}\right)x\right)$$

$$\leq \frac{s}{t}\varphi(x + th) + \left(1 - \frac{s}{t}\right)\varphi(x),$$

$$\Rightarrow \frac{\varphi(x + sh) - \varphi(x)}{s} \leq \frac{\varphi(x + th) - \varphi(x)}{t},$$

$$\Rightarrow t \to \frac{\varphi(x + th) - \varphi(x)}{t} \text{ is nondecreasing.}$$

So, we can make the following definition.

Definition 6.59 Let $\varphi : X \to \overline{\mathbb{R}} = \mathbb{R} \cup \{+\infty\}$ be a convex function, $x \in \text{dom}\varphi$, and $h \in X$. The "directional derivative" of $\varphi(\cdot)$ at x in the direction h is defined by

$$\varphi'_+(x; h) = \lim_{t \to 0^+} \frac{\varphi(x + th) - \varphi(x)}{t}.$$

Remark 6.60 Evidently, if $\varphi(\cdot)$ is Gateaux differentiable at x, then

$$\varphi'_G(x)(\cdot) = \varphi'_+(x; \cdot).$$

Proposition 6.61 *If $\varphi : X \to \overline{\mathbb{R}} = \mathbb{R} \cup \{+\infty\}$ is a convex function and $x \in \text{dom}\varphi$, then $\varphi'_+(x; \cdot)$ is sublinear (i.e., positively homogeneous and subadditive).*

Proof Let $\lambda > 0$. We have

$$\varphi'_+(x; \lambda h) = \lim_{t \to 0^+} \frac{\varphi(x + t\lambda h) - \varphi(x)}{t}$$

$$= \lambda \lim_{t \to 0^+} \frac{\varphi(x + \lambda t h) - \varphi(x)}{\lambda t} = \lambda \varphi'(x; h),$$

$$\Rightarrow \varphi'_+(x; \cdot) \text{ is positively homogeneous.}$$

Next let $h_1, h_2 \in X$. We have

$$\varphi'_+(x; h_1 + h_2) = 2\varphi'_+\left(x; \frac{1}{2}(h_1 + h_2)\right)$$

$$= 2 \lim_{t \to 0^+} \frac{\varphi\left(\frac{1}{2}(x + th_1) + \frac{1}{2}(x + th_2)\right) - \varphi(x)}{t}$$

$$\leq 2 \lim_{t \to 0^+} \frac{\frac{1}{2}\varphi(x + th_1) + \frac{1}{2}\varphi(x + th_2) - \varphi(x)}{t}$$

$$= \varphi'_+(x; h_1) + \varphi'_+(x; h_2),$$

$$\Rightarrow \varphi'_+(x; \cdot) \text{ is subadditive.}$$

Therefore, $\varphi'_+(x; \cdot)$ is sublinear. $\qquad \square$

Remark 6.62 If we consider $\varphi'_-(x; h) = \lim\limits_{t \to 0^-} \frac{\varphi(x+th)-\varphi(x)}{t}$, then

$$\varphi'_-(x; h) = \lim_{t \to 0^+} \frac{\varphi(x + t(-h)) - \varphi(x)}{-t} = -\varphi'_+(x; -h).$$

On account of the sublinearity of $\varphi'_+(x; \cdot)$, we have

$$0 = \varphi'_+(x; 0) = \varphi'_+(x; h - h) \leq \varphi'_+(x; h) + \varphi'_+(x; -h),$$

$$\Rightarrow -\varphi'_+(x; -h) \leq \varphi'_+(x; h).$$

Corollary 6.63 *If* $\varphi : X \to \overline{\mathbb{R}} = \mathbb{R} \cup \{+\infty\}$ *is convex,* $x \in \mathrm{dom}\varphi$, *and*

$$\lim_{t \to 0} \frac{\varphi(x + th) - \varphi(x)}{t}$$

exists for every $h \in X$, *then* $\varphi(\cdot)$ *is Gateaux differentiable at* x *and*

$$\varphi'_G(x)(\cdot) = \varphi'_+(x; \cdot).$$

Proposition 6.64 *If* $U \subseteq \mathbb{R}^N$ *is open convex,* $\varphi : U \to \mathbb{R}$ *is a convex function, and all the partial derivatives of* $\varphi(\cdot)$ *at* $x \in U$ *exist, then* φ *is Frechet differentiable at* x.

Proof From Theorem 6.56, we know that $\varphi(\cdot)$ is locally Lipschitz, while from Corollary 6.63 we know that $\varphi(\cdot)$ is Galeaux differentiable at x. Arguing by contradiction, suppose that $\varphi(\cdot)$ is not Frechet differentiable at x. Then we can find a sequence $\{h_n\}_{n\in\mathbb{N}} \subseteq X$ with $\|h_n\| \to 0$ such that

$$\mu_n = \frac{\varphi(x + h_n) - \varphi(x) - \varphi'_G(x)(h_n)}{\|h_n\|} \nrightarrow 0, \ as \ n \to \infty. \tag{6.42}$$

Let $\widehat{h}_n = \frac{h_n}{\|h_n\|}$, $n \in \mathbb{N}$. Then $\|\widehat{h}_n\| = 1$ and $h_n = t_n \widehat{h}_n$ with $t_n = \|h_n\|$ for all $n \in \mathbb{N}$. We may assume that $\widehat{h}_n \to \widehat{h}$ in $\mathbb{R}^N$ and $\|\widehat{h}\| = 1$. We have

$$|\mu_n| = \left| \frac{\varphi(x + t_n\widehat{h}_n) - \varphi(x)}{t_n} - \varphi'_G(x)(\widehat{h}_n) \right|$$

$$\leq \left| \frac{\varphi(x + t_n\widehat{h}) - \varphi(x)}{t_n} - \varphi'_G(x)(\widehat{h}) \right| + \frac{|\varphi(x + t_nh_n) - \varphi(x + t_n\widehat{h})|}{t_n}$$

$$+ |\varphi'_G(\widehat{h} - \widehat{h}_n)|$$

$$\leq \left| \frac{\varphi(x + t_n \widehat{h}) - \varphi(x)}{t_n} - \varphi_G'(x)(\widehat{h}) \right| + \left[k + \|\varphi_G'(x)\|_* \right] \|\widehat{h}_n - \widehat{h}\| \to 0,$$

which contradicts (6.42).

Therefore, $\varphi(\cdot)$ is Frechet differentiable at $x \in U$. $\qquad\qquad\qquad\qquad\square$

An interesting byproduct of the above proof is the following proposition.

Proposition 6.65 *If $\varphi : \mathbb{R}^N \to \mathbb{R}$ is Lipschitz continuous on some neighborhood of $x \in \mathbb{R}^N$, then $\varphi \cdot$ is Frechet differentiable at x if and only if it is Gateaux differentiable at x.*

Proposition 6.66 *If $\varphi : X \to \overline{\mathbb{R}} = \mathbb{R} \cup \{+\infty\}$ is Gateaux differentiable at every $x \in \mathrm{dom}\, \varphi = \{u \in X : \varphi(u) < +\infty\}$, then the following statements are equivalent:*

(a) $\varphi(\cdot)$ *is convex.*
(b) $\langle \varphi'(x), u - x \rangle \leq \varphi(u) - \varphi(x)$ *for all* $x, u \in \mathrm{dom}\, \varphi$.
(c) $\langle \varphi'(x) - \varphi'(u), u - x \rangle \geq 0$ *for all* $\mathrm{dom}\, \varphi$.

Proof (a)$\Rightarrow$(b) The convexity of $\varphi(\cdot)$ implies that

$$\varphi(x + t(u - x)) \leq (1 - t)\varphi(x) + t\varphi(u) \quad for\ all\ x, u \in \mathrm{dom}\, \varphi,\ all\ t \in [0, 1],$$

$$\Rightarrow \frac{\varphi(x + t(u - x)) - \varphi(x)}{t} \leq \varphi(u) - \varphi(x) \quad for\ all\ t \in (0, 1],$$

$$\Rightarrow \langle \varphi_G'(x), u - x \rangle \leq \varphi(u) - \varphi(x).$$

(b)$\Rightarrow$(c) By hypothesis, we have

$$\langle \varphi_G'(x), u - x \rangle \leq \varphi(u) - \varphi(x), \qquad\qquad\qquad\qquad (6.43)$$

$$\langle \varphi_G'(u), x - u \rangle \leq \varphi(x) - \varphi(u). \qquad\qquad\qquad\qquad (6.44)$$

Adding (6.43) and (6.44), we obtain

$$\langle \varphi_G'(x) - \varphi_G'(u), x - u \rangle \geq 0 \text{ for all } x, u \in \mathrm{dom}\varphi.$$

(c)$\Rightarrow$(a) Let $x, u \in \mathrm{dom}\, \varphi$, and let $\vartheta(t) = \varphi((1 - t)x + tu)$ for all $t \in \mathbb{R}$. Evidently, $\vartheta(\cdot)$ is differentiable and

$$\vartheta''(t) = \varphi_G' (x + t(u - x)) (u - x). \qquad\qquad\qquad\qquad (6.45)$$

For $0 \leq s < t \leq 1$, we have

$$(\vartheta'(t)-\vartheta'(s))(t-s) = \langle \varphi_G'(x + t(u - x)) - \varphi_G'(x + s(x - u)), x - u \rangle (t-s)^2 \geq 0.$$

$\Rightarrow \vartheta'$ is nondecreasing on $[0, 1]$.

So, by the mean value theorem, we have

$$\frac{\vartheta(t) - \vartheta(0)}{t} \leq \frac{\vartheta(1) - \vartheta(t)}{1 - t} \text{ for all } t \in (0, 1)$$

$$\Rightarrow \vartheta(t) \leq (1 - t)\vartheta(0) + t\vartheta(1) \text{ for all } t \in [0, 1]$$

$$\Rightarrow \varphi(\cdot) \text{ is convex.}$$

$\square$

Recall that $a : X \to \mathbb{R}$ is an "affine" functional if

$$a(x) = \langle x^*, x \rangle + c \text{ for all } x \in X, \text{ with } x^* \in X^*, c \in \mathbb{R}.$$

The graph of an affine functional is a hyperplane in $X \times \mathbb{R}$. Then the geometric meaning of statement (b) in Proposition 6.66 is that the affine functional

$$a(u) = \langle \varphi_G'(x), u - x \rangle + \varphi(x) \text{ for all } u \in X$$

is a minorant of $\varphi(\cdot)$ and $a(x) = \varphi(x)$. Hence, the graph of $a(\cdot)$ is a hyperplane supporting epi φ at $(x, \varphi(x))$. This observation leads to the following definition.

Definition 6.67 Let $\varphi : X \to \bar{\mathbb{R}} = \mathbb{R} \cup \{+\infty\}$ be a convex function. The "subdifferential" of $\varphi(\cdot)$ at $x \in \text{dom } \varphi$ is defined to be the set

$$\partial \varphi(x) = \{x^* \in X^* : \langle x^*, u - x \rangle \leq \varphi(u) - \varphi(x) \text{ for all } u \in \text{dom}\varphi\}.$$

Each $x^* \in \partial\varphi(x)$ is said to be a "subgradient" of φ at x. We say that $\varphi(\cdot)$ is subdifferentiable at x, if $\partial\varphi(x) \neq \emptyset$.

Remark 6.68 From Proposition 6.66, we see that if $\varphi(\cdot)$ is a convex function that is Gateaux differentiable at x, then $\varphi(\cdot)$ is subdifferentiable at x (i.e., $\partial\varphi(x) \neq \emptyset$). In general, the subdiffernetial $\partial\varphi(x)$ may be empty.

Example 6.69 Consider the convex function $\varphi : \mathbb{R}^N \to \bar{\mathbb{R}} = \mathbb{R} \cup \{+\infty\}$ defined by

$$\varphi(x) = \begin{cases} -[1 - |x|^2]^{\frac{1}{2}}, & \text{if } |x| \leq 1 \\ +\infty, & \text{otherwise.} \end{cases}$$

This function is differentiable at every $x \in B_1 = \{u \in \mathbb{R}^N : |u| < 1\}$. So, by Remark 6.68, $\partial\varphi(x) \neq \emptyset$ for all $x \in B_1$. On the other hand, $\partial\varphi(x) = \emptyset$ for all $x \in \partial B_1 = \{u \in \mathbb{R}^N : |u| = 1\}$.

In general, we have the following result concerning the subdifferentiability of convex functions.

Proposition 6.70 *If $\varphi : X \to \overline{\mathbb{R}} = \mathbb{R} \cup \{+\infty\}$ is a convex function that is continuous at $x \in X$, then $\partial \varphi(x) \subseteq X$ is nonempty, bounded, w^*-closed, and convex (i.e., $\partial \varphi(x) \in P_{bfc}(X^*)$).*

Proof It is clear from Definition 6.67 that $\partial \varphi(x) \subseteq X^*$ is closed and convex. The continuity of $\varphi(\cdot)$ at $x \in X$ implies that we can find $\delta > 0$ such that

$$\|u - x\| \leq \delta \Rightarrow |\varphi(u) - \varphi(x)| \leq 1. \tag{6.46}$$

If $x^* \in \partial \varphi(x)$, then we choose $h \in X$ such that $\|h\| = 1$. Namely, $h \in \partial B_1$) such that $\langle x^*, h \rangle = \|x^*\|$. We have

$$\delta \langle x^*, h \rangle = \langle x^*, \delta h \rangle \leq \varphi(x + \delta h) - \varphi(x) \leq 1$$

$$\Rightarrow \|x^*\|_* \leq \frac{1}{\delta}$$

$$\Rightarrow \partial \varphi(x) \text{ is bounded and so } w^*\text{-compact (being } w\text{-closed).}$$

Finally, we show that $\partial \varphi(x) \neq \emptyset$. Note that int epi $\varphi \neq \emptyset$ (it contains the open set $B_\delta(x) \times (\varphi(u) + 1, +\infty)$, see (6.46)). Also epi φ is convex. So by Theorem 3.41, we can find $x^* \in X^*$ and $\mu \in \mathbb{R}$, $(x^*, \mu) \neq (0, 0)$ such that

$$\langle x^*, x \rangle + \mu \varphi(x) \leq \langle x^*, u \rangle + \mu \lambda \text{ for all } (u, \lambda) \in \text{epi} \varphi. \tag{6.47}$$

If $\mu = 0$, then $\langle x^*, u - x \rangle \geq 0$ for all $u \in \text{dom} \varphi$, and since $x \in \text{int dom} \varphi$ (being a point of continuity of φ), we conclude that $x^* = 0$, a contradiction. Therefore, $\mu \neq 0$, and we can always assume that $\mu = 1$. From (6.47), we have

$$\langle x^*, x \rangle + \varphi(x) \leq \langle x^*, u \rangle + \lambda \text{ for all } (u, \lambda) \in \text{epi} \varphi.$$

Taking $\lambda = \varphi(u)$, we conclude that

$$\langle -x^*, u - x \rangle \leq \varphi(u) - \varphi(x) \text{ for all } u \in \text{dom} \varphi,$$

$$\Rightarrow -x^* \in \partial \varphi(x) \neq \emptyset.$$

$\square$

From the above proof, it is worth mentioning the following observation.

Proposition 6.71 *If $\varphi : X \to \overline{\mathbb{R}} = \mathbb{R} \cup \{+\infty\}$ is convex and continuous at a point, then $\text{intdom} \varphi \neq \emptyset$, $\text{int epi} \varphi \neq \emptyset$, and every continuity point of $\varphi(\cdot)$ is in $\text{int dom} \varphi$; moreover, $\text{int dom} \varphi \subseteq \text{dom}(\partial \varphi)$.*

So, summarizing the continuity properties of a convex function, we can state the following theorem.

Theorem 6.72 *If $\varphi : X \to \overline{\mathbb{R}} = \mathbb{R} \cup \{+\infty\}$ is a convex function, then the following statements are equivalent:*

(a) $\varphi(\cdot)$ *is bounded above in a neighborhood of a point.*
(b) $\text{int epi} \, \varphi \neq \emptyset$.
(c) $\text{intdom} \, \varphi \neq \emptyset$ *and* $\varphi \mid_{intdom\varphi}$ *is continuous.*

Moreover, if these statements hold, then

$$\text{int epi} \, \varphi = \{(x, \lambda) \in X \times \mathbb{R} : x \in \text{intdom} \, \varphi, \varphi(x) < \lambda\}.$$

In fact, if we assume that $\varphi(\cdot)$ is also lower semicontinuous, then we can remove the condition that $\varphi(\cdot)$ is bounded above in a neighborhood of a point.

Proposition 6.73 *If $\varphi : X \to \overline{\mathbb{R}} = \mathbb{R} \cup \{+\infty\}$ is convex and lower semicontinuous, then $\varphi \mid_{intdom\varphi}$ is continuous.*

Proof Let $x_0 \in \text{intdom} \, \varphi$. Translating φ by $-x_0$ if necessary, we may assume that $x_0 = 0$. Let $\varrho > 0$ such that $B_\varrho \subseteq \text{dom} \, f$. The convexity and lower semicontinuity of $\varphi(\cdot)$ imply that the set

$$C = \{x \in X : \varphi(x) \leq \varphi(0) + 1\}$$

is convex and closed, with $0 \in C$. We claim that $X = \bigcup_{n \in \mathbb{N}} nC$. To this end, let $x \in X, x \neq 0$ and consider the function $\vartheta(t) = \varphi(tx)$ for all $t \in \mathbb{R}$. Evidently, $\vartheta(\cdot)$ is convex and $\left(-\frac{\varrho}{\|x\|}, \frac{\varrho}{\|x\|}\right) \subseteq \text{dom} \, \varphi$ and so $\vartheta \mid_{\left(-\frac{\varrho}{\|x\|}, \frac{\varrho}{\|x\|}\right)}$ is continuous. Therefore, we can find $\delta \in \left(0, \frac{\varrho}{\|x\|}\right)$ such that

$$|\vartheta(t) - \vartheta(0)| = |\varphi(tx) - \varphi(0)| \leq 1 \quad \text{for all } |t| \leq \delta,$$

$$\Rightarrow tx \in C \quad \text{for all } |t| \leq \delta,$$

$$\Rightarrow x \in nC \text{ with } n \in \mathbb{N}, n \geq \frac{1}{\delta}.$$

This proves that $X = \bigcup_{n \in \mathbb{N}} nC$.

From Proposition 1.173 and Theorem 1.174, it follows that there exists $n_0 \in \mathbb{N}$ such that $\text{int}(n_0 C) \neq \emptyset$, hence $\text{int} C \neq \emptyset$. On account of this fact, we can apply Theorem 6.71 and conclude that $\varphi \mid_{intdom\varphi}$ is continuous. $\square$

Remark 6.74 We mention that on account of Mazur's Theorem (see Theorem 3.105), a convex function $\varphi : X \to \overline{\mathbb{R}} = \mathbb{R} \cup \{+\infty\}$ is lower semicontinuous if and only if it is weakly lower semicontinuous.

Example 6.75 The function $\varphi : L^1[0, 1] \to \overline{\mathbb{R}}_+ = [0, +\infty]$ defined by

$$\varphi(u) = \begin{cases} \|u\|_2^2, & \text{if } u \in L^2[0, 1] \\ +\infty, & \text{otherwise} \end{cases}$$

is convex and lower semicontinuous, but $\text{intdom}\,\varphi = \emptyset$.

Continuing with the subdifferential map, we extend Proposition 6.66.

Proposition 6.76 *If* $\varphi : X \to \overline{\mathbb{R}} = \mathbb{R} \cup \{+\infty\}$ *is convex, then the subdifferential multifunction* $\partial\varphi : X \to 2^{X^*}$ *is monotone in the sense that if* $x, u \in \text{dom}(\partial\varphi)$ *and* $x^* \in \partial\varphi(x), u^* \in \partial\varphi(u)$, *then* $0 \le \langle x^* - u^*, x - u \rangle$.

Remark 6.77 The subdifferential is one on the basic operators of monotone type, and so, we will encounter it again in Chap. 7.

Proposition 6.78 *If* $\varphi : X \to \overline{\mathbb{R}} = \mathbb{R} \cup \{+\infty\}$ *is convex and* $U \subseteq \text{intdom}\,\varphi$ *open, then* $\varphi\,|_U$ *is* k-*Lipschitz (with* $k > 0$*) if and only if*

$$\partial\varphi(x) \subseteq k\overline{B}_1^* \text{ for all } x \in U, \text{ with } \overline{B}_1^* = \{x^* \in X^* : \|x^*\|_* \le 1\}.$$

Proof $\Rightarrow$ Let $x \in U$ and $\varrho > 0$ such that $\overline{B}_\varrho(x) \subseteq U$. For $x^* \in \partial\varphi(x)$, we have

$$\|x^*\|_* = \frac{1}{\varrho} \sup \left[\langle x^*, h \rangle : \|h\| = \varrho \right]$$

$$\le \frac{1}{\varrho} \sup \left[\varphi(x + h) - \varphi(x) : \|h\| = \varrho \right] \quad (\text{since } x^* \in \partial\varphi(x))$$

$$\le \frac{1}{\varrho} k\|h\| = k$$

$\Rightarrow \partial\varphi(x) \subseteq k\overline{B}_1^*$.
$\Leftarrow$ Let $x, u \in U$ and $x^* \in \partial\varphi(x)$. We have

$$\varphi(x) - \varphi(u) \le \langle x^*, x - u \rangle \le \|x^*\|_* \|x - u\| \le k\|x - u\|.$$

Interchanging the roles of x and u in the above argument, we conclude that $|\varphi(x) - \varphi(u)| \le k\|x - u\|$ for all $x, u \in U$, and thus $\varphi\,|_U$ is k-Lipschitz. $\quad\square$

Corollary 6.79 *If* $\varphi : X \to \overline{\mathbb{R}} = \mathbb{R} \cup \{+\infty\}$ *is convex, then the subdifferential multifunction* $\partial\varphi\,|_{\text{intdom}\,\varphi}$ *is locally bounded.*

Proposition 6.80 *If* $\varphi : X \to \overline{\mathbb{R}} = \mathbb{R} \cup \{+\infty\}$ *is convex,* $x \in \text{intdom}\,\varphi$, *and* $x^* \in X^*$, *then the following statements are equivalent:*

(a) $x^* \in \partial\varphi(x)$.
(b) $\langle x^*, h \rangle \le \varphi'_+(x; h)$ *for all* $h \in X$.

(c) $-\varphi'_+(x; -h) \le \langle x^*, h \rangle \le \varphi'_+(x; h)$ *for all* $h \in X$.

Proof (a)$\Rightarrow$(b) Let $h \in X$ and choose $\varrho > 0$ such that $x + \varrho h \in \mathrm{intdom}\,\varphi$. For $0 < t < \varrho$, we have $\langle x^*, h \rangle = \frac{1}{t} \langle x^*, th \rangle \le \frac{\varphi(x+th)-\varphi(x)}{t}$. Let $t \to 0^+$ to obtain $\langle x^*, h \rangle \le \varphi'_+(x; h)$ for all $h \in X$.

(b)$\Rightarrow$(a) Let $u \in \mathrm{intdom}\,\varphi$. We have

$$\langle x^*, u - x \rangle \le \varphi'_+(x; u - x) \le \varphi(u) - \varphi(x),$$

$$\Rightarrow x^* \in \partial\varphi(x).$$

(b)$\Leftrightarrow$(c) The equivalence of these two statements follows from the fact that

$$-\langle x^*, h \rangle = \langle x^*, -h \rangle \le \varphi'_+(x; -h) \quad \text{for all } h \in X.$$

$\square$

Proposition 6.81 *If* $\varphi : X \to \overline{\mathbb{R}} = \mathbb{R} \cup \{+\infty\}$ *is convex that is finite and continuous at* $x \in X$, *then* $\varphi'_+(x; h) = \sigma(h, \partial\varphi(x))$ *for all* $h \in X$.

Proof Let $x^* \in \partial\varphi(x)$. Then

$$\langle x^*, h \rangle \le \frac{\varphi(x + th) - \varphi(x)}{t} \quad \text{for all } h \in X \,(\text{see Definition 6.67}),$$

$$\Rightarrow \sigma(h, \partial\varphi(x)) \le \varphi'_+(x; h) \quad \text{for all } h \in X. \tag{6.48}$$

Let $s(h) = \varphi'_+(x; h)$ for $h \in X$. Then $s(\cdot)$ is sublinear continuous, and we have $s(h) \le \varphi(x + h) - \varphi(x)$ for all $h \in X$. Hence, $\partial s(0) \subseteq \partial\varphi(x)$, and so

$$s(h) = \varphi'_+(x; h) \le \sigma(h, \partial\varphi(x)) \quad \text{for all } h \in X,$$

$$\Rightarrow \varphi'_+(x; h) = \sigma(h, \partial\varphi(x)) \quad \text{for all } h \in X \,(\text{see (6.48)}).$$

$\square$

Next we show that essentially the case of Gateaux differentiability of $\varphi(\cdot)$ is the same as that of the subdifferential being a singleton.

Proposition 6.82 *If* $\varphi : X \to \overline{\mathbb{R}} = \mathbb{R} \cup \{+\infty\}$ *is convex, then:*

(a) *Gateaux differentiability of* $\varphi(\cdot)$ *at* $x \in X$ *implies*

$$\partial\varphi(x) = \{\varphi'_G(x)\}.$$

(b) *If* $\varphi(x) \in \mathbb{R}$, $\varphi(\cdot)$ *is continuous at* x, $\partial\varphi(x) = \{x^*\}$, *then* $\varphi(\cdot)$ *is Gateaux differentiable at* $x \in X$ *and*

$$x^* = \varphi'_G(x).$$

Proof

(a) Since $\varphi(\cdot)$ is Gateaux differentiable at x, we have

$$- \varphi'_+(x; -h) = \varphi'_+(x; h) = \langle \varphi'_G(x), h \rangle \quad \text{for all } h \in X. \tag{6.49}$$

Using Proposition 6.80 and (6.49), for any $x^* \in \partial\varphi(x)$, we have

$$\langle x^*, h \rangle = \langle \varphi'_G(x), h \rangle \quad \text{for all } h \in X,$$
$$\Rightarrow x^* = \varphi'_G(x) \text{ and so } \partial\varphi(x) = \{\varphi'_G(x)\}.$$

(b) Arguing by contradiction, suppose that $\varphi(\cdot)$ is not Gateaux differentiable at x. Then there exists $h \subset X$ such that

$$\varphi'_+(x; h) \neq -\varphi'_+(x; -h). \tag{6.50}$$

On account of Proposition 6.81, we can find $x_1^*, x_2^* \in \partial\varphi(x)$ such that

$$\langle x_1^*, h \rangle = \varphi'_+(x; h) \text{ and } \langle x_2^*, h \rangle = -\varphi'_+(x; -h),$$
$$\Rightarrow \langle x_1^* - x_2^*, h \rangle = \varphi'_+(x; h) + \varphi'_+(x; -h) \neq 0 \text{ (see (6.50))},$$
$$\Rightarrow \partial\varphi(x) \text{ is not a singleton, a contradiction.}$$

$\square$

Now we will prove some calculus rules for the convex subdifferential.

Proposition 6.83 *If $\varphi, \psi : X \to \overline{\mathbb{R}} = \mathbb{R} \cup \{+\infty\}$ are convex functions, $\operatorname{dom} \varphi \cap \operatorname{dom} \psi \neq \emptyset$, and one of them is continuous at a point of the intersection, then $\partial(\varphi + \psi)(x) = \partial\varphi(x) + \partial\psi(x)$ for all $x \in X$.*

Proof From the definition of the subdifferential (see Definition 6.67), we have

$$\partial\varphi(x) + \partial\psi(x) \subseteq \partial(\varphi + \psi)(x) \quad \text{for all } x \in X. \tag{6.51}$$

We need to show that the opposite inclusion is also true. Let $x^* \in \partial(\varphi + \psi)(x)$. We have

$$\langle x^*, u - x \rangle \leq (\varphi + \psi)(u) - (\varphi + \psi)(x) \quad \text{for all } u \in X.$$

We introduce the following two sets:

$$C = \{(u, \lambda) \in X \times \mathbb{R} : \varphi(u) - \varphi(x) \leq \lambda + \langle x^*, u - v \rangle\},$$

$$D = \{(u, \lambda) \in X \times \mathbb{R} : \psi(u) - \psi(x) \leq \lambda\}.$$

Both sets are convex, $\text{int} C \neq \emptyset$, and $\text{int}\, C \cap D = \emptyset$. So, by Theorem 3.71, we can find $(u^*, \mu) \in X^* \times \mathbb{R}$, $(u^*, \mu) \neq (0, 0)$ such that

$$\psi(x) - \psi(u) \leq \mu - \langle u^*, u \rangle \leq \varphi(u) - \varphi(x) - \langle x^*, u - x \rangle \quad \text{for } u \in X. \qquad (6.52)$$

Choosing $u = x$, then we have $\mu = \langle u^*, x \rangle$. Hence,

$$\langle \widehat{u}^*, u - x \rangle \leq \varphi(u) - \varphi(x) \quad \text{for all } u \in X, \text{ with } \widehat{u}^* = x^* - u^*,$$
$$\Rightarrow \widehat{u}^* \in \partial\varphi(x).$$

Also from (6.52), it is clear that $u^* \in \partial\psi(x)$. So, finally, we have

$$x^* = u^* + \widehat{u}^* \in \partial\varphi(x) + \partial\psi(x),$$
$$\Rightarrow \partial(\varphi + \psi)(x) \subseteq \partial\varphi(x) + \partial\psi(x),$$
$$\Rightarrow \partial(\varphi + \psi)(x) = \partial\varphi(x) + \partial\psi(x). \qquad \qquad \square$$

Proposition 6.84 *If Y is another Banach space, $\varphi : X \to \overline{\mathbb{R}} = \mathbb{R} \cup \{+\infty\}$ is convex, $A \in \mathscr{L}(Y, X)$, and φ is continuous at some point of $R(A)$, then $\partial(\varphi \circ A)(y) = A^*\partial\varphi(A(y))$ for all $y \in Y$.*

Proof Evidently, $(\varphi \circ A)(\cdot)$ is convex on Y. Let $x^* \in \partial\varphi(A(y))$. We have

$$\langle x^*, u - A(y) \rangle_X \leq \varphi(u) - \varphi(A(y)) \quad \text{for all } u \in X,$$

$$\Rightarrow \langle A^*(x^*), v - y \rangle \leq \varphi(A(v)) - \varphi(A(y)) \quad \text{for all } v \in Y,$$

$$\Rightarrow A^*(x^*) \in \partial(\varphi \circ A)(y) \quad \text{for each } x^* \in \partial\varphi(A(y)),$$

$$\Rightarrow A^*\partial\varphi(A(y)) \subseteq \partial(\varphi \circ A)(y) \quad \text{for all } y \in Y. \qquad (6.53)$$

Let $y^* \in \partial(\varphi \circ A)(y)$. We have

$$\langle y^*, v - y \rangle_Y \leq \varphi(A(v)) - \varphi(A(y)) \quad \text{for all } v \in Y.$$

Consider the set

$$E = \{(A(y), \langle y^*, v - y \rangle_Y + \varphi(A(y))) : y \in Y\}.$$

This is an affine set and $E \cap \text{int}\,\text{epi}\,\varphi = \emptyset$. So by Theorem 3.71, we can find $(x^*, \mu) \in X^* \times \mathbb{R}$, $(x^*, \mu) \neq (0, 0)$ such that

$$-\langle x^*, A(v)\rangle_X + \langle y^*, v - y\rangle_Y + \varphi(A(y)) = \mu \text{ for all } v \in Y.$$

Let $v = y$. We obtain $\mu = -\langle x^*, A(y)\rangle_X + \varphi(A(y))$; thus

$$\langle x^*, A(v - y)\rangle_X = \langle y^*, v - y\rangle_Y \text{ for all } v \in Y,$$

$$\Rightarrow y^* = A^*(x^*). \tag{6.54}$$

Therefore,

$$\langle x^*, u - A(x)\rangle_X \leq \varphi(u) - \varphi(A(y)) \text{ for all } u \in X,$$

$$\Rightarrow x^* \in \partial\varphi(A(y)),$$

$$\partial(\varphi \circ A)(y) \leq A^*\vartheta_\varphi(A(y)) \ (see \ (6.54)),$$

$$\partial(\varphi \circ A)(y) = A^*\partial\varphi(A(y)) \ (see \ (6.53)). \qquad \square$$

Remark 6.85 From the above, we see that in general we have

$$A^*\partial\varphi(A(y)) \subseteq \vartheta(\varphi \circ A)(y) \text{ for all } y \in Y.$$

The following function is important in optimization theory.

Definition 6.86 For a Banach space X and $A \subseteq X$, set

$$i_A(x) = \begin{cases} 0 & \text{if } x \in A \\ \infty & \text{if } x \notin X \setminus A \end{cases}$$

and call this function the "indicator function" of A, which is convex if A is.

The proof of the next proposition is an easy consequence of the definitions.

Proposition 6.87 *For Banach space X and a closed and convex cone $K \subseteq X$, we have $\partial d_K(x) \subseteq \partial i_K(x) \cap \overline{B}_1^{X^*}$ and $\partial i_K(x) \subseteq \partial i_K(0)$.*

Using the subdifferential, we can generalize the classical Fermat's rule for minimizers.

Proposition 6.88 *If $\varphi : X \to \overline{\mathbb{R}} = \mathbb{R} \cup \{+\infty\}$ is convex, then $x_0 \in \text{dom } \varphi$ is a minimizer of $\varphi(\cdot)$ if and only if $0 \in \partial\varphi(x_0)$.*

Proof $\Rightarrow$ We have $0 \leq \varphi(u) - \varphi(x_0)$ for all $u \in X$ and so $0 \in \partial\varphi(x_0)$.

$\Leftarrow$ From the definition of the subdifferential, we have $0 \leq \varphi(u) - \varphi(x_0)$ for all $u \in \operatorname{dom}\varphi \Rightarrow \varphi(x_0) \leq \varphi(u)$ for all $u \in X$. $\square$

Remark 6.89 The convexity of φ implies that it has only global minimizers.

The next proposition will be used in the examples that follow, but it is also useful in many other situations.

Proposition 6.90 *If $\varphi : X \to \overline{\mathbb{R}} = \mathbb{R} \cup \{+\infty\}$ is convex and lower semicontinuous, then φ is minorized by a continuous affine function.*

Proof Let $x \in \operatorname{dom}\varphi$, and for $\varepsilon > 0$, note that $(x, \varphi(x) - \varepsilon) \notin \operatorname{epi}\varphi$. Since $\operatorname{epi}\varphi \subseteq X \times \mathbb{R}$ is convex, closed, by Theorem 3.71, we can find $(x^*, \mu) \in X^* \times \mathbb{R}, (x^*, \mu) \neq (0, 0)$ and $\delta > 0$ such that

$$\langle x^*, u \rangle + \mu\lambda \leq \langle x^*, x \rangle + \mu[\varphi(x) - \varepsilon] - \delta \text{ for all } (u, \lambda) \in \operatorname{epi}\varphi. \tag{6.55}$$

Since $\lambda > \varphi(x)$ can be arbitrary, from (6.55), we infer that $\mu \leq 0$. If $\mu = 0$, then $\langle x^*, \mu \rangle \leq \langle x^*, x \rangle - \delta$ for all $u \in \operatorname{dom}\varphi$, a contradiction. So, $\mu < 0$. We have

$$\langle x^*, u \rangle + \mu\varphi(u) \leq \langle x^*, x \rangle + \mu[\varphi(x) - \varepsilon] - \delta \text{ for all } u \in \operatorname{dom}\varphi$$

$$\Rightarrow \varphi(u) \geq -\frac{1}{\mu}\langle x^*, u \rangle + \frac{1}{\mu}\langle x^*, x \rangle + \varphi(x).$$

The function $\alpha(u) = -\frac{1}{\mu}\langle x^*, u \rangle + \frac{1}{\mu}\langle x^*, x \rangle + \varphi(x), u \in \operatorname{dom}\varphi$, is affine continuous and minorizes φ. $\square$

Now we present some examples of subdifferentials.

Example 6.91

(a) Consider a nondecreasing function $f : R \to R$. We know that the set of discontinuity points of $f(\cdot)$ is at most countable. We define

$$f_-(x) = \lim_{u \to x^-} f(u) \text{ and } f_+(x) = \lim_{u \to x^+} f(u).$$

We set $\varphi(x) = \int_{x_0}^x f(s)ds$ for some $x_0 \in R$, all $x \in R$. Evidently, φ is convex, and we can easily check that

$$\partial\varphi(x) = [f_-(x), f_+(x)] \text{ for all } x \in \mathbb{R}.$$

(b) Suppose (Ω, Σ, μ) is a finite measure space and $f : \overline{\mathbb{R}} \to \overline{\mathbb{R}} = \mathbb{R} \cup \{+\infty\}$ is convex and lower semicontinuous. We consider the integral functional $\varphi : L^1(\Omega) \to \overline{R} = R \cup \{+\infty\}$ defined by

$$\varphi(u) = \begin{cases} \int_\Omega f(u)d\mu & \text{if } f(u) \in L^1(\Omega) \\ +\infty & \text{otherwise} . \end{cases}$$

First we show that $\varphi(\cdot)$ is convex and lower semicontinuous. Convexity is clear from the convexity of f. For the lower semicontinuity, suppose that $\{u_n\}_{n \in N} \subseteq$ dom φ and assume that

$$u_n \to u \text{ in } L^1(\Omega) \text{ and } \varphi(u_n) \leq \lambda \text{ for all } n \in N, \text{ with } \lambda \in R. \qquad (6.56)$$

Since f is convex and lower semicontinuous, by Proposition 6.90, it is minorized by a continuous affine function. So, we can use the extended Fatou's lemma (see Proposition 2.149) and obtain

$$\int_\Omega \lim_{n \to \infty} \inf f(u_n)d\mu \leq \lim_{n \to \infty} \inf \int_\Omega f(u_n)d\mu,$$

$$\Rightarrow \int_\Omega f(u)d\mu \leq \lambda (see\ (6.56))$$

$$\Rightarrow \varphi(\cdot) \text{ is lower semicontinuous} (see\ Remark\ 1.177).$$

Next we show that

$$\text{``} u^* \in \partial\varphi(u) \text{ if and only if } u^*(w) \in \partial f(u(w)) \ \mu \cdot a.e. \text{''} \qquad (6.57)$$

$$\Rightarrow: \int_\Omega u^*(v - u)d\mu \leq \varphi(v) - \varphi(u) \text{ for all } v \in L^1(\Omega),$$

$$\Rightarrow \int_\Omega [f(v) - f(u) - u^*(v - u)]d\mu \geq 0. \qquad (6.58)$$

Let $A \in \Sigma$ and define $w(\omega) = \begin{cases} u(\omega) & \text{if } \omega \in \Omega \backslash A. \\ v(\omega) & \text{if } \omega \in A. \end{cases}$ Then $w \in L^1(\Omega)$, and from (6.58), we have

$$\int_A [f(v) - f(u) - u^*(v - u)]d\mu \geq 0,$$

$$\Rightarrow u^*(\omega)(v - u)(w) \leq f(v(\omega)) - f(u(\omega)) \ \mu \ a.e. \text{ for every } v \in L^1(\Omega)$$

$$\Rightarrow u^*(\omega) \in \partial f(u(\omega)) \ \mu \cdot a.e.$$

$\Leftarrow$: We have

$$\Rightarrow u^*(\omega)(v - u)(w) \le f(v(\omega)) - f(u(\omega)) \ \mu \ a.e. \ for \ all \ v \in L^1(\Omega)$$

$$f(u(\cdot)) \in L^1(\Omega), \ that \ is, \ u \in \text{dom} \ \varphi.$$

Moreover, integrating over Ω, we conclude that $u^* \in \partial\varphi(u)$.

(c) Let $f : \mathbb{R}_+ \to \mathbb{R}_+$ be a convex function with $f(0) = 0$, and consider the function $\varphi : X \to \mathbb{R}_+$ defined by $\varphi(u) = f(\|u\|)$ for all $u \in X$. We have

$$u^* \in \partial\varphi(u) \ if \ and \ if \ only \ \|u^*\|_*\|u\| = \langle u^*, u \rangle \ and \ \|u^*\|_* \in \partial f(\|u\|). \tag{6.59}$$

We show the equivalence in (6.59).

$\Rightarrow$: For $u \ne 0$, we have

$$\langle u^*, v - u \rangle \le \varphi(v) - \varphi(u) \ for \ all \ v \in X. \tag{6.60}$$

Let $e \in X$, with $\|e\| = 1$, and set $v = \|u\|e$. From (6.60), we have

$$\|u\|\langle u^*, e \rangle - \langle u^*, u \rangle \le \varphi(\|u\|e) - \varphi(u) = f(\|u\|) - f(\|u\|) = 0,$$

$$\Rightarrow \|u\|\langle u^*, e \rangle \le \langle u^*, u \rangle.$$

Since $e \in \partial B_1$ is arbitrary, we infer that

$$\|u\|\|u^*\|_* = \langle u^*, u \rangle. \tag{6.61}$$

Also for every $\vartheta \in \mathbb{R}$, we have

$$\langle u^*, \frac{\vartheta u}{\|u\|} \rangle - \langle u^*, u \rangle \le \varphi(\frac{\vartheta u}{\|u\|}) - \varphi(u),$$

$$\Rightarrow \|u^*\|_*[\vartheta - \|u\|] \le f(\vartheta) - f(\|u\|) \ (see \ (6.61)),$$

$$\Rightarrow \|u^*\|_* \in \partial f(\|u\|).$$

If $u = 0$, then in (6.60), let $v = \vartheta e$ with $\|e\| = 1$. We have

$$\langle u^*, \vartheta e \rangle \le \varphi(\vartheta e) = f(\vartheta),$$

$$\Rightarrow \vartheta \langle u^*, e \rangle \leq f(\vartheta),$$

$$\Rightarrow \vartheta \|u^*\|_* \leq f(\vartheta) \ (since \ e \in \partial B_1 \ is \ arbitrary),$$

$$\Rightarrow \|u^*\|_* \in \vartheta f(o).$$

$\Leftarrow$: We have

$$\langle u^*, v - u \rangle \leq \|u^*\|_* [\|v\| - \|u\|] \leq f(\|v\|) - f(\|u\|) = \varphi(v) - \varphi(u) \ for \ all \ v \in X,$$

$$\Rightarrow u^* \in \vartheta \varphi(u).$$

So, we have proved (6.59).

In particular, let $f(t) = \frac{1}{2}t^2 \ t \geq 0$. Then

$$\partial \varphi(x) = \mathcal{F}(x) = \{x^* \in X^* : \langle x^*, x \rangle = \|x^*\|_*^2 = \|x\|^2\} \ for \ all \ x \in X. \quad (6.62)$$

This is known as the duality map, and it is important in the study of the geometry of the Banach spaces and in evolution equations. We see that $\mathcal{F}(\cdot)$ has nonempty, closed, and convex values. If X^* is strictly convex (see Definition 3.24(a)), then $\mathcal{F}(\cdot)$ is single-valued. If X is a Hilbert space, then the duality map $\mathcal{F} : X \to X^*$ is single-valued and

$$\langle \mathcal{F}(x), u \rangle = (x, u) \ for \ all \ x, u \in X$$

with $(\cdot, \cdot)$ denoting the inner product in X. So $\mathcal{F}(\cdot)$ is the "Riesz map" for X (see Theorem 3.181). Next we establish some useful properties of $\mathcal{F}(\cdot)$ as we strengthen the conditions on the spaces X and X^*.

Definition 6.92 A map $f : X \to X^*$ is said to be "demicontinuous," if

$$u_n \to u \ in \ X \Rightarrow f(x_n) \xrightarrow{w} f(x) \ in \ X^*.$$

Proposition 6.93 *If X is reflexive and X^* is strictly convex, then $\mathcal{F}(\cdot)$ is demicontinuous, and for all $u, x \in X$, we have*

$$\langle \mathcal{F}(x) - \mathcal{F}(u), x - u \rangle \geq 0$$

that is $\mathcal{F}(\cdot)$ is monotone (see Sect. 7.2).

Proof For $x, u \in X$, we have

$$\langle \mathcal{F}(x) - \mathcal{F}(u), x - u \rangle \geq \|x\|^2 + \|u\|^2 - 2\|x\|\|u\| = (\|x\| - \|u\|)^2 \geq 0;$$

hence, $\mathcal{F}(\cdot)$ is indeed monotone.

Suppose $u_n \to u$ in X. Then $\|\mathcal{F}(x_n)\|_* = \|u_n\| \to \|u\|$ as $n \to \infty$. Since $\{\mathcal{F}(u_n)\}_{n \in N} \subseteq X^*$ is bounded and X^* is reflexive (since X is), we may assume that $\mathcal{F}(u_n) \xrightarrow{w} u^*$ in X^*. For all $h \in X$, we have

$$\langle u^*, h \rangle = \lim_{n \to \infty} \langle \mathcal{F}(u_n), h \rangle \leq \lim_{n \to \infty} \|u_n\|\|h\| = \|u_n\|\|h\|. \tag{6.63}$$

Also we have

$$\langle u^*, u \rangle = \lim_{n \to \infty} \langle \mathcal{F}(u_n), u_n \rangle = \lim_{n \to \infty} \|u_n\|^2 = \|u_n\|^2. \tag{6.64}$$

From (6.63) and (6.64), we infer that $\mathcal{F}(u) = u^*$ and so $\mathcal{F}(\cdot)$ is demicontinuous.
□

Proposition 6.94 *If X is a reflexive Banach space, X^* is strictly convex, and $\psi :$ $X \to R$ is the norm function, that is, $\psi(x) = \|x\|$ for all $x \in X$, then $\psi(\cdot)$ is Gateaux differentiable at every $x \in X \backslash \{0\}$ and*

$$\psi'(x) = \frac{\mathcal{F}x}{\|x\|} \text{ for all } x \in X \backslash \{0\}.$$

Proof Note that $\psi(x) = (2\varphi(x))^{1/2}$ for all $x \in X$, with $\varphi(x) = \frac{1}{2}\|x\|^2$. Then from proposition 6.10, we have that $\psi(\cdot)$ is Gateaux differentiable at $x \in X \backslash \{0\}$, and we have $\psi'(u) = \frac{1}{2}(2\varphi(u))^{-\frac{1}{2}}2\varphi'(u) = (2\varphi(u))^{-\frac{1}{2}}\mathcal{F}(u) = \frac{\mathcal{F}(u)}{\|u\|}$. □

Proposition 6.95 *If X is a reflexive Banach space and X^* is uniformly convex, then $\mathcal{F} : X \to X^*$ is uniformly continuous on bounded sets.*

Proof Let $\partial B_1 = \{u \in X : \|u\| = 1\}$. We start by showing that $\mathcal{F}(\cdot)$ is uniformly continuous on ∂B_1. Arguing by contradiction, suppose that $\mathcal{F}|_{\partial B_1}$ is not uniformly continuous. Then we can find $\{u_n\}_{n \in N}, \{v_n\}_{n \in N} \subseteq \partial B_1$ and $\varepsilon > 0$ such that

$$\|u_n - v_n\| \to 0 \text{ as } n \to \infty, \ \|\mathcal{F}(u_n) - \mathcal{F}(v_n)\|_* \geq \varepsilon \text{ for all } n \in N. \tag{6.65}$$

We have

$$\|\mathcal{F}(u_n) + \mathcal{F}(v_n)\|_*\|u_n\| \geq \langle \mathcal{F}(u_n) + \mathcal{F}(v_n), u_n \rangle$$

$$= \|u_n\|^2 + \|v_n\|^2 + \langle \mathcal{F}(v_n), u_n - v_n \rangle$$

$$\geq \|u_n\|^2 + \|v_n\|^2 - \|v_n\|\|v_n - u_n\|,$$

$$\Rightarrow \|\frac{1}{2}(\mathcal{F}(u_n) + \mathcal{F}(v_n))\|_* \geq 1 - \frac{1}{2}\|v_n - u_n\| \text{ (recall } u_n, v_n \in \partial B_1)$$

$$\Rightarrow \|\mathcal{F}(u_n) + \mathcal{F}(v_n)\|_* \to 2 \text{ as } n \to \infty,$$

$$\Rightarrow \|\mathcal{F}(u_n) - \mathcal{F}(v_n)\|_* \to 0 \text{ as } n \to \infty \text{ (see Definition 3.240}(c)),$$

which contradicts (6.65) (recall that $\|\mathcal{F}(u_n)\|_* = \|\mathcal{F}(v_n)\|_* = 1$ for all $n \in N$). Therefore, $\mathcal{F}|_{\partial(B_1)}$ is uniformly continuous.

Next note that $\mathcal{F}(\lambda y) = \lambda \mathcal{F}(y)$ for all $y \in X$, all $\lambda > 0$. For all $u, v \in X, u \neq 0, v \neq 0$, we have

$$\|\mathcal{F}(u) - \mathcal{F}(v)\|_* = \|\|u\|\mathcal{F}(\frac{u}{\|u\|}) - \|v\|\mathcal{F}(\frac{v}{\|v\|})\|_*$$

$$\leq \|u\|\|\mathcal{F}(\frac{u}{\|u\|}) - \mathcal{F}(\frac{v}{\|v\|})\|_* + |\|u\| - \|v\||\|\mathcal{F}(\frac{v}{\|v\|})\|_*,$$

and this on account of the uniform continuity of $\mathcal{F}(\cdot)$ on ∂B_1 implies the uniform continuity of $\mathcal{F}(\cdot)$ on bounded sets in X. $\qquad\square$

Remark 6.96 Also the map $\psi(u) = \|u\|$ is Frechet differentiable at every $u \neq 0$, and the Frechet derivative $\psi'(u) = \frac{\mathcal{F}(u)}{\|u\|}$, $u \neq 0$ (see Proposition 6.94) is uniformly continuous on closed bounded sets $D \subseteq X$ such that $0 \notin D$.

We conclude this section with a Gateaux differentiability theorem for convex functions on a Banach space. The result is known as "Mazur's Theorem."

Theorem 6.97 *If X is a separable Banach space, $U \subseteq X$ is open convex, and $\varphi : U \to R$ is continuous and convex, then $\varphi(\cdot)$ is Gateaux differentiable on a dense G_δ-set in U.*

Proof For $h \in X$ and $n \in N$, let

$$V_n(h) = \{x \in U : \sup_{0 < t < \delta} \frac{\varphi(x + th) + \varphi(x - th) - 2\varphi(x)}{t} < \frac{1}{n}\}$$

for some $\delta = \delta(x, n) > 0$. The continuity of φ implies that for each $k \in N$ the set

$$V_n^k(h) = \{x \in U : \frac{\varphi(x + \frac{1}{k}h) + \varphi(x - \frac{1}{k}h) - 2\varphi(x)}{\frac{1}{k}} < \frac{1}{n}\}$$

is open and $V_n(h) = \cup_{k \in N} V_n^k(h)$. Hence, $V_n(h)$ is open.

We show that $V_n(h)$ is dense in U. Arguing by contradiction, suppose that we do not have density of $V_n(h)$ in U. This means that we can find $x_0 \in U$ and $r > 0$ such that

$$V_n(h) \cap B_r(x_0) = \emptyset.$$

Then $t \to \varphi(x + th)$ is not differentiable on $(-r, r)$, a contradiction since this is a convex function on $\mathbb{R}$.

Since X is separable, we can find $\{h_m\}_{m \in \mathbb{N}} \subseteq \partial B_1$ dense. Then $\varphi(\cdot)$ is Gateaux differentiable at x in the direction $h_m, m \in \mathbb{N}$. So $\varphi(\cdot)$ is Gateaux differentiable on $\cap_{n,m \in \mathbb{N}} V_n(h_m)$. Each $V_n(h_m)$ is open dense, and so by Baire's Theorem (see Theorem 1.174 and Proposition 1.173), we have that $\cap_{n,m \in N} V_n(h_m) \subseteq U$ is dense and G_δ.   □

If we require this property of the Frechet derivative, then we have a particular class of Banach spaces, known as "Asplund spaces." A Banach space with a separable dual is Asplund (see Asplund [14]).

6.3 Convex Functions–Duality Theory

In this section, we introduce the notion of conjugate function and develop the duality theory for convex functions. Duality is an important feature of "Convex Analysis."

So let X be a normed space and X^* is topological dual. By $\langle \cdot, \cdot \rangle$, we denote the duality brackets for the pair (X, X^*). Again we consider in general $\overline{\mathbb{R}} = \mathbb{R} \cup \{+\infty\}$-valued functions, and as before, we always assume that they are not identically ∞.

Definition 6.98 Let $\varphi : X \to \overline{\mathbb{R}} = R \cup \{+\infty\}$. The "conjugate" (or "polar") function $\varphi^* : X^* \to \overline{\mathbb{R}} = \mathbb{R} \cup \{+\infty\}$ is defined by

$$\varphi^*(x^*) = \sup[\langle x^*, x \rangle - \varphi(x) : x \in \operatorname{dom} \varphi].$$

To conjugate of $\varphi^*(\cdot)$ is the function $\varphi^{**} : X \to \overline{\mathbb{R}}$ defined by

$$\varphi^{**}(x^*) = \sup[\langle x^*, x \rangle - \varphi^*(x^*) : x^* \in \operatorname{dom} \varphi^*].$$

Remark 6.99 From the above definitions, we see that we always have

$$\varphi(x) + \varphi^*(x^*) \geq \langle x^*, x \rangle \tag{6.66}$$

$$\varphi^*(x^*) + \varphi^{**}(x) \geq \langle x^*, x \rangle \ for \ all \ x \in X, \ x^* \in X^*. \tag{6.67}$$

Inequality (6.66) is known as "Young's inequality." Indeed, if $\varphi(t) = \frac{1}{p}t^p, t \geq 0$, then $\varphi^*(\tau) = \frac{1}{p'}\tau^{p'}, \tau \geq 0$, $(\frac{1}{p} + \frac{1}{p'} = 1)$, and from (6.66), we recover the classical Young inequality

$$t\tau \leq \frac{1}{p}t^p + \frac{1}{p'}\tau^{p'} \; for \; all \; t, \tau \geq 0.$$

If we consider all continuous affine minorants of $\varphi(\cdot)$, we have

$$\langle x^*, x \rangle - \mu \leq \varphi(x) \; for \; all \; x \in X,$$

$$\langle x^*, x \rangle - \varphi(x) \leq \mu \; for \; all \; x \in X.$$

So, the smallest $\mu \in R$ is $\varphi^*(x^*)$, and so $x \rightarrow \langle x^*, x \rangle - \varphi^*(x^*)$ is the biggest continuous affine minorant of $\varphi(\cdot)$.

Example 6.100

(a) $\varphi(x) = \frac{1}{p}\|x\|^p$ for all $x \in X$, with $1 < p < \infty$. Then

$$\varphi^*(x^*) = \frac{1}{p'}\|x^*\|_*^{p'} \; for \; all \; x^* \in X^*, \; \frac{1}{p} + \frac{1}{p'} = 1.$$

(b) $A \subset X$ nonempty and $\varphi(x) = i_A(x) = \begin{cases} 0 & if \; x \in A \\ +\infty & if \; x \notin A \end{cases}$ (the "indicator function" for the set A, see Definition 6.86). Then

$$i_A^*(x^*) = \sigma(x^*, A) = \sup[\langle x^*, \alpha \rangle : \alpha \in A] \; for \; all \; x^* \in X^*$$

(the support function of A).

(c) $\varphi(x) = \|x\|$ for all $x \in X$. Then

$$\varphi^*(x^*) = i_{\overline{B}_1}^*(x^*) = \begin{cases} 0 & if \; \|x^*\|_* \leq 1 \\ +\infty & if \; 1 < \|x^*\|_*. \end{cases}$$

(d) $f : \mathbb{R}_+ \rightarrow \overline{\mathbb{R}}$ and $\varphi(x) = f(\|x\|)$ for all $x \in X$. Then

$$\varphi^*(x^*) = \sup[\langle x^*, x \rangle - \varphi(x) : x \in X] = \sup[\langle x^*, x \rangle - f(\|x\|) : x \in X]$$

$$= \sup_{t \geq 0} \sup_{\|x\|=t} (\langle x^*, x \rangle - f(t))$$

$$= \sup_{t \geq 0}(t\|x^*\|_* - f(t)) = f^*(\|x^*\|_*) \; for \; all \; x^* \in X^*.$$

In the next proposition, we list some basic properties at the conjugate function that are easy consequences of Definition 6.98.

Proposition 6.101

(a) $\varphi(x) + \varphi^*(x^*) \geq \langle x^*, x \rangle$ for all $x \in X, x^* \in X^*$ (Young's inequality, see also Remark 6.99).

(b) $\varphi^*(0) = -\inf_{x \in X} \varphi$.

(c) $\varphi \leq \psi \Rightarrow \psi^* \leq \varphi^*$.

(d) $(\sup_{i \in I} \varphi_i)^* \leq \inf_{i \in I} \varphi_i^*$ for any index set I.

(e) $(\lambda \varphi)^*(x^*) = \lambda \varphi^*(\frac{1}{\lambda} x^*)$ for all $\lambda > 0$, all $x^* \in X^*$.

(f) If $\varphi_0(x) = \varphi(x - x_0)$ for all $x \in X$, some $x_0 \in X$, then $\varphi_0^*(x^*) = \varphi^*(x^*) + \langle x^*, x_0 \rangle$ for all $x^* \in X^*$.

Definition 6.102 By $\Gamma_0(x)$, we denote the cone of all functions $\varphi : X \to \overline{\mathbb{R}}$ (always not identically $+\infty$) that are convex and lower semicontinuous.

Then we have the following result.

Proposition 6.103 If $\varphi \in \Gamma_0(x)$, then $\varphi^* : X^* \to \overline{R} = R \cup \{+\infty\}$ is convex, w^*-lower semicontinuous, and not identically $+\infty$.

Proof From Definition 6.98 and Proposition 6.90, we have that $\varphi^*(\cdot)$ is $\overline{\mathbb{R}}$-valued and not identically $+\infty$. Also if $x^*, u^* \in X^*$ and $0 \leq t \leq 1$, then

$$\varphi^*((1-t)x^* + tu^*) = \sup[\langle (1-t)x^* + tu^*, x \rangle - \varphi(x) : x \in X]$$

$$\leq (1-t) \sup_{x \in X} [\langle x^*, x \rangle - \varphi(x)] + t \sup_{x \in X} [\langle x^*, x \rangle - \varphi(x)]$$

$$= (1-t)\varphi^*(x^*) + t\varphi^*(u^*);$$

hence, φ^* is convex.

Next let $\{x_\alpha^*\}_{\alpha \in D} \subseteq X^*$ be a net such that

$$\varphi^*(x_\alpha^*) \leq \lambda \ (\lambda \in R), x_\alpha^* \xrightarrow{w^*} x^* \ in \ X^*,$$

$$\Rightarrow \langle x_\alpha^*, x \rangle - \varphi(x) \leq \lambda \text{ for all } \alpha \in D, \ all \ x \in X \ (see \ (6.66)),$$

$$\Rightarrow \langle x^*, x \rangle - \varphi(x) \leq \lambda \text{ for all } x \in X,$$

$$\Rightarrow \varphi^*(x^*) \leq \lambda.$$

This proves that $\varphi(\cdot)$ is w^*-lower semicontinuous. $\square$

Using (6.66) (Young's inequality), we have

$$\varphi^{**}(x) = \sup_{x^* \in X^*} [\langle x^*, x \rangle - \varphi^*(x^*)] \leq \sup_{x^* \in X^*} [\langle x^*, x \rangle - \langle x^*, x \rangle + \varphi(x)] = \varphi(x).$$

So, we always have

$$\varphi^{**} \leq \varphi. \tag{6.68}$$

It is natural to ask under what conditions we have equality in (6.68).

Theorem 6.104 $\varphi \in \Gamma_0(x)$ *if and only if* $\varphi = \varphi^{**}$.

Proof $\Rightarrow$ From (6.68), we have $\varphi^{**} \leq \varphi$. We will show that the opposite inequality is also true. We need to consider only points $x \in \text{dom}\,\varphi^{**}$. So, suppose for some $x \in \text{dom}\,\varphi^{**}$, we have $\varphi(x) > \varphi^{**}(x)$. Let $m = \frac{1}{2}[\varphi(x) - \varphi^{**}(x)] > 0$. Note that $(x, \varphi(x) - m) \notin \text{epi}\,\varphi$, and so by Theorem 3.71, we can find $(\widehat{x}^*, \mu) \in X \times \mathbb{R}$, $(\widehat{x}^*, \mu) \neq (0, 0)$ such that

$$\sup[\langle x^*, x \rangle + \mu\lambda : (x, \lambda) \in \text{epi}\,\varphi] < \langle x^*, x \rangle + \mu(\varphi(x) - m). \tag{6.69}$$

Evidently, $\mu \leq 0$ (since $\lambda \geq \varphi(x)$ can be arbitrarily big). If $\mu = 0$, then $x^* = 0$ too, and we have a contradiction. Therefore, $\mu < 0$, and we can assume that $\mu = -1$. Then from (6.69), we have $x^* \neq 0$ and

$$\sup[\langle x^*, x \rangle - \varphi(x) : x \in \text{dom}\,\varphi] = \varphi^*(x^*) < \langle x^*, x \rangle - \varphi(x) + m,$$

thus $\varphi(x) < \varphi^{**}(x)$ (recall the definition of m), a contradiction.

$\Leftarrow$ From Proposition 6.103, we know that $\varphi^{**} \in \Gamma_0(X)$. Hence, $\varphi \in \Gamma_0(X)$. $\square$

From Theorem 6.104 and Proposition 6.103, we have that φ^{**} is the biggest element in the cone $\Gamma_0(X)$ minorizing φ. Also note that $\varphi^* = \varphi^{***}$.

Next we will see how conjugate functions are related to subdifferentials.

Theorem 6.105 *If* $\varphi : X \to \overline{\mathbb{R}}$ *is convex, then* $x^* \in \partial\varphi(x)$ *if and only if* $\varphi(x) + \varphi^*(x^*) = \langle x^*, x \rangle$.

Proof $\Rightarrow$ Since $x^* \in \partial\varphi(x)$, we have

$$\langle x^*, u - x \rangle \leq \varphi(u) - \varphi(x) \text{ for all } u \in X,$$

$$\Rightarrow \langle x^*, u \rangle - \varphi(u) \leq \langle x^*, x \rangle - \varphi(x) \text{ for all } u \in X,$$

$$\varphi^*(x^*) + \varphi(x) = \langle x^*, x \rangle \text{ (see (6.66))}.$$

$\Leftarrow$ We have

$$\langle x^*, x \rangle - \varphi(x) = \varphi^*(x^*) \geq \langle x^*, u \rangle - \varphi(u) \ for \ all \ u \in X,$$

$$\Rightarrow \varphi(u) - \varphi(x) \geq \langle x^*, u - x \rangle \ \text{for all } u \in X,$$

$$x^* \in \partial \varphi(x).$$

$\square$

Remark 6.106 From the above theorem, it follows that

$$\partial \varphi(x) = \{x^* \in X : \varphi(x) + \varphi^*(x^*) = \langle x^*, x \rangle\} \ for \ all \ x \in X.$$

Next we examine the relation between $\partial \varphi(\cdot)$ and $\partial \varphi^*(\cdot)$.

Theorem 6.107 *If $\varphi : X \to \overline{\mathbb{R}}$ is convex, then: (a) $x^* \in \partial \varphi(x) \Rightarrow x \in \partial \varphi^*(x^*)$. (b) If $\varphi \in \Gamma_0(x)$, then $x^* \in \partial \varphi(x) \Leftrightarrow x \in \partial \varphi^*(x^*)$.*

Proof (a) For any $u^* \in X^*$, we have

$$\langle u^*, x \rangle - \varphi(x) \leq \varphi^*(u^*) \ \text{(see (6.66))},$$

$$\Rightarrow \langle u^* - x^*, x \rangle + \varphi^*(x^*) \leq \varphi^*(u^*) \ \text{(see Theorem 6.105)},$$

$$\Rightarrow x \in \partial \varphi^*(x^*).$$

(b) On account of (a), we need to show only the $\Leftarrow$ part. Since $\varphi \in \Gamma_0(x)$, Theorem 6.104 implies that $\varphi = \varphi^{**}$. From (a), we have

$$x \in \partial \varphi^*(x^*) \Rightarrow x^* \in \partial \varphi^{**}(x) = \partial \varphi(x).$$

$\square$

Proposition 6.108 *If $\varphi : X \to \overline{\mathbb{R}}$ is convex and $\partial \varphi(x) \neq 0$, then $\varphi(x) = \varphi^{**}(x)$.*

Proof Let $x^* \in \partial \varphi(x)$. Then from Theorem 6.105, we have

$$\varphi(x) + \varphi^*(x^*) = \langle x^*, x \rangle,$$

$$\Rightarrow \varphi(x) \leq \varphi^{**}(x),$$

$$\Rightarrow \varphi(x) = \varphi^{**}(x) \ (see \ (6.68)).$$

$\square$

Proposition 6.109 *If $\varphi : X \to \overline{\mathbb{R}}$ is convex and $\varphi(x) = \varphi^{**}(x)$, then $\partial\varphi(x) = \partial\varphi^{**}(x)$.*

Proof Let $x^* \in \partial\varphi(x)$. Then using Theorem 6.105 and Proposition 6.108, we have

$$\langle x^*, x \rangle = \varphi(x) + \varphi^*(x^*) = \varphi^{**}(x) + \varphi^{***}(x);$$

thus $x^* \in \partial\varphi^{**}(x)$ and

$$\partial\varphi(x) \subseteq \partial\varphi^{**}(x).$$

Now let $x^* \in \partial\varphi^{**}(x)$. As above, we have

$$\langle x^*, x \rangle = \varphi^{**}(x) + \varphi^{***}(x^*) = \varphi(x) + \varphi^*(x^*),$$

hence $x^* \in \partial\varphi(x)$ and

$$\partial\varphi^{**}(x) \subseteq \partial\varphi(x). \tag{6.70}$$

Now from (6.69) and (6.70), we conclude that $\partial\varphi(x) = \partial\varphi^{**}(x)$. $\qquad\square$

Proposition 6.110 *If $\varphi : X \to \overline{\mathbb{R}}$ is convex, then φ is k-Lipschitz $\Longleftrightarrow$ dom $\varphi^* \subseteq k\overline{B}_1^*$.*

Proof $\Rightarrow$ Let $u^* \in X^*$, and assume that $\|u^*\|_* > k$. Then

$$\sup[\langle u^*, x \rangle - \varphi(x) : x \in X] = +\infty$$

and so $u^* \notin \text{dom } \varphi^*$. $\qquad\square$

$\Leftarrow$ Arguing by contradiction, suppose that $\varphi(\cdot)$ is not k-Lipschitz. Then we can find $x, u \in X$ such that $\varphi(x) - \varphi(u) > \vartheta \|x - u\|$ with $\vartheta > k$; hence,

$$\varphi(x) > \varphi(u) + \vartheta \|x - u\|.$$

By Theorem 3.71 (Separation Theorem), we can find $x^* \in X^*$ such that

$$\langle x^*, y - x \rangle \leqslant \varphi(y) - [\varphi(u) + \vartheta \|x - u\|] \quad \text{for all } y \in X;$$

hence, $x^* \in \text{dom } \varphi^*$ and $\|x^*\|_* \geqslant \vartheta$, a contradiction.

6.4 Infimal Convolution–Regularization–Coercivity

In this section, we introduce the operation of infimal convolution of two functions, and we use it to determine certain regular approximations of the original function. Then we examine the properties of coercive functions. As always, all $\overline{\mathbb{R}}$-valued functions are not identically $+\infty$.

Definition 6.111 Let $\varphi, \psi : X \to \overline{\mathbb{R}} = \mathbb{R} \cup \{+\infty\}$. The "infimal convolution" of φ, ψ, denoted by $\varphi \square \psi$, is defined by

$$(\varphi \square \psi)(x) = \inf[\varphi(u) + \psi(x - u) : u \in X].$$

It is clear the similarity with the classical integral convolution. Hence, the name "infimal convolution."

The next theorem shows that the operation of infimal convolution is dual to the operation of addition. For this, we will need the next proposition.

Proposition 6.112 *If $\varphi, \psi : X \to \overline{\mathbb{R}} = \mathbb{R} \cup \{+\infty\}$ are convex functions such that there exists $x_0 \in \mathrm{dom}\,\psi$ at which $\varphi(\cdot)$ is continuous, then*

$$\inf[\varphi(x) + \psi(x) : x \in X] = \max[-\varphi^*(x^*) - \psi^*(x^*) : x^* \in X^*].$$

Proof Note that

$$-\varphi^*(x^*) - \psi^*(-x^*) \leqslant \varphi(x) + \psi(x) \text{ for all } x \in X, \text{ all } x^* \in X^*,$$

$$m^* = \sup[-\varphi^*(x^*) - \psi^*(-x^*) : x^* \in X^*] \leqslant \inf[\varphi(x) + \psi(x) : x \in X] = m.$$
(6.71)

By hypothesis, $x_0 \in \mathrm{dom}\,\varphi \cap \mathrm{dom}\,\psi$ and so $-\infty \leqslant m < +\infty$. Hence, if $m = -\infty$, then the result is proved. So, we assume that $m \in \mathbb{R}$. Since $\varphi(\cdot)$ is continuous at x_0, we have that $D = \mathrm{int}\,\mathrm{epi}\,\varphi \neq \emptyset$ (see Theorem 6.72). Also let $C = \{(u, \vartheta) \in X \times \mathbb{R} : \vartheta \leqslant m - \psi(u)\} \neq \emptyset$. Both these sets D, C are convex and $D \cap C = \emptyset$. Since D is open, we can find $(\widehat{x}^*, \mu) \in X^* \times \mathbb{R}, (\widehat{x}^*, \mu) \neq (0,0)$, and $\eta \in \mathbb{R}$ such that

$$\langle \widehat{x}^*, x \rangle + \mu\lambda \leqslant \eta \leqslant \langle \widehat{x}^*, u \rangle + \mu\vartheta \text{ for all } (x, \lambda) \in D, (u, \vartheta) \in C. \quad (6.72)$$

Since λ can grow to $+\infty$, we see that $\mu \leqslant 0$. If $\mu = 0$, then

$$\langle \widehat{x}^*, x \rangle \leqslant \langle \widehat{x}^*, x_0 \rangle \text{ for all } x \in X, \Rightarrow \widehat{x}^* = 0, \text{ a contradiction.}$$

Hence, $\mu < 0$, and we can always take $\mu = -1$. Then from (6.72), we have

$$-\langle \widehat{x}^*, u \rangle + \vartheta \leqslant -\eta \text{ for all } (u, \vartheta) \in C,$$

$$\Rightarrow -\langle \widehat{x}^*, u \rangle + m - \psi(u) \leqslant -\eta \text{ for all } u \in X. \tag{6.73}$$

Also again from (6.72), we have

$$\langle \widehat{x}^*, x \rangle - \lambda \leqslant \eta \text{ for all } (x, \lambda) \in \overline{D} = \text{epi } \varphi. \tag{6.74}$$

From (6.73), we have

$$\psi^*(-\widehat{x}^*) \leqslant -\eta - m, \tag{6.75}$$

while from (6.74), we have

$$\varphi^*(\widehat{x}^*) \leqslant \eta. \tag{6.76}$$

Using (6.76) in (6.75), we obtain

$$m \leqslant -\varphi^*(\widehat{x}^*) - \psi^*(-\widehat{x}^*) \leqslant m^*, \Rightarrow m = m^* \ (see\ (6.71)). \qquad \square$$

Remark 6.113 This proposition is a special case of the so-called Fenchel Duality Theory (see Ekeland–Temam [102], Section III, 1).

Using this proposition, we can show that the operation of infimal convolution is dual to the operation of addition.

Theorem 6.114 *If* $\varphi, \psi : X \to \overline{\mathbb{R}} = \mathbb{R} \cup \{+\infty\}$ *are convex functions and there exists* $x_0 \in \text{dom } \psi$ *such that* $\varphi(\cdot)$ *is continuous at* x_0, *then* $(\varphi + \psi)^* = \varphi^* \square \psi^*$.

Proof For every $x^* \in X^*$, we have

$$-(\varphi + \psi)^*(x^*) = \inf[h(x) - \psi(x) : x \in X],$$

where h(x)=$\varphi(x) - \langle x^*, x \rangle$. Then we apply Proposition 6.112 and have

$$-(\varphi + \psi)^*(x^*) = -\min[h^*(u^*) + \psi^*(-u^*) : u^* \in X^*]$$
$$= -\min[\varphi^*(x^* - u^*) + \psi^*(u^*) : u^* \in X^*]$$
$$(see\ Proposition\ 6.101)$$
$$= -(\varphi^* \square \psi^*)(x^*). \qquad \square$$

Directly from the definitions, we have:

Proposition 6.115 *If* $\varphi, \psi : X \to \overline{\mathbb{R}} = \mathbb{R} \cup \{+\infty\}$ *are convex functions, then* $(\varphi \square \psi)^* = \varphi^* + \psi^*$.

Suppose that $\psi_\lambda(x) = \frac{1}{2\lambda} \|x\|^2$ for all $x \in X, \lambda > 0$. Then we define

$$\varphi_\lambda(x) = (\varphi \square \psi_\lambda)(x) = \inf[\varphi(u) + \frac{1}{2\lambda} \|u - x\|^2 : u \in X].\tag{6.77}$$

The function $\varphi_\lambda(\cdot)$ is known as the "Moreau–Yosida regularization" of $\varphi(\cdot)$. The next theorem justifies this name.

Theorem 6.116 *If X is reflexive and X and X^* are both strictly convex and $\varphi \in \Gamma_0(X)$, then $\varphi_\lambda(\cdot)$ is convex, continuous, Gateaux differentiable, and $\varphi_\lambda(x) \to \varphi(x)$ for all $x \in X$ as $\lambda \to 0^+$; moreover, if X is a Hilbert space, then $\varphi_\lambda(\cdot)$ is Frechet differentiable.*

Proof Fix $x \in X$ and consider the convex function $g_x : X \to \overline{\mathbb{R}} = \mathbb{R} \cup \{+\infty\}$ defined by

$$g_x(u) = \varphi(x) + \frac{1}{2\lambda} \|x - u\|^2.$$

Then using Proposition 6.83 and (6.62), we have

$$\partial g_x(u) = \partial \varphi(u) + \frac{1}{\lambda} \mathcal{F}(x - u).$$

So, $x_\lambda \in X$ is a minimizer of $g_x(\cdot)$ if and only if it solves the operator inclusion

$$0 \in \partial \varphi(x) + \frac{1}{\lambda} \mathcal{F}(x - u).\tag{6.78}$$

Note that $g_x(u) \to +\infty$ and then the reflexivity of X and Proposition 1.180 imply that $g_x(\cdot)$ has a minimizer and so (6.78) admits a solution $x_\lambda \in X$, which on account of the strict convexity of X is unique. Consider the map $J_\lambda : X \to X$ defined by

$$J_\lambda(x) = x_\lambda.$$

Hence, from (6.78), we have

$$\varphi_\lambda(x) = \varphi_\lambda(J_\lambda(x)) + \frac{1}{2\lambda} \|x - J_\lambda(x)\|^2 \quad \text{for all } x \in X.\tag{6.79}$$

Also since $\mathcal{F}$ is single-valued, we have

$$x_\lambda = \mathcal{F}^{-1}(\lambda x_\lambda^*) + x \quad \text{with } x_\lambda^* \in \partial \varphi(x_\lambda).$$

$$\Rightarrow x_\lambda = J_\lambda(x) \to x \quad \text{as } \lambda \to 0^+.\tag{6.80}$$

From (6.79) and (6.80), it follows that

$$\varphi(x) \leqslant \liminf_{\lambda \to 0^+} \varphi(J_\lambda(x)) \leqslant \liminf_{\lambda \to 0^+} \varphi_\lambda(x) \leqslant \varphi(x),$$

$$\Rightarrow \varphi_\lambda(x) \to \varphi(x) \ as \ \lambda \to 0^+.$$

Clearly, $\varphi_\lambda \in \Gamma_0(X)$, and since $\mathrm{dom}\,\varphi_\lambda = X$, from Theorem 6.72, we conclude that φ_λ is continuous. It remains to show that $\varphi_\lambda(\cdot)$ is Gateaux differentiable. Let $(\partial\varphi)_\lambda(x) = \frac{1}{\lambda}\mathcal{F}(x - x_\lambda)$, for all $x \in X$. Since X^* is strictly convex, $(\partial\varphi)_\lambda(\cdot)$ is single-valued, it is monotone and bounded (maps bounded sets to bounded sets), and it is demicontinuous. Using (6.79), we have

$$\varphi_\lambda(u) - \varphi_\lambda(x) \leqslant \langle J_\lambda(u) - J_\lambda(x), (\partial\varphi)_\lambda(u) \rangle$$

$$+ \frac{1}{2\lambda}[\|u - J_\lambda(u)\|^2 - \|x - J_\lambda(x)\|^2]$$

$$\leqslant \langle (\partial\varphi)(u), u - x \rangle;$$

hence, for all $x, u \in X, \lambda > 0$, we have

$$\varphi_\lambda(u) - \varphi_\lambda(x) - \langle (\partial\varphi)_\lambda(x), u - x \rangle \leqslant \langle (\partial\varphi)_\lambda(u) - (\partial\varphi)_\lambda(x), u - x \rangle$$

$$\Rightarrow \lim_{t \to 0^+} \frac{\varphi_\lambda(x + th) - \varphi_\lambda(x)}{t} \leqslant \langle (\partial\varphi)_\lambda(x), h \rangle \ \text{for all} \ h \in X$$

(since $(\partial\varphi)_\lambda(\cdot)$ is demicontinuous)

$$\Rightarrow (\varphi_\lambda)'_+(x; h) \leqslant \langle (\partial\varphi)_\lambda(x), h \rangle \ \text{for all} \ h \in X,$$

$$\Rightarrow (\varphi_\lambda)'_+(x; h) = \langle (\partial\varphi)(x), h \rangle \ \text{for all} \ h \in X.$$

Hence, $\varphi_\lambda(\cdot)$ is Gateaux differentiable and $(\varphi_\lambda)'_G(x)(\cdot) = (\partial\varphi)_\lambda(x)$. Finally, if X is a Hilbert space, then by Theorem 3.181, we can have $X^* = X$ and so $\mathcal{F} = id_X$. Then $(\partial\varphi)_\lambda(\cdot)$ is $\frac{2}{\lambda}$-Lipschitz and

$$|\varphi_\lambda(u) - \varphi_\lambda(x) - \langle (\partial\varphi)_\lambda(x), u - x \rangle| \leqslant \frac{2}{\lambda}\|u - x\|^2 \ for \ all \ x, u \in X,$$

$$\Rightarrow \varphi_\lambda(\cdot) \ \text{is Frechet differentiable.} \qquad \square$$

Remark 6.117 In Sect. 7.2, we will revisit the operators $J_\lambda(\cdot)$ and $(\partial\varphi)_\lambda(\cdot)$ in the more general framework of maximal monotone operators.

Lemma 6.118 *If* $\varphi : X \to \mathbb{R}$ *is convex, bounded on bounded sets,* $\varphi_n \leqslant \varphi$ *for all* $n \in \mathbb{N}$, *and* $\varphi_n^* \to \varphi^*$ *uniformly on bounded subsets of* $\mathrm{dom}\,\varphi^*$, *then* $\varphi_n \to \varphi$ *uniformly on bounded subsets of* X.

Proof Let $B \subseteq X$ be bounded and $\varepsilon > 0$. Since by hypothesis $\varphi(\cdot)$ is bounded on bounded sets, then $\partial\varphi : X \to 2^{X^*}$ maps bounded sets to bounded sets (see

Problem 6.12). So, we can find $n_0 \in \mathbb{N}$ such that

$$\varphi_n^*(u^*) \leqslant \varphi^*(u^*) + \varepsilon \text{ for all } u^* \in \partial\varphi(B), \text{ all } n \geqslant n_0.$$

Then for $u \in B$, $u^* \in \partial\varphi(x)$, and $n \geqslant n_0$, we have

$$\varphi^*(u^*) = \langle u^*, u \rangle - \varphi(u) \text{ (see Theorem 6.105)}$$
$$\geqslant \varphi_n^*(u^*) - \varepsilon \text{ (see (6.80))}$$
$$\geqslant \langle u^*, u \rangle - \varphi_n(u) - \varepsilon \text{ (see (6.66))},$$

$$\Rightarrow \varphi(u) - \varepsilon \leqslant \varphi_n(u) \leqslant \varphi(u) \text{ for all } u \in B, \ n \geqslant n_0,$$

$$\Rightarrow \varphi_n \to \varphi \text{ uniformly on bounded subsets of } X. \qquad \square$$

Using this lemma in the case of the Moreau–Yosida regularizations $\{\varphi_{\lambda_n}\}_{n \in \mathbb{N}}$ with $\lambda_n = \frac{1}{n}$, $n \in \mathbb{N}$, we obtain:

Theorem 6.119 *If $\varphi : X \to \mathbb{R}$ is convex and bounded on bounded sets, then $\varphi_{\lambda_n} \to \varphi$ as $n \to \infty$ uniformly on bounded sets in X.*

Proof From Proposition 6.115, we have

$$\varphi_{\lambda_n}^* = \varphi^* + \frac{1}{2n} \|\cdot\|_*^2 \text{ for all } n \in \mathbb{N}.$$

So, we can apply Lemma 6.118 and conclude that $\varphi_n \to \varphi$ uniformly on bounded subsets of X. $\qquad \square$

Let us recall the definition of a coercive function.

Definition 6.120 Let $\varphi : X \to \overline{\mathbb{R}} = \mathbb{R} \cup \{+\infty\}$.

(a) We say that $\varphi(\cdot)$ is "coercive," if $\lim_{\|x\| \to \infty} \varphi(x) = +\infty$.
(b) We say that $\varphi(\cdot)$ is "strongly coercive," if

$$\lim_{\|x\| \to \infty} \frac{\varphi(x)}{\|x\|} = +\infty.$$

The next proposition provides alternative equivalent characterizations of coercivity.

Proposition 6.121 *If $\varphi : X \to \overline{\mathbb{R}} = \mathbb{R} \cup \{+\infty\}$ is convex and lower semicontinuous at a point in its effective domain $\operatorname{dom}\varphi$, then the following statements are equivalent:*

(a) *φ is coercive.*
(b) *We can find $\vartheta > 0$ and $\eta \in \mathbb{R}$ such that*

$$\varphi(x) \geqslant \vartheta \, \|\cdot\| + \eta \ \text{for all } x \in X.$$

(c) $0 < \liminf_{\|x\| \to \infty} \frac{\varphi(x)}{\|x\|}$.

(d) *The sublevel sets* $L_\lambda = \{u \in X : \varphi(u) \leqslant \lambda\}$, $\lambda \in \mathbb{R}$, *are bounded.*

Proof (a)$\Rightarrow$(b) First assume that $\varphi(0)=0$ (i.e., $0 \in \mathrm{dom}\,\varphi$). By hypothesis, we can find $M > 0$ such that $\|x\| \geqslant M \Rightarrow \varphi(x) \geqslant 1$. For $\|x\| \geqslant M$, using the convexity of $\varphi(\cdot)$, we have

$$\varphi\Big(\frac{M}{\|x\|}x\Big) \leqslant \frac{\|x\| - M}{\|x\|}\varphi(0) + \frac{M}{\|x\|}\varphi\Big(\frac{M}{\|x\|}x\Big) = \frac{M}{\|x\|}\varphi(x),$$

$$\Rightarrow 1 \leqslant \frac{M}{\|x\|}\varphi(x) \ \text{(since } \Big\|\frac{M}{\|x\|}x\Big\| = M),$$

$$\Rightarrow \frac{\|x\|}{M} \leqslant \varphi(x) \ \text{if } \|x\| \geqslant M.$$

The lower semicontinuity at some point in $\mathrm{dom}\,\varphi$ implies

$$-\beta \leqslant \varphi(x) \ \text{ for all } \ x \in M B_1$$

$$\Rightarrow \vartheta \, \|x\| + \eta \leqslant \varphi(x) \ \text{ for all } \ x \in X \ with \ \vartheta = \frac{1}{M}, \ \eta = -\beta.$$

In the general case, let $x_0 \in \mathrm{dom}\,\varphi$ and set $\varphi(x) = \varphi(x + x_0) - \varphi(x_0)$. Then $\widehat{\varphi}(0) = 0$, and so we have the result for $\widehat{\varphi}(\cdot)$. By appropriately modifying $\eta \in \mathbb{R}$, we have the result also for $\varphi(\cdot)$.

(b)$\Rightarrow$(c) Clear.

(c)$\Rightarrow$(d) By hypothesis, there exist M, $\vartheta > 0$ such that

$$\varphi(x) \geqslant \vartheta \, \|x\| \ \text{for all } \|x\| \geqslant M. \tag{6.81}$$

Since $\varphi(\cdot)$ is lower semicontinuous at some point in $\mathrm{dom}\,\varphi$, we have $\varphi(x) \geqslant -M$ for some $M > 0$, all $x \in M\overline{B}_1$. This fact combined with (6.81) implies that every sublevel set L_λ is bounded.

(d)$\Rightarrow$(a) Clear. $\square$

Proposition 6.122 *If* $\varphi : X \to \overline{\mathbb{R}} = \mathbb{R} \cup \{+\infty\}$, *then* $\vartheta \, \|x\| + \eta \leqslant \varphi(x)$ *for all* $x \in X$, *some* $\vartheta > 0$, $\eta \in \mathbb{R}$ *if and only if*

$$\varphi^*(x^*) \leqslant -\eta \ \text{for all } x^* \in \vartheta \, \overline{B^*}_1.$$

Proof Just note that $\varphi \geqslant \vartheta \, \|\cdot\| + \eta$ iff $\varphi^* \leqslant 1_{\vartheta \overline{B^*}_1} - \eta$ (see Example 6.100(c)). $\square$

Combining Propositions 6.121, 6.122 with Theorem 6.72, we have the following result, known in the literature as the "Moreau–Rockafellar Theorem."

Theorem 6.123 *If $\varphi : X \to \overline{\mathbb{R}} = \mathbb{R} \cup \{+\infty\}$ is convex and lower semicontinuous at some point in* $\operatorname{dom} \varphi$, *then $\varphi(\cdot)$ is coercive iff $\varphi^*(\cdot)$ is continuous at 0.*

Another result in the same vein is the following theorem.

Theorem 6.124 *If $\varphi \in \Gamma_0(X)$, then $\varphi(\cdot)$ is continuous at 0 iff the sublevels sets of $\varphi^*(\cdot)$ are w^*-compact.*

Proof From Theorem 6.104, we have that $\varphi(\cdot)$ is continuous at 0 iff $\varphi^{**}(\cdot)$ is and this by Theorem 6.123 is equivalent to saying that $\varphi^*(\cdot)$ is coercive. The coercivity of $\varphi^*(\cdot)$ is equivalent to the boundedness of the sublevel sets of $\varphi^*(\cdot)$ (see Proposition 6.121). But these sets are w^*-closed (see Proposition 6.103); hence, they are w^*-compact (Alaoglu's Theorem, see Theorem 3.108). $\qquad\square$

Our last result concerns strongly coercive functions.

Theorem 6.125 *If $\varphi \in \Gamma_0(X)$, then:*

(a) φ *is strongly coercive iff φ^* is bounded on bounded sets.*
(b) φ *is bounded on bounded sets iff φ^* is strongly coercive.*

Proof

(a) $\Rightarrow$ By hypothesis, given any $\vartheta > 0$, we can find $\eta \in \mathbb{R}$, such that

$$\varphi(x) \geqslant \vartheta \, \|x\| + \eta \quad \text{for all } x \in X;$$

hence, φ^* is bounded on bounded sets (see Proposition 6.122).
$\Leftarrow$ Let $\vartheta > 0$. By hypothesis $\varphi^*(x^*) \leqslant \beta$ for some $\beta \in \mathbb{R}$, all $\|x^*\|_* \leqslant \vartheta$. Then $\varphi(x) \geqslant \vartheta \, \|x\| - \beta$ for all $x \in X$ (see Proposition 6.122). Hence,

$$\liminf_{\|x\| \to \infty} \frac{\varphi(x)}{\|x\|} \geqslant \vartheta.$$

Since $\vartheta > 0$ is arbitrary, we conclude that $\varphi(\cdot)$ is strongly coercive.
(b) From (a), we know that φ^* is strongly coercive iff φ^{**} is bounded on bounded sets. But from Theorem 6.104, we know that $\varphi^{**} = \varphi$. $\qquad\square$

6.5 Locally Lipschitz Functions

In this section, we discuss Lipschitz and locally Lipschitz functions. Such functions provide the right class of functions to extend the subdifferential theory of convex analysis (see Theorem 6.56). Also Lipschitz functions are a basic tool in geometric measure theory.

Definition 6.126 Let (X, d_X) and (Y, d_Y) be metric spaces:

(a) We say that $\varphi : X \to Y$ is "Lipschitz" if for some $k > 0$ we have

$$d_Y(\varphi(u), \varphi(x)) \leqslant k d_X(u, x) \text{ for all } u, x \in X. \tag{6.82}$$

We denote the space of Lipschitz functions by $\mathrm{Lip}(X, Y)$. The infimum of all k's satisfying (82) is called the "Lipschitz constant" of φ and is denoted by $\mathrm{Lip}(\varphi)$. Also, if $k \geq \mathrm{Lip}(\varphi)$, we often say that φ is "k-Lipschitz."

(b) We say that $\varphi : A \to Y$ is "locally Lipschitz," if every point in A has a neighborhood on which φ is Lipschitz. We denote the space of locally Lipschitz functions by $\mathrm{Lip}_{loc}(X, Y)$.

Remark 6.127 The Lipschitz condition is a purely metric condition and a typical example of a Lipschitz function and is the distance function. So, let (X, d_X) be a metric space and $C \subseteq X$ nonempty. For every $x \in X$, we define

$$d(x, C) = \inf[d_X(x, c) : c \in C].$$

Then we can easily see that $d(\cdot, C)$ is 1-Lipschitz. Note that 1-Lipschitz maps are also known as "nonexpansive" maps.

The following propositions are straightforward consequences of Definition 6.126.

Proposition 6.128 *If X is a metric space and $\varphi_\alpha : X \to Y$ ($\alpha \in J$) are k-Lipschitz functions, then so is $x \to \inf_{x \in J} \varphi_\alpha(x)$ and also $x \to \sup_{\alpha \in J} \varphi_\alpha(x)$ if it is finite at a point.*

Proposition 6.129 *If X, Y, V are metric spaces, $\varphi \in \mathrm{Lip}(X, Y)$ and $\psi \in \mathrm{Lip}(Y, V)$. Then, $\psi \circ \varphi \in \mathrm{Lip}(X, V)$ and $\mathrm{Lip}(\psi \circ \varphi) \leqslant \mathrm{Lip}(\psi) \mathrm{Lip}(\varphi)$.*

$\mathbb{R}$-valued Lipschitz functions have nice extension properties. The next theorem is known in the literature as "McShane's Extension Theorem."

Theorem 6.130 *If (X, d) is a metric space, $A \subseteq X$ is nonempty, and $\varphi \in \mathrm{Lip}(A, \mathbb{R})$, then there exists $\widehat{\varphi} \in \mathrm{Lip}(X, \mathbb{R})$ such that $\widehat{\varphi}|_A = \varphi$ and $\mathrm{Lip}(\widehat{\varphi}) = \mathrm{Lip}(\varphi)$.*

Proof Let $k \geqslant \mathrm{Lip}(\varphi)$, and let

$$\widehat{\varphi}(x) = \inf[\varphi(u) + k d(u, x); u \in X]. \tag{6.83}$$

Note that each of the functions $\widehat{\varphi}_x(u) = \varphi(u) + k d(u, x)$ is k-Lipschitz and so by Proposition 6.128 $\widehat{\varphi}(\cdot)$ is k-Lipschitz too. Moreover, $\widehat{\varphi}|_A = \varphi$. □

Remark 6.131 The function $\widehat{\varphi}(\cdot)$ defined by (6.83) is the biggest possible Lipschitz extension of $\varphi(\cdot)$. The smallest Lipschitz extension is given by

$$\widetilde{\varphi}(x) = \sup[\varphi(u) - kd(u, x); u \in A].$$

On account of this theorem, we see that when dealing with $\mathbb{R}$-valued Lipschitz functions, we can always assume that they are defined on the whole metric space X. Finally, we mention that the result is also true more generally for Hölder continuous functions of order $0 < \alpha \leqslant 1$. Recall that if (X, d_X), (Y, d_Y) are metric spaces, then a function $\varphi : X \to Y$ "Hölder continuous of order $\alpha \in (0,1]$," if

$$d_Y(\varphi(u), \varphi(x)) \leqslant kd_X(u, x)^\alpha \text{ for all } u, x \in X, some \ k > 0.$$

Evidently, a Lipschitz continuous function is Hölder continuous of order 1. Theorem 6.130 can be extended to vector valued function. The result is known as "Kriszbraun's Extension Theorem," and its proof can be found in Federer [108] (Theorem 2.10.43, p.201).

Theorem 6.132 *If $A \subseteq \mathbb{R}^N$ and $\varphi \in \mathrm{Lip}(A, \mathbb{R}^m)$, then there exists $\widehat{\varphi} \in \mathrm{Lip}(\mathbb{R}^N, \mathbb{R}^m)$ such that $\widehat{\varphi}|_A = \varphi$ and $\mathrm{Lip}(\widehat{\varphi}) = \mathrm{Lip}(\varphi)$.*

Remark 6.133 This theorem is not true for general finite dimensional Banach spaces (see Federer [108], Example 2.10.44, p. 202).

Also from the theory of Sobolev spaces (see Remark 4.71), we have the following result.

Proposition 6.134 *If $\Omega \subseteq \mathbb{R}^N$ is a bounded domain with a Lipschitz boundary, then $W^{1,\infty}(\Omega) = \mathrm{Lip}(\Omega, \mathbb{R}) \cap L^\infty(\Omega)$.*

$\mathbb{R}^m$-valued locally Lipschitz functions are differentiable almost everywhere. This is the famous "Rademacher's Theorem."

Theorem 6.135 *If $\varphi \in \mathrm{Lip}_{loc}(\mathbb{R}^N, \mathbb{R}^m)$, then $\varphi'_F(\cdot)$ exists at almost all $x \in \mathbb{R}^N$.*

Proof We may assume that $m = 1$ and $\varphi \in \mathrm{Lip}(\mathbb{R}^N, \mathbb{R}^m)$ since differentiability is a local property. Let $x, h \in \mathbb{R}^N$, with $|h| = 1$. We set $\vartheta(t) = \varphi(x + th)$ for all $t \in \mathbb{R}$. Since φ is Lipschitz continuous, then so is $\vartheta(\cdot)$, and so it is differentiable for almost all $t \in \mathbb{R}$ (clearly a Lipschitz function is absolutely continuous, see Definition 4.120). Let $N_h = \{x \in \mathbb{R}^N : \varphi'_G(x; h) \text{ fails to exist}\}$. We have

$$N_h = \left\{ x \in \mathbb{R}^N : \liminf_{t \to 0} \frac{\varphi(x + th) - \varphi(x)}{t} < \limsup_{t \to 0} \frac{\varphi(x + th) - \varphi(x)}{t} \right\}.$$

$$\Rightarrow N_h \in B(\mathbb{R}^N).$$

Note that for every line $L = \{x = x_0 + th : x_0 \in \mathbb{R}^N, t \in \mathbb{R}\}$, we have that $N_h \cap L$ has zero 1-Hausdorff measure. Hence, $\lambda^N(N_h) = 0$ (with $\lambda^N(\cdot)$ denoting the Lebesgue measure on $\mathbb{R}^N$). For L a line in the direction of h, we have for all $\vartheta \in C_c^\infty(\mathbb{R}^N)$,

$$\int_L \varphi'(x; h)\vartheta \, dx = -\int_L \varphi(x)(\vartheta'(x), h)_{\mathbb{R}^N} dx,$$

$$\Rightarrow \int_{\mathbb{R}^N} \varphi'(x; h)\vartheta \, dx = -\sum_{k=1}^N \int_{\mathbb{R}^N} \varphi(x)\frac{\partial \vartheta}{\partial x_k}(x)h_k dx$$

$$= \sum_{k=1}^N \int_{\mathbb{R}^N} \frac{\partial \varphi}{\partial x_k}(x)\vartheta(x)h_k dx$$

$$= \int_{\mathbb{R}^N} (\varphi'(x), h)_{\mathbb{R}^N}\vartheta(x)dx,$$

$$\Rightarrow \varphi'(x; h) = (\varphi'(x), h)_{\mathbb{R}^N} \ for \ a.a \ x \in \mathbb{R}^N.$$

From the separability of $\mathbb{R}^N$, we can find

$$\{\widehat{h}_n\}_{n\in\mathbb{N}} \subseteq \partial B_1^N = \left\{h \in \mathbb{R}^N : |h| = 1\right\} \ \text{dense}, \ D \subseteq \mathbb{R}^N \ \text{with} \ \lambda^N(D) = 0$$

such that

$$\varphi'(x; \widehat{h}_n) = (\varphi'(x), \widehat{h}_n)_{\mathbb{R}^N} \ \text{for all} \ x \in \mathbb{R}^N \setminus D, n \in \mathbb{N}.$$

Let $x \in \mathbb{R}^N \setminus D, h \in \partial B_1^N, t > 0$. We set

$$d(x, h, t) = \frac{\varphi(x + th) - \varphi(x)}{t} - (\varphi'(x), h)_{\mathbb{R}^N}.$$

Using the fact that $\varphi \in \mathrm{Lip}(\mathbb{R}^N, \mathbb{R})$, we see that

$$|d(x, h, t) - d(x, h', t)| \leqslant k[N + 1]|h - h'| \ \text{with} \ k = \mathrm{Lip}(\varphi).$$

We can find $n_0 \in \mathbb{N}$ such that for $h \in \partial B_1^N$ we have

$$|h - \widehat{h}_n| < \frac{\varepsilon}{2^{(N+1)k}} \ \text{for some} \ n \in \{1, \cdots, n_0\},$$

$$\Rightarrow d(x, h_n, t) < \frac{\varepsilon}{2} \ \text{for all} \ 0 < t < \delta, \ n \in \{1, \cdots, n_0\},$$

$$\Rightarrow d(x, h, t) < \frac{\varepsilon}{2} + \frac{\varepsilon}{2} = \varepsilon \ \text{for all} \ h \in \partial B_1^N, \ t \in (0, \delta),$$

$$\Rightarrow \varphi'(x; h) = (\varphi'(x), h)_{\mathbb{R}^N} \ \text{for all} \ x \in \mathbb{R}^N \setminus D, \ h \in \partial B_1^N,$$

$$\Rightarrow \varphi'_F(\cdot) \ \text{exists on} \ \mathbb{R}^N \setminus D \ (\text{see Proposition 6.4}). \qquad \square$$

Remark 6.136 A generalization of Theorem 6.134, due to Stepanov [263], says that a function $\varphi : \mathbb{R}^N \to \mathbb{R}^m$ is almost everywhere differentiable on the set

$$\left\{ x \in \mathbb{R}^N : \limsup_{u \to x} \frac{|\varphi(u) - \varphi(x)|}{|u - x|} < \infty \right\}.$$

The fact that a convex function is continuous in the interior of its effective domain provides a good starting point for an extension of the subdifferential theory.

So, let X be a Banach space, and let $\varphi : X \to \mathbb{R}$ be a locally Lipschitz function (i.e., for every $x \in X$, we can find $U \in \mathcal{N}(x)$ such that $|\varphi(v) - \varphi(u)| \leqslant k_U \|v - u\|$ for all $v, u \in U$, see Definition 6.126(b)). For such a function, the directional derivative need not exist. However, the locally Lipschitz property leads to the following substitute of it.

Definition 6.137 Let $\varphi \in \text{Lip}_{loc}(X, \mathbb{R})$. The "generalized directional derivative" of φ at $x \in X$ in the direction $h \in X$ is defined by

$$\varphi^0(x; h) = \limsup_{\substack{u \to x \\ t \to 0^+}} \frac{\varphi(u + th) - \varphi(u)}{t} = \inf_{\varepsilon, \delta > 0} \sup_{\substack{\|u - x\| \leqslant \varepsilon \\ 0 < t \leqslant \delta}} \frac{\varphi(u + th) - \varphi(u)}{t}.$$

This is a fruitful notion with good properties.

Proposition 6.138 *If $\varphi \in \text{Lip}_{loc}(X, \mathbb{R})$, then:*

(a) *For every $x \in X$, the function $h \to \varphi^0(x; h)$ is sublinear and Lipschitz.*
(b) *The function $(x,h) \to \varphi^0(x; h)$ is upper semicontinuous.*
(c) *$\varphi^0(x; -h) = (-\varphi)^0(x; h)$ for all $x, h \in X$.*

Proof (a) It is clear that from Definition 6.137 that $\varphi^0(x; \cdot)$ is positively homogeneous. Let $h, h' \in X$. We have

$$\varphi^0(x; h + h') = \limsup_{\substack{u \to x \\ t \to 0^+}} \frac{\varphi(u + t(h + h')) - \varphi(u)}{t}$$

$$= \limsup_{\substack{u \to x \\ t \to 0^+}} \frac{\varphi(u + t(h + h')) - \varphi(u + th') + \varphi(u + th') - \varphi(u)}{t}$$

$$\leqslant \varphi^0(x; h) + \varphi^0(x; h'),$$

$\Rightarrow \varphi^0(x; \cdot)$ is sublinear. Also note that for $t > 0$ small, we have $\frac{\varphi(u+th)-\varphi(u)}{t} \leqslant k\|h\|$, thus $\varphi^0(x; h) \leqslant k\|h\|$, and therefore

$$|\varphi^0(x; h)| \leq k\|h\|, \text{ for all } h \in X. \tag{6.84}$$

Hence, $\varphi^0(x; \cdot)$ is Lipschitz.

(b) Let $(x_n, h_n) \to (x, h)$ in $X \times X$, $v_n \in X$, and $t_n > C$ such that $\|v_n\| + t_n \le \frac{1}{n}$ and

$$\varphi^0(x_n; h_n) \le \frac{\varphi(x_n + v_n + t_0 h_n) - \varphi(x_n + v_n)}{t_n} + \frac{1}{n}$$

$$= \frac{\varphi(x_n + v_n + t_n h_n) - \varphi(x_n + v_n + t_n h)}{t_n}$$

$$+ \frac{\varphi(x_n + v_n + t_n h_n) - \varphi(x_n + v_n)}{t_n} + \frac{1}{n}.$$

Thus, $\limsup_{n \to \infty} \varphi^0(x_n; h_n) \le \varphi^0(x; h)$ since $\varphi \in \mathrm{Lip}_{loc}(X, \mathbb{R})$, and $(x, h) \to \varphi^0(x, h)$ is upper semicontinuous.

(c) We have

$$\varphi^0(x, -h) = \limsup_{u \to x, t \to 0^+} \frac{\varphi(u - th) - \varphi(u)}{t}$$

$$= \limsup_{v \to x, t \to 0^+} \frac{(-\varphi)(v + th) - (-\varphi)(v)}{t} \, (\mathrm{set}\, v = u - th)$$

$$= (-\varphi)^0(x; h) \text{ for all } x, h \in X.$$

$\square$

Using the generalized directional derivative and in particular the fact that $\varphi^0(x; \cdot)$ is sublinear continuous, we can define the "generalized (or Clarke) subdifferential."

Definition 6.139 If $\varphi \in \mathrm{Lip}_{loc}(X, \mathbb{R})$, then the "generalized or Clarke subdifferential" of $\varphi(\cdot)$ at $x \in X$ is defined by

$$\partial \varphi(x) = \left\{ x^* \in X^* : \langle x^*, h \rangle \le \varphi^0(x; h) \text{ for all } h \in X \right\}.$$

Remark 6.140 Proposition 6.90 guarantees that $\partial \varphi(x) \ne \emptyset$. Also Proposition 6.138(a) implies that $\partial \varphi(x)$ is convex, closed, and bounded. Therefore, for every $x \in X$, $\partial \varphi(x) \subseteq X^*$ is w^*-compact and convex. Note that $\varphi^0(x; \cdot) = \sigma(\cdot, \partial \varphi(x))$.

Proposition 6.141 *If* $\varphi \in \mathrm{Lip}_{loc}(X, \mathbb{R})$, $x_n \to x$, $x_n^* \xrightarrow{w^*} x^*$ *in* X^* *and* $x_n^* \in \partial \varphi(x_n)$ *for all* $n \in \mathbb{N}$, *then* $x^* \in \partial \varphi(x)$.

Proof By Definition 6.139, we have

$$\langle x_n^*, h \rangle \le \varphi^0(x_n; h) \text{ for all } h \in X, \text{ all } n \in \mathbb{N},$$

$$\Rightarrow \langle x^*, h \rangle \le \varphi^0(x; h) \text{ for all } h \in X \text{ and thus } \Rightarrow x^* \in \partial \varphi(x).$$

$\square$

Proposition 6.142 *If $\varphi \in \text{Lip}_{loc}(X, \mathbb{R})$, then:*

(a) *If $\varphi(\cdot)$ is Gateaux differentiable at $x \in X$, then $\varphi'(x) \in \partial\varphi(x)$.*
(b) *If $\varphi \in C^1(X, \mathbb{R})$, then $\partial\varphi(x) = \{\varphi'(x)\}$ for all $x \in X$.*

Proof

(a) $\langle \varphi'_G(x), h \rangle \leq \varphi^0(x; h)$ for all $h \in X$ (see Definition 6.137),

$$\Rightarrow \varphi'_G(x) \in \partial\varphi(x).$$

(b) Since $\varphi \in C^1(X, \mathbb{R})$, we have

$$\langle \varphi'(x), h \rangle = \varphi^0(x; h) \text{ for all } h \in X,$$

$$\Rightarrow \partial\varphi(x) = \{\varphi'(x)\} \text{ (see Definition 6.139).}$$

$\square$

Remark 6.143 In contrast to the convex case (see Proposition 6.82), Gateaux differentiability of $\varphi \in \text{Lip}_{loc}(X, \mathbb{R})$, for $x \in X$, does not imply that $\partial\varphi(x)$ is a singleton. Consider the function $\varphi \in \text{Lip}_{loc}(\mathbb{R}, \mathbb{R})$ defined by

$$\varphi(x) = \begin{cases} x^2 \sin \frac{1}{x} & \text{if } x \neq 0 \\ 0 & \text{if } x = 0. \end{cases}$$

Then $\varphi^0(0; h) = |h|$ and $\partial\varphi(0) = [-1, 1]$. Also $\varphi'(0) = 0$.

Definition 6.144 A function $\varphi : X \to \mathbb{R}$ is said to be "strictly differentiable" at x, if for every $x \in X$, we can find $\varphi'_s(x) \in X^*$ such that

$$\langle \varphi'_s(x), h \rangle = \lim_{\substack{x' \to x \\ t \to 0^+}} \frac{\varphi(x'th) - \varphi(x't}{.}$$

Remark 6.145 If $\varphi \in C^1(X, \mathbb{R})$, then $\varphi(\cdot)$ is strictly differentiable at every $x \in X$. If $\varphi(\cdot)$ is strictly differentiable, then $\varphi \in \text{Lip}_{loc}(X, \mathbb{R})$. In Proposition 6.142(b), we can replace the hypothesis that $\varphi \in C^1(X, \mathbb{R})$, by the weaker condition that φ is strictly differentiable.

Proposition 6.146 *If $\varphi \in \text{Lip}_{loc}(X, \mathbb{R})$ is also convex, then the convex subdifferential and the generalized subdifferential coincide.*

Proof Given $\varepsilon > 0$, we can find $\delta > 0$ such that

$$\frac{\varphi(u + th) - \varphi(u)}{t} \leq \frac{\varphi(x + \lambda h) - \varphi(x)}{\lambda} + \varepsilon, \text{ for } |t - \lambda| \leq \delta, \|u - x\| \leq \delta,$$

$$\Rightarrow \sup_{\|u-x\| \leq \delta} \frac{\varphi(u + (\lambda + \delta)h) - \varphi(u)}{\lambda} \leq \frac{\varphi(x + \lambda h) - \varphi(x)}{\lambda} + \varepsilon$$

$$\Rightarrow \varphi^0(x; h) \leq \varphi'_+(x; h) + \varepsilon.$$

Let $\varepsilon \downarrow 0$ to obtain $\varphi^0(x; h) \leq \varphi'_+(x; h)$ for all $h \in X$. Since the opposite inequality is always true, we conclude that

$$\varphi^0(x; \cdot) = \varphi'_+(x; \cdot) \text{ and so } \partial\varphi(x) = \partial_c\varphi(x) \text{ for all } x \in X,$$

with $\partial_c\varphi(x)$ denoting in this case the convex subdifferential. $\qquad\square$

In case $X = \mathbb{R}^N$, we can have a more intuitive characterization of the generalized subdifferential.

Theorem 6.147 *If $\varphi \in \text{Lip}_{loc}(\mathbb{R}^N, \mathbb{R})$ and $N \subseteq \mathbb{R}^N$ is a Lebesgue-null set, then $\partial\varphi(x) = \text{conv}\left\{\lim_{n\to\infty} \varphi'(x_n) : x_n \to x, x_n \notin N, x_n \in D_\varphi\right\}$, where*

$$D_\varphi = \text{set of differentiability points of } \varphi \text{ (recall that } \lambda^N(\mathbb{R}^N \setminus D_\varphi) = 0$$

with λ^N being the Lebesgue measure on $\mathbb{R}^N$, see Theorem 6.135).

Proof Note that $\partial\varphi(\cdot)$ maps bounded sets to bounded sets (see (6.84)). So, $\{\varphi'(x_n)\}_{n\in\mathbb{N}} \subseteq \mathbb{R}^N$ is bounded, and we may assume that $\varphi'(x_n) \to u^*$ in $\mathbb{R}^N$. Proposition 6.141 implies $u^* \in \partial\varphi(x)$. Hence,

$$S(x) = \text{conv}\left\{\lim \varphi'(x_n) : x_n \to x, x_n \notin N, x_n \in D_\varphi\right\} \leq \partial\varphi(x). \tag{6.85}$$

For $h \neq 0$, let $\vartheta_h = \lim_{\substack{x' \to x \\ x' \notin N \cup (\mathbb{R}^N \setminus D_\varphi)}} (\varphi'(x'h)_{\mathbb{R}^N}$. Given $\varepsilon > 0$, we can find $\delta > 0$, such that

$$(\varphi'(x'), h)_{\mathbb{R}^N} \leq \vartheta_h + \varepsilon \text{ for all } \|x'x\| \leq \delta, x' \in N \cup (\mathbb{R}^N \setminus D_\varphi).$$

For $0 < t < \frac{\delta}{2|h|}$ and for λ^N-a.a $x' \in \mathbb{R}^N$, $|x'x| \leq \frac{\delta}{2}$, we have

$$\varphi(x' + th) - \varphi(x'int_0^t(\varphi'(x' + sh), h)_{\mathbb{R}^N} ds \leq t(\vartheta_h + \varepsilon),$$

$$\Rightarrow \varphi^0(x; h) \leq \vartheta_h + \varepsilon,$$

$$\Rightarrow \varphi^0(x; h) \leq \sigma(h, s(x)) \text{ for all } h \in \mathbb{R}^N,$$

$$\Rightarrow \partial\varphi(x) \subseteq S(x),$$

$$\Rightarrow \partial\varphi(x) = S(x) \text{ see (6.85)}.$$

□

Corollary 6.148 *If $\varphi \in \text{Lip}_{loc}(\mathbb{R}^N, \mathbb{R})$, then for all $x \in \mathbb{R}^N$, we have*

$$\varphi^0(x; h) = \lim_{\substack{x' \to x \\ x' \notin N \cup (\mathbb{R}^N \setminus D_\varphi)}} (\varphi'(x'), h)_{\mathbb{R}^N}.$$

Next we present some calculus rules for the generalized subdifferential. First a definition.

Definition 6.149 Let $\varphi \in \text{Lip}_{loc}(\mathbb{R}^N, \mathbb{R})$. $\varphi(\cdot)$ is said to be "regular at x," if:

(a) The directional derivative $\varphi'_+(x; h)$ in the sense of Definition 6.59 exists for all $h \in X$.
(b) $\varphi'_+(x; h) = \varphi^0(x; h)$ for all $h \in X$.

Remark 6.150 Continuous convex function and strictly differentiable functions (see Definition 6.144) are regular of every $x \in X$. Recall strictly differentiable functions are locally Lipschitz. Also $C^1 - functions$ are strictly differentiable (see Remark 6.145).

Theorem 6.151 *If X, Y are Banach spaces, $\psi \in C^1(X, Y)$, and $\varphi \in \text{Lip}_{loc}(Y, \mathbb{R})$, then $\vartheta = \varphi \circ \psi \in \text{Lip}_{loc}(X, \mathbb{R})$ and*

$$\partial\vartheta(x) \subseteq \partial\varphi(\psi(x)) \circ \psi'(x) \text{ for all } x \in X;$$

equality holds if $\varphi(\cdot)$ is regular at $\psi(x)$ and then $\vartheta(\cdot)$ is also regular at x.

Proof From Proposition 6.129, we have that $\vartheta \in \text{Lip}_{loc}(X, \mathbb{R})$. Given $\varepsilon > 0$, we can find $0 < d < \varepsilon$ such that if $\|h' - h\|_Y < \delta$, then

$$\varphi^0(\psi(x); h') \leq \varphi^0(\psi(x); h) + \varepsilon. \tag{6.86}$$

Also from the definition of the generalized directional derivative (see Definition 6.133), we can find $0 < \eta, \mu$ such that

$$\|y - \psi(x)\|_Y < \eta, 0 < t < \mu, \|h' - h\|_Y < \delta$$

imply that

$$\frac{\varphi(y + th') - \varphi(y)}{t} \leq \varphi^0(\psi(x); h') + \varepsilon \leq \varphi^0(\psi(x); h) + 2\varepsilon. \tag{6.87}$$

Let $h = \psi'(x)(v)$, $v \in X$. Since $\psi \in C^1(X, Y)$, we can find $0 < \tau < \mu$ such that $\|u - x\|_X < \tau, 0 < t \leq \tau$, imply

$$\left\| \frac{\psi(u+th) - \psi(u)}{t} - \psi'(x)(h) \right\|_Y < \delta, \quad \|\psi(u) - \psi(x)\|_Y < \eta.$$

In (6.87), we set $y = \psi(x)$ and

$$h' = \frac{\psi(x+th) - \psi(v)}{t}.$$

Then

$$\frac{\vartheta(u+tv) - \vartheta(u)}{t} \le \varphi^0(\psi(x); \psi'(x)(v)) + 2\varepsilon.$$

If $\|u - x\|_X < \tau, o < t \le \tau$. Hence,

$$\vartheta^0(x; v) \le \varphi^0(\psi(x); \psi'(x)(v)), \tag{6.88}$$

$$\Rightarrow \partial\vartheta(x) \subseteq \partial\varphi(\psi(x)) \circ \psi'(x) \text{ for all } x \in X. \tag{6.89}$$

If φ is regular at $\psi(x)$, then using Definition 6.148, we obtain

$$\varphi^0(\psi(x); \psi'(x)(v)) \le \vartheta^0(x; v) \text{ for all } v \in X,$$

$$\Rightarrow \varphi^0(\psi(x); \psi'(x)(v)) = \vartheta^0(x; v) \text{ for all } v \in X \text{ (see (6.88))}.$$

Therefore, equality holds in (6.89), and clearly, $\vartheta(\cdot)$ is regular at $x \in X$. $\qquad \square$

Remark 6.152 Using the adjoint of $\psi'(x) \in \mathcal{L}(X, Y)$, it is more convenient to write (6.89) as

$$\partial\vartheta(x) = \psi'(x)^* \partial\varphi(\psi(x)) \text{ for all } x \in X.$$

We mention that Theorem 6.151 remains true if $\psi : X \to Y$ is only strictly differentiable (see Clarke [68], Theorem 2.3.10,P.45).

A useful consequence of Theorem 6.151 is the following corollary.

Corollary 6.153 *If X, Y are Banach spaces, $X \hookrightarrow Y$ continuously and densely,*

$$\varphi \in \text{Lip}_{loc}(Y, \mathbb{R}) \text{ and } \psi = \varphi|_X,$$

then $\partial\psi(x) = \partial\varphi(x)$ for all $x \in X$ (this equality means that every element of $\partial\psi(x)$ admits a unique extension to an element of $\partial\varphi(x)$).

Proof Let i$\in \mathcal{L}(X, Y)$ be the embedding map, and apply Theorem 6.151 with $\psi = i$.   □

Theorem 6.154 *If $T = [0, h]$, $\psi \in C^1(T, X)$, and $\varphi \in \mathrm{Lip}_{loc}(X, \mathbb{R})$, then $\vartheta = \varphi \circ \psi$ is differentiable a.e. on T and*

$$\vartheta'(t) \leq max[\langle x^*, \psi'(t) \rangle : x^* \in \partial\varphi(\psi(t))] \text{ for all } a.a t \in T.$$

Proof Since ϑ is locally Lipschitz on T, it is differentiable a.e. on T. Let $t_0 \in T$ be a point of differentiability of ϑ. Then, with $\frac{\varepsilon(t)}{t} \to 0$ as $t \to 0^+$,

$$\vartheta'(t_0) = \lim_{t \to 0} \frac{\varphi(\psi(t_0 + t)) - \varphi(\psi(t_0))}{t}$$

$$= \lim_{t \to 0} \frac{\varphi(\psi(t_0) + t\psi'(t_0) + \varepsilon(t)) - \varphi(\psi(t_0))}{t}$$

$$= \lim_{t \to 0} [\frac{\varphi(\psi(t_0) + t\psi'(t_0)) - \varphi(\psi(t_0))}{t}$$

$$+ \frac{\varphi(\psi(t_0) + t\psi'(t_0) + \varepsilon(t)) - \varphi(\psi(t_0) + t\psi'(t_0))}{t}$$

$$= \lim_{t \to 0} \frac{\varphi(\psi(t_0) + t\psi'(t_0)) - \varphi(\psi(t_0)}{t}$$

$$\leq \varphi^0(\psi(t_0); \psi'(t_0)) = max[\langle x^*, \psi'(t_0) \rangle : x^* \in \partial\varphi(\psi(t))].$$ □

We have a version of Fermat's rule for local extrema (local minima or local maxima).

Proposition 6.155 *If $\varphi \in \mathrm{Lip}_{loc}(X, \mathbb{R})$ and x is a local extremum of φ, then $0 \in \partial\varphi(x)$.*

Proof Since $\partial(-\varphi) = -\partial\varphi$ (see Proposition 6.138(c)), it suffices to consider the case where x is a local minimizer. Then from Definition 6.137, we have

$$0 \leq \varphi^0(x; h) \text{ for all } h \in X,$$

$$\Rightarrow 0 \in \partial\varphi(x) \text{ (see Definition 6139)}.$$ □

Using the generalized subdifferential, we can have a mean valued theorem for locally Lipschitz functions, known in the literature as "Lebourg's Mean Value Theorem."

Theorem 6.156 *If $\varphi \in \mathrm{Lip}_{loc}(X, \mathbb{R})$ and $x, u \in X$, then we can find $0 < t_0 < 1$ and $x^* \in \partial\varphi((1 - t_0)x + t_0u)$ such that:*

Proof Let $\vartheta : \mathbb{R} \to \mathbb{R}$ be defined by

$$\vartheta(t) = \varphi((1-t)x + tu) + t(\varphi(x) - \varphi(u)), \quad t \in \mathbb{R}.$$

This is a locally Lipschitz function. Also $\vartheta(0) = \vartheta(1) = \varphi(x)$. Hence, there is $t_0 \in \mathbb{R}$ that is a local extremum of $\vartheta(\cdot)$ and so $0 \in \partial\vartheta(t_0)$ (see Proposition 6.155). Also, we have

$$\partial\vartheta(t_0) \subseteq \langle \partial\varphi((1 \cdot t_0)x + t_0 u), u - x \rangle + \varphi(x) - \varphi(u),$$

$$\Rightarrow \varphi(u) - \varphi(x) \in \langle \partial\varphi((1-t_0)x + t_0 u), u - x \rangle. \qquad \square$$

6.6 Generalizations

In this section, we present very briefly some generalizations of the subdifferential theory beyond the realm of convex and locally Lipschitz functions. So, let X be a Banach space and $\varphi : X \to \overline{\mathbb{R}} = \mathbb{R} \cup \{+\infty\}$ a lower semicontinuous function.

Definition 6.157 Let $x \in \text{dom } f$. The "subdifferential" of φ at x is the set

$$\partial^-\varphi(x) = \left\{ x^* \in X^* : \liminf_{x \to 0} \frac{\varphi(x+h) - \varphi(x) - \langle x^*, h \rangle}{\|h\|} \geq 0 \right\}.$$

Remark 6.158 It is clear that if φ is convex, then $\partial^-\varphi(x)$ coincides with the convex subdifferential. Some of the important properties of the convex subdifferential also hold in this general case.

Proposition 6.159 *Suppose* $\varphi : X \to \overline{\mathbb{R}}$ *is lower semicontinuous and* $x \in \text{dom } \varphi$. *Then:*

(a) $\partial^-\varphi(x)$ *is closed and convex.*
(b) $o \in \partial^-\varphi(x)$ *if* x *is a local minimizer of* $\varphi(x)$.
(c) $\partial^-\varphi(x) \subseteq \{\varphi'_G(x)\}$ *if* $x \in \text{int dom } \varphi$ *and* φ *is Gateaux differentiable at* x.

Proof (a) Convexity of $\partial^-\varphi(x)$ is immediate from definition. To show the closedness of $\partial^-\varphi(x)$, we consider a sequence $\{x_n^*\} \subseteq \partial^-\varphi(x)$ and assume that $x_n^* \to x^*$ in X^*. We have

$$\liminf_{h \to 0} \frac{\varphi(x+h) - \varphi(x) - \langle x^*, h \rangle}{\|h\|}$$

$$= \liminf_{h \to 0} \left[\frac{\varphi(x+h) - \varphi(x) - \langle x_n^*, h \rangle}{\|h\|} + \frac{\langle x_n^* - x^*, h \rangle}{\|h\|} \right]$$

$$\geq \liminf_{h \to 0} \left[\frac{\varphi(x+h) - \varphi(x) - \langle x_n^*, h \rangle}{\|h\|} - \|x_n^* - x^*\|_* \right]$$

$$\geq 0.$$

Thus, $x^* \in \partial^- \varphi(x)$ and so $\partial^- \varphi(x)$ is closed.

(a) Just notice that $\varphi(x) \leq \varphi(x+h)$ for any $\|h\|$ small.
(b) Let $x^* \in \partial^- \varphi(x)$. We have for $y \in X$

$$\liminf_{t \to 0^+} \frac{\varphi(x+ty) - \varphi(x) - t \langle x^*, y \rangle}{t} \geq 0.$$

Thus,

$$\langle x^*, y \rangle \leq \liminf_{t \to 0^+} \frac{\varphi(x+ty) - \varphi(x)}{t} = \langle \varphi_G'(x), y \rangle.$$

Since $y \in X$ is arbitrary, we concluded that $x^* = \varphi_G'(x)$.   □

Remark 6.160 In part (c), the inclusion can be strict, namely, $\partial^- \varphi(x) = \emptyset$. Let $\varphi : \mathbb{R}^2 \to \mathbb{R}$ be defined by $\varphi(x, y) = -x$ if $y = x^2, x > 0$, and $\varphi(x, y) = 0$ otherwise. This function is Gateaux differentiable at $(0, 0)$, $\varphi_G'(0, 0) = (0, 0)$ but

$$\liminf_{(x,y) \to (0,0)} \frac{\varphi(x, y) - \varphi(0, 0) - ((0, 0), (x, y))_{\mathbb{R}^2}}{|(x, y)|}$$

$$= \liminf_{(x,y) \to (0,0)} \frac{\varphi(x, y)}{|(x, y)|} \leq \lim_{\delta \downarrow 0} \frac{-\sqrt{2}/2\delta}{\delta} = -\frac{\sqrt{2}}{2},$$

and thus $\partial^- \varphi(0, 0) = \emptyset$.

It can be shown that $x^* \in \partial^- \varphi(x)$ if and only if there exists a Frechet differentiable $f(x) = \varphi(x)$, with $x^* = f_F'(x)$, and there exits $\mathcal{U}$, a neighborhood of x, such that $f \leq \varphi$ on $\mathcal{U}$.

Using this fact, we obtain at once the following result.

Proposition 6.161 *If $\varphi : X \to \overline{\mathbb{R}}$ is lower semicontinuous and Frechet differentiable at $x \in \text{int dom } \varphi$, then $\partial^- \varphi(x) = \{\varphi_F'(x)\}$.*

We introduce a counterpart of the generalized directional derivative of a locally Lipschitz function (see Definition 6.137).

Definition 6.162 The "subderivative" of φ at $x \in \text{dom } \varphi$, denoted by $\varphi_0(x; \cdot)$, is defined by

$$\varphi_0'(x; h) = \liminf_{\substack{h' \to h, \\ t \downarrow 0}} \frac{\varphi(x + th') - \varphi(x)}{t} = \lim_{\delta \downarrow 0} \inf_{h' \in B_\delta(h)} \frac{\varphi(x + th') - \varphi(x)}{t}.$$

Proposition 6.163 *If* $\varphi : \mathbb{R}^N \to \overline{\mathbb{R}}$ *is lower semicontinuous and* $x \in \mathrm{dom}\, \varphi$, *then*

$$\partial^- \varphi(x) = \left\{ x^* \in \mathbb{R}^N : \langle x^*, h \rangle \leq \varphi_0'(x; h) \text{ for all } h \in \mathbb{R}^N \right\}.$$

Proof We have $\liminf_{h \to 0} \frac{\varphi(x+h) - \varphi(x) - \langle x^*, h \rangle}{\|h\|} \geq 0$ for $x^* \in \partial^- \varphi(x)$. Let $y \in \mathbb{R}^N$ and set $h = ty'$. We have

$$\liminf_{y' \to y, t \downarrow 0} \frac{\varphi(x + ty') - \varphi(x) - \langle x^*, ty' \rangle}{t} \geq 0.$$

Hence, $\varphi_0'(x; y) \geq \langle x^*, y \rangle$.

Next, let $x^* \in \mathbb{R}^N$ such that $\langle x^*, h \rangle \leq \varphi_0'(x; h)$ for all $h \in \mathbb{R}^N$. Suppose $x^* \notin \partial^- \varphi(x)$. Then we can find $\{h_n\} \subseteq \mathbb{R}^N$ such that $h_n \to 0$ and

$$\lim_{n \to \infty} \frac{\varphi(x + h_n) - \varphi(x) - \langle x^*, h_n \rangle}{\|h_n\|} < 0.$$

Let $y_n = \frac{h_n}{\|h_n\|}$ and $t_n = \|y_n\|$. We may assume that $y_n \to y$. Then, we have the following contradiction:

$$\varphi_0'(x; h) = \liminf_{h' \to h, t \downarrow 0} \frac{\varphi(x) + th') - \varphi(x)}{t}$$
$$\leq \lim_{n \to \infty} \frac{\varphi(x + t_n h_n) - \varphi(x)}{t_n}$$
$$< \langle x^*, h \rangle. \qquad \square$$

Example 6.164 Let X be strictly convex, reflexive, and $C \subseteq X$ be closed and convex. We define

$$d_C(x) = \inf[\|x - u\| : u \in C],$$

the distance function from C. If $x \notin C$, then there exists a unique $\widehat{u} \in C$ such that $d_C(x) = \|x - \widehat{u}\|$. We have for all $h \in X$

$$(d_C)_0'(x; h) = \frac{\langle x - \widehat{u}, h \rangle}{\|x - \widehat{u}\|}.$$

If $X = \mathbb{R}^N$, then d_C is differentiable at every $x \in C$ and $(d_C)'_F(x) = \frac{x-\widehat{u}}{\|x-\widehat{u}\|}$. If $x \in \text{int}\, C$, then again d_C is differentiable at x and $(d_C)'_F(x) = 0$.

The following property of the subdifferential is an easy consequence of Definition 6.518.

Proposition 6.165

(a) *If $\varphi : X \to \overline{\mathbb{R}}$ is lower semicontinuous and $\psi : X \to \mathbb{R}$ is Frechet differentiable at $x \in \text{dom}\, \varphi$, then*

$$\partial^-(\varphi + \psi)(x) = \partial^- \varphi(x) + \psi'_F(x).$$

(b) *If $\varphi, \psi : X \to \overline{\mathbb{R}}$ are lower semicontinuous and $x \in \text{dom}\, \varphi \cap \text{dom}\, \psi$, then*

$$\partial(\varphi + \psi) = \partial^- \varphi(x) + \partial \psi(x).$$

Now we introduce another kind of subdifferential.

Definition 6.166 Let $\varphi : X \to \overline{\mathbb{R}}$ be lower semicontinuous and $x \in \text{dom}\, \varphi$. The "proximal subdifferential" of φ at x, denoted by $\partial_p \varphi(x)$, is defined by

$$\partial_p \varphi(x) = \{x^* \in X^* : \varphi(u) - \varphi(x) + \tau \|u - x\|^2 \geq \langle x^*, u - x \rangle$$

$$\text{for all } u \in \mathcal{U} = \mathcal{U}(x, x^*) \in \mathcal{N}(x), \text{ with } \tau = \tau(x, x^*) > 0\}.$$

Proposition 6.167 $\partial_p \varphi(x) = \partial^- \varphi(x)$ *if φ is convex.*

Proof It is clear from Definition 6.167 that $\partial^- \varphi(x) \subseteq \partial_p \varphi(x)$. Now let $x^* \in \partial \varrho \varphi(x)$, and consider the convex function $\psi(u) = \varphi(u) + \tau \|u - x\|^2 - \langle x^*, u \rangle$. By Definition 6.167, we have that x is a local minimizer of ψ; hence, $0 \in \partial \psi(x)$. Since the subdifferential at 0 of $\| \cdot \|^2$ is $\{0\}$, we conclude that $x^* \in \partial^- \varphi(x)$ (see Proposition 6.83). $\square$

Remark 6.168 The above proof suggests that the geometric interpretation of the proximal subdifferential is the collection of all slopes at x of locally supporting parabolas to epi φ, namely, the functions $u \to \varphi(x) + \langle x^*, u - x \rangle - \tau \|u - x\|^2$.

Proposition 6.169 *If $\varphi : X \to \overline{\mathbb{R}}$ is lower semicontinuous and Gateaux differentiable at x, then*

$$\partial_p \varphi(x) \subseteq \{\varphi'_G(x)\}.$$

Proof Let $x^* \in \partial_p \varphi(x)$ and fix $h \in X$. In Definition 6.167, let $u = x + th$. We have for $t > 0$

$$\frac{1}{t}[\varphi(x + th) - \varphi(x)] \geq \langle x^*, h \rangle - \tau t \|h\|^2.$$

We pass to the limit as $t \to 0^0$ to obtain

$$\langle x^*, h \rangle \leq \langle \varphi'_G(x), h \rangle$$

for all $h \in X$. Thus, $x^* = \varphi'_G(x)$. □

Remark 6.170 The inclusion may be strict. Consider $\varphi : \mathbb{R} \to \mathbb{R}$, defined by $\varphi(x) = -|x|^{3/2}$. Then $\partial_p \varphi(0) = \emptyset$. It is easy to see from Definition 6.167 that there exists a C^2-function $f : X \to \mathbb{R}$ such that $f(x) = \varphi(x)$, $f'_F(x) = x^* \in \partial_p \varphi(x)$ and $f \leq \varphi$ in a neighborhood of x.

Minor changes in the proof of Proposition 6.160 lead to the following result.

Proposition 6.171 *If $\varphi : X \to \overline{\mathbb{R}}$ is lower semicontinuous and $x \in$ dom φ, then:*

(a) $\partial_p \varphi(x) \subseteq X^*$ *is convex.*
(b) $0 \in \partial_p \varphi(x)$ *is x if a local minimizer of φ.*

Remark 6.172 The proximal subdifferential $\partial_p \varphi(x)$ may fail to be closed. Consider $\varphi : \mathbb{R} \to \mathbb{R}$, defined by $\varphi(x) = x^{4/3}$ if $x < 0$ and $\varphi(x) = x$ if $x \geq 0$. Then $\partial_p \varphi(0) = (0, 1]$.

For the proximal subdifferential, we have the following sum rule.

Proposition 6.173 *If $\varphi : X \to \overline{\mathbb{R}}$ is lower semicontinuous, $\psi : X \to \mathbb{R}$ is Frechet differentiable in a neighborhood of $x \in$ dom φ with ψ'_F Lipschitz near x, then*

$$\partial_p(\varphi + \psi)(x) = \partial^- \varphi(x) + \{\psi'_F(x)\}.$$

Proof Since ψ'_F is locally Lipschitz at x, there exist $\delta > 0$ and $c > 0$ such that

$$\|\psi'_F(y) - \psi'_F(u)\|_* \leq c\|y - u\|$$

for any $y, u \in B_\delta(x)$.

For any $v \in B_\delta(x)$, using the mean value theorem, we can find $u \in B_\delta(x)$ such that $\psi(v) = \psi(x) + \langle \psi'_F(u), v - x \rangle$, and thus

$$|\psi(v) - \psi(x) - \langle \psi'_F(x), v - x \rangle = |\langle \psi'_F(u) - \psi'_F(x), v - x \rangle$$

$$\leq c\|u - x\|\|v - x\|$$

$$\leq c\|v - x\|^2.$$

Now let $x^* \in \partial_p(\varphi + \psi)(x)$. Then for some $\tau > 0$ and a neighborhood $\mathcal{U}$ of x, we have

$$\varphi(u) + \psi(u) - \varphi(x) - \psi(x) + \tau\|u - x\|^2 \geq \langle x^*, u - x \rangle$$

for all $u \in \mathcal{U}$. Thus,

$$\varphi(u) - \varphi(x) + [\tau + c]\|u - x\|^2 \geq \langle x^* - \psi_F'(x), u - x \rangle$$

for all $u \in \mathcal{U} \cap B_\delta(x)$; hence, $x^* - \psi_F'(x) \in \partial_p \varphi(x)$.

Conversely, if $u^* \in \partial_p \varphi(x)$, for some $\tau > 0$, and a neighborhood $\mathcal{U}$ of x, we have

$$\varphi(u) - \varphi(x) + \tau \|u - x\|^2 \geq \langle u^*, u - x \rangle$$

for all $u \in \mathcal{U}$. Hence,

$$\varphi(u) + \psi(u) - \varphi(x) - \psi(x) + [\tau + c]\|u - x\|^2 \geq \langle u^* + \psi_F'(x), u - x \rangle$$

for all $u \in \mathcal{U} \cap B_\delta(x)$. Therefore, $u^* + \psi_F'(x) \in \partial_p(\varphi_\psi)(x)$.   □

Setting $\varphi \equiv 0$ in the above proposition, we obtain:

Corollary 6.174 *If $\psi : X \to \mathbb{R}$ is Frechet differentiable in a neighborhood of x and ψ' is Lipschitz near x, then*

$$\partial_p \psi(x) = \{\psi_F'(x)\}.$$

Finally, we introduce a third subdifferential by applying a closure operation to the proximal subdifferential.

Definition 6.175 Let $\varphi : X \to \overline{\mathbb{R}}$ be lower semicontinuous and $x \in \operatorname{dom}\varphi$. "Limiting subdifferential" of φ at x, denoted by $\partial_L \varphi(x)$, is defined by

$$\partial_L(x) = \{x^* \in X^* : \text{ there exist } x_n \in \operatorname{dom}\varphi \text{ and } x^* \in \partial_p \varphi(x)$$

$$\text{such that } x_n \to x, x_n^* \to x^*, \text{ and } \varphi(x_n) \to \varphi(x)\}.$$

Remark 6.176 If φ is continuous at x, then the requirement of $\varphi(x_n) \to \varphi(x)$ is automatically true. We have

$$\partial_p \varphi(x) \subseteq \partial^- \varphi(x) \subseteq \partial_L \varphi(x).$$

The function $\varphi(x) = -|x|$ satisfies $\partial_p \varphi(0) = \partial^- \varphi(0) = \emptyset$, but $\partial_L \varphi(0) = \{-1, 1\}$.

Proposition 6.177 *(a) If φ is Lipschitz near x, then $\emptyset \neq \partial_L \varphi(x) \subseteq \partial_C \varphi(x)$ (= the Clarke subdifferential):*

(a) $\partial_L \varphi(x) = \{\varphi_F'(x)\}$ *if $\varphi(x)$ is C^1 near x.*
(b) $\partial_L \varphi(x) = \partial \varphi(x)$ (= *the convex subdifferential) if φ is convex.*

For the limiting subdifferential, we have the following "sum rule" (see Mordukhovich [198]).

Proposition 6.178 *If $\varphi, \psi : X \to \overline{\mathbb{R}}$ are lower semicontinuous and one of them is Lipschitz in a neighborhood of $x \in \operatorname{dom} \varphi \operatorname{dom} \psi$, then*

$$\partial_L(\varphi + \psi)(x) \subseteq \partial_L\varphi(x) + \partial_L\psi(x).$$

Remark 6.179 The inclusion can be strict. Let $\varphi, \psi : \mathbb{R} \to \mathbb{R}$ be defined by $\varphi(x) = |x|$ and $\psi(x) = -|x|$, respectively. Then

$$\partial_L(\varphi + \psi)(0) = \{0\}, \partial_L\varphi(0) + \partial_L\psi(0) = [-1, 1] + [-1, 1].$$

6.7 Remarks

6.1 Expositions of differential calculus in Banach spaces can be found in the books of Berger [31], Cartan [59], Denkowski–Migorski–Papageorgiou [81], Dieudonné [87], Gasinski–Papageorgiou [121], and Zeidler [290]. We mention that the first notion of derivative, for functions defined on an infinite dimensional space (function space), was produced by Volterra and soon thereafter came the more general definitions due to Frechet (1906) and Gateaux (1922).

6.2 The systematic study of convex functions starts with the works of Fenchel [110, 111]. His work was continued by Rockafellar [233]. Detailed expositions of the subject together with applications in optimization and optimal control can be found in the books of Barbu–Precupanu [28], Borwein–Vanderwerff [40], Ekeland–Temam [102], Gasinski–Papageorgiou [121], Giles [124], Ioffe–Tikhomirov [148], Papageorgiou–Winkert [216], Phelps [221] and Rockafellar [236]. More on integral functionals determined by a measurable convex integrand (a special case is presented in example 6.91(b)) can be found in Rockafellar [238]. For the duality map, we refer to Zeidler [292] (Proposition 32.22, p. 861). Theorem 6.97 is a classical result due to Mazur [190]. Extensions to Frechet differentiability were obtained by Asplund [14]. More on the subject can be found in the books of Borwein–Vanderwerff [40], Deville–Godefroy–Hare-–Zizler [82], Giles [124], and Phelps [221]. For the so-called "approximate subdifferential", see Borwein [39].

6.3 Duality is in the core of convex analysis. The conjugate function (also known as polar function or Fenchel transform) is an extension of the classical Legendre transform (Legendre (1786)). Related to Theorem 6.104 is the following correspondence established by Hörmander [144].

Theorem 6.180 *If X is a locally convex space, then there is a bijective correspondence between nonempty, closed, convex sets, and sublinear w^*-lower semicontinuous functions on X^* with values in $\overline{\mathbb{R}} = \mathbb{R} \cup \{+\infty\}$, the correspondence being*

$$P_{fc}(x) \ni C \to \sigma(\cdot; c) = \text{ the support function of } C.$$

Let (Ω, Σ, μ) be a σ-finite measure space and X a separable Banach space. We consider a $\Sigma \otimes B(X)$-measurable function (an integrand) $\varphi : \Omega \times X \to \overline{\mathbb{R}}$, and we assume that $\varphi(\omega, \cdot) \in \Gamma_0(X)$ for μ-a.a $\omega \in \Omega$. Such an integrand is called "normal integrand." We introduce the integral functional $I_\varphi : L^1(\Omega, X) \to \mathbb{R}^* = \mathbb{R} \cup \{\pm\infty\}$ defined by

$$I_\varphi(u) = \begin{cases} \int_\Omega \varphi(\omega, u(\omega)) d\mu & \text{if } \varphi(\cdot, u(\cdot))^+ \in L^1(\Omega) \\ +\infty & \text{otherwise.} \end{cases}$$

Proposition 6.181 *If $\varphi : \Omega \times X \to \overline{\mathbb{R}} = \mathbb{R} \cup \{+\infty\}$ is a normal integrand, then:*

(a) $\varphi^* : \Omega \times X^* \to \overline{\mathbb{R}} = \mathbb{R} \cup \{+\infty\}$ *defined by*

$$\varphi^*(\omega, x^*) : \sup[\langle x^*, x \rangle - \varphi(\omega, x) : x \in X]$$

is a normal integrand on $\Omega \times X^*_{\omega^*}$.
(b) *If I_φ is finite at $u_0 \in L^1(\Omega, X)$, then $(I_\varphi)^* = I_{\varphi^*}$.*

6.4 The operation of infimal convolution is due to Moreau [199]. There is an integral version of this operation. So, let (Ω, Σ, μ) be a finite, complete, nonatomic measure space and $\varphi : \Omega \times \mathbb{R}^N \to \mathbb{R}$ an integrand.

Definition 6.182 The "inf-convolution integral" of φ with respect to μ is the function $\oint_\Omega \varphi_\omega d\mu : \mathbb{R}^N \to \mathbb{R}$ defined by

$$(\oint_\Omega \varphi_\omega d\mu)(x) = \inf[\lambda \in \mathbb{R} : (x, \lambda) \in \int_\Omega \text{epi } \varphi(\omega, \cdot) d\mu].$$

Remark 6.183 In the above definition,

$$\int_\Omega \text{epi } \varphi(\omega, \cdot) d\mu = \left\{ \int_\Omega (u(\omega), \lambda(\omega)) d\mu : (u, \lambda) \in S^1_{\text{epi } \varphi(\cdot, \cdot)} \right\}.$$

If for all $u \in L^1(\Omega, \mathbb{R}^N)$, $I_\varphi(\cdot)$ exists (with possibly infinite value), then

$$(\oint_\Omega \varphi_\omega d\mu)(x) = \inf[I_\varphi(u) : u \in L^1(\Omega, \mathbb{R}^N), \int_\Omega u d\mu = x].$$

The Moreau–Yosida regularization is important in evolution equations (see Barbu [27] and Brezis [49]) and variational convergence (see Dal Maso [74]).

6.5 Lipschitz and locally Lipschitz play an important role in modern geometric measure theory. They are the smooth functions of metric spaces. For more on this subject, we refer to the books of Ambrosio–Tilli [11], Burago–Ivanov [56], Heinonen [137], and Shioya [253]. The subdifferential theory of locally Lipschitz functions (which extends the convex subdifferential theory) is due to Clarke [68, 69]. Theorem 6.156 was proved by Lebourg [177].

Finally, we mention that theorem 6.135 (Rademacher's theorem) can be extended to Lipschitz functions defined on an infinite dimensional Banach space. First a definition:

Definition 6.184 Let $(G, +)$ be an Abelian Polish group and d a compatible invariant metric on G. A universally measurable set $C \subseteq G$ is "Haar-null" if there exists a probability measure μ on G (not unique), such that $\chi_C * \mu = 0$. Here

$$(\chi_C * \mu)(x) = \int_C \chi_C(x + u) d\mu(u).$$

Remark 6.185 Every translation of C is μ-null too. The measure $\mu(\cdot)$ is known as the test measure.

The following extension of Rademacher's Theorem can be found in Christensen [67].

Theorem 6.186 *If X is a separable Banach space, Y is a Banach space with the RNP, and $\varphi \in \mathrm{Lip}_{loc}(X, Y)$, then there exists a universally measurable set $D \subseteq X$ with $X \setminus D$ Haar-null such that $\varphi(\cdot)$ is Gateaux differentiable on D.*

6.6 For the extensions of the subdifferential theory beyond the convex and Clarke's theories, we refer to the books of Clarke [69], Mordukhovich [198], and Rockafellar–Wets [239].

6.8 Problems

Problem 6.1 Let X, Y be Banach spaces and $\varrho : X \to Y$ is compact (see Definition 3.146), which is Frechet differentiable at $x \in X$. Show that $f'(x) \in \mathcal{L}_c(X, Y)$.

Problem 6.2 Let X be Banach space, $U \subseteq X$ open, $f : U \to \mathbb{R}$ a Frechet differentiable, and $f'(u) = 0$ for some $u \in U$. Show that:

(a) If f is twice differentiable at u and

$$f''(u)(x, x) \geq c\|x\|^2 \text{ for some } c > 0, \text{ all } x \in X,$$

then u is a strict local minimizer of f.

(b) If f is twice differentiable at u and there exists a ball $B_r(u) \subseteq u$, such that $f''(v)(x, x) \geq 0$ for $v \in B_r(u)$ and $x \in X$, then u is a local minimizer of f.

Problem 6.3 Let X, YV, W be Banach spaces, $\alpha : X \times Y \to V$ is bilinear,

$$\|\alpha(x, y)\| \leq c\|x\|_X \|y\|_Y \text{ for some } c > 0, \text{ all } x \in X, y \in Y,$$

and $f : W \to X, g : W \to Y$ are Frechet (resp., Gateaux) differentiable maps. Show that

$$\xi(w) = \alpha(f(w), g(w))$$

is Frechet (resp., Gateaux) differentiable.

Problem 6.4 Let H be a Hilbert space with inner product $(\cdot, \cdot)$ and $A \in \mathcal{L}(H)$. We consider the function $f : H \to \mathbb{R}$ defined by

$$f(u) = (A(u), u) \text{ for all } u \in H.$$

Show that $f \in C^1(H, \mathbb{R})$.

Problem 6.5 Let $f : L^p(\mathbb{R}^N) \to \mathbb{R}$ be defined by

$$f(u) = \|u\|_p^p \text{ for all } u \in L^p(\mathbb{R}^N).$$

Show that f is Frechet differentiable and find its derivative.

Problem 6.6 Let $\Omega \subseteq \mathbb{R}^N$ be a bounded domain with Lipschitz boundary $\partial\Omega$, $1 < p < \infty$, and $f : W_0^{1,p}(\Omega) \to \mathbb{R}$ is defined by

$$f(u) = \frac{1}{p}\|Du\|_p^p \text{ for all } u \in W_0^{1,p}(\Omega).$$

Show that $f \in C^1(W_0^{1,p}(\Omega))$ and find $f'(u)$.

Problem 6.7 Suppose X is a Banach space, $f : X \to \mathbb{R}$ is a Gateaux differentiable function such that

$$\langle f'(u), u \rangle > 0 \text{ for all } u \in X, \|u\| = \varrho,$$

and $f(u_0) = \min[f(u) : \|u\| \le \varrho]$. Show that $\|u_o\| < \varrho$.

Problem 6.8 Given a Banach space and a function $f : X \to \mathbb{R}$. We say that f is ϑ-homogeneous ($\vartheta \in \mathbb{R}$) if $f(tu) = t^\vartheta f(u)$ for all $u \in X \setminus 0, t > 0$. Suppose that $f : X \to \mathbb{R}$ is Frechet differentiable. Show that f is ϑ-homogeneous iff $\vartheta\varphi(u) = \langle \varphi'(u), u \rangle$ for all $u \ne 0$.

Problem 6.9 Suppose that $f \in C^1(\mathbb{R}^N, \mathbb{R}^N)$ and satisfies

$$(f'(u)h, h)_{\mathbb{R}^N} \ge c|h|^2 \text{ for all } u, h \in \mathbb{R}^N \text{ and some } c > 0.$$

Show that $f(\cdot)$ is a C^1-diffeomorphism.

Problem 6.10 Let X be a reflexive Banach space, H a Hilbert space, $X \hookrightarrow H$ continuously and densely, and $\varphi \in \Gamma_0(X)$ satisfies

$$c\|u\|^2 \le \varphi(u) \text{ for all } u \in X \text{ and some } c > 0.$$

We define $\widehat{\varphi} : H \to \overline{\mathbb{R}} = \mathbb{R} \bigcup \{+\infty\}$ by

$$\widehat{\varphi}(u) = \begin{cases} \varphi(u) & \text{if } u \in X \\ +\infty & \text{if } u \in H \setminus X. \end{cases}$$

Show that $\widehat{\varphi} \in \Gamma_0(H)$.

Problem 6.11 Let H be a Hilbert space, $C \subseteq H$ a nonempty, closed, and convex set, and $\varphi : H \to \mathbb{R}$ is defined by

$$\varphi(u) = \frac{1}{2}d(u, C)^2 \text{ for all } u \in H.$$

Show that φ is convex.

Problem 6.12 Let X be a Banach space and $\varphi : X \to \overline{\mathbb{R}}$ a convex function. Show that the following statements are equivalent:

(a) $\varphi(\cdot)$ is Lipschitz on bounded subset of X.
(b) $\partial\varphi$ maps bounded sets in X into bounded sets in X^*.

Problem 6.13 Let X be a Banach space and $\varphi : X \to \mathbb{R}$ a continuous convex function. Show that $0 \in \text{int } \partial\varphi(x)$ iff there exist $\varepsilon > 0$ such that $\varphi(x + h) \ge \varphi(x) + \varepsilon\|h\|$ for all $h \in X$.

Problem 6.14 Let $\Omega \subseteq \mathbb{R}^N$ be a bounded domain with Lipschitz boundary, $p > N$, and $u \in W^{1,p}(\Omega)$. Show that $u(\cdot)$ is differentiable at a.a $z \in \Omega$.

Problem 6.15 Let X be a Banach space, and consider the function $\varphi : X \to \mathbb{R}$ defined by

$$\varphi(u) = \frac{1}{2}\|u\|^2 \text{ for all } u \in X.$$

Show that $\varphi(\cdot)$ is Frechet differentiable iff the duality map $\mathcal{F}(\cdot)$ is single-valued and continuous.

Problem 6.16 Let K be a compact metric space and $\xi : C(K) \to \mathbb{R}$ defined by

$$\xi(u) = \max[u(t) : t \in K].$$

Show that $\xi(\cdot)$ is continuous convex and that

$$\mu \in \partial\xi(u) \text{ iff } \mu \geq 0, \langle\mu, i_0\rangle = 1, \operatorname{supp}\mu = \{t \in K : u(t) = \xi(u)\}$$

with $i_0(t) = 1$ for all $t \in K$.

Problem 6.17 Let X be a Banach space and $C \subseteq X$ nonempty closed. Show that $d(x, C) = \max[\langle x^*, x\rangle - \sigma(x^*, C) : \|x^*\|_* \leq 1]$.

Problem 6.18 Let H be a Hilbert space and $C \subseteq H$ nonempty, closed, and convex. Let $\varphi(u) = d(u, C)$ for all $u \in H$ and consider $x \notin C$. Show that $\varphi(\cdot)$ is Frechet differentiable at x and

$$\varphi'(x) = \frac{x - p_C(x)}{\|x - p_C(x)\|}$$

with $p_C(\cdot)$ being the metric projection map (see Theorem 3.175).

Problem 6.19 Let $\varphi \in \Gamma_0(\mathbb{R}^N)$ admit a unique minimization at x_0 and define

$$\vartheta(t) = \inf[\varphi(x) - \varphi(0) : |x - x_0| = t].$$

Show that $\varphi^*(0) + (x^*, x_0)_{\mathbb{R}^N} \leq \varphi^*(x^*) \leq \varphi^*(0) + (x^*, x_0)_{\mathbb{R}^N} + \vartheta(|x^*|)$ for all $x^* \in \mathbb{R}^N$ and that φ^* is differentiable at $x^* = 0$.

Problem 6.20 Let X be a Banach space and $\varphi \in \Gamma_0(X)$. Show that $\varphi(\cdot)$ is continuous at the origin iff φ^* has w^*-compact sublevel sets.

Problem 6.21 Let X be a Banach space, $C \subseteq X$ a nonempty, closed, convex set and define $\varphi_C(x) = d(x, C)$. Show that if $x \notin C$ and $x^* \in \partial\varphi_C(x)$, then $\|x^*\|_* = 1$.

Problem 6.22 Let X be a Banach space and $\varphi \in \text{Lip}_{loc}(X, \mathbb{R})$. Is φ Lipschitz on every bounded subset of X? Justify your answer.

Problem 6.23 Let X be a Banach space with strictly convex dual X^*. Suppose that $C \subseteq X$ is nonempty, closed, convex and $\varphi_C(x) = d(x, C)$. Show that if $x \notin C$, then $\partial\varphi_c(x)$ is a singleton.

Problem 6.24 Let $\varphi : \mathbb{R}^N \to \overline{\mathbb{R}} = \mathbb{R} \cup \{+\infty\}$ be a convex and $x_0 \in \text{dom}\,\varphi$. Show that $\varphi(\cdot)$ is Frechet differentiable at x_0 iff it is Gateaux differentiable at x_0 iff $\frac{\partial\varphi}{\partial Z_k}(x_0)$ exists for all $k \in \{1, \ldots, N\}$.

Problem 6.25 Let $M(n)$ be the space of all $n \times n$ real matrices and $\text{GL}(n) \subseteq M(n)$ the collection of all the invertible ones. Show that $\text{GL}(n) \subseteq M(n)$ is open and $f(A) = A^{-1}$ is Frechet differentiable and find its derivative.

Chapter 7
Nonlinear Operators

In this chapter we introduce and study some classes of nonlinear operators which we encounter often in applications. In Sect. 7.1, we consider compact maps, potential maps, nonlinear Fredholm maps, and proper maps. In Sect. 7.2 we consider monotone and maximal monotone operators. Such maps under coercivity conditions exhibit remarkable surjectivity properties. So, they lead to existence results for boundary value problems. A very characteristic example of a maximal monotone operator is the subdifferential $\partial\varphi(\cdot)$ of a function $\varphi \in \Gamma_0(X)$ (see Chap. 6). In Sect. 7.3 we introduce a broader class of nonlinear operators, which still has useful surjectivity properties. This is the class of pseudomonotone maps. We also deal with generalized pseudomonotone maps and maps of type $(S)_+$. The latter property is useful in verifying the compactness condition for the energy (Euler) functional, in the treatment of a boundary value problem using critical point theory (variational approach).

7.1 Compact and Fredholm Maps

In Sect. 3.7 we introduced compact maps (see Definition 3.146) and indicated that this is the class of maps to which finite dimensional results can be readily extended. For easy reference, let us recall the definition of compact maps.

Definition 7.1 Let X and Y be Banach spaces and $D \subseteq X$ nonempty. A map $f : D \to Y$ is said to be "compact" if it is continuous and maps bounded sets in D to relatively compact subsets of Y. We denote the set of compact maps by $K(D, Y)$. Related is also the notion of "finite rank map." We say that $f : D \to Y$ is "finite rank," if it is continuous and bounded (maps bounded sets to bounded sets) and takes values in a finite dimensional subspace of Y. We denote the space of "finite rank maps" by $K_f(D, Y)$. Evidently $K_f(D, Y) \subseteq K(D, Y)$. If $D = X$, the space

© The Author(s), under exclusive license to Springer Nature Switzerland AG 2022
S. Hu, N. S. Papageorgiou, *Research Topics in Analysis, Volume I*, Birkhäuser
Advanced Texts Basler Lehrbücher, https://doi.org/10.1007/978-3-031-17837-5_7

of linear, compact (resp., linear, finite rank) operators from X into Y is denoted by $\mathscr{L}_c(X, Y)$ (resp., $\mathscr{L}_f(X, Y)$).

We know that the elements of $K(D, Y)$ can be approximated by those in $K_f(D, Y)$. More precisely we have the following (see Theorem 3.151):

Theorem 7.2 *If X and Y are Banach spaces, $D \subseteq X$ is nonempty bounded and $f : D \to Y$, then the following statements are equivalent:*

(a) $f \in K(D, Y)$.
(b) Given $\varepsilon > 0$, we can find $f_\varepsilon \in K_f(D, Y)$ such that $\|f(x) - f_\varepsilon(x)\|_Y < \varepsilon$, for any $x \in D$; moreover $f_\varepsilon(D) \subseteq \overline{\mathrm{conv}} f(D)$

Remark 7.3 In fact, every $f \in K(D, Y)$ can be represented by a uniformly convergent series of maps in $K_f(D, Y)$, that is, $f(x) = \sum\limits_{n \geq 1} g_n(x)$ with $g_n \in K_f(D, Y)$ and $\|g_n(x)\|_Y \leq \frac{\varepsilon}{2^n}$ for all $x \in D$.

An important consequence of Theorem 7.2 is the following generalization of the "Tietze Extension Theorem" (see Dugundji [90] and Granas-Dugundji [128], Theorem 7.4, p.163 and Theorem 3.154).

Theorem 7.4 *If X and Y are Banach spaces, $D \subseteq X$ is nonempty and closed, and $f \in K(D, Y)$, then there exists $\widehat{f} \in K(X, Y)$ such that $\widehat{f} \subseteq \overline{\mathrm{conv}} f(D)$.*

Remark 7.5 As we already indicated in Theorem 3.151, the above extension is true for every $f \in C(D, Y)$, which admits a continuous extension $\widehat{f} \in C(X, Y)$ such that $\widehat{f} \subseteq \overline{\mathrm{conv}} f(D)$. In this more general form, the theorem implies that in a Banach space, every closed, convex set is a retract Recall that if X is a Banach space and $D \subseteq X$ is nonempty, then we say that D is a "retract," if there exists a continuous map $r : X \to D$ (called the "retraction" of X on D), if $r|_D = id$. Equivalently, $D \subseteq X$ is retract, if $id|_D$ admits a continuous extension on all of X.

Proposition 7.6 *If X and Y are Banach spaces, $U \subseteq X$ is nonempty and open, and $f \in K(U, Y)$ is Frechet differentiable in U, then $f'(x) \in \mathscr{L}_c(X, Y)$ for all $x \in U$.*

Proof We argue by contradiction. So, suppose that for some $x \in Uf'(x) \in \mathscr{L}(X, Y)$ is not compact. This means that $f'(x)(\partial B_1^X) \subseteq Y$ is not relatively compact. So, we can find $v_n \subseteq \partial B_1^X = \{x \in X : \|x\|_X = 1\}$ and $\varepsilon > 0$ such that

$$\|f'(x_0)(v_n - v_m)\|_Y \geq \varepsilon \quad \text{for all } n, m \in N, \ n \neq m. \tag{7.1}$$

Then on account of the Frechet differentiability of f, for $t > 0$ small, we have

$$\|f'(x)(v_n - v_m)\|_Y \leq \|f(x + tv_n) - f(x) - tf'(x)v_n\|_Y$$
$$+ \|f(x + tv_m) - f(x) - tf'(x)v_m\|_Y$$
$$+ \|f(x + tv_n) - f(x + tv_m)\|_Y,$$

and thus $t\varepsilon - \varepsilon(t) \leq \|f(x + tv_n) - f(x + tv_m)\|_Y$ with $\frac{\varepsilon(t)}{t} \to 0$ as $t \to 0^+$, see (7.1). Hence $\{(x + tv_n)\}_{n \in N} \subseteq f(U) \subseteq Y$ has no convergent subsequence. This contradicts the compactness of $f(\cdot)$. □

Next we identify another class of maps which arise often in applications and are generalizations of self-adjoint operators.

Definition 7.7 Let X be a Banach space, X^* its dual, $U \subseteq X$ nonempty and open, and $f \in C(U, X^*)$. We say that f is a "potential map," if we can find $\varphi \in C^1(U)$ such that $\varphi' = f(x)$ for all $x \in U$.

A straightforward criterion for potentiality of f is given by the next proposition, the proof of which is left as an exercise (see Problem 7.27).

Proposition 7.8 *If $f \in C^1(U, X^*)$, then f is a potential map if and only if*

$$\langle f'(x)h_1, h_2 \rangle = \langle f'(x)h_2, h_1 \rangle$$

for all $x \in U$, all $h_1, h_2 \in X$.

Theorem 7.9 *If X is a Banach space, $U \subseteq X$ is nonempty, open, and convex with $0 \in U$, and $f \in C^1(U, X^*)$, then the following statements are equivalent:*

(a) f is a potential operator.
(b) $\int_{x_0}^{x} f(u)du$ is independent of the path joining $x_0, x \in U$.

Proof $(a) \Rightarrow (b)$ By hypothesis, we have $f = \varphi'$ with $\varphi \in C^1(U)$. Let C be a curve in U (simple and rectifiable). We have

$$\int_C f(s(t))ds(t) = \int_C \varphi'(s(t))ds(t)$$

$$= \int_0^1 \frac{d}{dt}\varphi(s(t))dt$$

$$= \varphi(s(1)) - \varphi(s(0))$$

$$= \varphi(x) - \varphi(x_0).$$

$(b) \Rightarrow (a)$ Define $\varphi(x) = \int_0^1 \langle f(sx), x \rangle ds$. We have

$$\varphi(x+h) - \varphi(x) = \int_0^1 \langle f(s(x+h)), h\rangle ds + \int_0^1 \langle f(s(x+h)) - f(sx), xh\rangle ds.$$
(7.2)

Notice that

$$
\begin{aligned}
\int_0^1 \langle f(s(x+h)) - f(sx), x\rangle ds &= \int_0^1 \left[\int_0^s \frac{d}{dt}\langle f(sx+th), x\rangle dt \right] ds \\
&= \int_0^1 \int_0^s \langle f'(sx+th)h, x\rangle dt ds \\
&= \int_0^1 \int_t^1 \langle f'(sx+th)x, h\rangle ds dt \\
&= \int_0^1 \langle f(x+th) - f(tx+th), h\rangle dt
\end{aligned}
$$
(7.3)

(interchanging the order of integration and using Proposition 7.8)

Using (7.3) in (7.2), we obtain

$$\varphi(x+h) - \varphi(x) = \int_0^1 \langle f(x+th), h\rangle dt,$$

$$\Rightarrow |\varphi(x+h) - \varphi(x) - \langle f(x), h\rangle| = \varepsilon(\|h\|_X) \text{ with } \frac{\varepsilon(\|h\|_X)}{\|h\|_X} \to 0$$

$$\Rightarrow \varphi'(x) = f(x). \qquad \square$$

In Hilbert spaces, the property of being a potential operator is preserved under the action of self-adjoint operators.

Proposition 7.10 *If H is Hilbert space, $U \subseteq H$ is nonempty and open, $f \in C(H, H)$ is a potential map, and $A \in \mathcal{L}(H)$ is self-adjoint, then $g = A \circ f \circ A$ is potential map too.*

Proof Let $\varphi \in C^1(U)$ such that $\nabla \varphi = f$. We set $\widehat{\varphi}(x) = (\varphi \circ A)x$. Then

$$\nabla \widehat{\varphi}(x) = (A \circ f \circ A)(x), \text{ for all } x \in H. \qquad \square$$

Proposition 7.11 *If X is a Banach space, $f \in C(X, X^*)$ is potential map with potential function φ (that is, $\varphi' = f$), and $\varphi(0) = 0$, then*

(a) $\varphi(x) = \int_0^1 \langle f(sx), x\rangle ds$.

(b) *If f is completely continuous (see Definition 3.146), then φ is sequentially continuous from X with the w-topology into $\mathbb{R}$.*

(c) *φ is convex if and only if $\langle f(x) - f(u), x - u\rangle \geq 0$, for all $x, u \in X$.*

Proof

(a) Let $\widehat{\varphi}(x) = \int_0^1 \langle f(sx), x \rangle ds$ for for all $x \in X$. We have

$$\frac{d}{dt}\widehat{\varphi}(x + th) \mid_{t=0} = \langle f(x), h \rangle \text{ for every } h \in X, \Rightarrow \widehat{\varphi}'(x) = f(x), \text{ for all } x \in X.$$

Since $\widehat{\varphi}(0) = 0$, we infer that $\widehat{\varphi} = \varphi$.

(b) Let $x_n \xrightarrow{w} x$ in X. We have

$$\varphi(x_n) - \varphi(x) = \int_0^1 \langle f(x + s(x_n - x)), x_n - x \rangle ds$$

$$= \int_0^1 \langle f(x + s(x_n - x)) - f(x), x_n - x \rangle ds + \langle f(x), x_n - x \rangle,$$

$$\Rightarrow \lim_{n \to \infty} [\varphi(x_n) - \varphi(x)] = 0 \text{ (recall that } f(\cdot) \text{ is completely continuous)}.$$

(c) $\Rightarrow$ The convexity of $\varphi(\cdot)$ implies that

$$\varphi(x + t(u - x)) \leq (1 - t)\varphi(x) + t\varphi(u) \text{ for all } u, x \in X, \text{ all } t \in [0, 1],$$

$$\Rightarrow t\varphi(x) + \varphi(x + t(u - x)) - \varphi(x) \leq t\varphi(u),$$

$$\Rightarrow \varphi(x) + \frac{\varphi(x + t(u - x)) - \varphi(x)}{t} \leq \varphi(u),$$

$$\Rightarrow \varphi(x) + \langle \varphi'(x), u - x \rangle \leq \varphi(u). \tag{7.4}$$

Interchanging the roles of x and u in the above argument, we have

$$\varphi(u) + \langle \varphi'(u), x - u \rangle \leq \varphi(x). \tag{7.5}$$

Adding (7.4) and (7.5), we conclude that

$$0 \leq \langle \varphi'(x) - \varphi'(u), x - u \rangle = \langle f(x) - f(u), x - u \rangle \text{ for all } u, x \in X.$$

$\Leftarrow$ Without any loss of generality, we may assume that $\varphi(0) = 0$. Then from (a), we have

$$\varphi(tu - (1-t)x) = \varphi(x) + t \int_0^1 \langle f(x + st(u-x)), u-x \rangle ds$$

(using the monotonicty hypothesis)

$$= \varphi(x) + t[\varphi(u) - \varphi(x)]$$

$$= (1-t)\varphi(x) + t\varphi(u) \text{ for all } u, x \in X,$$

$\Rightarrow \varphi(\cdot)$ is convex.   □

Next we extend the notion of a Fredholm operator (see Definition 3.161).

Definition 7.12 Let X and Y be Banach spaces, $U \subseteq X$ is a nonempty domain (that is, open and connected). A map $f \in C^1(U, Y)$ is said to be a "Fredholm map" if $f'(x) \in \mathcal{L}(X, Y)$ is a Fredholm operator (see Definition 3.161). Then the "Fredholm index" of f is defined by ind $f = \text{ind } f'(x) = \dim \ker f'(x) - \text{codim } f'(x)$ for all $x \in U$.

Remark 7.13 The above notion of index is well defined, that is, it is independent of $x \in U$. Indeed, note that since $f \in C^1(U, Y)$, $x \to \text{ind } f'(x)$ from U into Z is continuous and because U is connected it must be constant. Hence, ind $f = \text{ind } f'$ is independent of $x \in U$. A diffeomorphism $f : X \to Y$ is a Fredholm map of index zero. Also, if f is a Fredholm map and $g \in C^1(U, Y)$ is compact, then on account of Proposition 7.6, $f + g$ is a Fredholm map and $\text{ind}(f + g) = \text{ind } f$.

First we deal with linear Fredholm maps (that is, with Fredholm operators, see Definition 3.161)

We start with a Lemma, which characterizes operators with a closed range and finite dimensional kernel. These are essential properties in our effort to recognize Fredholm operators (see Definition 3.161).

Lemma 7.14 *If X and Y are Banach spaces and $A \in \mathcal{L}(X, Y)$, then the following statements are equivalent:*

(a) A has closed range and a finite dimensional kernel.
(b) There exists a Banach space V, $K \in \mathcal{L}_c(X, V)$ and $c > 0$ such that

$$\|x\|_X \leq c[\|A(x)\|_Y + \|K(x)\|_V]$$

for all $x \in X$

Proof $(a) \Rightarrow (b)$ Let $m = \dim \ker A$ and we consider a basis $\{e_k\}_{k=1}^m$ for ker A. We can find $\{e_i^*\}_{i=1}^m \subseteq X^*$ such that

$$\langle e_i^*, e_k \rangle = \delta_{ik} = \begin{cases} 1 & if \quad i = k, \\ 0 & if \quad i \neq k, \end{cases}$$

$i, k \in 1, 2, \ldots, m$ (see Proposition 3.55). Let $V = \ker A$ and consider $K \in \mathscr{L}(X, Y)$ defined by

$$K(x) = \sum_{k=1}^{m} \langle e_k^*, x \rangle e_k.$$

Since V is finite dimensional, K is compact and $K|_V = id|_V$. So the operator $L : X \to Y \times X$ defined by $L(x) = (A(x), K(x))$ is 1-1 and $R(L) = R(A) \times V$ is a closed subspace of $Y \times V$. Hence by Remark 3.146 from Sect. 3.7, we can find $c > 0$ such that

$$\|x\|_X \leq c[\|A(x)\|_Y + \|K(x)\|_V], \text{ for all } x \in X. \tag{7.6}$$

$((b) \Rightarrow (a))$ First we show that $\dim \ker A < \infty$. If suffices to show that bounded sets are relatively compact. So, let $\{x_n\}_{k \in \mathbb{N}} \subseteq V = \ker A$ be bounded. Since $K \in \mathscr{L}_c(X, V)$, we can find a subsequence $\{x_{n_k}\}_{k \in \mathbb{N}}$ such that $\{K(x_{n_k})\}_{k \in \mathbb{N}} \subseteq V$ is Cauchy. We have $A(x_{n_k}) = 0$ for our $k \in \mathbb{N}$. Then from (7.6), we have

$$\|x_{n_k} - x_{n_i}\|_X \leq c \|K(x_{n_k} - x_{n_i})\|_V \text{ for all } k, i \in \mathbb{N},$$

$$\Rightarrow \{x_{n_k}\}_k \subseteq X \text{ is Cauchy,}$$

$$\Rightarrow x_{n_k} \to x \in V.$$

This proves that V is finite dimensional.

Next we show that $R(A) \subseteq Y$ is closed. We claim that

$$\inf[\|x + v\|_X : v \in \ker A] \leq \widehat{c} \|A(x)_Y\| \text{ for all } x \in X, \text{ some } \widehat{c} > 0. \tag{7.7}$$

We argue indirectly. So, suppose we could find $\{x_n\}_n \in \mathbb{N}$ such that

$$\inf[\|x_n + v\|_X : v \in \ker A] > n \|A(x_n)\|_Y \text{ for all } x \in X. \tag{7.8}$$

Without any loss of generality, we may assume that

$$\inf[\|x_n + v\|_X : v \in \ker A] = 1, 1 \leq \|x_n\| \leq 2, \text{ for all } n \in \mathbb{N}. \tag{7.9}$$

From (7.8) and (7.9), we have $\|A(x_n)\|_Y < \frac{1}{n}$ for all $n \in \mathbb{N}$. Hence,

$$A(x_n) \to 0, \text{ in } Y \text{ as } n \to \infty.$$

Also $\{x_n\}_{n \in \mathbb{N}} \subseteq X$ is bounded (see (7.9)). Since K is compact, we see that for some subsequence $\{x_{n_k}\}_{k \in \mathbb{N}}$, we have that $\{K(x_{n_k})\}_{n \in \mathbb{N}} \subseteq V$ is Cauchy and so $x_{n_k} \to x$ and clearly $x \in \ker A$. So, from (7.9), we have

$$1 = \inf[\|x_{n_k} + v\| : V \in \ker A] \leq \|x_{n_k} - x\| \text{ for all } k \in \mathbb{N},$$

which is a contradiction since $x_{n_k} \to x$. Therefore, (7.9) is true and it follows that $R(A) \subseteq Y$ is closed.   □

Now we are ready for the theorem that characterizes linear Fredholm operators.

Theorem 7.15 *If X and Y are Banach spaces and $A \in \mathcal{L}(X, Y)$, then the following statements are equivalent:*

(a) A is a Fredholm operator (see Definition 3.161).
(b) There exists $L \in \mathcal{L}(X, Y)$ such that

$$id_X - LA \in \mathscr{L}_c(X, X) \text{ and } id_Y - AL \in \mathscr{L}_c(Y, Y).$$

Proof $(a) \Rightarrow (b)$. Let $V = \ker A$ and $Y_0 = R(A)$. Then by Proposition 3.77, we can find $V_1 \subseteq X$ and $Y_1 \subseteq Y$ closed linear subspaces such that

$$X = V \oplus V_1 \text{ and } Y = Y_0 \oplus Y_1.$$

Hence $A_1 = A \mid_{V_1} \in \mathscr{L}(V_1, Y_1)$ is a bijection. Corollary 3.83 implies that $A^{-1} \in \mathscr{L}(Y_1, V_1)$. Let $L : Y \to X$ be defined by

$$L(y_0 + y_1) = A_1^{-1}(y_1) \text{ for all } y_0 \in Y_0, y_1 \in Y_1.$$

We have

$$(A \circ L)(y_0 + y_1) = y_1 \text{ and } (L \circ A)(x_0 + x_1) = x_1,$$

$$\Rightarrow (id_Y - AL)(y_0 + y_1) = y_0 \text{ and } (id_X - LA)(x_0 + x_1) = x_0$$

for all $x_0 \in V, x_i \in V_1, y_0 \in Y_0, y_1 \in Y_1$. But V and Y_1 are finite dimensional and so

$$F = id_Y - AL \in \mathscr{L}_f(Y, Y) \subseteq \mathscr{L}_c(Y, Y)$$

and

$$K = id_X - LA \in \mathscr{L}_f(X, X) \subseteq \mathscr{L}_c(X, X).$$

$(b) \Rightarrow (a)$ We have

$$\|x\|_X = \|LA(x) + K(x)\|_X \leq c[\|A(x)\|_Y + \|K(x)\|_X]$$

for all $x \in X$ with $c = \max\{1, \|L\|_{\mathscr{L}}\}$. By Lemma 7.14, it follows that dim ker $A \leq \infty$ and $R(A) \subseteq Y$ is closed.

Moreover, $F^* \in \mathcal{L}_c(Y^*, Y^*)$ and

$$\|y^*\|_{Y^*} = \|L^*A^*(y^*) + F^*(y^*)\|_{Y^*} \leq \widehat{c}[\|A^*(y^*)\|_{Y^*} + \|F^*(y^*)\|_{Y^*}]$$

for all $y* \in Y^*$, and some $\widehat{c} > 0$, $\Rightarrow$: ker A^* is of finite dimensional,

$\Rightarrow$: $^{\perp}$ ker A^* is of finite codimension, hence closed. $\square$

Then using this theorem, we can characterize nonlinear Fredholm maps (see Definition 7.12).

Theorem 7.16 *If X and Y are Banach spaces, $U \subseteq X$ is nonempty open, and $f \in C^1(U, Y)$, then the following statements are equivalent:*

(a) $f(\cdot)$ is a Fredholm map.
(b) For each fixed $x \in U$, we can find $L_x \in \mathscr{L}(X, Y)$ such that $id_x - L_x f'(x) \in \mathscr{L}_c(X, X)$ and $id_Y - f'(x)L_x \in \mathscr{L}_c(Y, Y)$.

Definition 7.17 Let X and Y be Banach spaces, $U \subseteq X$ is nonempty open and $f \in C^1(U, Y)$. We say that $x \in U$ is a "regular point" of $f(\cdot)$, if $f'(x) \in \mathscr{L}(X, Y)$ is surjective. otherwise we say that $x \in U$ is a "singular point." If $x \in U$ is a regular(resp., singular) point, then $y = f(x)$ is a "regular value" (resp., "singular value").

Proposition 7.18 *If X, Y are Banach spaces, $U \subseteq X$ is nonempty open, $f \in C^1(U, Y)$ is a Fredholm map, and $S_f \subseteq X$ is the set of singular points of $f(\cdot)$, then $S_f \subseteq X$ is closed.*

Proof By Definition 7.17,

$$S_f = \{x \in U : f'(x) \text{ is not surjective }\}.$$

Let $\{x_n\}_{n \in N} \subseteq S_f$ and assume that $x_n \to x$. The continuity of the index of f under small perturbations implies that

$$\text{ind } f'(x_n) = \text{ind } f'(x) \text{ for all large } n \in \mathbb{N}.$$

Also, if $B \in \mathscr{L}(X, Y)$ and $\|B\|_{\mathscr{L}}$ is small, then

$$\dim \text{coker}(f'(x) + B) \leq \dim \text{coker } f'(x),$$

So, for large $n \in \mathbb{N}$, we have

$$\dim \text{coker } f'(x_n) = \dim \text{coker}[f'(x) + (f'(x_n) - f'(x))] \leq \dim \text{coker } f'(x).$$

$\Rightarrow f'(x) \in \mathscr{L}(X, Y)$ is not subjective and so $x \in S_f$. $\square$

Let X and Y be Banach spaces and $f \in C(X, Y)$. We know that f maps compact sets to compact sets. However, inverse image of a compact set need not be compact (think of a constant map). Functions that return compact sets to compact ones are important, because such maps restrict the size of the solution set of an abstract equation of the form $f(x) = y_0$.

Definition 7.19 Let X and Y be Banach spaces and $f \in C(X, Y)$. We say that $f(\cdot)$ is "proper," if the inverse image of compact set $K \subseteq Y$, $f^{-1}(K) \subseteq X$ is compact too.

Proposition 7.20 *If X and Y are finite dimensional Banach spaces and $f \in C(X, Y)$, then $f(\cdot)$ is proper if and only if $f(\cdot)$ is coercive (that is, $\|f(x)\|_y \to \infty$ as $\|x\|_* \to \infty$).*

Proof $\Rightarrow$ On account of the properness hypothesis, $f(\cdot)$ returns bounded sets to bounded sets. But this is nothing else but a restatement of the coercivity property.

$\Leftarrow$ Since $f(\cdot)$ is coercive, if $K \subseteq Y$ is compact, $f^{-1}(K) \subseteq X$ is bounded closed, hence compact (recall that X is finite dimensional). $\qquad\square$

Proposition 7.21 *If X and Y are Banach spaces and $f \in C(X, Y)$, then $f(\cdot)$ is proper if and only if $f(\cdot)$ is closed and $S_y = \{x \in X : f(x) = y\}$ is compact for every $y \in Y$.*

Proof $\Rightarrow$ Since $f(\cdot)$ is proper and $\{y\} \subseteq Y$ is compact, it follows that $S_y \subseteq X$ is compact. Next we show that $f(\cdot)$ is closed. So, let $C \subseteq X$ be closed and let $\{y_n\}_{n \in \mathbb{N}} \subseteq f(C)$ such that $y_n \to y$ in Y. We have $y_n = f(x_n)$ with $x_n \in C$. Note that $K = \overline{\{y_n\}}_{n \in \mathbb{N}} = \{y_n, y\}_{n \in \mathbb{N}}$ is compact. Hence $f^{-1}(K) \subseteq X$ is compact (since $f(\cdot)$ is proper). Therefore we may assume that $x_n \to x \in C$. We have

$$y_n = f(x_n) \to f(x) = y \in f(C),$$

$$\Rightarrow f \text{ is closed.}$$

$\Leftarrow$ Let $K \subseteq Y$ be compact. We need to show that $f^{-1}(K) \subseteq X$ is compact too. So, let $\{x_n\}_{n \in \mathbb{N}} \subseteq f^{-1}(K)$. We need to show that this sequence has a cluster point. We have $y_n = f(x_n) \in K$ and so we may assume that $y_n = f(x_n) \to y$ in Y. Let $C_n = \{x_m\}_{m \geqslant n}$ and $D = f^{-1}(y) \subseteq X$. By hypothesis, $f(\overline{C_n}) = \overline{f(C_n)}$ (since $f(\cdot)$ is closed) and $D \subseteq X$ is compact (by hypothesis). We have

$$\{y\} = \bigcap_{n \in \mathbb{N}} \overline{f(C_n)} = \bigcap_{n \in \mathbb{N}} f(\overline{C_n}).$$

So, if $x \in D$, then $f(x) \in \bigcap_{n \in \mathbb{N}} f(\overline{C_n})$ and this implies that the closed sets $B_n = D \cap \overline{C_n} \subseteq D, n \in \mathbb{N}$ are nonempty and closed. The family $\{B_n\}_{n \in \mathbb{N}}$ has the finite intersection property. Therefore $\bigcap\limits_{n\in\mathbb{N}} B_n \neq \emptyset$ and so $\{x_n\}_{n\in\mathbb{N}}$ has a cluster point $x \in D$. □

Proposition 7.22 *If X and Y are Banach spaces and $f \in C(X, Y)$, $f(\cdot)$ is coercive(that is, $\|f(x)\|_Y \to \infty$ as $\|x\|_X \to \infty$), and one of the following is true:*

(i) $f = f_o + g$ with f_o proper, g compact.

(ii) X is reflexive and $x_n \overset{\omega}{\longrightarrow} x$ in X, $f(x_n) \to y$ in Y imply $x_n \to x$ in X,

then $f(\cdot)$ is proper.

Proof First assume that (i) holds.

Let $y_n \to y$ in Y and $y_n = f(x_n)$ for all $n \in \mathbb{N}$. On account of the coercivity of $f(\cdot)$, we have $\{x_n\}_{n\in\mathbb{N}} \subseteq X$ bounded. Because $g(\cdot)$ is compact, we may assume that $g(x_n) \to v$ in Y. Then $f_o(x_n) = f(x_n) - g(x_n) \to y - v$ in Y since $f(\cdot)$ is proper. It follows that by passing to a suitable subsequence, we have $x_n \to x$ in X. Then $f(x) = y$ and so $f(\cdot)$ is proper.

Next we assume that (ii) holds.

In this case since $\{x_n\}_{n\in\mathbb{N}} \subseteq X$ is bounded and X is reflexive, we have that $\{x_n\}_{n\in\mathbb{N}}$ is relatively w-compact. So, by the Eberlein–Smulian theorem, we may assume that $x_n \overset{\omega}{\longrightarrow} x$ in X. Then by hypothesis, we also have $x_n \to x$ in X and so by the continuity of $f(\cdot)$, we infer that $f(x) = y$, which proves that $f(\cdot)$ is proper. □

The next proposition is a straightforward consequence of Definition 7.19.

Proposition 7.23 *If X and Y are Banach spaces and $f \in C(X, Y)$ is $1-1$, then the following statements are equivalent:*

(a) $f(\cdot)$ is proper.
(b) $f(\cdot)$ is closed.
(c) f is a bicontinuous map from X onto $f(X)$.

Another easy consequence of Definition 7.19 is the next proposition.

Proposition 7.24 *If X, Y, and V are Banach spaces, $f \in C(X, Y)$, and $g \in C(Y, V)$, then*

(a) If f and g are both proper, then so is $g \circ f$.
(b) If $g \circ f$ is proper, then f is proper.
(d) If $g \circ f$ is proper and f is surjective, then g is proper.

Remark 7.25 Suppose X, Y are Banach spaces, $f \in C(X, Y)$, and we study the equation

$$f(x) = y_0, \quad y_0 \in Y. \tag{7.10}$$

We approximate the forcing term (right hand side) by $\{y_n\}_{n\in\mathbb{N}}$, $y_n \to y_0$. Suppose that the approximating equation $f(x) = y_n$ has solutions $\{x_n\} \subseteq X$. If f is proper, then we can extract a subsequence $\{x_{n_k}\}$ such that $x_{n_k} \to x_0$ with x_0 a solution of (7.10). Therefore we see that the concept of properness is crucial in approximation theory.

7.2 Monotone Operators

We start by introducing our notation. Suppose X and Y are vector spaces and $A : X \to 2^Y$ is a multivalued map. We introduce the following sets:

$$D(A) = \{x \in X : A(x) \neq \emptyset\} \text{ (the domain of A)},$$

$$R(A) = \bigcup_{x \in D(A)} A(x) \text{ (the range of A)},$$

$$\text{Gr } A = \{(x, y) \in X \times Y : y \in A(x)\} \text{ (the graph of A)},$$

$$A^{-1}(y) = \{x \in X : (x, y) \in \text{Gr } A\} \text{ for all } y \in Y.$$

From now on, X is a Banach space. Additional hypotheses will be introduced as needed. By X^* we denote the topological dual and by $\langle \cdot, \cdot \rangle$ the duality brackets for the pair (X, X^*) .

Definition 7.26 Consider an operator $A : X \to 2^{X^*}$.

(a) We say that $A(\cdot)$ is "monotone" if

$$\langle x^* - u^*, x - u \rangle \geq 0 \text{ for all } (x, x^*), (u, u^*) \in \text{Gr } A.$$

(b) We say that $A(\cdot)$ is "strictly monotone" if it is monotone, and for $x \neq u$, we have

$$\langle x^* - u^*, x - u \rangle > 0, \text{ for all } (x, x^*), (u, u^*) \in \text{Gr } A.$$

(c) We say that $A(\cdot)$ is "uniformly monotone" if there exists $\xi : \mathbb{R}_+ \longrightarrow \mathbb{R}_+$, strictly increasing, continuous, $\xi(0) = 0, \xi(r) \to +\infty$ as $r \to +\infty$ and

$$\langle u^* - x^*, u - x \rangle \geqslant \xi(\|x - u\|)\|x - u\|, \text{ for all } (x, x^*), (u, u^*) \in \text{Gr } A.$$

If $\xi(x) = cr$ with $c > 0$, then we say that the operator $A(\cdot)$ is "strongly monotone."

(d) Suppose $A(\cdot)$ is single valued. We see that $A(\cdot)$ is "hemicontinuous" if for all $x, h \in X$

$$A(x + th) \xrightarrow{\omega} A(x) \text{ as } t \to 0$$

(that is, for all $x, h, u \in X$, the real function $t \to \langle A(x + th), u \rangle$ is continuous; so the function is directionally weakly continuous).

(e) Suppose $A(\cdot)$ is single valued. We say that $A(\cdot)$ is "demicontinuous," if $x_n \to x$ in X implies $A(x_n) \xrightarrow{w} A(x)$ in X^*

(f) We say that $A(\cdot)$ is " bounded," if it maps bounded sets in X to bounded sets in X^*.

(g) We say that $A(\cdot)$ is "maximal monotone" if Gr A is not properly contained in the graph of another monotone operator.

(h) We say that $A(\cdot)$ is "coercive" ("strongly coercive"), if

$$\inf \left[\|x^*\|_* : x^* \in A(x) \right] \to \infty \text{ as } \|x\| \to \infty$$

$$\left(\text{resp., } \frac{\inf \left[\langle x^*, x \rangle : x^* \in A(x) \right]}{\|x\|} \to \infty \text{ as } \|x\| \to \infty \right).$$

Remark 7.27 According to the above definition, $A : X \to 2^{X^*}$ is "maximal monotone" if and only if "$\langle x^* - u^*, x - u \rangle \geq 0$ for all $(u, u^*) \in$ Gr A implies $(x, x^*) \in$ Gr A." We mention that the above notion of monotonicity is not related to any kind of ordering on X and X^*. It is an abstract generalization of the observation that a function $f : \mathbb{R} \to \mathbb{R}$ is nondecreasing if and only if $(f(x) - f(x')) (x - x') \geq 0$ for all $x, x' \in \mathbb{R}$. We have the following implications:

"A is strongly monotone $\Rightarrow$ A is uniformly monotone

$\Rightarrow$ A is strictly monotone $\Rightarrow$ A is monotone."

Moreover, if $A(0) \subseteq X^*$ is bounded, then we have

"A uniformly monotone $\Rightarrow$ A is strongly coercive."

Indeed note that for all $u^* \in A(u)$, we have

$$\langle u^*, u \rangle = \langle u^* - v^*, u \rangle + \langle v^*, u \rangle \text{ with } v^* \in A(0))$$

$$\geq [\xi(\|u\|) - M]\|u\| \text{ with } \|v^*\|_* \leq M \text{ for all } v^* \in A(0)$$

$$\Rightarrow \frac{\inf \left[\langle u^*, u \rangle : u^* \in A(u) \right]}{\|u\|} \to \infty \text{ as } \|u\| \to \infty$$

$$\Rightarrow A \text{ is strongly coercive.}$$

Finally if $A(\cdot)$ is maximal monotone, then so is $A^{-1} : X^* \to 2^X$.

Example 7.28

(a) As increasing function $f : \mathbb{R} \to \mathbb{R}$ is a monotone operator and it is strictly monotone if it is strictly increasing. But $f(\cdot)$ is a maximal monotone operator if and only if it is continuous. A discontinuous increasing function has jump discontinuities, at most countable . If we fill in these jumps, the resulting multivalued map is maximal monotone. This illustrates the need to consider multivalued operators.

(b) Let $\mathscr{F} : X \to 2^{X^*} \setminus \{\emptyset\}$ be the duality map (see Example 6.91(c)). We have

$$\mathscr{F}(x) = \left\{ x^* \in X^* : \langle x^*, x \rangle = \|x^*\|_*^2 = \|x\|^2 \right\} \text{ for all } x \in X. \tag{7.11}$$

We claim that $\mathscr{F}(\cdot)$ is monotone. Indeed for $(x, x^*), (u, u^*) \in \mathrm{Gr}\,\mathscr{F}$, we have

$$\langle x^* - u^*, x - u \rangle \geqslant \|x\|^2 - 2\|x\|\|u\| + \|u\|^2 \text{ (see (7.11))}$$

$$= [\|(x)\| - \|u\|]^2 \geqslant 0,$$

$$\Rightarrow \mathscr{F} \text{ is monotone.}$$

We will see below that in fact $\mathscr{F}$ is maximal monotone. We will see later in this section that for every $\varphi \in \Gamma_0(x)$, the subdifferential operator is maximal monotone.

(c) Let $X = H$ be a Hilbert spare. We identify $H = H^*$ and consider a nonexpansive map $g : H \to H$, that is,

$$\|g(x) - g(u)\| \leqslant \|x - u\| \text{ for all } x, u \in H.$$

For $|\lambda| \leqslant 1$, let $A = id + \lambda g$. Then $A(\cdot)$ is monotone. Let $(\cdot, \cdot)_H$ denote the inner product of H. We have

$$(A(x) - A(u), x - u)_H$$

$$= (x + \lambda g(x) - u - \lambda g(u), x - u)_H$$

$$\geqslant \|x - u\|^2 - |\lambda|\|g(x) - g(u)\|\|x - u\|$$

$$= [\|x - u\| - |\lambda|\|g(x) - g(u)\|]\|x - u\| \geqslant 0,$$

$$\Rightarrow A(\cdot) \text{ is monotone.}$$

(d) Let $X = H$ be a Hilbert spare, $T = [0, b]$ and $x_0 \in H$. We define $A : L^2(T, H) \to L^2(T, H)$ by $A(u) = u'$ and $u(0) = x_0$ (the time derivative operator with initial condition $x_0 \in H$). Then

$$D(A) = \left\{ u \in W^{1,2}((0, b), H) : u(0) = x_0 \right\}.$$

This operator is monotone. To see this, let $x, u \in D(A)$. We have

$$(A(x) - A(u), x - u)_{L^2(T,H)} = \int_0^b (x'(t) - u'(t), x(t) - u(t))_H dt$$

$$= \int_0^b \frac{1}{2} \frac{d}{dt} \|x(t) - u(t)\|^2 dt$$

$$= \frac{1}{2} \|x(b) - u(b)\|^2 \geqslant 0,$$

$$\Rightarrow A(\cdot) \text{ is monotone}.$$

Proposition 7.29 *If $A : X \to 2^{X^*}$ is maximal monotone, then for every every $x \in D(A)$, $A(x)$ is nonempty, closed, and convex.*

Proof First we show that $A(x) \subseteq X^*$ is closed. So, let $\{x_n^*\}_{n \in \mathbb{N}} \subseteq A(x)$ and assume that $x_n^* \to x^*$ in X^*. We have

$$\langle x_n^* - u^*, x - u \rangle \geqslant 0 \text{ for all } (u, u^*) \in \text{Gr } A, \text{ and all } n \in \mathbb{N},$$

$$\Rightarrow \langle x^* - u^*, x - u \rangle \geqslant 0 \text{ for all } (u, u^*) \in \text{Gr } A,$$

$$\Rightarrow x^* \in A(x) \text{ (since } A \text{ is maximal monotone)}.$$

So $A(x) \subseteq X^*$ is closed.
Next we show that $A(x)$ is convex. So, let $x_1^*, x_2^* \in A(x)$ and set

$$x_t^* = t x_1^* + (1 - t) x_2^* \text{ with } 0 < t < 1.$$

We have for all $(u, u^*) \in \text{Gr } A$,

$$\langle x_t^* - u^*, x - u \rangle = t \langle x_1^* - u^*, x - u \rangle + (1 - t) \langle x_2^* - u^*, x - u \rangle \geqslant 0.$$

Thus, $x_t^* \in A(x)$ (since A is maximal monotone) and $A(x)$ is convex. $\qquad \square$

Remark 7.30 So, by Mazur's Theorem (see Theorem 3.105), a maximal monotone operator $A : X \to 2^{X^*}$ has w-closed values.

Definition 7.31

(a) Let $C \subseteq X$ be nonempty, $c \in C$. Recall (see Proposition 3.11) that the set C is said to be "absorbing," if for every $x \in X$, we can find $t_x > 0$ such that $t_x x \in C$

(that is, $X = \bigcup_{t>0} tC$). Now suppose that $C \subseteq X$ and $x \in C$. We say that x is on "absorbing point of C," if $\widehat{C} = C - x$ is an absorbing set.

(b) A monotone operator $A : X \to 2^{X^*}$ is said to be "locally bounded at $x \in D(A)$, if there exist $c > 0$ and $\delta > 0$ such that $\|x^*\|_* \le c$ for all $x^* \in A(u), u \in D(A) \cap \overline{B}_\delta(x) \left(\overline{B}_\delta(x) = \{u \in X : \|u - x\| \le \delta\}\right)$.

Remark 7.32 If $\operatorname{int} C \neq \emptyset$ and $x \in \operatorname{int} C$, then clearly x is an absorbing point of C. However, a set $C \subseteq X$ can have absorbing points even if $\operatorname{int} C = \emptyset$. To see this, let $C = \partial B_1 \cup \{0\}$, where $\partial B_1 = \{x \in X : \|x\| = 1\}$. Then clearly C is absorbing, but $\operatorname{int} C = \emptyset$.

Proposition 7.33 *If $A : X \to 2^{X^*}$ is a monotone operator and $x \in D(A)$ is an absorbing point of $D(A)$, then $A(\cdot)$ is locally bounded at x.*

Proof Without any loss of generality, we may assume that $x = 0$ and that $0 \in A(0)$. To see this, fix $x^* \in A(x)$ and replace $A(\cdot)$ by $\widehat{A}(u) = A(u + x) - x^*$. So, we need to show that $A(\cdot)$ is locally banded at $x = 0$. Consider the function

$$f(x) = \sup\left[\langle u^*, x - u \rangle : u \in D(A), \|u\| \le 1, u^* \in A(u)\right] \tag{7.12}$$

and let $L_1 = \{x \in X : f(x) \le 1\}$. The function $f(\cdot)$ is the supremum of affine continuous functions (see (7.12)). Hence $f(\cdot)$ is convex and lower semicontinuous and so the set $L_1 \subseteq X$ is closed convex, $0 \in L_1$. Since $0 \in A(0)$, we see that $f \ge 0$. Also from the monotonicity of $A(\cdot)$, we see that $f(0) \le 0$, hence $f(0) = 0$. Let $C = L_1 \cap (-L_1)$. This set is closed, convex, and symmetric. We will show that it is also absorbing and so a neighborhood of the origin. So, let $x \in X$. Since $D(A)$ is absorbing, we can find $t > 0$ such that $A(tx) \neq \emptyset$. Let $x^* \in A(tx)$. We have

$$\langle u^*, tx - u \rangle \le \langle x^*, tx - u \rangle \quad \text{for all } (u, u^*) \in \operatorname{Gr} A,$$
$$\Rightarrow f(tx) \le \sup\left[\langle x^*, tx - u \rangle : u \in D(A), \|u\| \le 1\right]$$
$$\le \langle x^*, tx \rangle + \|x^*\|_* < \infty.$$

We choose $s \in (0, 1)$ small so that

$$sf(tx) < 1. \tag{7.13}$$

Since $f(\cdot)$ is convex, we have

$$f(stx) \le sf(tx) + (1 - s)f(0) = sf(tx) < 1 \quad \text{(see (7.13))},$$
$$\Rightarrow stx \in L_1.$$

Therefore C is a neighborhood of the origin. So, we can find $\delta > 0$ such that

$f(x) \leq 1$ for all $x \in \overline{B}_{2\delta}(0)$

$$\Rightarrow \langle u^*, x \rangle \leq \langle u^*, u \rangle + 1 \text{ if } x \in \overline{B}_{2\delta}(0), u \in D(A), \|u\| \leqslant 1, u^* \in A(u).$$

So if $u \in D(A) \cap \overline{B}_{\delta}(0), u^* \in A(u)$, then

$$2\delta \left\| u^* \right\|_* = \sup \left[\langle u^*, x \rangle : \|x\| \leqslant 2\delta \right]$$

$$\leq \left\| u^* \right\|_* \|x\| + 1$$

$$\leq \delta \left\| u^* \right\|_* + 1,$$

$$\Rightarrow \left\| u^* \right\|_* \leqslant \frac{1}{\delta}.$$

$\square$

Proposition 7.34 *If X is reflexive and $A : X \to 2^{X^*}$ is a maximal monotone operator with int $D(A) \neq \emptyset$, then $A|_{\text{int } D(A)}$ is usc from X with the norm topology into X^* with the weak topology (denoted by X^*_ω).*

Proof On account of Proposition 7.33 and since X is reflexive, $A|_{\text{int } D(A)}$ is locally compact. So, according to Proposition 5.13, it suffices to show that $\operatorname{Gr} A|_{\text{int } D(A)} \subseteq X \times X^*_w$ is closed. By the Eberlein–Smulian theorem, let $\left\{ (x_n, x_n^*) \right\}_{n \in \mathbb{N}} \subseteq G_r A|_{\text{int } D(A)}$ and assume that $x_n \to x \in \text{int } D(A)$ in $X, x_n^* \xrightarrow{w} x^*$ in X^*.

We will show that $x^* \in A(x)$. Note that

$$0 \leqslant \langle x_n^* - u^*, x_n - u \rangle \text{ for all } n \in \mathbb{N}, \text{ all } (u, u^*) \in \operatorname{Gr} A,$$

$$\Rightarrow 0 \leqslant \langle x^* - u^*, x - u \rangle \text{ for all } (u, u^*) \in \operatorname{Gr} A,$$

$$\Rightarrow x^* \in A(x) (\text{ since } A(\cdot) \text{ is maximal monotone}),$$

$$\Rightarrow A|_{\text{int } D(A)} \text{ is use from } X \text{ into } X^*_w.$$

$\square$

Proposition 7.35 *If X is Banach space and $A : X \to 2^{X^*}$ is a maximal monotone operator, then $\operatorname{Gr} A$ is closed in $X \times X^*_{w^*}$, and in $X_w \times X^*$ (by $X^*_{w^*}$ we denote the space X^* equipped with the w^*-topology and by X_w the space X equipped with the w-topology).*

Proof Let $\left\{ (x_\alpha, x_\alpha^*) \right\}_{\alpha \in J} \subseteq \operatorname{Gr} A$, we have

$$x_\alpha \to x \text{ in } X \text{ and } x_\alpha^* \xrightarrow{w^*} x^* \text{ in } X^*. \tag{7.14}$$

For every $(u, u^*) \in \operatorname{Gr} A$, we have

$$0 \leq \langle x_\alpha^* - u^*, x_\alpha - u \rangle \text{ for all } \alpha \in J,$$
$$\Rightarrow 0 \leq \langle x^* - u^*, x - u \rangle \text{ (see (7.14)).} \tag{7.15}$$

Since $(u, u^*) \in \operatorname{Gr} A$ is arbitrary, (7.15) and the maximal monotonicity of $A(\cdot)$ imply that $(x, x^*) \in \operatorname{Gr} A$.

Similarly we show that $\operatorname{Gr} A \subseteq X_w \times X^*$ is closed. $\qquad\square$

Now we will identify a broad class of monotone operators that are maximal monotone.

Proposition 7.36 *If X is a reflexive Banach space and $A : X \to X^*$ is hemicontinuous monotone, then $A(\cdot)$ is maximal monotone.*

Proof Arguing by contradiction, suppose that $A(\cdot)$ is not maximal monotone. So, we can find $(x_0, x_0^*) \in X \times X^*$ such that

$$x_0^* \neq A(x_0) \text{ and } \langle x_0^* - A(u), x_0 - u \rangle \geq 0 \; for \; all \; u \in X. \tag{7.16}$$

For any $h \in X$, let $u_t = tx_0 + (1 - t)h$ for all $0 \leq t \leq 1$. Using this in (7.16), we have

$$\langle x_0^* - A(u_t), x_0 - h \rangle \geq 0 \text{ for all } h \in X, \text{ all } t \in [0, 1].$$

Passing to the limit as $t \to 1^-$ and using the hemicontinuity of $A(\cdot)$, we obtain

$$\langle x_0^* - A(x_0), x_0 - h \rangle \geq 0 \text{ for all } h \in X$$

$$\Rightarrow x_0^* = A(x_0),$$

a contradiction to (7.16). Hence $A(\cdot)$ is maximal monotone. $\qquad\square$

Remark 7.37 In fact the result is also true for multivalued monotone operators $A : X \to 2^{X^*} \setminus \{\varnothing\}$, if we replace the hemicontinuity requirement by the assumption that for every $x, h \in X$, the multifunction $t \to A(x + th)$ is use from $\mathbb{R}$ into $X_{w^*}^*$. As a consequence of this proposition, we see that all the monotone operators from Examples 7.28 are in fact maximal monotone.

The next result is a straightforward consequence of the definition of maximal monotonicity.

Proposition 7.38 *If X is a reflexive Banach space and $A : X \to 2^{X^*}$, then $A(\cdot)$ is maximal monotone if and only if $A^{-1}(\cdot)$ is.*

The importance of maximal monotone operators comes from the fact that under reasonable conditions exhibit surjectivity properties. Surjectivity results imply existence results.

The main surjectivity result will be obtained using the Galerkin method (that is, finite dimensional approximations). So, we start with a finite dimensional result.

Proposition 7.39 *If X is a finite dimensional Banach space, $C \subseteq X$ is nonempty, closed, and convex, $A : X \to 2^{X^*}$ is a monotone map with $D(A) \subseteq C$, and $F : C \to X^*$ is monotone, continuous, and strongly coercive, then we can find $x_0 \in C$ such that*

$$\langle x^* + F(x_0), x - x_0 \rangle \geq 0 \ for \ all \ (x, x^*) \in \mathrm{Gr}\, A.$$

Proof Fix $(\widehat{x}, \widehat{x}^*) \in \mathrm{Gr}\, A$ and replace A and F by

$$\widehat{A}(x) = A(x + \widehat{x}) - \widehat{x}^* \ and \ \widehat{F}(x) = F(x + \widehat{x}) - \widehat{x}^*.$$

Clearly $\widehat{A}$ and $\widehat{F}$ retain the properties of A and F, respectively, and in addition we have $(0, 0) \in \mathrm{Gr}\, \widehat{A}$. So, without any loss of generality, we may assume that $(0, 0) \in \mathrm{Gr}\, A$.

First we assume that $D(A) \subseteq X$ is bounded. Let $K = \overline{conv}\, D(A)$. Then $K \subseteq X$ is compact and convex. Suppose that the proposition is not true. Then for every $v \in K$, we can find $(x, x^*) \in \mathrm{Gr}\, A$ such that

$$\langle x^* + F(v), x - v \rangle < 0.$$

It follows that

$$K = \bigcup_{(x,x^*) \in \mathrm{Gr}\, A} \{v \in K : \langle x^* + F(v), x - v \rangle < 0\}.$$

Each set in the union is open and K is compact. So, we can find $\{(x_k, x_k^*)\}_{k=1}^n \subseteq \mathrm{Gr}\, A$ such that

$$K = \bigcup_{k=1}^n \{v \in K : \langle x_k^* + F(v), x_k - v \rangle < 0\}.$$

Let $\{\vartheta_k\}_{k=1}^n$ be a continuous partition of unity subordinate to this finite open cover. Then we introduce the map $\xi : K \to K$ defined by

$$\xi(v) = \sum_{i=1}^n \vartheta_i(v) x_i.$$

This map is continuous and K is compact. So, Brouwer's fixed point theorem gives $x_0 \in K$ such that $\xi(x_0) = x_0$. For every $v \in K$, we have

$$g(v) = \langle \sum_{k=1}^{n} \vartheta_k(v)x_k^* + F(v), \xi(v) - v \rangle$$

$$= \langle \sum_{k=1}^{n} \vartheta_k(v)x_k^* + F(v), \sum_{i=1}^{n} \vartheta_i(v)x_i - v \rangle$$

$$= \langle \sum_{k=1}^{n} \vartheta_k(v)x_k^* + F(v), \sum_{i=1}^{n} \vartheta_k(v)(x_i - v) \rangle$$

$$= \sum_{k,i=1}^{n} \vartheta_k(v)\vartheta_i(v)\langle x_k^* + F(v), x_i - v \rangle. \tag{7.17}$$

If $k = i$ and $\vartheta_k^2(v) \neq 0$, then $v \in K$ and so $\langle x_k^* + F(v), x_k - v \rangle < 0$. If $k \neq i$ and $\vartheta_k(v)\vartheta_i(v) \neq 0$, then $v \in \{v \in K : \langle x_k^* + F(v), x_k - v \rangle < 0\} \cap \{v \in K : \langle x_i^* + F(v), x_i - v \rangle < 0\}$. Using the monotonicity of $A(\cdot)$, we have

$$\langle x_k^* + F(v), x_i - v \rangle + \langle x_i^* + F(v), x_k - v \rangle$$

$$= \langle x_k^* + F(v), x_k - v \rangle + \langle x_i^* + F(v), x_i - v \rangle - \langle x_k^* - x_i^*, x_k - x_i \rangle < 0$$

$$\Rightarrow g(v) < 0 \text{ for all } v \in K \text{ (see (7.17))}.$$

On the other hand, we have

$$g(x_0) = \langle \sum_{k=1}^{n} \vartheta_k(x_0)x_k^* + F(x_0), \xi(x_0) - x_0 \rangle = 0,$$

a contradiction. This proves the proposition when $D(A) \subseteq X$ is bounded. Next we remove this restriction.

From the first part of the proof, we can find $x_n \in C$ such that

$$\langle x^* + F(x_n), x - x_n \rangle \geq 0 \text{ for all } (x, x^*) \in Gr(A|_{\overline{B}_n}) \tag{7.18}$$

(recall $\overline{B}_n = \{x \in X : \|x\| \leq n\}$). Recall that $(0, 0) \in Gr\, A$. So from (7.18), we have

$$\langle F(x_n), x_n \rangle \leq 0 \text{ for all } n \in \mathbb{N}. \tag{7.19}$$

From (7.19) and the strong coercivity of $F(\cdot)$, it follows that $\{x_n\}_{n \in \mathbb{N}} \subseteq X$ is bounded. So, we may assume that $x_n \to x_0$. We have

$$x_0 \in C \text{ and } \langle x^* + F(x_0), x - x_0 \rangle \geq 0 \text{ for all } (x, x^*) \in Gr\, A \text{ (see (7.18))}. \qquad \square$$

Using the Galerkin method, we can extend the above result to infinite dimensional Banach spaces.

Proposition 7.40 *If X is a reflexive Banach space, $C \subseteq X$ is nonempty, closed, and convex, $A : X \to 2^{X^*}$ is a monotone map with $D(A) \subseteq C$, and $F : C \to X^*$ is monotone, hemicontinuous, bounded, and strongly coercive, then there exists $x_0 \in C$ such that*

$$\langle x^* + F(x_0), x - x_0 \rangle \geq 0 \ for \ all \ (x, x^*) \in Gr\, A.$$

Proof Consider a directed family $\{X_\alpha\}_{\alpha \in J}$ of finite dimensional subspaces such that $X = \overline{\cup_{\alpha \in J} X_\alpha}$. Let $p_\alpha \in \mathcal{L}(X, X_\alpha)$ be the corresponding projection operator. Evidently $p_\alpha^* = i_\alpha \in \mathcal{L}(X_\alpha^*, X^*)$ is the embedding operator. We set

$$A_\alpha = p_\alpha^* \circ A \circ p_\alpha, \ F_\alpha = p_\alpha^* \circ F \circ p_\alpha \ \text{and} \ C_\alpha = C \cap X_\alpha \ \text{for all} \ \alpha \in J.$$

Then for each $\alpha \in J$, the triple $(A_\alpha, F_\alpha, C_\alpha)$ satisfies the requirements of Proposition 7.39 on the finite dimensional Banach space X_α. So, we can find $x_\alpha \in C_\alpha$ such that $\langle \overline{x}_\alpha^* + F(x_\alpha), \overline{x}_\alpha - x_\alpha \rangle_{X_\alpha} \geq 0$ for all $(\overline{x}_\alpha, \overline{x}_\alpha^*) \in Gr\, A_\alpha$, and thus

$$\langle x^* + F(x_\alpha), x - x_\alpha \rangle \geq 0 \ \text{for all} \ (x, x^*) \in Gr(A \circ p_\alpha). \tag{7.20}$$

Since we can always assume that $(0, 0) \in Gr\, A$, we have

$$\langle F(x_\alpha), x_\alpha \rangle \leq 0 \ \text{for all} \ \alpha \in J,$$

$$\Rightarrow \{x_\alpha\}_{\alpha \in J} \subseteq X \ \text{and} \ \{F(x_\alpha)\}_{\alpha \in J} \subseteq X^* \ \text{are bounded,}$$

since $F(\cdot)$ is strongly coercive and bounded.

Since the space X is reflexive, the sets $\{x_\alpha\}_{\alpha \in J}$ and $\{F(x_\alpha)\}_{\alpha \in J}$ are both relatively w-compact. So, by the Eberlein–Smulian theorem, we can find a subsequence $\{x_{\alpha_n}\}_{n \in \mathbb{N}} \subseteq \{x_\alpha\}_{\alpha \in J}$ such that

$$x_{\alpha_n} \xrightarrow{w} x_0 \ \text{in} \ X \ \text{and} \ F(x_{\alpha_n}) \xrightarrow{w} x_0^* \ \text{in} \ X^*. \tag{7.21}$$

From (7.20), we have

$$\limsup_{n \to \infty} \langle F(x_{\alpha_n}), x_{\alpha_n} \rangle \leq \langle x^*, x - x_0 \rangle + \langle x_0^*, x \rangle \ \text{for all} \ (x, x^*) \in Gr\, A. \tag{7.22}$$

By Zorn's lemma, we can always assume that $A(\cdot)$ is maximal on $D(A)$. We claim that we can find $(x_1, x_1^*) \in Gr\, A$ such that

$$\langle x_1^*, x_1 - x_0 \rangle + \langle x_0^*, x_1 \rangle \leq \langle x_0^*, x_0 \rangle. \tag{7.23}$$

We argue indirectly. So, if the claim is not true, we will have

$$\langle x^* + x_0^*, x - x_0 \rangle > 0 \quad \text{for all} \quad (x, x^*) \in \operatorname{Gr} A, \tag{7.24}$$

and hence $(x_0, -x_0^*) \in \operatorname{Gr} A$ since A is maximal monotone.

Therefore, in (7.24), we can use $(x_0, -x_0^*) \in \operatorname{Gr} A$ and we have a contradiction. So, (7.23) holds for some $(x_1, x_1^*) \in \operatorname{Gr} A$.

We use (7.23) in (7.22) and have

$$\limsup_{n \to \infty} \langle F(x_{\alpha_n}), x_{\alpha_n} \rangle \leq \langle x_0^*, x_0 \rangle,$$

$$\Rightarrow \limsup_{n \to \infty} \langle F(x_{\alpha_n}), x_{\alpha_n} - x_0 \rangle \leq 0 \quad \text{(see (7.21))}. \tag{7.25}$$

Let $x \in D(A)$ and set $x_t = t x_0 + (1 - t)x, t \in [0, 1]$. Evidently $x_t \in C$. Since $F(\cdot)$ is monotone, we have

$$\langle F(x_{\alpha_n}) - F(x_t), x_{\alpha_n} - x_t \rangle \geq 0$$

$$\Rightarrow t \langle F(x_{\alpha_n}), x_{\alpha_n} - x_0 \rangle + (1 - t) \langle F(x_{\alpha_n}), x_{\alpha_n} - x \rangle$$

$$\geq t \langle F(x_t), x_{\alpha_n} - x_0 \rangle + (1 - t) \langle F(x_t), x_{\alpha_n} - x \rangle$$

$$\Rightarrow \liminf_{n \to \infty} \langle F(x_{\alpha_n}), x_{\alpha_n} - x \rangle \geq \langle F(x_t), x_0 - x \rangle \quad \text{for} \quad t \in [0, 1] \quad \text{(see (7.23), (7.25))}.$$

We send $t \to 1^-$ and use the hemicontinuity of $F(\cdot)$. We obtain

$$\liminf_{n \to \infty} \langle F(x_{\alpha_n}), x_{\alpha_n} - x \rangle \geq \langle F(x_0), x_0 - x \rangle$$

$$\Rightarrow \liminf_{n \to \infty} \langle F(x_{\alpha_n}), x_{\alpha_n} \rangle \leq \langle F(x_0), x_0 - x \rangle + \langle x_0^*, x \rangle \quad \text{for} \quad x \in D(A) \text{ (see (7.21))}. \tag{7.26}$$

From (7.22) and (7.26), it follows that

$$\langle F(x_0), x_0 - x \rangle \leq \langle x^*, x - x_0 \rangle \text{ for all } (x, x^*) \in \operatorname{Gr} A. \qquad \square$$

This leads to the following basic surjectivity result.

Theorem 7.41 *If X is a reflexive Banach space, $C \subseteq X$ is nonempty, closed, and convex, $A : X \to 2^{X^*}$ is maximal monotone with $D(A) \subseteq C$, and $F : C \to X^*$ is monotone, hemicontinuous, bounded, and strongly coercive, then $A + F$ is surjective.*

Proof Let $x_0^* \in X^*$ and let $\widehat{A}(x) = A(x) - x_0^*$ for all $x \in D(A)$. Evidently $\widehat{A}(\cdot)$ is still maximal monotone and $D(\widehat{A}) = D(A) \subseteq C$. We apply Proposition 7.40 and obtain $x_0 \in C$ such that

$$\langle \widehat{x}^* + F(x_0), x - x_0 \rangle \geq 0 \ \text{ for all } \ (x, \widehat{x}^*) \in \text{Gr } \widehat{A},$$

$$\Rightarrow \langle x^* - (x_0^* - F(x_0)), x - x_0 \rangle \geq 0 \ \text{ for all } \ (x, x^*) \in \text{Gr } A. \tag{7.27}$$

From (7.27) and the maximal monotonicity of A, we infer that

$$(x_0, x_0^* - F(x_0)) \in \text{Gr } A, \ \Rightarrow x_0^* \in A(x_0) + F(x_0).$$

Since $x_0^* \in X^*$ is arbitrary, we conclude that

$$R(A + F) = X^* \ \text{(that is, } A + F \text{ is surjective).} \qquad \square$$

This theorem leads to a necessary and sufficient condition for surjectivity of a maximal monotone operator. In this direction, we will use the following powerful renorming theorem of Troyanski [269].

Theorem 7.42 *If X is a reflexive Banach space, then we can equivalently renorm X so that both X and X^* are locally uniformly convex.*

Remark 7.43 Since the notion of maximal monotonicity is not affected by equivalent renorming of the ambient space, the above theorem is very useful in the context of reflexive Banach spaces, since it allows to assume rich structure on X and its topological dual, without any cost.

We can use Theorem 7.41 to prove a result which characterizes maximal monotone operators.

Theorem 7.44 *If X is a reflexive Banach space normed so that both X and X^* are locally uniformly convex (see Theorem 7.42), $\mathcal{F} : X \to X^*$ is the corresponding duality map, and $A : X \to 2^{X^*}$ is a monotone operator, then $A(\cdot)$ is maximal monotone if and only if for every $\lambda > 0$ (resp., some $\lambda > 0$) $R(A + \lambda\mathcal{F}) = X^*$.*

Proof $\Rightarrow$ The duality map $\mathcal{F} : X \to X^*$ is single-valued, strictly monotone, demicontinuous (hence hemicontinuous too), bounded, and strongly coercive (see Proposition 6.93). Let $\lambda > 0$. From Theorem 7.41, we have

$$R(A + \lambda\mathcal{F}) = X^*.$$

$\Leftarrow$ Suppose $R(A + \lambda\mathcal{F}) = X^*$ for some $\lambda > 0$. Arguing by contradiction, assume that $A(\cdot)$ is not maximal monotone. Then we can find $(v, v^*) \in (X \times X^*) \setminus \text{Gr } A$ such that

$$\langle u^* - v^*, u - v \rangle \geq 0 \text{ for all } (u, u^*) \in \text{Gr } A. \tag{7.28}$$

Let $(x_0, x_0^*) \in \text{Gr } A$ such that

$$x_0^* + \lambda \mathcal{F}(x_0) = v^* + \lambda \mathcal{F}(v), \tag{7.29}$$

$$\Rightarrow \lambda \langle \mathcal{F}(x_0) - \mathcal{F}(v), x_0 - v \rangle = \langle v^* - x_0^*, x_0 - v_0 \rangle \leq 0 \text{ (see (7.28))}$$

$$\Rightarrow \langle \mathcal{F}(x_0) - \mathcal{F}(v), x_0 - v \rangle \leq 0,$$

$$\Rightarrow \|x_0\|^2 + \|v\|^2 \leq \langle \mathcal{F}(v), x_0 \rangle + \langle \mathcal{F}(x_0), v \rangle,$$

$$\Rightarrow \langle \mathcal{F}(v) - \mathcal{F}(x_0), v - x_0 \rangle \leq 0,$$

$$\Rightarrow \mathcal{F}(v) = \mathcal{F}(x_0) \text{ (by strict monotonicity)}$$

$$\Rightarrow v = x_0 \text{ (since } \mathcal{F}^{-1} \text{ is single-valued)}.$$

It follows that $v^* = x_0^*$ (see (7.29)) and so finally

$$(v, v^*) = (x_0, x_0^*) \in \text{Gr } A,$$

a contradiction. This proves that A is maximal monotone. □

Proposition 7.45 *If X is a reflexive Banach space, $A : X \to 2^{X^*}$ is a maximal monotone operator, and $F : X \to X^*$ is monotone, hemicontinuous, and bounded, then $x \to (A + F)(x)$ is maximal monotone.*

Proof According to Theorem 7.42, without any loss of generality, we may assume that both X and X^* are locally uniformly convex. The map $x \to B(x) + \mathcal{F}(x)$ is single-valued, strictly monotone, demicontinuous, and bounded. So, by Theorem 7.41, we have

$$R(A + B + \mathcal{F}) = X^*.$$

Thus $A + B$ is a maximal monotone (see Theorem 7.44). □

Proposition 7.46 *If X is a reflexive Banach space and $F : X \to X^*$ is monotone hemicontinuous, then $F(\cdot)$ is maximal monotone.*

Proof Let $(v_0, v_0^*) \in X \times X^*$ and assume that

$$\langle v_0^* - F(x), v_0 - x \rangle \geq 0 \text{ for all } x \in X. \tag{7.30}$$

We need to show that $v_0^* = F(v_0)$ (see Remark 7.27).

Let $u_t = tv_0 + (1 - t)x$ for all $t \in [0, 1]$. In (7.30), replace x by u_t. Then

$$t\langle v_0^* - F(tv_0 + (1-t)x), v_0 - x\rangle \text{ for all } t \in [0, 1], \text{ all } x \in X.$$

Let $t \to 1^-$. Then since $F(\cdot)$ is hemicontinuous, we obtain

$$\langle v_0^* - F(v_0), v_0 - x\rangle \geq 0 \text{ for all } x \in X,$$

$$\Rightarrow v_0^* = F(v_0). \qquad \square$$

The next result highlights the importance of maximal monotone operators.

Theorem 7.47 *If X is a reflexive Banach space and $A : X \to 2^{X^*}$ is a maximal monotone coercive operator, then $A(\cdot)$ is surjective.*

Proof Let $v_0^* \in X^*$. On account of Theorem 7.44, we can find (unique) $x_\lambda \in D(A)$ such that

$$x_\lambda^* + \lambda\mathcal{F}(x_\lambda) = v_0^* \text{ with } x_\lambda^* \in A(x_\lambda), \lambda \in [0, 1]$$

$$\Rightarrow v_0^* - \lambda\mathcal{F}(x_\lambda) = x_\lambda^*. \tag{7.31}$$

By replacing $A(\cdot)$ with $\widehat{A}(x) = A(x + x_0)$ for all $x \in X$ and some $x_0 \in D(A)$, we see that without any loss of generality, we may assume that $0 \in D(A)$. Let $x^* \in A(0)$. Then from (7.31) and the monotonicity of $A(\cdot)$, we have

$$0 \leq \langle v_0^* - \lambda\mathcal{F}(x_\lambda) - x^*, x_\lambda\rangle,$$

$$\Rightarrow \lambda\|x_\lambda\|^2 \leq \|v_0^* - x^*\|_*\|x_\lambda\|,$$

$$\Rightarrow \lambda\|x_\lambda\| \leq \|v_0^*\|_* + |A(0)| = c. \tag{7.32}$$

Note that

$$\|\lambda\mathcal{F}(x_\lambda)\|_* = \|v_0^* - x_\lambda^*\|_* \text{ (see (7.31))},$$

$$\Rightarrow \lambda\|x_\lambda\| = \|v_0^* - x_\lambda^*\|_* \leq c \text{ (see (7.32))}. \tag{7.33}$$

The coercivity of $A(\cdot)$ implies that $A^{-1}|_{\overline{B}^*_{c+\|v_0^*\|_*}}$ is bounded (recall $\overline{B}^*_{c+\|v_0^*\|_*} = \{u^* \in X^* : \|u^*\|_* \leq c + \|v_0^*\|_*\}$). From (7.33), we see that $\{x_\lambda^*\}_{\lambda \in (0,1]} \subseteq \overline{B}^*_{c+\|v_0^*\|_*}$ is bounded and so $\{x_\lambda\}_{\lambda \in (0,1]} \subseteq X$ is bounded (see (7.33)). It follows that

$$\lambda \mathcal{F}(x_\lambda) \to 0 \ in \ X^* \ as \ \lambda \to 0^+,$$

$$\Rightarrow x_\lambda^* \to v_0^* \ in \ X^* \ as \ \lambda \to 0^+ \ (\text{see } (7.31)). \tag{7.34}$$

Also, the reflexivity of X implies that we can find a sequence $\lambda_n \to 0^+$ such that

$$x_{\lambda_n} \xrightarrow{w} x \ in \ X. \tag{7.35}$$

Then from (7.34), (7.35), and Proposition 7.35, we infer that

$$(x, v_0^*) \in \mathrm{Gr}\, A, \quad \text{that is,} \quad v_0^* \in A(x).$$

Since $v_0^* \in X^*$ is arbitrary, we conclude that $A(\cdot)$ is surjective.   □

In the above proof, A^{-1} had an important role combined with the coercivity hypothesis on $A(\cdot)$. In fact there is the following surjectivity criterion in terms of $A^{-1}(\cdot)$ (see Zeidler [292], Theorem 32.G, p.886)

Theorem 7.48 *If X is a reflexive Banach space and $A : X \to 2^{X^*}$ is a maximal monotone operator, then $A(\cdot)$ is surjective if and only if $A^{-1}(\cdot)$ is locally bounded.*

The next result is an immediate consequence of Proposition 7.46 and Theorem 7.47.

Corollary 7.49 *If X is a reflexive Banach space and $A : X \to X^*$ is monotone, hemicontinuous, and coercive, then $A(\cdot)$ is surjective.*

In Proposition 7.35, we saw that the graph of a maximal monotone operator is closed in $X \times X_{w^*}^*$ and in $X_w \times X^*$. The next proposition complements this fact.

Proposition 7.50 *If X is a reflexive Banach space, $A : X \to 2^{X^*}$ is a maximal monotone operator and $\{(x_n, x_n^*)\}_{n \in \mathbb{N}} \subseteq \mathrm{Gr}\, A$ with*

$$x_n \xrightarrow{w} x \ in \ X, \ x_n^* \xrightarrow{w} x^* \ in \ X^*$$

and

$$\limsup_{n,m \to \infty} \langle x_n^* - x_m^*, x_n - x_m \rangle \le 0 \tag{7.36}$$

or

$$\limsup_{n \to \infty} \langle x_n^* - x^*, x_n - x \rangle \le 0, \tag{7.37}$$

then $(x, x^*) \in \mathrm{Gr}\, A$ and $\langle x_n^*, x_n \rangle \to \langle x^*, x \rangle$.

Proof Suppose that (7.36) holds. This condition combined with the monotonicity of $A(\cdot)$ implies that

$$\lim_{n,m\to\infty} \langle x_n^* - x_m^*, x_n - x_m \rangle = 0. \tag{7.38}$$

Since $\left\{ \langle x_n^*, x_n \rangle \right\}_{n\in\mathbb{N}} \subseteq \mathbb{R}$ is bounded, we can find a subsequence $n_k \to \infty$ such that

$$\langle x_{n_k}^*, x_{n_k} \rangle \to \eta \in \mathbb{R}.$$

Then we have

$$0 = \lim_{k\to\infty} \left[\lim_{l\to\infty} \langle x_{n_k}^* - x_{n_l}^*, x_{n_k} - x_{n_l} \rangle \right] = 2\eta - 2\langle x^*, x \rangle \ (\text{see (7.38)}),$$

thus

$$\eta = \langle x^*, x \rangle. \tag{7.39}$$

For every $(u, u^*) \in \operatorname{Gr} A, 0 \leq \langle x_{n_k}^* - u^*, x_{n_k} - u \rangle$ for $k \in \mathbb{N}$, thus

$$0 \leq \langle x^* - u^*, x - u \rangle \text{ for all } (u, u^*) \in \operatorname{Gr} A \ (\text{see (7.39)}),$$

hence $(x, x^*) \in \operatorname{Gr} A$ and clearly $\langle x_n^*, x_n \rangle \to \langle x^*, x \rangle$. If (7.37) holds, then

$$\limsup_{n\to\infty} \langle x_n^*, x_n \rangle \leq \langle x^*, x \rangle. \tag{7.40}$$

Again we have

$$0 \leq \langle x_n^* - u^*, x_n - u \rangle \text{ for all } (u, u^*) \in \operatorname{Gr} A, \text{ all } n \in \mathbb{N},$$

$$\Rightarrow 0 \leq \langle x^* - u^*, x - u \rangle \text{ for all } (u, u^*) \in \operatorname{Gr} A \ (\text{see (7.40)}),$$

$$\Rightarrow (x, x^*) \in \operatorname{Gr} A \ \big(\text{since } A(\cdot) \text{ is maximal monotone}\big).$$

and clearly $\langle x_n^*, x_n \rangle \to \langle x_*, x \rangle$. □

Using this proposition, we can produce some useful single-valued approximations of a maximal monotone operator.

The setting is the following. We have X a reflexive Banach space and $A : X \to 2^{X^*}$ a maximal monotone operator. Theorem 7.42 implies that without any loss of generality, we may assume that both X and X^* are locally uniformly convex.

Let $u \in X$ and $\lambda > 0$ and consider the operator inclusion

$$0 \in \lambda A(x) + \mathcal{F}(x - u). \tag{7.41}$$

The operator $x \to \lambda A(x) + \mathcal{F}(x - u)$ is maximal monotone (see Proposition 7.45), coercive, hence surjective (see Theorem 7.47). So problem (7.41) admits a solution $x_\lambda \in X$. Hence, we have

$$0 = \lambda x_\lambda^* + \mathcal{F}(x_\lambda - u) \text{ with } x_\lambda^* \in A(x_\lambda). \tag{7.42}$$

Suppose $v_\lambda \in X$ is another solution of (7.41). We have

$$0 = \lambda v_\lambda^* + \mathcal{F}(v_\lambda - u) \text{ with } v_\lambda^* \in A(v_\lambda). \tag{7.43}$$

From (7.42) and (7.43), we have

$$0 \leq \langle \mathcal{F}(x_\lambda - u) - \mathcal{F}(v_\lambda - u), x_\lambda - v_\lambda \rangle = -\lambda \langle x_\lambda^* - v_\lambda^*, x_\lambda - v_\lambda \rangle \leq 0,$$

thus

$$\langle \mathcal{F}(x_\lambda - u) - \mathcal{F}(v_\lambda - u), x_\lambda - v_\lambda \rangle = 0. \tag{7.44}$$

But $\mathcal{F}$ is strictly monotone (recall that X is locally uniformly convex), and hence it follows from (7.44) that $x_\lambda = v_\lambda$, that is, the solution of the operator inclusion (7.41) is unique. So, we can define two maps $J_\lambda, A_\lambda : X \to X^*$ by $J_\lambda(u) = x_\lambda$ (to every $v \in X$, assign the unique solution of (7.41)), and

$$A_\lambda(u) = \frac{1}{\lambda} \mathcal{F}(u - x_\lambda) \quad \text{(see (7.41))}, \ u \in X, \ \lambda > 0. \tag{7.45}$$

Definition 7.51 The operator $J_\lambda : X \to X^*$ is called the "resolvent operator" corresponding to A, while $A_\lambda : X \to X^*$ is the "Yosida approximation" corresponding to A.

In the next proposition, we state the main properties of these two operators.

Proposition 7.52 *If X is a reflexive Banach space and both X and X^* are locally uniformly convex, then*

(a) $A_\lambda(\cdot)$ is single-valued, monotone, demicontinuous, and bounded.
(b) $\| A_\lambda(x) \|_* \leq \inf \left[\|x^*\|_* : x^* \in A(X) \right]$.
(c) $J_\lambda : X \to X^*$ is bounded and

$$J_\lambda(x) \to x \text{ as } \lambda \to 0^+ \text{ for all } x \in \overline{\text{conv}} D(A).$$

(d) If $\lambda_n \to 0^+$, $x_n \to x$ in X, $A_{\lambda_n}(x_n) \xrightarrow{w} x^*$ in X^* and

$$\limsup_{n,m \to \infty} \langle A_{\lambda_n}(x_n) - A_{\lambda_m}(x_m), x_n - x_m \rangle \leq 0,$$

then $(x, x^*) \in \text{Gr } A$ and $\lim\limits_{n,m \to \infty} \langle A_{\lambda_n}(x_n) - A_{\lambda_m}(x_m), x_n - x_m \rangle = 0$.

(e) $A_\lambda(x) \to A^0(x)$ *for all* $x \in D(A)$ *as* $\lambda \to 0^+$ *with*

$$A^0(x) = \min \left[\|x^*\|_* : x^* \in A(x) \right]$$

(see Proposition 7.29 and recall that X^* is locally uniformly convex).

Proof (a) Clearly $A_\lambda(\cdot)$ is single valued. For every $x, u \in X$, we have

$$\langle A_\lambda(x) - A_\lambda(u), x - u \rangle = \langle A_\lambda(x) - A_\lambda(u), J_\lambda(x) - J_\lambda(u) \rangle$$
$$+ \langle A_\lambda(x) - A_\lambda(u), (x - J_\lambda(x)) - (u - J_\lambda(u)) \rangle.$$
$$(7.46)$$

From (7.41), we know that $A_\lambda(v) \in A(J_\lambda(v))$ for all $v \in X$. Hence, from (7.46), it follows that

$$\langle A_\lambda(x) - A_\lambda(u), x - u \rangle \geq 0, \Rightarrow A_\lambda(\cdot) \text{ is monotone.}$$

Let $(v, v^*) \in \text{Gr } A$. From (7.42), we have

$$0 = \lambda \langle x_\lambda^*, J_\lambda(u) - v \rangle + \langle \mathcal{F}(x_\lambda - u), J_\lambda(u) - v \rangle \text{ with } x_\lambda^* \in A(x_\lambda),$$
$$\Rightarrow \lambda \langle v^*, J_\lambda(u) - v \rangle + \langle \mathcal{F}(x_\lambda - u), J_\lambda(u) - v \rangle \leq 0 \text{ (since } A(\cdot) \text{ is monotone)},$$
$$\Rightarrow \langle \mathcal{F}(x_\lambda - u), J_\lambda(u) - v \rangle \leq \lambda \langle v^*, v - J_\lambda(u) \rangle.$$

Therefore, we have

$$\| J_\lambda(u) - u \|^2 \leq \|u - v\| \|J_\lambda(u) - u\| + \lambda \|v^*\|_* \|u - v\|$$
$$+ \lambda \|v^*\|_* \|J_\lambda(u) - u\|,$$
$$\Rightarrow J_\lambda(\cdot) \text{ is bounded and then so is } A_\lambda(\cdot).$$

Now we show the demicontinuity of $A_\lambda(\cdot)$. To this end, let $u_n \to u$ in X and let $x_n = J_\lambda(u_n)$ and $x_n^* = A_\lambda(u_n)$. From (7.42), (7.44), and (7.45), we have $\lambda x_n^* + \mathcal{F}(x_n - u_n) = 0$ for all $n \in \mathbb{N}$, thus

$$\langle \mathcal{F}(x_n - u_n) - \mathcal{F}(x_m - u_m), (x_n - u_n) - (x_m - u_m) \rangle + \lambda \langle x_n^* - x_m^*, x_n - x_m \rangle$$
$$= \langle \mathcal{F}(x_n - u_n) - \mathcal{F}(x_m - u_m), u_m - u_n \rangle.$$

Thus we obtain

$$\lim\limits_{n,m \to \infty} \langle \mathcal{F}(x_n - u_n) - \mathcal{F}(x_m - u_m), (x_n - u_n) - (x_m - u_m) \rangle = 0 \qquad (7.47)$$

and

$$\lim_{n,m}\langle x_n^* - x_m^*, x_n - x_m\rangle = 0. \tag{7.48}$$

From the boundedness of $J_\lambda(\cdot)$ and $A_\lambda(\cdot)$ and the reflexivity of X and X^*, it follows that we can find a subsequence $\{n_k\}$ of $\{n\}$ such that

$$x_{n_k} \xrightarrow{w} x \text{ in } X, \ x_{n_k}^* \xrightarrow{w} x^* \text{ and } \mathcal{F}(x_{n_k} - u_{n_k}) \xrightarrow{w} v^* \text{ in } X^*. \tag{7.49}$$

Using (7.47), (7.48), (7.49), and Proposition 7.50, we have

$$(x, x^*) \in \operatorname{Gr} A, \ v^* = \mathcal{F}(x - u) \text{ and } \lambda x^* + \mathcal{F}(x - u) = 0,$$

$$\Rightarrow x = J_\lambda(u), x^* = A_\lambda(u) \text{ and } J_\lambda(u_n) \to J_\lambda(u), \ A_\lambda(u_n) \to A_\lambda(u),$$

$$\Rightarrow A_\lambda(\cdot) \text{ is demicontinuous.}$$

From Proposition 7.46, it follows that $A_\lambda(\cdot)$ is maximal monotone.

(b) Let $(u, u^*) \in \operatorname{Gr} A$. From the monotonicity of $A(\cdot)$, we have

$$0 \le \langle u^* - A_\lambda(u), u - J_\lambda(u)\rangle \le \|u^*\|_* \|u - x_\lambda\| - \frac{1}{\lambda}\|u - x_\lambda\|^2$$

$$\Rightarrow \|u - x_\lambda\| \le \lambda\|u^*\|_*$$

$$\Rightarrow \|A_\lambda(u)\|_* \le \|u^*\|_* \quad \text{(see (7.45)).}$$

Since $(u, u^*) \in \operatorname{Gr} A$ is arbitrary, it follows that

$$\|A_\lambda(u)\|_* = \inf\big[\|u^*\|_* : u^* \in A(u)\big].$$

(c) Let $x \in \overline{\operatorname{conv}} D(A)$ and let $(v, v^*) \in \operatorname{Gr} A$. We have

$$0 \le \langle A_\lambda(x) - v^*, J_\lambda(x) - v\rangle$$

$$\Rightarrow \|J_\lambda(x) - x\|^2 \le \lambda\langle v^*, v - J_\lambda(x)\rangle + \langle \mathcal{F}(J_\lambda(x) - x), v - x\rangle.$$

Consider a sequence $\lambda_n \to 0^+$ such that

$$\mathcal{F}(J_{\lambda_n}(x) - x) \xrightarrow{w} y^* \text{ in } X^*.$$

Then we have

$$\limsup_{n\to\infty} \|J_{\lambda_n}(x) - x\|^2 \le \langle y^*, v - x\rangle. \tag{7.50}$$

Since $(v, v^*) \in \operatorname{Gr} A$ is arbitrary, it follows that (7.50) holds for all $v \in \overline{\operatorname{conv}} D(A)$ and so we can take $u = x$. Then we have that

$$J_{\lambda_n}(x) \to x \text{ in } X \text{ as } n \to \infty,$$

$$\Rightarrow J_\lambda(x) \to x \text{ in } x \text{ as } \lambda \to 0^+ \text{ for all } x \in \overline{\operatorname{conv}} D(A).$$

(d) Using the monotonicity of $A(\cdot)$ and the fact that $A_\lambda(u) \in A(J_\lambda(u))$ for all $u \in X$, we obtain

$$\langle A_{\lambda_n}(x_n) - A_{\lambda_m}(x_m), x_n - x_m \rangle$$

$$\geq \langle \frac{1}{\lambda_n} \mathcal{F}(x_n - J_{\lambda_n}(x_n)) - \frac{1}{\lambda_m} \mathcal{F}(x_m - J_{\lambda_m}(x_m)), \qquad (7.51)$$

$$(x_n - J_{\lambda_n}(x_n)) - (x_m - J_{\lambda_m}(x_m)) \rangle.$$

From (7.51) and the hypotheses, it follows that

$$\lim_{n,m \to \infty} \langle A_{\lambda_n}(x_n) - A_{\lambda_m}(x_m), x_n - x_m \rangle = 0,$$

and

$$\lim_{n,m \to \infty} \langle A_{\lambda_n}(x_n) - A_{\lambda_m}(x_m), J_{\lambda_n}(x_n) - J_{\lambda_m}(x_m) \rangle = 0.$$

Using Proposition 7.50, we infer that $(x, x^*) \in \operatorname{Gr} A$.

(e) From Proposition 7.29 and since X^* is reflexive and locally uniformly convex, it follows that $A^0(x)$ is well defined and uniquely defined for all $x \in D(A)$.

Let $x \in D(A)$ and suppose that $\lambda_n \to 0^+$ and $A_{\lambda_n}(x) \xrightarrow{w} x^*$ in X^*. From (d), we have $x^* \in A(x)$. Since $\|A_{\lambda_n}(x)\|_* \leq \|A^0(x)\|_*$ (see (b)), it follows that $x^* = A^0(x)$. So, we have

$$\|A_{\lambda_n}(x)\|_* \to \|A^0(x)\|_*,$$

$$A_{\lambda_n}(x) \xrightarrow{w} A^0(x) \text{ in } X^*.$$

But from Proposition 3.244, we know that a locally uniformly convex Banach space has the Kadec–Klee property. So, we conclude that

$$A_{\lambda_n}(x) \to A^0(x) \text{ in } X^* \text{ as } n \to \infty \text{ for all } x \in D(A). \qquad \square$$

Also using Proposition 7.50, we have the following proposition:

Proposition 7.53 *If X is a reflexive Banach space, $A : X \to 2^{X^*}$ is a maximal monotone operator, and $\{(x_n, x_n^*)\}_{n \in \mathbb{N}} \in \operatorname{Gr} A$ such that*

$$x_n \xrightarrow{w} x \text{ in } X, \ x_n^* \xrightarrow{w} x^* \text{ in } X^* \text{ and } \limsup_{n\to\infty}\langle x_n^*, x_n - x\rangle \leq 0,$$

then $(x, x^*) \in \operatorname{Gr} A$.

Remark 7.54 However, in general the graph of a maximal monotone operator $A :$ $X \to 2^{X^*}$ with X reflexive is not sequentially closed in $X_w \times X_w^*$.

Definition 7.55 Let H be a Hilbert space. We will say that H is a "pivot Hilbert space" if $H = H^*$ (see Theorem 3.181).

In pivot Hilbert spaces, Proposition 7.52 has a much simpler form:

Proposition 7.56 *If H is a pivot Hilbert space, then*

(a) $J_\lambda : H \to H$ *is nonexpansive, that is,*

$$\|J_\lambda(x) - J_\lambda(u)\| \leqslant \|x - u\| \text{ for all } x, u \in H.$$

(b) $A_\lambda : H \to H$ *is* $\frac{2}{\lambda} -$ *Lipschitz, that is,*

$$\|A_\lambda(x) - A_\lambda(u)\| \leqslant \frac{2}{\lambda}\|x - u\| \text{ for all } x, u \in H.$$

(c) $A_\lambda(x) \to A^0(x)$ *as* $\lambda \to 0^+$ *for all* $x \in D(A)$

Proof

(a) Since H is a pivot Hilbert space $\mathcal{F} = id$. For $x, u \in H$, we set

$$x_\lambda = (id + \lambda A)^{-1}(x), \quad u_\lambda = (id + \lambda A)^{-1}(u)$$

and have

$$x_\lambda - u_\lambda + \lambda(A(x_\lambda) - A(u_\lambda)) \ni x - u \quad \text{(see (7.42))}. \tag{7.52}$$

Taking inner product with $x_\lambda - u_\lambda$ and using the monotonicity of $A(\cdot)$, we obtain

$$\|x_\lambda - u_\lambda\|^2 \leqslant \|x - u\|\|x_\lambda - u_\lambda\|,$$
$$\Rightarrow \|x_\lambda - u_\lambda\| = \|J_\lambda(x) - J_\lambda(u)\| \leqslant \|x - u\|.$$

(b) Since $\mathcal{F} = id$, from (7.45), we have

$$\|A_\lambda(x) - A_\lambda(u)\| = \|\frac{1}{\lambda}(x - x_\lambda) - \frac{1}{\lambda}(x - u_\lambda)\| \leqslant \frac{2}{\lambda}\|x - u\| \quad \text{(see (b))}.$$

(c) Follows from Proposition 7.52 (e).

$\square$

Proposition 7.57 *If X is reflexive and $A : X \to 2^{X^*}$ is a maximal monotone operator, then both $\overline{D(A)}$ and $\overline{R(A)}$ are convex.*

Proof As before using Theorem 7.42, we may assume without any loss of generality that both X and X^* are locally uniformly convex. From proposition 7.52(c), we know that

$$J_\lambda(x) \to x \text{ in } X \text{ as } \lambda \to 0^+ \text{ for all } x \in \overline{\text{conv}}D(A).$$

But $J_\lambda(x) \in D(A)$ for all $\lambda > 0$, all $x \in X$. Therefore

$$\overline{\text{conv}}D(A) = \overline{D(A)}$$

$$\Rightarrow \overline{D(A)} \subseteq X \text{ is convex.}$$

Note that $R(A) = D(A^{-1})$ and A^{-1} is maximal monotone too. Therefore the convexity $\overline{R(A)}$ follows from the first part of the proof. $\qquad \square$

Next we look at the sum of maximal monotone operators. We want to find conditions that guarantee that maximal monotonicity is preserved by addition.

Theorem 7.58 *If X is a reflexive Banach space and $A, K : X \to 2^{X^*}$ are maximal monotone operators such that*

$$(\text{int } D(A)) \cap D(K) \neq \emptyset,$$

then $x \to (A + K)(x)$ is maximal monotone.

Proof As before, on account of Theorem 7.42, we may assume that both X and X^* are locally uniformly convex. In addition, by shifting things if necessary, we may assume that

$$0 \in (\text{int } D(A)) \cap D(K), \ (0,0) \in \text{Gr } A, \ (0,0) \in \text{Gr } K. \tag{7.53}$$

We will show that $R(A + K + \mathcal{F}) = X^*$, and this by Theorem 7.44 implies the maximal monotonicity of $A + K$.

For $\lambda > 0$, let $K_\lambda : X \to X^*$ be the Yosida approximation. It is monotone, demicontinuous, and bounded (see Proposition 7.52(a)). Similarly for the duality map, $\mathcal{F} : X \to X^*$ (see Proposition 6.93). Then Proposition 7.45 implies that $x \to (A + K_\lambda + \mathcal{F})(x)$ is maximal monotone. Also, since $(0,0) \in \text{Gr } A$, $(0,0) \in \text{Gr } K$, we see that $(A + K_\lambda + \mathcal{F})(\cdot)$ is coercive. Therefore by Theorem 7.47, we have $R(A + K_\lambda + \mathcal{F}) = X^*$. So, given $v^* \in X^*$, we can find $x_\lambda \in D(A)$ such that

$$v^* \in A(x_\lambda) + K_\lambda(x_\lambda) + \mathcal{F}(x_\lambda). \tag{7.54}$$

In fact the strict monotonicity of $\mathcal{F}(\cdot)$, implies the uniqueness of this solution x_λ. On (7.52), we act with x_λ and obtain

$$\|x_\lambda\| \leqslant \|v^*\|_* \quad (\text{recall } (0,0) \in \text{Gr } A, \ K_\lambda(0) = 0).$$

Since $0 \in \text{int } D(A)$ (see (7.53)), from Proposition 7.33, we know that we can find $\varrho > 0$ and $\widehat{c} > 0$ such that

$$\|x^*\|_* \leq \widehat{c} \text{ for all } x^* \in A(x), \ x \in D(A) \cap \overline{B}_\varrho. \tag{7.55}$$

We act on (7.54) with $x_\lambda - \varrho h$ where $\|h\| = 1$. Then

$$\langle h^*, x_\lambda - \varrho h \rangle + \langle \mathcal{F}(x_\lambda) + K_\lambda(x_\lambda) - v^*, x_\lambda - \varrho h \rangle \leqslant 0,$$

with $h^* \in A(\varrho h)$, for all $\|h\| = 1$,

$$\Rightarrow \|x_\lambda\|^2 - \varrho \langle K_\lambda(x_\lambda), h \rangle \leqslant \widehat{c}\Big[\varrho + \|x_\lambda\|\Big] + \|x_\lambda\|\Big[\varrho + \|v^*\|_*\Big] \quad (\text{see } (7.55)).$$

$$\Rightarrow \{x_\lambda\}_{\lambda \in (0,1]} \subseteq X \text{ and } \{K_\lambda(x_\lambda)\}_{\lambda \in (0,1]} \subseteq X^* \text{ are bounded.}$$

Because X is reflexive, we may assume that for some sequence $\lambda_n \to 0^+$, we have

$$x_n = x_{\lambda_n} \xrightarrow{w} x \text{ in } X, \ x_n^* \xrightarrow{w} x^*(x_n^* \in A(x_n)), \ K_{\lambda_n}(x_n) \xrightarrow{w} u^*, \ \mathcal{F}(x_n) \xrightarrow{w} y^*.$$

From the monotonicity of $A + \mathcal{F}$, we have

$$\langle K_{\lambda_n}(x_n) - K_{\lambda_m}(x_m), x_n - x_m \rangle \leqslant 0 \text{ for all } n, m \in \mathbb{N}.$$

Using Proposition 7.52 (d), we have

$$\lim_{n,m \to \infty} \langle K_{\lambda_n}(x_n) - K_{\lambda_m}(x_m), x_n - x_m \rangle = 0, \ (x, u^*) \in \text{Gr } K.$$

So, using (7.55), we obtain

$$\lim_{n,m \to \infty} \langle x_n^* + \mathcal{F}(x_n) - (x_m^* + \mathcal{F}(x_m)), x_n - x_m \rangle = 0, \ x_n^* \in A(x_n), x_m^* \in A(x_m).$$

The operator $A + \mathcal{F}$ is maximal monotone. So Proposition 7.50 implies

$$(x, x^* + u^*) \in \text{Gr}(A + \mathcal{F}),$$

$$\Rightarrow v^* \in A(x) + K(x) + \mathcal{F}(x),$$

$$\Rightarrow (A + K)(\cdot) \text{ is maximal monotone.} \qquad \square$$

Combining this theorem with Proposition 7.46, we obtain the following result.

Corollary 7.59 *If X is a reflexive Banach space, $A : X \to 2^{X^*}$ is a maximal monotone operator, and $K : X \to X^*$ is monotone demicontinuous, then $x \to (A + K)(x)$ is maximal monotone.*

We know that the subdifferential of a convex function is a monotone operator (see Proposition 6.76). Now we will show that the subdifferential of a convex and lower semicontinuous function is maximal monotone. This is one of the most important examples of a maximal monotone operator. It is the notion that links the theory of nonlinear operators with nonsmooth analysis.

Theorem 7.60 *If X is Banach space and $\varphi \in \Gamma_0(x)$ (see Definition 6.102), then $\partial\varphi : X \to 2^{X^*}$ is a maximal monotone operator.*

Proof We will do the proof for X being a reflexive Banach space, and for the general case we refer to Rockafellor [237].

We already know that $\partial\varphi(\cdot)$ is monotone (see Proposition 6.76). We need to show the maximality of $\partial\varphi(\cdot)$. According to Theorem 7.44, it suffices to show that $x \to \partial\varphi(x) + \mathcal{F}(x)$ is surjective. We know that

$$\mathcal{F}(x) = \partial\psi(x) \text{ with } \psi(x) = \frac{1}{2}\|x\|^2 \text{ for all } x \in X \quad \text{(see Example 6.91(c)).}$$

Let $v^* \in X^*$ and consider the function

$$g(x) = \varphi(x) + \psi(x) - \langle v^*, x \rangle \text{ for all } x \in X.$$

Evidently, $g \in \Gamma_0(X)$, and on account of Proposition 6.83, we have

$$\partial g(x) = \partial\varphi(x) + \partial\psi(x) - v^* = \partial\varphi(x) + \mathcal{F}(x) - v^*. \quad (7.56)$$

Since $\varphi \in \Gamma_0(X)$, by Proposition 6.90, $\varphi(\cdot)$ is minorized by a continuous affine function. So we can find $x^* \in X^*$ and $\eta \in \mathbb{R}$ such that

$$\varphi(x) \geq \langle x^*, x \rangle + \eta \text{ for all } x \in X.$$

Hence we have

$$g(x) \geq \frac{1}{2}\|x\|^2 - \|x^* + v^*\|_*\|x\| + \eta \text{ for all } x \in X,$$

$$\Rightarrow g(x) \to \infty \text{ as } \|x\| \to \infty.$$

Since X is reflexive and $g(\cdot)$ is weakly lower semicontinuous (being convex), by Proposition 1.180, we can find $x_0 \in X$ such that

$$g(x_0) = \min[g(x) : x \in X],$$

$$\Rightarrow 0 \in \partial g(x_0),$$

$$\Rightarrow v^* \in \partial\varphi(x_0) + \mathcal{F}(x_0) \quad \text{(see (7.56)).}$$

But $v^* \in X^*$ was arbitrary. Therefore we infer that

$$R(\partial \varphi + \mathcal{F}) = X^*,$$

$$\Rightarrow x \to \partial \varphi(x) \text{is maximal monotone.} \qquad \square$$

Proposition 7.61 *If X is a Banach space and $\varphi \in \Gamma_0(X)$, then $D(\partial \varphi)$ is dense in* domφ.

Proof From Theorem 7.60, we know that $\partial \varphi(\cdot)$ is maximal monotone. So, for every $u \in dom\varphi$ and $\lambda > 0$, the inclusion $0 \in \lambda \partial \varphi(x) + \mathcal{F}(x - u)$ has a unique solution $x_\lambda = J_\lambda(u) \in D(\partial \varphi)$. Acting on the inclusion with $x_\lambda - u$, we obtain for some $x^* \in \partial \varphi(x_\lambda)$ and all $\lambda > 0$

$$0 = \lambda \langle x^*, x_\lambda - u \rangle + \|x_\lambda - u\|^2 \geqslant \lambda \left[\varphi(x_\lambda) - \varphi(u) \right] + \|x_\lambda - u\|^2. \qquad (7.57)$$

Since $\varphi(\cdot)$ is minorized by an affine continuous function (see Proposition 6.90), from (7.57), we see that

$$\|x_\lambda - u\|^2 \to 0 \text{ as } \lambda \to 0^+.$$

But $x_\lambda \in D(\partial \varphi)$ and $u \in dom\varphi$ is arbitrary. It follows that

$$\overline{D(\partial \varphi)} = \overline{dom\varphi}. \qquad \square$$

In Sect. 6.4, for $\varphi \in \Gamma_0(X)$, we introduced the "Moreau–Yosida regularization" of $\varphi(\cdot)$ defined by

$$\varphi_\lambda(x) = \inf \left[\varphi(u) + \frac{1}{2\lambda} \mid u - x \|^2 : u \in X \right] \text{ for all } x \in X.$$

In Theorem 6.116, when X is reflexive with both X and X^* strictly convex, we proved some important properties of $\varphi_\lambda(\cdot)$, $\lambda > 0$. In the next proposition, we relate φ_λ with the resolvent $J_\lambda(\cdot)$ and the Yosida approximation $(\partial \varphi)_\lambda(\cdot)$ of the maximal monotone operator $\partial \varphi(\cdot)$.

Proposition 7.62 *If X is a reflexive Banach space with both X and X^* being strictly convex and $\varphi \in \Gamma_0(X)$, then $\varphi_\lambda(\cdot)$ is convex, continuous, and Gateaux differentiable*

(a) $\varphi_\lambda(x) = \varphi(J_\lambda(x)) + \frac{1}{2\lambda} \|J_\lambda(x) - x\|^2$ *for all $x \in X$, all $\lambda > 0$.*
(b) $\lim_{\lambda \to 0^+} \varphi_\lambda(x) = \varphi(x)$ *for all $x \in X$.*
(c) $\varphi(J_\lambda(x)) \leq \varphi_\lambda(x) \leq \varphi(x)$ *for all $x \in X$, all $\lambda > 0$.*
(d) $(\partial \varphi)_\lambda = \partial \varphi_\lambda$ *for all $\lambda > 0$.*

Moreover, if X is Hilbert space, then $\varphi_\lambda(\cdot)$ is Frechet differentiable.

Proof The only thing that we need to prove is (d). The rest can be found in Theorem 6.116 and its proof.

From (a), we have

$$\varphi_\lambda(u) - \varphi_\lambda(x)$$

$$\leqslant \langle (\partial\varphi)_\lambda(u), J_\lambda(u) - J_\lambda(x) \rangle + \frac{1}{2\lambda}\left[\|u - J_\lambda(u)\|^2 - \|x - J_\lambda(x)\|^2 \right]$$

$$= \langle (\partial\varphi)_\lambda(u), u - x \rangle + \langle (\partial\varphi)_\lambda(u), J_\lambda(u) - u \rangle + \langle (\partial\varphi)_\lambda(x), x - J_\lambda(x) \rangle$$

$$+ \frac{1}{2\lambda}\left[\|u - J_\lambda(u)\|^2 - \|x - J_\lambda(x)\|^2 \right]$$

$$\leqslant \langle (\partial\varphi)_\lambda(u), u - x \rangle \left(\text{recall } \partial\left(\frac{1}{2}\|v - J_\lambda(v)\|^2\right) = \mathcal{F}(v - J_\lambda(v)) \right).$$

Thus, $\varphi_\lambda(u) - \varphi_\lambda(x) - \langle (\partial\varphi)_\lambda(x), u - x \rangle \leqslant \langle (\partial\varphi_\lambda)(u) \cdot (\partial\varphi)_\lambda(x), x - u \rangle$ for all $u, x \in X$ all $\lambda > 0$, and

$$\lim_{t \to 0} \frac{\varphi_\lambda(x + th) - \varphi_\lambda(x)}{t} \leqslant \langle (\partial\varphi)_\lambda(x), h \rangle$$

for all $x, h \in X$ (since $(\partial\varphi)_\lambda(\cdot)$ is demicontinuous, see Proposition 7.52(a)). Therefore, $\nabla\varphi_\lambda = (\partial\varphi)_\lambda$ for all $\lambda > 0$. $\qquad\square$

Example 7.63 Suppose H is a Hilbert space and $C \subset H$ is nonempty, closed, and convex. Let $\varphi(x) = i_C(x) = \begin{cases} 0 & \text{if } x \in C \\ +\infty & \text{if } x \notin C \end{cases}$ (the indicator function of the set C). Then $\varphi \in \Gamma_0(H)$ and

$$\varphi_\lambda(x) = \frac{1}{2\lambda}\|x - p_C(x)\|^2 \text{ for all } x \in H, \text{ all } \lambda > 0,$$

with $p_C(\cdot)$ being the metric projection on the set C (see Proposition 3.175). Also $J_\lambda = p_C$ for all $\lambda > 0$.

We can have a characterization of those maximal monotone operators which are of the subdifferential type.

Definition 7.64 Let X be a Banach space and $A : X \to 2^{X^*}$. We say that $A(\cdot)$ is "cyclically monotone" if for all $\{(x_k, x_k^*)\}_{k=0}^m \subseteq \text{Gr } A$, we have

$$0 \leqslant \sum_{k=0}^n \langle x_k^*, x_k - x_{k+1} \rangle \quad \text{with} \quad x_{n+1} = x_0.$$

We say that A is "maximal cyclically monotone," if it is cyclically monotone and has no proper cyclically monotone extension in $X \times X^*$ (that is, Gr A is maximal with respect to inclusion among the graphs of cyclically monotone operators).

Using this notion, we can characterize those maximal monotone operators that are of the subdifferential type. The result is due to Rockafellar [233].

Theorem 7.65 *If X is a Banach space and $A : X \to 2^{X^*}$, then $A = \partial\varphi$ for some $\varphi \in \Gamma_0(x)$ if and only if A is maximal cyclically monotone.*

We conclude this section with some additional examples of maximal monotone operators of the subdifferential type (see also Examples 6.91).

Example 7.66

(a) Let H be a Hilbert space and $A : H \to H$ is linear, densely defined, self-adjoint, and $A \geq 0$. Such an operator admits a square root, that is, three exists unique self-adjoint linear operator $K : H \to H$ such that $K \geq 0$ and $K^2 = A$ We donate K by $A^{1/2}$ (see Kato [152], Theorem 3.35, p.281). We have that $A = \partial\varphi$, with $\varphi \in \Gamma_0(H)$ defined by

$$\varphi(x) = \begin{cases} 1/2 \left\| A^{1/2}(x) \right\|^2 & \text{if } x \in D\left(A^{1/2}\right) \\ +\infty & \text{otherwise.} \end{cases} \tag{7.58}$$

Note that $A(\cdot)$ is maximal monotone, since $R(I + A) \subseteq H$ is closed and dense, hence $R(I + A) = H$, and this by Theorem 7.44 implies the maximal monotonicity of $A(\cdot)$. We have

$$\varphi(x) - \varphi(u) = \frac{1}{2}\left[\left\| A^{1/2}(x) \right\|^2 - \left\| A^{1/2}(u) \right\|^2 \right]$$

$$\leqslant \langle A(x), x - u \rangle \text{ for all } x \in D(A), u \in D\left(A^{1/2}\right)$$

$\Rightarrow A \subseteq \partial\varphi$,

$\Rightarrow A = \partial\varphi$ (since as we proved earlier A is maximal monotone).

Also a linear, densely defined operator $A : H \to H$ which is of the subdifferential type, that is, $A = \partial\varphi$ is necessarily self-adjoint. To see this, let $A = \partial\psi$ with $\psi \in \Gamma_0(H)$. Then $A_\lambda = \nabla\psi_\lambda$ with $A_\lambda = \frac{1}{\lambda}[I - \lambda A]^{-1}$. We have

$$\frac{d}{dt}\psi_\lambda(tx) = t\langle A_\lambda(x), x \rangle \text{ for all } t \in [0, 1], \text{ all } x \in H.$$

$\Rightarrow \psi_\lambda(x) = \frac{1}{2}\{A_\lambda(x), x\}$ for all $x \in H$, all $\lambda > 0$.

$\Rightarrow \nabla\psi_\lambda(x) = A_\lambda(x) = \frac{1}{2}\left(A_\lambda + A_\lambda^*\right)(x)$ for all $x \in H$, all $\lambda > 0$.

$\Rightarrow A_\lambda = A_\lambda^*$ for all $\lambda > 0$.

$\Rightarrow A = A^*$ (let $\lambda \to 0^+$).

So, we have proved that a linear densely defined maximal monotone operator $A : H \to H$ is cyclically monotone (that is, of the subdifferential type) if and only if $A = A^*$ (that is, A is self-adjoint).

Using this fact, we see that if $\Omega \in \mathbb{R}^N$ is a bounded domain with C^2-boundary $\partial\Omega$, $H = L^2(\Omega)$, and $A : H \to H$ is defined by

$$A(u) = -\Delta u \text{ with } u \in D(A) = H_0^1(\Omega) \cap H^2(\Omega),$$

then $A(\cdot)$ is self-adjoint and $A = \partial\varphi$, with $\varphi \in \Gamma_0(H)$ being defined by

$$\varphi(u) = \begin{cases} \frac{1}{2}\|Du\|_2^2 & \text{if } u \in H_0^1(\Omega) \\ +\infty & \text{otherwise.} \end{cases}$$

(b) This example extends Example 6.91(b). So, suppose $f : \Omega \times \mathbb{R}^N \to \mathbb{R} \cup \{+\infty\}$ is a normal convex integrand such that

- $f(z, y) \geqslant (a(z), y)_{\mathbb{R}^N} + \eta(z)$ for $a.a$ $z \in \Omega$, all $y \in \mathbb{R}^N$, with $\alpha \in L^{p'}(\Omega, \mathbb{R}^N)$ and $\eta \in L^1(\Omega)$ $\left(\frac{1}{p} + \frac{1}{p'} = 1\right)$.
- There exists $v_0 \in L^p(\Omega, \mathbb{R}^N)$ such that

$$f(\cdot, v_0(\cdot)) \in L^1(\Omega).$$

We set $X = L^p(\Omega, \mathbb{R}^N)$ and consider the integral functional $I_f : X \to \overline{\mathbb{R}} = \mathbb{R} \cup \{+\infty\}$ defined by

$$I_f(u) = \begin{cases} \int_\Omega f(z, u(z))dz & \text{if } f(\cdot, u(\cdot)) \in L^1(\Omega) \\ +\infty & \text{otherwise,} \end{cases} \quad \text{for all } u \in L^p(\Omega, \mathbb{R}^N)$$

We have $I_f \in \Gamma_0(X)$ and

$$\partial I_f(u) = \left\{ u^* \in L^{p'}(\Omega, \mathbb{R}^N) : u^*(z) \in \partial f(z, u(z)) \text{ for a.a } z \in \Omega \right\}.$$

Moreover, we have

$$(I_f)_\lambda(u) = I_{f_\lambda}(u) \text{ for all } u \in L^p(\Omega, \mathbb{R}^N), \text{ all } \lambda > 0.$$

(c) Every monotone map on $\mathbb{R}$ is automatically cyclically monotone.

7.3 Operators of Monotone Type

In this section we introduce and study nonlinear operators of monotone type, primarily pseudomonotone operators. Such operators arise in many boundary value problems and they are an essential tool in their modern treatment, which uses energy-method techniques. On a theoretical level, these techniques are based on the reflexivity of the ambient spaces (in such spaces, bounded sets are relatively sequentially weakly compact by the James and Eberlein–Smulian theorems) and on the pseudomonotonicity properties of the driving differential operator. So, it is important to know the main properties of such operators.

The setting is the following: X is a reflexive Banach space (in most applications it is also separable), X^* is its topological dual (this is reflexive too and separable if X is so), and by $\langle \cdot, \cdot \rangle$ as always we denote the duality brackets for the pair (X, X^*). Also we have a multivalued man $A : X \to 2^{X^*}$.

Definition 7.67

(a) We say that the operator $A : X \to 2^{X^*}$ is "pseudomonotone," provided that

  (1) For every $x \in X$, $A(x) \in P_{wkc}(X^*)$.

  (2) A is usc from every finite dimensional subspace $Y \subseteq X$ into $X_w^* = $ the space X^* equipped with the weak topology.

  (3) If $x_n \xrightarrow{w} x$ in X and $x_n^* \in A(x_n)$ $n \in \mathbb{N}$ satisfy $\limsup_{n\to\infty}\langle x_n^*, x_n - x \rangle \le 0$, then for every $u \in X$, we can find $x^*(u) \in A(x)$ such that

$$\langle x^*(u), x - u \rangle \le \liminf_{n\to\infty}\langle x_n^*, x_n - u \rangle.$$

(b) We say that A is "generalized pseudomonotone," if for all sequences $x_n \xrightarrow{w} x$ in X, $x_n^* \xrightarrow{w} x^*$ in X^*, $x_n^* \in A(x_n)$ $n \in \mathbb{N}$, which satisfy

$$\limsup_{n\to\infty}\langle x_n^*, x_n - x \rangle \le 0,$$

we have $x^* \in A(x)$ and $\langle x_n^*, x_n \rangle \to \langle x^*, x \rangle$.

Proposition 7.68 *If $A : X \to 2^{X^*}$ is pseudomonotone, then $A(\cdot)$ is generalized pseudomonotone.*

Proof Consider a sequence $\{(x_n, x_n^*)\}_{n\in\mathbb{N}} \subseteq \operatorname{Gr} A$ such that

$$x_n \xrightarrow{w} x \text{ in } X, x_n^* \xrightarrow{w} x^* \text{ in } X^* \text{ and } \limsup_{n\to\infty}\langle x_n^*, x_n - x \rangle \le 0. \tag{7.59}$$

By Definition 7.67(a), for every $u \in X$, we can find $x^*(u) \in A(x)$ such that

$$\langle x^*(u), x - u \rangle \le \liminf_{n\to\infty}\langle x_n^*, x_n - u \rangle. \tag{7.60}$$

From (7.59), $\{(x_n^*, x_n)\}_{n\in\mathbb{N}} \subseteq \mathbb{R}$ is bounded. So, by passing to a suitable subsequence if necessary, we may assume that $\langle x_n^*, x_n \rangle \to \eta \in \mathbb{R}$, thus

$$\limsup\langle x_n^*, x_n - x \rangle = \eta - \langle x^*, x \rangle \le 0 \ (\text{see } (7.59)). \tag{7.61}$$

Also, from (7.60), we have $\langle x^*(u), x - u \rangle \leq \eta - \langle x^*, u \rangle$, thus

$$\langle x^*(u), x - u \rangle \leq \langle x^*, x - u \rangle \quad \text{for all } u \in X \quad \text{(see (7.61)).} \tag{7.62}$$

Suppose $x^* \notin A(x)$. Then by the strong separation theorem (see Theorem 3.69(b)), we can find $y \in X \setminus \{0\}$ such that

$$\langle x^*, y \rangle < \vartheta \leq \inf[\langle u^*, y \rangle : u^* \in A(x)]. \tag{7.63}$$

In (7.62), let $u = x - y$. Then, we have

$$\langle x^*(u), y \rangle \leq \langle x^*, y \rangle \text{ with } x^*(u) \in A(x). \tag{7.64}$$

Comparing (7.63) and (7.64), we have a contradiction. This proves that $x^* \in A(x)$. From (7.60) with $u = x$, we have

$$\langle x^*, x \rangle \leq \liminf_{n \to \infty} \langle x_n^*, x_n \rangle. \tag{7.65}$$

On the other hand, from (7.59), we have

$$\limsup_{n \to \infty} \langle x_n^*, x_n \rangle \leq \langle x^*, x \rangle. \tag{7.66}$$

Finally (7.65) and (7.66) imply that $\langle x_n^*, x_n \rangle \to \langle x^*, x \rangle$. So, we have proved that $A(\cdot)$ is generalized pseudomonotone. □

We can have a converse of the above result.

Proposition 7.69 *If $A : X \to P_{fc}(X^*)$ is bounded and generalized pseudomonotone, then $A(\cdot)$ is pseudomonotone.*

Proof We check (1), (2), and (3) in Definition 7.67(a).

The boundedness hypothesis on $A(\cdot)$, the convexity of the values of $A(\cdot)$, and the reflexivity of X imply that for all $x \in X$ $A(x) \in P_{wkc}(X^*)$.

Next we check (3). So, suppose $\{(x_n, x_n^*)\}_{n \in \mathbb{N}} \subseteq \text{Gr } A$ and assume that

$$x_n \xrightarrow{w} x \text{ in } X \text{ and } \limsup_{n \to \infty} \langle x_n^*, x_n - x \rangle \leq 0. \tag{7.67}$$

The boundedness of $A(\cdot)$ implies that $\{(x_n^*)\}_{n \in \mathbb{N}} \subseteq X^*$ is bounded and so we may assume that $x_n^* \xrightarrow{w} x^*$ in X^*. Then according to Definition 7.67(b), on account of (7.67), we have

$$x^* \in A(x) \text{ and } \langle x_n^*, x_n \rangle \to \langle x^*, x \rangle. \tag{7.68}$$

If (3) in Definition 7.67(a) is not true, then we can find $u \in X$ such that

$$\liminf_{n\to\infty}\langle x_n^*, x_n - u\rangle < \inf[\langle v^*, x - u\rangle : v^* \in A(x)],$$

$$\Rightarrow \liminf_{n\to\infty}\langle x_n^*, x_n - u\rangle < \langle x^*, x - u\rangle \quad \text{(see (7.68))},$$

$$\liminf_{n\to\infty}\langle x_n^*, x_n\rangle < \langle x^*, x\rangle, \quad \text{which contradicts (7.67)}.$$

Therefore (3) in Definition 7.67(a) holds.

Finally, we show (2) in Definition 7.67(a). So, let $Y \subseteq X$ be finite dimensional. Then according to Proposition 5.13 to show the upper semicontinuity of $A(\cdot)$ from Y into X_w^*, it suffices to show that $\operatorname{Gr} A|_Y \subseteq Y \times X_w^*$ is closed, which is true since $A(\cdot)$ is generalized pseudomonotone. □

Proposition 7.70 *If $A : X \to X^*$ is bounded and pseudomonotone, then $A(\cdot)$ is demicontinuous.*

Proof Consider a sequence $x_n \to x$ in X. Then, since $A(\cdot)$ is bounded, we have that $\{A(x_n)\}_{n\in\mathbb{N}} \subseteq X^*$ is relatively weakly compact, and hence we may assume that $A(x_n) \overset{w}{\longrightarrow} x^*$ in X^*. Then, we have $\lim_{n\to\infty}\langle A(x_n), x_n - x\rangle = 0$ and so from Proposition 7.69 and Definition 7.67(b), we have $A(x) = x^*$. Then by the Urysohn criterion for the convergence of sequences, for the initial sequence, we have $A(x_n) \overset{w}{\longrightarrow} A(x)$ in X^* and so $A(\cdot)$ is demicontinuous. □

Now, we introduce another class of nonlinear operators of monotone type.

Definition 7.71 An operator $A : X \to X^*$ is said to be type $(S)_+$, if for every sequence $x_n \overset{w}{\longrightarrow} x$ in X such that $\limsup_{n\to\infty}\langle A(x_n), x_n - x\rangle \leq 0$, we have $x_n \to x$ in X^*.

Using this notion, we can have a converse of Proposition 7.70.

Proposition 7.72 *If $A : X \to X^*$ is demicontinuous and of type $(S)_+$, then $A(\cdot)$ is pseudomonotone.*

Proof Let $x_n \overset{w}{\longrightarrow} x$ in X^* and assume that $\limsup_{n\to\infty}\langle A(x_n), x_n - x\rangle \leq 0$. Since $A(\cdot)$ is of type $(S)_+$, we have $x_n \to x$ in X. Hence by demicontinuity, we have $A(x_n) \overset{w}{\longrightarrow} A(x)$ in X^* and so for every $u \in X$, we have

$$\langle A(x), x - u\rangle = \lim_{n\to\infty}\langle A(x_n), x_n - u\rangle.$$

This verifies (3) in Definition 7.67(a). Properties (1) and (2) in that definition are clearly true. Hence $A(\cdot)$ is pseudomonotone. □

Next, we show that generalized pseudomonotonicity extends the notion of maximal monotonicity.

Proposition 7.73 *If $A : X \to 2^{X^*}$ is maximal monotone, then $A(\cdot)$ is generalized pseudomonotone.*

Proof Let $\{(x_n, x_n^*)\}_{n \in \mathbb{N}} \subseteq \operatorname{Gr} A$ and assume that $x_n \xrightarrow{w} x$ in X, $x_n^* \xrightarrow{w} x^*$ in X^* and

$$\limsup_{n \to \infty} \langle x_n^*, x_n - x \rangle \leq 0. \tag{7.69}$$

It follows from (7.69) that

$$\limsup_{n \to \infty} \langle x_n^* - x^*, x_n - x \rangle \leq 0$$

and so from Proposition 7.50, we have

$$x^* \in A(x) \text{ and } \langle x_n^*, x_n \rangle \to \langle x^*, x \rangle,$$

$\Rightarrow A(\cdot)$ is generalized pseudomonotone. $\qquad\qquad\square$

Pseudomonotonicity is preserved under addition.

Proposition 7.74 *If $A, K : X \to 2^{X^*}$ are two pseudomonotone maps, then so is $x \to S(x) = (A + K)(x)$.*

Proof Evidently, $S(\cdot)$ is $P_{wkc}(X^*)$-valued and it is usc from every finite dimensional subspace $Y \subseteq X$ into X_w^*. So, it remains to verify (3) in Definition 7.67(a). So, suppose $\{(x_n, x_n^*)\}_{n \in \mathbb{N}} \subseteq \operatorname{Gr} S$ and assume that

$$x_n \xrightarrow{w} x \text{ in } X \text{ and } \limsup_{n \to \infty} \langle x_n^*, x_n - x \rangle \leq 0. \tag{7.70}$$

We have $x_n^* = y_n^* + v_n^*$ with $y_n^* \in A(x_n)$, $v_n^* \in K(x_n)$ for all $n \in \mathbb{N}$. From (7.70), we have

$$\limsup_{n \to \infty} [\langle y_n^*, x_n - x \rangle + \langle v_n^*, x_n - x \rangle] \leq 0. \tag{7.71}$$

We will show that (7.71) implies that

$$\limsup_{n \to \infty} \langle y_n^*, x_n - x \rangle \leq 0 \text{ and } \limsup_{n \to \infty} \langle v_n^*, x_n - x \rangle \leq 0 \tag{7.72}$$

If (7.72) is not true, then at least one of the two inequalities is not true. Assume that the first fails (the proof is similar if the second fails). So, we have $0 < \limsup \langle y_n^*, x_n - x \rangle$. By passing to a subsequence if necessary, we can have $\langle y_n^*, x_n - x \rangle \to \eta > 0$ as $n \to \infty$, thus

$$\limsup_{n\to\infty} \langle v_n^*, x_n - x \rangle \leq -c < 0. \tag{7.73}$$

Let $u \in X$. Then (7.73) and the pseudomonotonicity of $K(\cdot)$ imply that there exists $v^*(u) \in K(x)$ such that

$$\langle v^*(u), x - u \rangle \leq \liminf_{n\to\infty} \langle v_n^*, x_n - u \rangle,$$

$$\Rightarrow 0 \leq \liminf_{n\to\infty} \langle v_n^*, x_n - x \rangle \leq 0 \quad \text{choosing } u = x. \tag{7.74}$$

Comparing (7.73) and (7.74), we have a contradiction. So, (7.72) holds, and then for every $u \in X$, we can find $y^*(u) \in A(x)$, $v^*(u) \in K(x)$ such that

$$\langle y^*(u), x - u \rangle \leq \liminf_{n\to\infty} \langle y_n^*, x_n - u \rangle, \quad \langle v^*(u), x - u \rangle \leq \liminf_{n\to\infty} \langle v_n^*, x_n - u \rangle. \tag{7.75}$$

Set $x^*(u) = y^*(u) + v^*(u) \in S(x) = (A + K)(x)$, and from (7.75), we have

$$\langle x^*(u), x - u \rangle \leq \liminf_{n\to\infty} \langle x_n^*, x_n - u \rangle,$$

and hence $S(\cdot) = (A + K)(\cdot)$ is pseudomonotone.   □

In the sequel we will prove a surjectivity result for pseudomonotone maps, which is the reason why this class of operators is so important in applications. The result that we will prove is the pseudomonotone counterpart of Theorem 7.47. As was the case with maximal monotone operators (see the proof of Proposition 7.40), the proof will be based on Galerkin (finite dimensional) approximations. For this reason, first we prove a finite dimensional surjectivity result.

Proposition 7.75 *If X is a finite dimensional Banach space and $F : X \to P_{kc}(X^*)$ is usc and strongly coercive, then $F(\cdot)$ is surjective.*

Proof If $y^* \in X^*$ and consider the multifunction $x \to F_{y^*}(x) = F(x) - y^*$, then $F_{y^*}(\cdot)$ shares the properties of $F(\cdot)$ and so it suffices to show that $0 \in R(F)$.

Arguing by contradiction, suppose that $0 \notin R(F)$. Then for every $x \in X$, we can find $h(x) \in X \setminus \{0\}$ such that

$$0 < \inf[\langle x^*, h(x) \rangle : x^* \in F(x)].$$

The strongly coercive property of $F(\cdot)$ implies that given $M > 0$, we can find $\varrho = \varrho(M) > 0$ such that

$$\langle x^*, x \rangle \geq M \|x\| \quad \text{for all } \|x\| \geq \varrho, \quad x^* \in F(x)$$

and

$$\langle x^*, x \rangle \geq M\varrho \text{ for all } \|x\| = \varrho, \quad x^* \in F(x).$$

Then for such an $x \in X$, we can take $u(x) = x$. For $x \neq 0$, we define

$$U(x) = \{v \in X : \inf[\langle v^*, x \rangle : v^* \in F(v)] > 0\}.$$

From Proposition 5.53, we can know that $v \to \inf[\langle v^*, x \rangle : v^* \in F(v)]$ is lower semicontinuous. It follows that $U(x)$ is open in X and so $\{U(x)\}_{x \in X \setminus \{0\}}$ is an open cover of X. Let $\{V_k\}_{k=1}^m$ be an open cover of $\overline{B}_\varrho = \{x \in X : \|x\| \leq \varrho\}$ such that for each $k \in \{1, ..., m\}$, we can find $x_k \in X$ such that $V_k \subseteq U(x_k)$, and if $V_k \cap \partial B_\varrho \neq 0$, then $x_k \in V_k \cap \partial B_\varrho$ and diam $V_k < \frac{\varrho}{2}$. Let $\{\vartheta_k\}_{k=1}^m$ be a continuous partition of unity subordinate to the cover $\{V_k\}_{k=1}^m$. Let $f : \overline{B}_\varrho \to X$ be defined by

$$f(x) = \sum_{k=1}^m \vartheta_k(x) x_k.$$

Clearly $f(\cdot)$ is continuous, and for every $k \in \{1, ..., m\}$ such that $\vartheta_k(x) > 0$ and for every $x^* \in F(x)$, we have $\langle x^*, x_k \rangle > 0$ (since $x \in V_k \subseteq U(x_k)$). Hence for every $x \in \overline{B}_\varrho$ and $x^* \in F(x)$, we have

$$\langle x^*, f(x) \rangle = \sum_{k=1}^m \vartheta_k(x) \langle x^*, x_k \rangle > 0$$

and

$$f(x) \neq 0 \text{ for all } x \in \overline{B}_r.$$

It follows that $d_B(f, B_r, 0) = 0$ (d_B being the Brouwer degree). On the other hand, if $x \in \partial B_\varrho$, $f(x)$ is a convex combination of the points $\{x_k\}_{k=1}^m \subseteq \partial B_\varrho$ and $\|x_k - x\| < \frac{\varrho}{2}$ for all $k \in \{1, ..., m\}$. Hence

$$\|f(x) - x\| \leq \frac{\varrho}{2} \text{ for all } x \in \partial B_\varrho.$$

Therefore $f(\cdot)$ is homotopic to $id(\cdot)$, and from the properties of the Brouwer degree, we have $d_B(f, B_\varrho, 0) = 1$, a contradiction. So, we conclude that $0 \in R(A)$, which implies the surjectivity of $F(\cdot)$. $\qquad \square$

Using this finite dimensional result, we can have the main surjectivity result for pseudomonotone operators. The result is the "pseudomonotone" counterpart of Theorem 7.47.

Theorem 7.76 *If $A : X \to 2^{X^*}$ is pseudomonotone and strongly coercive, then $A(\cdot)$ is surjective.*

Proof Let $\mathscr{S}$ be the family of finite dimensional subspaces of X endowed with the partial order defined by inclusion. Let $V \in \mathscr{S}$ and let $i_V : V \rightarrow X$ be the embedding of V into X. We have that $i_V^* : X^* \rightarrow V^*$ is the projection operator of X^* onto V^*. Let $A_V : V \rightarrow 2^{V^*}$ be defined by

$$A_V = i_V^* \circ A \circ i_V.$$

Evidently $A_V(x) \in P_{kc}(V)$ and it is usc (see Definition 7.67(a)(2)). For every $x_V^* \in A_V(x)$, we have $x_V^* = i_V^*(x^*)$ with $x^* \in A(x)$. Then

$$\langle x_V^*, x \rangle_V = \langle i^*(x^*), x \rangle_V = \langle x^*, i(x) \rangle.$$

So, $A_V(\cdot)$ is coercive.

As in the previous proof, to show the surjectivity of $A(\cdot)$, it suffices to show that $0 \in R(A)$. From Proposition 7.75, we can find $x_V \in V$ such that

$$0 \in A_V(x_V),$$

hence

$$0 = i_V^*(\widehat{x}_V^*) \quad \text{for some } \widehat{x}_V^* \in A(x_V).$$

The coercivity of $A(\cdot)$ implies that $\{x_V\}_{V \in \mathscr{S}} \subseteq X$ is bounded. Let

$$T_V = \bigcup_{V' \in \mathscr{S}, V \subseteq V'} \{x_{V'}\}.$$

We have $T_V \subseteq \overline{B}_M = \{x \in X : \|x\| \leq M\}$ for some big $M > 0$. The reflexivity of X implies that each $\overline{T}_V^w \subseteq X$ is w-compact. So, by the finite intersection property, we have

$$\bigcap_{V \in \mathscr{S}} \overline{T}_V^w \neq \emptyset.$$

Consider $x_0 \in \bigcap_{V \in \mathscr{S}} \overline{T}_V^w$ and $y \in X$. Choose $V \in \mathscr{S}$ such that $\{x_0, y\} \subseteq V$. Let $\{x_{V_k}\}_{k \in \mathbb{N}} \subseteq T_V$ be such that

$$x_{V_k} \xrightarrow{w} x_0 \quad in \quad X$$

by the Eberlern–Smulian theorem.

We know that $0 = i_{V_k}^*(\widehat{x}_{V_k}^*)$ with $\widehat{x}_{V_k}^* \in A(x_{V_k})$. So, we have

$$\langle x_{V_k}^*, x_{V_k} - x_0 \rangle = 0 \quad \text{for all } k \in \mathbb{N}.$$

On account of the pseudomonotonicity of $A(\cdot)$, we can find $x^*(y) \in A(x_0)$ such that

$$\langle x^*(y), x_0 - y \rangle \leq \lim_{k \to \infty} \inf \langle x^*_{V_k}, x_{V_k} - y \rangle = 0 \ \text{ for all } \ y \in X. \tag{7.76}$$

Suppose that $0 \notin A(x_0) \in P_{wkc}(x^*)$. So, we can find $y \in X$ such that

$$0 < \inf[\langle x^*, x_0 - y \rangle : x^* \in A(x_0)].$$

Therefore, we reach a contradiction (see (7.76)). This proves the surjectivity of $A(\cdot)$.
□

Let $\Omega \subseteq \mathbb{R}^N$ be a bounded domain with Lipschitz boundary $\partial\Omega$, $1 < p < \infty$ and consider the operator $A : W^{1,p}(\Omega) \to W^{1,p}(\Omega)^*$ defined by

$$\langle A(u), h \rangle = \int_\Omega |Du|^{p-2}(Du, Dh)_{\mathbb{R}^N} dz \ \text{ for all } \ u, h \in W^{1,p}(\Omega). \tag{7.77}$$

We recall the following elementary inequalities. Let $a, b \in \mathbb{R}^N$. We have

$$(|a|^{p-2}a - |b|^{p-2}b, a - b)_{\mathbb{R}^N} \geq \eta \begin{cases} |a-b|^p & \text{if } p \geq 2, \\ |a-b|^2[1 + |a| + |b|]^{p-2} & \text{if } 1 < p \leq 2. \end{cases} \tag{7.78}$$

Proposition 7.77 *The nonlinear operator* $A : W^{1,p}(\Omega) \to W^{1,p}(\Omega)^*$ *defined by (7.77) is maximal monotone and of type* $(S)_+$.

Proof From (7.78), we see that $A(\cdot)$ is monotone. Also $A(\cdot)$ is continuous. To see this, consider a sequence $\{u_n\}_{n \in \mathbb{N}} \subseteq W^{1,p}(\Omega)$ such that $u_n \to u$ in $W^{1,p}(\Omega)$. We have

$$\langle A(u_n) - A(u), h \rangle \leq \int_\Omega (|Du_n|^{p-2}Du_n - |Du|^{p-2}Du, Dh)_{\mathbb{R}^N} dz$$

$$\leq \||Du_n|^{p-2}Du_n - |Du|^{p-2}Du\|_{p'}\|Dh\|_p$$

$$\leq \||Du_n|^{p-2}Du_n - |Du|^{p-2}Du\|_p\|h\|_{1,p}$$

by Hölder's inequality, where $\|\cdot\|_{1,p}$ is the norm of $W^{1,p}(\Omega)$. So,

$$\|A(u_n) - A(u)\|_* \leq \||Du_n|^{p-2}Du - |Du|^{p-2}Du\|_{p'} \to 0.$$

$A(\cdot)$ is continuous.

Then Proposition 7.36 implies the maximal monotonicity of $A(\cdot)$.

Now we show that $A(\cdot)$ is of type $(S)_+$. So, suppose that $u_n \xrightarrow{w} u$ in $W^{1,p}(\Omega)$ and assume that

$$\lim_{n\to\infty} \sup \langle A(u_n), u_n - u \rangle \leq 0. \tag{7.79}$$

The monotonicity of $A(\cdot)$ and (7.79) imply that

$$\lim_{n\to\infty} \langle A(u_n), u_n - u \rangle = 0. \tag{7.80}$$

Let $\eta_n(z) = \langle |Du_n(z)|^{p-2} Du_n(z) - |Du(z)|^{p-2} Du(z), Du_n(z) - Du(z) \rangle_{\mathbb{R}^N}$. Evidently $\eta_n \in L^1(\Omega)$, $\eta_n(z) \geq 0$ for a.a $z \in \Omega$. From (7.80), we have

$$\int_\Omega \eta_n(z) dz \to 0,$$

hence

$$\eta_n \to 0 \quad in \quad L^1(\Omega).$$

Therefore, by passing to a suitable subsequence if necessary, we may assume that

$$\eta_n(z) \to 0 \text{ for a.a } z \in \Omega,$$
$$0 \leq \eta_n(z) \leq \vartheta(z) \text{ for a.a. } z \in \Omega, \text{ all } n \in \mathbb{N}, \tag{7.81}$$

with $\vartheta \in L^1(\Omega)$. Then

$$\vartheta(z) \geq \eta_n(z) \geq |Du_n(z)|^{p-1} + |Du(z)|^{p-1}$$
$$- |Du_n(z)|^{p-1} |Du(z)| - |Du(z)|^{p-1} |Du_n(z)| \tag{7.82}$$

for a.a $z \in \Omega$, all $n \in \mathbb{N}$. Hence we can find $D \subseteq \mathbb{R}^N$ Lebesgue-null such that $\{Du_n(z)\}_{n\in\mathbb{N}} \subseteq \mathbb{R}^N$ is bounded for all $z \in \Omega \setminus D$. By passing to a subsequence if necessary (in general this subsequence depends on $z \in \Omega \setminus D$), we have

$$Du_n(z) \to \xi(z) \text{ for all } z \in \Omega \setminus D.$$

Passing to the limit as $n \to \infty$, we obtain

$$(|\xi(z)|^{p-2} \xi(z) - |Du(z)|^{p-2} Du(z), \xi(z) - Du(z))_{\mathbb{R}^N} = 0 \tag{7.83}$$

for all $z \in \Omega \setminus D$ (see 7.81).

From (7.78) and (7.83), we infer that $\xi(z) = Du(z)$ for a.a $z \in \Omega$. So, by the Urysohn criterion for the convergence of sequences, we have that

$$Du_n(z) \to Du(z) \quad for \quad a.a \quad z \in \Omega. \tag{7.84}$$

From (7.82) and Hölder's inequality, we see that

$$\{|Du_n|^p\}_{n \in \mathbb{N}} \subseteq L^1(\Omega) \tag{7.85}$$

is uniformly integrable. Then on account of (7.84), (7.85), and Vitali's Theorem (see Theorem 2.147), we have

$$\|Du_n\|_p \to \|Du\|_p. \tag{7.86}$$

Since $Du_n \overset{w}{\to} Du$ in $L^p(\Omega, \mathbb{R}^N)$ and the Kadec–Klee property of the uniformly convex Banach space $L^p(\Omega, \mathbb{R}^N)$ (see Definition 3.243 and Proposition 3.244), we infer that

$$Du_n \to Du \quad in \quad L^p(\Omega, \mathbb{R}^N).$$

From the Sobolev embedding theorem (see Theorem 4.114), we have

$$u_n \to u \quad in \quad L^p(\Omega).$$

So, we conclude that $u_n \to u$ in $W^{1,p}(\Omega)$, and hence $A(\cdot)$ is of type $(S)_+$. □

Remark 7.78 If instead of $W^{1,p}(\Omega)$ we consider $W_0^{1,p}(\Omega)$, then the operator $A(\cdot)$ is strictly monotone if $1 < p < 2$ and strongly monotone if $p \geq 2$ (see (7.78)). In addition $A(\cdot)$ is also coercive (by Poincare's inequality).

7.4 Remarks

7.1 Compact operators were the first class of operators used to study nonlinear equations in infinite dimensional Banach spaces. The reason is that compact operators can be approximated by finite dimensional ones (see Theorem 7.2) and so finite dimensional results can be extended to infinite dimensional equations. This approximation result is due to Schauder and can be found in the seminal paper of Leray-Schauder [181] (see also Schauder [246, 247]). It extends to locally convex spaces (see Leray [180] and Nagumo [204]). The same is also true for the Dugundji extension theorem (see Theorem 7.4). We refer to the paper of Dugundji [90]. For potential maps (see Definition 7.7), the main references are the books of Vainberg [273],[274] (see also Berger [31]). For linear Fredholm operators, we refer to Kato [155] and for nonlinear ones to Berger [31]. Proper maps (see Definition 7.19) are discussed in Berger [31].

7.2 Monotone operators were introduced in the 60s in order to have a theoretical framework broader than that provided by compact operators. The systematic study of monotone operators started with the works of Kachurovski [158] and Minty [194]. Subsequent important contributions on the subject were made by Brezis [49], Browder [53], and Rockafellar [235], [237]. The local boundedness of a monotone operator (see Proposition 7.33) is due to Rockafellar [235] (see also Phelps [221]). We mention also the following genericity result due to Kenderov [159].

Proposition 7.79 *If X is a separable, reflexive Banach space and $A : X \to 2^{X^*}$ is a maximal monotone operator with int $D(A) \neq \emptyset$, then there is a G_δ-set $C \subseteq$ int $D(A)$ such that $A|_C$ is single valued and use from X into X^*.*

Theorems 7.41 and 7.47 are due to Browder [53], while Theorem 7.44 was proved by Rockafellar [237], as in Theorem 7.58.

7.3 Pseudomonotone operators were introduced by Brezis [48] using nets. The passage to sequences was made by Browder [53]. Theorem 7.76 is due to Browder-Hess [54]. More on Pseudomonotone operators can be found in the papers of Kenmochi [160].

Maximal monotone operators and operators of monotone type are discussed to different levels of generality in Barbu [27], Brezis [49], Browder [53], Gasinski-Popageorgiou [121], Hu-Popageorgiou [145], Papageorgiou-Winkert [216], Pascali-Sburlan [219], Roubiček [240], Showalter [254], and Zeidler [292].

Finally we mention the following result about the square root of a bounded linear operator on a Hilbert space which is monotone (positive).

Proposition 7.80 *If H is a Hilbert space, $A \in \mathcal{L}(H)$, and A is monotone (that is, $A \geq 0$), then there exists unique $L \in \mathcal{L}(H)$, $L \geq 0$, such that*

$$L^2 = A.$$

We say that L is the square root of A and write $L = A^{1/2}$. Moreover, L commutes with every element in $\mathcal{L}(H)$ which commutes with A.

7.5 Problems

Problem 7.1 Let $A : \mathbb{R}^N \to 2^{\mathbb{R}^N}$ be an use multifunction which is usc and for every $x \in \mathbb{R}^N$, $A(x) \subseteq \mathbb{R}^N$ is convex. Show that $A(\cdot)$ is maximal monotone.

Problem 7.2 Let X be a finite dimensional Banach space and $A : X \to 2^{X^*}$ is a monotone map such that $D(A) = X$. Show that $A(\cdot)$ in bounded (that is, maps bounded sets to bounded sets).

Problem 7.3 Let X be a reflexive Banach space and suppose that $A : X \to X^*$ is linear, demicontinuous, and monotone. Show that $A \in \mathcal{L}(X, X^*)$.

Problem 7.4 Let X be a reflexive Banach space and $A : X \to X^*$ is uniformly monotone and hemicontinuous. Show that $A(\cdot)$ is surjective.

Problem 7.5 Show that the duality map of a Banach space X is linear if and only if X is a Hilbert space.

Problem 7.6 Let X be a Banach space and $\mathcal{F}(\cdot)$ is its duality map. Show that X is reflexive if and only if $\mathcal{F}(\cdot)$ is surjective.

Problem 7.7 Let X be a reflexive Banach space, $A : X \to X^*$ is monotone hemicontinuous, and $K \subseteq X$ is w-closed and bounded. Show that $A(K) \subseteq X^*$ is closed.

Problem 7.8 Let X be a reflexive Banach space, $A : X \to X^*$ is everywhere defined, and for every $x, h \in X : t \to < A(x + th), h >$ is continuous on $\mathbb{R}$. Suppose that $< u^* - A(x), u - x >\geq 0$ for all $x \in X$. Show that $u^* = A(u)$.

Problem 7.9 Let X be a reflexive Banach space, $A : X \to X^*$ is everywhere defined, bounded, and coercive, and for every $x, h \in X : t \to < A(x + th), h >$ is continuous on $\mathbb{R}$ and uniformly (resp., strongly) monotone. Show $A^{-1} : X^* \to X$ is uniformly (resp., Lipschitz) continuous.

Problem 7.10 Let X be a reflexive Banach space and $A : X \to X^*$ is an everywhere defined monotone map such that for all $x, h \in X : t \to < A(x + th), h >$ is continuous on $\mathbb{R}$. Show that $A(\cdot)$ is demicontinuous.

Problem 7.11 Let X be a reflexive Banach space and $A : X \to 2^{X^*}$ is a maximal monotone map. Let $m(x) = inf[\|x^*\| : x^* \in A(x)]$ (as always $inf\emptyset = +\infty$). Show that the function $m : X \to \overline{\mathbb{R}} = \mathbb{R} \cup +\infty$ is lower semicontinuous.

Problem 7.12 Let (X, H, X^*) be an evolution triple (see Definition 4.201), $1 < p < \infty$, consider the operators $A_1, A_2 : L^p([0, b], X) \to L^{p'}([0, b], X^*)(\frac{1}{p} + \frac{1}{p'} = 1)$ defined by

$$
\begin{aligned}
A_1(u) = u' \quad &forall \quad u \in D(A_1) = \{u \in W_p(0, b) : u(0) = 0\}, \\
A_2(u) = u' \quad &forall \quad u \in D(A_2) = \{u \in W_p(0, b) : u(0) = u(b)\}.
\end{aligned}
\tag{7.87}
$$

Show that both maps $A_1(\cdot)$ and $A_2(\cdot)$ are maximal monotone.

Problem 7.13 Let $A : \mathbb{R}^N \to \mathbb{R}^N$ be an everywhere defined, surjective monotone map. Show that $A(\cdot)$ is coercive.

Problem 7.14 Let X be a reflexive Banach space, $A : X \to 2^{X^*}$ bounded, and has the properly that "If $\{(x_n, x_n^*)\}_{n \in \mathbb{N}} \subseteq \mathrm{Gr}\, A$, $x_n \xrightarrow{w} x$, $x_n^* \xrightarrow{w} x^*$ and $\limsup_{n \to \infty} \langle x_n^*, x_n - x \rangle \leq 0$, then $(x, x^*) \in \mathrm{Gr}\, A$." Show that if $Y \subseteq X$ is finite dimensional, then $A|_Y$ is usc from Y into X_w^*.

Problem 7.15 Let X be a Banach space and $A, F : X \to 2^{X^*}$ are maximal monotone with $D(A) = D(F) = X$. Is it true that $R(A + F) = R(A) + R(F)$? Justify your answer.

Problem 7.16 Let (Ω, Σ, μ) be a σ-finite measure space and H be a pivot Hilbert space, $A : H \to 2^H$ is a maximal monotone map with $(0, 0) \in \mathrm{Gr}\, A$, and $\widehat{A} : L^2(\Omega, H) \to 2^{L^2(\Omega, H)}$ is the realization of $A(\cdot)$ on the Hilbert space $L^2(\Omega, H)$, that is,

$$\widehat{A}(u) = \{h \in L^2(\Omega, H) : h(\omega) \in A(u(\omega)) \quad \mu - a.e.\}$$

for all $u \in D(\widehat{A}) = \{v \in L^2(\Omega, H) : S^2 A(v(\cdot)) \neq 0\}$. Show that $A(\cdot)$ is maximal monotone and find $A_\lambda(\cdot)$ for $\lambda > 0$. (If $\mu(\cdot)$ is finite, we can drop the requirement that $(0, 0) \in \mathrm{Gr}\, A$.)

Problem 7.17 Let H be a pivot Hilbert space and $C \subseteq H$ a nonempty bounded set. We introduce the multifunction $G : H \to 2^H$ defined by

$$G(x) = \{\widehat{c} \in C : \|x - \widehat{c}\| = \sup_{c \in C} \|x - c\|\}.$$

Show that $(-G)(\cdot)$ is monotone.

Problem 7.18 Let H be a pivot Hilbert space and $A \in \mathcal{L}(H)$ is monotone (that is, $A \geq 0$). Show that

$$ker\, A = ker\, A^* \quad and \quad \overline{R(A)} = \overline{R(A^*)}.$$

Problem 7.19 Let H be a pivot Hilbert space and $A \circ K \in \mathcal{L}(H)$ are self-adjoint such that $A \circ K = K \circ A$. Show that $A \cdot K$ is monotone (that is, $A \cdot K \geq 0$).

Problem 7.20 Let X, Y be Banach spaces and $f \in C(X, Y)$ a proper function. Show that given $y \in Y$ and $\varepsilon > 0$, we can find $\delta > 0$ such that

$$\|f(x) - y\|_Y \leq \delta \Rightarrow \|x - f^{-1}(y)\|_X \leq \varepsilon.$$

Problem 7.21 Let X, Y be Banach spaces, $U \subseteq X$ and $V \subseteq Y$ are nonempty open subsets, and $f \in C(X, Y)$ satisfies $f(U) = V$. Suppose $f|_U$ is locally invertible and proper. For $y_0 \in Y$, let

$$S_0 = \{x \in U : f(x) = y_0\}$$

and

$$\mu_0 = card\, S_0.$$

Show that μ_0 is finite.

Problem 7.22 Let X be a reflexive Banach space and $A : X \to 2^{X^*}$ a pseudomonotone map. Show that for every $h \in X$, $A_0(u) = A(u + h)$ is pseudomonotone too.

Problem 7.23 Let X be a reflexive Banach space, $A : X \to 2^{X^*}$ a pseudomonotone map, and $K : X \to X^*$ a strongly continuous map (that is, if $x_n \xrightarrow{w} x$ in X, then $K(x_n) \to K(x)$ in X^*). Show that $x \to (A + K)(x)$ is pseudomonotone.

Problem 7.24 Suppose $\varphi : \mathbb{R}^N \to \mathbb{R}^N$ is monotone, measurable. Let D_N denote the family of all Lebesgue-null sets in $\mathbb{R}^N$. We define

$$\widehat{\varphi}(x) = \cap_{\varepsilon > 0} \cap_{E \in D_N} \overline{conv}\varphi(\overline{B}_\varepsilon(x) \setminus E)$$

for all $x \in \mathbb{R}^N$. Show that $\widehat{\varphi} : \mathbb{R}^N \to \mathcal{L}^{\mathbb{R}^N}$ is maximal monotone.

Problem 7.25 Let H be a pivot Hilbert space and $A : H \to 2^H$. Show that $A(\cdot)$ is monotone if and only if $\|x - u + t(x^* - u^*)\| \geq \|x - u\|$ for all $x^* \in A(x)$, $u^* \in A(u)$, and all $t > 0$.

Problem 7.26 Let X be a reflexive Banach space and $A : X \to X^*$ is bounded, demicontinuous, strongly coercive, and a type $(S)_+$. Show that $A(\cdot)$ is surjective.

Problem 7.27 Prove Proposition 7.8

Chapter 8
Variational Analysis

In this chapter, we present some subjects related to variational analysis. In Sect. 8.1, we introduce some modes of set convergence, examine the relations between them, and discuss some of the hyperspace topologies associated with them. In Sect. 8.2, we introduce a notion of convergence of functions that is suitable for the study of the stability (sensitivity) analysis of variational problems. So, we introduce the notion of Γ-convergence (or epigraphical convergence) of functions and establish that this is the appropriate mode to discuss issues related to the stability of optimization problems. We also show that this functional convergence is closely related to the convergence of sets from Sect. 8.1, the link between the two being the epigraph of the function. We also introduce and study the associated G-convergence of operators. Finally, in Sect. 8.3, we present three basic variational principles: the Ekeland Variational Principle, the Caristi Fixed Point Theorem, and the Takahashi Variational Principle. We show that all three are equivalent.

8.1 Convergence of Sets

In this section, we introduce and study certain modes of convergence of sequences of nonempty sets. We also make an effort to associate these modes of convergence to certain hyperspace topologies. These convergences of sets arise in many applications, and so it is important to know their properties and how they are related to each other.

First, we introduce the Hausdorff convergence of sets. So, let (X, d) be a metric space and $P_f(X)$ the hyperspace of all nonempty, closed subsets of X. We know that on $P_f(X)$ we can define an extended (generalized) metric (that is, the metric can take the value $+\infty$), known as the Hausdorff metric (see Definition 5.39), which is denoted by $h(\cdot, \cdot)$. Recall that on $P_{bf}(X) = \{A \in P_f(X) : A \text{ is bounded}\}$, then $h(\cdot, \cdot)$ is a finite-valued metric.

© The Author(s), under exclusive license to Springer Nature Switzerland AG 2022
S. Hu, N. S. Papageorgiou, *Research Topics in Analysis, Volume I*, Birkhäuser
Advanced Texts Basler Lehrbücher, https://doi.org/10.1007/978-3-031-17837-5_8

Definition 8.1 Let (X, d) be a metric space and $\{A_n, A\}_{n \in \mathbb{N}} \subseteq P_f(X)$. We say that the $A_n's$ converge to A in the Hausdorff metric if and only if $h(A_n, A) \to 0$ as $n \to \infty$. We denote this convergence by $A_n \overset{h}{\to} A$.

Remark 8.2 We know that this is a topological notion since $(P_f(X), h)$ is a metric space that is complete if and only if X is complete (see Proposition 5.43). Note that the Hausdorff metric topology on $P_f(X)$ is the topology that $P_f(X)$ inherits from $C(X, \mathbb{R})$ endowed with the topology of uniform convergence under the identification $A \leftrightarrow d(\cdot, A)$. So $A_n \overset{h}{\to} A$ if and only if $d(\cdot, A_n) \overset{u}{\to} d(\cdot, A)$. Moreover, if X is a normed space and $\{A_n, A\}_{n \in \mathbb{N}} \subseteq P_{bfc}(X) = \{A \in P_{bf}(X) : A$ is also convex$\}$, then $A_n \overset{h}{\to} A$ if and only if $\sigma(\cdot, A_n) \to \sigma(\cdot, A)$ uniformly on $\overline{B}_1^* = \{x^* \in X^* \|x^*\|_* \leq 1\}$ (by Hörmander's formula, see Proposition 5.42(b)). Finally, we mention that $(P_f(X), h)$ is compact if and only if (X, d) is compact.

The excess and gap functions are defined by

$$h^*(A, C) = sup[d(a, C) : a \in A] = inf[\varepsilon > 0 : A \subseteq C_\varepsilon]$$

with $C_\varepsilon = \{x \in X : d(x, C) \leq \varepsilon\}$ and

$$G(A, C) = inf[d(a, c) : a \in A, c \in C] = inf[\varepsilon > 0 : A_\varepsilon \cap C \neq \emptyset].$$

Directly from the definition of the Hausdorff distance (metric), see Definition 5.39, we have:

Proposition 8.3 *If (X, d) is a metric space and $A, C \in P_f(X)$, then the functionals* $e_A : (P_f(X), h) \to \overline{\mathbb{R}}_+ : \mathbb{R}_+ \cup \{+\infty\}$, $e_C : (P_f(X), h) \to \overline{\mathbb{R}}_+$, *and* $G_A : (P_f(X), h) \to \overline{\mathbb{R}}_+$ *defined by*

$$e_A(C) = h^*(A, C), e_C(A) = h^*(C, A), G_A(C) = G(A, C)$$

are all Lipschitz continuous with Lipschitz constant 1.

Next we will introduce the so-called Kuratowski convergence of a sequence of nonempty sets.

Definition 8.4 Let (X, τ) be a Hausdorff topological space with τ denoting the topology, and let $\{A_n\}_{n \in \mathbb{N}} \subseteq 2^X \setminus \{\emptyset\}$. We define

$$\tau\text{-}\liminf_{n \to \infty} A_n = \{x \in X : x = \tau\text{-}\lim_{n \to \infty} x_n, x_n \in A_n \ n \in \mathbb{N}\}$$

(the Kuratowski limit inferior of $\{A_n\}_{n \in \mathbb{N}}$),

$$\tau\text{-}\limsup_{n\to\infty} A_n = \{x \in X : x = \tau\text{-}\lim x_{n_k}, x_{n_k} \in A_{n_k}, n_k < n_{k+1} \ k \in \mathbb{N}\}$$

(the Kuratowski limit superior of $\{A_n\}_{n\in\mathbb{N}}$).

We say that the sequence $\{A_n\}_{n\in\mathbb{N}}$ converges to $A \in 2^X$ in the "Kuratowski sense" if and only if $A = \liminf_{n\to\infty} A_n = \limsup_{n\to\infty} A_n$. Then we write $A_n \overset{K_\tau}{\to} A$.

Remark 8.5 If X is a metric space with metric $d(\cdot, \cdot)$, then

$$d\text{-}\liminf_{n\to\infty} A_n = \{x \in X : \lim_{n\to\infty} d(x, A_n) = 0\},$$

$$d\text{-}\limsup_{n\to\infty} A_n = \{x \in X : \liminf_{n\to\infty} d(x, A_n) = 0\}.$$

From Definition 8.4, it is clear that we always have $\tau\text{-}\liminf_{n\to\infty} A_n \subseteq \tau\text{-}\limsup_{n\to\infty} A_n$, and the inclusion can be strict. To see this, let $X = \mathbb{R}$ and $A_n = [-n, -\frac{1}{n}]$ if $n \in \mathbb{N}$ is even, and $A_n = [\frac{1}{n}, n]$ if $n \in \mathbb{N}$ is odd. Then $\liminf_{n\to\infty} A_n = \{0\}, \limsup_{n\to\infty} A_n = \mathbb{R}$. Also when the topology (resp., the metric) is clearly understood from the context, we drop the use of the letter τ (resp., d). If X is first countable, then we can have the following topological descriptions of the two Kuratowski limits:

$$\tau\text{-}\liminf_{n\to\infty} A_n = \bigcap_C c\ell[\bigcup_{n\in C} A_n : C \subseteq \mathbb{N} \text{ cofinal}]$$

$$\tau\text{-}\limsup_{n\to\infty} A_n = \bigcap_{k\geq 1} \overline{\bigcup_{n\geq k} A_n}.$$

Evidently, then the two sets $\tau\text{-}\liminf A_n$, $\tau\text{-}\limsup A_n$ are closed sets (possibly empty). We can also have a topological version of Definition 8.4. So, let $\{A_i\}_{i\in D} \subseteq 2^X\setminus\{\emptyset\}$ be a net. Then

$$\tau\text{-}\liminf A_i = \{x \in X : A_i \cap U \neq \emptyset \text{ for } U \in \mathcal{N}(x) \text{ and } i \geq i_0\}$$

$$\tau\text{-}\limsup A_i = \{x \in X : A_i \cap U \neq \emptyset \text{ for } U \in \mathcal{N}(x) \text{ and } i \in C, C \subseteq D \text{ cofinal}\}.$$

If X is separable, then every $\{A_n\}_{n\in\mathbb{N}} \subseteq 2^X\setminus\{\emptyset\}$ has a K-convergent subsequence.

On $P_f(X)$, we define a topology with subbasis of the sets

$$U^- = \{A \in P_f(X) : A \cap U \neq \emptyset\}, U \in \tau\setminus\{\emptyset\},$$

$$V^+ = \{A \in P_f(X) : A \subseteq V\}, V \in \tau \setminus \{\emptyset\}, X \setminus V \in P_k(X).$$

This hyperspace topology is known as the "Fell topology," and it is denoted by τ_F. It is weaker than the Vietoris topology τ_V generated by the subbasis $\{U^-, V^+ : U, V \in \tau \setminus \{\emptyset\}\}$. The Fell topology has nice properties when X is locally compact, namely $(P_f(X), \tau_F)$ is completely regular if and only if X is locally compact. Then we have $A = \tau_F\text{-}\lim_{i \in D} A_i$ if and only if $A = K_\tau\text{-}\lim_{i \in D} A_i$. So, in applications where X is a normed space, the Kuratowski convergence is most useful if X is finite dimensional. If X is not locally compact, then the Kuratowski convergence is not topological. For this reason, in the context of infinite dimensional Banach spaces, we introduce the following convergence mode that mixes the norm and weak topologies on X.

Definition 8.6 Let X be a Banach space and $\{A_n\}_{n \in \mathbb{N}} \subseteq 2^X \setminus \{\emptyset\}$. We define

$$s\text{-}\liminf_{n \to \infty} A_n = \{x \in X : x = \lim x_n, x_n \in A_n \ n \in \mathbb{N}\},$$

$$w\text{-}\limsup_{n \to \infty} A_n = \{x \in X : x = w\text{-}\lim x_{n_k}, x_{n_k} \in A_{n_k}, n_k < n_{k+1} \ k \in \mathbb{N}\}.$$

We say that the sequence $\{A_n\}_{n \in \mathbb{N}}$ converges to $A \in 2^X$ in the "Mosco sense" if and only if $A = s\text{-}\liminf_{n \to \infty} A_n = w\text{-}\limsup_{n \to \infty} A_n$. Then we write $A_n \overset{M}{\to} A$.

Remark 8.7 Since $s\text{-}\liminf_{n \to \infty} A_n \subseteq w\text{-}\liminf_{n \to \infty} A_n$, $s\text{-}\limsup_{n \to \infty} A_n \subseteq w\text{-}\limsup_{n \to \infty} A_n$, we see that

$$\text{"}A_n \overset{M}{\to} A \text{ if and only if } A_n \overset{K_s}{\to} A \text{ and } A_n \overset{K_w}{\to} A.\text{"}$$

Also the Mosco convergence and the Hausdorff convergence are distinct. To see this, let $X = \ell^2$, and let $\{e_n\}_{n \in \mathbb{N}}$ be the standard orthonormal basis of ℓ^2. We set $A_n = \{te_n : 0 \le t \le 1\}$ and $A = \{0\}$. We see that $A_n \overset{M}{\to} A$ but $h(A_n, A) = 1$ for all $n \in \mathbb{N}$, and so we do not have h-convergence. On the other hand, if X is a reflexive Banach space and $A_n = A = \partial B_1 = \{x \in X : \|x\|_X = 1\}$ for all $n \in \mathbb{N}$, then $A_n \overset{h}{\to} A$, but since $\overline{\partial B_1}^w = \overline{B}_1 = \{x \in X : \|x\|_X \le 1\}$, we see that we cannot have Mosco convergence of the $A_n's$ to A.

Now we introduce a fourth mode of set of convergence. It is defined on the closed subsets of a metric space.

Definition 8.8 Let (X, d) be a metric space and $\{A_n, A\}_{n \in \mathbb{N}} \subseteq P_f(X)$. We say that the sequence $\{A_n\}_{n \in \mathbb{N}}$ converges to A in the "Wijsman sense" if and only if for all $x \in X, d(x, A_n) \to d(x, A)$. Then we write $A_n \overset{W}{\to} A$.

Remark 8.9 It is clear that "$A_n \overset{h}{\to} A \Rightarrow A_n \overset{W}{\to} A$." Also $A_n \overset{W}{\to} A \Rightarrow A_n \overset{K_d}{\to}$ A (see Remark 8.5). However, the reverse implication is not true. To see this, let $X = \ell^2$, $\{e_n\}_{n \in \mathbb{N}}$ be the standard orthonormal basis of ℓ^2 and $A_n = \{x \in X : x = te_1 + (1-t)e_n, 0 \le t \le 1\}$, $n \in \mathbb{N}$, and $A = \{e_1\}$. Then $A_n \overset{K_s}{\to} A$, but $d(0, A_n) = \frac{1}{2}\|e_1 + e_n\| = \frac{\sqrt{2}}{2}$ for all $n \in \mathbb{N}$, and so we cannot have Wijsman convergence. This mode of set convergence is topological, and to determine the topology, consider the identification of $A \in P_f(X)$ with $d(\cdot, A) \subset C(X, \mathbb{R})$ to embed $P_f(X)$ in $C(X, \mathbb{R})$ and then consider the topology of pointwise convergence for $C(X, \mathbb{R})$ restricted on $P_f(X)$ (i.e., the weak topology on $P_f(X)$ generated by the family $\{d(x, \cdot)\}_{x \in X}$). We denote this topology by τ_W, and in general, this topology is not first countable. So, sequences are not enough to describe it. However, if X is separable, then τ_W is metrizable and separable.

Finally, we introduce a fifth mode of set convergence.

Definition 8.10 Let X be a Banach space and $\{A_n, A\}_{n \in \mathbb{N}} \subseteq P_{f_c}(X)$. We say that the sequence $\{A_n\}_{n \in \mathbb{N}}$ converges to A in the "weak sense" if and only if for all $x^* \in X^*$, we have $\sigma(x^*, A_n) \to \sigma(x^*, A)$ as $n \to \infty$. Then we write $A_n \overset{w}{\to} A$.

Remark 8.11 If $\{A_n\}_{n \in \mathbb{N}} \subseteq P_{bfc}(X)$, then on account of the Hörmander formula (see Proposition 5.42(b)), we have "$A_n \overset{h}{\to} A \Rightarrow A_n \overset{w}{\to} A$." Of course, the reverse implication is not in general true. Simply consider a sequence in X that converges weakly but not strongly.

So, summarizing what we have seen for these modes of convergence, we can state the following proposition.

Proposition 8.12 *If X is a Banach space and $\{A_n\}_{n \in \mathbb{N}} \subseteq 2^X \setminus \{\emptyset\}$, then:*

(a) $A_n \overset{h}{\to} A \Rightarrow A_n \overset{K_s}{\to} A$ *and* $A_n \overset{W}{\to} A$.

(b) *If $\{A_n, A\}_{n \in \mathbb{N}} \subseteq P_{bfc}(X)$, then "$A_n \overset{h}{\to} A \Rightarrow A_n \overset{w}{\to} A$."*

(c) $A_n \overset{M}{\to} A$ *if and only if* $A_n \overset{K_s}{\to} A$ *and* $A_n \overset{K_w}{\to} A$.

(d) $A_n \overset{W}{\to} A \Rightarrow A_n \overset{K_s}{\to} A$.

Another byproduct of the above definitions and remarks is the next proposition.

Proposition 8.13 *If (X, d) is a metric space and $\{A_n\}_{n \in \mathbb{N}} \subseteq 2^X \setminus \{\emptyset\}$, then:*

(a) *The sets d-$\liminf\limits_{n \to \infty} A_n$, d-$\limsup\limits_{n \to \infty} A_n$ are closed (possibly empty).*

(b) *If $\{A_n\}_{n \in \mathbb{N}}$ is increasing, then d-$\liminf\limits_{n \to \infty} A_n = d$-$\limsup\limits_{n \to \infty} A_n = \overline{\bigcup\limits_{n \in \mathbb{N}} A_n}$.*

(c) *If $\{A_n\}_{n \in \mathbb{N}}$ is decreasing, then d-$\liminf\limits_{n \to \infty} A_n = d$-$\limsup\limits_{n \to \infty} A_n = \bigcap\limits_{n \in \mathbb{N}} \overline{A_n}$.*

Remark 8.14 Note that $\bigcap\limits_{k \ge 1} \overline{\bigcup\limits_{n \ge k} A_n} \subseteq d$-$\liminf A_n$ (see Remark 8.5).

To go beyond the above basic observations, we need to impose additional conditions on the sequence of sets.

Proposition 8.15 *If X is a Banach space, $\{A_n\}_{n\in\mathbb{N}} \subseteq P_{bfc}(X)$, and $A_n \overset{h}{\to} A$, then $A_n \overset{M}{\to} A$.*

Proof First note that $A \in P_{bfc}(X)$ (see Proposition 5.43(d)). Also on account of Proposition 8.12(a), we have that $s\text{-}\liminf_{n\to\infty} A_n = A$. Therefore, if we show that $w\text{-}\limsup_{n\to\infty} A_n \subseteq A$, we will reach the desired conclusion. So, let $x \in w\text{-}\limsup_{n\to\infty} A_n$. We can find a subsequence $\{n_k\}_{k\in\mathbb{N}}$ of $\{n\}$ such that $x_{n_k} \in A_{n_k}, x_{n_k} \overset{w}{\to} x$ in X. Since A is convex, $d(\cdot, A)$ is convex and so $d(x, A) \leq \liminf d(x_{n_k}, A)$. On the other hand, $d(x_{n_k}, A) \leq h(A_{n_k}, A) \Rightarrow \limsup d(x_{n_k}, A) = 0$. Therefore, $d(x, A) = 0$ and so $x \in A$, and we have proved the desired inclusion. It follows that $A_n \overset{M}{\to} A$.
□

It is clear that Hausdorff convergence implies Kuratowski convergence, but the reverse implication is not in general true as the following simple example illustrates.

Example 8.16 $X = \mathbb{R}$ and $A_n = [0, \frac{1}{n}] \cup [n, \infty)$. Then $A_n \overset{K}{\to} \{0\}$ but not in h-metric.

To have that K-convergence implies h-convergence, we need an additional compactness condition.

Proposition 8.17 *If (X, d) is a metric space, $\{A_n\}_{n\in\mathbb{N}} \subseteq P_f(X)$, $A_n \overset{K}{\to} A$, and $A_n \subseteq C \in P_k(X)$ for all $n \in \mathbb{N}$, then $A_n \overset{h}{\to} A$ as $n \to \infty$.*

Proof Evidently, $A \in P_k(X)$ (see Proposition 8.13(a)). Then we can find $x_n \in A$ such that $d(x_n, A_n) = \max[d(x, A_n) : x \in A] = h^*(A, A_n)$. We have that $\{x_n\}_{n\in\mathbb{N}} \subseteq C$, and so we can find a subsequence $\{x_{n_k}\}_{k\in\mathbb{N}}$ of $\{x_n\}_{n\in\mathbb{N}}$ such that $x_{n_k} \to x \in A$. Moreover, since by hypothesis $A_n \overset{K}{\to} A$, we can find $u_n \in A_n$ $n \in \mathbb{N}$ such that $u_n \to x$. We have

$$h^*(A, A_{n_k}) = d(x_{n_k}, A_{n_k}) \leq d(x_{n_k}, u_{n_k}) \to 0 \text{ as } k \to \infty.$$

Also let $v_n \in A_n$ such that

$$d(v_n, A) = \max[d(v, A) : v \in A_n] = h^*(A_n, A),$$

and we can find $\{v_{n_k}\}_{k\in\mathbb{N}}$ a subsequence of $\{v_n\}_{n\in\mathbb{N}}$ such that $v_{n_k} \to v \in A$. We have

$$h^*(A_{n_k}, A) = d(v_{n_k}, A) \leq d(v_{n_k}, v) \to 0 \text{ as } k \to \infty.$$

The Urysohn criterion for the convergence of sequences implies that

$$h^*(A, A_n) \to 0 \text{ and } h^*(A_n, A) \to 0 \text{ as } n \to \infty,$$

$$\Rightarrow A_n \xrightarrow{h} A \text{ as } n \to \infty. \qquad \square$$

Remark 8.18 We mention that the Hausdorff metric topology on $P_f(X)$ is the weakest topology on $P_f(X)$, for which for every $C \in P_f(X)$ the functions $A \to h^*(A, C)$ and $A \to h^*(C, A)$ are both continuous.

Now suppose that X is a Banach space and $\{A_n\}_{n \in \mathbb{N}} \subseteq 2^X \setminus \{\emptyset\}$. It is clear from Definition 8.6 that we always have

$$\sigma(x^*, w\text{-}\limsup_{n \to \infty} A_n) \leq \limsup_{n \to \infty} \sigma(x^*, A_n) \text{ for all } x^* \in X^*. \qquad (8.1)$$

It is useful to know when we can guarantee that in (8.1) equation holds. Again a kind of compactness condition is needed.

Proposition 8.19 *If X is a Banach space and $\{A_n\}_{n \in \mathbb{N}} \subseteq 2^X \setminus \{\emptyset\}$ and $A_n \subseteq W \in P_{wk}(X)$ for all $n \in \mathbb{N}$, then $w\text{-}\limsup_{n \to \infty} A_n \neq \emptyset$ and $\limsup_{n \to \infty} \sigma(x^*, A_n) = \sigma(x^*, w\text{-}\limsup_{n \to \infty} A_n)$ for all $x^* \in X^*$.*

Proof From the Eberlein–Smulian theorem (see Theorem 3.121), it follows that $w\text{-}\limsup_{n \to \infty} A_n \neq \emptyset$. Now let $x^* \in X^*$, and consider $u_n \in A_n$ such that

$$\sigma(x^*, A_n) - \frac{1}{n} \leq \langle x^*, u_n \rangle \quad n \in \mathbb{N}. \qquad (8.2)$$

Since $\{u_n\}_{n \in \mathbb{N}} \subseteq W \in P_{wk}(X)$, we may assume that $u_n \xrightarrow{w} u \in w\text{-}\limsup_{n \to \infty} A_n$. Then if in (8.2) we pass to the limit as $n \to \infty$, we obtain

$$\limsup_{n \to \infty} \sigma(x^*, A_n) \leq \langle x^*, u \rangle \leq \sigma(x^*, w\text{-}\limsup_{n \to \infty} A_n).$$

Combining this with (8.1), we conclude that equality holds. $\qquad \square$

Proposition 8.20 *If X is a Banach space, $\{A_n, A\}_{n \in \mathbb{N}} \subseteq 2^X \setminus \{\emptyset\}$ and for all $x^* \in X^*$, $\limsup_{n \to \infty} \sigma(x^*, A_n) \leq \sigma(x^*, A)$, then $w\text{-}\limsup_{n \to \infty} A_n \subseteq \overline{\text{conv}}A$.*

Next we observe some additional relations between the five modes of convergence introduced in the beginning of this section.

Proposition 8.21 *If X is a Banach space, $\{A_n\}_{n\in\mathbb{N}} \subseteq P_{fc}(X)$, $A_n \subseteq W \in P_{wk}(X)$ for all $n \in \mathbb{N}$, and $A_n \xrightarrow{K_w} A$, then $A_n \xrightarrow{w} A$.*

Proof We know that $A \in P_{fc}(X)$. Also given $x^* \in X^*$, we can find $x_n \in A_n$ $n \in \mathbb{N}$ such that $\langle x^*, x_n \rangle = \sigma(x^*, A_n)$. We may assume $x_n \xrightarrow{w} x \in A$, and so from Proposition 8.19, we have $\sigma(x^*, A_n) \to \sigma(x^*, A)$ and so $A_n \xrightarrow{w} A$. $\square$

Remark 8.22 The result fails if we drop the condition that $A_n \subseteq W \in P_{wk}(X)$ for all $n \in \mathbb{N}$. On the other hand, if in Proposition 8.21 the space X is separable, then $A_n \xrightarrow{K_w} A$ if and only if $A_n \xrightarrow{w} A$.

In general, the set $w\text{-}\limsup\limits_{n\to\infty} A_n$ is neither strongly closed nor weakly closed as the next examples show.

Example 8.23

(a) Let X be a Banach space with separable dual. We know that $(\overline{B}, w)$ is sequentially compact $(\overline{B}_1 = \{u \in X : \|u\|_X \leq 1\})$, and we can find $\{u_n\}_{n\in\mathbb{N}} \subseteq \partial B_1$ such that $u_n \xrightarrow{w} 0$. We define $A_n = \{u_k\}_{k=1}^n$. Then $w\text{-}\limsup\limits_{n\to\infty} A_n = \{u_k\}_{k\in\mathbb{N}} \cup \{0\}$, which is not strongly closed.

(b) Let $X = l^1$ and $A_n = \partial B_1$ for all $n \in \mathbb{N}$. We know that $0 \in \overline{\partial B_1}^w$ (see Proposition 3.112). But by the Schur property, no sequence of elements on ∂B_1 can converge weakly to 0. Hence, $w\text{-}\limsup\limits_{n\to\infty} A_n$ is not w-closed.

Proposition 8.24 *If (X, d) is a metric space, $\{A_n\}_{n\in\mathbb{N}} \subseteq 2^X \setminus \{\emptyset\}$, then for every $x \in X$, we have $\limsup\limits_{n\to\infty} d(x, A_n) \leq d(x, \liminf\limits_{n\to\infty} A_n)$.*

Proof We assume that $\liminf\limits_{n\to\infty} A_n \neq \emptyset$ (otherwise, the result is trivial since the right hand side is $+\infty$). Let $u \in \liminf\limits_{n\to\infty} A_n$. We can find $u_n \in A_n$ $n \in \mathbb{N}$ such that $u_n \to u$. We have $d(x, A_n) \leq d(x, u_n)$; hence, $\limsup\limits_{n\to\infty} d(x, A_n) \leq d(x, u)$. Since $u \in \liminf\limits_{n\to\infty} A_n$ is arbitrary, we infer that $\limsup\limits_{n\to\infty} d(x, A_n) \leq d(x, \liminf\limits_{n\to\infty} A_n)$. $\square$

Proposition 8.25 *If (X, d) is a metric space, $\{A_n, A\}_{n\in\mathbb{N}} \subseteq 2^X \setminus \{\emptyset\}$, and for all $x \in X$, we have $\limsup\limits_{n\to\infty} d(x, A_n) \leq d(x, A)$, then $A \subseteq \liminf\limits_{n\to\infty} A_n$.*

Proof For $u \in A$, we have $d(u, A_n) \to 0$ and so $u \in \liminf\limits_{n\to\infty} A_n$ (see Remark 8.5) and so $A \subseteq \liminf\limits_{n\to\infty} A_n$. $\square$

Then Propositions 8.20 and 8.25 lead to the following result.

Proposition 8.26 *If X is a Banach space, $\{A_n, A\}_{n\in\mathbb{N}} \subseteq P_{fc}(X)$, and we have $A_n \xrightarrow{W} A$, $A_n \xrightarrow{w} A$, then $A_n \xrightarrow{M} A$ as $n \to \infty$.*

We mention that if X is a locally compact, separable metric space, then $(P_f(X), \tau_F)$ with τ_F denoting the Fell topology (see Remark 8.5) is a Polish space (see Flachsmeyer [113]). Then we have:

Proposition 8.27 *If (X, d) is a locally compact, separable metric space, $\{A_n, A\}_{n \in \mathbb{N}} \subseteq 2^X \setminus \{\emptyset\}$, and we have:*

(i) $A \subseteq \liminf\limits_{n \to \infty} A_n$.

(ii) For every $K \in P_k(X)$, $\limsup\limits_{n \to \infty}(A_n \cap K) \subseteq A$; then $A_n \overset{K}{\to} A$.

Proof Condition (i) implies that $A_n \to A$ in the hyperspace topology generated by $\{U^- : U \subseteq X \text{ open}\}$. Next suppose that $K \in P_k(X)$ and assume that $A_n \cap K \neq \emptyset$ for an infinite number of indices $n \in \mathbb{N}$. Then $\limsup\limits_{n \to \infty}(A_n \cap K) \neq \emptyset$ and so $A \cap K \neq \emptyset$ (see (ii)). Therefore,

$$A \cap K = \emptyset \Rightarrow A_n \cap K = \emptyset \text{ for all } n \geq n_0,$$

$$\Rightarrow A = \tau_F\text{-}\lim A_n \text{ (see Remark 8.5)},$$

$$\Rightarrow A_n \overset{K}{\to} A \text{ as } n \to \infty. \qquad \square$$

Next we will focus on sequences of sets in a Banach space. We introduce the following class of subsets.

Definition 8.28 Let X be a Banach space. We define

$$\mathcal{A} = \{C \in P_{wf}(X) : \text{for all } r > 0, \ C \cap \overline{B_r} \in P_{wk}(X)\}$$

with $\overline{B_r} = \{x \in X : \|x\|_X \leq \xi\}$. Also we set

$$\mathcal{A}_c = \{C \in \mathcal{A} : C \text{ is convex}\}.$$

Remark 8.29 This family of subsets of X is closed under finite unions and arbitrary intersections. Also $\mathcal{A}$ contains all weakly closed and locally weakly compact subsets of X. If X is reflexive, then by Proposition 3.124, $\mathcal{A} = P_{wf}(X) = P_f(X_w)$.

Proposition 8.30 *If X is a Banach space, $\{A_n\}_{n \in \mathbb{N}} \subseteq 2^X \setminus \{\emptyset\}$, and $A_n \subseteq W \in \mathcal{A}$ for all $n \in \mathbb{N}$, then for every $x \in X$, we have*

$$d(x, w\text{-}\limsup\limits_{n \to \infty} A_n) \leq \liminf\limits_{n \to \infty} d(x, A_n).$$

Proof Let $x \in X$ and $\vartheta = \liminf\limits_{n \to \infty} d(x, A_n)$. If $\vartheta = +\infty$, then we are done. So, we assume that $\vartheta < +\infty$. Arguing indirectly, suppose that the assertion of the proposition is not true. This means that for some $x \in X$, we have

$$\vartheta < d(x, w\text{-}\limsup_{n \to \infty} A_n). \tag{8.3}$$

Let $\{n_k\}_{k \in \mathbb{N}}$ be a subsequence of $\{n\}$ such that

$$\vartheta = \lim_{k \to \infty} d(x, A_{n_k}) < \infty.$$

Consider $u_{n_k} \in A_{n_k}$ such that

$$\|x - u_{n_k}\| \le d(x, A_{n_k}) + \frac{1}{k} \text{ for all } k \in \mathbb{N}. \tag{8.4}$$

For $k \in \mathbb{N}$ big, we will have $u_{n_k} \in W \cap \overline{B}_{(\vartheta + \|x\| + 1)} \in P_{wk}(X)$. So, we may assume that $u_{n_k} \xrightarrow{w} u \in w\text{-}\limsup\limits_{n \to \infty} A_n \ne \emptyset$. We have by (8.3) and (8.4) that

$$\|x - u\|_X \le \liminf_{k \to \infty} \|x - u_{n_k}\| \le \vartheta < d(x, w\text{-}\limsup_{n \to \infty} A_n),$$

a contradiction. $\qquad\qquad\qquad\qquad\qquad\qquad\qquad\qquad\qquad\qquad\qquad\qquad\square$

A byproduct of the above proof is that $w\text{-}\limsup\limits_{n \to \infty} A_n \ne \emptyset$. So, we can state the following result.

Proposition 8.31 *If X is a Banach space, $\{A_n\}_{n \in \mathbb{N}} \subseteq 2^X \setminus \{\emptyset\}$, and for some $x \in X$, $\vartheta = \liminf\limits_{n \to \infty} d(x, A_n)$, then:*

(a) $w\text{-}\limsup\limits_{n \to \infty} A_n \ne \emptyset \Rightarrow \vartheta < +\infty.$
(b) If $A_n \subseteq W \in \mathcal{A}$ for all $n \in \mathbb{N}$ and $\vartheta < +\infty$, then $w\text{-}\limsup\limits_{n \to \infty} A_n \ne \emptyset.$

Next we will prove a result that is the counterpart for $w\text{-}\limsup\limits_{n \to \infty} A_n$ of Proposition 8.25. We need the following Lemma.

Lemma 8.32 *If X is reflexive with dual X^* that is locally uniformly convex, $x \in X \setminus \{0\}$, and $x_n \xrightarrow{w} x$ in X, then there exists $t \in (0, 1]$ such that*

$$\limsup_{n \to \infty} \|x - tx_n\| < \|x\|.$$

Proof Consider the duality map $\mathcal{F} : X \to X^*$ that is single-valued since X^* is locally uniformly convex. We show that $\mathcal{F}(\cdot)$ is continuous. To this end, let $x_n \to x$ in X. Then

$$\|\mathcal{F}(x_n)\|_*^2 = \|x_n\|^2 = \langle \mathcal{F}(x_n), x_n \rangle \quad \text{for all } n \in \mathbb{N}.$$

Since X^* is reflexive, by passing to a suitable subsequence if necessary, we have $\mathcal{F}(x_n) \xrightarrow{w} x^*$ in X^*. Then

$$\|x^*\|_* \leq \liminf_{n\to\infty} \|\mathcal{F}(x_n)\|_*$$

and

$$\lim_{n\to\infty} \|\mathcal{F}(x_n)\|_*^2 = \|x\|^2 = \langle x^*, x \rangle. \tag{8.5}$$

It follows that $\|\mathcal{F}(x_n)\|_* \to \|x^*\|_*$, and by the Kadec–Klec property (see Definition 3.242), we have $\mathcal{F}(x_n) \to x^* = \mathcal{F}(x)$ (see (8.5)) and so $\mathcal{F}(\cdot)$ is continuous.

Let $x \neq 0$, and suppose $x_n \xrightarrow{w} x$ in X. Choose $\varepsilon > 0$ so that $\varepsilon \sup_{n\geq 1} \|x_n\| < \|x\|^2$ and $t \in (0,1]$ such that $\|\mathcal{F}(x - tx_n) - \mathcal{F}(x)\|_* < \varepsilon$ for all $n \in \mathbb{N}$. (We use the continuity of $\mathcal{F}(\cdot)$ and the boundedness of $\{x_n\}_{n\in\mathbb{N}}$.) Consider a subsequence $\{x_{n_k}\}_{k\in N}$ of $\{x_n\}_{n\in N}$ such that

$$\lim_{k\to\infty} \|x - tx_{n_k}\| = \limsup_{n\to\infty} \|x - tx_n\|.$$

Recall that $\mathcal{F}(x) = \partial\varphi(x)$ with $\varphi(x) = \frac{1}{2}\|x\|^2$ (see Example 6.91(c)). So,

$$\frac{1}{2}[\|x - tx_{n_k}\|^2 - \|x\|^2] \leq - < \mathcal{F}(x - tx_{n_k}), tx_{n_k} >$$

$$= - < \mathcal{F}(x - tx_{n_k}) - \mathcal{F}(x), tx_{n_k} > - < \mathcal{F}(x), tx_{n_k} >$$

$$\leq t\varepsilon\|x_{n_k}\| - t < \mathcal{F}(x), x_{n_k} >,$$

$$\Rightarrow \lim_{k\to\infty} \frac{1}{2}[\|x - tx_{n_k}\|^2 - \|x\|^2] \leq t[\varepsilon \sup_{n\geq 1} \|x_n\| - \|x\|^2] < 0,$$

$$\Rightarrow \limsup_{n\to\infty} \|x - tx_n\| < \|x\|. \qquad \square$$

Using this lemma, we can prove the counterpart of Proposition 8.25 mentioned earlier.

Proposition 8.33 *If X is reflexive with dual X^* that is locally uniformly convex, $\{A_n, A\}_{n\in N} \subseteq P_{fc}(X)$, and for every $x \in X$, we have*

$$d(x, A) \leq \liminf_{n\to\infty} d(x, A_n),$$

then $w\text{-}\limsup_{n\to\infty} A_n \subseteq A$.

Proof Let $x \in w\text{-}\limsup_{n\to\infty} A_n$. We can find a subsequence $\{n_k\}$ of $\{n\}$ and $x_{n_k} \in A_{n_k}$ such that $x_{n_k} \xrightarrow{w} x$ in X. Translating things, we may assume that $x = 0$. Suppose $0 \notin A$ and let $u \in p(0, A) = \{u \in A : \|u\| = d(0, A)\}$. Then $u \neq 0$ and $u \in p(-tu, A)$ for every $t \geq 0$. So by hypothesis

$$d(-tu, A) \leq \liminf_{n\to\infty} d(-tu, A_n) \leq \liminf_{k\to\infty} d(-tu, A_{n_k})$$

$$\leq \liminf_{k\to\infty} \|x_{n_k} + tu\|,$$

$$\Rightarrow (1+t)\|u\| \leq \liminf_{k\to\infty} \|x_{n_k} + tu\| \quad \text{for all } t \geq 0,$$

$$\Rightarrow \|u\| \leq \liminf_{k\to\infty} \left\|\frac{1}{1+t}x_{n_k} + \frac{t}{1+t}u\right\| \quad \text{for all } t \geq 0,$$

and this contradicts Lemma 8.32. Hence, $0 \in A$ and so $w\text{-}\limsup_{n\to\infty} A_n \subseteq A$. $\qquad \square$

We know that weakly convergent sequences are bounded (see Proposition 3.101). Therefore, we have

$$w\text{-}\limsup_{n\to\infty} A_n = \cup_{k\geq 1} w\text{-}\limsup_{n\to\infty}(A_n \cap k\overline{B}_1).$$

From this, we infer the following result.

Proposition 8.34 *If X is a Banach space, $\{A_n\}_{n\in N} \subseteq 2^X \setminus \{\emptyset\}$, and either:*

(a) X^* is separable or
(b) X is separable and $A_n \subseteq W \in \mathcal{A}$ for all $n \in \mathbb{N}$,

then $w\text{-}\limsup_{n\to\infty} A_n = \cup_{k\geq 1} \cap_{m\geq 1} \overline{\cup_{n\geq 1} A_n \cap k\overline{B}_1}^w$.

Of course, if we are in a finite dimensional Banach space, then the situation is much simpler.

Proposition 8.35 *If X is a finite dimensional Banach space and*

$$\{A_n, A\}_{n\in N} \subseteq P_{fc}(x)$$

with A compact, then

$$A_n \xrightarrow{h} A \Leftrightarrow A_n \xrightarrow{K} A \Leftrightarrow A_n \xrightarrow{W} A \Leftrightarrow A_n \xrightarrow{w} A.$$

We also mention a classical compactness result that can be found in Kuratowski[167](p.164).

Theorem 8.36 *If (X, d) is a separable metric space and $\{A_n\}_{n\in N} \subseteq 2^X \setminus \{\emptyset\}$, then there exists a K-convergent subsequence whose limit may be the empty set.*

Without the separability hypothesis, the above theorem fails. The Wijsman convergence of sequences in $P_f(X)$ is equivalent to the K-convergence of the closed enlargements of the sets.

Proposition 8.37 *If X is a Banach space and $\{A_n, A\}_{n\in N} \subseteq P_f(X)$, then $A_n \overset{w}{\to} A$ if and only if for all $r > 0$ $\overline{(A_n)}_r \overset{k}{\to} \overline{A_r}$ (recall that for $C \in P_f(X)$, $C_r = \{x \in X : d(x, C) < r\}$).*

Proof $\Rightarrow$: Let $u \in \overline{A_r}$ and $\varepsilon > 0$. Then we can find $x \in A$ such that $\|u-x\| \leq r+\frac{\varepsilon}{2}$. Since by hypothesis we have $A_n \overset{w}{\to} A$, we can find $n_0 \in \mathbb{N}$ such that $d(x, A_n) \leq \frac{\varepsilon}{2}$ for all $n \geq n_0$. Let $x_n \in A$ such that $\|x_n - x\| \leq \frac{\varepsilon}{2}$. Then we have

$$\|u - x_n\| \leq r + \varepsilon \quad \text{for all } n \geq n_0,$$

$$\Rightarrow \overline{(A_n)}_r \cap \overline{B}_\varepsilon(u) \neq \emptyset \quad \text{for all } n \geq n_0, \tag{8.6}$$

$$\Rightarrow \overline{(A_r)} \subseteq \liminf_{n\to\infty} \overline{(A_r)}_r.$$

Next let $u \in \limsup_{n\to\infty} \overline{(A_n)}_r$. Then for each $\varepsilon > 0$, we have

$$\overline{(A_n)}_r \cap \overline{B}_\varepsilon(u) \neq \emptyset \quad \text{for infinitely many } n \in \mathbb{N},$$

$$\Rightarrow d(u, A) \leq \liminf_{n\to\infty} d(u, A_n) \leq r,$$

$$\Rightarrow u \in \overline{A_r}, \tag{8.7}$$

$$\Rightarrow \limsup_{n\to\infty} \overline{(A_n)}_r \subseteq \overline{A_r}.$$

From (8.6) and (8.7), we conclude that

$$\overline{(A_n)}_r \overset{K}{\to} \overline{A_r}. \quad \text{as } n \to \infty.$$

$\Leftarrow$ Let $u \in X$ such that $d(u, A) < r$. Choose ϑ such that $d(u, A) < \vartheta < r$. Then $u \in \overline{A_\vartheta}$, and by hypothesis, $\overline{A_\vartheta} = \liminf_{n\to\infty} \overline{(A_n)}_\vartheta$. Therefore, we can find $n_0 \in \mathbb{N}$ such that $\overline{(A_n)}_\vartheta \cap B_{r-\vartheta}(u) \neq \emptyset$ and $d(u, A_n) < r$ for all $n \geq n_0$. On the other hand, if $d(u, A_n) \leq \vartheta$ for an infinite number of $n \in \mathbb{N}$, then $u \in \limsup_{n\to\infty} \overline{(A_n)}_r \subseteq \overline{A_r}$ and so $d(u, A) \leq r$. Therefore, $d(u, A) > r$ implies $d(u, A_n) > r$ for all $n \geq n_0$, and so we can say that $A_n \overset{w}{\to} A$. $\qquad\square$

Now we turn our attention to the convergence of L^p-selectors of certain measurable multifunctions that leads to convergence theorems for set-valued integrals.

We start with a result that is very helpful in many situations since it provides information about the pointwise behavior of a weakly convergent sequence in the Lebesgue–Bochner space $L^p(\Omega, X)$, $1 \le p < \infty$.

Theorem 8.38 *If* (Ω, Σ, μ) *is a finite measure space,* X *is a Banach space,* $\{f_n, f\}_{n \in \mathbb{N}} \subseteq L^p(\Omega, X)(1 \le p < \infty)$, $f_n \xrightarrow{w} f$ *in* $L^p(\Omega, X)$, *and* $f_n(z) \in W(z) \in P_{wk}(X)$ *for* $\mu - a.a.$ $z \in \Omega$, *all* $n \in \mathbb{N}$, *then* $f(z) \in \overline{\mathrm{conv}}\ w\text{-}\limsup_{n \to \infty}\{f_n(z)\}$ *for* $\mu - a.a.$ $z \in \Omega$.

Proof From Mazur's Theorem (see Theorem 3.105), for every $k \in \mathbb{N}$, we have

$$f(z) \in \overline{\mathrm{conv}} \bigcup_{n \ge k} \{f_n(z)\} \quad \mu - a.e.$$

For every $k \in \mathbb{N}$, $x^* \in X^*$, and $z \in \Omega \setminus N$ with $\mu(N) = 0$, we have

$$\langle x^*, f(z) \rangle \le \sigma(x^*, \cup_{n \ge k}\{f_n(z)\}) = \sup_{n \ge k}\langle x^*, f_n(z)\rangle,$$

$$\Rightarrow \langle x^*, f(z)\rangle \le \liminf_{n \to \infty}\langle x^*, f_n(z)\rangle,$$

$$\Rightarrow \langle x^*, f(z)\rangle \le \sigma(x^*, w\text{-}\limsup_{n \to \infty}\{f_n(z)\}) \text{ (see Proposition 8.19)},$$

$$\Rightarrow f(z) \in \overline{\mathrm{conv}}\ w\text{-}\limsup_{n \to \infty}\{f_n(z)\} \text{ for } \mu - a.a.\ z \in \Omega. \qquad \square$$

We will use this theorem to prove Fatou-type Lemmata for set-valued integrals and for sets of L^p-selectors. This analysis is fruitful in the context of separable Banach spaces.

So, let X be a separable Banach space, and consider its dual X^* equipped with the Mackey topology $m = m(X^*, X)$. We know that X_m^* is separable, and so we can find $\{x_n^*\}_{n \in \mathbb{N}} \subseteq X^*$ such that $X^* = \overline{\{x_n^*\}_{n \in \mathbb{N}}}^m$. For every $x, u \in X$, we define

$$d(x, u) = \sum_{n \in \mathbb{N}} \frac{1}{2^n} \frac{|\langle x_n^*, x - u \rangle|}{1 + |\langle x_n^*, x - u \rangle|}.$$

This is a metric on X, and the d-metric topology on X is weaker the weak topology. The two coincide on relatively weakly compact subsets of X (see Theorem 3.117). Moreover, for the Borel algebras, we have that $B(X) = B(X_w) = B(X_d)$.

Lemma 8.39 *If* (Ω, Σ) *is a measurable space,* X *is separable. Banach space and* $F_n : \Omega \to 2^X \setminus \{\emptyset\}$ $n \in \mathbb{N}$ *are measurable multifunctions such that* $F_n(z) \subseteq W(z) \in \mathcal{A}$ *for all* $z \in \Omega$; *then* $z \to w - \limsup_{n \to \infty} F_n(z)$ *is measurable.*

Proof From Proposition 8.34, we know that

$$w\text{-}\limsup_{n\to\infty} F_n(z) = \cup_{k\geq 1} \cap_{m\geq 1} \overline{\cup_{n\geq m} F_n(z) \cap k\overline{B}_1}^w$$

$$= \cup_{k\geq 1} \cap_{m\geq 1} \overline{\cup_{n\geq m} F_n(z) \cap k\overline{B}_1}^d$$

$\Rightarrow z \to w\text{-}\limsup\limits_{n\to\infty} F_n(z)$ is measurable (see Propositions 5.60 and 5.61).

$\square$

Remark 8.40 If X is also reflexive, then we can have $W(z) = X$ for all $z \in \Omega$, and so $z \to w\text{-}\limsup\limits_{n\to\infty} F_n(z)$ is always measurable.

Now we can prove the first Fatou-type lemma for set-valued integrals.

Proposition 8.41 *If (Ω, Σ, μ) is nonatomic, σ-finite measure space, X is a separable Banach space, and $F_n : \Omega \to 2^X \setminus \{\emptyset\}, n \in N$, are graph measurable multifunctions such that $F_n(z) \subseteq W(z) \mu - a.e$ with $W : \Omega \to P_{wkc}(X)$ integrably bounded, then $w\text{-}\limsup\limits_{n\to\infty} \int_\Omega F_n d\mu \subseteq \int_\Omega w\text{-}\limsup\limits_{n\to\infty} d\mu$.*

Proof Let $x \in w\text{-}\limsup\limits_{n\to\infty} \int_\Omega F_n d\mu$. We can find a subsequence $\{n_k\}_{k\in\mathbb{N}}$ of $\{n\}$ and $x_{n_k} \in \int_\Omega F_{n_k} d\mu$ such that $x_{n_k} \xrightarrow{w} x$ in X. We have $x_{n_k} = \int_\Omega f_{n_k} d\mu$ with $f_{n_k} \in S^1_{F_{n_k}} \subseteq S^1_W$, and the latter is w-compact in $L^1(\Omega, X)$ (see Theorem 5.133). So, we may assume that $f_{n_k} \xrightarrow{w} f$ in $L^1(\Omega, X)$. Then Theorem 8.38 implies that $f(z) \in \overline{conv}\ w\text{-}\limsup\limits_{n\to\infty} F_n(z) = \overline{conv}\ w\text{-}\limsup\limits_{n\to\infty} \overline{F_n(z)}$, with $z \to \overline{F_n(z)}, \Sigma_\mu$-measurable for all $n \in \mathbb{N}$. Hence, Lemma 8.39 implies that $z \to w\text{-}\limsup\limits_{n\to\infty} F_n(z)$ is Σ_μ-measurable. Using Corollary 5.143 and Theorem 5.133, we conclude that

$$x = \int_\Omega f d\mu \in cl \int_\Omega w\text{-}\limsup_{n\to\infty} F_n d\mu,$$

$$\Rightarrow w\text{-}\limsup_{n\to\infty} \int_\Omega F_n d\mu \subseteq \overline{\int_\Omega w\text{-}\limsup_{n\to\infty} F_n d\mu}.$$

$\square$

We can have an analogous Fatou-type lemma for the s-liminf.

Proposition 8.42 *If (Ω, Σ, μ) is a σ-finite measure space, X is a separable Banach space, and $F_n : \Omega \to 2^X \setminus \{\emptyset\}, n \in \mathbb{N}$, are graph measurable multifunctions such that $\sup\limits_{n\in\mathbb{N}} d(0, F_n(\cdot)) \in L^1(\Omega)$, then*

$$\int_\Omega s\text{-}\liminf_{n\to\infty} F_n d\mu \subseteq s\text{-}\liminf_{n\to\infty} \int_\Omega F_n d\mu.$$

Proof Let $x \in \int_\Omega s\text{-}\liminf\limits_{n\to\infty} F_n d\mu$. Then $x = \int_\Omega f d\mu$ with $f \in S^1_{s\text{-}\liminf F_n}$. Consider the multifunction $G_n(\cdot)$ defined by

$$G_n(z) = \{x \in F_n(z) : \|f(z) - x\| \le d(f(z), F_n(z)) + \frac{1}{n}\}.$$

We see that $\mathrm{Gr}\, G_n \in \Sigma_\mu \otimes B(X)$. So, by Theorem 5.105 (the Yankov–von Neumann–Aumann Selection Theorem), we can find $f_n : \Omega \to X$ $n \in \mathbb{N}$, Σ-measurable function such that $f_n(z) \in G_n(z)$ $\mu - a.e.$ Then

$$\|f(z) - f_n(z)\| \to 0 \quad \mu - a.e \text{ as } n \to \infty.$$

Thus, $x_n = \int_\Omega f_n d\mu \to x = \int_\Omega f d\mu$, by the Lebesgue dominated convergence theorem. Therefore,

$$\int_\Omega s\text{-}\liminf_{n\to\infty} F_n d\mu \subseteq s\text{-}\liminf_{n\to\infty} \int_\Omega F_n d\mu. \qquad \square$$

We can have the corresponding results for the set of integrable selectors.

Proposition 8.43 *If (Ω, Σ, μ) is a σ-finite measure space, X is a separable Banach space, and $F_n : \Omega \to 2^X \setminus \{\emptyset\}$ $n \in \mathbb{N}$ are graph measurable multifunctions and $F_n(z) \subseteq W(z) \in P_{wk}(X)$ $\mu - a.e$ and $|W(z)| \le \varphi(z)$ $\mu - a.e$ with $\varphi \in L^1(\Omega)$, then $w\text{-}\limsup_{n\to\infty} S^1_{F_n} \subseteq \overline{\mathrm{conv}} S^1_{w\text{-}\limsup F_n}$.*

Proof On account of Proposition 5.170, we may assume that $z \to W(z)$ is a measurable multifunction. From Corollary 5.128, we know that for all $u^* \in L^\infty(\Omega, X^*_{w^*})$, we have

$$\sigma(u^*, S^1_{F_n}) = \int_\Omega \sigma(u^*(z), F_n(z)) d\mu \text{ for all } n \in \mathbb{N}$$

$$\Rightarrow \limsup_{n\to\infty} \sigma(u^*, S^1_{F_n}) \le \int_\Omega \limsup_{n\to\infty} \sigma(u^*, F_n) d\mu \text{ (by Fatou's lemma)}$$
$$= \int_\Omega \sigma(u^*, w\text{-}\limsup F_n) d\mu \text{ (see Proposition 8.19)}$$
$$= \sigma(u^*, S^1_{w\text{-}\limsup F_n}),$$

$$\Rightarrow w\text{-}\limsup_{n\to\infty} S^1_{F_n} \subseteq \overline{\mathrm{conv}} S^1_{w\text{-}\limsup F_n} \text{ (see Proposition 8.20).} \qquad \square$$

Proposition 8.44 *If (Ω, Σ, μ) is a σ-finite measure space, X is a separable Banach space, $F_n : \Omega \to 2^X \setminus \{\emptyset\}$ $n \in \mathbb{N}$ are graph measurable, and $\sup_{n\in\mathbb{N}} d(0, F_n(\cdot)) \in L^1(\Omega)$, then $S^1_{s\text{-}\liminf F_n} \subseteq s\text{-}\liminf_{n\to\infty} S^1_{F_n}$.*

Proof For every $u \in L^1(\Omega, X)$ and every $n \in \mathbb{N}$, we have

$$d(u, S^1_{F_n}) = \int_\Omega d(u, F_n) d\mu \text{ (see Theorem 5.125),}$$

$$\Rightarrow \limsup_{n\to\infty} d(u, S^1_{F_n}) \le \int_{\Omega} \limsup_{n\to\infty} d(u, F_n) d\mu \text{ (by Fatou's lemma)}$$

$$\le \int_{\Omega} d(u, s\text{-}\liminf F_n) d\mu \text{ (see Proposition 8.24)}$$

$$= d(u, S^1_{s\text{-}\liminf F_n}) \text{ (see Theorem 5.126)},$$

$$\Rightarrow S^1_{s\text{-}\liminf F_n} \subseteq s\text{-}\liminf_{n\to\infty} S^1_{F_n} \text{ (see Proposition 8.25)}. \qquad \Box$$

Remark 8.45 If μ is finite, then it suffices to assume that $\{d(0, F_n(\cdot))\}_{n\in\mathbb{N}} \subseteq L^1(\Omega)$ is uniformly integrable.

Combining Propositions 8.43 and 8.44, we have the following convergence result for the sets of integrable selectors and the set-valued integrals.

Theorem 8.46 *If (Ω, Σ, μ) is a σ-finite measure space, X is a separable Banach space, $F_n : \Omega \to P_{wkc}(X)$ $n \in \mathbb{N}$ are graph measurable multifunctions with $F_n(z) \subseteq W(z)$ μ-a.e with $W : \Omega \to P_{wkc}(X)$ on integrably bounded multifunction, and $F_n(z) \xrightarrow{M} F(z)$ $\mu - a.e$, then*

$$S^1_{F_n} \xrightarrow{M} S^1_F \text{ and } \int_{\Omega} F_n d\mu \xrightarrow{M} \int_{\Omega} F d\mu.$$

Corollary 8.47 *If (Ω, Σ, μ) is a σ-finite measure space, X is a finite dimensional Banach space, and $F_n : \Omega \to 2^X \setminus \{\emptyset\}$ $n \in \mathbb{N}$ are graph measurable multifunctions such that*

$$\sup_{n\in\mathbb{N}} |F_n(\cdot)| \in L^1(\Omega) \text{ and } F_n(z) \xrightarrow{K} F(z) \mu - a.e,$$

then $\int_{\Omega} F_n d\mu \xrightarrow{K} \int_{\Omega} F d\mu$.

A sequence $\{f_n\}_{\{n\in\mathbb{N}\}} \subseteq L^1(\Omega)$ that converges weakly but not strongly oscillates around its weak limit, and this prevents the strong convergence. So, in order to guarantee strong convergence, we need a condition that prohibits this oscillation pattern.

Proposition 8.48 *If (Ω, Σ, μ) is a finite measure space and $\{f_n\}_{n\in\mathbb{N}}$ such that*

$$f_n \xrightarrow{w} f \in L^1(\Omega) \text{ as } n \to \infty,$$

$$f(z) \le \liminf f_n(z) \mu - a.e, \tag{8.8}$$

then $f_n \to f$ in $L^1(\Omega)$.

Proof By replacing f_n by $f_n - f$, we may assume that $f = 0$. By the Dunford–Pettis Theorem (see Theorem 4.27), we know that $\{f_n\}_{n\in\mathbb{N}} \subseteq L^1(\Omega)$ is uniformly integrable. So, given $\epsilon > 0$, we can find $\delta > 0$ such that

$$A \in \Sigma, \mu(A) \leq \delta \Rightarrow \sup_{n\in\mathbb{N}} \int_A |f_n| d\mu \leq \epsilon. \tag{8.9}$$

For every $m \in \mathbb{N}$, let $A_m = \{z \in \Omega, \inf_{n\geq m} f_n(z) \geq -\epsilon\}$. If we choose m big, we can have $\mu(\Omega \setminus A_m) \leq \delta$ (see (8.8) and recall that we have assumed that $f = 0$). Let $k \in \mathbb{N}, k \geq m$ such that

$$|\int_{A_m} f_n d\mu| \leq \epsilon \text{ for all } n \geq k \text{ (since } f_n \overset{w}{\to} f = 0 \text{ in } L^1(\Omega)).$$

Therefore for $n \geq k$, we have

$$\int_\Omega |f_n| d\mu = \int_{A_m} |f_n| d\mu + \int_{\Omega \setminus A_m} |f_n| d\mu$$
$$\leq \int_{A_m} |f_n + \epsilon| d\mu + \int_{A_m} \epsilon d\mu + \int_{\Omega \setminus A_m} |f_n| d\mu$$
$$\leq |\int_{A_m} f_n d\mu| + 2\epsilon \mu(\Omega) + \epsilon \text{ (see (8.9))}$$
$$= 2\epsilon[1 + \mu(\Omega)],$$

$$\Rightarrow f_n \to f = 0 \text{ in } L^1(\Omega). \qquad \square$$

Remark 8.49 Evidently, we can replace (8.8) by

$$\limsup_{n\to\infty} f_n(z) \leq f(z) \ \mu - a.e.$$

For $\mathbb{R}^N$-valued functions, Proposition 8.48 takes the following form, see A. Visintin [277].

Proposition 8.50 *If (Ω, Σ, μ) is a finite measure space, $\{f_n\}_{n\in\mathbb{N}} \subseteq L^1(\Omega, \mathbb{R}^N)$ satisfy*

$$f_n \overset{w}{\to} f \text{ in } L^1(\Omega, \mathbb{R}^N)$$

$$f(z) \in ext \, [\overline{\text{conv}} \limsup_{n\to\infty} \{f_n(z)\}] \ \mu - a.e.,$$

then $f_n \to f$ in $L^1(\Omega, \mathbb{R}^N)$.

In the next section, we will study the variational convergence of functions.

8.2 Variational Convergence of Functions

In this section, we will develop a mode of functional convergence that is suitable for the sensitivity analysis of optimization problems and in the study of homogenization problems.

The setting is a metric space (X, d). In fact, the theory can be developed more generally in the context of a first countable topological space, but for the sake of simplicity, we choose to work with a metric space.

Definition 8.51 Let (X, d) be a metric space and $f_n : X \to \overline{\mathbb{R}} = \mathbb{R} \cup \{+\infty\}, n \in \mathbb{N}, f : X \to \overline{\mathbb{R}} = \mathbb{R} \cup \{+\infty\}$. We say that $\{f_n\}_{n\in\mathbb{N}}$"$\Gamma - converges$"(or"$epi - converges$") to f at $x \in X$, if the following two statements are satisfied:

(a) For every sequence $x_n \to x$ in X, we have

$$f(x) \leq \liminf_{n\to\infty} f(x_n).$$

(b) There exists a sequence $u_n \to x$ in X, and we have

$$\limsup_{n\to\infty} f(u_n) \leq f(x).$$

If statements (a) and (b) hold for all $x \in X$, then we say that $\{f_n\}_{n\in\mathbb{N}}$ "$\Gamma - converges$" (or "$epi - converges$") to f, and we write $f_n \overset{\Gamma}{\to} f$ (or $f_n \overset{e}{\to} f$) and $\Gamma\text{-}\lim_{n\to\infty} f_n = f$ (or $e\text{-}\lim_{n\to\infty} f_n = f$).

Remark 8.52 Evidently, statements (a) and (b) in the above definition are equivalent to (a) and

(b)' there exists a sequence $u_n \to x$ in X such that

$$f(x) = \lim_{n\to\infty} f(u_n).$$

In general, this is not a topological notion. For example, if $f_n = f : \Omega \to \overline{\mathbb{R}} = \mathbb{R} \cup \{+\infty\}$, then $f_n \overset{\Gamma}{\to} \operatorname{cl} f =$ the lower semicontinuous envelope of f (i.e., $\operatorname{cl} f$ is the biggest lower semicontinuous function minorizing f). It is easy to check that $f_n \overset{\Gamma}{\to} f$ if and only if $\operatorname{epl} f_n \overset{K}{\to} \operatorname{epi} f$ (see Definition 6.51). For this reason, this mode of convergence is also called "epi-convergence."

We define the following $\overline{\mathbb{R}} = \mathbb{R} \cup \{+\infty\}$-valued functions

$$(\Gamma\text{-}\liminf_{n\to\infty} f_n)(x) = \min[\liminf_{n\to\infty} f_n(x_n) : x_n \to x \text{ in } X],$$

$$(\Gamma\text{-}\limsup_{n\to\infty} f_n)(x) = \min[\limsup_{n\to\infty} f_n(x_n) : x_n \to x \text{ in } X].$$

It is easy to see from these definitions that the following proposition is true.

Proposition 8.53 *If (X, d) is a metric space and $f_n, f : X \to \overline{\mathbb{R}} = \mathbb{R} \cup \{+\infty\}$ $n \in \mathbb{N}$, then:*

(a) $x \to \Gamma\text{-}\liminf_{n\to\infty} f_n(x)$ *and* $x \to \Gamma\text{-}\limsup_{n\to\infty} f_n(x)$ *are lower semicontinuous.*

(b) $f_n \xrightarrow{\Gamma} f$ *if and only if* $\Gamma\text{-}\limsup_{n\to\infty} f_n \leq f \leq \Gamma\text{-}\liminf_{n\to\infty} f_n$.

Another way to describe these two limits is the following. Recall that if $x \in X$, by $\mathcal{N}(x)$, we denote the filter of neighborhoods of x

$$(\Gamma\text{-}\liminf_{n\to\infty} f_n)(x) = \sup_{U\in\mathcal{N}(x)} \liminf_{n\to\infty} \inf_{u\in U} f_n(u), \tag{8.10}$$

$$(\Gamma\text{-}\limsup_{n\to\infty} f_n)(x) = \sup_{U\in\mathcal{N}(x)} \limsup_{n\to\infty} \inf_{u\in U} f_n(u). \tag{8.11}$$

These expressions indicate the close connection of this mode of convergence with variational problems.

In general, Γ-convergence and pointwise convergence are distinct notions. We have already seen this earlier with the constant sequence $\{f_n = f\}$ with f not lower semicontinuous. Another example is the following.

Example 8.54

(a) Let $X = \mathbb{R}$, and consider the sequence $\{f_n(x) = nxe^{-n^2x^2}\}_{n\in\mathbb{N}}$. We have

$$(\Gamma\text{-}\liminf_{n\to\infty} f_n)(x) = \begin{cases} -\frac{1}{\sqrt{2e}}, & if \ x \neq 0 \\ 0, & if \ x = 0 \end{cases} \quad \text{and} \ (\lim f_n)(x) = 0, \ x \in \mathbb{R}.$$

So, both limits exist but are not equal.

(b) Let $X = \mathbb{R}$ consider the sequence $\{f_n(x) = sin(nx)\}_{n\in\mathbb{N}}$. Then

$$\Gamma\text{-}\lim f_n = -1,$$

but the pointwise limit does not exist.

Recall that if $C \subseteq X$, then the "indicator function" of C is defined by

$$i_C(x) = \begin{cases} 0, & if \ x \in C \\ +\infty, & if \ x \neq C. \end{cases}$$

Proposition 8.55 *If* $\{C_n\}_{n\in\mathbb{N}} \subseteq 2^X \setminus \{\emptyset\}$ *and we define*

$$C_* = K\text{-}\liminf_{n\to\infty} C_n, \quad C^* = K\text{-}\limsup_{n\to\infty} C_n,$$

then $i_{C_*} = \Gamma\text{-}\limsup_{n\to\infty} i_{C_n}, \ i_{C^*} = \Gamma\text{-}\liminf_{n\to\infty} i_{C_n}.$

Proof We prove the first equality, and the second can be proved in a similar way. So, let $f = \Gamma\text{-}\limsup_{n\to\infty} i_{C_n}$. Evidently, the range of f is in $\{0, 1\}$. Hence, we need to show that $f(x) = 0$ if and only if $x \in C_*$. We know that $x \in C_*$ if and only if for every $U \in \mathcal{N}(x)$ we have $C_n \cap U \neq \emptyset$ for all $n \geq n_0$. This is equivalent to $\inf_{u \in U} i_{C_n}(u) = 0$ for all $n \geq n_0$. Therefore, $x \in C_*$ if and only if $\limsup_{n\to\infty} \inf_{u \in U} i_{C_n}(u) = 0$ for all $U \in \mathcal{N}(x)$. We conclude that $f = i_{C_*}$ (see (8.2)), similarly for the other equality using (8.10). $\square$

Proposition 8.56 *If* $f_n : X \to \overline{\mathbb{R}} = \mathbb{R} \cup \{+\infty\} \ n \in \mathbb{N}$ *and we define*

$$f_* = \Gamma\text{-}\liminf_{n\to\infty} f_n \ and \ f^* = \Gamma\text{-}\limsup_{n\to\infty} f_n,$$

then $\operatorname{epi} f_* = K\text{-}\limsup \operatorname{epi} f_n$ *and* $\operatorname{epi} f^* = K\text{-}\liminf_{n\to\infty} \operatorname{epi} f_n$.

Proof Again we prove the first equality, the second following in a similar way. We know that $(x, \lambda) \in \operatorname{epi} f_*$ if and only if $f_*(x) \leq \lambda$. From (8.10), we see that $f_*(x) \leq \lambda$ if and only if for every $\epsilon > 0$ and $U \in \mathcal{N}(x)$, we have

$$\liminf_{n\to\infty} \inf_{u \in U} f_n(u) < \lambda + \epsilon. \tag{8.12}$$

From (8.12), we see that for every $\epsilon > 0$, every $U \in \mathcal{N}(x)$, and every $k \in \mathbb{N}$, we can find $n \geq k$ such that $\inf_{u \in U} f_n(u) < \lambda + \epsilon$. This is equivalent to

$$U \times (\lambda - \epsilon, \lambda + \epsilon) \cap \operatorname{epi} f_n \neq \emptyset.$$

The sets $U \times (\lambda - \epsilon, \lambda + \epsilon)$ with $U \in \mathcal{N}(x)$ and $\epsilon > 0$ form a local basis for (x, λ) in $X \times \mathbb{R}$ with the product topology. So, we conclude that

$$(x, \lambda) \in \operatorname{epi} f_* \text{ if and only if } (x, \lambda) \in K\text{-}\limsup_{n\to\infty} \operatorname{epi} f_n,$$

similarly for the second equality. $\square$

We have already seen that the Γ-convergence and the pointwise convergence are in general distinct notions. However, directly from the definition of Γ-convergence (see (8.10), (8.11)), we can have the following result.

Proposition 8.57 *If* $f_n : X \to \overline{\mathbb{R}} = \mathbb{R} \cup \{+\infty\}, n \in \mathbb{N}$, *then*

$$\Gamma\text{-}\liminf_{n\to\infty} f_n \leq \liminf_{n\to\infty} f_n \ and \ \Gamma\text{-}\limsup_{n\to\infty} f_n \leq \limsup_{n\to\infty} f_n;$$

therefore, if $f_\Gamma = \Gamma\text{-}\lim_{n\to\infty} f_n$ and $f = \lim f_n$, then $f_\Gamma \leq f$.

If we replace pointwise convergence by uniform convergence, then we can improve the above proposition.

Proposition 8.58 *If $f_n : X \to \overline{\mathbb{R}} = \mathbb{R} \cup \{+\infty\}$ $n \in \mathbb{N}$ and $f_n \overset{u}{\to} f$, then $\Gamma\text{-}\lim_{n\to\infty} f_n = cl\, f$ (= the lower semicontinuous envelope of f).*

Proof Let $U \subseteq X$ nonempty open. Then

$$\lim_{n\to\infty} \inf_{u\in U} f_n(u) = \inf_{u\in U} f(u),$$

$$\Rightarrow \sup_{U\in\mathcal{N}(x)} \lim_{n\to\infty} \inf_{u\in U} f_n(u) = \sup_{U\in\mathcal{N}(x)} \inf_{u\in U} f(u) = (cl\, f)(x) \text{ for all } x \in X$$

$$\Rightarrow \Gamma\text{-}\lim_{n\to\infty} f_n = cl\, f.$$

$\square$

Since the uniform limit of lower semicontinuous functions is lower semicontinuous, we have the following corollary.

Corollary 8.59 *If $\{f_n\}_{n\in\mathbb{N}} \subseteq \Gamma_0(X)$ and $f_n \overset{u}{\to} f$, then $\Gamma\text{-}\lim_{n\to\infty} f_n = f$.*

For monotone sequences of functions, the situation is better, and this explains the importance of monotone methods in optimization.

Proposition 8.60 *If $f_n : X \to \overline{\mathbb{R}} = \mathbb{R} \cup \{+\infty\}$, $n \in \mathbb{N}$, is increasing, then*

$$\Gamma\text{-}\lim_{n\to\infty} f_n = \Gamma\text{-}\lim_{n\to\infty} cl\, f_n = \sup_{n\in\mathbb{N}} cl\, f_n.$$

Proof Let $U \subseteq X$ be a nonempty open set. We have

$$\lim_{n\to\infty} \inf_{u\in U} f_n(u) = \sup_{n\in\mathbb{N}} \inf_{u\in U} f_n(u),$$

$$\Rightarrow \sup_{U\in\mathcal{N}(x)} \lim_{n\to\infty} \inf_{u\in U} f_n(u)$$

$$= \sup_{U\in\mathcal{N}(x)} \sup_{n\in\mathbb{N}} \inf_{u\in U} f_n(u)$$

$$= \sup_{n\in\mathbb{N}} \sup_{U\in\mathcal{N}(x)} \inf_{u\in U} f_n(u)$$

$$= \sup_{n\in\mathbb{N}} cl\, f_n.$$

$\square$

We know that if $\{f_n\}_{n\in\mathbb{N}} \subseteq \Gamma_0(X)$ is increasing, then $\sup_{n\in\mathbb{N}} f_n$ is lower semicontinuous too. So, as a consequence of Proposition 8.60, we have the following corollary, which gives a situation where Γ and pointwise convergence coincide.

Corollary 8.61 *If $\{f_n\}_{n\in\mathbb{N}} \subseteq \Gamma_0(X)$ is increasing and $f = \sup_{n\in\mathbb{N}} f_n$, then $f = \Gamma\text{-}\lim_{n\to\infty} f_n$.*

We have a corresponding result for decreasing sequences.

Proposition 8.62 *If* $f_n : X \to \overline{\mathbb{R}} = \mathbb{R} \cup \{+\infty\}$, $n \in \mathbb{N}$, *is a decreasing sequence and* $f_n(x) \to f(x)$ *for all* $x \in X$, *then* $\Gamma\text{-}\lim_{n\to\infty} f_n = clf$.

Proof Let $U \subseteq X$ be nonempty open. We have

$$\lim_{n\to\infty} \inf_{u\in U} f_n(u) = \inf_{n\in\mathbb{N}} \inf_{u\in U} f_n(u) = \inf_{u\in U} f(u),$$

$$\Rightarrow \Gamma\text{-}\lim_{n\to\infty} f_n = clf. \qquad \square$$

The next notion will allow us to have equivalence of Γ and pointwise convergence beyond monotone sequences. Using this notion, we will extend Proposition 8.58.

Definition 8.63 We say that a sequence $f_n : X \to \overline{\mathbb{R}} = \mathbb{R} \cup \{+\infty\}$, $n \in \mathbb{N}$, is "equi-lower semicontinuous at $x \in X$," if given any $\varepsilon > 0$, we can find $U \in \mathcal{N}(x)$ such that

$$f_n(x) - \varepsilon \le f_n(u) \text{ for all } u \in U, \text{ all } n \in \mathbb{N}.$$

We say that a sequence $\{f_n\}_{n\in\mathbb{N}}$ is "equi-lower semicontinuous," if it is equi-lower semicontinuous at every $x \in X$.

Proposition 8.64 *If* $f_n : X \to \mathbb{R}$, $n \in \mathbb{N}$, *is an equi-lower semicontinuous sequence, then*

$$\Gamma\text{-}\liminf_{n\to\infty} f_n = \liminf_{n\to\infty} f_n \text{ and } \Gamma\text{-}\limsup_{n\to\infty} f_n = \limsup_{n\to\infty} f_n.$$

Proof As before, we prove the first equality, the proof of the second being similar. On account of Proposition 8.57, it suffices to show that

$$\liminf_{n\to\infty} f(x) \le (\Gamma\text{-}\liminf_{n\to\infty} f_n)(x) \text{ for all } x \in X. \tag{8.13}$$

The equi-lower semicontinuity of $\{f_n\}_{n\in\mathbb{N}}$ implies that given $\varepsilon > 0$, we can find $U \in \mathcal{N}(x)$ $(x \in X)$ such that

$$f_n(x) - \varepsilon \le \inf_{u\in U} f_n(u) \text{ for all } n \in \mathbb{N},$$

$$\Rightarrow \liminf_{n\to\infty} f_n(x) - \varepsilon \le \sup_{U\in\mathcal{N}(x)} \liminf_{n\to\infty} \inf_{u\in U} f_n(u),$$

$$\Rightarrow \liminf_{n\to\infty} f_n(x) \le (\Gamma\text{-}\liminf_{n\to\infty} f_n)(x) \text{ (letting } \varepsilon \downarrow 0).$$

So (8.13) holds, and by Proposition 8.57, we conclude the equality must hold. Similarly, we show the other equality. $\qquad \square$

Corollary 8.65 *If* $f_n : X \to \overline{\mathbb{R}} = \mathbb{R} \cup \{+\infty\}$, $n \in \mathbb{N}$, *is equi-lower semicontinuous,* *then* $f_n \overset{\Gamma}{\to} f \Leftrightarrow f_n(x) \to f(x)$ *for all* $x \in X$.

If we are in a Banach space and the sequence $\{f_n\}_{n \in \mathbb{N}}$ consists of convex functions, then we can use Theorem 6.56 to have the following result.

Proposition 8.66 *If* X *is a Banach space,* $f_n : X \to \overline{\mathbb{R}} = \mathbb{R} \cup \{+\infty\}$, $n \in \mathbb{N}$ *is a sequence of convex functions that is equibounded above on* $U \in \mathcal{N}(x)$ *(i.e., there exists* $M > 0$ *such that* $\sup\limits_{n \in \mathbb{N}} \sup\limits_{u \in U} f_n(u) \leq M$*), then* $f_n \overset{\Gamma}{\to} f \Leftrightarrow f_n(x) \to f(x)$ *for all* $x \in X$.

Remark 8.67 If X is finite dimensional and $\{f_n, f\}_{n \in \mathbb{N}}$ are $\mathbb{R}$-valued, convex functions, then $f_n \overset{\Gamma}{\to} f \Leftrightarrow f_n(x) \to f(x)$ for all $x \in X$ (see Corollary 6.58).

The main interest for the notion of Γ-convergence comes from its variational nature. In the sequel, we develop these variational properties of Γ-convergence.

We start with a definition that provides a topological counterpart of the notion of coercivity.

Definition 8.68 Let (X, τ) be a Hausdorff topological space (τ denotes the topology of X).

(a) A function $f : X \to \overline{\mathbb{R}} = \mathbb{R} \cup \{+\infty\}$ is said to be "coercive" (resp., "sequentially coercive"), if for every $\lambda \in \mathbb{R}$, the set

$$L_\lambda = \overline{\{x \in X : f(x) \leq \lambda\}}^\tau$$

is τ-countably compact (resp., τ-sequentially compact).

(b) A sequence of functions $f_n : X \to \overline{\mathbb{R}} = \mathbb{R} \cup \{+\infty\}$, $n \in \mathbb{N}$ is said to be "equicoercive," if for every $\lambda \in \mathbb{R}$, there exists a τ-closed, τ-countably compact set $K_\lambda \subseteq X$ such that

$$\{x \in X : f_n(x) \leq \lambda\} \subseteq K_\lambda \text{ for all } n \in \mathbb{N}.$$

Remark 8.69 Note that if X is a reflexive Banach space with the weak topology, then we recover the usual notion of coercivity, namely

"$f(\cdot)$ is coercive if and only if $f(x) \to +\infty$ as $\|x\| \to \infty$."

If $\frac{f(x)}{\|x\|} \to +\infty$ as $\|x\| \to \infty$ (superlinear growth for $f(\cdot)$), then we say that $f(\cdot)$ is "strongly coercive" (see also Definition 7.26(b)).

Next we provide a necessary and sufficient condition in order to have equicoercivity.

Proposition 8.70 *If (X, τ) is a Hausdorff topological space, $f_n : X \to \overline{\mathbb{R}} = \mathbb{R} \cup \{+\infty\}$, and $n \in \mathbb{N}$ is a sequence of functions such that*

$$-\infty < \inf_{n \in \mathbb{N}} f_n(x) \text{ for all } x \in X,$$

then the sequence $\{f_n\}_{n \in \mathbb{N}}$ is equicoercive if and only if there exists a τ-lower semicontinuous coercive function φ such that $\varphi \leq f_n$ for all $n \in \mathbb{N}$.

Proof $\Rightarrow$: Let $\varphi_0(x) = \inf_{n \in \mathbb{N}} f_n(x)$ and $\varphi = \overline{\varphi_0}^{\tau}$ (the τ-lower semicontinuous regularization of φ_0). Since by hypothesis the family $\{f_n\}_{n \in \mathbb{N}}$ is equicoercive, we can find $\{K_\lambda\}_{\lambda \in \mathbb{R}}$, τ-closed and τ-countably compact sets in X such that $\{f_n \leq \lambda\} \subseteq K_\lambda$ for all $\lambda \in \mathbb{R}$, all $n \in \mathbb{N}$. Let $x \in \{\varphi \leq \lambda\}$ and $\varepsilon > 0$. Then we can find $n_\varepsilon \in \mathbb{N}$ such that $f_n(x) \leq \lambda + \varepsilon$ for all $n \geq n_\varepsilon$; hence, $x \in K_{\lambda + \varepsilon}$. Therefore, $\{\varphi \leq \lambda\} \subseteq \bigcap_{\varepsilon > 0} K_{\lambda + \varepsilon}$, and the latter is τ-closed and τ-countably compact. So, we conclude that φ is τ-lower semicontinuous and coercive such that $\varphi \leq f_n$ for all $n \in \mathbb{N}$.

$\Leftarrow$: For every $\lambda \in \mathbb{R}$ and every $n \in \mathbb{N}$, we have

$$\{f_n \leq \lambda\} \subseteq \{\varphi \leq \lambda\} = K_\lambda = \tau\text{-closed, } \tau\text{-countably compact,}$$

$$\Rightarrow \{f_n\}_{n \in \mathbb{N}} \text{ equicoercive (see Definition 8.68).} \qquad \square$$

Equicoercivity leads to variational stability.

Theorem 8.71 *If (X, d) is a metric space, $f_n : X \to \overline{\mathbb{R}} = \mathbb{R} \cup \{+\infty\}$, $n \in \mathbb{N}$, is an equicoercive family, $-\infty < \inf_{n \in \mathbb{N}} f_n(x)$ for all $x \in X$, and we set*

$$f_* = \Gamma - \liminf_{n \to \infty} f_n, \quad f^* = \Gamma - \limsup_{n \to \infty} f_n,$$

then both f_* and f^* are coercive and

$$\min_X f_* = \liminf_{n \to \infty} \inf_X f_n.$$

Moreover, if $f_n \overset{\Gamma}{\to} f$, then f is coercive and

$$\min_X f = \lim_{n \to \infty} \inf_X f_n.$$

Proof According to Proposition 8.70, there exists $\varphi : X \to \overline{\mathbb{R}} = \mathbb{R} \cup \{+\infty\}$ lower semicontinuous and coercive such that $\varphi \leq f_n$ for all $n \in \mathbb{N}$. Hence, $\varphi \leq f_* \leq f^*$, and so both functions f_* and f^* are coercive and of course lower semicontinuous (see Proposition 8.53(a)).

Consider the minimization problem $m_* = \inf_X f_*$, and let $\{x_n\}_{n\in\mathbb{N}}$ be a minimizing sequence for this problem. Then on account of the coercivity of f_*, $\{x_n\}_{n\in\mathbb{N}}$ has a cluster point $x \in X$, and we have

$$m_* \leq f_*(x) \leq \liminf_{n\to\infty} f_*(x_n) = m_*,$$

$$\Rightarrow f_*(x) = m_*, \tag{8.14}$$

$$\Rightarrow \liminf_{n\to\infty} \inf_X f_n \leq \min_X f_* \text{ (see (8.1))}.$$

We assume that $\liminf_{n\to\infty} \inf_X f_n < \infty$ (or otherwise there is nothing to prove). We can find a subsequence $\{n_k\}$ of $\{n\}$ and $\lambda \in \mathbb{R}$ such that

$$\lim_{k\to\infty} \inf_X f_{n_k} = \liminf_{n\to\infty} \inf_X f_n < \lambda.$$

We may assume that

$$\inf_X f_{n_k} < \lambda \text{ for all } k \in \mathbb{N}. \tag{8.15}$$

The equicoercivity of $\{f_n\}_{n\in\mathbb{N}}$ implies that we can find a set K_λ that is compact such that $\{f_{n_k} \leq \lambda\} \subseteq K_\lambda$ for all $k \in \mathbb{N}$ and the sets $\{f_{n_k} \leq \lambda\} \neq \emptyset$ (see (8.15)). It follows that

$$\inf_X f_{n_k} = \inf_{K_\lambda} f_{n_k}.$$

We can find $x_{n_k} \in K_\lambda$ such that $f_{n_k}(x_{n_k}) = \inf_X f_{n_k}$. On account of the compactness of K_λ, we may assume that $x_{n_k} \to x^*$ in X. Then

$$\widehat{f}_*(x_*) \leq \liminf_{n\to\infty} \inf_X f_n \text{ where } \widehat{f}_* = \Gamma \cdot \liminf_{k\to\infty} f_{n_k},$$

$$\Rightarrow \min_X \widehat{f}_* \leq \liminf_{n\to\infty} \inf_X f_n,$$

$$\Rightarrow \min_X f_* \leq \liminf_{n\to\infty} \inf_X f_n \text{ (since } f_* \leq \widehat{f}_*).$$

From (8.14), we conclude that

$$\min_X f_* = \liminf_{n\to\infty} \inf_X f_n.$$

Finally, suppose that $f_n \xrightarrow{\Gamma} f$. Then from (8.11), we have

$$\limsup_{n \to \infty} \inf_X f_n \leq \min_X f,$$

$$\Rightarrow \min_X f = \lim_{n \to \infty} \inf_X f_n.$$ $\square$

We will prove some more variational features of the Γ-convergence.

Definition 8.72 Let $f : X \to \overline{\mathbb{R}} = \mathbb{R} \cup \{+\infty\}$ be a nontrivial function (i.e., $f \not\equiv +\infty$). By M_f, we denote the set of minimizers of f, that is,

$$M_f = \{x \in X : f(x) = \inf_X f\}.$$

If the infimum is not attained, we look for ε-minimizers ($\varepsilon > 0$). So, we introduce the set

$$M_f^\varepsilon = \left\{x \in X : f(x) \leq \max\{\inf_X f + \varepsilon, -\frac{1}{\varepsilon}\}\right\}.$$

Remark 8.73 This definition of ε-minimizer is designed to incorporate also the case when $\inf_X f = -\infty$. Note that M_f can be empty, while $M_f^\varepsilon \neq \emptyset$ for all $\varepsilon > 0$.

Proposition 8.74 *If (X, d) is a metric space and $f_n \xrightarrow{\Gamma} f$ in X, then:*

(a) $K - \limsup_{n \to \infty} M_{f_n} \subseteq \bigcap_{\varepsilon > 0} K - \limsup_{n \to \infty} M_{f_n}^\varepsilon \subseteq M_f.$

(b) *If $\bigcap_{\varepsilon > 0} K - \limsup_{n \to \infty} M_{f_n}^\varepsilon \neq \emptyset$, then $M_f \neq \emptyset$ and $\min_X f = \limsup_{n \to \infty} \inf_X f_n.$*

(c) *If $\bigcap_{\varepsilon > 0} K - \liminf_{n \to \infty} M_{f_n}^\varepsilon \neq \emptyset$, then $M_f \neq \emptyset$ and $\min_X f = \lim_{n \to \infty} \inf_X f_n.$*

Proof

(a) Since $M_{f_n} \subseteq M_{f_n}^\varepsilon$ for all $n \in \mathbb{N}$, all $\varepsilon > 0$, the first inclusion is clear.

Let $x \in \bigcap_{\varepsilon > 0} K - \limsup_{n \to \infty} M_{f_n}^\varepsilon$. Then for every $\varepsilon > 0$, $U \in \mathcal{N}(x)$, and $k \in \mathbb{N}$, we can find $n \geq k$ such that $U \cap M_{f_n}^\varepsilon \neq \emptyset$. Hence,

$$\liminf_{n \to \infty} \inf_{x \in U} f_n(x) \leq \max\left\{\limsup_{n \to \infty} \inf_X f_n + \varepsilon, -\frac{1}{\varepsilon}\right\},$$

$$\Rightarrow f(x) \leq \limsup_{n \to \infty} \inf_X f_n.$$

On the other hand, from (8.11), we see that

$$\limsup_{n\to\infty} \inf_X f_n \leq \inf_X f,$$

$$\Rightarrow \min_X f = \limsup_{n\to\infty} \inf_X f_n.$$

(b) Follows from (a).
(c) We have (see the proof of (a))

$$f(x) \leq \liminf_{n\to\infty} \inf_X f_n \leq \inf_X f,$$

$$\Rightarrow \min_X f = \liminf_{n\to\infty} \inf_X f_n,$$

$$\Rightarrow \min_X f = \lim_{n\to\infty} \inf_X f_n \text{ (see (b))}. \qquad \square$$

When f is not identically $+\infty$, then using Proposition 8.74 and Definition 8.73, we easily establish the following theorem.

Proposition 8.75 *If (X, d) is a metric space, $f_n \xrightarrow{\Gamma} f$, $f \not\equiv +\infty$, and we consider the following statements:*

(i) $\bigcap_{\varepsilon>0} K - \limsup_{n\to\infty} M_{f_n}^\varepsilon \neq \emptyset$

(ii) $M_f \neq \emptyset$ and $\min_X \varphi = \limsup_{n\to\infty} \inf_X f_n$

(iii) $M_f = \bigcap_{\varepsilon>0} K - \limsup_{n\to\infty} M_{f_n}^\varepsilon$

(iv) $\bigcap_{\varepsilon>0} K - \liminf_{n\to\infty} M_{f_n}^\varepsilon \neq \emptyset$

(v) $M_f \neq \emptyset$ and $\min_X f = \lim_{n\to\infty} \inf_X f_n$

(vi) $M_f = \bigcap_{\varepsilon>0} K - \liminf_{n\to\infty} M_{f_n}^\varepsilon = \bigcap_{\varepsilon>0} K - \limsup_{n\to\infty} M_{f_n}^\varepsilon$

then the following implications hold:

(a) $(i) \Leftrightarrow (ii) \Rightarrow (iii)$
(b) $(iv) \Leftrightarrow (v) \Rightarrow (vi)$

Proof (a) Note that $(i) \Leftrightarrow (ii)$ was proved in Proposition 8.74(b). So we prove $(ii) \Leftrightarrow (i)$. Let $x \in M_f$. Then,

$$f(x) = \min_X f = \limsup_{n\to\infty} \inf_X f_n < +\infty.$$

Given $\varepsilon > 0$, we have

$$f(x) - \varepsilon \leq \inf_X f_n \text{ for infinitely many } n \in \mathbb{N}. \qquad (8.16)$$

From (8.11), we have for all $U \in \mathcal{N}(x)$

$$\limsup_{n\to\infty} \inf_U f_n < \max\left\{ f(x) + \frac{\varepsilon}{2}, -\frac{1}{\varepsilon} \right\},$$

$$\Rightarrow \inf_U f_n < \max\left\{ f(x) + \frac{\varepsilon}{2}, -\frac{1}{\varepsilon} \right\} \text{ for } n \in \mathbb{N} \text{ big,}$$

$$\Rightarrow \inf_U f_n < \max\left\{ f(x) + \varepsilon, -\frac{1}{\varepsilon} \right\} \text{ for infinitely many } n \in \mathbb{N}, \text{ see (8.16),}$$

$$\Rightarrow U \cap M_{f_n}^{\varepsilon} \neq \emptyset \text{ for infinitely many } n \in \mathbb{N},$$

$$\Rightarrow x \in \bigcap_{\varepsilon > 0} K - \limsup_{n\to\infty} M_{f_n}^{\varepsilon}.$$

This argument also shows that (i) and (ii) both imply (iii).
(b) In this case, (8.16) holds for all $n \in \mathbb{N}$ big. Arguing as above, we obtain

$$\inf_U f_n < \max\{ f(x) + \varepsilon, -\frac{1}{\varepsilon} \} \text{ for all } n \in \mathbb{N} \text{ big,}$$

$$\Rightarrow U \cap M_{f_n}^{\varepsilon} \neq \emptyset \text{ for all } n \in \mathbb{N} \text{ big, all } \varepsilon > 0,$$

$$\Rightarrow x \in \bigcap_{\varepsilon > 0} K - \liminf_{n\to\infty} M_{f_n}^{\varepsilon}.$$

This combined with (a) gives the desired implications.   □

Corollary 8.76 *If (X, d) is a metric space, $f_n \xrightarrow{\Gamma} f$, $x_n \in M_{f_n}^{\varepsilon_n}$, $n \in \mathbb{N}$ with $\varepsilon_n \downarrow 0$, and x is a cluster point of $\{x_n\}_{n\in\mathbb{N}}$, then $x \in M_f$ and $f(x) = \limsup_{n\to\infty} f_n(x_n)$; moreover, if $x_n \to x$ in X, then $x \in M_f$ and $f(x) = \lim_{n\to\infty} f_n(x_n)$.*

8.3   *G*-Convergence of Operators

The notion of *G*-convergence was introduced in order to give an answer to the following problem that we encounter in many applications. If $\{f_n\}_{n\in\mathbb{N}}$ is a sequence of functionals defined on a function space, does there exist a functional f such that the solutions to problems for $f_n(\cdot)$ converge toward the solutions of the corresponding problem with f? De Giorgl and Spagnolo, who initiated this line of research, established the variational nature of *G*-convergence, by establishing its connection to the convergence of the energy functionals (Γ-convergence).

Let X be a separable reflexive Banach space, and let $\| \cdot \|$ denote its norm. By X^*, we denote the dual of X, and by $\| \cdot \|_*$ its norm. We know that X^* is separable reflexive too, and by $\langle \cdot, \cdot \rangle$, we denote the duality brackets for the pair (X, X^*). We will consider in general multivalued operators $A : X \to 2^{X^*}$, and so we will identify them with their graph $\operatorname{Gr} A = \{(x, x^*) \in X \times X^* : x^* \in A(x)\}$. So, the product space $X \times X^*$ is important in our considerations. On $X \times X^*$, we consider the product topology $\tau = w \times s$ with w denoting the weak topology on X and s being the norm (strong) topology on X^*.

Definition 8.77 Consider a sequence of operators $A_n : X \to 2^{X^*}$. We say that the sequence is "G-convergent" to A, if

$$\operatorname{Gr} A_n \xrightarrow{K_\tau} \operatorname{Gr} A.$$

Remark 8.78 According to this definition, G-convergence is equivalent to the following two properties:

G1 If $x_n \xrightarrow{w} x$ in X, $x_n^* \to x^*$ and $x_n^* \in A(x_n)$ for all $n \in \mathbb{N}$, then $x^* \in A(x)$.
G2 If $x^* \in A(x)$, then we can find $x_n \xrightarrow{w} x$ in X, $x_n^* \to x^*$ in X^*, and $x_n^* \in A(x_n)$ for all $n \in \mathbb{N}$.

Directly from Definition 8.4, we have the following result.

Proposition 8.79 *If* $A_n \xrightarrow{G} A$, *then for any subsequence* $\{n_k\}$ *of* $\{n\}$, *we have* $A_{n_k} \xrightarrow{G} A$.

Remark 8.80 In fact, we have the Urysohn characterization of G-convergence, namely $A_n \xrightarrow{G} A$ if and only if every subsequence of $\{A_n\}_{n \in \mathbb{N}}$ has a further subsequence that is G-convergent to A. Note that if $A_n = A$ for all $n \in \mathbb{N}$ and $\operatorname{Gr} A$ a is sequentially closed in $X_w \times X^*$, then $A_n \xrightarrow{G} A$. Recall that if $A(\cdot)$ is maximal monotone, then $\operatorname{Gr} A$ is $\tau = w \times s$-closed in $X \times X^*$, and it is also τ-sequentially closed (recall that in general $\tau \subseteq \tau_{seq}$).

Proposition 8.81 *If* $A_n \xrightarrow{G} A$ *and* $K : X \to X^*$ *is sequentially continuous from* X_w *into* X^*, *then* $A_n + K \xrightarrow{G} A + K$.

Proof Let $x_n \xrightarrow{w} x$ in X, $x_n^* \to x$ in X^*, and $x_n^* \in A(x_n) + K(x_n)$ for all $n \in \mathbb{N}$. Then $K(x_n) \to K(x)$ in X^x and $x_n^* - K(x_n) \in A(x_n)$ for all $n \in \mathbb{N}$. Since $A_n \xrightarrow{G} A$, we infer that $x^* - K(x) \in A(x)$ and so $x^* \in A(x) + K(x)$. So, condition $(G1)$ in Remark 6.78 is satisfied.

Next let $x \in X$ and $x^* \in X^x$ such that $x^* \in A(x) + K(x)$. So $x^* - K(x) = u^* \in A(x)$, and since $A_n \xrightarrow{G} A$, we can find $x_n \in X$, $u_n^* \in X_n^*$, and $u_n^* \in A(x_n)$ $n \in \mathbb{N}$ such that $x_n \xrightarrow{w} x$ in X and $u_n^* \to u^*$ in X^*. Let $x_n^* = u_n^* + K(x_n)$, $n \in \mathbb{N}$. Then $x_n^* \to u^* + K(x)$ in X^*, and so condition $(G2)$ of Remark 6.78 is satisfied. We conclude that $A_n + K \xrightarrow{G} A + K$. $\square$

Let $1 < p < \infty$, and let p' be the corresponding conjugate exponent (i.e., $\frac{1}{p} + \frac{1}{p'} = 1$). Also, $m_1, m_2 \geq 0$ and $c_1, c_2 > 0$. We introduce the following class of operators.

Definition 8.82 By $M = M(m_1, m_2, c_1, c_2)$, we denote the set of all monotone operators $A : X \to 2^{X^*}$ such that for all $(x, x^*) \in \operatorname{Gr} A$, we have

$$\|x^*\|_*^{p'} \leqslant m_1 + c_1 \langle x^*, x \rangle, \tag{8.17}$$

$$\|x\|^p \leqslant m_2 + c_2 \langle x^*, x \rangle. \tag{8.18}$$

By $M_0 = M_0(m_1, m_2, c_1, c_2)$, we denote the subset of M consisting all maximal monotone elements of M.

Remark 8.83 Note that from (8.17), we have

$$\|x^*\|_*^{p'} \leq m_1 + \frac{c_1 \varepsilon}{p'} \|x^*\|_*^{p'} + \frac{c_1}{p\varepsilon} \|x\|^p \text{ (Young's inequality with } \varepsilon > 0)$$

$$\Rightarrow \|x^*\|_* \leqslant m_3 + c_3 \|x\|^{p-1} \text{ for some } m_3 \geqslant 0, c_3 > 0. \tag{8.19}$$

Moreover, from (8.18), we have

$$\|x\|^p - m_2 \leqslant \langle x^*, x \rangle. \tag{8.20}$$

Conversely, if (8.19) and (8.20) hold, then (8.17) and (8.18) are true too.

Suppose $A \in M_0$. Then A^{-1} is maximal monotone too, and from (8.17), we see that A^{-1} is coercive. Then by Theorem 7.47, $A^{-1}(\cdot)$ is surjective, that is, $R(A^{-1}) = D(A) = X$.

We have the following fundamental compactness result.

Theorem 8.84 *If $\{A_n\}_{n \in \mathbb{N}} \subseteq M$, then we can extract a subsequence $\{A_{n_k}\}_{k \in \mathbb{N}} \subseteq \{A_n\}_{k \in \mathbb{N}}$ such that $A_{n_k} \xrightarrow{G} A \in M$ as $k \to \infty$.*

Proof Recall that X^* is separable, and let $\{x_n^*\}_{n \in \mathbb{N}}$ be a dense sequence. Then the function $d : X \times X \to \mathbb{R}$ defined by

$$d(x, u) = \sum_{n \in \mathbb{N}} \frac{1}{2^n} \frac{|\langle x_n^*, x - u \rangle|}{\|x_n^*\|_*}$$

is a metric on X and on bounded sets (thus, reactively weakly compact sets, since X is reflexive); it generates the weak topology (see Theorem 3.19). The product topology $\widehat{\tau} = d \times s$ on $X \times X^*$ is separable metric. So, by passing to a subsequence if necessary, we can say that

$$\text{Gr } A_n \xrightarrow{K_{\widehat{\tau}}} \text{Gr } A \tag{8.21}$$

with $A : X \to 2^{X^*}$ an operator. We will show that

$$\text{Gr } A_n \xrightarrow{K_{\widehat{\tau}}} \text{Gr } A \quad (\tau = w \times s) \text{ and } A \in M. \tag{8.22}$$

First we show that

$$K_\tau - \limsup_{n \to \infty} A_n \subseteq A. \tag{8.23}$$

To this end, let $(x, x^*) \in K_\tau - \limsup_{n \to \infty} A_n$. Then we can find a subsequence $\{A_{n_k}\}_{k \in \mathbb{N}}$ of $\{A_n\}_{n \in \mathbb{N}}$ and $(x_k, x_k^*) \in \text{Gr } A_{n_k}, k \in \mathbb{N}$, such that

$$x_k \xrightarrow{w} x \text{ in } X, \quad x_k^* \to x^* \text{ in } X^*.$$

Since weakly convergent sequences are bounded and on bounded sets the w and d topologies coincide, we have

$$\left(x_k, x_k^*\right) \xrightarrow{\widehat{\tau}} \left(x, x^*\right) \text{ in } X \times X^*$$

$$\Rightarrow \left(x, x^*\right) \in \text{Gr } A \quad (\text{ see } (8.21)).$$

Therefore, (8.23) is true. Next we show that

$$A \subseteq K_\tau - \liminf_{n \to \infty} A_n. \tag{8.24}$$

To this end, let $(x, x^*) \in \text{Gr } A$. Then by (8.21), we can find $\left\{(x_n, x_n^*)\right\}_{n \in \mathbb{N}}$ such that

$$\left(x_n, x_n^*\right) \xrightarrow{\widehat{\tau}} \left(x, x^*\right) \text{ and } x_n^* \in A(x_n) \text{ for all } n \in \mathbb{N}.$$

Then from (8.20), it follows that $\{x_n\}_{n \in \mathbb{N}} \subseteq X$ is bounded and so $x_n \xrightarrow{w} x$, thus $\left(x_n, x_n^*\right) \xrightarrow{\tau} (x, x^*)$, and this proves (8.24). From (8.23)–(8.24), we infer that the convergence in (8.22) is true.

It remains to show that $A \in M$. Let $(x_n, x_n^*) \in \text{Gr } A$. From (8.22), we have that there exist $\{x_n\}_{n \in \mathbb{N}} \subseteq X, \left\{x_n^*\right\}_{n \in \mathbb{N}} \subseteq X^*$ such that

$$x_n \xrightarrow{w} x_n \text{ in } X, x_n^* \to x^* \text{ in } X^*, x_n^* \in A_n(x_n) \text{ for all } n \in \mathbb{N}.$$

Since $A_n \in M$, from (8.17), we have

$$\|x_n^*\|_*^{p'} \leqslant m_1 + c_1 \langle x_n^*, x_n \rangle \text{ for all } n \in \mathbb{N}$$

$$\Rightarrow \|x^*\|_*^{p'} \leq m_1 + c_1 \langle x_1^*, x \rangle.$$

Similarly, we show that

$$\|x\|^p \leqslant m_2 + c_2 \langle x^*, x \rangle.$$

Therefore from the monotonicity of $A(\cdot)$ and Definition 8.82, we conclude that $A \in M$.   □

Next we will show that maximal monotonicity is stable under G-convergence. First, a simple lemma.

Lemma 8.85 *If Y is a reflexive Banach space, $A : Y \to 2^{Y^*}$ is monotone, $K : Y \to Y^*$ is strictly monotone with $D(K) = Y$ and $R(A + K) = Y^*$, then $A(\cdot)$ is maximal monotone.*

Proof Suppose $0 \leq \langle u^* - y^*, u - y \rangle_Y$ for all $(y, y^*) \in \mathrm{Gr}\, A$. We can find $v_0 \in Y$ such that $u^* + K(u) \in A(v_0) + K(v_0)$. Then $h^* = u^* + K(u) - K(v_0) \in A(v_0)$, and so we have

$$0 \leqslant \langle u^* - h^*, u - v_0 \rangle_Y = \langle K(v_0) - K(u), u - v_0 \rangle_Y \leq 0$$

$$\Rightarrow u = v_0 \text{ (since } K \text{ is strictly monotone).}$$

It follows $u^* = h^* \in A(v_0) = A(u)$, which proves the maximality of $A(\cdot)$.   □

Now, we can state and prove the stability result for maximal monotonicity.

Proposition 8.86 *If $\{A_n\}_{n \in \mathbb{N}} \subseteq M_0$, $A_n \xrightarrow{G} A$, and there exists a strictly monotone map $K : X \to X^*$ that is continuous from X_w into X^*, then $A \in M_0$.*

Proof We already know that $A \in M$ (see Theorem 8.84). So, we need to show that A is maximal monotone.

Let K be as in the hypotheses. Proposition 7.45 says that $x \to (A_n + K)(x)$ is maximal monotone. Also, for every $x \in X$, we have

$$\langle A_n(x), x \rangle + \langle K(x), x \rangle \geq \|x\|^p - m_2 - \|K(0)\|_* \|x\| \text{ (see (8.20))} \tag{8.25}$$

$$\Rightarrow x \to (A_n + K(x)) \text{ is coercive (since } p > 1).$$

Then Theorem 7.47 implies that $(A_n + K)(\cdot)$ is surjective. Also Proposition 8.81 implies that

$$A_n + K \xrightarrow{G} A + K. \tag{8.26}$$

Let $x^* \in X^*$. We can find unique (due to the strict monotonicity of K) $x_n \in D(A_n)$, $n \in \mathbb{N}$, such that $x^* \in (A_n + K)(x_n)$. Evidently, $\{x_n\}_{n \in \mathbb{N}}$ is bounded, and so we may assume that $x_n \xrightarrow{w} x$ in X. On account of (8.17), we have $x^* \in (A+K)(x)$ and so $R(A + K) = X^*$. Then Lemma 8.85 implies $A \in M_0$. $\square$

This proposition leads to a stability result of maximal monotonicity with respect to G-convergence.

Theorem 8.87 *If $\{A_n\}_{n \in \mathbb{N}} \subseteq M_0$ and there exists a strictly monotone map $K :$ $X \to X^*$ that is continuous from X_w into X^*, then we can extract a subsequence $\{A_{n_k}\}_{k \in \mathbb{N}}$ of $\{A_n\}_{n \in \mathbb{N}}$ such that $A_{n_k} \xrightarrow{G} A \in M_0$.*

Remark 8.88 We need to elaborate on the hypothesis concerning the existence of the map $K(\cdot)$. Let Y be a reflexive Banach space and Y, Y^* are strictly convex. Suppose that $X \hookrightarrow Y$ compactly and densely. Then $Y^* \hookrightarrow X^*$ compactly and densely (see Lemma 4.200). Let $\mathscr{F} : Y \to Y^*$ be the duality map. Then $\mathscr{F}(\cdot)$ is strictly monotone and demicontinuous (see Proposition 6.93), and so $K = \mathscr{F}|_X$ has the desired properties.

Proposition 8.89 *If $\{A_n, A\}_{n \in \mathbb{N}} \subseteq M_0$ and are strictly monotone, then $A_n \xrightarrow{G} A$ if and only if for any $x^* \in X^*$, $A_n^{-1}(x^*) \xrightarrow{w} A^{-1}(x^*)$.*

Proof $\Rightarrow$ This is an immediate consequence of Definition 8.77 (see also Remark 8.78).

$\Leftarrow$ On account of Theorem 8.84, we may assume that $A_n \xrightarrow{G} \widehat{A} \in M$. Then $A^{-1}(x^*) \in \widehat{A}^{-1}(x^*)$ and so $\widehat{A}$ is a monotone extension of A, and by the maximality of A, we conclude that $\widehat{A} = A$. Therefore, $A_n \xrightarrow{G} A$ (see Remark 8.80). $\square$

Remark 8.90 Some authors define G-convergence using this proposition. Namely consider operators with single-valued inverse and require that for every $x^* \in X^*$ we have $A_n^{-1}(x^*) \xrightarrow{w} A^{-1}(x^*)$.

For single-valued operators, we introduce some other modes of convergence.

Definition 8.91 Let $\{A_n, A\}_{n \in \mathbb{N}}$ be single-valued operators from X into X^*.

(a) We say that the A_n's "converge uniformly" to A, denoted by $A_n \xrightarrow{u} A$ if and only if $d(A_n, A) \to 0$ with $d(\cdot, \cdot)$ being the metric defined by

$$d(A, E) = \sup[\frac{\|A(x) - E(x)\|_*}{1 + \|x\|^p} : x \in X].$$

(b) We say that the A_n's "converge pointwise" to A, denoted by $A_n \xrightarrow{p} A$ if and only if for all $x \in X$, $A_n(x) \to A(x)$ in X^*.

(c) We say that A_n "converges weakly" to A, denoted by $A_n \xrightarrow{w} A$ if and only if for all $x \in X$, $A_n(x) \xrightarrow{w} A(x)$ in X^*.

Evidently, $\xrightarrow{u} \Rightarrow \xrightarrow{p} \Rightarrow \xrightarrow{w}$.

Proposition 8.92 *If $\{A_n, A\}_{n \in \mathbb{N}}$ are single-valued operators from X into X^* that are uniformly monotone, then $A_n \xrightarrow{p} A \Rightarrow A_n \xrightarrow{G} A$.*

Proof Let $x^* \in X^*$. We can find unique $x_n, x \in X$ such that

$$A_n(x_n) = x^* \text{ for all } n \in \mathbb{N}, A(x) = x^*.$$

From (8.20), we see that $\{x_n\}_{n \in \mathbb{N}} \subseteq X$ is bounded. Also by hypothesis $A_n(x) \to A(x) = x^*$ in X^*. On account of the uniform monotonicity of $A_n(\cdot)$, we have

$$\langle A_n(x_n) - A_n(x), x_n - x \rangle \geq \xi(\|x_n - x\|)\|x_n - x\|$$
$$\Rightarrow x_n \to x, \text{ which implies } A_n \xrightarrow{G} A. \qquad \square$$

Remark 8.93 Weak and *G*-convergence are distinct notions. Consider the following example due to Spagnolo [260]. Let $X = H_0^1(0, 2\pi)$, $X^* = H^{-1}(0, 2\pi)$ and consider $a : \mathbb{R} \to \mathbb{R}$ a 2π-periodic, strictly monotone function such that $0 < c \leq a(t) \leq \widehat{c}$ for a.a. $t \in [0, 2\pi]$. We introduce the operator $A_n : X \to X^*, n \in \mathbb{N}$, defined by

$$A_n(u)(t) = (a(nt)u'(t))'$$

with the derivatives being defined in the weak sense.

Then we have

$$A_n \xrightarrow{w} \widehat{A} \text{ with } \widehat{A}(u) = \left(\frac{1}{2\pi} \int_0^{2\pi} a\, dt\right) u'' \text{ for all } u \in X,$$

$$A_n \xrightarrow{G} \widetilde{A} \text{ with } \widetilde{A}(u) = \left(\frac{1}{2\pi} \int_0^{2\pi} a^{-1} dt\right)^{-1} u'' \text{ for all } u \in X.$$

Next we focus on subdifferential operators. First it is straightforward to check that G-convergence preserves cyclic monotonicity.

Proposition 8.94 *If $\{A_n\}_{n \in \mathbb{N}}$ are cyclically monotone operators from X into 2^{X^*} and $A_n \xrightarrow{G} A$, then A is cyclically monotone too.*

Corollary 8.95 *If $A_n = \partial \varphi_n$ with $\varphi_n \in \Gamma_0(X), n \in \mathbb{N}$, and $A_n \overset{G}{\to} A$, then $A = \partial \varphi$ with $\varphi \in \Gamma_0(X)$.*

We conclude this section with a quick look at multivalued monotone operators on Sobolev spaces.

Definition 8.96 Let $\Omega \subseteq \mathbb{R}^N$ be a bounded domain with Lipschitz boundary $\partial \Omega$. By $M(\Omega, \mathbb{R}^N)$, we denote the space of multifunctions $a : \Omega \times \mathbb{R}^N \to 2^{\mathbb{R}^N}$ such that:

(a) For all $y \in \mathbb{R}^N$, $z \to a(z, y)$ is graph measurable.
(b) For a.a. $z \in \Omega$, $y \to a(z, y)$ is maximal monotone.
(c) There exist $m_1, m_2 \in L^1(\Omega)$ and constants $c_1, c_2 > 0$ such that

$$|y|^{p'} \leq m_1(z) + c_1(\xi, y)_{\mathbb{R}^N} \ (\frac{1}{p} + \frac{1}{p'} = 1),$$

$$|\xi|^p \leq m_2(z) + c_2(\xi, y)_{\mathbb{R}^N}$$

for $a.a. z \in \Omega$ and all $(y, \xi) \in Gra(z, \cdot)$.

Remark 8.97 As before, condition (c) is equivalent to

$$|y| \leq m_3(z) + c_3|\xi|^{p^{-1}},$$

$$(\xi, y)_{\mathbb{R}^N} \geq m_4(z) + c_4|\xi|^p$$

for $a.a. \ z \in \Omega$, some $c_3, c_4 > 0$, $m_3 \in L^{p'}(\Omega)$, $m_4 \in L^1(\Omega)$, and all $(y, \xi) \in Gra(z, \cdot)$. Note that on account of Proposition 7.29 for $a.a. \ z \in \Omega$, $a(z, \cdot)$ has closed and convex values.

Now given $a \in M(\Omega, \mathbb{R}^N)$, we introduce the operator $\widehat{A} : W_0^{1,p}(\Omega) \to W^{-1,p'}(\Omega) = W_0^{1,p}(\Omega)^*$ defined by, for all $u \in W_0^{1,p}(\Omega)$,

$$\widehat{A}(u) = \{-\operatorname{div} h : h \in S_{a(\cdot, Du(\cdot))}^{p'}\}.$$

Using Lemma 8.85, we can easily check that $\widehat{A}(\cdot)$ is maximal monotone. The next theorem due to Chiado Piat–Dal Maso–Defranceschi [65] (Theorem 3.11) relates the G-convergence of a sequence $\{a_n\}_{n \in \mathbb{N}} \subseteq M(\Omega, \mathbb{R}^N)$, with the corresponding sequence $\{\widehat{A}_n\}_{n \in \mathbb{N}}$.

Theorem 8.98 *If for a.a.* $z \in \Omega$, $a_n(z, \cdot) \xrightarrow{G} a(z, \cdot)$, *then* $\widehat{A}_n \xrightarrow{G} \widehat{A}$.

8.4 Variational Principles

We start with the very versatile Ekeland variational principle. When we look for the minima of a differentiable function, we check the points where the derivative vanishes, which leads to points that can support the epigraph of the function from below with a half space. In the absence of differentiability, we replace the half space with a cone that touches the epigraph from below at a single point. If the cone is wide (i.e., the $\varepsilon > 0$ in Theorem 8.99 below is small), then the point of contact is almost a true minimizer. This is the essence of the Ekeland variational principle, which proved to be a very helpful tool in many situations.

Theorem 8.99 *If* (X, d) *is a complete metric space,* $f : X \to \overline{\mathbb{R}} = \mathbb{R} \cup \{+\infty\}$ *is a lower semicontinuous function that is bounded from below,* $\varepsilon > 0$ *and* u *satisfy* $f(u) \le \inf_X f + \varepsilon$ *and* $\lambda > 0$, *then we can find* $v \in X$ *such that:*

(a) $d(u, v) \le \lambda$.
(b) $f(v) + \frac{\varepsilon}{\lambda} d(u, v) \le f(u)$.
(c) $f(x) + \frac{\varepsilon}{\lambda} d(v, x) > f(v)$ *for all* $x \in X \setminus \{v\}$.

Proof By replacing $d(\cdot, \cdot)$ with $\frac{1}{\lambda} d(\cdot, \cdot)$, we see that without any loss of generality, we may assume that $\lambda = 1$.

Inductively, we will produce a sequence $\{u_n\}_{n \in \mathbb{N}_0} \subseteq X$ such that $u_n \to u$ with $u_0 = u$. Suppose we have produced u_n, $n \in \mathbb{N}$. Let

$$D_n = \{x \in X : f(x) + \varepsilon d(x, u_n) \le f(u_n)\}.$$

We have two possibilities:

(i) $f(u_n) = \inf_{D_n} f$: In this case, we set $u_{n+1} = u_n$.
(ii) $\inf_{D_n} f < f(u_n)$. Then we choose $u_{n+1} \in D_n$ such that

$$f(u_{n+1}) < \inf_{D_n} f + \frac{1}{2}[f(u_n) - \inf_{D_n} f] = \frac{1}{2}[f(u_n) + \inf_{D_n} f] < f(u_n). \quad (8.27)$$

We show that $\{u_n\}_{n \in \mathbb{N}_0}$ is Cauchy. Indeed, if (i) occurs at a certain step, then $\{u_n\}_{n \in \mathbb{N}_0}$ is eventually constant, thus Cauchy. If (ii) occurs at all steps, then

$$\varepsilon d(u_{n+1}, u_n) \le f(u_n) - f(u_{n+1}), n \in \mathbb{N}. \quad (8.28)$$

Adding (8.28) from n to $m - 1 > n$, we obtain

$$\varepsilon d(u_m, u_n) \leq f(u_n) - f(u_m), n \in \mathbb{N}, n + 1 < m. \tag{8.29}$$

From (8.27), we know that $\{f(u_n)\}_{n \in \mathbb{N}}$ is decreasing and bounded below. Therefore, it is convergent. Then from (8.29), it follows that $\{u_n\}_{n \in \mathbb{N}}$ is Cauchy, and so we have $u_n \to v$ in X. From (8.29) with $n = 0$, we have

$$\varepsilon d(u_m, u) \leq f(u) - f(u_m) \text{ for all } m \in \mathbb{N},$$

$$\Rightarrow \varepsilon d(v, u) \leq f(u) - f(v) \tag{8.30}$$

(since $f(\cdot)$ is lower semicontinuous), which is statement (b).

We have

$$f(u) - f(v) \leq f(u) - \inf_X f < \varepsilon. \tag{8.31}$$

From (8.31) and (8.30), we infer statement (a).

From (8.29) with $n \in \mathbb{N}$ fixed, we let $m \to \infty$ and obtain

$$\varepsilon d(v, u_n) \leq f(u_n) - f(v) \text{ (since } f(\cdot) \text{ is lower semicontinuous)},$$
$$\Rightarrow v \in D_n \text{ for all }, n \in \mathbb{N}_0,$$
$$\Rightarrow v \in \cap_{n \in \mathbb{N}_0} D_n.$$

On the other hand, if $y \in \cap_{n \in \mathbb{N}_0} D_n$, then

$$\varepsilon d(y, u_{n+1}) \leq f(u_{n+1}) - f(y) \leq f(u_{n+1}) - \inf_{D_n} f. \tag{8.32}$$

From (8.27), we have

$$f(u_{n+1}) - \inf_{D_n} f \leq f(u_n) - f(u_{n+1}),$$
$$\Rightarrow \lim_{n \to \infty} [f(u_{n+1}) - \inf_{D_n} f] = 0,$$
$$\Rightarrow \varepsilon d(y, v) = 0 \text{ (see (8.32))}$$
$$\Rightarrow y = v.$$

Therefore, we have

$$\bigcap_{n \in \mathbb{N}} D_n = \{v\}. \tag{8.33}$$

It is easy to see that $\{D_n\}_{n \in \mathbb{N}}$ is decreasing. Hence, if $y \neq v$, then on account of (8.33), we have $y \notin D_n$ for all $n \in \mathbb{N}$ big. This means that

$$f(u_n) \le f(y) + \varepsilon d(y, u_n) \text{ for all, } n \ge n_0,$$
$$\Rightarrow f(v) \le f(y) + \varepsilon d(y, v),$$

and this inequality is strict if $y \ne v$. This proves statement (c). □

Remark 8.100 Note that in the above theorem, statements (a) and (b) are complementary. Namely if $\lambda > 0$ is big, then (a) gives little information on the whereabouts of v with respect to the approximate minimizer u, while statement (b) says that v is close to being a global minimizer of $f(\cdot)$. The situation is reversed of $\lambda > 0$ is small. We will check two special cases of importance: when $\lambda = 1$ (which means that we are indifferent on the whereabouts of v) and when $\lambda = \sqrt{\varepsilon}$ (which means that we split the difference).

Corollary 8.101 *If (X, d) is a complete metric space and $f : X \to \overline{\mathbb{R}} = \mathbb{R} \cup \{+\infty\}$ is lower semicontinuous and bounded below, then for any $\varepsilon > 0$, we can find $u_\varepsilon \in X$ such that*

$$f(u_\varepsilon) \le \inf_X f + \varepsilon \text{ and } f(u_\varepsilon) \le f(u) + \varepsilon d(u, u_\varepsilon) \text{ for all } u \in X.$$

Corollary 8.102 *If (X, d) is a complete metric space, $f : X \to \overline{\mathbb{R}} = \mathbb{R} \cup \{+\infty\}$ is lower semicontinuous and bounded below, and $\varepsilon > 0$, $u_\varepsilon \in X$ satisfy*

$$f(u_\varepsilon) \le \inf_X f + \varepsilon,$$

then we can find $v_\varepsilon \in X$ such that

$$f(v_\varepsilon) \le f(u_\varepsilon), d(v_\varepsilon, u_\varepsilon) \le \sqrt{\varepsilon}, f(v_\varepsilon) \le f(u_\varepsilon) + \sqrt{\varepsilon} d(u, u_\varepsilon) \text{ for all } u \in X.$$

If we assume more structure on X and φ, we can have interesting consequences of the Ekeland variational principle.

Theorem 8.103 *If X is a Banach space, $f : X \to \mathbb{R}$ is lower semicontinuous, bounded below, and Gateaux differentiable, and $\varepsilon > 0$, $u_\varepsilon \in X$ satisfy*

$$f(u_\varepsilon) \le \inf_X f + \varepsilon,$$

then we can find $v_\varepsilon \in X$ such that

$$f(v_\varepsilon) \le f(u_\varepsilon), \|v_\varepsilon - u_\varepsilon\| \le \sqrt{\varepsilon}, \|f'(v_\varepsilon)\|_X \le \sqrt{\varepsilon}.$$

Proof From corollary 8.102, we know that there exists $v_\varepsilon \in X$ such that

$$f(v_\varepsilon) \le f(u_\varepsilon), \|v_\varepsilon - u_\varepsilon\| \le \sqrt{\varepsilon} \text{ and } -\varepsilon\sqrt{\varepsilon}\|u - v_\varepsilon\| \le f(u) - f(v_\varepsilon) \quad (8.34)$$

for all $u \in X$.

In the third inequality in (8.34), we let $x = v_\varepsilon + th$ with $t > 0$ and $h \in \overline{B}_1^X$. We have

$$-\sqrt{\varepsilon} \leq \frac{f(v_\varepsilon + th) - f(v_\varepsilon)}{t}$$

$$\Rightarrow -\sqrt{\varepsilon} \leq \langle f'(v_\varepsilon), h \rangle \text{ for all } h \in \overline{B}_1^X$$

$$\Rightarrow \|f'(v_\varepsilon)\|_X \leq \sqrt{\varepsilon}. \qquad \square$$

This theorem leads to the existence of a minimizing sequence of almost critical points.

Corollary 8.104 *If X is a Banach space, $f : X \to \mathbb{R}$ is lower semicontinuous, bounded below, and Gateaux differentiable, then given a minimizing sequence $\{u_n\}_{n \in \mathbb{N}} \subseteq X$, we can find another minimizing sequence $\{v_n\}_{n \in \mathbb{N}} \subseteq X$ such that $f(v_n) \leq f(u_n)$, $\|v_n - u_n\| \to 0$ and $\|f'(v_n)\|_* \to 0$.*

In critical point theory, the following condition plays a central role.

Definition 8.105 Let X be a Banach space and $f \in C^1(X)$. We say that f satisfies the "C-condition," if every sequence $\{u_n\}_{n \in \mathbb{N}} \subseteq X$, such that $\{f(u_n)\}_{n \in \mathbb{N}} \subseteq \mathbb{R}$ is bounded and $(1 + \|u_n\|)f'(u_n) \to 0$ in X^*, admits a strongly convergent subsequence.

Remark 8.106 This is a compactness-type condition on the functional f. It compensates for the fact that the ambient space need not be locally compact (in most applications, it is infinite dimensional). Similar to the situation in "Degree Theory," the infinite dimensionality of X forces us to pass the burden of compactness from X to f.

Proposition 8.107 *If X is a Banach space and $f \in C^1(X)$ satisfies the "C-condition," then there exists $\widehat{u} \in X$ such that*

$$f(\widehat{u}) = \inf_X f.$$

Proof On account of Corollary 8.104, we can find a minimizing sequence $\{u_n\}_{n \in \mathbb{N}} \subseteq X$ such that $f'(u_n) \to 0$ in X^*. Since $f(\cdot)$ satisfies the "C-condition," we can assume that $u_n \to \widehat{u}$ in X. Then

$$f(u_n) \to f(\widehat{u}) = \inf_X f. \qquad \square$$

We present another important application of the Ekeland Variational Principle, this time on convex analysis. First we need to introduce the following "approximate" version of the subdifferential (see Definition 6.67).

Definition 8.108 Let X be a Banach space, $f \in \Gamma_0(X)$, and $u \in \text{dom} f$. For any $\varepsilon > 0$, we define the "ε-subdifferential $\partial_\varepsilon f(u)$" by

$$\partial_\varepsilon f(u) = \{x^* \in X^* : \langle x^*, x - u \rangle \leq f(x) - f(u) + \varepsilon \text{ for all } x \in X\}.$$

Clearly, if $0 < \varepsilon_1 < \varepsilon_2$, then $\partial_{\varepsilon_1} f(u) \subseteq \partial_{\varepsilon_2} f(u)$.

Remark 8.109 In contrast to the subdifferential, the ε-subdifferential is nonempty at every $u \in \text{dom} f$. Moreover, while ∂f is a local notion, $\partial_\varepsilon f$ is a global one in the sense that the behavior of f on all of X may be relevant for the construction of $\partial_\varepsilon f$.

We will also need the following result due to Borwein [39].

Proposition 8.110 *If X is a Banach space, $g \in \Gamma_0(X)$, $\xi \in C(X)$ is convex, and there exists $v \in X$ such that $g(v) = -\xi(v)$, then there exist $x^* \in X^*$ and $\vartheta \in \mathbb{R}$ such that*

$$-\xi(x) \leq \langle x^*, x \rangle + \vartheta \leq g(x) \text{ for all } x \in X, -x^* \in \partial g(v).$$

The next theorem relates the two subdifferentials.

Theorem 8.111 *If X is a Banach space, $f \in \Gamma_0(X)$, then given $u_0 \in \text{dom} f, \varepsilon > 0, \lambda > 0$ and $u_0^* \in \partial_\varepsilon f(u_0)$, we can find $u \in \text{dom} f$ and $u^* \in \partial f(u)$ such that*

$$\|u - u_0\| \leq \frac{\varepsilon}{\lambda}, \ \|u^\star - u_0^*\|_* \leq \lambda;$$

it follows that $D(\partial f)$ is dense $\text{dom} f = D(\partial_\varepsilon f)$.

Proof Let $g(x) = f(x) - \langle u_0^*, x \rangle, x \in X$. Then $g \in \Gamma_0(X)$ and $\text{dom} g = \text{dom} f$. Since $u_0^* \in \partial_\varepsilon f(u_0)$, we have $g(u_0) \leq \inf_X g + \varepsilon$. Invoking the Ekeland Variational Principle, we can find $v \in \text{dom} f$ such that

$$\lambda\|v - u_0\| \leq \varepsilon \text{ and } g(v) \leq g(x) + \lambda\|x - v\| \text{ for all } x \in X. \tag{8.35}$$

If we set $\xi(x) = \lambda\|x - v\| - g(v), x \in X$, then $\xi(\cdot)$ is continuous convex and $-\xi(x) \leq g(x)$ for all $x \in X$ (see (8.35)), $-\xi(v) = g(v)$. On account of Proposition 8.110, we can find $x^* \in X^*$ and $\vartheta \in \mathbb{R}$ such that

$$-\xi(x) \leq \langle x^*, x \rangle + \vartheta \leq g(x) \text{ for all } x \in X, x^* \in \partial g(v), x^* \in \partial h(v).$$

Let $u = v$ and $u^* = x^* + u_0^*$. These have the desired properties. □

Corollary 8.112 *If X is a Banach space, $f \in \Gamma_0(X)$, and $u_0 \in \text{dom} f, \varepsilon > 0$, then there exists $u \in D(\partial f)$ such that*

$$|f(u) - f(u_0)| < \varepsilon, \ \|u - u_0\| < \varepsilon.$$

Proof Let $u_0^* \in \partial_{\varepsilon/2} f(u_0)$ and $k = \max\{1, \|u_0^*\|_*\}$. By Theorem 8.111, we can find $u \in \mathrm{dom}\, f$ and $u^* \in X^*$ such that

$$u^* \in \partial f(u), \|u - u_0\| \leq \frac{\varepsilon}{2k} \text{ and } \|u^* - u_0^*\|_* \leq k. \tag{8.36}$$

Then we have

$$f(u) - f(u_0) \geq \langle u_0^*, u - u_0 \rangle - \frac{\varepsilon}{2} \geq -k \frac{\varepsilon}{2k} - \frac{\varepsilon}{2} = -\varepsilon \tag{8.37}$$

$$\|u^*\|_* \leq 2k \text{ (see(8.36))}.$$

Since $u^* \in \partial f(u)$ (see(8.36)), we have

$$f(u_0) - f(u) \geq \langle u^*, u_0 - u \rangle \geq -2k\|u - u_0\| = -\varepsilon. \tag{8.38}$$

From (8.37) and (8.38), we conclude that $|f(u) - f(u_0)| \leq \varepsilon$.   □

The next consequence of Theorem 8.111, and so of the Ekeland Variational Principle, is the following fundamental theorem from the Banach space theory, known as the "Bishop–Phelps Theorem."

Theorem 8.113 *If X is a Banach space, then the set of all elements of X^* that attain their norm on $\overline{B}_1^X = \{u \in X : \|u\| \leq \varepsilon\}$ is dense in X^*.*

Proof Let $u_0^\star \in X^*$ and $\varepsilon \in (0, 1)$. We can find $u_0 \in X$, $\|u_0\| = 1$ such that

$$\|u_0^*\|_* - \varepsilon \leq \langle u_0^*, u_0 \rangle.$$

Consider the continuous convex function $f(x) = \|u_0^*\|_\star \|x\|$ for all $x \in X$. We see that $u_0^* \in \partial_\varepsilon f(u_0)$. We apply Theorem 8.111 with $\lambda = \sqrt{\varepsilon}$. Then we can find $u \in X$ and $u^* \in \partial f(u)$ such that

$$\|u - u_0\| \leq \sqrt{\varepsilon} \text{ and } \|u^* - u_0^*\|_* \leq \sqrt{\varepsilon}.$$

The functional u^* attains its norm at $\frac{u}{\|u\|}$.   □

Proposition 8.114 *If X is a Banach space and $f : X \to \mathbb{R}$ is lower semicontinuous, bounded below, Gateaux differentiable and satisfies*

$$f(x) \geq c_1\|x\| - c_2 \text{ for all } x \in X, \text{ some } c_1, c_2 > 0,$$

then $f'(x)$ is dense in $c_1 \overline{B}_1^{X^*} = \{u^* \in X^* : \|u^*\|_* \leq c_1\}$.

Proof Let $u^* \in c_1 \overline{B}_1^{X^*}$, and let $g(u) = f(u) - \langle u^*, u \rangle$ for all $u \in X$. Evidently, $g(\cdot)$ is lower semicontinuous, bounded below, and Gateaux differentiable. Applying

Theorem 8.103, we can find $v_\varepsilon \in X$ such that $\|g'(v_\varepsilon)\|_* \leq \varepsilon$. Hence,

$$\|f'(v_\varepsilon) - x^*\|_* < \varepsilon$$

$$\Rightarrow f'(X) \text{ is dense in } c_1\overline{B}_1^{X^*}. \qquad \square$$

Proposition 8.115 *If X is a Banach space and $f : X \to \mathbb{R}$ is lower semicontinuous, Gateaux differentiable, and satisfies*

$$\vartheta(\|u\|) \leq f(u) \text{ for all } u \in X$$

with $\vartheta : \mathbb{R}_+ \to \mathbb{R}$ continuous and $\frac{\vartheta(t)}{t} \to +\infty$ as $t \to +\infty$, then $f'(x)$ is dense in X^*.

Next we present another remarkable result of nonlinear analysis that is actually equivalent to the Ekeland Variational Principle. This is the "Caristi Fixed Point Theorem."

Theorem 8.116 *If (X, d) is a complete metric space, $f : X \to \overline{\mathbb{R}} = \mathbb{R} \cup \{+\infty\}$ is lower semicontinuous and bounded below, and $F : X \to 2^X \setminus \{\emptyset\}$ is a multifunction such that*

$$f(v) \leq f(u) - d(u, v) \text{ for all } u \in X, \text{ all } v \in F(u),$$

then we can find $u_0 \in X$ such that $u_0 \in F(u_0)$.

Proof According to Corollary 8.101, there exists $u_0 \in X$ such that

$$f(u_0) < f(x) + d(x, u_0) \text{ for all } x \neq x_0 \ (\varepsilon = 1). \qquad (8.39)$$

We claim that $u_0 \in F(u_0)$. Arguing by contradition, suppose that $u_0 \notin F(u_0)$. Then from the hypothesis and (8.39), we have

$$f(v) \leq f(u_0) - d(u_0, v) < f(v) \text{ for all } v \in F(u_0)$$

a contradiction. $\square$

Remark 8.117 Let F be single-valued and a contraction, that is,

$$d(F(v), F(u)) \leq kd(v, u) \text{ for all } v, u \in X, \text{ with } k < 1.$$

Let $f(x) = \frac{1}{1-k}d(x, F(x))$. Then this $f(\cdot)$ satisfies the requirements of Theorem 8.116, and so we produce $u_0 \in X$ such that $F(u_0) = u_0$. So, Theorem 8.116 gives the existence part in the Banach fixed point theorem. However, Banach's theorem includes much more information.

In fact, the Ekeland Variational Principle (in the form of Corollary 8.101) and the Caristi Fixed Point Theorem are equivalent.

Proposition 8.118 *Corollary 8.101 $\Longleftrightarrow$ Theorem 8.116.*

Proof $\Longrightarrow$ We have seen this in the proof of Theorem 8.116.

$\Longleftarrow$ Arguing by contradiction, suppose that there is no $u_\varepsilon \in X$ such that $f(u_\varepsilon) < f(u) + \varepsilon d(u, u_\varepsilon)$ for all $u \neq u_\varepsilon$. Consider the multifunction $F(u) = \{v \in X : f(u) \geq f(v) + \varepsilon d(v, u), v \neq u\}$. We see that $F(u) \neq \emptyset$ for all $u \in X$. We apply Theorem 8.116 and obtain $u_0 \in X$ such that $u_0 \in F(u_0)$, which cannot happen (see the definition of $F(\cdot)$). $\qquad\square$

We present another variational principle known in the literature as the "Takahashi Variational Principle."

Theorem 8.119 *If (X, d) is a complete metric space and $f : X \to \overline{\mathbb{R}} = \mathbb{R} \cup \{+\infty\}$ is lower semicontinuous, bounded below, and for an $u \in X$ with $f(u) > \inf_X f$, there exists $v \in X$, $v \neq u$ such that $f(v) + d(u, v) \leq f(u)$, then we can find $u_0 \in X$ such that $f(u_0) = \inf_X f$.*

Proof Arguing by contradiction, suppose that $\inf_X f < f(u)$ for all $u \in X$. Consider the multifunction $F(u) = \{v \in X : f(v) + d(u, v) \leq f(u), v \neq u\}$. Then $F(u) \neq \emptyset$ for all $u \in X$, and by Theorem 8.116, we can find $\widehat{u} \in X$ such that $\widehat{u} \in F(\widehat{u})$, a contradiction. So, there exist $u_0 \in X$ such that $f(u_0) = \inf_X f$. $\qquad\square$

Proposition 8.120 *Theorem 8.116 $\Leftrightarrow$ Theorem 8.119.*

Proof $\Rightarrow$ See the proof of Theorem 8.16.

$\Leftarrow$ Again we proceed by contradiction. So, suppose that $u \notin F(u)$ for all $u \in X$. From the property of $F(\cdot)$ (see Theorem 8.116), we see that for every $u \in X$, we can find $v \neq u$ such that

$$f(v) + d(u, v) \leq f(u).$$

This is precisely the hypothesis in Theorem 8.119. So, we can find $u_0 \in X$ such that $f(u_0) = \inf_X f$. Let $v \in F(u_0)$, $v \neq u_0$ and

$$f(v) + d(v, u_0) \leq f(u_0)$$
$$\Rightarrow 0 < d(v, u_0) \leq f(u_0) - f(v) \leq 0$$

a contradiction. So, $F(\cdot)$ has a fixed point. $\qquad\square$

8.5 Remarks

(8.1) The Hausdorff metric topology and the associated mode of convergence are the most well studied hyperspace notions. The Kuratowski convergence of sets started with Hausdorff [136] and was popularized and used extensively by Kuratowski [167]. The associated Fell topology (see Remark 8.5) was introduced by Fell [109]. Earlier important work was done by Mosco [200], [201] in order to study stability problems of variational inequalities and the continuity properties of the operation of convex conjugation. His idea to mix in the definition of the Kuratowski limits the norm and weak topologies, turned out to be very fruitful. The W-convergence (see Definition 8.8) and the weak convergence (see Definition 8.10) were introduced by Wijsman [283] in connection with problems of mathematical statistics.

The next theorem summarizes some important compactness properties of the K-convergence (see also Remark 8.5).

Theorem 8.121

(a) If (X, τ) is a second countable Hausdorff space and $\{A_n\}_{n \in \mathbb{N}} \subseteq 2^X \setminus \{\emptyset\}$, then we can find a K_τ-convergent subsequence.
(b) If (X, d) is a metric space, $\{A_n\}_{n \in \mathbb{N}} \subseteq P_f(X)$, and $A_n \subseteq K \in P_k(X)$ for all $n \in \mathbb{N}$, then we can find $A \in P_k(X)$ and a subsequence $\{A_{n_k}\}_{k \in \mathbb{N}}$ such that

$$A_{n_k} \xrightarrow{K} A.$$

Remark 8.122 It is easy to see that (b) follows from (a). Part (b) is often used in the literature and is known as "Blaschke's Theorem."

More on the convergence of sets can be found in the books of Aubin–Frankowska [16], Hu–Papageorgiou [145], Klein–Thompson [162], and Papageorgiou–Kyritsi [215]. For more on the corresponding hyperspace topologies, we refer to the books of Beer [29] and Klein–Thompson [162].

(8.2) The Γ-convergence of functions was introduced by De Giorgi [79]. Earlier related notion can be found in the seminal works of Mosco [200], [201]. It turns out that the Mosco convergence of functions is equivalent to the Γ-convergence of functions (sequential) in both the norm and weak topologies. The related G-convergence of operators started with the work of Spagnolo [260].

More on these subjects can be found in the books of Braides [45], Dal Maso [74], Pankov [214] and in the paper of Zhikov–Kozlov–Oleinik [294]. For the G-convergence of monotone operators on Sobolev spaces, we refer to the papers of Chiado Piat–Dal Maso [65] and Defranceschi [78].

(8.3) Theorem 8.99 is due to Ekeland [99]. It turned out that the theorem has many applications in different parts of mathematical analysis and its applications; see the works of Ekeland [100], [101]. Theorem 8.116 is due to Caristi [58], while Theorem 8.119 is due to Takahashi [264].

8.6 Problems

Problem 8.1 Let X be a Banach space and $\{A_n\}_{n\in\mathbb{N}} \subseteq P_{fc}(X)$ an increasing (resp., decreasing) sequence. Show that

$$A_n \xrightarrow{M} \overline{\bigcup_{n\geq 1} A_n} \quad (resp\ A_n \xrightarrow{M} \bigcap_{n\geq 1} A_n).$$

Problem 8.2 Let X be a Banach space and $\{A_n\}_{n\in\mathbb{N}} \subseteq P_f(X)$. Show that the set $w - \limsup_{n\to\infty} A_n$ need not be closed or w-closed.

Problem 8.3 Let X be a separable Banach space, $\{A_n\}_{n\in\mathbb{N}} \subseteq P_{fc}(X)$, and

$$A_n \subseteq W \subseteq P_{wk}(X) \ for\ all\ n \in \mathbb{N}.$$

Show that we can extract a subsequence $\{A_{n_k}\}_{k\in\mathbb{N}}$ of $\{A_n\}_{n\in\mathbb{N}}$ such that $A_{nk} \xrightarrow{w} A \in P_{wkc}(X)$.

Problem 8.4 Let X be a Banach space, $\{A_n, A\}_{n\in\mathbb{N}} \subseteq P_{fc}(X)$ and assume that $A_n \xrightarrow{w} A$. Show that we can find a sequence $\{C_n\}_{n\in\mathbb{N}}$ consisting of convex combinations of the $A_n's$, such that $C_n \xrightarrow{h} A$.

Problem 8.5 Let (X, d) be a metric space, $\{A_n\}_{n\in\mathbb{N}} \subseteq P_f(X)$ and are connected, and $A_n \xrightarrow{w} A \in P_k(X)$. Show that A is connected too.

Problem 8.6 Show that the result of Problem 8.5 fails if A is not compact.

Problem 8.7 Let X be a Banach space, $\{A_n, A\}_{n\in\mathbb{N}} \subseteq P_{fc}(X)$, and suppose that $A_n \xrightarrow{w} A$. Show that $A = \bigcap_{k\geq 1} \overline{conv} \bigcup_{n\geq k} A_n$.

Problem 8.8 Let X be a Banach space, $\{A_n, A\}_{n\in\mathbb{N}} \subseteq P_{fc}(X)$, and $A_n \xrightarrow{M} A$. Show that for every $x \in X$, $w - \limsup_{n\to\infty} p(x, A_n) \subseteq p(x, A)$, where for every $C \subseteq X$ nonempty $p(x, C) = \{c \in C : \|x - c\| = d(x, C)\}$ (the metric projection multifunction).

Problem 8.9 Let X be a Banach space, $\{A_n\}_{n\in\mathbb{N}} \subseteq P_{fc}(X)$, $A_n \subseteq W \in P_{wk}(X)$ for all $n \in \mathbb{N}$, and assume that $x_n \to x$ in X and $A_n \xrightarrow{M} A$. Show that $d(x_n, A_n) \to d(x, A)$.

Problem 8.10 Let H be a Hilbert space and $\{V_n\}_{n\in\mathbb{N}}$ a sequence of nontrivial closed subspaces of H. Show that

$$s - \liminf_{n\to\infty} V_n^{\perp} = w - \limsup V_n^{\perp}.$$

Problem 8.11 Let X be a finite dimensional Banach space $\{A_n, C_n\}_{n\in\mathbb{N}} \subseteq P_{fc}(X)$, and assume that

$$A_n \xrightarrow{K} A, \ C_n \xrightarrow{K} C \ \text{and} \ \text{int}\, A \cap C \neq \emptyset.$$

Show that $A_n \cap C_n \xrightarrow{K} A \cap C$.

Problem 8.12 Let X be a Banach space, $\{A_n\}_{n\in\mathbb{N}} \subseteq P_{fc}(X)$, $A_n \xrightarrow{M} A, C \in P_{fc}(X)$, and $A \cap \text{int}\, C \neq \emptyset$. Show that $A_n \cap C \xrightarrow{M} A \cap C$.

Problem 8.13 Show that the result of Problem 8.12 fails if $\text{int}\, C = \emptyset$.

Problem 8.14 Suppose H is a Hilbert space, $C \subseteq H$ is nonempty, bounded, closed, convex, and $\{L_n\}_{n\in\mathbb{N}} \subseteq \mathcal{L}(H)$ a sequence of self-adjoint operators such that $L_n(u) \to u$ in H for all $u \in H$. Show that $L_n(C) \xrightarrow{M} C$.

Problem 8.15 Show that the result of Problem 8.14 fails if C is not bounded.

Problem 8.16 Suppose (Ω, Σ, μ) is a complete measure space, X a Banach space with a separable dual, and $F_n : \Omega \to 2^X \setminus \{\emptyset\}$ are measurable multifunctions. Show that $\omega \to w - \limsup_{n\to\infty} F_n(\omega)$ is measurable.

Problem 8.17 Let (X, d) be a metric space, $f_n \xrightarrow{\Gamma} f$, and $g_n \to g$ uniformly on compact sets. Show that $f_n + g_n \xrightarrow{\Gamma} f + g$.

Problem 8.18 Consider the sequence $f_n(x) = tan^{-1}(nx + (-1)^n)$ for all $x \in \mathbb{R}$. Show that $\{f_n\}_{n\in\mathbb{N}}$ Γ-converges and find its Γ-limit.

Problem 8.19 Let (X, d) be a metric space, $f_n \xrightarrow{\Gamma} f$, and $g \in C(X, \mathbb{R})$. Show that $\max\{f_n, g\} \xrightarrow{\Gamma} \max\{f, g\}$ and $\min\{f_n, g\} \xrightarrow{\Gamma} \min\{f, g\}$.

Problem 8.20 Let (X, d) be a metric space, $K \in P_k(X)$, and $f_k = \Gamma - \liminf_{n \to \infty} f_n$. Show that $\min_K f_k \leq \liminf_{n \to \infty} \inf_K f_n$.

Problem 8.21 Suppose $A_n \overset{G}{\to} A$ and each A_n is odd (resp., homogeneous). Show that A is odd (resp., homogeneous).

Problem 8.22 Let X be a separable Banach space, Y a reflexive Banach space, and $\{A_n\}_{n \in \mathbb{N}} \subseteq \mathcal{L}(X, Y)$ that is norm bounded. Show that we can extract a subsequence $\{A_{n_k}\}_{k \in \mathbb{N}}$ of $\{A_n\}_{n \in \mathbb{N}}$ and find $A \in \mathcal{L}(X, Y)$ such that

$$A_{n_k}(x) \overset{w}{\to} A(x) \text{ for all } x \in X.$$

Problem 8.23 Let (X, d) be a metric space, $f_n : X \to \overline{\mathbb{R}} = \mathbb{R} \cup \{+\infty\}, n \in \mathbb{N}$, a sequence of ϑ-homogeneous (resp., convex) functions, and $f_n \overset{\Gamma}{\to} f$. Show that f is ϑ-homogeneous (resp., convex).

Problem 8.24 Let $\Omega \subseteq \mathbb{R}^N$ be abounded domain with Lipschitz boundary and $A_n : \Omega \to \mathbb{R}^{N \times N}$ is measurable with symmetric values and $\eta_1 id \leq A_n \leq \eta_2 id, \eta_1, \eta_2 > 0$. Show that we can find a subsequence $\{A_{n_k}\}_{k \in \mathbb{N}}$ of $\{A_n\}_{n \in \mathbb{N}}$ and A such that

$$\int_\Omega (A(z)Du, Du)_{\mathbb{R}^N} dz = \Gamma - \lim_{n \to \infty} \int_\Omega (A_n(z)Du, Du)_{\mathbb{R}^N} dz$$

in $L^2(\Omega)$ for all $u \in H^1(\Omega)$.

Problem 8.25 Show that the Ekeland Variational Principle holds if and only if (X, d) is a complete metric space.

Problem 8.26 Let X be a Banach space and $f \in \Gamma_0(X)$. Show that for every $\varepsilon > 0$ and every $x \in \mathrm{dom} f, \partial_\varepsilon f(x) \neq \emptyset$.

References

1. R. Adams-J. Fournier: "Sobolev Spaces" (Second edition), Elsevier/Academic Press, Amsterdam, 2003
2. N. Akhiezer-I. Glazman: "Theory of Linear Operators in Hilbert Spaces" slowromancapi@, slowromancapii@ F.Ungar Publishing Co, New York, (1961, 1963)
3. L. Alaoglu: "Weak topologies of normed linear spaces" Ann Math 42 (1940), 252–267
4. P. Alexandrov: "Sur les Ensembles de la Première Classe ef Les Espaccs Abstraits" CRAS Paris 5. 178 (1924), 185–187
5. P. Alexandrov: "Über die metrisation der im kleinen kompakten tepologische räume" Math. Ann. 92 (1924), 294–301
6. P. Alexandrov: "Über stetige abkildung kompakter räume" Math. Ann. 96 (1926), 555–571
7. P. Alexandrov-P. Urysohn:"Zür theorie der topologischen räume" Math. Ann. 92 (1924), 258–266
8. C. Aliprantis-K. Border: "Infinite Dimensional Analysis" (Third Edition) Springer-Verlag, Berlin, 2006
9. L. Ambrosio-P. Tilli: "Topics on Analysis in Metric Spaces" Oxford Univ. Press., Oxford, UK, 2004
10. W. Arveson: "An Invitation to C^*-Algebra" Springer-Verlag, New York 1976
11. R. Ash: "Real Analysis and Probability" Academic Press, New York, 1972
12. E. Asplund: "Frechet differentiability of convex functions" Acta Math 121 (1968), 31–47
13. H. Attouch: "On the maximal monotonicity of the sum of two maximal monotone operators" Nonlin. Anal. 5 (1981) 143–147
14. H. Attouch: "Variational Convergence for Functions and Operators" Pitman, London, 1984
15. J-P. Aubin-A. Cellina: "Differential Inclusions" Springer-Verlag, Berlin, 1984
16. J-P. Aubin-H. Frankowska: "Set-Valued Analysis" Birkhäuse, Boston, 1991
17. R.J. Aumann: "Borel structures for function spaces" Illinois Jour. Math 5 (1961), 641–630
18. R.Aumann: "Integral of set-valued functions" J. Math. Anal. Appl. 12 (1965), 1–12
19. R. Aumann: "Measurable utility and the measurable choice theorem" in Actes Collog. Internat. du CNRS, Aix-en-Provence 1967, Paris (1969), 13–26
20. R. Baire: "Sur les operations de variables réelles" Ann. Mat. Pura Appl. 3 (1899) 1–123
21. S. Banach: "Sur les operahons dans les ensembles abstraits et leur application aux equations integrales" Fund Math. 3 (1922), 133–181
22. S. Banach: "Sur les fonctionelles lineaires I, II" slowromancapi@, slowromancapii@, Studia Math 1 (1929), 211–216 and 223–239
23. S. Banach: "Über die Bairesche kategorie gewisser funktionenmengen" Studia Math. 3 (1931), 174–179

© The Author(s), under exclusive license to Springer Nature Switzerland AG 2022
S. Hu, N. S. Papageorgiou, *Research Topics in Analysis, Volume I*, Birkhäuser
Advanced Texts Basler Lehrbücher, https://doi.org/10.1007/978-3-031-17837-5

24. S. Banach: "Theory of Linear Operations" North Holland, Amsterdam, 1987
25. S. Banach-J. H. Steinhaus: "Sur le principe de la condensation de singularités" Fund. Math. 9 (1927), 50–61
26. S. Banach-A. Tarski: "Sur la decomposition des ensembles de points en parties respectivement congruentes" Fund. Math 6 (1924), 244–277
27. V. Barbu: "Nonlinear Semigroups and Differential Equations in Banach Spaces" Noordhoff International Publishing, The Netherlands, Leyden, 1976
28. V. Barbu-Th. Precupanu: "Convexity and Optimization in Banach Spaces" (Second edition), D. Reidel Publishing co, Dordrecht, The Netherlands, 1986
29. G. Beer: "Topologies on Closed and Closed Convex Sets" Kluwer Academic Publishers, Dordrecht, The Netherlands, 1993
30. C. Berge: "Espaces Topologiques, Fonctions Multivogues" (Second Edtion), Dunod, Paris, 1966
31. M. Berger: "Nonlinearity and Functional Analysis" Academic Pres. New York, 1977
32. P. Billingsley: "Convergence of Probability Measures" (Second edition), Wiley, New York, 1999
33. G. Birkhoff: "Moore-Smith convergence in general topology" Ann Math 38 (1937), 39–56
34. S. Bochner: "Integration von funktionen deren werte die elemente eines vectorraumes sind" Fund. Math 20 (1933), 262–276
35. V. I. Bogachev: "Measure Theory" Vols. I, II, Springer-Verlag, Berlin, 2007
36. V. I. Bogachev: "Weak Convergence of Measures" Mathematical Surveys and Monographs, Vol. 234, American Math. Society, Providence, R.I., 2018
37. B. Bollobas: "Linear Analysis" (Second edition), Cambridge Univ. Press Cambridge, UK, 1999
38. V. Boltyanski-G. Gamkrelidze-L.S. Pontryagin: "Theory of optimal processes Maximum Principle" lzv. Akad. Nauk USSR 24 (1960) 3–42
39. J.M. Borwein: "A note on ϵ-subgradients and maximal monotonicity," Pacific J. Math., 103 (1982), 307–314
40. J.M. Borwein- J.D. Vanderwerff: "Convex Functions: Constructions, Characterizations and Counterexamples" Cambridge Univ. Press, Cambridge, UK, 2010
41. N. Bourbaki: "General Topology: Part 1" Addison-Wesley, Reading, 1966
42. N. Bourbaki: "Integration. Chapters 7–9" Springer, Berlin, 2004
43. N. Bourbaki: "Topologie Generale: Chapitres 5 à 10" Springer-Verlag, Berlin, 2007
44. R. Bourgin: "Geometric Aspects of Convex Sets with the Radon-Nikodym Property" Lecture Notes in Math 1667, Springer-Verlag, Berlin, 1983
45. A. Braides: "Γ-Convergence for Beginners" Oxford Univ. Press, Oxford UK, 2002
46. A. Bressan: "Directionally continuous selections and differential inclusions" Funkc. Ekvac. 31 (1988), 459–470
47. A. Bressan-G. Colombo: "Extensions and selections of maps with decomposable values" Studia Math 90 (1988), 69–85
48. H.Brezis: "Equations et inequations non lineaires dans les espaces vectoriels en dualité " Ann. Inst. Fourier 18 (1968), 115–175
49. H. Brezis: "Operateurs Maximaux Monntones" North-Holland, Amsterdam, 1973
50. H. Brezis: "Functional Analysis, Sobolev Spaces and Partial Differential Equations" Springer, New York, 2011
51. H. Brezis-E. Lieb: "A relation between pointwise convergence of functions and convergence of functionals" Proc. Amer. Math. Soc. 88 (1983), 486–490
52. J.K. Brooks-R.V. Chacon: "Continuity and compactness of measures" Adv. Math. 37 (1980), 16–26
53. F. Browder: "Nonlinear Operators and Nonlinear Equations of Evolution in Banach Spaces" Proc. Symposia in Pure Math, Vol.17, Part2 Amer Math. Soc, Providence, R.I, 1976
54. F. Browder-P. Hess: "Nonlinear mappings of monotone type in Banach spaces" J. Funct. Anal. 11 (1972), 251–294

55. A.Brown-C.Pearcy, "Introduction to Operator Theory I: Elements of Functional Analysis," Springer, New York (1977)
56. D. Burago-S. lvanov: "A Course in Metric Geometry" American Math. Soc, Providence, R.I, 2001
57. C. Caratheodory: "Vorlesungen Über Reele Funktionen" Teubner, Leipzig (second edition), 1972
58. J. Caristi: "Fixed point theorems for mappings satisfying inwardness conditions" Trans. Amer. Math. Soc 215 (1976), 241–251
59. H. Cartan: "Calcul Differentiel" Hermann, Paris. 1967
60. E. Casas-L. Fernandez: "A Green's formula for quasilinear elliptic operators" J. Math. Anal. Appl. 142 (1989), 62–73
61. C. Castaing: "Sur les multiapplications mesurables" Revue Francaise Inform Rech. Operat. 1 (1967), 91–120
62. C. Castaing- M. Valadier: "Convex Analysis and Measurable Multifunctions" Lecture Notes in Math, Vol. 508, Springer-Verlag, Berlin, 1977
63. R.E. Castillo-H. Rafeiro: "An Introductory Course in Lebesgue Space" Canadian Math. Society, Springer, Switzerland, 2016
64. A. Cellina: "The role of approximation in the theory of multivalued mappings" in "Differential Games and Related Topies" eds H.W. Kuhn-G.P. Azego, North-Holland, Amsterdam (1971), 209–220
65. V. Chiado Piat-G. Dal Maso-A. Ddfranceschi: "G-convergence of monotone operators" Ann. Inst. H Poincare-Analyse Nonlineaire 7 (1990), 123–160
66. G. Choquet: "Lectures on Analysis I, II, III," Benjamin, Reading, Mass, 1969.
67. J.P.R. Christensen: "Topology and Borel Strucyure" North-Hlland, Amsterdam, 1974
68. F.H. Clarke: "Optimization and Nonsmooth Analysis" Wiley-Inter-Science, New York, 1983
69. F.H. Clarke: "Functional Analysis, Calculus of Variations and Optimal Control" Springer-Verlag, London, 2013
70. J.A. Clarkson: "Uniformly convex spaces" Trans. Amer. Math. Soc. 40 (1936), 396–414
71. D.L. Cohn: "Measure Theory" Birkhäuser, Boston, 1993
72. C. Constantinescu: "Spaces of Measures" De Gruyter, Berlin, 1984
73. D.V. Cruz Uribe-A. Fiorenza: "Variable Lebesgue Spaces. Foundations and Harmonic Analysis" Birkhäuser/ Springer, New York, 2013
74. G. Dal Maso: "An Introduction to Γ-Convergence" Birkhäuser, Boston, 1993
75. P.J. Daniell: "A general form of integral" Ann. Math 19 (1918), 279–294
76. M. M. Day: "Normed Linear Spaces" (Third edition), Springer-Verlag, New York, 1973
77. G. Debreu: "Integrantion of correspondences" in Proceedings of the Fifth Berkeley Symp on Math statistics and Probability (1965/66), Vol. slowromancapii@, Univ. of California Press, Berkeley (1967), 351–372
78. A. Defranceschi: "G-convergence of cyclically monotone operators" Asymp. Anal. 2 (1989), 21–37
79. E. De Giorgi: "Sulla convergenza di alcune successions di alcune del tipo dell area" Rend. Mat. 8 (1975), 277–294
80. C. Dellacherie: "Quelques examples familiers en probabilités des ensembk analytiques non boreliens" Seminaire de Probabilités Univ. Strasbourg (1976–1977), Lecture Notes in Math, Vol. 649 (1978), 746–756
81. Z. Denkowski-S. Migorski-N. S. Popageorgiou: "An Introduction to Nonlinear Analysis: Theory" Kluwer Academic. Publishers, Boston, 2003
82. R. Deville-G. Godefroy-V. Zizler: "Smoothness and Renorming in Banach Spaces" Pitman, New York, 1993
83. L. Diening-P. Harjulehto-P. Hästo- M. Ruzička: "Lebesgue and Sobolev Spaces with Variable Exponents" Lecture Notes in Math, Vol. 2017, Springer, Heidelberg, 2011
84. J. Diestel: "Sequences and Series in Banach Spaces" Springer-Verlag, New York, 1984
85. J. Diestel-J. Uhl: "Vector Measures" American Math. Society, Providence, R.I, 1977

86. J. Dieudonné: "Une generalisation des espaces compacts" J. Math. Pures Appl. 23 (1944) 65–76

87. J. Dieudonné: "Foundations of Modern Analysis" Academic Press, New York, 1969

88. J. Dieudonné-L. Schwartz: "La dualité dans les espaces $\mathscr{F}$ ei $(\mathscr{LF})$" Ann. Inst. Fourier, Grenoble 1 (1949), 61–101

89. R. Dudley: "Real Analysis and Probability" Wadsworth and Books/Cole Pacific Grove, CA., 1989

90. J. Dugundji: "An extension of Tietze's theorem" Pacific J. Math. 1 (1951) 353–367

91. J. Dugundji: "Topology" Allyn and Bacon, Boston, 1966

92. N. Dunford: "Uniformity in linear spaces" Trans. Amer. Math. Soc 37 (1935), 305–356

93. N. Dunford: "Integration in general analysis" Trans. Amer. Math. Soc. 37 (1935), 441–453

94. N. Dunford-B. J. Pettis: "Linear operations on summable functions" Trans. Amer. Math. Soc. 47 (1940), 323–392

95. N. Dunford-J. Schwartz: "Linear Oprators Part I: General Theory" Wiley Interscience, New York, 1958

96. W.F. Eberlein: "Weak compactness in Banach spaces I" Proc. Nat. Acac Sci USA 33 (1947), 51–53

97. M. Edelheit: "Eur theorie der konvexen mengen in linearen normierten räumen" Studia Math 6 (1936), 104–111

98. D.T. Egorov: "Sur les suites de fonctions mesurables" CRAS Paris, t. 152 (1911), 244–246

99. I. Ekeland: "On the variational principle" J. Math. Anal. Appl 47 (1974) 324–353

100. I. Ekeland: "Nonconvex minimization problems" Bull Amer. Math Soc(NS) 1 (1979), 443–474

101. I. Ekeland: "The ε-variational principle revisited" in "Methods of Nonconvex Analysis" ed. A. Cellina, Lecture Notes in Math, Vol. 1446, Springer-Verlag, Berlin (1989),1–15

102. I. Ekeland-R. Temam: "Convex Analysis and Variational Problems" North-Holland, Amsterdam, 1976

103. P. Enflo: "A counterexample in the approximation problem in Banach spaces" Acta Math. 130 (1973), 309–327

104. R. Engelking: "Outline of General Topology" North Holland, Amsterdam, 1968

105. L.C. Evans-R.F. Gariepy: "Measure Theory and Fine Properties of Functions" CRC Press, Boca Raton, FL, 1992

106. M. Fabian-P. Habala-P. Hajek-V. Montesinos Santalucian-J. Pelant-V. Zizler: "Functional Analysis and Infinite Dimensional Geometry" Canadian. Math. Soc, Springer, New York, 2001

107. Ky Fan: "Fixed points and minimax theorems in locally convex spaces" Proc. Nat. Acad. USA, 38 (1952) 121–126

108. H. Federer: "Geometric Measure Theory" Springer-Verlag, New York, 1969

109. J. Fell: "A Hausdorff topology for the closed subsets of a locally compact non-Hausdorff space" Proc. Amer. Math Soc. 15 (1962), 472–476

110. W. Fenchel: "On conjugate convex functiona" Canadian Jour. Math. 1 (1949), 73–77

111. W. Fenchel: "Convex Cones, Sets and Functions" Lecture Notes, Princeton Univ, Princeton, N. T, 1951

112. E. Fischer: "Sur le convergence en moyenne" CRAS Paris, t. 144 (1907), 1022–1024

113. J. Flachsmeyer: "Verschiedene topologisierungen in $rte\ddot{x}ta$um der aogeschlossenen teilmengen" Math. Nachr. 26 (1964), 321–333

114. K. Floret: "Weakly Compact Sets" Lecture Notes in Math, Vol. 801, Springer-Verlag, Berlin, 1980

115. G.B. Folland: "Real Analysis" (2nd edition), Wiley, New York, 1999)

116. M. Frechet: "Sur quelques points du calcul fonctionnel" Rendicanti del Circolo Mat. di Palermo 22 (1906), 1–74

117. M. Frechet: "Sur les ensembles de fonctions et les operations lineares" CRAS Paris, t. 144 (1907), 1414–1416

118. M. Frechet: "Des familles et fonctions additives d'ensembles abstraits" Fund. Math 5 (1924), 206–251

119. A. Fryszkowski: "Continuous selections for a class of nonconvex multivalued maps" Studia Math. 76 (1983), 163–174.

120. A. Fryszkowski: "Fixed Point Theory for Decomposable Sets" Kluwer Academic Publishers, New York, Springer, Dordrecht, 2005

121. L. Gasinski-N.S. Papageorgiou: "Nonlinear Analysis" Chapman Hall/CRC, Boca Raton, Fl, 2006

122. L. Gasinski-N.S. Papageorgiou: "Exercises in Analysis. Part 2: Nonlinear Analysis" Springer, Cham, 2016

123. L. Gasinski-N.S. Papageorgiou: "Convergence theorems for adapted sequences of random sets" Stoch. Anal. Appl 37 (2019), 189–218

124. J.R. Giles: "Convex Analysis with Application in Differentiation of Convex Functions" Pitman, Boston, 1982

125. D.C. Gillespie-W.A. Hurwitz: "On sequences of continuous functions having continuous limits" Trans. Amer. Math. Sor. 32 (1930), 527–543

126. H.H. Goldstine: "Weakly complete Banach spaces" Duke Math. Jour. 4 (1938), 125–131

127. L. Gorniewice: "Topological Fixed Point Theory of Multivalued Mappings" (Second edition), Springer, Dordrecht, The Netherlands, 2006

128. A. Granas-J. Dugundji: "Fixed Point Theory" Springer-Verlag, New York, 2003

129. G. Haddad-J.M. Lasry: "Periodic solutions of functional differential inclusions and fixed points of δ-selectionable correspondences" J. Math. Anal. Appl. 96 (1983), 295–312

130. H. Hahn: "Über folgen linearen operationen" Monatsh Math Physik 32 (1922), 3–88

131. H. Hahn: "Über linearer gleichungsysteme in linearer räumen" J. Reine Angew. Math. 157 (1927), 214–229

132. H. Hahn: "Über die multiplikation total-additiver mengenfunktionen" Annali Scu. Norm. Pisa 2 (1933), 429–452

133. P. Halmos: "The range of a vector measure" Bull. Amer. Math. Soc. 54 (1948), 493–510

134. P. Halmos: "Introduction to Hilbert Spaces" Chelsea Publ. Co, New York, 1957

135. P. Halmos: "Measure Theory" Springer-Verlag, New York, 1974

136. F. Hausdorff: "Grundzüge der Mengenlehre" Chelsea Publishing Company New York, 1949

137. J. Heinonen: "Lectures on Analysis in Metric Spaces", Springer-Verlag, New York, 2001

138. E. Helly: "Über linearer funktionaloperationen" Sifzber. Kous. Akad. Wiss. Math-Noturwiss. Kl. Wien. 125 (1912), 265–297

139. E. Hewitt-K. Stromberg: "Real and Abstract Analysis" Graduate Texts in Math, Vol. 18, Springer-Verlag, New York, 1975

140. F. Hiai-H. Umegaki: "Integrats, conditional expectations and martingales of multivalued functions" J. Multiv. Anal 7 (1977), 149–182

141. E. Hille-R. Phillips: "Functional Analysis and Semigroups" AMS Collog. Publ, Vol 31, American Math. Soc, Providence, R.I, 1957

142. C. Himmelberg: "Measurable relations" Fund. Math. 87 (1975), 53–72

143. L. Hörmander: "Sur la fonction d'appui des ensembles convexes dans un espace localement convexe" Ark. Mat. 3 (1955), 181–186

144. L. Hörmander: "The Analysis of Linear Partial Differential Operators I" Springer, Berlin (1983s)

145. S. Hu-N.S. Papageorgious: "Handbook of Multivalues Analysis. Vol I: Theory" Kluwer Academic Publishers, Dordrecht, The Netherlands, 1997

146. S. Hu-N.S. Papageorgious: "Handbook of Multivalued Analysis. Volume II: Applications" Kluwer Academic Publishers, Dordrecht, The Netherlands, 2000

147. A and C Ionescu Tulcea: "Topics in the Theory of Lifting" Springer, Berlin, 1969

148. A. Ioffe-V. Tichomirov: "Theory of Extremal Problems" North-Holland, Amsterdam, 1979

149. M. Jacobs: "Measurable multivalued maps and Lusin's theorem" Trans. Amer. Math. Soc 134 (1968), 471–481

150. R.C. James: "A non-reflexive Banach space isometric with its second conjugate," Proc. Nat Acad. Sci USA, 37 (1951), 174–177

151. R.C. James: "Weakly compact sets" Trans. Amer. Math. Soc. 113 (1964), 129–140

152. R.C. James: "Reflexivity and the sup of linear functionals" Israel J. Math. 13 (1973), 289–300

153. R. Kachurovski: "On monotone operators and convex functionals" Uspekhi Mat. Nauk 15 (1960), 213–215

154. S. Kakutani: "Concrete representation of abstract (M)-spaces" Ann. Math. 42 (1941), 994–1024

155. T. Kato: "Perturbation Theory for Linear Operators" (Second edition) Springer-Verlag, Berlin, 1976

156. J. Kelley: "Convergence in topology" Duke Math. Jour 17 (1950), 277–283

157. J. Kelley: "General Topology" Springer-Verlag, New York, 1955

158. J. Kelley-I. Namioka: "Linear Topological Spaces" Springer-Verlag, New York, 1976

159. P. Kenderov: "The set-valued monotone mappings are almost everywhere single valued" C. R. Acad. Bulgare Sci. 27 (1974), 1173–1175

160. N. Kenmochi: "Pseudomonotone operators and nonlinear elliptic boundary value problems" J. Math. Soc. Japan 27 (1975), 121–149

161. H.A. Klei: "A compactness criterion in $L^1(E)$ and Radon-Nikodym theorems for multimeasures," Bull. Sci. Math., 112 (1988), 306–324

162. E. Klein-A. Thompson: "Theory of Correspondence" Wiley, New York, 1984

163. O. Kovačik-J. Rakosnik: "On spaces $L^{p(x)}$ and $W^{1,p(x)}$" Czechoslovak Math. Jour. 41 (1991), 592–618

164. M. Krasnoselskii: "Topological Methods in the Theory of Nonlinear Integral Equations" MacMillan, New York, 1964

165. C.S. Kubrusly: "Elements of Operator Theory," Birkhäuser, Boston, (2001)

166. A. Kufner-O. John-S. Fučik: "Function Spaces" Noordhoff International Publishing, Leyden, The Netherlands, 1977

167. K. Kuratowski: "Topology: Volume I" Academic Press, New York, 1966

168. K. Kuratowski: "Topology: Volume II" Academic Press, New York, 1968

169. K. Kuratowski-C. Ryll Nardzewski: "A general theorem on selection" Bull. Acad. Polon. Sci. Ser Math, Asfron, Phys. 13 (1965), 397–403

170. K. Kuratowski-W. Sierpinski: "Le théorème de Borel-Lebesgue dans la theorie des ensembles abstraits" Fund. Math 2 (1921), 172–178

171. M. Lavrentiev: "Contribution ò la theorie des ensembles homeomorphes" Fund. Math. 6 (1924), 149–160

172. A Le: "Eigenvalue problems for the p-Laplacian," Nonlin. Anal. 64 (2006), 1057–1099

173. H. Lebesgue: "Integrale, longeur, aire" Annali Mat Pura Appl 7 (1902), 231–359

174. H. Lebesgue: "Lecons sur l'Integration et la Recherche des Fonctions Primitives" Gouthier-Villars, Paris (1904)

175. H. Lebesgue: "Sur les fonctions représentables analytiquement" Journal de Mathematiques 1 (1905), 139–216

176. H. Lebesgue: "Sur l'integration des fonctions discontinues" Ann. Sci Ecole Norm. Sup. 27 (1910), 361–450

177. G. Lebourg: "Valeur moyenne pour gradient généralisé" CRAS Paris, t. 281, 795–797

178. S. Leese: "Multifunctions of Souslin type" Bull. Australian Math. Soc 11 (1974), 395–41

179. G. Leoni: "A First Course in Sobolev Spaces" Graduate Studies in Math, Vol 105, American Math. Society, Providnece, R. I, 2009.

180. J. Leray: "La theorie des points fixes et ses applications en analyse" in Proceedings Intern. Cong. Math, Vol 2, Cambridge, UK (1950), 202–208

181. J. Leray-J. Schauder: "Topologie et equations fonctionelles" Ann. Sci. Ecole Nocm. Sup. 51 (1934), 45–78

182. V. Levin: "Borel selections of many valued maps," Siberian Math Jour., 19 (1979), 434–438

183. E. Lindelöf: "Sur quelques points de la theorie des ensembles" CRAR Paris, t. 13 (1903), 697–700

184. J. Lindenstrauss: "A short proof of Lyapunov's convexity theorem" J. Math. Mech 15 (1966), 971–972

185. N. Lusin: "Sur les propriétés des fonctions mesurables" CRAS Paris, t. 156 (1912), 1688–1690

186. M. Marcus-V. Mieel: "Every superposition operator mapping one Sobolev space into another is continuous" J. Funct. Anal. 38 (1979) 217–229

187. G.W. Mackey: "The Theory of Unitary Group Representations" The University of Chicago Press, Chicago, 1976

188. A. Markov: "On mean values and extreme densities" (Russian) Mat. Sbornik N S 4 (1938), 165–191

189. S. Mazurkiewicz: "Über Bordsche mengen" Bull. del Academie dcs Scieices, Cracovie (1916), 490–494

190. S. Mazur: "Über konvexe mengen in linearen normierten räumen" Studia Math. 4 (1933), 70–84

191. E.J. McShane: "Linear functionals on certain Banach spaces" Proc. Amer. Math. Soc. 1 (1950), 402–408

192. R.E. Megginson: "An Introduction to Banach Space Theory" Springer-Verlag, New York, 1998

193. E. Michael: "Continuous selections I" Ann. Math. 63 (1956), 361–382

194. G. Minty: "Monotone nonlinear operators in a Hilbert space" Duke Math. Jour 29 (1962), 341–346

195. E.H. Moore: "Definition of Limit in General Integral Analysis" Proc. Nat Aca Sci, USA 1 (1915), 628–632

196. R.L. Moore: "An extension of the theorem that no countable point set is perfect" Proc. Nat. Acad. Sci. USA 10 (1924), 168–170

197. R.L. Moore: "Concerning upper semicontinuous collections of continua" Monatshefte für Math. Physik 36 (1929), 81–88

198. B.S. Mordukhovich: "Variational Analysis and Generalized Differentiation I – Basic Theory" Springer-Verlag, Berlin, 2006

199. J.J. Moreau: "Fonctionelles Convexes" College de France, Seminaire surles equations aux dccivées partielles, Paris, 1967

200. U. Mosco: "Convergence of convex sets and of solutions of variational inequalities" Adv. Math 3 (1969), 510–585

201. U. Mosco: "On the continuity of the Young-Fenchel transform" J. Math. Anal. Appl. 35 (1971), 518–535

202. J. Munkres: "Topology. A First Course" Prentice-Hall Inc, Englewood Cliffs New Jersey, 1975

203. J. Nagata: "Modern General Topology" North Holland, Amsterdam, 1968

204. M. Nagumo: "A theory of degree of mapping based on infinitesimal analysis" Amer. J. Math 73 (1951), 485–496

205. H. Nakano: "Modulared Semi-Ordered Linear Spaces" Maruzen Co.Ltd, Tokyo, 1950

206. J.von Neumann: "Zur algebra der funktionaloper ationen und theorie denormalen operatoren" Math. Ann 102 (1929–1930), 370–427

207. J.von Neumann: "Allgemeine eigenwerttheorie Hermitescher funkfional operatoren" Math. Ann 102 (1929–1930), 19–131

208. J.von Neumann: "On rings of operators, reduction theory" Ann. Math 50 (1949), 401–485

209. L. Neustadt: "Optimization. A Theory of Necessary Conditions" Princeton Univ. Press, Princeton, N. J, 1976

210. J. Neveu: "Discrete Parameter Martingales" North Holland, Amsterdam, 1975

211. O.M. Nikodym: "Sur une generalisation des mesures de M J.Radon" Fund. Math 15 (1930), 131–179

212. C. Olech: "Decomposability as a subsitute of convexity" in "Proceedings of the Conference on Multifunctions and Integrals" ed. G. Salinettl, Lecture Notes in Math, Vol. 1091, Springer-Verlag, Berlin, 1984

213. W. Orlicz: "Über konjugierte exponentenfolgen" Studia Math 3 (1931), 200–211
214. A. Pankov: "G-Convergence and Homogenization of Nonlinear Partial Differential Opera-tors" Springer, Dordrecht, 1997
215. N.S. Papageorgiou-S.T. Kyritsi: "Handbook of Applied Analysis" Springer, Dordrecht, 2009
216. N.S. Papageorgiou-P. Winkert: "Applied Nonlinear Functional Analysis" DeGruyter, Berlin, 2018
217. K.P. Parthasarathy: "Probability Measures on Metric Spaces" Academic Press, New York, 1967
218. T. Parthasarathy: "Selection Theorems and Applications" Lecture Notes in Math, Vol. 263, Springer-Verlag, Berlin, 1972
219. D. Pascali-S. Sburlan: "Nonlinear Mappings of Monotone Type" Sijthoff and Noordhoff, The Netherlands, 1978
220. B.J. Pettis: "On integration in vector spaces" Trans. Amer. Math. Soc. 44 (1938), 277–304
221. R.R. Phelps: "Convex Functions, Monotone Operators and Differentiability" Lecture Notes in Math, Vol 1364, Spriner-Verlag, Berlin, 1989
222. L. Pick-A. Kufner-O. John-S.Fučik: "Function Spaces" DeGruyter Berlin, 2013
223. J. Radon: "Theorie and Anwendungen der absolut additiven Mengenfunktionen" Sitzungsber. Akad Wiss Wien Abt slowromancapii@ a (1913), 1295–1438
224. H. Radström: "An embedding theorem for spaces of convex sets" Proc. Amer. Math. Soc. 3 (1952), 165–169
225. M. Reed-B. Simon: "Functional Analysis" Academic Press, New York 1972
226. F. Riesz: "Sur les ensembles de fonctions" CRAS Paris t.143 (1906), 738–741
227. F. Riesz: "Sur une espèce de geometrie analytique des systemes defonctions sommables" CRAS Paris, t.144 (1907), 1409–1411
228. F. Riesz: "Sur les suites defonctions mesurables" CRAS Paris, t148 (1909), 396–397 and 405–406
229. F. Riesz: "Sur les operations fonctionelles lineaires" CRAS Paris, t.149 (1909), 974–977
230. F. Riesz: "Untersuchungen über systeme integrierbarer funktionen" Math. Ann. 69 (1910), 449–497
231. F. Riesz: "Les Systemes d'Equations Lineaires a une Infinite d'Inconnues" Gauthier Vdlars, Paris, 1913
232. F. Riesz: "Über lineare funktionalgleichungen" Acta Math. 41 (1931) 71–98
233. R.T. Rockafellar: "Extension of Fenchel's duality theorem for convex functions" Duke Math. J. 33 (1966), 81–89
234. R.T. Rockafellar: "Measurable dependence of convex sets and functions on parameters" J. Math. Anal. Appl. 28 (1969), 4–25
235. R.T. Rockafellar: "Local boundedness of nonlinear monotone operators" Mich. Math. Jour. 16 (1969), 397–407
236. R.J. Rockafellar: "Convex Analysis" Princeton University Press, Princeton, N.J, 1970
237. R.T. Rockafellar: "On the maximum monotonicity of subdifferential mappings" Pacific J. Math. 33 (1970), 209–216
238. R.T. Rockafellar: "Integral functionals, normal integrands and measurable selections" in "Nonlinear Operators and the Calculus of Variations" ed. I. Waelbroek, Lecture Notes in Math, Vol.543, Springer-Verlag New York (1976), 157–207
239. R.T. Rockafellar-R. Wets: "Variational Analysis" Spring-Verlag, Berlin, 1998
240. T. Roubiček: "Nonlinear Partial Differential Equations with Applications" Birkhöuser-Springer, Basel, 2005
241. H.L. Royden: "Real Analysis" (Second edition), MacMillan Publishing Comp, New York, 1968
242. W. Rudin: "Functional Analysis" McGraw-Hill, New York, 1973
243. W. Rudin: "Real and Complex Analysis" (Second Edition), McGraw-Hill New York, 1974
244. M-F. Saint Beuve: "On the extension of von Neumann-Aumann theorem" J. Funct. Anai 17 (1974), 112–129
245. S. Saks: "Theory of the Integral" Hafner, New York, 1937

246. J. Schauder: "Zur theorie stetiger abbildungen in funktionenräumen" Math. Z. 26 (1927), 47–65
247. J. Schauder: "Über die umkehrung linearer, stetiger funktionaloperationen" Studia Math 2 (1930), 1–6
248. J. Schauder: "Über lineare, vollstetige funktionaloperationen" Studia Math. 2 (1930), 183–196
249. H.H. Schaefer: "Topological Vector Spaces" Springer-Verlag, New York 1971
250. E. Schmidt: "Über die auflosung linearer gleichungen mit unendlich vielen unbekannten" Rend. Circolo Mat. Palermo 25 (1908) 53–77
251. J. Schwartz: "A note on the space L_p^*" Proc. Amer. Math Soc 2 (1951), 270–275
252. L. Schwartz: "Radon Measures on Arbitrary Topological Spaces and Cylindrical Measures" Tata institute of Fundamental Research, Studies in Math, No. 6, Oxford University Press, 1973
253. T. Shioya: "Metric Measure Geometry" European Math. Soc., Zurich, 2016
254. R. Showalter: "Monotonic Operators in Banach Space and Nonlinear Partial Differential Equations" Math Surveys and Monographs, Vol. 49, Amer. Math Soc, Providence, R.I 1997
255. W. Sierpinski: "Sur un problème concernant les ensembles mésurables superficiellment" Fund. Math. 1 (1920), 112–115
256. I. Singer: "Best Approximation in Normed Linear Spaces by Elements of Linear Subspaces" Springer-Verlag, Berlin, 1970
257. V.L. Smulian: "On the principle of inclusion in the space of type (B)" Rec. Math. Moscou NS 5 (1939), 317–328
258. V.L. Smulian: "Über lineare topologische räume" Mat. Sbornik N.S. 7 (49) (1940), 425–448
259. M.Y.Souslin: "Sur une definiyion des ensembles mesurables B sans nombres transfinis" CRAS Paris, t.164 (1917), 88–91
260. S. Spagnolo: "Sul limite delle soluzioni di problemi di Couchy relativi all' equazioni del colore" Ann. Scuola Norm. Sup. Pisa 21 (1967), 657–699
261. S.M. Srivastava: "A Course on Borel Sets" Springer-Verlag, New York 1998
262. H. Steinhaus: "Additive und stetige funktionaloperationen" Math. Z 5 (1919), 186–221
263. W. Stepanov: "Über totale differenzierbarkeit" Math. Ann. 90 (1923), 318–320
264. W. Takahashi: "Existence theorems generalizing fixed point theorems of multivalued mappings" in "Fixed point Theory and Applications" eds J. Baillon-M.Thera, Pitman, Harlow (1991), 397–406
265. H. Tietze: "Über Funktionen die auf einer abgeschlossenen Menge stetig stind" J.für die Reine und Angenw. Math 145 (1915), 9–14
266. H. Tietze: "Beiträge zur allgemeinen Topologie I" Math. Ann 88 (1923), 290–312
267. L. Tonelli: "Fondamenti di Calcolo delle Variazioni" Zanichelli, Bologna, 1921–1923
268. F. Treves: "Topological Vector Spaces, Distributions and Kernels" Academic Press, New York, 1967
269. S.L. Troyanski: "On equivalent norms and minimal systems in nonseparable Banach spaces" Studia Math 43 (1972), 125–138
270. A. Tychonov: "Über die topologische erweiterung von räumen" Math. Ann 102 (1930), 544–561
271. P. Urysohn: "Über die Mächtigkeit der zusammenhängenden Mengen" Math. Ann 94 (1925), 262–295
272. P. Urysohn: "Zum metrisation problem" Math. Ann. 94 (1925), 309–315
273. M. Vainberg: "Variational Methods for the Study of Nonlinear Operators' Holden-Day, San Francisco, 1964
274. M. Vainberg: "Variational Method and Method of Monotone Operators in the Theory of Nonlinear Equations" Halsted Press, New York, 1973
275. N.N. Vakhania-V.J.Tarieladze-S.A.Chobanyan: "Probability distributions in Banach spaces' Mathematics and its Applications, Vol. 14, D.Reidel Publishing Co, Dordrecht, The Netherland 1987
276. L. Vietoris: "Stetige Mengen" Monatsh. Math. 31 (1921), 173–204

277. A. Visintin: "Strong convergence results related to strict convexity" Comm. PDE. 9 (1987), 439–466
278. G. Vitali: "Sull' integrazione per serie" Rend. Circ. Mat. Palecmo 23 (1907), 137–155
279. D. Wagner: "Survey of measurable selection theorems" SIAM J.Control Optim 15 (1977), 859–903
280. D. Wagner: "Survey of measurable selection theorems: an update" in "Measure Theory in Oberwolfach 1979" ed D.Közlow, Lecture Notes in Math, Vol. 794, Springer-Verlag, Berlin 1980
281. J. Weidmann: "Linear Operators in Hilbert Spaces" Springer-Verlag, New York, 1980
282. N. Wiener: "Limit in terms of continuous transformation" Bull Soc Math France 50 (1922), 119–134
283. R. Wijsman: "Convergence of sequences of convex sets, cones and functions slowroman-capii@ Trons. Amer. Math. Soc 123 (1966), 32–45
284. A. Wilansky: "Modern Methods in Topological Vector Spaces" McGraw Hill, New York, 1978
285. S. Willard: "General Topology" Addison-Wesley, Reading, Massachusetts, 1970
286. V. Yankov: "On the uniformization of A-sets" (Russian), Dokl. Akad. Nauk USSR 30 (1941), 591–592
287. K. Yosida: "Functional Analysis" (Fifth edition), Springer-Verlag, Berlin 1978
288. D. Yost: "There can be no lipschitz version of Michael's selection theorem" in "Proceedings of the Analysis Conference, Singapore 1986" eds. S.Choy-J.Jesudason-P.Lee, North-Holland, Amsterdam (1988), 295–299
289. Z. Zalcwasser: "Sur ane propriété des champs des fonctions continues" Studia Math 2 (1930), 63–67
290. E. Zeidler: "Nonlinear Functional Analysis and its Applications Vol. I" Springer-Verlag, New York, 1985
291. E. Zeidler: "Nonlinear Functional Analysis and its Applications Vol. III" Springer-Verlag, New York, 1986
292. E. Zeidler: "Nonlinear Functional Analysis and its Applications". Volumes II/A, II/B, Springer-Verlag, New York, 1990
293. V.V. Zhikov: "On variational problems and nonlinear elliptic equations with nonstandard growth conditions" J.Math. Sci. 173 (2011), 463–570
294. V.V. Zhikov-S.M. Kozlov-O.A Oleinik: "G-convergence of parabolic operators" Russian Math. Surveys 36 (1981), 9–60
295. W.P. Ziemer: "Weakly Differentiable Functions" Springer-Verlag, New York, 1989

Index

© The Author(s), under exclusive license to Springer Nature Switzerland AG 2022

S. Hu, N. S. Papageorgiou, *Research Topics in Analysis, Volume I*, Birkhäuser

Advanced Texts Basler Lehrbücher, https://doi.org/10.1007/978-3-031-17837-5

Printed in the United States
by Baker & Taylor Publisher Services